# The Original Survey

# The Original Survey

## Recognition and Significance

**Donald A. Wilson**

**CRC Press**
Taylor & Francis Group
Boca Raton  London  New York

CRC Press is an imprint of the
Taylor & Francis Group, an **informa** business

First edition published 2022
by CRC Press
6000 Broken Sound Parkway NW, Suite 300, Boca Raton, FL 33487-2742

and by CRC Press
2 Park Square, Milton Park, Abingdon, Oxon, OX14 4RN

**Library of Congress Cataloging-in-Publication Data**

Names: Wilson, Donald A., 1941- author.
Title: The original survey : recognition and significance / Donald A. Wilson.
Description: First edition. | Boca Raton : CRC Press, 2022. | Includes
bibliographical references and index.
Identifiers: LCCN 2021022148 (print) | LCCN 2021022149 (ebook) | ISBN
9780367405472 (hbk) | ISBN 9781032116785 (pbk) | ISBN 9781003032557
(ebk)
Subjects: LCSH: Surveying--United States--History. | Land tenure--United
States--History. | Boundaries (Estates)--United States. | Surveying--Law
and legislation--United States.
Classification: LCC TA592.6.U6 W55 2022 (print) | LCC TA592.6.U6 (ebook)
| DDC 624.029--dc23
LC record available at https://lccn.loc.gov/2021022148
LC ebook record available at https://lccn.loc.gov/2021022149

ISBN: 978-0-367-40547-2 (hbk)
ISBN: 978-1-032-11678-5 (pbk)
ISBN: 978-1-003-03255-7 (ebk)

DOI: 10.1201/9781003032557

Typeset in Times
by KnowledgeWorks Global Ltd.

*This work is dedicated to my late wife and life partner,*
*Christine, who was always there when*
*I needed assistance with a writing, and who, as an expert title examiner,*
*knew the significance and value of an original survey.*
*Also to the original surveyors of long ago, who pushed their way into*
*unknown territories enduring the hardships of the wilderness far*
*from home ...*
*Knowingly or unknowingly, they set the stage for everything*
*that came later, and we owe them a debt of gratitude.*

# Contents

# SECTION II   Original Survey

# SECTION III   Locating Original Surveys and Related Information

# Preface

In the logical scheme of things, it would seem a book on Original Surveys should precede a book on the retracement of them, but it did not work out that way. Specializing my entire career on survey retracement, as well as coming to the realization that even with all of the seminars, lectures, and various writings, there has never been a full-length treatment published on boundary retracement. Such an important topic deserved a study and comprehensive review of its various facets and resulting court decisions.

The central theme is, and always has been, following the footsteps of the original surveyor. In studying, as well as experiencing, various retracement projects, it became apparent that there had never been a full treatment of original surveys either. It is one thing to expect, and also require by law, that an original survey in every case is the guiding factor. Then the questions quickly arise: What is it in each case? Is there more than one? Where do I find it? How do I interpret it? And, of course, if there never has been a traditional survey, what is the proper procedure to follow?

Thus a book devoted to the original survey was born. It has now been created, and even though the two companion volumes were created in what some might say out of order, I hope that they are useful and fit the need regardless.

**Donald A. Wilson**
*Newfields, New Hampshire*

# About This Book

## THE ORIGINAL SURVEY – WHY YOU NEED IT AND HOW TO FIND IT

Sometimes people do things because it is the law; other times because it is the right thing to do. Ideally, the two would coincide.

### THE ORIGINAL SURVEY – ITS RECOGNITION AND SIGNIFICANCE

Since the "sole duty, function and power" of the retracement surveyor is to "follow the footsteps" of the *original* surveyor,[1] it becomes of utmost importance to begin with the original survey. It follows that it is also axiomatic that such original survey must be correctly identified. Finding the original survey, interpreting and locating it on the ground can be unique challenges. Occasionally, there is no original survey as such, but the courts have ruled on how to deal with such an anomaly. It seems that we always assume that somewhere there is an original survey to be found, but it turns out that is not the case, if we depend on the usual connotation of the word. There is, however, somewhere, an original creation of the title. If not, then there is no need to consider a survey process.

Often the practice by the modern "retracement" surveyor is to review the current available descriptions, maps, plans, and so-called surveys in the neighborhood, then search for any markers in the vicinity. Any that are found are often accepted at face value as long as they "look ok" and fit within reason with the documents collected. Those not found need to be set in proper position, which is often done according to the measurements found in various documents, occasionally by other creative endeavors. Relying on these measurements, some of which are mere estimates, likely results in something other than a true survey, and probably is no better than a "best guess." At any rate, it is not adhering to the rule of law to follow the original surveyor, or the original survey. Consequently, because rule of law was not followed, such practice may constitute either a breach of contract, or a violation of one or more laws.

The basic flaw in comparing locations of evidence on the ground with written descriptions is that the latter are not trustworthy. That is why work on the ground is paramount (right, wrong, good, or bad as Judge Cowart articulated in *Tyson v. Edwards*[2]) and controls all other elements. As stated in *Outlaw v. Gulf Oil Corporation*,[3] field notes are a written description of what the surveyor did on the ground, and the plat is a resulting picture of the survey done on the ground. To proceed one more step, the written description in a document (or the "legal description" as some phrase it), is derived from the plat or some other source, and is secondary at best, potentially making it even more untrustworthy. Numbers on a plat are most often adjusted values with whatever accumulated and compensating errors they contain, which are not identifiable. Being fourth down on the list, and dependent on the other three, their reliability is highly suspect. While many rely on descriptions, plats, and even field notes, they do not control and must yield to the superior element – that of the original work on the ground. Or in other words, the original survey, in whatever form it may take.

---

[1] *Rivers v. Lozeau*, 539 So.2d 1147 (Fla.App. 5 Dist. 1989).
[2] 433 So.2d 549 (Fla.App. 5 Dist. 1983).
[3] 137 S.W.2d 787 (Tex.Civ.App. 1940).

**Author's Note:** Since it is common knowledge that numbers on a plat are derived from some (probably unknown) adjustment routine, which survey may also contain compensating (random) errors, it is difficult to understand why a retracement surveyor would attempt to use such information as being evidence superior to other available items. In the extreme, such practice may lead to additional markers being placed based on this faulty, unreliable information, leading to what is commonly known as "pincushion corners."

# Acknowledgments

Works of this nature do not happen by themselves, nor are they accomplished without a team of assistants regardless of whether an author or a reader is conscious of that fact. To Gary Kent, a modern leader in our profession who, with his very busy schedule, takes the time and interest to not only read, but study what I write, offering many helpful suggestions. Gary is a true friend and I am grateful to know him as a fellow professional.

Kris Kline, fellow author, seminar provider, associate and close friend. Kris has enhanced my recent writings with a careful eye for words continually offering new insights, examples, and legal principles. His input truly helps me produce a better piece of writing.

John "Steve" Parrish has always been very supportive of my work, and always willing to offer information and advice when asked. His understanding of public land surveys and western surveys in general is difficult to match and his experience unique. I much appreciate his input and value his friendship.

My sincere thanks goes to not only these three surveyors, but to others who have assisted me along the way, far too numerous to mention. Generally unappreciated, perhaps, are the students and audiences who come to learn, receive credit, or be entertained, but who also keep a presenter on the straight and narrow, whether in a classroom, with an email, or in studying words and pictures on a page – without this focus, such endeavors become much less meaningful.

# About the Author

**Donald A. Wilson** recently established a new company called Donald Wilson Consulting, LLC. He was previously president of Land & Boundary Consultants and has been in practice for over 60 years, consulting to groups throughout the United States and Canada. He is both a licensed land surveyor and professional forester, having conducted more than 500 programs on a variety of topics, including description interpretation, boundary evidence, law, title problems, and forensic procedures. Mr. Wilson has more than 200 technical publications in several areas, and has been involved with over 60 books, which include titles on Maine history and several books on fishing. Besides being co-author of *Evidence and Procedures for Boundary Location* and *Boundary Control and Legal Principles*; he is author of *Deed Descriptions I Have Known But Could Have Done Without*; *Easements and Reversions*; *Interpreting Land Records*; and *Forensic Procedures for Boundary and Title Investigation*. His latest titles are *Easements Relating to Land Surveying and Title Examination; Boundary Retracement: Processes and Procedures*; and a co-authorship entitled, *Land Tenure, Boundary Surveys, and Cadastral Systems*. Don is an instructor for RedVector's online professional courses and a regular presenter at the University of New Hampshire's Professional Development Program. He is lead instructor in Surveyors Educational Seminars. In his professional practice, Don has testified numerous times, in a variety of courts, on boundary and title matters.

# Section I

## Land Titles and Surveys

# 1 Introduction

## PART I: WHAT IT IS

> The rights of the parties in the land in controversy, must be determined by the original survey, made under the authority of the government,[1] and returned to the office of the surveyor general; and according to that survey, the disputed township line was a straight one, running through the entire length, from corner to corner; and it cannot now be varied to suit the interests or caprice of purchasers. To permit it to the done would be to annul the authority of all the public surveys; obliterate the lines of demarcation between the property of man and man, and open wide the door to confuse, harassing, and endless litigation.
>
> ***M'Clintock v. Rogers*, 11 Ill. 279 (Ill. 1849)**

Inherent in court decisions is the protection of the rights of the parties, already established, which cannot later be altered without a proper process. Put succinctly, it is the maintenance of the sanctity of titles and the stability of their boundaries. In fact, it has been declared that "*all* the rules of law adopted for guidance in locating boundary lines have been to the end that the steps of the surveyor who originally projected the lines on the ground may be retraced as nearly as possible"[2] (emphasis supplied).

It is interesting that the term *original survey* is used throughout the court system, and in retracement manuals and discussions, but generally not defined, rather taken for granted that everyone knows what it is.[3] However, it has been found that many confuse original surveys and first surveys, in particular calling any first location of boundary an original survey, when in fact, many are not – the boundaries are actually defined by existing features, which already exist.

Take for example a treaty with a description drawn from faulty and incomplete, not to mention conflicting, knowledge of land features. Distances when presented do not harmonize, some features cited do not exist, or are far from where thought to be, and geographical lines such as grants and lines of latitude, when marked, are erroneous and misleading. Such was common with early grants and titles, leading to such disputes as between the Calvert family and Penn family, eventually resulting in the survey and location of what has become to be known as the Mason-Dixon line. This line, by the way, is the result of a *first* survey, not an *original* survey, since it involved the location of an existing line, one which had previously been defined, but not located. Mason's and Dixon's task was to mark, for the very first time, the existing line on the surface of the earth.

## THE RIVERS DECISION

The decision in what has become known as the "Rivers case" is one of the most well-reasoned and explained concerning the distinction between original surveys and retracement surveys. Judge Cowart succinctly explained the differences between the two and how each is applied, as

---

[1] Since this is a case within the PLSS, "government" signifies the U.S. government and its authoritative departments charged with the subdivision of townships and otherwise surveying public lands. Government, however, in the general sense, should rightly include "the sovereign," since outside of the PLSS, titles are traceable back to a sovereign entity, a state, a governing nation, or, on occasion, the representative of a king.

[2] *Rivers v. Lozeau*, Fla.App. 5 Dist., 539 So.2d 1147 (Fla., 1989).

[3] See Chapter 3 for a discussion of what is an *original survey*.

DOI: 10.1201/9781003032557-2

3

well as putting in perspective a number of other survey issues while providing a summarized history of the rectangular survey process. An examination of what happened in this case and why it is important to be familiar with it deserves attention.

Judge Cowart discussed several points of interest to the practicing surveyor. After defining property descriptions and their component parts dependent upon the lines which are dependent upon their terminal points, he undertook to explain the duties of a land surveyor:

LAND SURVEYORS. Although title attorneys and others who regularly work with them develop expertise as to land descriptions, the only professional authorized to locate land lines on the ground is a registered land surveyor. In fact, the definition of a legally sufficient real property description is one that can be located on the ground by a surveyor. However, in the absence of statute, a surveyor is not an official and has no authority to establish boundaries; like an attorney speaking on a legal question, he can only state or express his professional opinion as to surveying questions. In working for a client, a surveyor basically performs two distinctly different roles or functions:

First, the surveyor can, in the first instance, lay out or establish boundary lines within an original division of a tract of land which has theretofore existed as one unit or parcel. In performing this function, he is known as the "original surveyor" and when his survey results in a property description used by the owner to transfer title to property [6] that survey has a certain special authority in that the monuments set by the original surveyor on the ground control over discrepancies within the total parcel description and, more importantly, control over all subsequent surveys attempting to locate the same line.

Second, a surveyor can be retained to locate on the ground a boundary line which has theretofore been established. When he does this, he "traces the footsteps" of the "original surveyor" in locating existing boundaries. Correctly stated, this is a "retracement" survey, not a resurvey, and in performing this function, the second and each succeeding surveyor is a "following" or "tracing" surveyor and his sole duty, function and power is to locate on the ground the boundaries corners and boundary line or lines established by the original survey; he cannot establish a new corner or new line terminal point, nor may he correct errors of the original surveyor. He must only track the footsteps of the original surveyor. The following surveyor, rather than being the creator of the boundary line, is only its discoverer and is only that when he correctly locates it.

THE FACTS. The controversy in this case involves the correct location of the line between two parcels of land lying within the 40 acre quarter-quarter section described as the Southeast 1/4 of the Southwest 1/4 of Section 15, Township 14 South, Range 24 East, in Marion County, Florida. In 1964 Joseph Rizzo and his wife owned that portion of this quarter-quarter section that is in question. The U.S. Forestry Service owns the land to the north. At that time, the Rizzos retained a surveyor, Moorhead Engineering, to survey their land and to establish certain internal land lines dividing it into parts. Moorhead undertook to locate and monument Rizzos' external boundary lines and corners and to establish and monument the terminal points of certain internal division lines.

In 1969, the Rizzos conveyed to Marcus E. Brown and wife by deed containing the following land description:

The North 400.00 feet of SE 1/4 of SW 1/4 of Section 15, Township 14 South, Range 24 East, Marion County, Florida.

The west, north, and east lines of the Brown parcel followed the outer or external boundary lines of the property owned by the Rizzos. The south line of the Brown parcel did not follow any internal line established by the Moorhead survey. Mr. Rizzo showed Marcus Brown the monuments Moorhead had set as being the north corners of this quarter-quarter section and certain other Moorhead monuments which the Rizzos told Marcus Brown were 33 feet south of the south line of the parcel the

Rizzos conveyed to Brown. Later in 1977 or 1978, Marcus Brown measured 33 feet north of the Moorhead monuments shown him by Mr. Rizzo and placed a metal rod at the point Mr. Rizzo had told him was his south boundary line. Marcus Brown conveyed this property by the same description to George Brown who conveyed by the same description to appellees Raymond S. Lozeau and his wife.

In 1975, the Rizzos conveyed a parcel of their remaining land to Paul W. Adams and wife, which parcel was described by reference to the boundary lines of this quarter-quarter section with the north line of the property conveyed being described as:

thence N 89 53'01"'E. along a line 400.00 feet south of and parallel to the North line of said SE 1/4 of SW 1/4 a distance of 1327.04 feet to a point on the East line of said SE 1/4 of SW 1/4;

Using substantially the same land description, the Adamses conveyed to Daniel E. Reader and wife, who conveyed to appellants Harold J. Rivers and wife.

In 1982 the U.S. Bureau of Land Management did a 'dependent resurvey' of the lands of the U.S. Forestry Service which retraced the lines of the original government survey and identified, restored, and remonumented the original position of the corners of the original U.S. government survey. This remonumenting of the original government survey, along with a 1986 survey by Whit Holley Britt, made obvious to all the true location of the north line of this quarter-quarter section on the ground and that the Moorhead monuments intended to denote that line were actually located 28.71 feet north of the true location of that line as it was originally established by the official U.S. government survey and reestablished by the 1982 government 'dependent survey.'

Appellees Lozeaus brought this action in ejectment and for declaratory judgment against the appellants Riverses who had possession of the south 28.71 feet of the north 400 feet measured from the north line of the quarter-quarter section according to the U.S. government (and Britt) surveys. The Lozeaus argued that they acquired legal title to the disputed land by virtue of the 1969 deed from Rizzo to Marcus Brown and the successive conveyances to them. The Riverses argued that Moorhead was the original surveyor and that his monuments on the ground controlled the location of the land subsequently conveyed by Rizzo, notwithstanding that "later" surveys, i.e., the government survey of 1982 and the 1986 Britt survey, may show the Moorhead monuments to have been in error. After a non-jury trial, the trial court found that the property descriptions of the parties overlapped and ordered that the exact dimensions of the overlap be established and the overlapping property split evenly between the plaintiffs and defendants. The Riverses appeal and the Lozeaus cross-appeal.

After discussing the historical aspects of the title and the history of the rectangular survey system, the court proceeded to discuss land descriptions and the role of the surveyor. Afterwards, a discussion of the basic problems in this case were analyzed.

ORIGINAL LAND LINES. When there is a boundary dispute caused by an ambiguity in the property description in a deed, it is often stated that the courts seek to effectuate the intent of the parties. This is not an accurate notion. The intent of the parties to a contract for the sale and purchase of land, both the buyer and the seller, may be relevant to a dispute concerning that contract, but in a real sense, the grantee in a deed is not a party to the deed, he does not sign it and his intent as to the quality of the legal title he receives and as to the location and extent of the land legally conveyed by the deed is quite immaterial as to those matters. The owner of a parcel of land, being the grantee under a patent or deed, or devisee under a will or the heir of a prior owner, has no authority or power to establish the boundaries of the land he owns; he has only the power to establish the division or boundary line between parcels when he owns the land on both sides of the boundary line he is establishing. In short, an original surveyor can establish an original boundary line only for an owner who owns the land on both sides of the line that is being established and that line becomes an authentic original line only when the owner makes a conveyance based on a description of the surveyed line and has good legal title to the land described in his conveyance.

THE LAW APPLIED TO THE FACTS OF THIS CASE. In establishing the internal lines within Rizzo's subdivision, Moorhead acted as an "original surveyor" but in attempting to locate and monument Rizzo's external boundary lines which are described by reference to the federal rectangle system of surveying, Moorhead was a "following surveyor" and not only failed to properly find the northern boundary of this quarter-quarter section where it was located by the original government surveyor (and also re-established by an authorized federal government resurvey) but to evidence his erroneous opinion as to the true line, the Moorhead surveyor placed monuments 28.71 feet north of the true north line of this quarter-quarter section. From the time the federal government granted this quarter-quarter section to the original grantee down to the Rizzos, the title conveyed was to a tract of land located according to the original government survey and by the deed from the Rizzos to Brown, and subsequent deeds, the Lozeaus acquired title to the north 400 feet of this quarter-quarter section according to the true boundary line established by the original government surveyors. This is true regardless of the fact that Mr. Rizzo showed Marcus Brown the erroneous monuments set by the Moorhead surveyors and regardless of where anyone erroneously thought or believed the correct location of this land boundary line to be. Neither the title to land nor the boundaries to a deeded parcel move about from time to time based on where someone, including a particular surveyor, might erroneously believe the correct location of the true boundary line to be. In 1975, the Rizzos conveyed to appellant Rivers' predecessor in title property the northern boundary of which is defined as being 400 feet south of, and parallel to, the north line of this quarter-quarter section. Regardless of any assertion that this conveyance was made relying on the Moorhead survey, the description itself does not describe the line in question by reference to the survey or monuments set by the Moorhead surveyor. On the contrary, that description adopts by reference the true north line of this quarter-quarter section which is necessarily controlled by the location of that line as established by the original government survey. Even if the description in the subsequent deed is considered to overlap the south 28 feet of the property previously conveyed by the Rizzos to Lozeaus' predecessor in title (which it does not), it is quite immaterial because, at the time of the conveyance to Paul W. Adams, Mr. and Mrs. Rizzo did not own that south 28 feet, they having previously conveyed legal title to it to Marcus Brown, Lozeaus' predecessor in title. All else argued in this case is immaterial. The Lozeaus are entitled to prevail in this controversy. All legal theories that could change the result in this case, such as those relating to adverse possession, title by acquiescence, estoppel, lack of legal title, etc., were neither asserted, nor argued, nor material in this case. This case is reversed and remanded with instructions that the trial court enter a judgment in favor of the appellees Raymond S. Lozeau and wife, in accordance with the land description as controlled by the official U.S. government survey.

FOOTNOTE in decision:[6] This is a most important qualification.

This case is one of the best examples demonstrating the result of not recognizing an original survey to be followed as a retracement, combined with an original creation of lines and why the two are not the same. This decision is indispensable to the practicing surveyor, in that first, it emphasizes that the original survey controls from the time it is created and has a certain special authority in that the monuments set by the original surveyor on the ground control over discrepancies within the total parcel description and, more importantly, control over all subsequent surveys attempting to locate the same line. Second, the decision explains that a recovery of an existing line is a retracement survey, and in performing this function, the second and each succeeding surveyor is a "following" or "tracing" surveyor and his sole duty, function, and power is to locate on the ground the boundaries corners and boundary line or lines established by the original survey; he cannot establish a new corner or new line terminal point, nor may he correct errors of the original surveyor. He must only track the footsteps of the original surveyor. The following surveyor, rather than being the creator of the boundary line, is only its discoverer and is only that when he correctly locates it.

Inherent in this decision are two principles: the importance of doing quality work when accomplishing an original survey, and understanding the original survey when doing a retracement thereof. These are the proper procedures, and, as the court stated, a retracement surveyor only has authority to discover the location of the original survey, and only when done correctly.

Also noteworthy is the court's analysis of the title aspects, both historical and modern. The discussion elaborated, putting it all into perspective, by stating, "real property descriptions are controlled by the descriptions of their boundary lines which are themselves controlled by the terminal points or corners as established on the ground by the original surveyor creating those lines. A property description that refers to, and adopts by reference, the description of a boundary line is DEPENDENT upon the proper location of the adopted line, which is dependent upon the location of the terminal points of the adopted line, which are dependent on their location on the ground as established by the original surveyor creating that adopted line."

One of the biggest problems facing the retracement surveyor today is being able to recognize when a previous *retracement* surveyor, not the *original* surveyor, has not made the appropriate identification of the type of survey they are working with. Frequently, a simple problem has been compounded, and likely led to some type of dispute as to boundary location. Such dispute may or may not have been recognized by the property owners until a proper retracement has been conducted at some point in the future, which may be days, weeks, or even years later. The more time that passes, the more problems that potentially arise, such as improvements mis-located, fences erected in the wrong place, and a host of other difficulties. In addition, after a period of time, unwritten rights, such as adverse possession, acquiescence, estoppel, or other unwritten doctrines, may have been perfected which eventually will require some further resolution to create a record title satisfactory to the marketable title requirements.

There are two aspects of a boundary, that of identification and that of location, the latter being in the realm of the land surveyor. The age-old maxim, sprinkled throughout court decisions in all jurisdictions is, "what are boundaries is a matter of law, and where boundaries are is a matter of fact." What a boundary is depends on its identification and where it is, location on, in, or above the earth, depends on facts. In the absence of appropriate facts, an investigator must rely on evidence. One must never lose sight of the fact that what a boundary is must be determined before its location can be determined, otherwise one can't know what one is looking for. Another important characteristic which the surveyor must never lose sight of is that not all boundaries are marked, or obvious, yet still exist. What surveyors wrestle with on a daily basis is corners and lines once marked, that are no longer marked, or obvious, yet corners and lines still exist, and are where the original survey, or the creation of title, established them. Therein lies the true challenge.

## PART II: THE COURT SYSTEM

Likely the first analysis of the concept was made by Judge George Mortimer Bibb (1776–1859), who was Chief Justice of the Kentucky Court of Appeals as well as a United States Senator. While compiling the very first volume of Kentucky Reports containing the case of *Beckley v. Bryan and Ransdale*,[4] his discussion in *Bryan, & c. v. Beckley*[5] emphasized that "existing lines and corners are to govern." In this decision he also sets out a series of rules for retracement surveys. In addition, Judge Bibb decided two other cases of significance to the land surveyor, *Finnie v. Clay*,[6] having to do with magnetic bearings, and *Vance v. Marshall*,[7] which had to do with proper surveying procedures.

---

[4] 1 Ky (Ky. Dec.) 91 (1801).
[5] 16 Ky (Litt Sel Case 91 (1809).
[6] 5 Ky (2 Bibb) 351 (1811).
[7] 6 Ky (3 Bibb) 148 (1813).

Chief Justice John Marshall[8] is quoted as having said that, "The most material and most certain calls shall control those which are less material and less certain." In this case it is laid down as a prime rule that the "Footsteps of the surveyor must be followed, and the above rules are found to afford the best and most unerring guides to enable one to do so."[9] *W. M. Ritter Lumber Co. v. Montvale Lumber Co.*[10]

As stated by the Texas court in *Goodson v. Fitzgerald*,[11] "the purpose of the inquiry, and the end to which all evidence is addressed, in a boundary suit is to find the footsteps of the original surveyor." This same court followed up 35 years later by stating, "… the testimony related to the search for the footsteps of the surveyors." As we said in *Turnbow v. Bland*,[12] "In all of the cases, from *Stafford v. King*,[13] on down, the law has been that search must be made for the footsteps of the original surveyor and, when found, the case is solved."[14] The court also cited *Taylor v. Higgins Oil & Fuel Co.*,[15] where it is said: "Every rule of evidence laid down for guidance in boundary questions is for the purpose of ascertaining the true location of the line in dispute, by which is meant the place at which the original surveyor ran the line."

To distinguish, the Kentucky case of *Rowe v. Kidd*[16] dealt with a description that was a combination of surveyed and protracted (nonsurveyed) lines, which contained significant mistakes with both courses and distances. After extensive analysis of both the description and its governing legal principles, the court included in its written decision:

> In determining the location of an actual survey, the fundamental principle is that it is to be located where the surveyor ran it. As it has been put, the thing to be done is to track the surveyor. This being so, it is where he ran and not where his certificate says he ran that governs. If there is a conflict between where he ran and where he thus says he ran, the latter must yield. In *Dimmitt v. Lashbrook*, 2 Dana, 1, Judge Robertson said:

> "When a line is actually run, it must be, as so run, the true boundary."

> Where, however, there was no actual running, i.e., where the running was mental only, the certificate with such exception is always conclusive. Whilst here, as in the case of an actual survey, it is where the surveyor ran which is the determining consideration, there is no other evidence of where he so did than his certificate, apart from the plat. But it is to be construed in the light of then existing conditions and what he may be known to have done. Suppose, then, in a given case, according to the certificate, the surveyor ran from a certain point a certain course and distance to a certain thing, and a line from that point to that thing is not according to such course and distance, and the plat sheds no light, and there are no marks left by the surveyor, what is to be done? As it is not possible for the surveyor both to have run from such point to such thing, and from such point such course and distance, he must have made a mistake either in saying that he ran from such point to such thing, or that he ran therefrom such course and distance, and the thing to be done is to ascertain which of the two statements was a mistake. Two alternative cases present themselves for consideration. One is where the surveyor actually ran the line; the other where he protracted it, i.e., ran it mentally. And I take it that, in the absence of persuasive evidence that he did not actually run it, it is to be taken that he did. If, then, it is a case of actual running, and the thing to which he says he ran is a visible thing, either natural or artificial, he must have known whether or not he ran to such thing. He could not have

---

[8] Fourth Chief Justice of the U.S. Supreme Court (1801–1835).
[9] The rules cited are (1) natural objects, (2) artificial marks, (3) course and distance.
[10] 169 N.C. 80, 85 S.E. 438 (1915).
[11] 40 Tex. Civ. 619, 90 S.W. 898 (Tex., 1905).
[12] 149 S.W.2d 604 (1941).
[13] 30 Tex. 257, 94 Am.Dec. 304.
[14] *Hart v. Greis*, 155 S.W.2d 997 (Texas, 1941).
[15] 2 S.W.2d 288 *(Tex.Civ.App.*, 1928).
[16] 249 F. 882 (E.D.Ky., 1916).

been mistaken in thinking that he ran thereto if he did not. The only possible room for mistake in such case is in making notes of what he did or transcribing the notes into his survey, and there would be little chance of his making a mistake in this particular. In such case, therefore, that is, where the line was actually run, as the call is to run to a visible thing, the mistake is to be taken to have been made in the statement as to the course and distance. Such mistakes are readily made. So it is that we have the rule, as to which there is no question, that where the call is for a natural object or a marked corner or line, the call for course and distance must give way to the call for such object, corner, or line. How, then, is it in case of actual running and the thing to which he says he ran is an invisible thing, i.e., an ideal or open line of another survey? In such case it is possible for him to have made a mistake in saying that he ran to such thing, if he did not run such line of the other survey, or otherwise know definitely where it was. If he did run it or otherwise knew, the likelihood of mistake is greatly lessened. Though, if he did not run it, the chance of mistake is greater, can it be said that the chance of mistake is greater than the chance of mistake in saying that he ran the course and distance? Possibly it is. If, then, there is a rule that, in a call for an ideal line of a survey, such call must give way to that for course and distance, here is a case for its application. And such case is where the line was actually run, and that of the survey called for was not run and he did not otherwise know where it was. But as, in such a case, it is well-nigh impossible to say whether the line of such survey was run, to see where it was located, or he otherwise knew, it is questionable, to say the least, whether any distinction should be made, where the line has been actually run, between a call for a visible thing and one for an invisible thing, and whether or not in both cases the call for course and distance should not yield to the call for the thing.

Take, then, the other case, i.e., where there is evidence which persuades that the line was not actually run. If it was protracted, i.e., mentally run, is there any reason for a distinction between the call for a visible thing and one for an invisible thing? If the call is for a visible thing, as the surveyor did not run the line he was not at it. The fact that it was visible was of no value to him in knowing whether he ran to the thing. Unless he mentally ran to the thing, he would not have called for it. The chance for mistake in the course and distance to the thing over what it would have been had he actually run the line was much greater, and in some cases it would be marvelous whether he could give it correctly. There is therefore no reason for making a distinction, where the call is for a visible thing, between a case where the line was actually run and where it was not. And so the rule is that in case of such call always, i.e., whether the line was run or not, the call for course and distance yields to the call for the thing. If, then, where the line was not actually run, and the call is for a visible thing, the call for course and distance yields to the call for the thing, there is no conceivable reason why, in such a case, and the call is for an invisible thing, the rule should not be exactly the same. In such a case the call for a visible thing has no advantage, as a corrective, because of the visibility of the thing, over a call for an invisible thing. In neither case was the surveyor at the thing to determine whether he had gone to it or not. In the latter, as in the former, case, he would not have called for the thing if he had not mentally run the line to it. Such is the meaning of the call. It can have no other meaning. And the possibility of mistake in giving the course and distance is as great where the call is for an invisible thing as in a call for a visible one. I therefore see no reason why the rule should not be that in all cases, whether the line was actually run or was not, and whether the thing called for was visible or invisible, that the call for the course and distance yields to the call for the thing.

There is reason, however, for an exception to this rule. This exception comes in only where the line was not actually run, and it makes no difference whether the thing called for was visible or invisible. The position thus taken that, if the line was not actually run, the call for the thing, whether visible or invisible, takes precedence over the call for the course and distance, is based on the assumption that the surveyor does not think that the thing called for is in a particular place. The exception comes in where he does think that it is in a particular place and hence mentally runs the line to that place. If it turns out that the thing called for at that place was not in fact there, i.e., the surveyor was mistaken in thinking that it was there, the call for the thing should give way to the call for course and distance; for the surveyor mentally ran the line to where he thought the thing was and not to where it in fact was. And the cases of *Mercer v. Bate* and *Ralston v. McClurg* were simply applications of this exception. In each case the line was not actually run, and the surveyor thought the thing called for was in a particular place, as to which he was mistaken. It was held in each case that the thing called for should give way to course and distance, because he had mentally run the line to where he thought the thing was and not where it actually was. It made no difference in the position there taken as to whether the thing called for was visible or invisible. A detailed consideration of these two cases will make this good.

Judge Robertson made a point of the fact that Mercer's upper or closing line was an ideal or open line.

It is a doctrine never heretofore applied in adjudicating on boundary in this country, so far as we know. It was not advanced by the learned counsel of the appellants, so that it might have met a reply. Its consequences are not easily foreseen, and as there are many lines left open, owing to the witchery in which surveyors acted and the dangers that surrounded them, it may be fatal; for it is a doctrine that will rend bounds and limits, hereafter to be fixed by a series of adjudications.

There was no occasion in the case for pointing out any distinction between a call for marked and visible lines and ideal lines. It is true that [a] line was not a marked line, but an ideal one. But it could have made no difference if it had been a marked one. The surveyor had actually run the line called for according to the course and distance called for, and as we have seen that it is where the surveyor ran, actually or mentally, that determines where the survey should be located. The fact that he thought that such line was the line called for, but was mistaken in so thinking, cannot change the fact as to where he ran, and hence does not allow the line to be located in accordance with the line called for, and that whether such line is a marked or an ideal line.

In summary, there can be only one original boundary survey and description; all subsequent surveys are retracements, or in a rare case, a resurvey. Original surveys have the following attributes:[17]

- They are without error.
- They are unassailable through the courts.
- They are permanent even if no evidence is found to identify them.
- They can be retraced and redefined by retracements, but never changed.

Synthesizing all this down to one principle, *An original survey creates boundary lines; it does not ascertain them.*[18] The principle is the subject of the following decisions:

A survey of public lands does not ascertain boundaries; it creates them. *Robinson v. Forrest,* 29 Cal. 317; *Sawyer v. Gray,* 205 F. 160.

### *WILLIAM ROBINSON v. W. G. FORREST*, 29 Cal. 317 (1865):

Neither a private survey nor one made under the authority of the State will answer this purpose, but it must be made under the authority of the United States. Even after a principal meridian and a base line have been established, and the exterior lines of the townships have been surveyed, neither the sections nor their subdivisions can be said to have any existence until the township is subdivided into sections and quarter sections by an approved survey. The lines are not ascertained by the survey, but they are created, and although a surveyor may, in advance of the making of the subdivision of the township, by the Deputy of the United States Surveyor-General, run lines with the greatest practicable exactness from the corners established on the exterior lines of the township, to ascertain the bounds of any given quarter-quarter section, still when the survey comes to be made under the

---

[17] Principle 11, Section 1.16, *Brown's Boundary Control and Legal Principles.*
[18] *Robinson v. Forrest,* 29 Cal. 317, 325; *Sawyer v. Gray,* 205 F. 160, 163.

direction of the Surveyor-General, the difference between the two surveys may be such that the forty acre lot, which, under the private and theoretically the more accurate survey, appeared to fall within the lands listed to the State, will be excluded from the list, or *vice versa*.

### *SAWYER et al. v. GRAY et al.*, No. 1,696., United States District Court, W.D. Washington, Southern Division, 205 F. 160 (W.D. Wash. 1913):

The township plat having been filed in April, 1901, the lands in dispute were unsurveyed at the time of the first alleged selection, in March, 1900. The government survey creates, not merely identifies, sections of land. There were no such lands as those described in the first application at the time of selection. *U.S. v. Curtner (C.C.)* 38 F. 1; *So. Pac. R. Co. v. Burlingame, 5 Land Dec. Dept. Int.* 415, and cases cited; *Robinson v. Forrest*, 29 Cal. 317; *Middleton v. Low*, 30 Cal. 596; *Bullock v. Rouse*, 81 Cal. 590, 22 P. 919; *Smith v. City of Los Angeles*, 158 Cal. 702, 112 P. 307.

# 2 The Elements of Title and Its Significance

While the farmer holds the title to the land, actually, it belongs to all the people because civilization itself rests upon the soil.

**Thomas Jefferson**

Sometimes title and boundary are confused. In some manner, a title contains, and is dependent on, boundaries; without title there are no boundaries. And without boundaries, no one can possibly understand what and where the title is situated, nor its extent, in size or in content. In addition, many surveyors shy away from title aspects, without giving a thought to their inspecting the title every time they read a title document, e.g., a deed.[1] Title and boundary are inter-related, each being dependent on the other, as the Tennessee court aptly stated.

## TITLE TO LAND CANNOT EXIST WITHOUT BOUNDARY

In the Tennessee case of *Moore v. Brannan*,[2] the court stated, "Complainants" bill was dismissed by the Chancellor because complainants had failed to locate their lands on the ground. More specifically, the Chancellor dismissed the bill on the ground that complainants had failed to locate Barrell Grant No. 5196 from which Grant complainants deraigned their title.

Our Supreme Court approved the following rule of law as applicable to such factual situations as we have in the case on trial; that is, when the controlling question is the location of the land on the ground. *Southern Coal & Iron Co. v. Schwoon*, 145 Tenn. 191, 239 S.W. 398, 403: See Figure 2.1.

The burden of the evidence on this point is upon the complainants. The rule of law which is to guide us in determining whether the complainants have met this burden is aptly stated in 1 Meigs' Digest, p. 540, as follows: [see following decision for text of the statement].

The cited case, *Southern Coal*, went a bit more in depth, by stating, "2. *The question of location of the complainants' land* is common to both cases and to all defendants. If the complainants have failed to show that their title papers cover the land in dispute, then they must fail altogether. We may therefore conveniently consider this phase of the case first.

Grant No. 4936 and the other title papers of the complainants describe the land sued for in case No. 15802 as follows:

On the waters of Collins river, beginning on a hickory tree, Peter Yates' southeast corner of his 5,000-acre entry; thence south, crossing the left-hand fork of Collins river at 1,000 poles, in all 1,100 poles, to a stake on the bluff; thence west 734 poles to a stake; thence with the mountain north 18 degrees west 760 poles to a stake; thence north 44 degrees east 440 poles to a black oak, Peter Yates'

---

[1] Additional documents, such as plans, may also be title documents, particularly if they are referenced directly or indirectly in title deeds. It is widely held that references in a description are as much a part of the description as if copied therein.

[2] 42 Tenn.App. 542, 304 S.W.2d 660 (1957).

DOI: 10.1201/9781003032557-3

corner; thence with his several lines south 45 degrees east 120 poles to a white oak; thence east 180 poles to a dogwood; thence north 70 degrees east 434 poles to the beginning.

Grant No. 4940 and the other title papers of the complainants describe the land sued for in case No. 15806 as follows:

Situated in Grundy county, Tennessee, on the waters of Collins river, beginning on the southeast corner of an entry made in the name of Peter Yates on a hickory, thence running east with Elias Mayo's survey 900 poles to two Spanish oaks; thence north with the line of John Rogers and Sterling Savage 896 poles to a hickory; thence west 900 poles to a hickory; thence south with said Yates survey 896 poles to the beginning.

Thus it will be seen that the northeast corner of grant No. 4936 and the southwest corner of grant No. 4940 are the same, and that this common corner is the southeast corner of another grant by No. 4191, issued to Samuel Edmondson, known in this record as the Peter Yates grant, by reason of the entry upon which it is based having been made in the name of Peter Yates, and by which name we shall have occasion later on to refer to it.

The location of complainants' grants with reference to each other and to the Peter Yates grant is shown by the diagram appended.

There is no difficulty whatever about the relative location of these grants, but the difficulty lies in locating them upon the ground.

It is admitted by the complainants that they have no other way of locating their lands on the ground except than by first locating the corner of the Peter Yates grant, from which point both of complainants' grants start, and it is conceded by the defendants that, if the complainants have successfully shown the location of the southeast corner of the Peter Yates grant to be as claimed by them, they have successfully located the lands described in the grants which form the basis of their title. The location on the ground claimed by the complainants as being the hickory or beginning corner of both

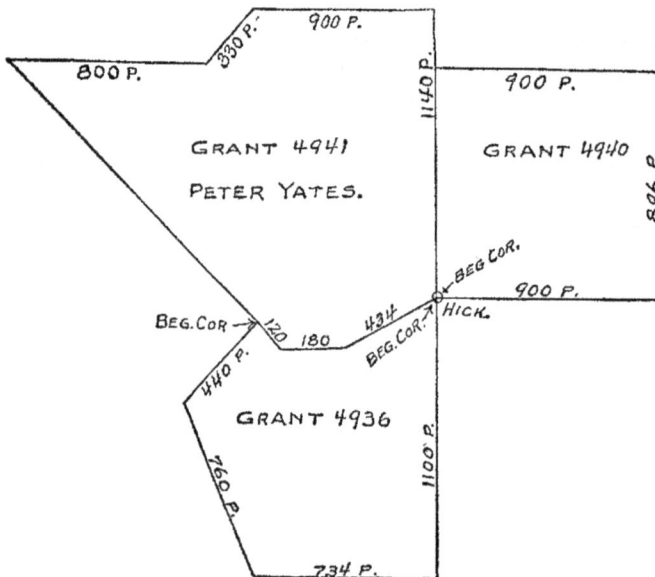

**FIGURE 2.1** Diagram contained in *Southern Coal & Iron Co. v. Schwoon.*

grants, and as being the southeast corner of the Peter Yates grant and of the black oak or southwest corner of the Peter Yates grant, and the lines between these corners which constitute the dividing line between grant No. 4936 and the Peter Yates grant, are fully shown in the record and well understood by both sides; their position with reference to the streams, roads, and other natural objects is shown upon the maps and need not be stated more particularly here.

The burden of the evidence on this point is upon the complainants. The rule of law which is to guide us in determining whether the complainants have met this burden is aptly stated in 1 Meigs' Digest, p. 540, as follows:

Title to land cannot exist without boundary; the plaintiff must show a marked boundary, or some proof from which boundary can be ascertained, before he can say to the defendant, even though he is a naked possessor: 'I have a better right to possess this particular piece of land than you have.' But then, in order to establish boundary, it is not indispensably necessary that some particular corner or marked line should be proven to exist. If it be proven to have existed, or any monument, corner, or marks from which the boundaries called for in a grant or deed can be satisfactorily ascertained, according to any easy and natural interpretation, it is sufficient. Nevertheless a grant can be lost for uncertainty in its boundary, there being no means by which its boundary can be defined. But not if by any reasonable means the intention of the contracting parties can be ascertained.

**Author's Note:** The importance of this case cannot be over-stressed. Not only does it emphasize the necessity of boundary identification, but also makes a point of stating that without the ability to satisfactorily show the existence of a boundary, a grant may fail.

## TITLE IS KEY

The basic understanding of any endeavor comes from its very foundation. The foundation of a boundary stems from its title, and the origin of the latter.

Every title has one or more boundaries, and parcels of land, easements, airspace, and water all have lines around them, confining their identity and separating them from an abutting title. Boundaries, therefore, are only fully identifiable through an understanding of the title. To rely on the senses, hearsay, opinions, and other less-than-adequate sources of information is to invite being misled and coming to an incorrect conclusion or opinion. Numerous examples throughout the text illustrate this and the courts, at all levels and jurisdictions, are replete with examples of misunderstandings.

Many, often the inexperienced, attempt to find a deed to provide the source of title, or more likely a description, to answer the object of their search. As will be seen, deeds, although often the most common source of information (even though often unreliable and misleading[3]), are not the only information to be sought. The major problem with this fallacy is that many give up when a deed is not found, often arriving at an incorrect conclusion, more likely an unfounded or misguided assumption. An obvious example is the person searching for a deed to a roadway and not finding one since deeded ways are much in the minority, conclude that it is not a legal way, or else, if it is, it must have been created through long-term use (prescription). There are numerous alternatives, often taking the searcher to a wide range of sources, and sometimes relying on rules of law. The bottom line is that they cannot know what it is that they have.

---

[3] Descriptions, unappreciated by many, are not to identify property; they are to furnish the means to identify. Descriptions are not to identify land but to furnish the means of identification. *City of North Mankato v. Carlstrom,* 2 N.W.2d 130, 212 Minn. 32 (1942).

## TITLE, RIGHTS, AND INTERESTS IN LAND

> Definition of Title: The right to or ownership of property. The word is used to designate the means by which an owner of lands has the just possession of his property, the legal evidence of his ownership, or the means by which his right to the property has accrued.

**Patton on Titles, § 1**

## THE CREATION OF TITLE

The creation of title sets at least some of the boundaries of the tract or right described. Along with the title there may be one boundary created, two, or more to almost any extent, although the more boundary definitions included, the more cumbersome to deal with the description becomes. Following are several examples of a single, as well as multiple boundaries, defining a title. To find the source of the title and its accompanying boundaries requires knowledge of where it arose. The following summary describes the various ways this can come about. Chapter 8 contains examples of likely sources containing information in the form of available records and other evidence.

## MEANS OF ACQUIRING OR TRANSFERRING TITLE TO, OR RIGHTS IN, LAND[4]

### TITLE BY TREATY*

Treaties are a result of a process of negotiation between nation-states over a period of time. Where boundaries are involved, they may include exploratory surveys followed by location and monumentation of any results. They are a form of agreement, and will have extensive amounts of associated correspondence and other documentation, for example, maps, opinions, reports, and descriptions. Likely, perhaps years later, they will be memorialized in a book or story. Treaties may accept existing boundaries, create new boundaries, or both.

### TITLE BY PUBLIC GRANT (E.G., PATENT FROM THE SOVEREIGN)

A public grant traditionally emanates from a sovereign entity. Extreme care must be exercised in searching for and tracking them, as the basic requirement is knowing the correct sovereign entity at the time of the original conveyance and creation of the title. Laws and requirements have varied over time, and may be dependent to the nature of the grant. Generally, there will be formal paperwork involved, with a record filed in one or more appropriate repositories.

Important examples are the boundaries established by sovereign nations prior to the United States becoming a sovereign entity. Such boundaries were established by other countries – England, France, Spain, Mexico, and Russia, in addition to the Netherlands and Sweden. Most of these were for large tracts of land, creating exterior boundaries, some of which were disputed, but all of which set the stage for future, subdivided, boundaries.

A familiar example is the Mason-Dixon line about which much has been written. The Grant to Lord Calvert by King Charles I and the grant to William Penn by King Charles II overlapped to great extent. Eventually, an agreement was reached whereby the accepted boundary would be an existing line of latitude, which was afterward located and monumented by the two English astronomers, Charles Mason and Jeremiah Dixon. The survey team spent five years locating 233 miles of boundary, placing stones every mile and what were called "crown stones" at five-mile intervals.[5]

---

[4] Taken in part from *Brown's Boundary Control & Legal Principles*, 7th Edition.
[5] Danson, Edwin. *Drawing the Line: How Mason and Dixon Surveyed the Most Famous Border in America.* Hoboken: John Wiley & Sons. (2001).

## Title by Private Grant (Such as by Deed)*

This is by far the most common type of conveyance of land and/or rights. Because of this, many parties are misled into believing that a deed is essential to prove any kind of transfer, and some-times conclude that, without such an instrument, no estate or rights have passed, or perhaps worse yet, did pass, but for an entirely different reason. This is a very dangerous notion. If placed on record, deeds are generally found in a designated repository, often a local courthouse, but other repositories are known to exist. However, two things must constantly be kept in mind: not all deeds are put on the public record, and not all descriptions are found in deeds.

## Title by Will (from the Decedent)*

The word "will" has been defined to mean the legal expression of a person's mind or wishes as to the disposition of his or her property, to be performed or take effect after death.[6] It is frequently termed a person's "last will and testament."

Not all decedents pass with a will, and in the absence thereof titles pass automatically *to the legal heirs at the time of death* by process of intestate (without a will) succession. Copies of wills, when probated, are usually found in probate records of an appropriate court. Where land is divided among devisees, wills may offer original boundary creations with their accompany-ing descriptions. Occasionally, a will is not filed but exists in a hidden, unusual, hopefully safe, location. Depending on the jurisdiction, the point in time and the circumstances, they are likely no less valid.

## Title by Descent (Intestate Succession)*

Legal heirs are defined by statute, which varies from state to state and from time to time. The law existing at the time of a person's death must be examined to determine who is a legal heir and therefore entitled to a share of the decedent's estate. Extra care must be exercised when more than one country is involved, such as when a person residing in one country holds title to real property in another country. Since no paperwork is involved, a process is demanded whereby the appropriate persons are identified. Sometimes, when complex or far-reaching, it is necessary to involve genealogical experts to undertake family searches in order to determine relationships and legal recipients.

## Title by Involuntary Alienation (Bankruptcy or Foreclosure)*

Encumbrances on land, such as mortgages, may expire according to their own terms, whereupon a designated process is executed to recover the value of the encumbrance. Either state, federal, or private holders may be involved, and records must be searched accordingly through the appropri-ate agency or repository. However, not all of them get recorded each time. Mortgage discharges are a common culprit of failure of recordation, and when in existence, are frequently found in private files.

## Title by Adverse Possession or Unwritten Agreement*

Possession may ripen into title through unwritten means, resulting in a transfer of title or of rights. Generally, there is a governing statute, usually termed statute of limitations for recovery of

---

[6] Black's Law Dictionary.

property. The statute, along with accompanying case law, will recite the requirements for transfer of title, along with any exceptions. While there is no direct paperwork involved, there may be related information in the form of other documents or affidavits, and, to perfect *marketable title*, a court procedure may have resulted. With the latter, there should exist, through some court proceeding, a decree as to the title, which may be a recorded writing or otherwise, depending on the nature of the action.

### TITLE BY EMINENT DOMAIN (PUBLIC TAKING WITH COMPENSATION)*

Certain agencies, governmental, state, and private, may have the power of eminent domain to acquire land and rights through a legal process and by paying the appropriate parties for the deprivation of their property, or its use. Such compensation is guaranteed in both federal and state constitutions. There should be detailed paperwork available filed in appropriate locations, which may be somewhat variable, depending on the nature of the taking, and the agency involved. Some takings are challenged, in which case there should be additional records as a result of the challenge and its result. Takings may result in either fee title or easement. Absent a statute allowing a fee taking, the taking results in an easement interest.

### TITLE BY ESCHEAT (PROPERTY REVERTING TO THE STATE)■

Escheat is a common law process whereby a person owning property dies without a will and without heirs (properly termed "without issue"), thereby having no successor of their estate, their property automatically vests in the state in which the land is situated. Extensive research is sometimes necessary to prove the non-existence of both a will and legal heirs. Occasionally, a subtle clue may arise whereby the recipient of the estate later conveys all or part of it.

### TITLE BY DEDICATION (E.G., EASEMENTS FOR PUBLIC USE)♦

Dedication is an extremely complex process of devoting private land or rights to public use. There are several possible procedures involved whereby a dedication may be statutory and thereby falling under the requirements of the current statute at the time, or it may fall under the common law of dedication. In addition, dedications may be express, such as with a writing which often is in the form of a map or subdivision plat, or they may be implied, wherein there likely will be no formal records, although meeting minutes of approval bodies may provide some insight. There may or may not be an acceptance, completing the process; although, when occurring, it may be either express or implied as well. Both dedication and acceptance are required for private property to become public. In addition, although sometimes fee title is involved, most dedications result in easement rights.

### TITLE WITH THE ELEMENT OF ESTOPPEL ENTERING♦

Estoppel is a legal bar arising through deed, a record, or by conduct. It prevents a person from doing or acting contrary to a previous act or deed whereby, if permitted, would result in harm to another person.

### TITLE THROUGH ACCRETION■

Accretion is a natural process whereby land accumulates in a gradual and imperceptible manner, resulting in an addition of soil to the parcel it attaches to. Accretions may be by water, which is the most common, or by wind, and sometimes a combination of both. They may also be natural, artificial, or natural due to an artificial interference. No formal records exist of such processes, although photographs, various accounts, parol testimony, and scientific studies may offer insight as to the ultimate result and effect of such processes, as well as their extent.

Accretion, which is "gradual and imperceptible," must not be confused with avulsion, which is sudden and abrupt generally resulting from a catastrophic weather phenomenon. There have been a few instances whereby property boundaries have been affected by a combination of both processes, the difficulty being separating them, since accretion results in a change of boundary, and therefore title, whereas avulsion does not.

## TITLE BY PAROL GIFT (FOLLOWED BY ADVERSE POSSESSION OR ACTS OF PARTIES)[7]■

Notwithstanding the Statute of Frauds designates that transfer of land, among other things, must be in writing to be enforceable, some jurisdictions have allowed the parol transfer of land. This is particularly prevalent in some jurisdictions, whereby family members may retain property within the family under the proper circumstances.

## TITLE THROUGH OPERATION OF LAW■

Certain transfers are valid under the law without any paperwork, or records. Reversion rights, determinable fees, expiration of easements, and other estates for years, as well as escheat fall under this category. The events happen automatically under their own terms, or according to law, without any further clarification or recordation (or as sometimes stated, "without office found").

## TITLE BY CUSTOM■

Creation of rights through the doctrine of custom was elaborated by Sir William Blackstone in his *Commentaries on the Laws of England*,[8] an influential 18th-century treatise on the common law of England. "Custom has often been declared to be the source of all law, except what rests upon statutes." Blackstone wrote, "this unwritten or common law is properly distinguishable into three kinds: 1. General customs; which are the universal rule of the whole kingdom, and form the common law, in its stricter and more usual signification. 2. Particular customs; which for the most part affect only the inhabitants of particular districts. 3. Certain particular laws; which by custom are adapted and used by some particular courts, of pretty general and extensive jurisdiction."[9]

---

[7] For extensive discussion and a collection of decisions, see 43 ALR2d 6, *Adverse Possession Under Parol Gift of Land*, and additional related ALR's.

[8] Originally published by the Clarendon Press at Oxford, 1765–1769, the work is divided into four volumes, on the rights of persons, the rights of things, of private wrongs, and of public wrongs.

The *Commentaries* were long regarded as the leading work on the development of English law and played a role in the development of the American legal system. They were in fact the first methodical treatise on the common law suitable for a lay readership since at least the Middle Ages. The common law of England has relied on precedent more than statute and codifications and has been far less amenable than the civil law, developed from the Roman law, to the needs of a treatise. The *Commentaries* were influential largely because they were in fact readable, and because they met a need.

The *Commentaries* are often quoted as the definitive pre-Revolutionary source of common law by United States courts. Opinions of the Supreme Court of the United States quote from Blackstone's work whenever they wish to engage in historical discussion that goes back that far, or farther (for example, when discussing the intent of the Framers of the Constitution).

[9] Gray, John Chipman. *The Nature and Sources of the Law.*

## EASEMENTS BY CUSTOM

An easement by virtue of custom is a legal right acquired by the operation of law through continuous use of land for a specific purpose over a long period of time. An easement right acquired by custom lacks a dominant tenement and is in generally in favor of the inhabitants of a particular locality.

Custom differs from prescription, which is personal and is annexed to the person of the owner of a particular estate; while the other is local, and relates to a particular district. The distinction has been thus expressed: "While prescription is the making of a right, custom is the making of a law."[10] Bouvier's Law Dictionary defines custom as "such a usage as by common consent and uniform practice has become the law of the place, or of the subject matter to which it relates."

The early New Hampshire decision of *Perley v. Langley*[11] provides a good explanation of how the doctrine of custom operates, explaining the difference between custom and prescription:

> Whether rights are holden as a custom, or as a prescription, depends on the manner in which they are holden – whether by a local usage, or as a personal claim, or as dependent on a particular estate.

> All rights which may be holden as a custom, may be holden as a prescription, but the reverse is not true.

> If the rights are common to any manor, district, hundred, parish or county, as a local right, they are holden as a custom.

> If the same rights are limited to an individual and his descendants, to a body politic and its successors, or are attached to a particular estate, and are only exercised by those who have the ownership of such estate, they are holden as a prescription.

Custom appears in tort, property, commercial, and constitutional law.[12]

### TITLE BY PRIOR APPROPRIATION

Another of Blackstone's concepts crafted to apply to running water, which he called *transient property*. Since it is always moving, as water touches the land it flows by, the respective landowners have the right to make use of it temporarily, so long as they do not infringe on another's rights, either upstream or downstream. Common uses are for domestic purposes, as well as irrigation, mill and dam operation, canals and flood control.

Key to symbols in the above categories:

- ▪ No written record
- ◆ May or may not be a record
- • Not always found in the usual locations

## TITLE AND ITS ACCOMPANYING BOUNDARIES

An original creation of title will establish one or more boundaries.[13] The question then becomes, what constitutes the original creation of the title, and what is its description defining its limits.

---

[10] Lawson, Usages & Cust. 15, note 2.

[11] 7 N.H. 233 (N.H., 1834).

[12] "Custom's jurisprudential foundations ... are premised on such diverse justifications as the functional reciprocity of obligations, the rationality and utility of customary rules, and the positive act of law-making by a community of legal actors." David J. Bederman, *Custom as a Source of Law.*

[13] Title to land cannot exist without boundary, 1 Meigs' Digest, p. 540, quoted in *Moore v. Brannan* 42 Tenn.App. 542, 304 S.W.2d 660 (Tenn.App. 1957).

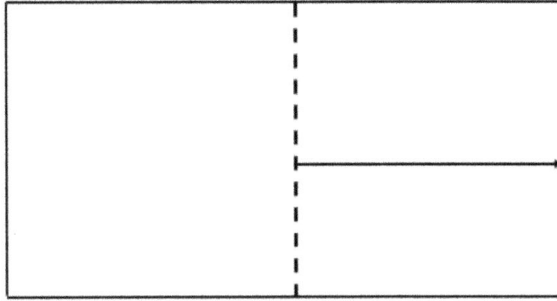

FIGURE 2.2   Area on one side of a described line (one boundary).

The following serves to illustrate the creation of titles with their accompanying boundaries:

**Example of a lot conveyed from corner of parent parcel.**[14] See Figure 2.3.

*Osteen v. Wynn*, 131 Ga. 209, 62 S.E. 37 (Ga 1908). A description of land, located in a county which is laid off by government survey into square land lots, as all of a named lot except 50 acres in the southeast corner, means all the land included in the lot except 50 acres, located by taking the southeast corner as a base point from which two sides of the excepted land shall extend equal distances so as to include by parallel lines 50 acres, and is sufficiently definite.

**Principle. If a given quantity of land is excepted from a corner of a tract, it must, as a general rule, be laid off in a square form. Lego v. Medley, 79 Wis. 211, 48 N.W. 375 (Wis. 1891)**

......except one acre from the southeast corner of the southwest quarter of the southwest quarter of said section, town, and range, together with the buildings thereon.

That a square acre laid off from the southeast corner of said land would not include the said dwelling-house; and that an oblong square acre laid off from said corner, having for its southern boundary the center of said highway, would include said dwelling-house, but would not include all the other

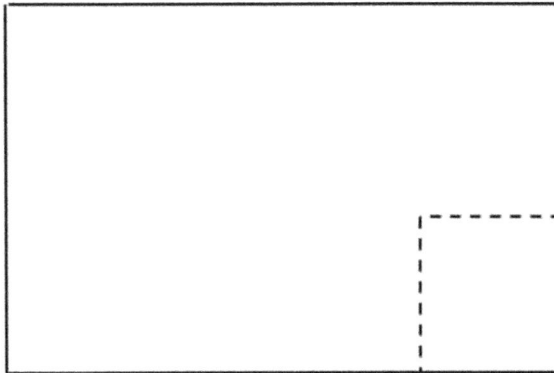

FIGURE 2.3   Subdivision in a corner of parent tract (two boundaries). Sometimes a special case with a unique court interpretation.

---

[14] For further discussion and examples, see 139 ALR 1180, *Validity and effect of deed which identifies tract conveyed only by reference to its area and a specified corner or other point of a larger tract from which it is to be taken.*

buildings referred to as used in connection therewith; but an acre so laid off from said corner, excluding the highway, that is, taking for the corner the point where the east boundary of said land intersects with the highway, would include all of said buildings; said acre would be sixteen rods long on the south boundary, and ten rods wide on the east boundary.

That said conveyance from Rose Medley to Richard P. Medley should be construed with reference to the circumstances attending the transaction, the situation of the parties, state of the property, the location of said dwelling-house, and other buildings, and the existence of the highway; and, having regard for these circumstances, the court holds that it was the evident intention of the parties, by the language used in said conveyance, that the acre excepted should be laid off from the southeast corner of said west half of the southwest quarter in said section nine (9), excluding the highway, so as to include said dwelling-house and said outbuildings used in connection therewith as the same were located at the time of the execution of said conveyance, which said acre, as near as can be determined from the testimony, is bounded as follows: Beginning at a point where the east boundary line of the southwest quarter of the southwest quarter of section nine (9), town thirty (30), range six (6), intersects with the highway on the south side of said land; thence west along the said highway sixteen rods; thence, at right angles, north ten rods, to the said east boundary line of said land; thence at right angles south to the place of beginning.

There are many different ways in which an improvement called for in a location, may be included in a survey; and the most proper way of doing it must be settled by the other parts of the location; if the call to include the improvement is accompanied with no other call which explains or restricts this call, the survey should be made in a square, with the lines running to the cardinal points, and the improvement in the center of that square; where the call is to include the improvement and to run a particular course for quantity, the survey ought to be in a square with the improvement in the center of the boundary opposite to the course called for, and the improvement to be barely included in the survey. See the case Miller's heirs v. Fox's heirs, 1 Ky. 100.

Then from the adjudication of the late supreme court in the case of Smith v. Grimes (1 Ky. 35), and from a principle which that court and the court of appeals have generally regarded, viz: that all surveys should be square when a compliance with the entries on which they are founded do not necessarily require them to be otherwise. Kenton's survey on his location with the commissioners ought to have been a square (unless the distance of his improvement from Preston's survey had made it necessary to extend it further), and have so adjoined Preston's survey as that the lines at right angles thereto, would be at equal distances from the improvement. But agreeably to Kenton's entry with the surveyor, and his location with the commissioners taken as one entry, and rejecting the call for Douglass' line, his survey should begin at Preston's south corner, and extend along that survey north-westwardly so far, that a line at right angles thereto would just include the said improvement, if so doing would stretch the survey beyond a square, which happens to be the case.

*—McConnell v. Kenton*, 1 Ky. 257 (1799)

## ESTABLISHING RIGHTS OF PARTIES

Referenced several times throughout the book in other decisions, *M'Clintock v. Rogers*[15] stresses several important points regarding original surveys.

The rights of the parties in the land in controversy, must be determined by the original survey, made under the authority of the government, and returned to the office of the surveyor general; and according to that survey, the disputed township line was a straight one, running through the entire length, from corner to corner; and it can not now be varied to suit the interests or caprice of purchasers. To

---

[15] 11 Ill. 279 (Ill. 1849).

permit it to be done would be to annul the authority of all the public surveys; obliterate the lines of demarcation between the property of man and man, and open wide the door to confused, harassing, and endless litigation.

It is a question of fact to be determined, how was this line originally run by the surveyors of the United States government. If that fact can be ascertained, it must determine the case.

A mathematical line is defined to be length without breadth: it exists only in the mind. A surveyor's line has a local habitation: it consists of a series of marked or established points on the ground, approaching a right line[16], according to the skill and correctness of the surveyor and perfection of his instruments. An actual township line consists of a series of section corners, never in point of fact falling in a right line from township corner to township corner. When these corners are gone, we resort to the next best evidence to ascertain where they were originally placed.

When the location of a lost corner is sought, by approaching it from others which are established, we must be guided by the best lights at command, however imperfect and unsatisfactory they may be.

There is, not infrequently, more than one original survey affecting a parcel. For example, a section line that is coincident with a township line, the question being, which one came first? It is about establishing the rights of the parties. Referenced several times throughout the book with other scenarios, the early decision of *M'Clintock v. Rogers*[17] stresses several important points regarding original surveys. "The rights of the parties in the land in controversy, must be determined by the original survey, made under the authority of the government, and returned to the office of the surveyor general; and according to that survey, the disputed township line was a straight one, running through the entire length, from corner to corner; and it can not now be varied to suit the interests or caprice of purchasers. To permit it to be done would be to annul the authority of all the public surveys; obliterate the lines of demarcation between the property of man and man, and open wide the door to confused, harassing, and endless litigation."

In this case, a later surveyor trying to honor, or survey, a section line without taking into consideration the township line it is coincident with may draw the incorrect conclusion, and be controlled by the wrong evidence. Such an approach is flirting with disastrous consequences. These situations are relatively numerous, as will be witnessed throughout the text.

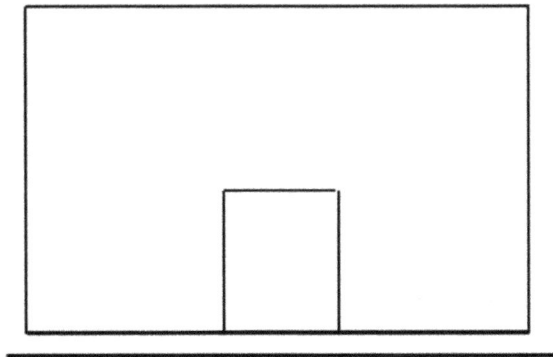

**FIGURE 2.4**   Subdivision in middle on a road (three boundaries + the road line).

---

[16] **Author's Note:** A right line is a straight line from point to point. Where the line of a survey was not in fact marked, or the evidence of where it was actually run has become extinct a right line from corner to corner must govern. *M'Clintock v. Rogers*, 11 Ill. 279, Ill. 1849).

[17] 11 Ill. 279 (Ill. 1849).

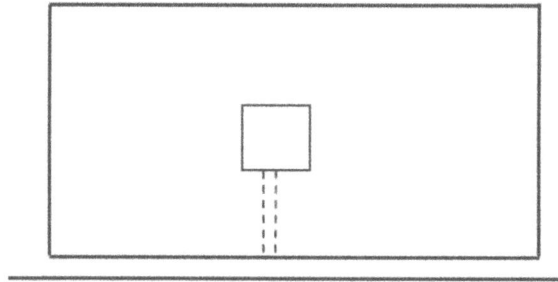

**FIGURE 2.5** Subdivision in middle with a right-of-way (4 boundaries + the r/w and its boundaries – it also may be an original survey regarding location and r/w limits). With the right-of-way, there are 2 possibilities, r/w created with the dominant tract, r/w created separate from the dominant tract, both however are separate titles. Another consideration is whether the right-of-way is fee or easement.

## HOW BOUNDARIES ARE ESTABLISHED

Boundaries are not established by survey, regardless of what some of the sources imply, or simply state. They may be the result of the survey process, but a surveyor has no authority, unless otherwise conferred, to establish a boundary. Even in the case of a subdivision creating new lines, the operating mechanism, as emphasized in *Rivers v. Lozeau*,[18] requires a transfer of title from the owner to create a lot, or one or more boundary lines.

As discussed in *Boundaries and Surveys*,[19] a boundary is an invisible line of division between two contiguous parcels of land [or between two *estates* in land]. Continuing, it was noted that boundaries may originate, be fixed or be varied by statutory authority, by proved acts of the respective owners (as by plans and deeds, possession, estoppel, or by agreement), or by the courts exercising statutory or inherent jurisdiction.[20]

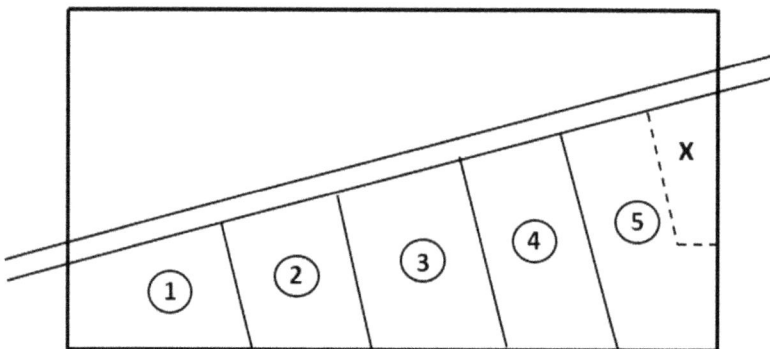

**FIGURE 2.6** More than one original survey for a parcel. Parcel **x** has several original surveys making it up, in order, the perimeter lines, the highway right-of-way, the subdivision south of the highway and the dashed lines creating it.

---

[18] 539 So.2d 1147 (Fla.App. 5 Dist. 1989), stating "this is a most important qualification."

[19] *Boundaries and Surveys, §1*. Lambden & de Rijcke. The Carswell Company Limited. 1985.

[20] Note that a surveyor establishing a boundary is not included. Surveyors do not establish boundaries, since (1) boundaries are discovered and located by surveyors, and, although lots are created on a subdivision plan, lines do not become established until there is a transfer of title, that is, a formal act by the parties involved in the transfer. Government surveys, not surveys conducted by private individuals, create, rather than merely identify, boundaries. *Cox v. Hart,* 260 U.S. 427, 43 S.Ct. 154, 67 L.Ed. 332 (1922).

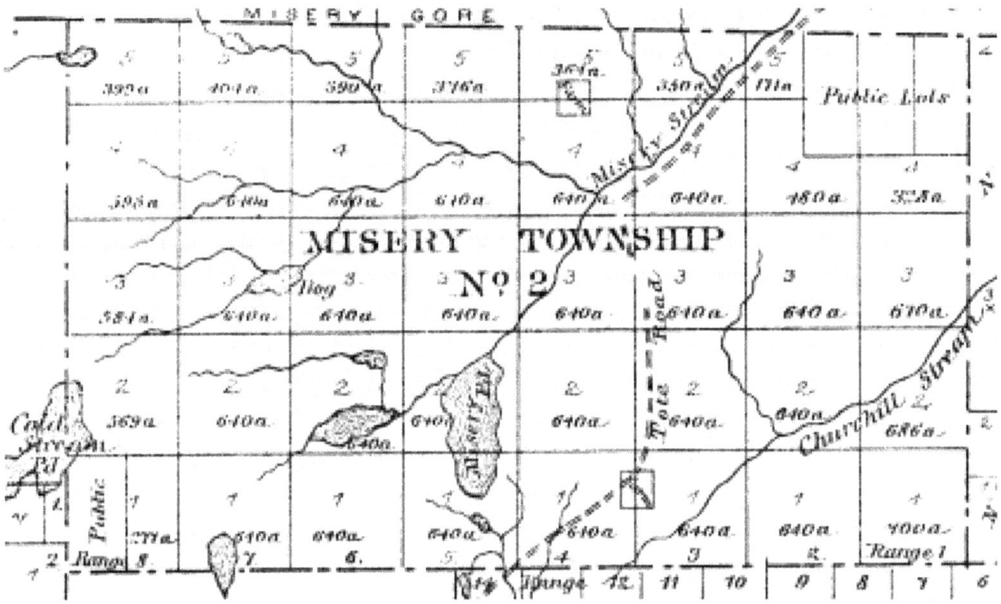

**FIGURE 2.7** An original township division into lots in Maine (then Massachusetts). Note the two farms which were grants to individuals prior to the regular lot division; also, in the northeast and southwest corners are standard Public Lots, for school and church purposes. These grants take priority over the regular lots created for sale at a later time. (Atlas of Somerset County, Maine. Geo. N. Colby & Co. 1883.)

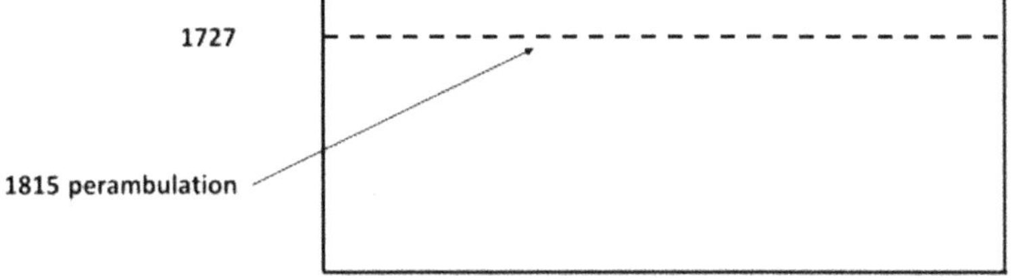

**FIGURE 2.8** Town divided on two separate occasions. Original town divided into two parts in 1727 (the 1727 boundary); subject town severed from northerly town in 1849. Note particularly that subject town has three original boundaries – east and west from the original parent town; the southerly boundary created in 1727, and the northerly boundary created in 1849. (Researching this evolution results in an original survey of the southerly boundary in 1815 as part of the required statutory perambulation.)

Note that a boundary is an invisible line. Clarification should be made that it should not be restricted to parcels of land, since it can also be a line of division between any two estates in land. The common and obvious example is the line between a fee title and an encumbering easement in or on the land.

A boundary is defined by two end points, generally known as corners, which are also invisible. While many believe a piece of physical evidence is a boundary (such as a fence), many also believe that a piece of evidence is the corner (such as an iron pipe, a stake, or a stone). The Tennessee decision of *Christian v. Gernt*[21] defined a corner as "the intersection of two converging lines or surfaces, an angle, whether internal or external; as the corner of a building; the four corners of a square; the corner of two streets."[22] Court decisions and literature speak of a corner being set, or marked with physical evidence, where, technically, the evidence as set *intends* to mark the *position* of the corner. The evidence *may* be at the position of the corner or it may be offset, as an accessory, or a witness, to the corner. A piece of evidence once set at the position of a corner may also have been disturbed, moved or even obliterated while the corner itself, its position, may still be determined with the aid of evidence since it never moves once created and positioned. Later discussion in Chapter 9 provides appropriate court decisions with discussion(s).

## THE ELEMENTS OF THE DEFINITION

Examining each of these elements individually provides insight to the processes creating boundaries, i.e., lines of separation between individual estates, including but not limited to, what are known as property boundaries.

The **origin** of a boundary is the key to its creation and subsequently to the foundation of its history through time.[23]

### ORIGINATION

A title along with its defining boundaries originates through a formal process, usually governed by law, at a point in time. This is known as its creation.

**Fixing** a boundary may be a matter of going through an appropriate process. While title holders may respect a certain line between them, without a formal process actually establishing the line as a boundary under the requirements of law, it is merely a line of convenience between the respective parties.[24]

Lines may be **varied** through designated processes, which include boundary line adjustments, exchange of title documents such as deeds, and long-term occupation.

---

[21] 64 S.W. 399 (Tenn.Ch., 1900).

[22] Citing Cent. Dict. 1269

[23] *Harvey v. Inhabitants of Town of Sandwich, 152 N.E. 625, 256 Mass. 379 (1926)*: Ancient deeds are to be construed as written in light of the then use of properties conveyed and adjacent land, and cannot be cut down by vagueness in subsequent conveyances.

[24] Courts have also rejected application of the agreed-boundary doctrine where parties not engaged in a boundary dispute are merely mistaken about the true boundary. "[W]here a boundary fence is built on what the parties suppose to be the true line, and there is no dispute, and it is not known that there is any uncertainty, the establishment of a boundary by implied consent and acquiescence does not follow therefrom." (*Kirkegaard v. McLain, supra,* 199 Cal.App.2d at p. 491 ["mere acquiescence in what adjoining landowners mistakenly believe to be the true line, without any notion on their part of fixing a disputed or uncertain boundary, does not amount to an agreement fixing a disputed or uncertain boundary line"]; *Pra v. Bradshaw* (1953) 121 Cal.App.2d 267, 269 ["An agreement or acquiescence in a wrong boundary when the true boundary is known, or can be ascertained from the deed, is treated in both law and equity as a mistake, and neither party is stopped from claiming the true line"].) *ROBBY G. HOGGATT et al., Plaintiffs and Respondents, v. KAZMER LEONARD PEZDEK et al.,* Defendants and Appellants. D056322 California Court of Appeal, Fourth District, First Division, February 1, 2011.

Depending on the jurisdiction, and the point in time, certain **statutes** may provide **authority** to create one or more boundaries. Eminent domain proceedings, probate partitions and highway laws fall into this category.

**Acts of owners** include agreements of various types, adverse relationships, artificial accretion, and conveying land parcels from subdivisions of larger parcels, are examples.

**A court having inherent jurisdiction** over not only the geographical region, but also the nature of the issue, may create a line, and therefore its location, which may be either temporary or permanent.

**Author's Note:** While surveyors may mathematically define a point, or a position or location, surveyors, by themselves, contrary to the belief of some, have no authority to create a boundary or a corner.[25] Any points shown on a subdivision plat do not become actual (legal) locations unless and until such survey results in a property description used by the owner to transfer title.[26] This is an extremely important concept to never lose sight of. Many, if not most, boundaries already exist, and it is the responsibility of anyone who follows the creator to find them – what they are, and where they exist. As the court stated in *Rivers*,[27] the following surveyor, [or anyone else, for that matter] rather than being the creator of the boundary line, is only its discoverer and is only that when he correctly locates it.

Sometimes a surveyor, along with a court as well, will state that a boundary was established, when in fact, the boundary had already been established at some point in the past, and the surveyor rendered an opinion as to its location, and marked it, or the court decided its proper location from the evidence submitted. This needs to be considered very carefully when deciding which rules to apply. One should ask the question, is this an original survey, a first survey, a [proper] resurvey, or a retracement survey? The rules are different for each.

A word of caution: it is sometimes easy to mistake an original survey for a first survey (or vice versa). If the boundary exists and is later surveyed (located) for the first time, that survey is not an original, since the line already exists, from a prior creation. It is a first survey, in that it was surveyed (located) for the first time through, or by, formal process.

## WHAT DOES IT MEAN TO ESTABLISH A BOUNDARY?

### ESTABLISH

To make stable or firm; to fix immovably or firmly; settle.[28] To settle or fix firmly; place on a permanent footing; found; create; put beyond doubt or dispute; prove; convince.[29]

Note that the list says nothing about a surveyor establishing a boundary. Surveyors do not "establish" [title] boundaries, except in a few, specified, circumstances. As Judge Cowart stated in *Rivers*, a surveyor has two functions, laying out or establishing boundary lines within an original division of a tract of land, which has theretofore existed as one unit or parcel, and retained to locate on the ground a boundary line, which has theretofore been established.

As emphasized, when the survey results in a property description used by the owner to transfer title, then the survey takes on special authority. There is no difference between the private actions as described above and the subdivision of public lands for which separation lines are created by the U.S. Congress and later marked on the ground by federal surveyors. The lines do not

---

[25] Unless provided with special authority, on a case-by-case basis.

[26] The court in *Rivers v. Lozeau* noted that this statement is a very important qualification.

[27] Ibid.

[28] Webster's Dictionary.

[29] Black's Law Dictionary.

become permanent until the plat is appropriately filed, after which a title transfer takes place in some form of public grant, usually a patent.

ORIGINAL LAND LINES. The owner of a parcel of land, being the grantee under a patent or deed, or devisee under a will or the heir of a prior owner, has no authority or power to establish the boundaries of the land he owns; he has only the power to establish the division or boundary line between parcels when he owns the land on both sides of the boundary line he is establishing. In short, an original surveyor can establish an original boundary line only for an owner who owns the land on both sides of the line that is being established and that line becomes an authentic original line only when the owner makes a conveyance based on a description of the surveyed line and has good legal title to the land described in his conveyance.

## CONNECTION OF TITLE AND BOUNDARY

There always is some act connecting boundary location and title. Since they are so inter-connected, to fully understand one, it is necessary to understand the other.

**In summary, a boundary does not become fixed (established) until there is a transfer of title – as pointed out in *Rivers*, "this is a most important qualification." With a Federal land patent, a boundary does not become fixed until the plat is approved and filed.**[30]

Robert J. Griffin[31] took the next step as far as the significance of the establishment of a boundary by stating, as Principle Number 1, **Location of a Boundary Line is Determined as of the Time of its Creation.** He stated that, "a boundary once established should remain fixed in its original position through any series of mesne conveyances." Continuing, "A grantee who purchases the entire extent of particular lands owned by the grantor determines the boundaries of his purchase as of the time that the particular parcel was carved out of some larger tract. He takes to the bounds of the estate of his grantor, who in turn took to the limits of his grantor's estate, etc., to the time of the creation of the boundary."[32]

There is nothing that can serve to alter a parcel boundary once established. Resurveys may replace them, agreements, by law, only serve to "fix that which is uncertain" (if a change takes place it becomes contrary to the Statute of Frauds because it actually results in a transfer of a small sliver of land from one person to another, which requires a writing in proper form) and, while overall ownership boundaries may change with subsequent acquisitions and sales, a parcel line remains fixed forever from its moment of creation.

The Pennsylvania decision in *Roth v. Halberstadt*[33] contained the statement:

the primary function of a court faced with a boundary dispute is to ascertain and effectuate the intent of the parties at the time of the original subdivision. To achieve this result, our courts have employed certain rules of construction that are commonly thought to provide the best indication of that intent. As a general rule, where there is a conflict between courses and distances or quantity of land and natural or artificial monuments, the monuments prevail. Dallas Borough Annexation Case, 169 Pa. Super. 129, 82 A.2d 676 (1951). Moreover, natural monuments normally take preference over artificial marks or monuments. Dallas Borough Annexation Case, supra. These rules, however, are not imperative but are merely aids in construction that must yield to a contrary showing." *Baker v. Roslyn Swim Club,* 206 Pa. Super. 192, 213 A.2d 45(1965); *Walleigh v. Emery,* 193 Pa. Super. 53, 163 A.2d 665 (1960).

---

[30] *United States v. Curtner, et al.*, United States Circuit Court, N.D. California, 38 F. 1 (N.D.Cal. 1889).

[31] Retracement & Apportionment as Surveying Methods for Re-establishing Property Corners. *Marquette Law Review*, Vo. 43, Issue 484, Article 5. Spring 1960.

[32] Retracement and Apportionment as Surveying Methods for Re-establishing Property Corners. *Marquette Law Review*, Vol. 43, Issue 484, 1960.

[33] 258 Pa.Super. 401, 392 A.2d 855 (Pa.Super. 1978).

## SURVEY AND TITLE RELATED

The decision rendered in the Mississippi case of *Newman, Appellant, v. Foster's Heirs*[34] explained the relationship between a survey map and its accompanying patent. It referenced an earlier decision wherein a dotted line of partition was shown on the original plat of survey. This court stated:

> that decision was, that this line being part of the map and so appearing upon its face, could not be corrected after a sale of the land, though made in mistake. The rule on this subject is well settled, and is uniform in all the cases which have been adjudicated where boundary was to be ascertained. The survey is to be taken as part of the patent. *It is the source of title*, is a matter of record, and may, therefore, be resorted to in order to control the calls of the patent. A consequence of this principle is, that if the plat and certificate of survey show an artificial or natural boundaries, though they may vary from the course or distance called for, they will nevertheless be taken as the true boundaries of the tract, if they can be well ascertained as described in the grant. 1 Marshall, 96; Pirtle's Digest, 125. The reason and policy of this doctrine is well explained in the case of *Hubert and Wife v. Wise and Others*, 3 Call. Rep. 238. Judge Pendleton says, 'the marked trees upon the land remain invariable, and are to govern as to the boundary. Such lines, therefore, when proved, are never suffered to be departed from. If the true line according to course and distance, called for in the plat and certificate of survey, depart from the line proved to be actually run, and evidenced by marked trees or other natural or artificial monuments, the latter must prevail;' and he gives the reason, the liabilities to mistakes by the surveyor, sometimes putting north for south, east for west, or in copying the descriptions into the patent. It would be highly detrimental, therefore, if a mistake in the calls of a patent might not be corrected by reference to the plat and certificate of survey.

---

[34] 4 Miss. 383 (Miss. 1839).

# 3 Types of Surveys

To survey land means to ascertain corners, boundaries, divisions, with distances and directions, and not necessarily to compute areas included within defined boundaries; such computation being merely a matter of mathematics.

*Kerr v. Fee*, **161 N.W. 1079; 179 Iowa 545 (1917)**

## WHAT IS AN ORIGINAL SURVEY, AND WHAT MAKES IT UNIQUE?

It is original, because it relates to the first time a title line or lines are created, and establishing one or more rights in real property. If it is found to be incorrect, and should be revised, a *resurvey*[1] may be in order. When an existing boundary is found on the ground, having been previously established (created) such search is a *retracement* – a retracing of a line already in existence, having been created by some procedure in the past. When a boundary has been created, but without benefit of a technical survey, the first time it is located, measured, and probably marked in some fashion, such endeavor is known as a *first survey*. Each of these has different characteristics, different purposes, and yield a different ultimate result, although, in theory, a retracement locates a line in the precise position that it was previously placed, in most instances an original line, created by the original actions of the respective parties.

For purposes of this treatment, an original survey equates to the origin of title, which may be a parcel, one or more lines, a space, or a right. The original survey of a tract equates to the description of the tract. The creation of one line, equates to the survey of that one line.

In *Rowe, et al v. Kidd, et al.,*[2] Judge Cochran presented an extensive review and rationale of applicable precedent.

In *Morgan v. Renfro*, Judge O'Rear said:

In all such cases, where it comes to locate again the survey so made, the object is to reproduce if possible, or as near as may be, what was originally done in appropriating the land by the survey. * * * There are several means which may be adopted. The rule is to prefer the best evidence. Therefore marked corners, i.e., those clearly identified, and which are notorious objects, are seized upon as the most satisfactory; then natural objects not marked, such as a stream, a ridge, a cliff, or the like, for they, while not so exact, are nevertheless reasonably sure to afford satisfactory evidence of the location of the patent at or near that point; then calls for the lines of other patents which are of record, and which are susceptible of definite and certain location; then courses; and then distances, in the order named.

In determining the location of an actual survey, the fundamental principle is that it is to be located where the surveyor ran it. As it has been put, the thing to be done is to track the surveyor. This being so, it is where he ran and not where his certificate says he ran that governs. If there is a conflict between where he ran and where he thus says he ran, the latter must yield. In *Dimmitt v. Lashbrook*,[3] 2 Dana, 1, Judge Robertson said:

---

[1] The term resurvey is frequently used in place of the term retracement. As distinguished here, and again later, they are not the same and fulfill different functions. The reader and the researcher are cautioned to have an understanding which they are dealing with.

[2] 249 F. 882 (E.D.Ky. 1916).

[3] Kentucky, 1834.

DOI: 10.1201/9781003032557-4

"When a line is actually run, it must be, as so run, the true boundary."

In reflecting upon this I have been puzzled to reconcile it with the rule that parol evidence is inadmissible to vary a writing. I have reached the conclusion that there is no conflict here, because such a case does not come within the rule. The making of the certificate is not contemporaneous with the making of the survey, though made from notes taken at the time thereof. When it is made the making of the survey is a thing of the past. It is a statement as to how the surveyor actually ran the lines. Just as, then, parol evidence is admissible as to what took place at a meeting of a board of directors or a town council by one who was present and heard what took place, and the minutes of the meeting made up from notes then taken, are not conclusive, so testimony of one, who was present at the survey, as to how the lines were actually run is admissible and the surveyor's certificate is not conclusive. The determining thing, then, in locating an actual survey, is to ascertain where the surveyor actually ran and not where he says he ran in his certificate. It is merely evidential as to where he actually ran. But, in the case of ancient surveys, no parol evidence as to where he actually ran the lines is obtainable, because of the death of all of those who may have been present at the survey. In such cases, in determining where the surveyor ran the lines, we are limited necessarily to his certificate except so far as may be affected by accompanying plat, and such marks as he may have made at the time of the survey, and, in the absence of such marks, to his certificate, except so far as it may be affected by the plat. In the latter case, of necessity, the certificate with such exception is conclusive. Where, however, there was no actual running, i.e., where the running was mental only, the certificate with such exception is always conclusive. Whilst here, as in the case of an actual survey, it is where the surveyor ran which is the determining consideration, there is no other evidence of where he so did than his certificate, apart from the plat. But it is to be construed in the light of then existing conditions and what he may be known to have done. Suppose, then, in a given case, according to the certificate, the surveyor ran from a certain point a certain course and distance to a certain thing, and a line from that point to that thing is not according to such course and distance, and the plat sheds no light, and there are no marks left by the surveyor, what is to be done? As it is not possible for the surveyor both to have run from such point to such thing, and from such point such course and distance, he must have made a mistake either in saying that he ran from such point to such thing, or that he ran therefrom such course and distance, and the thing to be done is to ascertain which of the two statements was a mistake. Two alternative cases present themselves for consideration. One is where the surveyor actually ran the line; the other where he protracted it, i.e., ran it mentally. And I take it that, in the absence of persuasive evidence that he did not actually run it, it is to be taken that he did. If, then, it is a case of actual running, and the thing to which he says he ran is a visible thing, either natural or artificial, he must have known whether or not he ran to such thing. He could not have been mistaken in thinking that he ran thereto if he did not. The only possible room for mistake in such case is in making notes of what he did or transcribing the notes into his survey, and there would be little chance of his making a mistake in this particular. In such case, therefore, that is, where the line was actually run, as the call is to run to a visible thing, the mistake is to be taken to have been made in the statement as to the course and distance. Such mistakes are readily made. So it is that we have the rule, as to which there is no question, that where the call is for a natural object or a marked corner or line, the call for course and distance must give way to the call for such object, corner, or line. How, then, is it in case of actual running and the thing to which he says he ran is an invisible thing, i.e., an ideal or open line of another survey? In such case it is possible for him to have made a mistake in saying that he ran to such thing, if he did not run such line of the other survey, or otherwise know definitely where it was. If he did run it or otherwise knew, the likelihood of mistake is greatly lessened. Though, if he did not run it, the chance of mistake is greater, can it be said that the chance of mistake is greater than the chance of mistake in saying that he ran the course and distance? Possibly it is. If, then, there is a rule that, in a call for an ideal line of a survey, such call must give way to that for course and distance, here is a case for its application. And such case is where the line was actually run, and that of the survey called for was not run and he did not otherwise know where it was. But as, in such a case, it is well-nigh impossible to say whether the line of such survey was run, to see where it was located, or he otherwise knew, it is questionable, to say the least, whether any distinction should be made, where the line has been actually run, between a call for a visible thing and one for an invisible thing, and whether or not in both cases the call for course and distance should not yield to the call for the thing.

Take, then, the other case, i.e., where there is evidence which persuades that the line was not actually run. If it was protracted, i.e., mentally run, is there any reason for a distinction between the call for a visible thing and one for an invisible thing? If the call is for a visible thing, as the surveyor did not run the line he was not at it. The fact that it was visible was of no value to him in knowing whether he ran to the thing. Unless he mentally ran to the thing, he would not have called for it. The chance for mistake in the course and distance to the thing over what it would have been had he actually run the line was much greater, and in some cases it would be marvelous whether he could give it correctly. There is therefore no reason for making a distinction, where the call is for a visible thing, between a case where the line was actually run and where it was not. And so the rule is that in case of such call always, i.e., whether the line was run or not, the call for course and distance yields to the call for the thing. If, then, where the line was not actually run, and the call is for a visible thing, the call for course and distance yields to the call for the thing, there is no conceivable reason why, in such a case, and the call is for an invisible thing, the rule should not be exactly the same. In such a case the call for a visible thing has no advantage, as a corrective, because of the visibility of the thing, over a call for an invisible thing. In neither case was the surveyor at the thing to determine whether he had gone to it or not. In the latter, as in the former, case, he would not have called for the thing if he had not mentally run the line to it. Such is the meaning of the call. It can have no other meaning. And the possibility of mistake in giving the course and distance is as great where the call is for an invisible thing as in a call for a visible one. I therefore see no reason why the rule should not be that in all cases, whether the line was actually run or was not, and whether the thing called for was visible or invisible, that the call for the course and distance yields to the call for the thing.

There is reason, however, for an exception to this rule. This exception comes in only where the line was not actually run, and it makes no difference whether the thing called for was visible or invisible. The position thus taken that, if the line was not actually run, the call for the thing, whether visible or invisible, takes precedence over the call for the course and distance, is based on the assumption that the surveyor does not think that the thing called for is in a particular place. The exception comes in where he does think that it is in a particular place and hence mentally runs the line to that place. If it turns out that the thing called for at that place was not in fact there, i.e., the surveyor was mistaken in thinking that it was there, the call for the thing should give way to the call for course and distance; for the surveyor mentally ran the line to where he thought the thing was and not to where it in fact was. And the cases of *Mercer v. Bate* and *Ralston v. McClurg* were simply applications of this exception. In each case the line was not actually run, and the surveyor thought the thing called for was in a particular place, as to which he was mistaken. It was held in each case that the thing called for should give way to course and distance, because he had mentally run the line to where he thought the thing was and not where it actually was. It made no difference in the position there taken as to whether the thing called for was visible or invisible. A detailed consideration of these two cases will make this good.

Translate this to the modern survey. The plat and resulting deeds are equivalent to the certificate. They are where someone, including the surveyor, stated the line was run. As Judge Cocoran stated, "where the surveyor actually ran is evidential and he also reflects on previous cases where lines were protracted," or as he stated, "mental."

## FEDERAL DEFINITION

While the term "original survey" is found within many decisions, in practically all jurisdictions, few definitions, totally or partially, appear. A thorough study of many decisions will provide the researcher with a confirmation of the obvious conclusion. However, it is with caution that one does not confuse the original survey with other types, particularly a first survey, since the rules for each are different.

The Supplement[4] to the Manual of Instructions dwells on retracement and following the footsteps of the original survey; "the original survey controls" is sprinkled throughout court decisions.

## RESURVEY[5]

A number of instances are found where the terms "resurvey" and "retracement survey" are used interchangeably, however, they are not the same.[6] A resurvey is properly defined as the reestablishment or restoration of land boundaries and subdivisions by the re-running and re-marking of the lines that were represented in the field note record and on the plat of the previous official survey. *Glossaries of BLM Surveying and Mapping Terms, U.S.D.I., BLM, 1980.*

In the Public Land Survey System (PLSS), there is a provision in the Manual of Surveying Instructions for conducting a resurvey. In the Manual, the concept involves re-running and re-marking existing boundaries created by the official survey (i.e., the original). Chapter V in the 2009 edition of the Manual, is entitled *Principles of Resurveys.*

Both Dependent and Independent Resurvey procedures are outlined in detail in this section.

## RESURVEY DEFINITION

A reconstruction of land boundaries and subdivisions accomplished by re-running and re-marking the lines represented in the fieldnote record or on the plat of a previous official survey. The resurvey includes a field-note record describing the technical manner in which the resurvey was made, full reference to recovered evidence of the previous survey, surveys, or resurveys, a complete description of the work performed and monuments established, and a plat that represents such resurvey. The resurvey, like an original survey, is subject to approval of the directing authority and official filing.

A *dependent resurvey* is a retracement and reestablishment of the lines of the original survey or of a prior resurvey in their true original positions according to the best available evidence[7] of

---

[4] Restoration of Lost or Obliterated Corners and Subdivisions of Sections of the Bureau of Land Management (1963 ed.).

[5] See Chapter 8 concerning unsurveyed lands for insight as to where an actual resurvey may be necessary.

[6] The researcher must take great care when studying or relying on court decisions, as most of the time a decision will state "resurvey" when it is actually discussing a retracement survey. Judge Cowart makes that plain in the *Rivers* decision. Judge Lamar elaborates on resurveying a township in *Cragin v. Powell*, which is the leading authority on the subject, and quoted extensively.

[7] The determination of the best available evidence of the original survey involves consideration of both direct and collateral evidence. Direct evidence from the record of the original survey should lead to the adoption of certain points as existent corners, while both direct and collateral evidence may lead to the adoption of other points as obliterated corners.

the positions of the original corners. Section 5.1, *Manual of Surveying Instructions, U.S.D.I., B.L.M., 2009.*

An Independent Resurvey is a retracement and reestablishment in reliance on evidence of the original survey in order to give official recognition and respect to all alienated lands within its scope, and where applicable, it also includes an establishment of new section lines, and often new township lines, independent of and without reference to the corners of the original survey. *Manual of Surveying Instructions, U.S.D.I., B.L.M., 2009.*

## TRACT SEGREGATION: TRACT SURVEY AND DESCRIPTION OF ALIENATED LANDS

Section 5.62 of the Manual begins the description of alienated lands that will be dependently resurveyed. The special instructions must designate the sections. It states, "where there is acceptable evidence of the original survey, the identification of the areas that have been disposed of must be the same as would ordinarily be derived by the regular subdivision of the section. Areas to be segregated by survey and described as tracts are (1) those areas that cannot be so identified, nor confirmed satisfactorily, (2) those areas where correction of conveyance document appears not to be an available remedy, and (3) those areas where the disposals are found to be in conflict by overlap. Every corner of these tracts is marked by angle-point monumentation, and a tie is made from each tract to a corner of the resurvey."

The succeeding sections elaborate on the details for the subsequent survey procedures by discussing various aspects of the process. The Manual stresses that before tract segregations are made and before new section lines are run, it is necessary to insure that the discrepancies are such that no adequate or satisfactory basis can be shown for the restoration of the former section line boundaries as a whole. The plan of the independent resurvey will be such that no lines, monuments, or plat representations will duplicate the description of any previous section where disposals have been made. With the filing of the resurvey, the record-field notes, and plat representing the prior survey are canceled, and must not be used for any future disposals, leases, and other land or resource transactions.

Section 5.65 details the proper procedures to be followed in accomplishing the resurvey process. Section 5.66 et seq. is devoted to the projection of new lines to complete the final work.

## CONCERNING RESURVEYS

In all cases, bona fide rights of purchasers and landowners must be honored and preserved whenever any type of survey is performed. Chapter VI of the Manual is devoted to this protection. In addition, the following two decisions address this concept in detail.

*Cragin v. Powell*[8] is a United States Supreme Court decision written by Justice Lamar that addresses *resurvey v. retracement survey*. In this case, the surveyor attempted to conduct a resurvey to correct "mistakes" he encountered. However, the court ultimately stated that lines once established by proper authority may not be altered, except by proper authority and then only where private rights are not interfered with. Since the U.S. Congress set the original boundaries, only the U.S. Congress may change them, not a surveyor and not a court of law.

The 1878 Michigan case of *Diehl v. Zanger,*[9] reported the same conclusion in the private sector. Its discussion included, "This litigation grows out of a new survey recently made by the city surveyor. This officer, after searching for the original stakes and finding none, has proceeded to take measurements according to the original plat, and to drive stakes of his own. According to this

[8] 128 U.S. 691, 9 S.Ct. 203, 32 L.Ed. 566 (La., 1888).
[9] 39 Mich. 601 (1878).

survey the practical location of the whole plat is wrong, and all the lines should be moved between four and five feet to the east. The surveyor testifies with positiveness and apparently without the least hesitation that 'the fences and buildings on all the lots are not correctly located' and there is of course an opportunity for forty-eight suits at law and probably many more than that."

When an office proposes thus dogmatically to unsettle the landmarks of a whole community, it becomes of the highest importance to know what has been the basis of his opinion. The record in this case fails to give any explanation, but the reasonable inference is that the surveyor has reached his conclusion by first satisfying himself what was the initial point of Mr. Campau's survey, and then proceeding to survey out the plat anew with that as his starting point. Of course by this method if no mistake is made, there is no difficulty in ascertaining with positive certainty where, according to Mr. Campau's plat, the original street and lot lines ought to have been located; and apparently the surveyor has assumed that that was all he had to do.

Nothing is better understood than that few of our early plats will stand the test of a careful and accurate survey without disclosing errors. This is as true of the government surveys as of any others, and if all the lines were now subject to correction on new surveys, the confusion of lines and titles that would follow would cause consternation in many communities. Indeed the mischiefs that must follow would be simply incalculable, and the visitation of the surveyor might well be set down as a great public calamity.

But no law can sanction this course. The surveyor has mistaken entirely the point to which his attention should have been directed. The question is not how an entirely accurate survey would locate these lot, but how the original stakes located them.

No rule in real estate law is more inflexible than that monuments control course and distance,—a rule that we have frequent occasion to apply in the case of public surveys, where its propriety, justice and necessity are never questioned. But its application in other cases is quite as proper, and quite as necessary to the protection of substantial rights.

A lesson may be taken from this decision, in that too frequently a surveyor will: (1) rely solely on the plat with its inherent errors as being the best evidence, or (2) assume they have the authority to "fix" or "correct" the believed inadequacies of the original work. As has been stated numerous times, in this text and throughout court decisions, the only authority the following surveyor has is to locate the original work, as Judge Cowart stated in *Tyson v. Edwards*,[10] "right, wrong, good, or bad."

## Retracement Survey[11]

A survey that is made to ascertain the direction and length of lines, and to identify the monuments and other marks of an established prior survey. *Manual of Surveying Instructions, U.S.D.I., B.L.M., 2009.*

## Local Survey

Any survey, retracement, or remonumentation of township, section, subdivision-of-section, or special survey lines that is not an official survey. *Manual of Surveying Instructions, U.S.D.I., B.L.M., 2009.*

---

[10] 433 So.2d 549; Fla.App. 5 Dist., 1983.
[11] See Wilson, Donald A. *Boundary Retracement: Processes and Procedures* for an extensive treatment of retracement surveys and their accompanying supporting legal principles. CRC Press, 2017.

## Independent Survey

This survey does not appear to have been made from any known government corner and is called by one of the witnesses "an independent survey" – that is to say a survey not dependent upon any known government corner. *Bentley v. Jenne*, 33 Wyo. 1, 236 P. 509 (Wyo., 1925).

We have already mentioned the fact that subsequent to the commencement of this action, a new survey was made by the government of the United States, whereby Township 57, Range 74, was divided into lots. This survey does not appear to have been made from any known government corner and is called by one of the witnesses "an independent survey" – that is to say a survey not dependent upon any known government corner. Respondents filed a supplemental petition setting forth the making of this survey. A demurrer was filed to such supplemental petition and overruled. Upon the trial of the case all evidence in connection with this resurvey was objected to, but the objections were overruled. Counsel for appellant have devoted a great deal of their argument to this subject, but we fail to see the importance thereof. We concede that a resurvey cannot disturb the title which parties have acquired up to the time that it is made. We do not, however, understand that respondents claim any rights under any new survey. The only purpose, apparently, of introducing any evidence in regard thereto, was to settle all future questions as to the actual location and description of the Hamilton land, and to locate and describe it in accordance with the new survey. It would seem that the description of the land according to the new survey, simply gives the land as described in the patent a new name; instead of calling it "X," it is now called "Y." We can see no possible harm in doing so.

## Indefinite Survey

A survey which describes one of the lines as running to a stake on line of another parcel is not invalid for indefiniteness, although there were sever parcels which fitted description of one referred to, but it must be ascertained from all circumstances which parcel was intended.

That there are two things which answer to a call of a survey, and it cannot be determined definitely which is the one called for, does not invalidate the survey, the rule being that the one which is most against the survey is to be taken.[12]

## No Survey

In some instances, there is no "original survey." For example, the designation in a treaty of a line of latitude incorporates an existing line as part of its description. That line has previously been created, and when "surveyed," or located and perhaps monumented, the survey becomes the "first survey," in that it is the first time the line has actually been "surveyed" (located, measured, and possibly monumented). Other examples are protracted lines, lines of agreement, Congressional lines in the rectangular system (PLSS) along with other designated land parcels, and miscellaneous pre-designated lines that are incorporated in title and boundary descriptions. Eventually, many of these, some out of necessity, are surveyed.

A protracted line is not a surveyed line. Early grants were common where part of the tract was located on the ground, the rest being protracted on paper. Each is identified, and retraced, by its own rules. It is important to remember that protracted lines are not part of the survey,[13] and

---

[12] See *Rowe v. Kidd*, D.C. Ky, 249 F. 882, affirmed 259 F. 127, 170 C.C.A. 195; C.J.S. §49a. p. 599.

[13] Protraction is the method whereby boundary lines are created without being run on the ground. They may be drawn on a plat, or merely written out in a land description. See *Ralston v. M'Clurg*, 39 Ky. 338 (Ky.App. 1840).

Also see Wilson, Donald A. *Boundary Retracement: Processes and Procedures* (CRC Press), Chapter 4, for an extensive treatment of protracted boundaries.

therefore should not be treated as surveyed lines. Obviously, that means that most of the traditional rules do not apply.

A recent decision where one such situation was treated by the Montana court may be found in *Larsen v. Richardson*.[14] Two surveyors had different opinions as to the location of the boundary line between their respective clients. Both testified that they used the so-called "bibles" of surveying and both understood that it is a retracement surveyor's duty to "follow the footsteps of the original surveyor." However, in this instance, there was no original survey to follow. The court concluded that they would follow the words of the original deed writers. The intent of the parties was put into words, the transaction completed, and the title established – without a survey.

There does not have to be a survey to create a title; and by definition, as demonstrated in previous discussions, where there is a title (to something) there is necessarily a location and a confinement, i.e., boundaries (of some sort). Therefore, once a title has been created, one or more boundaries have been established.

## THE FLAWED SURVEY

The Tennessee case of *Dillehay v. Gibbs*[15] provides insight from the court system when it has less than adequate evidence to work with. The opening words of the Appeals Court were, "This case essentially boils down to which of two flawed surveys the trial court most credited – Surveyor A's, cobbled together from ancient deeds with little apparent connection to the land and a disclaimer as to its veracity, or Surveyor B's, shot from the ground on an old fence with slight support from the underlying deeds." However, the bare facts in this case present a result that should have resulted in an obvious, well-founded conclusion. Since the court did not have that information, it is not part of the decision.

The two current descriptions of the parties are merely by abutting calls with acreage estimates provided. Tracing the two descriptions back to 1878 results in two parcels created from a common source tract, divided by a line described by metes and bounds calls for monumentation that was identifiable at the time of the litigation.

> The evidence on either side was problematic and not particularly compelling. All three surveyors were reluctant to establish an exact boundary line and noted the inherent difficulties in doing so based on boundary deeds. Surveyor A used the old Boze and Richardson deeds to draw a line. However, he refused to call his line the boundary line. Moreover, he never shot his line from the ground, and his straight, compass-point lines do not appear to match the contours of the properties. Plaintiff's second surveyor, Surveyor C, although his line more closely followed the natural contours of the land, testified that he did not survey the line himself but rather prepared a trial exhibit showing where everyone else had purported the line to be. [Surveyor B's] line appears to be based largely on the location of the barbed-wire fence.[16]

> This case essentially boils down to which of two flawed surveys the trial court most credited – Surveyor A's, cobbled together from ancient deeds with little apparent connection to the land and a disclaimer as to its veracity, or Surveyor B's, shot from the ground on an old fence with slight support from the underlying deeds. It appears from the record that Surveyor B was the only surveyor to establish a boundary line within a reasonable degree of surveying certainty. He provided detailed reasons supporting his decision and extensive critiques of the other surveyors' methods. In his expert opinion, Surveyor B believed that the location of the barbed-wire fence on a steep, wooded slope indicated that it was placed there to serve as a boundary line. He testified that the other fences in the

---

[14] 361 Mont. 344, 260 P.3d 103 (2011).

[15] No. M2010-01750-COA-R3-CV; Court of Appeals of Tennessee, Nashville; June 16, 2011.

[16] Surveyor B testified that he did not know who erected the fence, nor what its purpose was.

disputed area, including the woven-wire fence upon which the A and C lines relied, were in locations indicating their service as containment fences for farm animals.

However, examining the deeds in the chain of title brings us back in time to the division deed in 1878 where in the dividing line between the two farms (Dillehay & Gibbs properties) originated, or was established (created) for the first time. Its description was stated as follows:

> containing by survey One hundred and nineteen acres, 3 R., 34 poles to be the same more or less and bounded as follows to wit, Beginning on a chestnut on top of the ridge Smiths and Climen's corner running thence north 120 poles to a beech thence East 53 poles to a stake thence South 56 poles to a stake on the East Side of a branch thence East 36 poles to an oak Stump thence South 57 poles to a Sourwood thence East 26 poles to a Stake thence South 45 poles to a beech on the South Side of the branch in the head of a hollow thence down the hollow with its meanders west 39-1/4 poles to a rock on the west side of the branch thence North 5° west 6 poles near a rock Spring thence north 78° west 26 poles to a Sugar tree thence with the right hand brink of the hill with a marked line Northwardly in all 130 poles to a Stake thence North 82° west 5 poles to a Stake Thence South 39-1/4° west 11 poles to the Beginning.

The description was created by survey, and while the perimeter of the source tract was created and described some years prior, the issue in the case was only the dividing line between the two farms, and could be placed on the ground relying on the calls in the description. The fact is, in 2011, most, if not all of the called-for monumentation was still in existence. The ultimate problem with the survey(s) was due to the inability of the surveyors involved to find and locate that line, as reflected by their testimony and the decision from the Court of Appeals.

In using the fence, the Gibbs property comes out 20 acres over the specified acreage, while the Dillehay property comes out 20 acres under its specified acreage.

## NO SURVEY

There may not be a survey simply because one was never done, or else some surveying work was done, but it either did not meet the requirements of a legal survey, or else is deficient, substandard, or unlawful[17] in some fashion that it does not legally qualify as being a bona fide survey. As such, it begs the question whether it has any value, or should not be considered to control any title elements or survey considerations. As one individual stated, "it is nothing more than an attempt at a survey that failed."[18]

## FRAUDULENT SURVEYS

### What Is a Fraudulent Survey?

Obviously one that was not done in accordance with law, i.e., by a non-licensed individual. In fact, the result is probably not a survey at all. But also a surveyor, who does not follow the basic rules of law. Certainly there are matters of judgement, and being human, we are all subject to making mistakes. However, violations of ethics, ignoring standards and legal requirements, are inexcusable, and subject to sanction. Others are blatant violations, knowingly and purposely, for personal gain. Those border on criminal activity.

---

[17] The obvious situation is where work was done by an unlicensed individual, yet relied on by clients and other parties. In the extreme, it is part of a transaction instrument, perhaps even on record, giving rise to a situation that may void either the transaction or the title.

[18] Attributed to George F. Butts, LS, former Staff Surveyor, Green Mountain National Forest.

Fraudulent surveys, at times and locations, were rampant in some parts of the of the PLSS. However, fraudulent surveys may appear anywhere. Two very real examples can be common. First, in many of the early grants, some of which were very large, it was common to protract many of the lines never locating them on the ground. This was also not uncommon within eastern, particularly New England, townships where a grant perimeter was surveyed, then subdivided on paper in some fashion. Second, modern day practices sometimes took advantage of measurements and error theory to "fudge" figures, usually to somehow derive an extra land parcel, large or small, where in reality that amount of land did not physically exist even though survey results attempt to convince that it did.

The following are some well-documented examples in the PLSS.

## DEFINITION

When part of the public domain (PLSS) the term "fraudulent survey" has been defined by the Commissioner of the General Land Office.

ANNUAL INSTRUCTIONS TO SURVEYORS GENERAL OF THE UNITED STATES DEPARTMENT OF THE INTERIOR, GENERAL LAND OFFICE

**Washington, June 26, 1880**

- - The laws regulating the surveys of public lands assign to each subdivisional landmark a particular position with relation to township boundaries, each landmark, properly located, constituting a corner monument to four principal subdivisions of two or more sections. The manual of 'Instructions to Surveyors General,' with supplements, fully describe the means and methods to be employed by the deputy surveyor in determining the true and legal positions of the land corners. The manual - - - makes ample provision for seemingly unavoidable deviations from true lines. When it is found that the limits thus provided in the manual have been exceeded by the deputy surveyor, the original survey so made should be treated as fraudulent, the returns of the same should be canceled, and the deputy, in case such result arises solely from lack of skill on his part, should be barred from further contracts. Where such erroneous surveys are attributable to willful negligence upon the part of the deputy, the act is fraudulent, and approval of the returns must not only be refused by you, but the particulars of the case must be promptly forwarded to this office for such action as may be proper under existing laws.

In view of the serious results which usually attend the acceptance of erroneous or fraudulent surveys, inflicting injury upon one or other of the contracting parties, or upon the occupants of lands so surveyed, I shall feel obliged to hold you to strict account for the manner in which you exercise your knowledge and judgement of the experience, ability, integrity, and habits of the several deputies to whom you may intrust the work of prosecuting the public land surveys of your district.

## THE BENSON SYNDICATE

Widely known in the western United States, this group was what has been called an organized crime organization which received contracts from the General Land Office to perform land surveys of the public lands.

The organization was led by, as well as named after, John A. Benson (1845–1910), who had been a school teacher and a county surveyor, later becoming a reputable deputy surveyor, Mineral Surveyor and Civil Engineer. The sphere of operation was from the Rocky Mountain to the Pacific Ocean, being most active in California and headquartered in San Francisco. It operated from 1875 to 1898, peaking from 1883 to 1886. Just in California, at least 40 individuals were

known to be involved. The operation generated false demand for public land surveys using fictitious land patent applications, followed by contracting with the GLO (General Land Office) for
surveying the lands. Those surveys were then fraudulently executed, and were known to be substandard, incomplete, or outright fictitious. They were made under contract to individual deputy
surveyors, some of whom were unaware that their names were involved, as they had been tricked
into signing blank papers which later became contracts and other legal documents without their
knowledge.

In addition to this fraud, the Syndicate had people with either minimal experience or no
experience, or lacking proper qualifications as deputy surveys, performing the work with the
contracting surveyor being physically present. This practice was clearly unlawful. Commonly, an
area would only be surveyed to the extent necessary to create reasonable, but fictional, plats and
field notes for the remainder of the area. At other times, entire contracted area, often consisting of
several townships, were fabricated by syndicate members at the San Francisco office, with little
or no actual work on the ground.

The organization eventually infiltrated into high levels of the government, and syndicate members in influential positions along with members of congress made the group's schemes possible.
In California, at least two Surveyors General in the 1880s approved numerous fraudulent survey
results and approved requests for government payment that were 200–700% of the originally
estimated survey cost, which the government paid.

The Report for the year 1887 from the Commissioner of the General Land Office stated, "In
April last the United States grand jury at San Francisco returned forty one indictments for perjury and conspiracy in connection with fraudulent surveys of the public lands. The operations of
this syndicate were not confined to California, but extended to the States of Nevada, Oregon and
Colorado, and the Territories of Arizona, New Mexico, Idaho, Montana, Utah, and Washington.
The principal portion of the contracts were let in the names of thirty-four alleged deputy surveyors, of whom three were 'dummies', two were intimately connected with the surveyor-general
who approved the bulk of the contracts, and the remainder were relatives, partners, associates,
or employees of the head of the syndicate. There was paid to the order or purported order, of
these thirty-four pretended deputy surveyors upwards of $1,000,000, all of which went into the
treasury of the syndicate."

It was the completely implausible survey results, along with sworn testimony of disenchanted
employees or associates, that led to the recognition of the widespread fraud of Benson's group.
Beginning about 1886, contracts held by certain surveyors thought to be aligned with Benson
were not paid by the government, leading to various lawsuits. In 1887, 41 federal indictments
for conspiracy and perjury were brought against Benson and several others. Trials did not
occur until 1892, and when they did all were found not guilty based on legal technicalities.
Because the word was out, and surveyors had difficulty getting work, Benson proposed what
became known as the "Benson Compromise" in 1895, where the proposal, to the California
Surveyor General, stated that the survey work would be corrected or completed on several
contracts. The compromise was accepted by the government, but little of the work was ever
reported to have been accepted as valid.

In California, government estimates indicate that nearly a thousand townships might have
been involved, comprising an area of approximately 20 million acres, or about 20% of all land
in the state.[19] Special agents were sent by the government to investigate fraudulent practices. By
1887, their work had been so thorough that the unlawful practices of the Benson Syndicate were
effectively terminated.

---

[19] Uzes, Francois D. *Chaining the Land*. Sacramento: Landmark Enterprises. 1977.

## The Oregon Land Fraud Scandal

Taking place in the early 20th century, U.S. government land grants in Oregon were illegally obtained with the assistance of public officials. Most of Oregon's U.S. congressional delegation received indictments in the case against them.

In 1870, the Oregon and the U.S. government granted the Oregon and California Railroad 3.7 million acres of land to build a line, from Portland south to California. The land, granted in a checkerboard pattern along both side of the railroad's right-of-way, was then sold to settlers in parcels of 160 acres at the low price of $2.50 per acre to encourage people to settle along the line, fostering development accepted by the government, but little of the work was ever reported to have been accepted as valid.

Much of the land being unfit for development, few settlers took advantage, but since it was covered with prime timber, timber companies were ready buyers at higher prices. In order to bypass the requirements of the original grant, the president of the Southern Pacific Railroad, the then owner of the Oregon and California Railroad, hired former surveyor Stephen A. Douglas Puter to round up people from saloons in Portland's waterfront district, take them to the land office, have them register a parcel as a settler, then transfer it to Puter's people. The accumulated parcels were then sold in large blocks to the highest bidder in the timber industry. Harriman, the president, had a dispute with Puter and fired him. When a lumber company bookkeeper exposed the scheme, Puter turned on his former boss and, testifying against him, wrote a scathing exposé about the scheme while in prison, which became the first 25 chapters in *Looters of the Public Domain*.

In the beginning of the lawsuits, more than 1,000 indictments were issued, but the U.S. Attorney narrowed the list to the 35 most egregious offenders, including one U.S. senator and two representatives.

## The Yazoo Land Scandal

After the Revolutionary War, Georgia, like most states, had a huge debt due to the commitments made toward the war efforts. At this time, Georgia claimed all land west of the state to the Mississippi River, which then included what is now Alabama, Mississippi, part of Florida's panhandle, and part of Louisiana. The governor at the time, George Mathews, along with the Georgia General Assembly, created a massive real estate fraud scheme, wherein Georgia politicians sold large tracts in what was known as the Yazoo lands, to political insiders at very low prices.

The first attempt for Georgia to settle was a 1784 proposal to establish Houstoun County in the Muscle Shoals area. This never happened because the major proponents instead became interested in an effort to establish the State of Franklin in present-day eastern Tennessee.

Next, in 1785, Governor George Mathews endorsed the Bourbon County Act, organizing Bourbon County, Georgia in the area east of the Mississippi and south of the Yazoo River. The area included Natchez and was in the area claimed by Spain. Because of pressure from the federal government due to the Spanish claim and because of unresolved claims by Native American tribes, Georgia dissolved Bourbon County in 1788.

Then, in 1789, the first Yazoo Act gave the land to three companies formed as a secret society, The South Carolina Yazoo Company, The Virginia Yazoo Company, and the Tennessee Company. In this transaction, the Governor at the time signed a deal to sell 20,000,000 acres to the companies. The deal fell through when the state demanded payment in gold or silver instead of currency. Although the law enabling the sales was overturned by reformers, its ability to do so was challenged in the courts, the case making its way to the U.S. Supreme Court. In the landmark decision of *Fletcher v. Peck*[20] in 1810, the Court ruled that the contracts were binding and the

---

[20] 10 U.S. 87, 3 L.Ed. 162 (1810).

state could not retroactively invalidate the earlier land sales. In 1802, because of the controversy and the litigation, Georgia ceded all of its claims to lands west of its current border to the U.S. government. The claims involving the land purchases were not completely resolved until legislation was passed in 1814, establishing a claims-resolution fund.

In 1794 the Yazoo Act resurfaced with many of the same participants. Four companies, the Georgia Company, the Georgia-Mississippi Company, the Upper Mississippi Company, and the new Tennessee Company would buy the land from Georgia, then resell it to investors. The companies persuaded the Georgia state assembly to sell more than 40,000,000 acres. A number of Georgia politicians received bribes in exchange for votes to pass the act. One, in opposition to the act, resigned his Senate seat and returned to Georgia. He was elected by the Georgia house and worked to dismantle the act. He led the passing of the Rescinding Act, which nullified the Yazoo Land Act. Subsequent hearings implicated many of the elected officials involved in the case. Eventually, Georgia and the U.S. government came to an agreement, whereby the claims against Georgia would be discontinued in exchange for assistance in removing the Creek and Cherokee from western Georgia.

## PINE BARRENS SPECULATION

At the same time, and often confused with the Yazoo scandal, was the Pine Barrens speculation. From 1789 to 1796, Georgia governors, while in office, made gifts of land grants covering more than three times as much land as Georgia then contained. In all they conveyed 29,097,866 acres in counties that contained only 8,717,960 acres.

In Montgomery County alone, with an area of 407,680 acres, three men received land grants totaling 2,664,000 acres. All the grants in Montgomery County totaled 7,436,995 acres.

## CONCERN FOR THE SURVEYOR

Surveyors may encounter situations like the foregoing, and knowing the history of fraudulent surveys, and the areas in which they occur, one can be on guard for such anomalies. The resolution of such problems depends on the nature of each and is best left to the court system since land titles are likely affected. Maps exist, some of which are a matter of public record, others may be found in archives and historical collections. Field notes exist and will be uncovered. Markers will be found in the ground, with confusing relationship to titles and descriptions, or for which no explanation whatever is available.

**Author's Note:** While little may be found regarding the ultimate result of a fraudulent survey, some things may be contemplated. One might be, does it do anything at all, besides confound a person's regular procedures, or cause concern for a duped buyer? More importantly, if it is indeed not a survey, then it could not have boundaries. The immediate thought is that even if difficult, ambiguous or questionable, is that there are boundaries, whether described or just nebulous, but boundaries exist nevertheless. Surveyors deal with such titles daily, but yet with ability and perseverance, aided by the court system and rules based in law, overcome these obstacles, and prove that seemingly bogus information is, in fact, adequate, and vice-versa. Returning to the title itself, the rules are clear even though the supporting information is scant: without a bona fide title, there can be no boundaries. If a title does not exist, neither can its elements.

## ERRONEOUS SURVEY

All measurements are prone to error, therefore it is more than safe to assume that surveys and surveyed lines contain error, not all of which may be identified, or compensated for. Any known closure error may be adjusted by some selected standard rule, depending on the character of the work. There are legitimate errors due to a variety of human and natural influences, and there are also errors due to carelessness of observers,[21] sometimes classed as blunders. Erroneous surveys are different from fraudulent surveys, although they may appear to be the same due to the similarity of results and remaining evidence. However, it is important to make the appropriate identification, as fraudulent surveys may be subject to correction or replacement, while erroneous surveys are not likely to be.

## OVERVIEW OF EARLY SURVEYS IN HAWAII

Equally applicable to early colonial surveys reported by West Virginia and Kentucky courts as well as others, Alexander[22] reported on erroneous, or perhaps sloppy, early surveys of lands in Hawaii.

## THE GREAT MAHELE OF 1848

As early as 1842, government lands began to be set apart, and a treasury board was appointed to administer those lands in which it had an undivided interest and undefined claim in all land in the Kingdom until the *Mahele*. The great mass of lands consisted of those surrendered and made over to the government by King Kamehameha III, enumerated by name in the Act of 1848. A Land Office was created in 1846, authorizing the minister of the interior to sell or lease government land. From 1850 to 1860, nearly all desirable government land was sold, generally to natives, and surveyed at the expense of the purchaser.

Thus, the traditional system of land tenure in Hawaii ended in 1848, as King Kamehameha III was persuaded by foreigners to institute the Great *Mahele* (division), allowing lands to be bought and sold.[23]

According to Alexander (1882), the early surveys, those made under the Land Commission (est. 1862), commonly known as *kuleana* surveys, had the same defects as the first surveys in most new countries. There was a lack of proper supervision, no Bureau of Surveying, and the president of the Land Commission was overwhelmed with work such that there was no time for supervision. In addition, there was little money to pay for the work and little time to accomplish it.

No uniform rules or instructions were given to the surveyors, who were "practically irresponsible." Few of them were regarded as thoroughly competent surveyors, while some were not only incompetent but "careless and unscrupulous." The result was that almost every possible method of measurement was adopted. Some used 50-foot chains, while others used 4 pole (66-foot) chains divided into links; some attempted to survey by the true meridian, others by the average magnetic meridian, while most made no allowance for local variations of the needle. Some recorded surveys have been found to have been made with a ship's compass and even a pocket compass. Few made the effort to mark corners or to note topographic features.

In addition, rarely was one section or district assigned to one surveyor. It has been reported that over a dozen were employed in the surveying of Waikiki, for instance, not one of whom knew what the other surveyors had done, or made any attempt to make his surveys agree with theirs where they adjoined one another. As one might expect, overlaps and gaps became rule rather than the exception, so that it became generally impossible to put these old surveys together correctly

---

[21] *Stafford v. King*, 30 Tex. 257, 94 Am. Dec. 304 (1867).
[22] Alexander, William DeWitt (1882).
[23] Ibid.

on paper without determining their true relative locations through actual measurements on the ground.

Consequently, the land grant and tenure systems of what is now the State of Hawaii is a mixture of conveyances and processes over a period of time. Despite the plan to confirm titles and organize lands, the net result is a variety of categories of land tenure and land records. While many surveys were performed, and are available, they often do not provide the most desirable definition of claims.[24]

---

[24] Cole, George M. & Donald A. Wilson. Land Tenure, Boundary Surveys, and Cadastral Systems (2017).

# Section II

Original Survey

# 4 The Original Survey

Learn the duty as well as taste the pleasure of original work.

**Robert James Graves**

## CREATING THE FOOTSTEPS TO BE FOLLOWED

Once again, to emphasize, *an original survey creates boundary lines; it does not ascertain them.*[1] Of all the different types of surveys, besides an original survey, only a resurvey can create a boundary, and only under certain specified circumstances. However, it is not quite that simple, in that, with a resurvey, a boundary that had once been created but was improperly surveyed had to be corrected. It is more about a survey process than it is about a boundary creation. As stated in the previous chapter, *Cragin v. Powell*[2] analyzes the concept nicely. Since a title boundary is not created until title is transferred by the appropriate parties, it does not become a true entity without a confirmation. Mere lines on paper and words in a document are only *proposed* boundaries, awaiting the final step that gives them status. In other words, a specified required process must be finalized.

In the case of a resurvey, substituting an entirely new survey for the original survey when the latter is inadequate, or grossly in error, the first consideration is whether any title has transferred, thereby creating rights in private parties. The court, Justice Lamar, was clear and detailed about the fact that a new survey to replace a previous survey cannot be done if it would disturb the rights of others created under the original survey. No entity but the United States Congress can change an original survey (in the PLSS). The same concept applies to non-GLO (General Land Office) surveys as well, in that once rights have been created from original boundaries created, they may only be changed with the involvement of all parties affected.

## THE SURVEY

Probably the best designation of "survey" is found in the Texas case of *Outlaw v. Gulf Oil Corporation*,[3] wherein it is stated, "survey is the substance and consists of the actual acts of the surveyor." Since some confuse other aspects of the procedure as the survey, or a part thereof, the court went on to elaborate: "a 'map' is a picture of a survey,[4] 'field notes' constitute a written description thereof, and the 'survey' is the substance and consists of the actual acts of the surveyor, and, if existing actual monuments are on the ground evidencing such acts, such monuments control because they are the best evidence of what the surveyor actually did in making the survey and are part at least of what the surveyor did."

---

[1] *Robinson v. Forrest,* 29 Cal. 317, 325; *Sawyer v. Gray,* 205 F. 160, 163.

[2] 128 U.S. 691, 9 S.Ct. 203, 32 L.Ed. 566 (La., 1888).

[3] 137 S.W.2d 787 (Tex.Civ.App., 1940).

[4] This has often been emphasized in court decisions, and is frequently misunderstood by the average person. The common misconception is that people believe that a piece of paper (the plat) is their survey, when it is actually not, and sometimes far from it.

DOI: 10.1201/9781003032557-6

Interestingly, this explanation began much earlier, as found in the Maine case of *Bean v. Batchelder*,[5] wherein the court gave instructions to the jury that the lines run by the surveyor on the surface of the earth, as and for the boundaries of the lot, would still be boundaries of that lot, if their locality could be found. "The plan was merely a picture. The survey was the substance. The plan was not made to show where the lots were to be hereafter located, or how they were to be hereafter bounded. It was made as evidence of where they had before been located and bounded. The lot actually surveyed, bounded by the lines actually run, was the lot intended to be conveyed. The plan was named in the deed, rather as a picture, indicating the location and lines of the lot. Still the actual boundaries, rather than the pictured boundaries, were to be sought for. The picture might not be wholly accurate."

Judge Joe A. Cowart, Jr. added a philosophical perspective to this concept in the Florida case of *Tyson v. Edwards*,[6] wherein part of a dispute was whether the plat or the monumentation on the ground should govern lot location. The court brilliantly stated that reasoning "lies in the historic development of the concept of land boundaries and of the profession of surveying. Man set monuments as landmarks before he invented paper and even today the true survey is what the original surveyor did on the ground by way of fixing boundaries by setting monuments and running lines ('metes and bounds'), and the paper 'survey' or plat of survey is intended only as a map of what is on the ground."

## WHAT IS AN ORIGINAL SURVEY?

A definition of an original survey is "A cadastral survey which creates land boundaries and marks them for the first time." *Glossaries of BLM Surveying and Mapping Terms* [Prepared by the Cadastral Survey Training Staff, Denver Service Center. USDI-BLM (1980)].

The term "original survey" refers to the official government survey performed under the laws of the federal government by its official agency. *Cox v. Hart*, 260 U.S. 427, 43 S.Ct. 154, 67 L.Ed. 332 (1922); *Block v. Howell*, 346 N.W.2d 441 (S.D., 1984). A subsequent survey by a private individual or non-government entity is more accurately described as a retracing or resurvey. *Block*, 346 N.W.2d 441; *Randall v. Burk Tp.*, 4 S.D. 337, 57 N.W. 4 (1893).

An examination of *Titus v. Chapman*[7] summarizes the principles involved:

Government surveys, not surveys conducted by private individuals, create, rather than merely identify, boundaries." *Cox v. Hart*, 260 U.S. 427, 43 S.Ct. 154, 67 L.Ed. 332, (1922). The term "original survey" refers to the official government survey performed under the laws of the federal government by its official agency. *See Id.; Block v. Howell*, 346 N.W.2d 441, (S.D.1984); Walter G. Robillard & Lane J. Bouman, *Clark on Surveying and Boundaries* § 4.12 (5th ed 1976).

[18.] A subsequent survey by a private individual or non-government entity is more accurately described as a retracing or resurvey. *Block,* 346 N.W.2d at 444; *Randall v. Burk Tp.,* 4 S.D. 337, 57 N.W. 4, (1893). In a retracing or resurvey, a surveyor must "take care to observe and follow the boundaries and monuments as run and marked by the original survey." *Block,* 346 N.W.2d at 444. Boundaries as established by original government surveys are unchangeable and must control disputes. *Christianson v. Daneville Tp.,* 61 S.D. 55, 58, 246 N.W. 101, (1932).

[19.] Original monuments, those located by the original surveyor, mark true corners. *Lawson v. Viola Tp.,* 50 S.D. 555, 210 N.W. 979, (1926). "Where the location of the original monument can be found, or can be established by evidence, such location shall be held to be the true corner, regardless

---

[5] 78 Me. 184, 3 A. 279, Me., 1886.
[6] 433 So.2d 549, Fla.App. 5 Dist. 1983.
[7] 2004 SD 106, 687 N.W.2d 918 (S.D. 2004).

of the fact that resurveys may show that it should have been located elsewhere." *Id.* (citing *Byrne v. McKeachie,* 34 S.D. 589, 149 N.W. 552 (1914); *Hoekman v. Iowa Civil Township,* 28 S.D. 206, 132 N.W. 1004 (1911); *Randall,* 4 S.D. 337, 57 N.W. 4; *Beardsley v. Crane,* 52 Minn. 537, 54 N.W. 740 (1893); *Ogilvie v. Copeland,* 145 Ill. 98, 33 N.E. 1085 (1893); *Nesselroad v. Parrish,* 59 Iowa 570, 13 N.W. 746 (1882). Where the original monument is obliterated, that is it cannot be located nor established by evidence, then a corner can be established by a new survey. *Lawson,* 50 S.D. at 558, 210 N.W. at 980 (citing *Randall,* 4 S.D. at 355, 57 N.W. at 10; *Washington Rock Co. v. Young,* 29 Utah 108, 80 P. 382 (1905)). Only upon obliteration of an original corner may a new survey be made from points that can be determined in accordance with the original surveyors field notes. *Id.* However, if the point at which an original monument was located can be ascertained by the court, the line as indicated by the government survey prevails. *Dowdle v. Cornue,* 9 S.D. 126, 68 N.W. 194 (1896).

SDCL 43-18-7 provides:

In retracing lines or making the survey the surveyor shall take care to observe and follow the boundaries and monuments as run and marked by the original survey, but shall not give undue weight to partial and doubtful evidence or appearances of monuments, the recognition of which shall require the presumption of marked errors in the original survey, and he shall note an exact description of such apparent monuments.

## HOW ARE ORIGINAL SURVEYS CREATED?

### The Colonial System

Some would characterize the so-called metes and bounds areas as an indiscriminate system, perhaps even haphazard. However, nothing could be further from the truth, since even though many land grants followed the topography, mostly deliberately, and perhaps had no specific order to the granting of tracts of land, systems were developed based on need. When one contemplates the early descriptions, calls were for known entities: lines of latitude, water bodies, particularly rivers, ridge and mountain crests, features that were tangible. Later, regular descriptions attempted to attach some order to the surveying process by using directions and distances as dividing lines to place straight lines on a rough and curved surface.

Longlots, townsites, rectangular, and other regular procedures were used depending on the design of the subdividers, the number of shares needed, the size and shape of the parcel being subdivided, as well as other considerations that arose from time to time. As examples, see Figures 4.1 and 4.2.

The fascinating part of the development of rectangular systems is that the underlying concept was a design of the Romans known as centuriation. Regular lines of division into squares and rectangles of varying sizes were laid out in orderly fashion, often based on the cardinal directions. Thomas Jefferson learned of this system while working in Europe and, adopting the concept, attempted to convince colonists in Virginia to lay out lands in such fashion. That failing, he tried to convince the powers in the Federal Government that an orderly system was to their advantage. Part of his plan was to adopt the metric system of measurement, which most were not comfortable with, having already developed numerous grants, descriptions and surveys based on rods and links[8] (which were not always consistent in themselves) and using the Gunter chain developed by a mathematician in 1620 which made computation of areas convenient and simple. With a few modifications, much of Jefferson's plan was adopted and the first layout, that of the Seven Ranges, began in 1785.

---

[8] As an example, when Tennessee separated from North Carolina, nearly the entire state was laid out in rectangular fashion. The system did not last long, as people reverted to the traditional metes and bounds system. A few areas persist today, described in rectangular fashion.

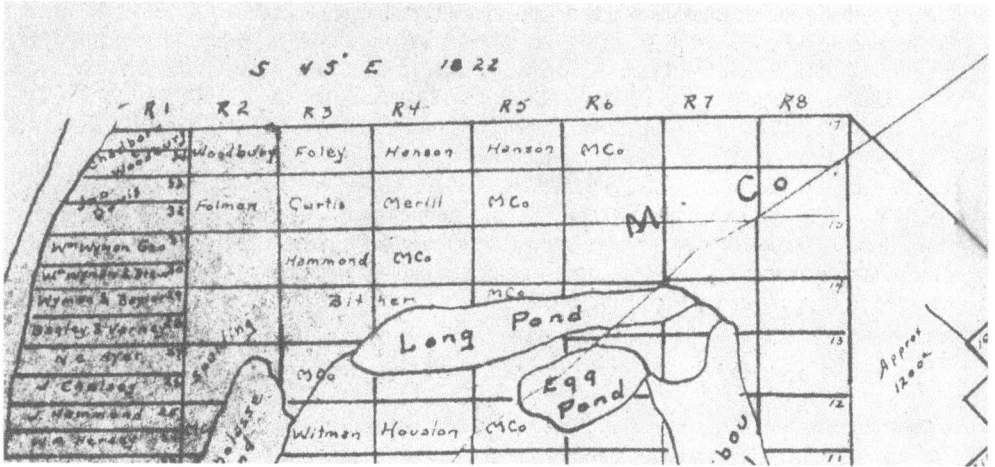

**FIGURE 4.1** Section of lotting plan of Lincoln, Maine. Longlots, designated as "river lots" on left side, rectangular lots and ranges on right side, each 100 acres, measuring 100 × 160 rods. There are two different surveying (cadastral) systems laid out in this town, possibly at two different times by two different surveyors. At the least, each has its own sets of errors.

**FIGURE 4.2** Township C in southwestern Maine, a nearly triangular township. Many lots are square, while the majority are what are termed fractional in the Public Land Survey System (PLSS).

Before 1785 when the PLSS began with its first surveys, there were no formal instructions for original surveys, other than word of mouth or a writing from a proprietorship, or a large land holder-claimant such as a king. Designs were created on a case-by-case basis, leaving one to conclude that the entire lotting system was actually a series of *special instructions* based on specific needs.

As an example, King George III administered Royal Commissions, Instructions, Laws and Orders to Benning Wentworth, newly appointed Governor of the Province of New Hampshire from 1741 to 1766.

Instructions to Benning Wentworth Esq$^r$ His Majesty's Governor and Commander in Chief in and over the Province of New Hampshire in New England in America[9]:

First. With these Instructions You will receive His Majesty's Commission under the Great Seal of Great Britain constituting you Governor and Commander in Chief in and over the Province of New Hampshire, within His Majesty's Dominion of New England in America; You are therefore to fit yourself with all convenient Speed, and to repair to the said Province of New Hampshire, and being arrived there, You are to take upon you the Execution of the Place and Trust His Maj$^{ty}$ has reposed in You and forthwith to call together the Members of His Majesty's Council in that Province Viz$^t$

38. Townships. And whereas it has been found by long Experience that the settling Planters in Townships hath redounded very much to their Advantage not only with respect to the assistance they have been able to afford each other in their civil Concerns; But likewise with regard to the Security they have thereby acquired against the Insults & Incursions of Neighboring Indians or other Enemies. His Majesty has therefore thought it for his Service that Townships should be settled on the Frontiers of your Province & that each Township may consist of about twenty thousand Acres of Land but not to exceed six miles square, & in each such Township a proper Place shall be laid out for the Seite of the Town itself, where any Planter besides fifty Acres of land for each Person in his Family shall have set out a Lot or Footland for a Town House, and that no Town be set out or any such Lands or Lots granted until there be fifty or more Families ready to begin the Settlement, & that so soon as any such Township has got one hundred or more Families settled therein it shall have & enjoy all the Immunities & Privileges as do of right belong to any other Parish or Township in the said Province.

**Author's Note:** As early as 1741, the Kingdom of Great Britain was thinking in terms of townships being 6 miles square. According to Truesdell, W.A., *The Rectangular System of Surveying* (1908), Bennington, Vermont (chartered 1749) was the first township to be laid out six miles square with boundaries north, south, east, and west.

40. Grants of Land. Whereas great Inconveniences have arisen in many of His Majesty's Colonies in American from the granting excessive Quantities of Land to particular Persons which they have cultivated, and have thereby prevented others more industrious from improving the same; To prevent the like Inconveniences in the Province under your Government in all Grants of Land to be made y you by & with the Advice & Consent of His Majesty's Council there; Your are to take especial Care that no Grants be made to any Person but in Proportion to his Ability to cultivate the same and that proper Clauses be inserted for vacating the said Grants on Failure of Cultivation or Payment of Quit-Rents reserved thereon. And you are hereby directed not to grant to any Person more than fifty Acres for every Man, Woman or Child of w$^{ch}$ the Grantee's Family shall consist at the time such Grant shall be made.

72. Harbors to be surveyed. You shall cause a Survey to be made of all the considerable landing Places & Harbours in the said Province and with the Advice of his Majewsty's Council there, erect

---

[9] Laws of New Hampshire. Volume Two: Province Period, 1702–1745.

in any of them such Fortifications as shall be necessary for the Security and Advantage of the said Province which shall be done at the publick Charge of the Country; And You are accordingly to move the General Assembly to the passing of such Acts as may be requisite for the carrying on of that Work in which his Majesty doubts not of their cheerful Concurrence from the common Security & Benefit they will receive thereby.

74. To send a map. You shall transmit to his Majesty & to His Commissioners for Trade and Plantations by the first Opportunity a Map with the exact Description of the whole Province of New Hampshire with the several Plantations upon it & of the Fortifications. See Figure 4.3.

## THE PUBLIC LAND SURVEY SYSTEM (PLSS)

Beginning in 1785, after several trials and false starts, Congress adopted a standard system for laying out public lands. Through the years, the instructions were changed to meet specific needs and refined to become very detailed in order to allow for every contingency, the latest detailed Manual of Instructions being the edition of 2009.

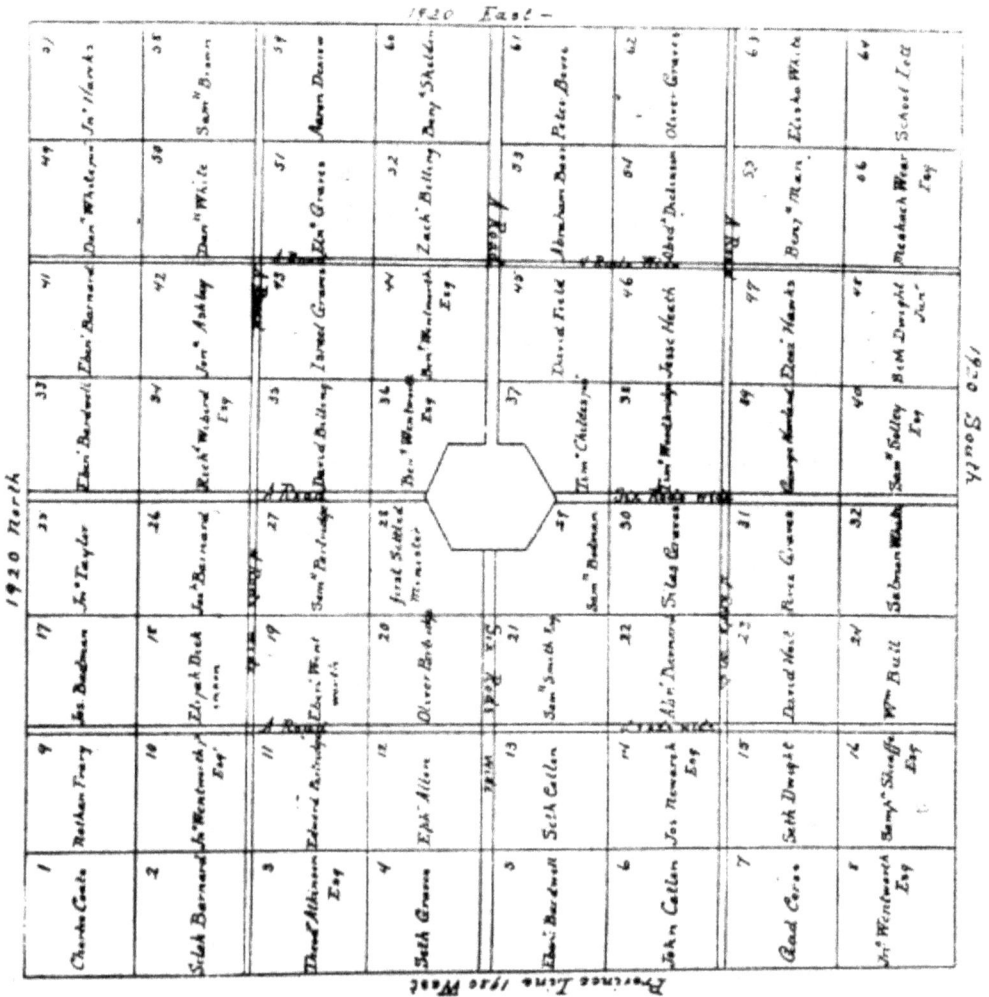

FIGURE 4.3  A governor Wentworth Township and its subdivision into lots.

## MANUAL OF INSTRUCTIONS

With the several trials, and subsequent modifications, by 1851 it was realized that a manual of instructions was necessary in order to achieve consistency within the system. What is now known as the Oregon Manual was the first attempt to provide uniform instructions to all surveyors. An updated manual quickly followed in 1855, and each succeeding edition provided instruction on proper procedure for making original surveys.

## THE SUPPLEMENT TO THE MANUAL

Titled *Restoration of Lost or Obliterated Corners & Subdivision of Sections, a guide for surveyors*, this booklet is a brief (in comparison to the Manual) pamphlet, originally produced in 1883, to offer guidance to surveyors and others attempting to locate and follow the original surveys done in accordance with Manual in existence at the time of the survey. The Supplement has been cited in court decisions and has been reprinted several times, so is readily available to anyone for guidance in locating original survey evidence.

## SPECIAL INSTRUCTIONS

*Special Instructions* have been issued a number of times over time for governing specific projects. These should be consulted whenever gathering information on original surveys, as there may be specific rules, tasks and contingencies required beyond what appears in the *Manual* itself. *Special Instructions* cite their pertinent authority, the appropriation or other funding, the nature of the work to be performed, and specific instructions for the survey that may not be covered in current *Manual of Surveying Instructions*.

All original surveying, as well as subdivision of townships, prior to the creation of the PLSS was done under special instructions. Each subdivision was somewhat different, done by a variety of surveyors, each having their own equipment without much standardization, using their own techniques, and accumulating their own sets of errors. Following these surveys demands expertise and an understanding of how each was done. In order to work with such a conglomeration, to be successful in following footsteps, it is necessary to determine each surveyor's system, or method of operation. Some rules were set by law, others created on a need basis. Fortunately, very wise judges such as George Mortimer Bibb (1776–1859), Chief Justice John Marshall (1755–1835), and their contemporaries set down guidelines for others to follow, inserting some consistency into the systems. These leaders where followed by other very capable jurists with familiar names: Story, Lamar, Lumpkin, Taylor, and Cooley, among others.

## WHAT IS NOT AN ORIGINAL SURVEY

As important as knowing exactly what is an original survey, is knowing what is not an original survey, as there appears to be widespread confusion as reflected by not only practice, but also within various court proceedings at all levels. There are a number of scenarios that may be encountered where a retracement surveyor did not uncover the original survey, but relied instead on something else. While in some cases, this approach may produce the same results as a full investigation to the origin of survey and title, it would seem that: (1) this retracement person is incurring quite a risk concerning whether the information being relied upon is in fact correct, and (2) putting a tremendous amount of faith in someone else's determination or

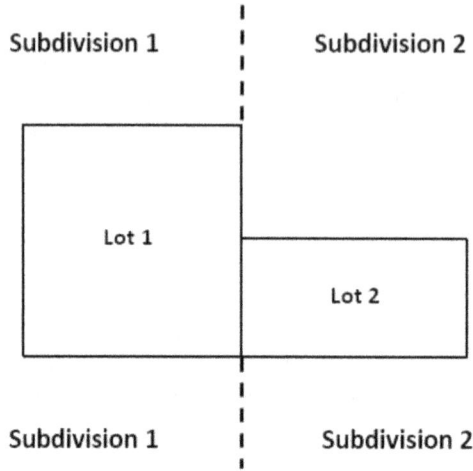

**FIGURE 4.4**   One ownership consisting of two lots from two different subdivisions.

opinion, probably without checking the background of the source. This practice may invite considerable liability. Figure 4.6 provides an example with a disastrous result.

- Tracing records back to the previous [purported] "survey," not an original, actually someone's retracement, at best.
- Not recognizing that lines are a combination of several creations; even the original creation of a lot in a subdivision may have one or more lines coincident with a previously created "surveyed' boundary. This frequently happens with parcels that have at least one line coincident with the perimeter, surveyed by another party or sometimes the same surveyor at a different time and under different conditions. That necessitates the examination of two different surveys, each with their own characteristics and being two separate, independent, retracements, each with its own characteristics (Figures 4.4 and 4.5).

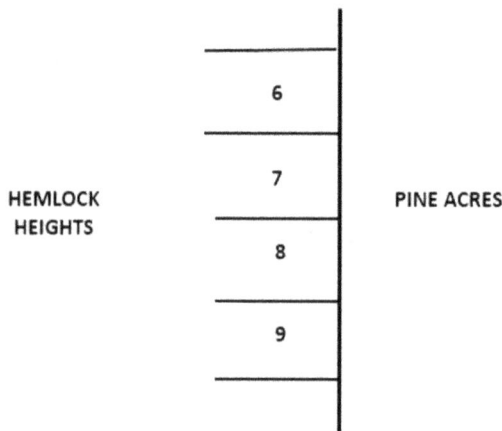

**FIGURE 4.5**   Lots 6–9 each have two original boundaries: one from Pine Acres subdivision, the other from the Hemlock Heights subdivision.

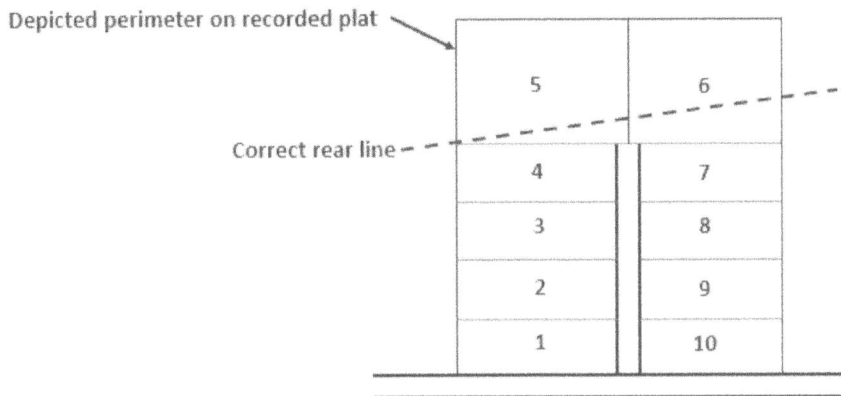

**FIGURE 4.6** An attempt at a subdivision using, without checking, a previous perimeter survey, which turned out to be incorrect.

- Holding someone else's retracement, such as subdividing someone else's perimeter without a detailed investigation of its validity.[10]
- Selecting an intermediate description in the chain of title, not tracing back to the beginning, the creation of the title.

### EXAMPLE

Following a chain of title and description back in time to its origin resulted in the following misleading and incorrect information. Beginning in 1903 with the earliest description, a parcel described with the following information was created: an abutting call of Mary A. Eaton, directions of South about 3 degrees West, North about 50 degrees West and an area of 100 acres. Going forward, the next description, in 1906, reported South 30 degrees West and "over 100 acres." The next description was in a 1913 conveyance, which reported South 30 degrees West and "more than 100 acres." The next description, in 1916, contained South 30 degrees West, North 50 degrees West, and "160 acres." Then, in 1920, the description read Mary A. Eaton as abutter, South 3 degrees West, North 56 degrees West, and "100 acres." Finally, in 1938, the description which carries forward to present day reads, Mary A. Easton, North 50 degrees West, South 3 degrees West, but without a recitation of area. The entire chain of title spans over 100 years, the parcel never change in size or shape the entire time, yet the descriptions for the same tract of land changed several times, misleading a reader and, if the lot had been staked, it could easily been done incorrectly, further misleading owners and abutting owners. There are ten deeds in this chain of title, nine of which contain an incorrect description with one or more mistakes.

This example illustrates why it is vitally important to read every conveyance in a chain of title from origin to present day, then following up to account for any changes. Randomly selecting a conveyance can very easily lead a reader to an incorrect opinion as to the property, and result in a surveyor producing incorrect results.

- Not separating a combo of original + already existing lines (e.g., the Rivers case)
- Not recognizing all of the parts, such as a right-of-way to the subject tract – appurtenant easements travel with the land, whether mentioned or not, and are part of the title

---

[10] Done properly, the checking process may often involve more than doing the entire work over again. For example, most times inherent errors cannot be identified, there is no check on whether any critical evidence has been overlooked either in the records or on the ground, nor is there insight as to any adjustment procedures performed in order to remove allowable errors so as to produce a mathematically closed survey figure.

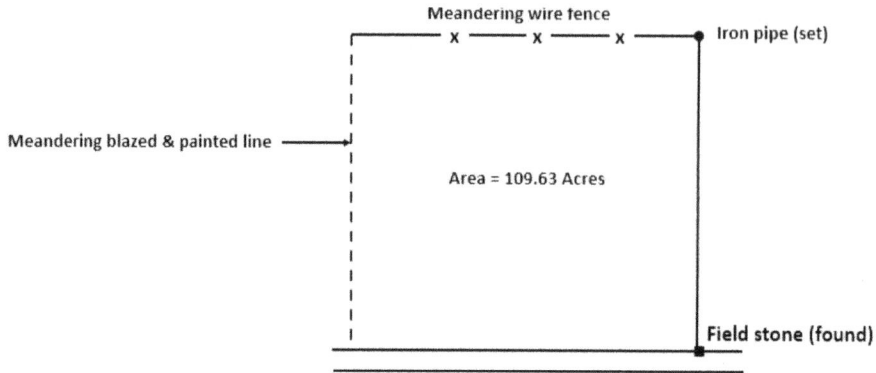

**FIGURE 4.7** Survey plan on record composed of undocumented markers and (possible) evidence, but without verification or documentation. Written description: southerly by the highway; westerly by land of Jones, northerly by the Smith Farm; easterly by owners unknown, containing 100 acres, more or less.

- Not recognizing the significance of an abutting tract (senior deed, senior survey or a monument controlling the location (called for, in existence, identifiable)
- Not examining and comparing all the evidence – monuments, plats, field notes, descriptions (deeds, etc.). Sometimes critical evidence is extremely difficult to find, causing projects to take much longer than predicted. Two extreme examples of this problem are presented in Chapter 8.
- Relying on numbers in the [current] description, without knowing their origin or reliability and what adjustments have been performed. This is a leading cause of "pincushion corners."
- Giving up too soon and executing a boundary line agreement. This is frequently a violation of statutory and longtime case law requirements of a line or corner being unknown or unascertainable.
- Not giving the title due consideration. e.g., "all of such-and-such parcel, excepting, etc., etc., etc."
- Treating protracted lines as part of the survey as if done using the same procedures and adjustments. As detailed in Chapter 6 and emphasized by the courts, they are independent, in that protracted lines are not part of the survey.
- Not satisfying the legal requirements of what a survey is. Read and understand not only the current state or federal statute, but be intimately familiar with pertinent case law.
- Blanket measurements without any legal or title basis – fences, ground features, area controlling size and extent of parcel. Generally this amounts to nothing more than guesswork, even if it is an "educated guess." As a summary of the foregoing, refer to Figure 4.7.

## THE SIGNIFICANCE OF THE ORIGINAL SURVEY

**Author's Note:** Why the original survey and not something else? Simply, because "something else" is likely just someone else' opinion, without foundation of title. The original survey is the basis for the original title, and vice versa. In other words, in most cases the original survey is the creation and definition of the original source of title.

As discussed in Chapter 1, the decision in the case of *Rivers v. Lozeau*[11] puts in perspective the relationship between the establishment of boundaries and land surveying:

ORIGINAL LAND LINES. The owner of a parcel of land, being the grantee under a patent or deed, or devisee under a will or the heir of a prior owner, has no authority or power to establish the boundaries of the land he owns; he has only the power to establish the division or boundary line between parcels when he owns the land on both sides of the boundary line he is establishing. In short, an original surveyor can establish an original boundary line only for an owner who owns the land on both sides of the line that is being established and that line becomes an authentic original line only when the owner makes a conveyance based on a description of the surveyed line and has good legal title to the land described in his conveyance.

First, the surveyor can lay out or establish boundary lines within an original division of a tract of land which has existed as one unit or parcel. In performing this function, he is known as the "original surveyor" and *when his survey results in a property description used by the owner to transfer title to property*[12] that survey has a certain special authority in that the monuments set by the original surveyor on the ground control over discrepancies within the total parcel description and more importantly, control over all subsequent surveys attempting to locate the same line.

This is why it is important to do good work. Careful measurements, proper adjustments, keeping good records, and documenting, everything for others to follow (and theoretically "get it right").

One of the most important decisions of guidance for the following surveyor is from the federal court system.[13]

## I. GENERAL SURVEYING PRACTICES
### a. Background Research

The court will first set forth basic principles of surveying based on its review of relevant treatises and case law as well as the expert testimony offered at trial by the parties.[21] A surveyor should strive first to locate and examine all historical records, deeds, prior surveys, maps and drawings in preparation for conducting an original survey. *See, generally*, CURTIS BROWN ET AL., BOUNDARY CONTROL AND LEGAL PRINCIPLES 371–74 (3rd ed. 1986) [hereinafter "BOUNDARY CONTROL"]; WALTER G. ROBILLARD & LANE J. BOUMAN, CLARK ON SURVEYING AND BOUNDARIES § 4 (5th ed. 1987) (hereinafter CLARK ON SURVEYING) If the surveyor is not performing an original survey then the surveyor must also carefully review the original survey, as well as subsequent surveys or drawings. The purpose of thoroughly researching the history of a parcel of land is to ensure that the surveyor will be able to incorporate the most complete and accurate data into his or her survey. If a surveyor does not complete such research, the surveyor might perform the survey without having the benefit of essential information. For instance, the surveyor might not adequately search for crucial monuments or might misinterpret other field or documentary evidence. BROWN, BOUNDARY CONTROL at 371. In addition, if a surveyor knows that his or her survey will be used in a particular manner, a surveyor should review relevant documents and field surveys of adjacent parcels of land to ensure that his or her particular survey will be reliable and consistent with other existing surveys, so as to discourage litigation. *Id.* at 374.

### C. *The Centrality of the Original Survey*

Since the physical position of monuments referenced in a conveyance reflect the original boundaries of a particular parcel, a subsequent surveyor must attempt to conform his or her survey as closely as

---

[11] Fla. App. 5 Dist., 539 So.2d 1147 (Florida, 1989).

[12] This is a most important qualification.

[13] *Newfound Management v. Sewer*, 885 F.Supp. 727 (U.S. Dist., 1995).

possible to the prior surveyor's work. Hence treatises and courts frequently recite an admonishing maxim, namely that a surveyor must follow in the footsteps of the original surveyor. *See* Rudolph Galiber's Testimony (Tr. 1B, p. 35.), Marvin Berning's Testimony (Tr. 2, p. 112–114). The purpose and result of this principle is to give effect to the intentions of the parties at the time of the survey as well as ensuring the continuity of boundaries over time. Accordingly, "[t]he general rule governing the determination of boundary lines by resurvey is that the intent of the new survey should be to ascertain where the original surveyors placed the boundaries," not to determine new modern boundaries. *Thein v. Burrows*, 13 Wash.App. 761, 537 P.2d 1064 (Wash.App.1975); *see, also, U.S. v. Champion Papers, Inc.*, 361 F.Supp. 481 (S.D.Tex.1973) (boundary dispute involving 135-year-old survey resolved by the court's attention to totality of the evidence including evidence of the parties' intentions).

## WHAT IF THE ORIGINAL MONUMENTS ARE TRULY GONE?

Remember, a corner is an invisible point, which may or may not be defined by, or related to, some type of monument.[14] The Maine court dealt with this problem on more than one occasion, for different reasons.

In the case of *Lloyd v. Benson*,[15] the court stated, "When a monument referenced in a deed is missing, it does not lose its significance as a monument if its original location can be determined. We have previously expressed this principle in both positive and negative terms. For example, in *Theriault v. Murray*, 588 A.2d 720, 722 (Me. 1991), we stated: 'The physical disappearance of a monument does not end its use in defining a boundary if its former location can be ascertained.'"

In contrast, in *Milligan v. Milligan*, 624 A.2d 474, 478 (Me. 1993), we stated, "[t]he physical disappearance of a monument terminates its status as a boundary marker *unless* its former location can be ascertained through extrinsic evidence." (Emphasis added.) We concluded in *Milligan* that "[b]ecause the unrebutted testimony in this case was that no pin … was ever located at [the terminus described in the deed], the pin could no longer be considered a monument. *Id.*"

Regardless of whether expressed in the positive or negative, the principle remains the same: the location of a monument that is described in a deed, but is missing from the face of the earth, can be established through extrinsic evidence. *See Hennessy*, 2002 ME 76, 796 A.2d at 48.

Once so established, the monument has the same legal significance as if it were not missing. *Theriault*, 588 A.2d at 722 (stating that if the locations of missing monuments can be determined, the "monuments as a matter of law must prevail over the deed's course and distance calls").

If [a corner] is truly gone, then resort must of necessity to appropriate procedures. However, decisions state that a corner must be so lost that it cannot be replaced with ANY evidence (See detailed discussion in Chapter 9).

Further emphasis of missing monuments/existing corners is found in the following decisions from other jurisdictions:

"The courses and distances in a deed always give way to the boundaries found upon the ground, *or supplied by the proof of their former existence*, where the marks or monuments are gone" (emphasis supplied). *Cullacott v. Cash Gold & Silver Mining Co.*, 8 Colo. 179, 183, 6 P. 211, 214 (1885) (citing *Lodge v. Barnett*, 46 Pa. St. 477 (Pa. 1864)); 12 Am.Jur.2d Boundaries § 74.

Like the South Dakota case of *Titus v. Chapman*,[16] the original monuments, *or the places where they stood*, govern over unknown markers with no known basis and official maps with

---

[14] When called for, identifiable, and in existence at the time called for.

[15] 2006 ME 129, 910 A.2d 1048 (Me. 2006).

[16] 2004 SD 106, 687 N.W.2d 918 (S.D. 2004).

protracted, unsurveyed, lines. The map is nothing more than a picture by a draftsperson and the protracted lines are not part of the survey.

These decisions underscore the definition that a corner is a point and not a physical entity, by stating the monument, or the place where it stood (the point, or the position). Several cases have stressed this, such as those above from the Maine court, by stating, two different ways, that so long as the corner can be determined, or located, it remains controlling over lesser, conflicting, calls.

## UNSURVEYED LAND

This anomaly is addressed in the federal system in the case of *Cox v. Hart*.[17] Later treatment in Chapter 8 details with several decisions on point discussing the problems concerning this category with their ultimate resolutions.

"5. Public lands lose their status as 'surveyed lands' and become 'unsurveyed' when the lines and marks of the original survey have become obliterated for practical purposes and when, for that reason, a resurvey has been directed by an act of Congress. P. 436. 270 F. 51 affirmed."

**Unsurveyed Lands** are lands not yet surveyed. Fractional section surveys, for example, leave unsurveyed land within a section. The land is known to exist, but the survey has either not yet been authorized or has not yet, for whatever reason, been completed. **Completion Surveys** are executed to finish a partially subdivided township or section, or to finish parts of boundaries of townships or sections that are unsurveyed. An **Extension Survey** is executed to add to an existing survey. An "extension survey" does not, however, complete a survey of boundaries of townships or sections or the subdivision of a township or a section, it only adds to an existing partial survey.[18]

Even though the eastern states and colonies (prior 1776) developed considerable case law as a solid foundation, later decisions frequently reviewed, and sometimes relied on as precedent, decisions from states in the PLSS after its creation in 1785. The Tennessee decision of *Wood v. Starko*,[19] one of the most significant decisions to come from eastern courts, stated the following:

> While these general rules [as reviewed in this decision – DAW] apparently have their origin in surveys reflecting government grants, such rules are equally applicable to private surveys.

### Quoting *Staub v. Hampton*, 117 Tenn. 706, 101 S.W. 776 (1907)

In Staub, by the court, "It is insisted for defendants that all of the foregoing [reviewed] were cases in which the survey was made by a public surveyor, as a preliminary to the issuance of a grant by the state, and that the rule which gives commanding importance to the survey and the marks made upon the ground pursuant thereto, in arriving at the intention of the parties as to the identity of the land sold and purchased, in case of an apparent conflict between the written conveyance and such actual survey and marking, applies only to such public surveys and to grants by the state, and not to deeds between private persons. Before entering upon a consideration of the point it is due that we should quote from our cases the rule as it has been heretofore applied." The court went on to analyze a number of local cases in which deeds were from private persons in support of their decision.

---

[17] 260 U.S. 427, 43 S.Ct. 154, 67 L.Ed. 332 (1922).
[18] Glossaries of BLM Surveying and Mapping Terms. Cadastral Survey Training Staff. 1980.
[19] 197 S.W.3d 255 (Tenn.App. 2006).

In *Lodge v. Barnett*,[20] it is said: "On this point there is nothing more fixed or better ascertained than the law of this state."

Staub continued to say, "It is a familiar principle of our system, and one in reason applicable to this species of title, as well as any other, that it is the work on the ground, and not on the diagram returned, which constitutes the survey, the latter being but evidence (and by no means conclusive) of the former ... It is conceded that the patent may be rectified by the return of survey; and why not the return of survey by the lines on the ground, and particularly the numbered tree, which is the foundation of the whole?" In the latter case, Kennedy, J. said: "That the original lines as found marked on the ground must govern, in determining the location and extent of the survey, is a well-established rule, in general applicable to all cases." .... "We know, in point of fact, that the marks made on the ground at the time of making the survey are the original, and therefore the best evidence of what is done in making it; that everything that is committed to paper afterwards in relation to it intended and ought to be, as it were, a copy of what was done, and ought to appear on the ground, in the doing of which errors may be committed, which renders it less to be relied on than the work as it appears by the marks made on the ground" 46 Pa., [477,] 484, 485 [1864].

## PRESUMPTION THAT A SURVEYOR DID THEIR JOB FAITHFULLY

This is an important presumption, as it provides a baseline from which to work. If we presume that work was done in accordance with a standard, or by following certain rules or instructions, we can put our faith in the work, and focus on other aspects.[21] Applying later rules of retracement, or merely identifying certain aspects of the survey, prevents the necessity of wide scale searching based on unknowns or according to logical theories, which may or may not be successful. The Texas court summarized the principle: until otherwise shown, it is presumed the surveyor did his duty in making a survey.[22]

## LACK OF SIGNIFICANCE OF A FLAWED SURVEY

As noted in Chapter 3, if it is not a true and valid survey, perhaps it does not do, or control, anything. If it is void, or voidable as a purported survey, perhaps it is a nullity and does not control anything, particularly anything sourced from a different title. The Wyoming court made a point in *Hagerman v. Thompson*[23] that the three mineral surveys in question were resurveys and biding on no one, "unless one of them perchance should ultimately in a proper proceeding be found to be correct."

## DECISIONS CONDEMNING NON-USE OF ORIGINAL SURVEY

It is a well settled principle that if lands are granted according to an official plat of the survey of such lands, the plat itself, with all its notes, lines, descriptions, and landmarks, becomes as much a part of the grant or deed as if such descriptive features were written out upon the face of the deed or grant itself, and controls so far as limits are concerned. *Nicolin v. Schneiderhan*, 37 Minn. 63, 33 N.W. 33; *Turnbull v. Schroeder*, 29 Minn. 49, 11 N.W. 147; *Fox v. Union Sugar Refinery*, 109 Mass. 292; *Jefferis v. East Omaha Land Co.*, 134 U.S. 178, 10 S.Ct. 518, 33 L.Ed. 872. Monuments are the best evidence of the lines and corners actually made by a survey, and, when ascertained, are satisfactory and conclusive evidence of the location of the lines as originally run. 8 Am. Jur., Boundaries, p. 787,

---

[20] 46 Pa. 477.

[21] We should be able to expect that the original surveyor did what he said he did, barring certain legitimate exceptions.

[22] *Phillips v. Ayres*, 45 Tex. 601 (1876).

[23] 235 P.2d 750 (Wyo., 1951).

§ 59, states: "* * * Monuments set at the time of an original survey on the ground and named or referred to in the plat are the best evidence of the true line."

A resurvey that changes lines and distances and purports to correct inaccuracies or mistakes in the old plat is not competent evidence of the true line fixed by the original plat. 8 Am.Jur., Boundaries, p. 819, § 102; *Cragin v. Powell*, 128 U.S. 691, 9 S.Ct. 203, 32 L.Ed. 566, following the rule of *City of Racine v. Emerson*, 85 Wis. 80, 55 N.W. 177, 39 Am.St. Rep. 819, where the court stated: "* * * A resurvey must agree with the old survey and plat to be of any use in determining it. * * * Resurveys for the lawful purpose of determining the lines of an old survey and plat are generally very unreliable as evidence of the true lines. The fact, generally known and quite apparent in the records of courts, is that two consecutive surveys by different surveyors seldom, if ever, agree; and the greater number of surveys, the greater number of differences and disagreements will occur. When two surveys disagree, the correct one cannot be determined by still another survey. It follows that resurveys are of very little use in such a case as this, except to confuse it."

# 5A Necessity of the Original Survey

A boundary line once established should remain fixed in its original position through any series of mesne conveyances.[1]

**Robert J. Griffin**

The Wyoming court emphasized the principle of a retracement necessarily reproducing the original survey. In *Hagerman v. Thompson*,[2] the court stated "Three surveys in question in this case were resurveys, binding on no one, unless one of them perchance should ultimately in a proper proceeding be found to be correct." This court had no way of knowing which one of them was correct, if any. The court continued, "it is a well-known fact that surveyors are apt to differ from each other, and surveyors employed by the United States government are not immune from the frailties of their profession. Which one of the resurveys is correct is a question of fact." Without supporting evidence, it is not possible to evaluate an existing survey. Subsequent surveys must follow the original survey as precisely as possible, from which came the establishment of the title. The court therefore, in this case, was unable to either evaluate or accept the three surveys presented in evidence.

## EFFECT OF SURVEYING WITHOUT CONSIDERING THE ORIGINAL

First and foremost, such practice ignores the basic requirement of following the footsteps of the original surveyor. Second, such surveyor is accomplishing two things: violating a strong and basic rule of surveying as well as law, and likely creating future problems for landowners and surveyors. Such performance may be a breach of ethics, and more importantly, it may be a negligent act.

The foregoing decision, along with others, states that such survey work is binding on no one "unless perchance it should be found to be correct." In short, such surveying practice should be questioned, and examined for correctness. One might even go so far as to fail to give it any credence as being "surveying type work" but not a "survey."

For certain, the following surveyor has not followed the rule, that is, has failed in the duty of following proper procedure set by law. In doing so, others should seriously consider whether it was indeed a survey, and whether it enjoys any status whatever. An attempt at a survey that fails, is certainly not a survey.

## FIRST STATEMENT OF PRINCIPLE CONTAINING THE CONTROL OF THE ORIGINAL SURVEY

While, with all of the emphasis in relatively recent decisions on "following the footsteps of the original surveyor," it would seem that the sanctity of the original survey would be entrenched in the Public Land Survey System (PLSS) and the very early decisions from

---

[1] First Principle: **Location of a Boundary Line Is Determined as of the Time of Its Creation.** Griffin, Robert J. Retracement & Apportionment as Surveying Methods for Re-establishing Property Corners. *Marquette Law Review*, Vol. 43, Issue 4, Article 5. Spring 1960.

[2] 235 P.2d 750 (Wyo., 1951).

DOI: 10.1201/9781003032557-7

the PLSS states. It would appear, however, that is not the case, in that early decisions from Maine seem to set the precedent, even though not referenced in later decisions.

An early decision, which may be the first of its kind in the PLSS, is *M'Clintock v. Rogers*.[3] This is a case worth studying for a variety of reasons, but noteworthy here is the emphasis on the concept of rights: "The rights of the parties in the land in controversy, must be determined by the original survey, made under the authority of the government and returned to the office of the surveyor general; and according to that survey, the disputed township line was a straight one, running through the entire length, from corner to corner; and it cannot now be varied to suit the interests or caprice of purchasers. To permit it to be done would be to annul the authority of all the public surveys; obliterate the lines of demarcation between the property of man and man, and open wide the door to confused, harassing, and endless litigation."

Continuing, stated by J. Grimshaw, for defendant in error, "It is a question of fact to be determined, how was this line originally run by the surveyors of the United States government. If that fact can be ascertained, it must determine the case. The actual survey must be deemed part and parcel of the description of lot. A proprietary surveyor held not to be an agent of people whose lands he surveyed, but of proprietor alone. Such surveyor could not bar title by his return, which was vested by actual survey. The courses and distances on the ground are the true survey."[4]

Turning to Maine, an earlier case decided by Justice Weston, is that of *Brown v. Gay*.[5] This is one of Maine's earliest land cases, the state having been created in 1820, out of Massachusetts. The problem arose from the deeds calling for lots on a survey plan, but the work on the ground differed. Parties were in possession according to the information on the plan.

The opinion stated, "The rights of the parties do not depend upon their respective possessions; but upon a sound construction of the deeds and of the plan, which forms a part of them. The original locations by the surveyor, as far as they can be found, are to be sustained; and if any variance appears to exist between them and the plan, the locations actually made control the plan."

Justice Weston elaborated more fully in the subsequent decision of *Esmond v. Tarbox*.[6] Again incorporating the original survey plan, the court found, "This is an exceedingly plain case. The deeds of both the demandant and tenant, refer to *Adam's* plan. And that is *Hobart's* plan, upon a reduced scale. *Hobart* surveyed the ground and set up monuments. *Adams* made no survey of the land in question; but adopted *Hobart's* survey and monuments, as he was directed to do. The jury have, by their verdict, established the line between these parties, according to these monuments; the location of which was proved to their satisfaction. The plan and the monuments do not exactly coincide; but this is no uncommon case; and where a difference is found to exist, it has been long the settled practice, both of Massachusetts and of this State, to give effect to the latter, rather than the former.

"The monuments adopted, or placed upon the face of the earth, are the best evidence of the lines and corners actually made by the survey. Of this the plan is intended to be an accurate delineation. The survey is the original work, and the plan is derived from it, and intended to represent it. If it fails to do so, the survey, if it can be ascertained, and not the erroneous delineation of it, is to govern. – Purchasers look to actual monuments, which they are, or should be, careful to preserve; and public policy, as well as principles of law, requires that their titles and possessions should be protected and secured by them.

"It makes no difference that the plan referred to was made by one man, and the survey by another; or that a plan upon a larger scale intervened. Both were intended to be coincident, and derived from

---

[3] 11 Ill. 279 (Ill., 1849). Cited by *Browning & Bushnell*, for plaintiff in error.
[4] Citing *Lilly v. Kitzmiller*, 1 Yeates 29; *Yoder v. Fleming*, 2 Yeates 311. See act of 1800, Land Laws.
[5] 3 Me. 126 (Me., 1824).
[6] 7 Me. 61 (Me., 1830).

one source, the survey. The legal construction of what is done in these cases, is not affected by the number of agents employed. One may make the survey, and locate the monuments, and another may delineate the plan from his field book or minutes, and the actual survey will be equally conclusive, as if all had been done by the same hand."

Some early decisions from the Commonwealth of Kentucky[7] are also noteworthy.

1. After an entry has been once surveyed, and the survey recorded, the warrant merges in the survey; and a second, or re-survey, is illegal.

5. A survey variant from the entry, is not void, but voidable. *Estill v. Hart's Heirs*, 3 Ky. 567, (1808). The positions taken in argument, by Hughes, are so fully embraced by the opinion of the court that it is deemed unnecessary to insert them. So much of the argument for the appellees as related to their right to make a re-survey, was as follows:

CLAY for appellees – By the laws of Virginia, prior to 1779, relating to land claims, a warrant was a sufficient authority to a surveyor to make a survey, and entries were not required. Since 1779, warrants are required to be entered, and give no authority to make a survey until entered. It is, therefore, the entry which gives the authority. The warrant is merged in it; and the authority derived under the entry, is to make a survey in pursuance of the entry, not to survey generally. A survey made, but not in pursuance of the entry, is a mere nullity.

**Author's Note:** As stated earlier, verified by this decision, surveying that does not strictly adhere to the rules, may be nothing other than an exercise in futility. Surveyors should consider such an outcome when evaluating previous work by others.

The supreme court of the United States decided in the case of *Mason v. Wilson*, 1 Cranch, 45, that "such survey conveyed no right to the land, and was not an appropriation. This court, in the case of *Greenup and Keene v. Kenton and Frazier*, ante, 14, decided, that a patent did appropriate land; and, in the case of *Bradford v. Patterson*, ante, 108, that neither the registry nor any intermediate act between the entry and patent, gave validity to a survey variant from the entry; or, in other words, appropriated land. Such a survey, is, therefore, without authority or validity. It is a void attempt at executing a power, and cannot hinder a proper execution of the power. No matter how far the steps toward procuring the patent may have gone, if the patent has not issued, it cannot destroy the right of properly executing the power given by the entry.

"It is true, that the court, in the case of *Jones's Heirs v. Taylor's Heirs, etc.*, Pr. Dec. 71, decided, that the surveyor should have the warrant, as his authority, when he made the survey; and in this case, the warrant was in the register's office, at the time our last survey was made. But that case is clearly distinguishable from this. That was a survey made prior to 1779; when the warrant, and the warrant only, gave the right to survey. Ours was a survey, made under the acts of 1779, under the authority of the entry which remained in the books uncomplied with, and after the warrant had been properly deposited with the surveyor, and then was transferred into the hands of the officers designated by the law for the final reception of it, for the purpose of perfecting the claim to the land.

"It has been every day's practice, for the principal surveyor, when his deputy has returned to him a plat and certificate, if he finds it defective, or the survey not made conformable to law or so erroneous that the courses will not meet, or the like, to send it back and have a new survey made. The propriety of this proceeding has never been questioned.

---

[7] It is necessary to examine the law of the jurisdiction at the time to determine whether something is actually a bona fide survey or not.

"If the surveyor have the power thus to correct his mistakes of admeasurement, etc., there is the same reason for his correcting his mistakes in giving form to the survey, or a construction to the entry. That the mistakes I have mentioned are corrected before the plat and certificate is admitted to record, and what I contend for is afterward, can make no difference according to the principle of the decisions I have cited; for neither recording or registering give a right to the land surveyed.

"But if there could be doubts as to the right of an adult to make a re-survey, there can be none as to the rights of these infants. If an act be done for infants, it will bind them, if for their benefit; but it will not, if it be not for their benefit. The erroneous survey of 1784 was not for their benefit; and therefore, cannot bind them. They had a right to avoid what was thus improperly done in their names. The executors, if under the will they had a right to have surveys made, it must have been to make legal surveys, or surveys in conformity to the entries, not surveys which would destroy their claims.

"Mrs. Hart had only a life estate in the lands, and could not do an act which would bind the inheritance. She could not direct the surveying, without the assent and concurrence of the guardians of the infants; which it does not appear was obtained. This survey was, therefore, without the authority, in fact, of the persons entitled to the land; and on this account also, not binding. *Estill v. Hart's Heirs*, 3 Ky. 567, (1808).

"For the appellees it was argued that they had a right to re-survey. First. Upon general principles. Secondly. Because of their infancy at the time the first surveys were executed. Thirdly. Because of the fraud of Daniel Boone in misdirecting the surveys for the purpose of saving his own claim. Fourth. That the re-survey shall hold, at least, against the purchasers from said Boone, who are charged with notice of Boone's fraud, before they purchased.

"Upon general principles, we consider the surveyor, by virtue of his office, as having no authority to lay off a particular tract of land, for any individual. To do so, he must have a special precept or authority. In case of four hundred acres, granted by the act of 1779, for settlement rights, the certificate of the commissioners is considered as a sufficient warrant of authority to the surveyor to admit an entry and afterward to lay off the land. In all other cases, a warrant duly obtained from the proper authority, was required by that act. When the special authority was once executed, it merged in the survey, and could not authorize the surveyor again and again to make surveys, as the whims, the frauds, or the better judgment of the owner of the warrant, or certificate of authority, should dictate.

"That mistakes and uncertainty in surveys, and in the description of the boundaries, would in some instances, render re-surveys desirable, the legislature contemplated, in enacting the original land law of 1779. But to guard against the mischiefs which might result from such re-surveys, they were, under certain regulations, to be permitted by the county courts; who were to hear the petition after notice had been previously given, and thereupon, in their discretion, empower the surveyor of their county to re-survey such lands, at the charge of the party, according to his directions and the original or authentic title papers, taking care not to intrude upon the possessions of any other person, and to return a fair plat and certificate of such re-survey into the said court, to be examined and compared with the title papers; and if the court shall certify that, in their opinion, such re-survey is just and reasonable, the party may return the same, together with his material title papers into the land office, and demand the register's receipt for them," etc.

"The salutary provisions of this act; the notice to be given to the owners of adjoining claims; and advertisement at the door of the court-house, on two several court days, of the time of the intended application, form a striking contrast with the mere protest of the acting executor, and the re-survey thereon, as used in the present instance.

"The legislature foresaw that many frauds and inconveniences, as well upon the commonwealth as upon individuals, would enter in if the door was open to re-surveys at the will of every individual. The mischiefs which have resulted from a removal of one of the checks against impositions, that is to say, from the authority given by the act of 1784, to the register, to receive a plat and certificate of survey, and to issue a grant thereon, without the original warrant of survey, ought to warn the judge against removing others (if it were in his power and within his province), by judicial opinions.

"The land law contains many expressions which, so far from doing away the general principle that a special authority once executed is merged and at an end, confirms and establishes the principle with respect to the permission and authority to surveyors to lay off tracts of land for individuals. The land law of 1779 declares that warrants "shall be always good and valid, until executed by actual survey, or exchanged in the manner herein (by the act) after directed."

Again, where the lands on which "any warrant is located shall be sufficient to satisfy such warrant, the party may have such warrant exchanged by the register of the land office for others of the same amount in the whole, but divided as best may answer the purposes of the party, or entitle him to so much land elsewhere as will make good the deficiency."

> Every person having a land warrant, founded on any of the before mentioned rights, and being desirous of locating the same on any particular waste and unappropriated lands, shall lodge such warrant with the chief surveyor of the county wherein the said lands, or the greater part of them, lie; who shall give a receipt for the same if required.

No surveyor was permitted "to admit an entry for any land without a warrant from the register of the land office, except in the particular case of certificates from the commissioners of the county, for tracts of land not exceeding four hundred acres, allowed in consideration of settlements," etc.

The plat and certificate of survey is to express, among other things, "the nature of the warrant and rights on which such survey was made." The warrant was to be delivered to the party, with the plat and certificate of survey, and to be returned to the register's office. Even where lands, surveyed by virtue of any warrant, were adjudged to another, upon a caveat, the party thus losing his land could not re-locate or survey by virtue of the old warrant; but was put to the necessity of suing out a new warrant from the register, for the quantity of land so granted to another, "reciting the original warrant and rights, and the particular cause of granting the new warrant. And to prevent confusion and mistakes in the application, exchange, or removal of warrants, the register of the land office is hereby directed and required to leave a sufficient margin in the record books of his office, and whenever any warrant shall be exchanged, renewed, or finally carried into execution by a grant, to note the same in the margin opposite to such warrant, with folio references to the grant, or other mode of application; and also, to note in the margin opposite to each grant, the warrant or warrants, and survey, on which such grant is founded, with proper folio references," etc. "All persons, as well foreigners as others, shall have a right to assign or transfer warrants, or certificates of survey, for lands."

"Such are the expressions, and such is the general view of the land law respecting warrants; from which it seems to be the fair understanding of the act, that a special authority or warrant to the surveyor was in every case necessary. That in the case of settlement rights, the certificate of the commissioners was the warrant; in others, the warrant from the register. That without such warrant or special authority, the surveyor was not permitted to make a survey for an individual, any more than any private person was permitted to lay off a tract of land. That the law considered every warrant as merged in the survey, when made a matter of record, at least. Whether that survey was in conformity with the entry, or variant from it, cannot prevent the merger of the warrant. A survey variant from an entry was not void, but voidable only. If not caveated, a grant issued upon it, which grant would hold the land against all claimants, except those who could show a prior title in law or equity; whereas, if such survey was void ab initio, and not merely voidable, no subsequent act could give it validity; the maxim being, *quod ab initio non valet, in tractu temporis non convalescit.*

"Whatever doubts might be raised, as to the particular time at which the warrant shall be said to be merged in the survey; whether from the time it is approved by the chief surveyor and recorded, or from the time it is delivered out to the owner, or from the end of three months after making the survey; we conceive the case clear, that after registering, the warrant was no longer an authority to any surveyor to receive an entry or make another survey. The warrant, in fact, then is, or ought to be with the register; and we know of no authority in law which could justify the register in delivering out the original warrant, lodged with a survey made by virtue of such warrant.

"The infancy of the owner of such warrant, is nowhere made an exception by the land law to this general principle. Indeed, infants are not (in the general) excepted out of the several forfeitures, conditions, and limitations, prescribed to other holders of land claims; but in this instance, there is no pretense for making an exception in favor of the complainants. The widow having a life estate in the lands, the guardian and curator, by testament of the infants, the adult executrix appointed, who does not seem to have renounced the trust or interest given by the will, did procure the survey to be made in 1784. Another executor arrived at full age and died in 1786. Nathaniel Hart, another executor, arrived at full age in 1791, and yet no protest was made against the survey, or other survey executed, until 1798; fourteen years after the first survey had been registered.

If such resurveys were indulged, it would make strange confusion in neighborhoods and among the owners of adjacent claims. *Estill v. Hart's Heirs*, 3 Ky. 567, (1808)."

## SURVEY FAILURE IN THE PLSS

Just as in colonial states, failure in the PLSS may lead to incomplete results as described in the case of *United States v. Curtner et al.*[8]

"The congressional acts of 1862 and 1864 granting aid in the construction of a railroad and telegraph line to the Pacific Ocean, etc., operated as a present grant of land to the railroad company, upon conditions subsequent, which could only be defeated by breach of conditions, and divestiture of title thereupon, by proper legal proceedings on behalf of the United States."

The lands granted were the odd-numbered sections within 20 miles of the line of the road, such as were public lands at the date of the act, not sold, reserved, or otherwise disposed of by the United States; and such odd-numbered sections within the same limits as were public lands, to which a pre-emption or homestead claim had not attached at the time the line of the road was definitely fixed.

"No right other than that of the railroad company could be acquired or initiated in any of said odd sections of land, after the filing in the local land-office of the district, on January 30, 1865, of the order of withdrawal provided for in section 7 of the act of July 1, 1862.

"The filing of the map of the general route and the withdrawal thereupon protected the lands against the acquisition of any right by any other parties until the line should become 'definitely fixed,' when the grant became specific by attaching itself to every odd section within the prescribed limits.

"State selections of lieu lands for school purposes made upon lands unsurveyed by the United States are utterly void.

"All the state selections shown in the bill being upon lands unsurveyed by the United States at the date of selection, in townships 2 S., 1 E., and 3 S., 3 E., Mt. Diablo B. and M., were therefore void.

---

[8] United States Circuit Court, N.D. California, 38 F. 1 (N.D.Cal. 1889).

"*Lands are not surveyed lands by the United States until a certified copy of the official plat of survey has been filed in the local land-office.*

"The state selections in question were also void, for the reason that the act of 1853, under which these selections were made, excepted from selection by the state in lieu of school sections lost, "lands reserved by competent authority" and, "lands claimed under any foreign grant or title," and "mineral lands."

"No right of any kind had attached to these lands when they were withdrawn for the purposes of the railroad grant on January 30, 1865, that, under the recent decision of the United States supreme court, in *U.S. v. McLaughlin,* 8 Sup.Ct.Rep. 1177, could prevent that grant from attaching. It was, therefore, the first grant to attach, and by performance of the conditions subsequent the title of the company became absolute.

"The selections in question were excepted from confirmation by the act of 1866, (14 St. 218;) but had it been otherwise, it was not in the power of congress at that time to divest the right of the company.

"The act of March 1, 1877, (19 St. 267,) for like reasons, cannot affect the rights of the railroad company. At the date of this confirmatory act, seven years after the title of this company became perfect, the United States had no interest whatever in the land upon which the act could operate.

"Parties purchasing under state locations in township 2 S., 1 E., since June 10, 1865, had official record notice of the right of the railroad company; for the map filed in the office of the register of the local land-office had distinctly indorsed upon it in red ink the following, viz: 'The odd-numbered sections on this plat are granted to the Western Pacific Railroad.'

"The statute of limitations does not run against the United States; and the cause of action here was not stale, the company having been, from the first, active in pursuing its right before the department of the interior.

"The government is not without interest in this action, being responsible to the company for the land or its full value, by reason of the statutory grant and contract in the congressional acts of 1862 and 1864.

"The Mexican grant called 'Las Pocitas,' was a float – a grant of two leagues within exterior boundaries embracing ten or more leagues, which two leagues so granted were confirmed and patented to the claimants, and the odd-numbered sections outside of the two leagues granted and confirmed, but inside of the exterior boundaries, passed to the railroad company.

"The prior decision, in *Newhall v. Sanger,* 92 U.S. 761, by the United States supreme court, materially limited in its operation by the recent decision in U.S. v. McLaughlin.

"This is a bill in equity, filed by the attorney general on behalf of the United States, at the request of the secretary of the interior, to obtain a decree of the court vacating and annulling the listing over to the state of certain lands selected by the state, in lieu of sections 16 and 36, as was supposed, in pursuance of the act of congress on the subject, adjudging such listing to be unauthorized and void, annulling and vacating the patents issued to purchasers by the state, after such selecting and listing, and decreeing that no title to the lands passed thereby to the patentees. The grounds of the bill are, that the listing over to the state was by mistake and without authority of law; the lands having been granted to the Central Pacific Railroad Company before any right could have attached in favor of the state, and were therefore, not subject to selection by the state under the said acts. After a contest continued for many years, the secretary of the interior has finally decided that the lands in question belong to the railroad company, and that it is entitled to a patent, that they were listed to the state by mistake, without authority of law, and that the listing is void. But the department refuses to complicate matters by issuing patents. According to the view of the secretary of the interior, the United States are under obligation to convey a clear title to the railroad company, and they are unable to do so by reason of the mistake of the officers of the government, in unlawfully listing the lands to the state; and, consequently, that it is the duty of the government to have the prior listing to the state annulled, and the patents issued thereon declared to be unauthorized and void by a decree of the court, before issuing patents to the party entitled. For these reasons, and upon these grounds, this bill has been filed by the attorney general, at the request of the secretary of the interior.

"The lands in question are odd sections, lying within the 20-mile limit of the grant of lands made to the Central Pacific Railroad Company, to aid in the construction of its road, by the act of congress of July 1, 1862, and the act of 1864 amending said act. 12 St.p. 492, § 3; 13 St.p. 358, § 4. Part of the lands lie in township 3 S., range 3 E., Mt. Diablo Base and Meridian, and a part in township 2 S., range 1 E. The lands in township 3, range 3, were surveyed in the field in August, 1862, and sectionized, and a plat thereof was made and approved by the surveyor general of California, December 24, 1862, but a duly-certified copy of the plat was not filed in the land-office of the district till June 4, 1869. The certified copy of the plat then filed is regarded by the department as the official plat, and the date of its filing, June 4, 1869, as the date of the survey. On December 28, 1865, a plat of the township, approved by the surveyor general December 18, 1865, was filed in the district land-office, but this plat is not regarded by the department as official, or as indicating the date of the official survey. Township 2 S., range 1 E., was first surveyed in the field in March, 1865, and an approved plat thereof first filed in the district land-office June 10, 1865. In accordance with the provisions of said acts of 1862 and 1864, the railroad company filed in the department of the interior, on December 8, 1864, its map designating the general route of the road, and on December 23, 1864, the secretary of the interior, in pursuance of the provisions of said acts, issued an order withdrawing the said lands for the distance of 25 miles on each side of the line of said road so designated, 'from sale, location, pre-emption and homestead. A map, showing distinctly the lands so withdrawn, accompanied said order. Said order of withdrawal and map were received and filed in the district land-office, and went into effect, at latest, on January 30, 1865. This action was before any of the lands in township 2, range 1, had been surveyed in the field, and before any plat recognized by the department as official, of the lands surveyed in township 3, range 3, had been filed, but after this latter township had been actually surveyed in the field. The road having been fully completed and accepted by the president, the railroad company filed its map of definite location on February 1, 1870. In 1839 the Mexican governor, Alvarado, made a grant of land called 'Las Pocitas,' to one Livermore and another, who presented it to the board of land commissioners for confirmation, and it was confirmed by the board, February 14, 1854. The decree is in the words following, to-wit:

> The lands of which confirmation are hereby made of 'Las Pocitas,' are bounded and described as follows, to-wit: On the north by the Lomas de las Cuevas; on the east by the Sierra de Buenos Ayres; on the south by the dividing line of the establishment of San Jose; and on the west by the rancho of Don Jose Dolores Pacheco, containing in all two square leagues, a little more or less. Reference for further description to be had to the map marked 'C,' and filed in the cause.

"The exterior boundaries contained from 10 to 12 leagues. The district court, on appeal, affirmed the decree of the board, February 18, 1859, and the supreme court of the United States finally confirmed the grant on appeal in January, 1861.[1] The final decree of confirmation is in the words following:

> The land of which confirmation is hereby made is known as 'Las Pocitas,' and is bounded and described as follows, to-wit: On the north by the Lomas de las Cuevas; on the east by the Sierra de Buenos Ayres; on the south by the dividing line of the establishment of San Jose; and on the west by the rancho of Don Jose Dolores Pacheco, containing in all two square leagues, provided that quantity be contained within the boundaries named, and if less than that quantity be contained therein, then the less quantity is hereby confirmed. Reference for further description to be had to the map marked 'C,' filed in this case.

"After confirmation by the board, and before the appeal, at the request of Livermore, then the owner of the grant, on April 5, 1854, William J. Lewis, a deputy-surveyor, was directed by the United States surveyor general of California to make a survey. He was directed to notify any adjoining claimants who might be interested, of the time and place when any line would be run; to note any objections, and report any protest that might be made. He made the survey in accordance with the instructions. Livermore being present, and pointing out his corners and boundaries; and the deputy-surveyor reported that the owner, Livermore, 'expressed himself entirely satisfied with the boundaries as I surveyed them, and as represented in the accompanying map. 'He reports that he has no doubt that 'the survey as made fulfills the intentions of the Mexican grant, as derived from the terms of the grant. 'The neighboring owners were notified, and were also present with Livermore, and pointed out their boundaries; and they, as well as Livermore, were satisfied. This survey was approved by the surveyor general June 19, 1854. It embraced over four – nearly five – square leagues of land, more than double the amount afterwards stated in the decree of confirmation by the supreme court, but did not include any of the lands now in controversy. An appeal having afterwards been taken by the United States from the decree of confirmation, nothing further was done under this survey. The final decree of confirmation by the supreme court in January, 1861, limited the amount to two square leagues, by striking out the words 'more or less,' in the decree of the board, and adding other words indicating the purpose; the language of the final decree being 'containing in all two square leagues, provided that quantity is contained within the boundaries named,' etc. In 1858, pending the appeal. Livermore died. The claim having been finally confirmed in 1861, Mr. Dyer, a deputy-surveyor, in 1865, under instructions dated September 21, 1865, made a survey, which embraced ten square leagues instead of two, to which the quantity was limited by the terms of the final decree. This survey embraced the entire Lewis survey, and extended far beyond it, in nearly all directions, and especially to the south-east and north-west. It also embraced the lands in controversy in this suit, at the two extremities of the survey, in the longest direction of the survey. The survey was approved by the surveyor general of California on February 8, 1867. On July 30, 1868, the secretary of the interior set aside this survey as being 'clearly wrong,' and directed the commissioner to return it to the 'surveyor general, with instructions to reduce the quantity of land to two square leagues. 'A new survey was made by Dyer, deputy-surveyor, by which the land was reduced to two square leagues, all of which lies within the boundaries of the Lewis survey, but does not cover one-half of that survey. None of the lands in controversy are within the two square leagues, or even within the boundaries of the Lewis survey. This last survey of two square leagues was approved by the surveyor general May 11, 1870, by the commissioners of the general land-office, March 1, 1871, and by the acting secretary of the interior on June 6, 1871, by which it became final. The land was patented in accordance with this survey, and the patent accepted by the claimant. Between May 15, 1863, and May 16, 1864, after actual survey in the field, but before the survey had been officially adopted or recognized by the secretary of the interior, and before it had been approved by the surveyor general, and filed in the district land-office, the state of California, by its locating agent, made selections and locations of all the lands now in controversy in township 3, range 3, in part satisfaction of the grant to the state, of lands in lieu of sections 16 and 36, under the act of March 3, 1853, (10 St.p. 246, §§ 6, 7.) Between February 17, 1864, and February 9, 1866, the state had issued its certificates of purchase to the several purchasers thereof, the first payments of the purchase money having been made. The selections, apparently, at their respective dates were by the register of the land-office entered in his office.

"A portion of these lands was certified over to the state by the land department at Washington, approved by the secretary of the interior on November 15, 1871, and the remainder on March 24, 1873, and they were afterwards patented to the purchasers by the state.

"The lands in controversy situate in said township 2, range 1, were selected in advance of any survey in the field by the United States surveyor general, upon surveys made by the county surveyors of the state, between July 28, 1862, and July 20, 1863. Certificates of sale were issued to purchasers by the state for a part between March 2, 1863, and January 25, 1864, and for the remainder, between February 20, and March 14, 1865. These selections were entered by the register of the land-office on June 12, 1865. A part was certified over to the state by the secretary of the interior on September 8, 1870, and the rest on March 11, 1871. These lands were also afterwards patented to the purchasers by the state. The listings over to the state were all after the final approval of the two square league survey of the Rancho Las Pocitas, which was on June 6, 1871; also after the filing of the map of general route of the road by the railroad company in December, 1864, and the withdrawal by the secretary of the interior in January, 1865; as well as after the filing of the map of the definite location of the Western Pacific Railroad Company, on February 1, 1870. But the surveys and selections and issue of certificates of purchase by the state were before the said dates of June 6, 1871, and February 1, 1870. The Western Pacific Railroad was completed in accordance with the terms of the several acts of congress relating to the subject, on or before December 29, 1869, and the company thereby became entitled to the lands granted. A contest thereupon immediately arose before the department of the interior, between the railroad company and the settlers who settled subsequently to the grants on the odd sections, as to what lands were included by the grant, and this was supposed to depend upon the exterior boundaries of the Las Pocitas grant. This matter was earnestly litigated before the department, a test case, (Arthur St. Clair v. The Western Pacific Railroad Company,) having been made by stipulation with the settlers, until January, 1874, when it was decided in favor of the railroad company. Soon thereafter, on May 12, 1874, the land agent of the company presented a list of lands for which the company claimed patents, including the lands in controversy, when it was discovered that the latter had been listed over to the state by mistake, upon the state selections hereinbefore referred to, as indemnity lands for losses of sections 16 and 36 granted for school purposes, and that they were claimed by purchasers from the state. The claim of the company for patents to these lands was vigorously prosecuted by the company, with varying results; until it was finally determined by the secretary of the interior, upon petition for reconsideration by the company, filed April 22, 1880, that the company was entitled to the lands; but he declined to complicate matters by issuing patents until the question of right should be settled by the courts. Thereupon, and for the purpose of having the question authoritatively adjudicated, upon his request the bill in this case was filed by the attorney general on July 23, 1883. Upon the allegations of the bill, a demurrer was interposed, on the ground, among others, that the cause of action was barred by the statute of limitations; and if the statute of limitations does not run against the United States, then that the cause of action is stale, and it would be inequitable to enforce it at this late day. The demurrer was overruled, since the statute does not run against the United States, and the railroad company had, from the first, been active in pursuing its right before the department. The delay was entirely owing to the course of procedure in the department, and the large amount of other similar business incident to the administration of its affairs. *U.S.* v. *Curtner*, 11 Sawy. 411, 26 F. 296. Since the decision on the demurrer, the supreme court has decided the case of *U.S.* v. *Beebe* 127 U.S. 338, 8 S.Ct. 1086, in which it is held that, after a lapse of 45 years, a suit in the name of the United States to cancel a patent obtained by fraud, and in which the United States has no interest, is barred–the suit being affected by the laches of those whose interests it asserts. The point is, therefore, now again made at the hearing, and this case is relied on as determining the question. We do not think it reaches the case. There has, certainly, been no laches here on the part of the railroad company. It has been pressing its claim earnestly before the department from the first, and it could not go any faster than the business and course of procedure of the department permitted. The company could not sue the government. Besides, we do not think the government is wholly without interest. If these lands are within the statutory grant, the company has earned them by a full performance of its part of the statutory contract, and an absolute indefeasible right to a patent, unincumbered by any cloud, has vested. The government, in that case, is legally bound to make a good title. It is legally liable to perform its part of the contract, and issue the patent as required by the statute. The United States are, therefore, responsible to the railroad company for the land, or its full value of the land in setting aside the listing and patents resulting from their mistakes, or having them judicially adjudged inoperative and void, in order that they may relieve themselves from their liability. For these reasons, we do not think the decision relied on reaches the case.

"As we have seen from the facts stated, the lands in question are odd sections within the limits prescribed by the act of 1862, granting lands to aid the construction of the Western Pacific Railroad. The Mexican grant called 'Las Pocitas' was a float – a grant of two leagues within exterior boundaries embracing ten or more leagues, unlocated both at the date of the act of 1862, and at the time when the claims of the state to the land in question were initiated. After the rights of both parties, whatever they were, had attached, this grant was finally located and patented so as to exclude the lands in controversy. There was then ample land other than these lands to satisfy this float, both at the time of the passage of the act of 1862, and at the time when the right of the railroad company attached to the particular odd sections, and became specific and indefeasible. In *U.S. v. McLaughlin,* 127 U.S. 428, 8 S.Ct. 1177, (decided at the last term of the supreme court,) it was held, after the most mature consideration, that, in case of a floating Mexican grant of a specific quantity of land within large exterior bounds, the lands within such exterior boundaries are public lands, subject to a railroad grant, there being sufficient left to satisfy the float; and that the said act of 1862 took effect upon the odd sections of land within such exterior boundaries as were not finally required to satisfy the float; thus very materially limiting the operation of the prior decision in Newhall v. Sanger. That is precisely this case; and the same act of 1862 granted to the same company all the odd sections within the exterior boundaries of the Las Pocitas grant, not embraced within the two leagues as it was finally located, 'not sold, reserved, or otherwise disposed of by the United States, and to which a pre-emption or homestead claim may not have attached at the time the line of said road is definitely fixed.' 12 St. 492, § 3. The lands in question are odd sections within the prescribed limits, and are not embraced in the Las Pocitas grant as finally patented. These lands, therefore, upon completion of the road, passed to the railroad company, unless some one of the rights specified in the statute had attached before the attaching of the right of the company. Section 7 of the act provides that the 'said company shall designate the general route of said road, as near as may be, and shall file a map of the same in the department of the interior, whereupon the secretary of the interior shall cause the lands within fifteen miles of said designated route or routes to be withdrawn from pre-emption, private entry, or sale. '*Id.* 493. This map of general location was filed in the office of the secretary of the interior on December 8, 1864, and on December 23, 1864, the secretary issued an order in pursuance of the acts of congress, as they then were, withdrawing for 25 miles on each side of the designated line 'from sale, location, pre-emption and homestead,' and forwarded it, together with a map showing the location and lands withdrawn, to the register of the land-office of the district embracing the lands where it was received, filed and promulgated on January 30, 1865; from which date at the latest, no right other than that of the railroad company, could be acquired or initiated in any of said odd sections of land. If, then, no right of the kind specified by the statute had legally attached to the lands in question before the 30th of January, 1865, none could thereafter attach in favor of the state by selection, listing over by the land department, or otherwise, nor could congress even authorize any subsequent legal transfer of title. The grant to the railroad company was a present grant upon conditions subsequent, which could only be defeated by breach of condition and its divestiture of title thereupon, by proper legal proceedings on behalf of the United States. The

filing of the map of the general route, and withdrawal thereupon from sale, pre-emption, etc., protected the lands against the acquisition of any other right by any other parties until the line should
become 'definitely fixed,' when the grant would become specific by attaching itself to every odd
section within the prescribed limits, and could not thereafter be changed. *U.S. v. McLaughlin,*
12 Sawy. 191, 30 F. 147; *Buttz v. Railroad Co.,* 119 U.S. 55, 7 S.Ct. 100; *Railroad Co. v. Orton, 6
Sawy.* 198, and cases there cited; *Denny v. Dodson,* 32 F. 899; *Schulenberg v. Harriman,* 21 Wall.
44; *Railway Co. v. Railroad Co.,* 97 U.S. 491.

"The only remaining question, therefore, is: Had any such right, as is excepted by the statute,
legally attached in favor of the state in the lands in question, or any of them, on January 30,
1865? It is not pretended that any other right than that under the state selection had attached. It
has been settled by numerous decisions in the state of California, and affirmed by the United
States supreme court, that the state could acquire no right whatever by a selection of lieu lands
made before the lands have been surveyed by the United States; and that a selection made upon
unsurveyed lands is utterly void. *Grogan v. Knight,* 27 Cal. 516; *Railroad Co. v. Robinson,* 49
Cal. 446, 448; Chant v. Reynolds, *Id.* 217; Young g. Shinn, 48 Cal. 26; *Hastings v. Devlin,* 40 Cal.
358; *Toland v. Mandell,* 38 Cal. 31, 41; *Aurrecoechea v. Sinclair,* 60 Cal. 549; Collins. Bartlett,
44 Cal. 371, 380; *Smith v. Athern,* 34 Cal. 506; *Aurrecoechea v. Bangs,* 114 U.S. 383, 5 S.Ct. 892;
*Barnard's Heirs v. Ashley's Heirs,* 18 How. 46. None of the lands in question situate in township
2 S., range 1 E., as we have seen, were surveyed in the field by authority of the United States till
the month of March, 1865, and the approved plats were not filed in the district land-office till
June 10, 1865. The applications of the state locating agent to locate all said lands in township 2
S. were made and entered in the office of the register of the land-office on the 12th and 13th of
June, 1865; the register having refused to recognize applications made in 1862 and 1863 upon
surveys made under authority of the state. As we have seen, the acts of the state in making selections prior to the United States survey in March, 1865; and the filing of the plat in the land-office
in June, were utterly void, and no rights attached to the lands or any of them by virtue of those
acts performed before said survey in March. On January 30th, at latest, the grant to the railroad
company attached in such manner that it could not be thereafter limited or divested; and the
absolute right to the lands by the completion of the road and filing the map of definite location
indefeasibly vested in the company. There can be no doubt, therefore, that the complainant should
have a decree that they are entitled to the lands in said township 2 S. The lands in question lying
in township 3 S., stand in no different situation from those in township 2 S., except that they were
surveyed in the field by the United States deputy-surveyor in August, 1862, and a plat thereof was
made and approved by the surveyor general on December 24, 1862; but a certified copy was not
filed in the office of the register of the land-office of the district embracing the lands until June 4,
1869. This plat (so filed in 1869) is regarded by the interior department as official, and the survey
as made of the date of filing. A plat approved by the surveyor general December 18, 1865, however, was filed in the district land-office on December 28, 1865, this being the first plat filed in
that office; but this map is not regarded by the interior department as official, as it had not at that
time been approved and adopted by the department. Were it otherwise, this filing was too late.
Unless the actual survey in the field, and making and approving a plat by the surveyor general
without filing it, or a certified copy of it, in the local land-office, places the lands in the category
of surveyed lands in contemplation of law, then these lands were also selected before they were
surveyed by the United States, and the selections were void. The interior department did not
regard the survey as official until the certified copy of the official plat was filed by direction of
the department in the local land-office, June 4, 1869. Whether this is to be regarded as the date of
the survey or not, we are satisfied that the lands could not be regarded as legally surveyed in such
sense as to open them to selection, location, sale, or other disposition till the approved copy of the
plat was filed on December 28, 1865. This is the earliest date at which they could be considered

open to selection, if open to selection then. The land-office was the place for the disposition and record of the public lands; and until they had an authentic official plat of the surveys of the public land, it would be impracticable to keep a record of them or of their disposition. If we are correct in this view, then no valid selection could be made, at the earliest, till December 28, 1865, and this was several months after the grant to the railroad company had indefeasibly attached.

"On another ground the state selections in question are clearly void, and no interest attached to the lands selected in favor of the state. By the express terms of the act of 1853, under which the selections were made, 'lands reserved by competent authority,' 'lands claimed under any foreign grant, or title, and the mineral lands,' are excepted from the operation of the act. Consequently, neither such 'reserved lands,' lands claimed under Mexican grants, nor mineral lands could be legally selected in lieu of school sections lost, or otherwise disposed of. And this was manifestly the view of congress, for when it passed the act of 1866, to quiet titles in California by confirming void selections, it also expressly excepted from confirmation 'any land held or claimed under any valid Mexican or Spanish grant.' 14 St.p. 218, § 1. Those selections of lands so claimed under Spanish grants were void, and created no right whatever in the state, is directly decided and settled by the supreme court of the United States in cases arising under this very grant, Las Pocitas, upon locations made in 1863, at the same time and in the same manner as the lands now in question were selected and located. *Aurrecoechea v. Bangs,* 114 U.S. 382, 5 S.Ct. 892, and *Huff v. Doyle,* 93 U.S. 558. These cases are controlling. The lands were claimed under the Las Pocitas grant, at the time of their selection, location and sale by the state, and they were afterwards in fact included in one of the surveys upon the final decree of confirmation; but that survey was set aside, and they were finally excluded by the survey which became final in the year 1871. The supreme court held that no valid selection could be made by the state until the grant was finally located. No right of any kind then had attached to these lands when they were withdrawn for the purposes of the railroad grant on January 30, 1865, that under the recent decision of the supreme court in *U.S. v. McLaughlin,* could prevent that grant from attaching. It was, therefore, the first grant to attach, and, by performance of the conditions subsequent, the title of the company became absolute. The selections in question were excepted from confirmation by the act of 1866; but had it been otherwise, as we have seen, it was not in the power of congress at that time to divest the right of the company. The act of March 1, 1877 (19 St. 267,) for like reasons, cannot affect the rights of the railroad company. The right of the company had not only attached, but by the performance of the required conditions within the prescribed time, and of the filing of the map of definite location, the grant had become specific on February 1, 1870, and the title of the company had become absolute and indefeasible. At the date of this confirmatory act, therefore, seven years afterwards, the United States had no interest whatever in the land upon which the act could operate.

"This case affords another instance of hardship arising from the ill-advised efforts of the state to prematurely select the lands to which it was entitled, without regard to the existing laws of the United States. But with respect to the particular lands now in question, the parties purchasing in township 2 S., 1 E., since June 10, 1865, had official record notice of the right of the railroad company, for the map filed in the office of the register of the land-office had distinctly indorsed upon it, in red ink, the following: 'The odd-numbered sections on this plat are granted to the Western Pacific Railroad. See letter of instructions dated December 23, 1864. 'It follows from these views that there must be a decree in favor of the United States, adjudging that the listing to the state of the lands in controversy was unauthorized and void, and that the patents issued by the state upon such listing to purchasers from her passed no title to them in the lands patented, and enjoining them from claiming, in any way or form, title to such lands, or to any part of them, under the said patents, and that the title to the lands passed to the Central Pacific Railroad Company by the acts of congress of July 1, 1862, and of July 2, 1864, the said company having complied with the conditions of the grant to it, and constructed the road and telegraph line designated therein; and that the said company is entitled to a patent of the United States for such lands. No costs will be allowed to the complainants."

Demonstrated by the foregoing decisions, from two different land systems, that if rules and proper procedures are not followed, the results and efforts are likely to be void or ineffectual.

One of the earliest decisions is *Bryan v. Beckley*[9] decided in 1809 by the Kentucky court, presenting not only the principle, but the reasoning behind it. This decision stated, among a series of principles, that existing lines and corners are to govern, however variant from the courses called for. Judge George Mortimer Bibb wrote in his decision, "as to those lines which are extant, or whose bearings are ascertained by existing corners, they are to govern, however variant from the courses called for. It is to the lost lines and corners that these rules are to be applied; where the ancient boundaries are visible and identified, resort to courses and to distances is unnecessary. Visible and actual boundaries, as rules to govern the property of men, are far preferable to ideal lines and corners; but when these actual landmarks fail, we must resort to the next best evidences, courses and distances, as producing a reasonable degree of certainty, and a necessary security against the acts of the fraudulent and depraved, and against time and the elements."

## THE COURT SYSTEM AND A STANDARD

First, the definition of a standard is "that which is established by authority, custom, or general consent, as a model or example; criterion; or test."[10] The significance of a standard is that it sets a bar, which is the minimum requirement everyone is supposed to meet. In negligence actions, one of the first considerations is whether a standard exists within a profession, or calling. This is quickly followed with whether that standard, if it exists, has been breached. The duty owned in a professional endeavor, if not met, may easily result in negligence. Therefore, any professional should pay very close attention to standards, and the extent of their duty in any professional undertaking.

The recent case of *Newfound Management v. Sewer,*[11] set a standard for the practicing surveyor. Perhaps, more importantly, it succinctly summarized from a long line of related decisions, what a surveyor does, or should be doing:

> After a surveyor has completed a comprehensive review of *all available records, deeds and prior surveys*, the surveyor begins the field survey. Once in the field, the surveyor has a *duty* to make a diligent *search for all monuments referenced directly or indirectly* in the deed or property description that either occur naturally or were put in place by prior surveyors or other persons. (Emphasis supplied.)

## CONCLUSIONS OF LAW

This case requires the court to resolve legal issues which primarily relate to: 1) identifying and locating real property, and 2) the conveyance or ownership of real property. To identify or locate parcels of property the court must look at the original documents, such as deeds or survey reports, defining a parcel. By analyzing legal descriptions within deeds, comparing surveys and surveyor's reports, the court evaluates the legal description, weighing its contents, and related surveying documents to locate the parcel on the ground. To determine whether property has been conveyed or transferred properly, the court considers whether the parties have transferred title to land pursuant to the Virgin Islands' recording statutes. The legal analysis required to determine

---

[9] 16 KY 91 (1809).
[10] Webster's New Collegiate Dictionary (1950).
[11] 885 F.Supp. 727 (U.S.Dist., 1995).

questions of ownership and title in this case is somewhat more discrete than the analysis necessary to identify a parcel's location.

Because there are numerous parcels at issue, repetitive legal questions arise concerning the location of property on the ground and the evaluation of surveying practices. To avoid duplicative analysis, in determining each parcel's boundaries, and to focus the reader's attention on the detailed facts at issue, the court has set forth the generally applicable law regarding these common underlying issues in the following preliminary sections. Specifically, these sections summarize law relating to locating property, surveying concepts, and the surveying traditions in the Virgin Islands.

## I. GENERAL SURVEYING PRACTICES

### A. *Background Research*

The court will first set forth basic principles of surveying based on its review of relevant treatises and case law as well as the expert testimony offered at trial by the parties.[21] A surveyor should strive first to locate and examine all historical records, deeds, prior surveys, maps and drawings in preparation for conducting an original survey. *See, generally,* CURTIS BROWN ET AL., BOUNDARY CONTROL AND LEGAL PRINCIPLES 371-74 (3rd ed. 1986) [hereinafter "BOUNDARY CONTROL"L"]; WALTER G. ROBILLARD & LANE J. BOUMAN, CLARK ON SURVEYING AND BOUNDARIES § 4 (5th ed. 1987) (hereinafter CLARK ON SURVEYING) If the surveyor is not performing an original survey, then the surveyor must also carefully review the original survey, as well as subsequent surveys or drawings. The purpose of thoroughly researching the history of a parcel of land is to ensure that the surveyor will be able to incorporate the most complete and accurate data into his or her survey. If a surveyor does not complete such research, the surveyor might perform the survey without having the benefit of essential information. For instance, the surveyor might not adequately search for crucial monuments or might misinterpret other field or documentary evidence. BROWN, BOUNDARY CONTROL at 371. In addition, if a surveyor knows that his or her survey will be used in a particular manner, a surveyor should review relevant documents and field surveys of adjacent parcels of land to ensure that his or her particular survey will be reliable and consistent with other existing surveys, so as to discourage litigation. *Id.* at 374.

### B. *Field Surveys*

After a surveyor has completed a comprehensive review of all available records, deeds and prior surveys, the surveyor begins the field survey. Once in the field, the surveyor has a duty to make a diligent search for all monuments referenced directly or indirectly in the deed or property description that either occur naturally or were put in place by prior surveyors or other persons. *Id.* at 371.

"Monuments have special significance because monuments indicate the location of property at issue on the ground. The search for monuments must continue until the monuments are located or until there is an explanation for their absence. *Id.* at 372. If necessary, the surveyor should consult former surveyors, landowners, residents, or other knowledgeable parties to determine monument sites or obtain other information tending to show where a piece of property should be located. *Id.* Testimony of neighbors and informed residents concerning boundaries is an important source of information for resurveys. *Maplesden v. U.S.,* 764 F.2d 1290, 1291 (9th Cir. 1985). As stated in one treatise, "[a] diligent, thorough, and complete search for all evidence is the fundamental essence of land surveying." BROWN, BOUNDARY CONTROL at 372. Through these investigative efforts, the surveyor attempts to reach his or her goal: the "location of land boundaries in accordance with the best available evidence" even though the best evidence may be "mere hearsay or reputation." *Id.* at 372-3; *see* Part II(B) *infra* on determining the order of importance of conflicting descriptive elements in a conveyance.

### C. *The Centrality of the Original Survey*

Since the physical position of monuments referenced in a conveyance reflect the original boundaries of a particular parcel, a subsequent surveyor must attempt to conform his or her survey as closely as possible to the prior surveyor's work. Hence treatises and courts frequently recite an admonishing maxim, namely that a surveyor must follow in the footsteps of the original surveyor. *See* Rudolph Galiber's Testimony (Tr. 1B, p. 35.), Marvin Berning's Testimony (Tr. 2, p. 112–114). The purpose and result of this principle is to give effect to the intentions of the parties at the time of the survey as well as ensuring the continuity of boundaries over time. Accordingly, "[t]he general rule governing the determination of boundary lines by resurvey is that the intent of the new survey should be to ascertain where the original surveyors placed the boundaries," not to determine new modern boundaries. *Thein v. Burrows,* 13 Wash.App. 761, 537 P.2d 1064, 1065 (Wash.App.1975); *see, also, U.S. v. Champion Papers, Inc.,* 361 F.Supp. 481 (S.D.Tex.1973) (boundary dispute involving 135-year-old survey resolved by the court's attention to totality of the evidence including evidence of the parties' intentions).

## II. DETERMINING THE INTENT OF PARTIES TO A CONVEYANCE

While a surveyor must aspire to walk in the exact steps of an original surveyor, sometimes a surveyor may be unable to find monuments placed by the original surveyor because the monuments may have been obliterated or lost. When a surveyor is unable to follow the precise "footsteps" of his or her predecessor, then a surveyor must attempt to track the original surveyor's work using whatever recoverable evidence that exists. *See, generally,* ROBILLARD, CLARK ON SURVEYING § 14 (section on tracking a survey); 11 C.J.S. § 61. Ultimately, a surveyor may only be able to "say with a great degree of certainty, 'this is where the surveyor walked.'" *See,* BROWN, BOUNDARY CONTROL at 294.

A. *Original Survey Lines or Lines of Possession?*

When a surveyor has difficulty retracing the original surveyor's steps, either because field evidence is missing or conflicting, certain principles guide the evaluation of existing field evidence. First, because original lines control other information contained in the conveyance, a surveyor should determine whether or not a line of possession, such as a fence, marks the location of the original survey line. *See* ROBILLARD, CLARK ON SURVEYING § 16.17. For instance, if the possession line is marked by an old boundary fence erected at approximately the same time as the original surveyor ran the lines, the fence may memorialize the survey line itself. BROWN, BOUNDARY CONTROL at 372. A surveyor's determination that a line of possession corresponds with an original survey line should be made according to the best evidence available which may include testimony of residents and the evaluation of the age of fencing or other natural monuments. *Id.* In addition, where surveyors disagree on the location of property lines and where a true survey line may be uncertain, monuments, such as fences which mark a possession line and which were established soon after the original survey, will control. *Id.* at 89 and 93.

"In the context of a surveyor's inability to locate original monuments or the original survey lines, lines of possession may become significant precisely because they give effect to the conveyer's intentions. This is particularly true when a conveyance contains a written statement describing these intentions. Accordingly, where a deed contains such a recitation of the parties' intentions, a surveyor should compare all of the conflicting descriptive elements, such as lines of possession, monuments, and acreage, and give the most weight to the element or elements which best effectuates the intentions of the parties to the deed. *See* BROWN, BOUNDARY CONTROL at 82.

B. *An Ordering System*

1. In The Virgin Islands

When a legal description in a deed is ambiguous, territorial law guides a surveyor's efforts to resolve unclear terms. *See* 28 V.I.CODE ANN. tit. 28 § 47 (Supp.1978)[23]; *see, generally, M.B.M. Inc. v. George,* 655 F.2d 530, (3d Cir. 1981) ("sufficient circumstances" supported the proper construction of the conveyance, including evidence of the grantor's intent, so that the trial court did need not resort to the statute to interpret the deed); *Roebuck v. Hendricks,* 255 F.2d 211, 211-12 (3d Cir. 1958) (a legal description referring to the conveyance of an undivided parcel of property merely in terms of acreage was overly vague and unenforceable). When a surveyor construes a description in a conveyance, the surveyor will look first to permanent or ascertainable monuments or boundaries, then to lines, angles, and finally surface area. Permanent or ascertainable monuments or boundaries are paramount over lines, angles or surface area in that order. These Code provisions have guided the court's consideration of legal descriptions contained in conveyances and its review of surveyors' interpretations of these same descriptions.

"Territorial law parallels general surveying conventions. In general, if the boundaries of a parcel are unclear, the deed or other historical documents include conflicting descriptive elements, and there is no one element which expresses the concerned parties' intent, then a surveyor may turn to a widely accepted ordering system. Courts and surveyors use this ordering system as a means of weighing and choosing between different descriptive elements. For example, the relative importance of conflicting elements is, in descending order, a) original surveyed lines, b) natural monuments, c) artificial monuments, d) metes and bounds descriptions, e) courses and distances, and f) quantity and acreage. *See, generally,* 11 C.J.S. *Boundaries* § 49-57; *see also Kruger & Birch v. Du Boyce,* 241 F.2d 849, 853 (3d Cir. 1957).

2. Monuments

Existing and undisturbed monuments called for in a conveyance are afforded the most weight and are given precedence over distance, direction or area. BROWN, BOUNDARY CONTROL at 89. In turn, natural monuments like a well-known tree or large rock, take precedence over artificial monuments like a fence, stake or ditch, because they are fixed and naturally occurring. Since monuments or objects afford greater certainty than computations of courses or distances, the "true intention of the parties will more probably be ascertained by adopting the call for natural monuments." *Kruger & Birch,* 241 F.2d at 853; *see also U.S. v. Doyle,* 468 F.2d 633, (10th Cir. 1972).[24] If a monument is obliterated the testimony of residents, witnesses, or other surveyors may reestablish its original location. BROWN, BOUNDARY CONTROL at 87.

3. Acreage

In contrast to monuments, the most credited elements of a description according to the canons of construction, quantitative elements, such as stated acreage, have the least relative importance. "In the determination of boundaries of land, quantity or area has been variously declared, with qualifications ..., the least certain or reliable element of description, ... without weight or effect, ... the last element to be resorted to." 11 C.J.S. *Boundaries* § 57 (1938). When a conveyance includes a description of acreage combined with the words "more or less," the element is a recognized approximation. As such, acreage is less reliable as a means of determining the location and boundaries of the described property when more substantial evidence exists.

"Consequently, if an individual purchases property from a seller who intended to convey certain defined property described in the conveyance by metes and bounds and approximate acreage "more or less," the stated acreage loses its authoritative value if a subsequent survey shows that the property is larger or smaller than the stated acreage. *Thorp v. Smith,* 344 F.2d 452, (3d Cir. 1965). The purchaser's holdings are limited by the seller's intent to convey the certain parcel as described by metes and bounds. The stated acreage does not entitle the purchaser to any more or less property. In *Thorp,* the court stated the "phrase 'more or less" indicated merely that 13 acres was the approximate and not the precise area of the parcel of land which was conveyed by designation. While a survey may demonstrate that a stated acreage amount is incorrect, a survey may not carve out (or eliminate) parcels of adjacent land and add them to the first parcel to increase (or diminish) the acreage to conform to the quantitative description in the deed. Similarly, in *Pendall v. Virgin Islands Title & Trust Company,* 6 V.I. 105, 279 F.Supp. 733, (D.C.V.I.1968), where plots of land were pointed out to the purchaser and described by metes and bounds on a survey plan, the deed's description by metes and bounds prevailed over an inconsistent reference to acreage.

4. The Cumulative Weight of the Evidence

Even though monuments usually control other inferior descriptive elements, occasionally, upon examination of all of the different elements, a surveyor may conclude he or she should follow the inferior elements called for in a conveyance rather than a particular monument. Surveyors should be sensitive to the weight of the evidence when all the relevant elements are considered. For instance, a surveyor may locate property according to the distances and area described in a deed rather than relying on a monument because the distances and area taken together seem to better reflect the original intentions of the parties to the conveyance. Better surveying practice requires a surveyor to evaluate initially all of the available evidence, even if ambiguous, regardless of its character. Then the surveyor should draw his or her conclusions based on the most persuasive information, rather than blindly relying on an abstract ordering scale to evaluate evidence on his or her behalf. BROWN, BOUNDARY CONTROL at 88-9."

## WHY IT IS NEEDED

The reasons for going back to the beginning owe their basis to the fundamentals of proper practice. In addition to the identification of title and people's rights, there are a number of other reasons why it is necessary to trace a property back to its origin. These may be thought of as safeguards for the surveyor, in that they will help avoid the misidentification and exclusion of important evidence which will in turn assist in avoiding mistakes. They are such checks on one's work that it is an amazement why every retracement surveyor does not include them as routine procedures.

**Author's Note:** Think of it as a comfort level, or a security blanket.

- First and foremost, to trace back to the creating document, whether it be an original survey, or the source of title. As explained in detail in Chapter 4 (see Figure 4.4), there may be more than one original document to consider, perhaps several.

In an extreme case, which happens more often than one might imagine, there may be no basis of title. See case discussion later in this chapter.

Without it, one cannot know what one began with nor can one tell if later descriptions and documents are accurate or even relevant.

- **To search specifically for an original survey**, or to determine whether the parcel in question has indeed been surveyed (located and/or measured). It is necessary to compare original calls with today's information to determine whether the current information is reliable, incorrect, or misleading. Since the rule is to follow the original survey, there must be a determination of whether there is an original survey in order to follow it. As treated in some detail, if there is no original survey, there are appropriate allowances for the correct procedure. The procedures applied to an original survey are not the same as applied to other foundations of original title.
- **To check for transcription errors**. Without comparing the current description with the original, there is no way of knowing if the current one is accurate, or even the right parcel of land. The more descriptions are copied from deed to deed, the more chance there is for errors in copying. Common problems include transposition of numbers, spelling, and punctuation errors, and words and even whole lines or sentences inadvertently omitted. Occasionally it will be encountered where mathematical conversions have been

made in an effort to "update" or "upgrade" the description, and are found to have been done incorrectly. In addition, some transcribers having difficulty reading or attaching meaning to certain words, phrases, and handwriting style, will often insert their interpretation of what they are copying. This is a very dangerous practice and leads to erroneous and misleading alterations to sound descriptive information. These changes need to be found, identified, and corrected.

Reading handwriting from early documents has always been a problem, and people have often interpreted handwritten words incorrectly. Words in a deed are presumed to have a purpose, and every word must be given meaning, treating none as surplusage.[12] Obviously, to do that, one must have the correct word, otherwise the entire meaning of a document may be changed, affecting the true intent of the parties, a paramount consideration in all jurisdictions.

- **To determine the base year for the computation of change in magnetic declination.**
  At times, the only way corners can be found is through an investigation into "following the footsteps" of the original surveyor, or a previous surveyor. Most early surveys were magnetic surveys, and changes over time must be accounted for. Without using the earliest date attainable, conversion is, at best, only a guess, and may contain considerable error thereby leading to some arbitrary point a considerable distance from the correct one. Depending on location, even an error of 50 years could mean a difference of several degrees, resulting in the failure to find an existing corner, or setting a new one in the wrong place. An error of 1 degree will result in a distance error of 92 feet in a distance of 1 mile. This is obviously significant within the PLSS, such as following a section line, but also translates to significant error in other systems. For example, an error of one degree can mean a 10-foot difference in a distance of 500 feet. Consider an error of several degrees, placing the retracement surveyor substantial distance from the designated corner. In many areas, with extensive brush and marsh, the chances of success in finding an existent corner can be significantly diminished.

The court system has dealt with this issue numerous times, always with the same result: changes in magnetic meridian must be taken into account. There are numerous early decisions from the mid-Atlantic states, including one from the Supreme Court of United States, authored by Chief Justice John Marshall.[13]

The case of *M'Clintock v. Rogers*[14] previously mentioned, contains an interesting statement by Justice Caton, who delivered the opinion. "Unless the compass is beyond the influence of disturbing causes, and the surveyor is very careful in adjusting it properly, and noting minutely the variation at which the line is run – and we know the date of the survey – so that the increase or decrease of the variation since, can be added or deducted, no surveyor can ever feel confident that he is running even very near to the line traversed by his predecessor, and by whose minutes he is working."

In the early case of *Bryan v. Beckley*[15] previously noted, Judge Bibb wrote:

The variation of the magnetic meridian from the true meridian is recognized by the statutes, and by the former opinion of this court. That such variation was eastwardly of the true meridian at the time of the original survey, (in 1774), that it had progressed eastwardly from that time until the time of

---

[12] *Woods v. Seymour,* 183 N.E. 458, 350 Ill. 493 (1932).
[13] *M'Iver's Lessee v. Walker,* 17 U.S. 444, 4 Wheat. (U.S.) 444, 4 L.Ed. 611 (1815).
[14] 11 Ill. 279 (Ill., 1849).
[15] 16 Ky 91 (1809).

making the survey, preparatory to the decree now complained of, is one of those principles acknowledged by scientific men, which this court are bound to notice as relative to surveys, as much as they would be bound to notice the laws of gravitation, the descent of the waters, the diurnal revolution of the earth, or the change of seasons, in cases where they would apply. The particular degree of variation is difficult to be ascertained, and the method of ascertaining the variation to be allowed, was directed in the former opinion. From the acknowledged eastwardly variation at the time of the original survey, compared with the survey upon which the decree complained of is predicated, it is evident that the courses of Beckley's military survey, as inserted in his grant, reciting the survey, were ascertained by the magnetic needle without regard to the true meridian; and such we believe was the prevailing practice of the country at the time of that survey. Perhaps not a single instance of a contrary practice has occurred in original surveys; at least, not an instance to the contrary has fallen under the notice of this court. Applying these observations to the case before us, it is apparent, that a gross error was committed in running the existing line of Beckley's survey; for the line, as now reported, is south 69 1-2 degrees east, which would make the variation of the magnet westwardly instead of eastwardly. This mistake is farther proved, by adverting to the course of the line A B, as now reported; for it is only south 43 degrees west, instead of 45 degrees, as called for in the grant, that is to say, a variation of two degrees eastwardly, had intervened since 1774, which, it is believed, is about the average variation which has been found by those who have traced a considerable number of the surveys of 1774, and compared the variations generally in those surveys with the variations of the magnet about the year 1802. But whatever might have been the variation in 1802, the time when the preparatory report was decreed upon by the court, whether two degrees eastwardly, or more or less, it is evident that the circuit court erred in communicating to the line B C, the mistake, and westwardly variation of the existing line E A, which is contrary to the first and sixth principles recited. For it is evident that the course of the line B C, has been made to depart from the course called for in the grant, and without necessity. Neither is this departure owing to a due allowance for the magnetic variation; but an undue and improper westwardly variation has been given to it, instead of an eastwardly. The circuit court have no doubt fallen into this error by attending to the direction, that this line should be parallel to the given or existing line, without examining that direction in conjunction with the principles of the decree; but nakedly and abstractedly, as well from those principles as from the additional light afforded by the report to which that direction has been applied. According to the grant, those lines appeared parallel; the court of appeals, not informed of the mistake committed in the existing line, directed the corresponding line to be renewed by running a parallel. But it is evident from the first, second, third, and fifth principles extracted from that decree, that if the court had been informed of the mistake existing in that line, they could not have directed a parallel to it, without doing violence upon the very principles which they meant to preserve undefiled. The mind cannot for a moment assent to the proposition that by any rational construction of the former opinion, the court have intended, by introducing the word "parallel," to produce a departure from the course of that line as expressed in the grant, after due allowance was made for the variation of the magnet. If any intention of departing from course, except from inevitable necessity, had been intended, the direction to connect the extremities of the two lines, the one of 500 poles, and the other of 460 poles, would have been nugatory; for a line run from those extremities to connect them, would have closed the survey. But so to connect those lines, did not comport with the views and intention of the court. Such a connection might have, and in fact would have, produced a departure from the patent course expressed for that line. For it is a truth, that the courses and distances named in the grant, without any variation, without mistake in any one instance, without any mistake in course, and tried upon a perfect plane, will not close the survey. To guard against such an event, and against mistakes, the court, therefore, added that these extremities were to be connected by a line "parallel to the line which connects the two first mentioned corners"; which provision was only tantamount to a direction to observe the patent course, for in the patent those lines were apparently parallel, each to the other. The opinion does not notice any mistake or departure in the given line. The very decretal order, declares that Beckley was to have a decree for so much of Bryan's settlement survey, as should be found to be within Beckley's military survey, when its lost lines and corners are run and fixed conformably to the exception contained in the aforegoing opinion; "and that exception expressly declares, that the two lost lines are to run from the corners extant, and extended on the course and to the distances called for in the plat and certificate of survey, making a proper allowance on each line for the unevenness of the ground over which it passes," and also to preserve the course of the other lost line, as called for in the same." Hence it may be affirmed with confidence, that a departure from course is repelled by every part of the former decision. The transfusing a mistake in one course into another course expressed in the patent, and thereby producing a departure from that other course, as is done by the decree complained of, was contrary to the manifest spirit and intention of the former decree of this court in the premises.

*Bryan v. Beckley,* **16 Ky. 91 (1809)**

This decision is so well reasoned and explained that the principle has never been altered over time and through the examination of countless retracements of early surveys, the vast majority of which were run by the magnetic needle.

**Author's Note:** It is nothing short of astonishing the lack of understanding and application of the principle of correcting for the variation in the magnetic needle over time. Lines defined by magnetic direction at a point in time do not change their position, even though their definition continually changes due to the movement of the magnetic pole. This has been a scientific fact for a long time,[16] and has been documented in scientific as well as surveying literature many times. As is pointed out several times, directions and distances are somewhat unreliable since they inherently contain error, some of which cannot be identified or accounted for, especially with distance measurement. However, changes in magnetic readings are well documented and simply determined. There should be no reason for a retracement surveyor not to routinely factor in this difference. If done correctly, they would likely find an increase in their success rate of finding original evidence.

## Some Supporting Court Decisions Demonstrate This as a Requirement in State Law

Perhaps the earliest court decision to discuss magnetic variation and the importance of allowing for it in following or retracing a previous survey, particularly an original survey, is the early Kentucky decision of *Beckley v. Bryan.*[17] This decision analyzed, in the eyes of the law, the difficulties in following an original survey: "It is equally necessary to premise, that the variation of the magnetic needle from the true meridian, which is different at different times, and at all times hard to be ascertained with accuracy – the inadvertent deviations of surveyors from the courses they meant to pursue – the carelessness and the involuntary mistakes of chain carriers, and the greater unevenness of the ground whereon one line of a survey is run than that of another, for which it is nearly impossible to make an accurate allowance, are the principal sources of the difficulties arising on this question, for the removal of which our land law contains no provision, only as to that arising from the variations of the magnetic needle, which, it is believed, has not generally been carried into effect with any reasonable degree of precision. Be this as it may, from this provision it is clear that the legislature considered the courses specified in a plat and certificate of survey as the principal guide to recover lost lines and corners. To confirm this position, it might be observed that for any other purpose the provision is wholly unnecessary. If the lines and corners of all survey could be shown, the true meridian and the variations of the magnetic needle from it would be of no importance in adjusting claims to land. Considerable aid may also be derived from the distances called for in plats and certificates of survey, yet, when a departure from either course or distance becomes necessary, reason as well as law seems to suggest that the

---

[16] The first extensive study of the earth's magnetic field, resulting in the first declination chart, was a three-year endeavor across the width of the Atlantic Ocean by Edmond Halley (1656–1742), published in 1701.

[17] 2 Ky. 91 (Ky.App. 1801).

distances, taken in our mode of mensuration, ought to yield, as being much the most uncertain of the two. Indeed, it never has been held, that in laying off vacant land purchased from government, that a line established by a surveyor can be altered on account of its being longer or shorter than the distance specified in the plat and certificate he has returned.

"It will, however, be understood that in all cases where lost corners and lines are to be renewed, due allowances must be made for the variation of the magnetic needle from the true meridian. Where it shall be found that the direction of the land law, in this respect, was accurately complied with at making the survey, a legal standard is fixed. But where it was wholly omitted, or carelessly done, the defect may be rationally supplied by ascertaining from other lines of that or the neighboring surveys what was the difference of the variation, at or about the time the lost lines were made, from the variation at the time being."

Followed by *Bryan v. Beckley*[18] a few years later, Judge Bibb emphasized the following points:

- Allowances to be made for variation of the needle.
- Court bound to take notice that there is a magnetic variation from the true meridian.
- Surveyors generally took their courses from the magnetic meridian.

The variation of the magnet since the original survey, is the allowance to be made.

"The variation of the magnetic meridian from the true meridian is recognized by the statutes, and by the former opinion of this court. That such variation was eastwardly of the true meridian at the time of the original survey, (in 1774), that it had progressed eastwardly from that time until the time of making the survey, preparatory to the decree now complained of, is one of those principles acknowledged by scientific men, which this court are bound to notice as relative to surveys, as much as they would be bound to notice the laws of gravitation, the descent of the waters, the diurnal revolution of the earth, or the change of seasons, in cases where they would apply. The particular degree of variation is difficult to be ascertained, and the method of ascertaining the variation to be allowed, was directed in the former opinion. From the acknowledged eastwardly variation at the time of the original survey, compared with the survey upon which the decree complained of is predicated, it is evident that the courses of Beckley's military survey, as inserted in his grant, reciting the survey, were ascertained by the magnetic needle without regard to the true meridian; and such we believe was the prevailing practice of the country at the time of that survey. Perhaps not a single instance of a contrary practice has occurred in original surveys; at least, not an instance to the contrary has fallen under the notice of this court. Applying these observations to the case before us, it is apparent, that a gross error was committed in running the existing line of Beckley's survey; for the line, as now reported, is south 69 1-2 degrees east, which would make the variation of the magnet westwardly instead of eastwardly. This mistake is farther proved, by adverting to the course of the line A B, as now reported; for it is only south 43 degrees west, instead of 45 degrees, as called for in the grant, that is to say, a variation of two degrees eastwardly, had intervened since 1774, which, it is believed, is about the average variation which has been found by those who have traced a considerable number of the surveys of 1774, and compared the variations generally in those surveys with the variations of the magnet about the year 1802. But whatever might have been the variation in 1802, the time when the preparatory report was decreed upon by the court, whether two degrees eastwardly, or more or less, it is evident that the circuit court erred in communicating to the line B C, the mistake, and westwardly variation of the existing line E A, which is contrary to the first and sixth principles recited. For it is evident that the course of the line B C, has been made to depart from the course called for in the grant, and without necessity. Neither is this departure owing to a due allowance

---

[18] 16 Ky. 91 (Ky.App. 1809).

for the magnetic variation; but an undue and improper westwardly variation has been given to it, instead of an eastwardly. The circuit court have no doubt fallen into this error by attending to the direction, that this line should be parallel to the given or existing line, without examining that direction in conjunction with the principles of the decree; but nakedly and abstractedly, as well from those principles as from the additional light afforded by the report to which that direction has been applied. According to the grant, those lines appeared parallel; the court of appeals, not informed of the mistake committed in the existing line, directed the corresponding line to be renewed by running a parallel. But it is evident from the first, second, third and fifth principles extracted from that decree, that if the court had been informed of the mistake existing in that line, they could not have directed a parallel to it, without doing violence upon the very principles which they meant to preserve undefiled. The mind cannot for a moment assent to the proposition that by any rational construction of the former opinion, the court have intended, by introducing the word "parallel," to produce a departure from the course of that line as expressed in the grant, after due allowance was made for the variation of the magnet. If any intention of departing from course, except from inevitable necessity, had been intended, the direction to connect the extremities of the two lines, the one of 500 poles, and the other of 460 poles, would have been nugatory; for a line run from those extremities to connect them, would have closed the survey. But so to connect those lines, did not comport with the views and intention of the court. Such a connection might have, and in fact would have, produced a departure from the patent course expressed for that line. For it is a truth, that the courses and distances named in the grant, without any variation, without mistake in any one instance, without any mistake in course, and tried upon a perfect plane, will not close the survey. To guard against such an event, and against mistakes, the court, therefore, added that these extremities were to be connected by a line "parallel to the line which connects the two first mentioned corners"; which provision was only tantamount to a direction to observe the patent course, for in the patent those lines were apparently parallel, each to the other. The opinion does not notice any mistake or departure in the given line. The very decretal order, declares that Beckley was to have a decree for so much of Bryan's settlement survey, as should be found to be within Beckley's military survey, when its lost lines and corners are run and fixed conformably to the exception contained in the aforegoing opinion; "and that exception expressly declares, that the two lost lines are to run from the corners extant, and extended on the course and to the distances called for in the plat and certificate of survey, making a proper allowance on each line for the unevenness of the ground over which it passes," and also to preserve the course of the other lost line, as called for in the same." Hence it may be affirmed with confidence, that a departure from course is repelled by every part of the former decision. The transfusing a mistake in one course into another course expressed in the patent, and thereby producing a departure from that other course, as is done by the decree complained of, was contrary to the manifest spirit and intention of the former decree of this court in the premises."

A few years later, the same court said the following[19]: "We are of opinion, that a correct exposition of the former opinion and decree of this court, requires that Beckley's line, which is designated as running N. 20 E. 460 poles, should be run by making a due allowance for the variation of the magnetic, from the true meridian, from the time the survey was made until the present. This variation is reported by the surveyor to be three degrees and thirty minutes; and, consequently, the course of that line should have been N. 16 1-2 E. according to the present magnetic meridian, instead of N. 20 E. as the court below decreed it should be."

The importance of identifying the date of the original survey for the magnetic changes to be made is emphasized in the decision found in *Bodley v. Hernden*[20]:

---

[19] *Bryan v. Beckley*, 16 Ky. 100 (Ky.App. 1814).
[20] 10 Ky. 21 (Ky.App. 1820).

A line run shortly after the date of an entry with an intent to execute a survey thereon, and pursued when the survey was afterwards actually made, is evidence of the magnetic variation at the date of the entry, and shall form the boundary, regardless of a subsequent variation or the true meridian.

The court below adopted the present magnetic line, and decreed accordingly, disregarding the other modes of surveying; to reverse which decree, this writ of error is prosecuted.

If the true meridian was the true position, which ought to have been given to the original survey, it is evident, that Bodley can have no claim to run to that line, because his old survey did not go to it, and he cannot be entitled to a decree for any land which his ancient survey did not include. If, however, the magnetic line at the date of the entry is to govern as was intimated by this court in the case of Vance against Marshall, 3 Bibb, 148, then his old line may set up a formidable claim, when opposed by the present magnetic line; and if this ancient survey was made according to that line, no good reason can be given why he should not be allowed to recover to it. It seems evident, from the facts in the cause, that the old line is truly shown from A. to 14 running north, and that the line east from 14 to 21 is at right angles thereto. It is proved that Tebbs, in the year 1784 started from the beginning and made a considerable part of the survey; but from some cause or other not explained did not finish it. In the year 1796, Payne made the survey and swears that in doing so he pursued the old marked line then existing on the ground, and according to them the survey has been reported in the cause. To impeach this, the defendant's counsel caused the surveyor to state sundry declarations of Payne, the original surveyor. The report of the surveyor is not evidence of these statements, nor can it prove that these statements were ever made. It is true that the report of the surveyor made under an order in a cause is evidence; but it is only so with regard to such facts as he can report officially; such as the existence of objects found by him existing on the ground, and their identity as shown to him by others. But he is not authorized to report the sayings and doings of others, touching the merits of the controversy, and thereby introduce them into the cause as good evidence. The same may be said with regard to his reporting the acts of Fulton, another surveyor, and the fact stated by him, that the compass which he used in making the survey, was the same which made the original. These facts must be disregarded in giving a decision in this cause. It is plain that the old line run in 1784, only one year after the entry was made, varied eastwardly from the true meridian. It is also shown that the variation has progressed still more eastwardly since; for the present magnetic line presents a still greater variation. As it was usual in this country to make original surveys by the magnetic meridian, as is evidenced by most of the plats and certificates and numerous resurveys which have come before this court, we think it fair to presume, in addition to the facts already stated that the line of 1784 came nearer the magnetic line at the time the entry was made than any other; and this having been adopted by the surveyor, Payne, when he completed the survey, we conceive that Bodley is entitled to the full benefit of this line now; and that the court ought to have decreed in his favor according to that position given on the plat, as the one which he, at least, was entitled to.

The New Jersey court in the case of *State v. Schanck*[21] analyzed a highway layout in terms of the surveying done, raising questions about the work, and suggesting possible interpretations.

The reason is, that the return made by the surveyors is vague and uncertain. It is insisted to be so, first, because the several courses of the road set forth in the return are said to be as the magnetic needle of the compass of Leonard Walling, the practical surveyor engaged in laying out the road, pointed on the 17th and 18th of August, 1824, the days on which they were run out. In this mode of expression we see nothing of uncertainty. It is no more than a designation of the instrument by which the observation of the courses was made. It is admitted that it would have been correct for the surveyors to have said, we run N. 35 [degree] 43 E. Can it then be uncertain to say, what is really the fact, and what is always understood, though not always expressed, as the needle pointed which was used by us? The remark made by the counsel of the defendant in certiorari is just. It really adds to

---

[21] 9 N.J.L. 107 (N.J. 1827).

the certainty of the return. Let us suppose it to be requisite some twenty years hence, to ascertain the exact location of these courses when the monuments mentioned may be gone. What variation shall the surveyor allow? What allowance shall he make for the known difference of instruments? His answers to these questions must be conjectural, as these returns are commonly drawn. But in the present case, he will resort to some line in the neighborhood known also to have been run by the compass of Leonard Walling, and his answers will then approach, if not attain, to certainty. Second: The return was said to be uncertain because the route had been run on a subsequent day by James H. Newell and James Robinson, deputy surveyors, with their compasses, who agreed in their observations, and found a small difference of course on the first line, which, followed out through the courses of a road of several miles, found at the termination a departure of about four chains. By the depositions of Leonard Walling, and of Thomas Debow, another surveyor, it appears they also, at a subsequent day, again run out the road, each with his own compass. They agreed, and found the original survey, as they believed, entirely correct. Who then can say the courses contained in the return are incorrect? Is it shewn that the instruments of Newell and Robinson were more sure and perfect than those of Walling and Debow? Or that the former were more skillful and experienced artists than the latter? The difference between them is a very few minutes, and is nothing more than the variance common among instruments, and very common among surveyors. If a return is to be set aside because an instrument and surveyor, not shewn to be in the slightest degree superior, nay, not even shewn to be equal, in accuracy and skill, to the original instrument and surveyor, differ somewhat in the survey, is it not manifest that the act for laying out roads is repealed without the will or intervention of the legislature?

Stressing the magnitude and effect of a change between two surveys over time, the New Jersey court in *Scott v. Yard*,[22] had this to say:

The proofs, I think, also make it tolerably clear that the surveyor who located the Bell tract meant to run the first line of that tract precisely on the same course on which the closing line of the Hartshorne tract had been run. The Bell survey, as already stated, was made sixty-seven or sixty-eight years after the Hartshorne survey *had been made*. This lapse of time had made a considerable change in the course which the magnetic needle would indicate, placed at exactly the same point where it had been placed in 1745. There is a difference of three degrees and fifteen minutes between the courses on which the two surveys were made. Is this variation so great, having regard to the period of time which had elapsed between the two surveys, as to make it safe to assume that the surveyor who made the last survey intended to establish a new and independent line, and not to make his line coincident with the one previously established? On the argument both parties made use of the report of the United States Coast Survey for 1886, showing the secular variation of the magnetic declination in the United States for the period intervening between 1745 and 1813. If we take the three points given in this report nearest to the land in dispute, namely, New York, Philadelphia and Hatborough, Pennsylvania, and calculate the observed variation for the period in question, it will be found that it was two degrees and five minutes at New York, three degrees and forty minutes at Philadelphia and three degrees and seven minutes at Hatborough. So that, according to the data furnished in this report, a course which, in 1745, stood at New York, Philadelphia and Hatborough, south, fourteen degrees east, would, in 1813, stand at New York, south, sixteen degrees and five minutes east; and at Philadelphia, south, seventeen degrees and forty minutes east; and at Hatborough, south, seventeen degrees and seven minutes east. The last line of the Hartshorne tract was run, it will be remembered, in 1745, on a course south, fourteen degrees east, and the first line of the Bell tract was run in 1813 on a course south, seventeen degrees and fifteen minutes east. The difference in the course shown by the two surveys may, I think, be reasonably and satisfactorily explained by the change which sixty-seven years had made in the position of the magnetic needle, to say nothing of the imperfect character of the instruments in use both in 1745 and 1813, nor of the careless manner in which both surveys were probably made.

---

[22] 46 N.J.Eq. 79, 18 A. 359 (N.J.Super.Ch. 1889).

**Author's Note:** This is always a common consideration when comparing two surveys, or aiding one, to rely on another made at a different time. The question is whether they can be considered together as reported, or if they must be treated differently, in order to put them both in the same time frame.

Emphasizing again the difference in variation between localities, the Vermont court[23] reported: "It is common knowledge among surveyors that in ascertaining the locality of an old line an allowance must be made for the variation of the magnetic needle, and that that allowance is not a matter of exact computation. The variation is different in different localities. In running this line from the conceded point the parties came out twenty-two inches apart on Pearl street; hence both relied upon the fence as indicating the true division line."

Considering both magnetic variation and distance measurement, and factoring in a subsequent agreement, the West Virginia court in *Warren v. Boggs*,[24] made the following analysis. The conclusion in the decision was to following the original survey.

> Plaintiffs insist that the surveys should be made according to the courses, lines, distances, and corners denoted as of the date of the compromise agreement of May 12, 1913, allowing for the magnetic variation from that date, or, if the survey be made as of the date of the original survey in 1886, and the magnetic variation calculated from that date, then surface measurement should be invoked, the method of measurement then used, as expressly stated in the original deed from Smith and Wells to Hoff. Defendant asserts that the survey should now be made with magnetic variation calculated from 1886, and by horizontal measurement, under section 2, c. 67, Code (sec. 3685). It appears that all of the surveyors calculated the magnetic variation from 1886, each using practically the same degree of variation when running the lines. If the survey is to be made as of that date, and we think it proper to do so, the same measurement should be used as was then used–surface measurement. In that way only could the footsteps of the original surveyors be followed. In that way only can the boundaries be determined and fixed as they were then ascertained and fixed. Any other method will bring about a different result. That is fully demonstrated, for, by using horizontal measurement, the side lines have been shortened about 6 poles and the eastern line lengthened by 1 1/2 poles. These surveyors should, as nearly as possible, do exactly what the surveyor then did.

> Was it the intention of the parties, when they made the compromise agreement in 1913, that the 18-acre tract should be run out and located by the survey and deed as of 1886? Such seems to be the contention of defendant, for his surveyors have calculated and used the magnetic variation from that date. It follows that the surface measurement as set out in that deed must also be adopted. The beneficial portions of that deed cannot be taken by either party and the portions which are not beneficial ignored. The agreement refers specifically to the deed of 1886, and the courses, distances, and corners are copied "as therein bounded and described." Horizontal measurement should govern unless surface measurement is contracted for. Smith and Wells contracted for surface measurement with Hoff, and the parties here have contracted for the location of the land just as the original deed stipulated, by specific reference thereto.

In *Scott v. Jessee*,[25] the Virginia court considered a "rule of thumb" for considering magnetic variation, depending on the region. Customary allowances in measurements are sometimes the prevailing rule.

> In our view of the case, therefore, starting at the white oak (H on the sketch) in the relocation of this disputed line, the courses and distances of the Gabriel Jessee survey should have been followed throughout, with due and proper allowance for the variation of the magnetic needle with regard to the time between the original survey and its relocation.

---

[23] *Stiles v. Estabrook*, 66 Vt. 535, 29 A. 961 (Vt. 1894).
[24] 90 W.Va. 329, 111 S.E. 331 (W.Va. 1922).
[25] 143 Va. 150, 129 S.E. 333 (Va. 1925).

In renewing lost corners, allowance should be made for variation in the magnetic needle since the date of the original survey, and the courses and distances should not be departed from except when necessary. In making measurements allowance should be made for unevenness of the ground.

**9 Corpus Juris, p. 163, sec. 16(b)**

There was some difference of opinion between the surveyor who testified on the one side for the plaintiffs and the surveyors who testified on the other side for the defendants as to the allowance for variation of the magnetic needle. Mr. Dorton, when he undertook to locate the line in 1921, states that he used thirty minutes variation west, and he was locating the disputed line of a survey made in 1854, but admitted that the general rule in that community was to allow one degree for every twenty years. The surveyor, Taylor, made the same allowance. Baldwin, however, using the Thompson title bond dated 1848 when he made his survey in 1921, allowed a variation to the west of three degrees and twenty-one minutes from the original calls.

The Massachusetts court in *Crawford v. Roloson*[26] distinguished between elements of law and elements of fact, magnetic variation being of the latter.

We are unable to perceive any question of law in the various requests for rulings presented by the petitioner. They all appear to us to relate to matters of fact. Engineering theories as to the extent of the "variation of the compass, or magnetic, north from the true north," and as to the correct method of calculation of courses contained in instruments of divers dates and used with respect to land geographically distant from Boston, where it seems to be assumed that observations have been made, present no question of law. Where a line or lines described by specified courses and distances may fall upon the face of the earth or may be delineated upon a plan, is not a question of law but a matter of fact. Counsel for the petitioner has argued his contentions with great earnestness. But it would serve no useful purpose to examine his contentions one by one, because they are all resolved into disputes of fact and do not present any question of law.

Proper procedure is summarized in the West Virginia case of *State v. West Virginia Pulp & Paper Co.*[27] As recently as 1930, the court is relying on the standards reported in *Bryan v. Beckley* (1809), previously noted.

In restoring lost lines of an old survey, it is the general practice to begin at an established corner and to locate such lines according to the courses and distances mentioned in the survey. Allowances should be made for the variation of the magnetic needle from the true meridian as well as for the greater precision of modern surveying.

Allowances should be made for the variation of the magnetic needle from the true meridian as well as for the unevenness of the ground over which each line passes. *Bryan v. Beckley, Litt. Sel. Cas. (Ky.)* 91, 100, 12 Am.Dec. 276.

The case of *Parker v. T. O. Sutton and Sons*[28] notes the court's taking judicial notice of a scientific fact as well as what is necessary to follow original surveyors' footsteps at two different times.

The Hereford Survey was patented in 1948. R. A. Ellis surveyed the Hereford Survey July 25, 1942, using a variation of 9 degrees 20 minutes East, being the compass declination from true North Ellis used for magnetic North. The surveyor Owens made his survey in 1961 and used a variation of 8 degrees East for his compass declination from true North.

---

[26] 262 Mass. 527 (1928).
[27] 108 W.Va. 553, 152 S.E. 197 (W.Va. 1930).
[28] 384 S.W.2d 433 (Tex.Civ.App. – Beaumont 1964).

The variation from magnetic North and true North constantly changes according to the evidence in this case. Actually, this is a scientific fact commonly known, of which this court takes judicial notice. The surveyor Owens testified that in order to follow the footsteps of surveyors in running out old surveys of the age of the Hereford Survey he had found an 8 degree declination to fit their lines.

The more recent Maine decision of *Milliken v. Buswell*[29] relied on precedent discussed in earlier cases:

In *M'Iver's Lessee v. Walker*, 1815, 9 Cranch 173, 13 U.S. 173, 3 L.Ed. 694, Mr. Chief Justice Marshall said:

It is, undoubtedly, the practice of surveyors, and the practice was proved in this cause (the surveyors in the instant case confirmed the practice), to express in their plats and certificates of survey, the course which are designated by the needle; and if nothing exists to control the call for course and distance, the land must be bounded by the courses and distances of the patent, according to the magnetic meridian.

We need not presently decide whether in the construction of deeds the use of magnetic north instead of 'true' north is to be presumed as the New Hampshire Court ruled in *Wells v. Jackson Iron Manufacturing Company*, 1866, 47 N.H. 235.[30] But, in weighing evidence of recent surveys of ancient lines, consideration must be given to the variation of the needle in determining a magnetic course. *Brooks v. Tyler*, 1829, 2 Vt. 348.

*Brooks v. Tyler*, referenced above, stated as follows:

The declaration, after giving the *terminus a quo*, describes the land by courses and distances only, without any reference to the lines of the lot of which the land is alleged to be a part, or to any certain or natural monuments; and whether or not the defendant was in possession of any part of the land described, could be determined only by actual survey. When there is nothing stated to control the courses and distances, the lines must be run by the needle. – (*McIver v. Walker*, 13 U.S. 173, 3 L.Ed. 694, 9 Cranch 173.) – It is true, that the magnetic course is subject to variation and uncertainty; and in weighing evidence of recent surveys of ancient lines, regard may be had to the variation of the needle. But the declaration in this case contains no reference to the lines of the lot, or to any ancient survey; and we cannot understand the courses stated in the declaration, to mean courses as designated by the needle forty or fifty years ago, instead of courses designated by it now. If the defendant was in possession of land, belonging to the plaintiff as a part of lot no. 100, the plaintiff might have given a description, either by the lines of the lot, by metes and bounds, or by such courses and distances, as would, according to the course of the needle, have included the land of which the defendant was in possession. But as it was admitted on the trial, that the courses and distances stated in the declaration, if run according to the present direction of the needle, would not include any land of which the defendant was in possession, and there being no other description given, we are of opinion that the judgment of the court below was right and must be affirmed.

In summary, three important cases emphasizing the basic principles are as follows:

- In running line established in 1885, allowance must be made for variation of needle. *McCourry et al. v. McCourry*, 105 S.E. 166 (N.C., 1920).

---

[29] 313 A.2d 111 (Me. 1973).

[30] Reference may be had to this case for an extensive treatment and discussion of magnetic directions. The principles analyzed in this case are further discussed in Wilson, Donald A., *Boundary Retracement: Processes and Procedures*, CRC Press, 2017.

- Nothing else appearing, the calls in a deed must be followed as of the date thereof. Where it clearly appears upon the face of the deed or where the evidence shows, that a line as established on a prior date was adopted and was copied in the deed according to the courses and distances thereof, it is necessary to take into consideration the variations of the magnetic needle in locating the same. *Greer v. Hayes,* 216 N.C. 396 (1939).
- **Principle:** The importance of this cannot be over-stressed. While variation of the magnetic needle changes with the years, a line which has been run on the ground never changes. *Giles v. DiRobbio,* 186 Md. 258, 46 A.2d 611 (Md. 1946).

Unless otherwise required, such as by statute, it has long been the practice of surveyors to reference directions to the magnetic meridian, particularly in the early surveys. The majority of original surveys were magnetic surveys. In later years, the BLM Instructions required that original surveys be according to the true meridian. This has been common knowledge within the court system since the earliest boundary decisions. For example, detailed previously, *allowance must be made for the magnetic meridian.*[31]

In *M'Iver's Lessee v. Walker,*[32] Mr. Chief Justice Marshall said:

> It is, undoubtedly, the practice of surveyors, and the practice was proved in this cause [the surveyors in the instant case confirmed the practice], to express in their plats and certificates of survey, the courses which are designated by the needle, and if nothing exists to control the call for course and distance, the land must be bounded by the courses and distances of the patent, according to the magnetic meridian.

And in New Hampshire, the courses in a deed are to be run according to the magnetic meridian, unless there be something in the instrument showing that a different mode is intended.[33] The same could be said for any of the states, particularly those surveys prior to, and outside of, the Public Land Survey System (Public Domain).

To follow this procedure, it is necessary to determine, as closely as possible, the date of the original definition of a particular line. Any substitution of the prescribed procedure may easily result in an erroneous determination of the position of a designated (called-for) corner. This is likely a leading cause of "pincushion corners," along with errors in distance measurement using a variety of instruments and techniques, none of which duplicate those of the original survey, and therefore result in not following original footsteps.

- **To determine the order of conveyancing and senior rights.** Without knowledge of the precise date a parcel was created (or its title established), it is impossible to determine how it fits in the order of conveyancing, and the priority it should be given, if any, for reasons of apportionment when there is excess or deficiency, or when allowing for errors in measurement.

It must be remembered that measurements in documents of any kind, are only means to any end – to lead one from corner to corner, or, in some cases, monument to monument. If the corner can be located, or an original monument is found, and in place, it controls and any measurements are secondary, and must yield. Some retracement people have a difficult time with this concept, and believing that the original scheme or the intentions (not *intent*[34]) is what the parties meant

---

[31] *Bryan, & c. v. Beckley,* 16 Ky. 91 (1809).
[32] 9 Cranch 173, 13 U.S. 173, 3 L.Ed 694 (1815).
[33] *Wells v. Jackson Iron Mfg. Co.,* 47 N.H. 235 (N.H., 1866).
[34] Parties' intention and the intent of the parties, which by law is the controlling rule, are not the same thing. See following section for a detailed discussion of the role intent plays.

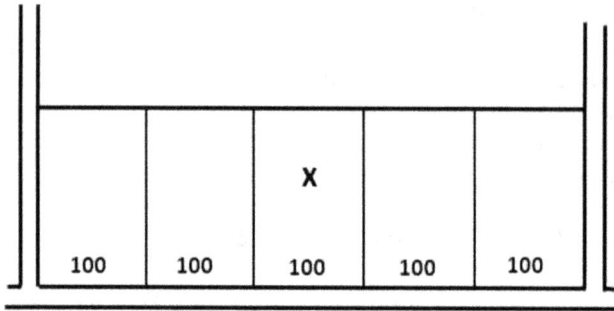

**FIGURE 5.1** Example of several possible outcomes to any one lot in a subdivision, contingent on a lot creation and its subsequent conveyancing.

and therefore should control. There are many situations where a corner, or a found monument, does not quite fit the reported measurement, and another is set to "correct" what is believed to be an error, or mistake. This practice is too prevalent and is contrary to the rule that corners created and markers set by the original survey control.

It must also be kept in mind that apportionment, known as proportioning in the PLSS, is a rule of last resort and is only appropriate when retracement fails.[35] If one knows and understands not only the rules, but also the proper procedures, used in retracement, the retracement will seldom, if ever, fail to reveal the position of a corner, or at least a rational opinion thereof.

A typical situation is presented in Figure 5.1.

In the above diagram, a 500-foot lot situated between two streets had been subdivided into five 100-foot lots, and all five subsequently sold by the subdivider. Each deed description calls for having 100-feet frontage on the street. However, the original 500-foot measurement was incorrect, and it is found that the block containing the five lots has only 490 feet, or a shortage of 10 feet. Some might be tempted to proportion the shortage such that each lot bears a deficiency of 2 feet. In some cases, with 100-foot zoning, the lots would only have 98 feet of frontage, rendering them non-buildable. There is only one out of six possible scenarios where this is possible.

Focusing on the middle lot (**x**) consider a subsequent survey resulting in setting markers at the corners. Keeping in mind recorded deeds state that each of the lots have 100 feet of frontage on the street.

The following scenarios are possible:

1. One or more original markers are found; they must be accepted regardless of measurements, so long as they have not been disturbed.
2. No original markers exist, and may never have been set. The question then becomes where to measure from to locate the lot in question. Given that the block has a 10-foot shortage, a decision must subsequently be made as to where to place any error(s).
3. The middle lot was conveyed first, making it senior in title, and therefore entitled to its entire 100 feet of frontage, leaving any shortage to be applied elsewhere.
4. The middle lot was conveyed last, being a remainder, and therefore can only have 90 feet of frontage since that is all that is left.

---

[35] See Griffin, *Retracement & Apportionment as Surveying Methods for Re-establishing Property Corners. Marquette Law Review*, Vol. 43, Issue 4, Article 5. Spring 1960.

5. The five lots were created simultaneously, giving them equal standing and sharing equally in any excess or deficiency. All lots would therefore be entitled to 98 feet of frontage.
6. Three lots were sold sequentially, leaving 200 feet by description, but actually only 190 feet. Then it is a question of how the remaining 200 feet is conveyed: sequentially, such that one of the lots is a remainder having 90 feet of frontage, or subdivided with the remaining 2 lots created simultaneously, each having a frontage of 95 feet, sharing equally in the deficiency.

With such a situation, it is tempting to divide any excess or deficiency proportionately through the subdivision of the parent tract. But, once again, (1) that is appropriate *only when retracement* fails to produce the location of corners, and (2) if the original *corners* cannot be located. Far too many situations exist created by lack of knowledge, impatience, or inexperience where apportionment was selected to make the difficulty disappear.

The basic problem then becomes that the lot staker has been preceded by a previous surveyor who has not followed any of the basic rules and has set one or more markers, thereby misleading the neighborhood, likely resulting in improvements either encroaching on someone else or in violation of setback requirements, and ultimately creating unforeseen problems for the next surveyor who is anticipating a simple lot stake-out.

- **To gain insight as to the intention**[36] of the parties at the time of original description. Many court decisions stress that it is the court's responsibility to determine the intent of the parties under the conditions and circumstances at the time of the document.[37] Names, conditions, laws, and customs change. While these problems will be discussed in some detail later, the chain of title example in Chapter 4 containing description changes demonstrates what can happen when the incorrect date is selected, or when situations are not placed in their appropriate time frame.

**Example:** During the trial of a title case, one of the descriptions stated that it was "4 acres, west of the railroad." It doesn't take a genius to realized that the call leaves a lot of latitude in that going west could cover a lot of territory. However, the testimony from a title examiner on the opposite side of the case was, under oath, that the parcel had to be located near the railroad. That, by itself, removed the parcel from the locus in dispute, which was later proved by additional evidence to the court' satisfaction.

In the New York case of Dolphin Lane, several attempts were made in court to resolve a dispute, to no avail. Finally, the court demanded a complete chain of title and a thorough study back to the original proprietors, stating, "the past must be explored to understand the present."

---

[36] "Intention" as applied to the construction of a deed is a term of art, and signifies a meaning of the writing. *26 C.J.S., § 83.*

Intent of the grantor as spelled out in the deed itself must be interpreted, not the grantor's intent in general, or even what he may have intended. *Wilson v. DeGenaro,* 415 A.2d 1334, 36 Conn. Sup. 200 (1970).

[37] The courts, in their endeavor to arrive at meaning of description in deed, should assume the position of the parties, consider circumstances of transaction and read and interpret words of instrument in light of those circumstances. *Brown v. Windland,* 457 S.W.2d 840, 240 Ark. 6 (1970).

Construction of a writing is for the purpose of determining the true intent of the parties to it, and, to this end, the subject-matter, the situation of the parties, and the surrounding circumstances at the time the writing was executed, should be taken into consideration. *Curtis v. Meadows,* 99 S.E. 286 (W.Va., 1919).

Fundamental rule of construction is that purpose or intent of a written instrument is to be determined from language used in light of circumstances under which it was written. *Phipps v. Leftwich,* 222 S.E.2d 536 (Va., 1976).

As is sometimes worded, *to determine the intent of the surveyor.*[38] Insight as to the significance of this may be found in the case of *Outlaw v. Gulf Oil Corporation*,[39] a case that appears several times because of its significance to surveying and to the philosophy of retracement.

> In the case of *Blackwell v. Coleman County*, 94 Tex. 216, 59 S.W. 530, it is said: 'In determining the location of land in such cases the courts seek to ascertain the true intention of the parties concerned in the survey. But the intention referred to is not that which exists only in the minds of the surveyor. It is defined as that which may 'be gathered from the language of the grant' or as 'the intention apparent on the face of the grant' (*Hubert v. Bartlett's Heirs*, 9 Tex. 97), or 'the legal meaning of the language of the patent when considered in the light shed upon it by the acts constituting the survey' (*Robertson v. Mosson*, 26 Tex. 248; *Robinson v. Doss*, 53 Tex. 496; *Brown v. Bedinger*, 72 Tex. 247, 10 S.W. 90; *Richardson v. Powell*, 83 Tex. 588, 19 S.W. 262). When reference is made in the decisions to the intention of the surveyor, the purpose deduced from what he did in making the survey and description of the land is meant, and not one which has not found expression in his acts. Grants are issued by the state and accepted by the grantees upon the acts done by the surveyor in identifying and describing the lands, and the rights of both are to be determined by the legal effect of those acts, and not by intentions which cannot be deduced from a construction of the descriptions in the grants with the aid of the facts constituting the surveys upon which they are based.' In short, it is the expressed intention of the grant that is the material inquiry.

> The inquiry * * * is directed to the survey as made, and not as it should have been made. *Forbis v. Withers*, 71 Tex. 302, 9 S.W. 154.

**Author's Note:** There sometimes is a serious misunderstanding as to what is meant by the "intent of the parties." This intent is the guiding principle of the court system, and the controlling consideration in the interpretation of documents. As previously noted, too often parties, as well as retracement surveyors, will speculate as to what the parties meant to do. What the parties meant to do is not to be considered and is merely an opinion, often without sound basis. In following this practice, when two people have opposing opinions, whether landowners or surveyors, there is a potential for needless litigation. The rules are clear; the court decisions are abundant. If all parties would go back to basics and apply the appropriate principles, there would be considerably less disagreement, and, as a result, less litigation.

While the *Outlaw* decision succinctly spells out the court's position on the word "intent," Texas is by no means the only court system to follow this rule. In fact, it is a universal rule, and a few examples are appropriate here.

While in a case of ambiguity "… it is the duty of the court to place itself as nearly as possible in the situation of the parties at the time the instrument was made, that it may gather their intention from the language used, viewed in the light of the surrounding circumstances", yet the rule "that the construction of a written document is the ascertainment of the intention of the parties… does not authorize the use of evidence of matter not proper for consideration in the interpretation of a writing."

What is to be determined is the meaning of the deed, and not the parties' understanding of the meaning. What they in fact intended cannot control or affect the language used when its meaning is ascertained. The process of construction builds upon the language to develop the intention, and not upon the intention to interpret the language."[40]

---

[38] The intention of the parties is considered to be essentially the same as that of the surveyor. The surveyor's intention is to be ascertained by scrutinizing what he actually did in making the survey as reflected by his field notes and the attending totality of circumstances of the survey. *U.S. v. Champion Papers, Inc.*, 361 F.Supp. 481 (1973).

[39] 137 S.W.2d 787, (1940).

[40] *Smart v. Huckins*, 82 NH 342 (1926). This decision is one of the most important for the practicing surveyor to be aware of.

Intent of the grantor as spelled out in the deed itself must be interpreted, not the grantor's intent in general, or even what he may have intended.[41]

The intention sought in the construction of a deed is that expressed in the deed, and not some secret, unexpressed intention, even though such secret intention be that actually in mind at the time of execution.[42]

It is not what the parties meant to say, but the meaning of what they did say that is controlling.[43]

## SURVEYOR'S INTENTION

Orn, in *Vanishing Footsteps of the Original Surveyor*,[44] elaborated on intent by discussing the surveyor's intention.

> The well-recognized rules of construction to be applied in locating the boundaries of a grant are designed to carry out the intention of the parties. They are controlling only when they enable the court to arrive at that intention. The intention of the surveyor is frequently referred to as being the same as the intention of the parties. In ascertaining the surveyor's intention the inquiry is not the intention which exists in his mind, but the intention which may be deduced from what he did in making the survey as reflected by his field notes.[45] Moreover, the court is not concerned with some secret intention of the surveyor not expressed in his field notes ad in his acts, but is it concerned with his purpose, what he called for in his field notes, and all of the attending circumstances.[46]

> As a general rule, when effect is given to the calls in accordance with their comparative dignity, effect is likewise being given to the intention of the surveyor. But, should the evidence clearly show that the intention of the surveyor will be defeated if the calls are followed in accordance with their grade, the call which will defeat the intention will be rejected as false regardless of the comparative dignity of the conflicting calls.[47]

- **To examine the abutting tracts for consistency.** While searching an abutting parcel is a separate review, it is important to include the process. There are several reasons for doing this: abutting tracts, when called for and identifiable, may be classed as monuments, therefore of the highest dignity in the overall position of conflicting evidence. Abutting parcels may verify consistency with the subject tract, or may provide additional information, even valuable information not available with the subject parcel. And, a verification whether there are any gaps or overlaps between the subject tract and its abutting parcels, all of which, when mentioned, provide notice of their existence.

The Tennessee case of *Burton v. Duncan*[48] presented an interesting dilemma whereby the court discovered a parcel they termed "no-man's land." The two parties were both claiming what they deemed to be an overlap of their lands, only to find out that neither had title to a tract lying between them. This the court termed "no-man's land" and stated "We recognize that our holding creates a so-called 'no-man's land' concerning the sliver of property at issue in this action. We

---

[41] *Wilson v. DeGenaro*, 415 A.2d 1334, 36 Conn. Sup. 200 (1970).

[42] *Sullivan v. Rhode Island Hospital Trust Co.*, 185 A. 148, 56 R.I. 253 (1936).

[43] *Urban v. Urban*, 18 Conn. Sup. 83 (1952).

[44] *Baylor Law Review*, Vol. IV, No. 3, Spring 1952, pp. 274–295.

[45] *Finberg v. Gilbert*, 104 Tex. 539, 141 S.W. 82 (1911).

[46] *Masterson v. Ribble* 78 S.W. 358 (Tex.Civ.App. 1904).

[47] *Gill v. Peterson*, 126 Tex. 216, 86 S.W.2d 629 (1935); *Livingston Oil & Gas Co. v. Shasta Oil Co.*, 114 S.W.2d 378 (Tex. Civ.App. 1938); *Miller v. Southland Life Ins. Co.*, 68 S.W.2d 558 (Tex.Civ.App. (1934).

[48] *Charles C. Burton v. Bill J. Duncon et al.;* No. M2009-00569-COA-R3-CV; Court of Appeals of Tennessee, Nashville. April 28, 2010.

regret that fact, but it was the duty of the court to determine the boundary line of the Duncans' property and the boundary line of Mr. Burton's property. We have done that. The fact that the disputed property does not lie within the litigants' respective boundaries is problematic. The record suggests there may be others who own or have a claim to the disputed property, but the record is not sufficient to permit us to determine who owns the disputed property. Moreover, the resolution of that issue may require that others, who are not parties to this action, are essential parties to such a determination; accordingly, they may be indispensable parties, which prevents this court from making that determination. Accordingly, we make no decision concerning who may own the disputed property. Our decision is limited to the determination of the respective boundary lines of Mr. Burton's and the Duncans' property."

An abutting call may qualify as a monument, to which all other calls will yield. In order for an abutting call to be classed as a monument, it must be: (1) called for, (2) existing at the time, and (3) identifiable. What constituted an abutter at the time of the original survey or description likely is not the same as the abutter existing later in time. Without analyzing the description of the parcel called for is not examining, or satisfying, the call in the description, and if equivalent to a monument, not satisfying the call for the monument.[49]

## Principle

In the absence of calls for other monuments, calls for established and well-known adjoiners will generally control other conflicting calls, unless manifestly erroneous.[50] The reason for is, where they are certain, they are monuments of the highest dignity. Therefore, if called for in an original description, they are original monuments.[51] To have controlling effect, they must be established and well known, and must be called for in the conveyance.[52]

- **To understand the basis of the title, which creates the boundaries.** As discussed in Chapter 2, the basis of title is what creates original boundaries, and the subject tract itself, whether the entirety of the subject tract, or one of several parcels that make it up. The source of the title is what *established* the boundaries.

A recent decision that answers a title question rather than a survey question serves to illustrate the importance of the creating document in answering the question of what exactly it is one is dealing with. In the case of *Brucker v. Burgess and Town Council of Carlisle*,[53] the argument concerned the title to a town square created over 200 years prior. In the words of the court, "there is no doubt that the mere fact of the use of [the Square] by the public for now more than 200 years is sufficient to raise a conclusive presumption of an original grant for the purpose of a public square; such is an ancient and well-established principle of the law. Nor can it be denied that, where such a dedication has been established and the public has accepted it, there cannot be any diversion of such use from a public to a private purpose, and it is also true that, where a dedication is for a limited or restricted use, and diversion therefrom to some purpose other than the one designated is likewise forbidden."

- **To understand if there was any title to begin with.** Several recent court cases have concluded that early creation of rights as well as examining original documents have resulted in there being no basis of title or in the claims of the parties.

---

[49] *United Fuel & Gas Co. v. Snyder*, 135 S.E. 164, 162 W.V. 75.
[50] 11 C.J.S. Boundaries, § 53.
[51] Ibid.
[52] *Coleman County v. Stewart.*, 65 S.W. 383 (Tex.Civ.App., 1901).
[53] 376 Pa. 330; 102 A.2d 418 (Pa., 1954).

As has been stated, one must know what one started with to understand what one ends up with. Or in the words of the New York court, "the past must be explored to understand the present."[54]

In the Maryland case of *Ski Roundtop v. Wagerman*,[55] the court analyzed the evidence and came to the conclusion that neither party had title to the parcel in dispute. It went further by stating, "a requisite for valid title to real property is an original conveyance from the State. Absent such a conveyance, one purporting to transfer an ownership interest in such property transfers nothing, and no quantity of successive transfers by deed nor the mere passage of time will metamorphose good title from void title." In this case, 100 or more years of records did not demonstrate evidence of a claim, since there was no claim to begin with. The court elaborated further and emphasized one cardinal point in stating that the intent of the original parties or the surveyor is of paramount consideration. Further, it seems that the 60-year rule of thumb for title searches, as practiced in Maryland, may not be sufficient. The Appeals Court found that despite successive deeds in the Wagerman's chain of record title from 1812, title was void because an original land patent was lacking.

The court closed by referencing a previous decision, *Maryland Coal and Realty Co. v. Eckhart*,[56] quoting, *"an albatross in the wake of every title searcher is the ominous question of whether he has gone back far enough in the chain of title."* Had the parties, in their title examinations, gone back in the records to the original conveyances, they would have known that their clients, in fact neither party, had a claim of title to the parcel in question.

- **To search for easements or other encumbrances**, which may add additional rights and boundaries, sometimes accomplished with survey information. Easements may add additional boundaries, corners, and information. Occasionally an easement boundary will be coincident with a boundary of the subject tract, adding additional information and demanding additional examination.
- **To look for possible agreements that may affect the existence or location of another boundary with its accompanying corners.** As provided for by law, unwritten means affecting title and boundary location may occur under the appropriate circumstances. Such may have an effect on, even alter, an original description, or it may refine it. When a difference is discovered between an original boundary location and an agreed upon boundary, which in itself may be legitimate, there is an immediate question of which is the controlling line. This is a matter of law, and the area between the two lines is a matter of title, possibly affected by the Statute of Frauds.[57]
- **To comply with the notice requirement.**

As investigation progresses, boundary and title information come to the surveyor's attention. The United States District Court stated in the case of *Newfound Management Corp. v. Sewer*[58] succinctly spelled out the surveyor's duty in the investigative process[59]:

"After a surveyor has completed a comprehensive review of *all*

*available records, deeds and prior surveys*, the surveyor begins

---

[54] *Dolphin Lane Associates, Ltd. v. Town of Southampton*, 339 N.Y.S.2d 966, 72 Misc.2d 868 (1971).
[55] 79 Md.App. 357, 556 A.2d 1144 (1989).
[56] 25 Md.App. 605, 337 A.2d 150 (1975).
[57] This is the common designation of a very celebrated English Statute (29 Car. II. Ch. 3) passed in 1677, entitled an *Act For Prevention of Frauds and Perjuries*. It has been adopted, in a more or less modified form, in nearly all of the United States.
[58] 885 F.Supp. 727 (U.S.Dist., 1995).
[59] Most, if not all, states, as well as the ACSM-ALTA standards for survey list these as requirements.

the field survey. Once in the field, the surveyor has a duty to make

a diligent search for *all monuments referenced directly or indirectly*

in the deed or property description that either occur naturally or were

put in place by prior surveyors or other persons." *Emphasis supplied.*

**Author's Note:** Suggested practice is to not take shortcuts, and to follow up on any and all leads, at least to the point of asking appropriate questions and examining any evidence that is reasonably available. As will be seen, people are *chargeable by law* to inquire of anything they have any sort of notice about. The duty extends to considering both direct and implied knowledge, and presumes that the responsible party not only has knowledge, but also is aware of, and understands, all relevant facts related thereto. *Emphasis supplied.*

The Newfound Management case aforementioned should not only be read, but should be fully understood and taken seriously.

# 5B Notice

Implied actual notice is that which one who is put on a trail is in duty bound to seek to know, even though the track or scent lead to knowledge of unpleasant and unwelcome facts.

*Hopkins v. McCarthy,* **121 Me. 27, 29, 115 A. 513, 515 (1921)**

**NOTICE**: There are three kinds of notice: actual notice, constructive notice, and inquiry notice, defined as follows. Inquiry notice is also called imputed notice and implied notice.

## ACTUAL NOTICE

Actual notice may be defined as knowledge of the contents of a document or of other facts which may affect title to an interest in real property. If a person has actual notice of a recorded document, it is immaterial whether the document appears in the record chain of title. If a person has actual notice of an unrecorded document, it is immaterial that the document is not recorded. An investigator must consider the effect of any document of which he or she has actual notice in the preparation of his or her title opinion.

## CONSTRUCTIVE NOTICE

Constructive notice may be defined as being charged by law with notice of the effect on title to an interest in real property of the contents of a document or of other facts without knowledge of the document itself or the facts themselves. A document recorded in the real property records in the office of the county clerk and recorder is constructive notice of its existence and of its contents to all persons subsequently acquiring an interest in the real property affected by that document, even if the document is not properly indexed or copied in the records by the clerk and recorder. While the recording of a document is constructive notice to the persons subsequently acquiring an interest in the real property affected by that document, an investigator is only responsible for analyzing the effect on title of those recorded documents which would be revealed by a properly conducted search of the real property records by the investigator, or which are contained in the abstract of title examined.

## INQUIRY NOTICE

Inquiry notice may be defined as being charged by law with notice of the effect on title of facts that would have been revealed by an inquiry if known facts would cause a reasonable person to inquire. If a person acquiring an interest in real property has knowledge of facts which, in the exercise of common reason and prudence, ought to put him or her upon particular inquiry as to the effect of such facts on the title to such real property, he or she will be presumed to have made the inquiry and will be charged with notice of every fact which would in all probability have been revealed had a reasonably diligent inquiry been undertaken. Whether the known facts are sufficient to charge such person with inquiry notice will depend upon the circumstances of each case. If, in the course of a title examination, an investigator discovers a document which is not in the record chain of title but which sets forth or refers to facts (other than the existence of an unrecorded document) that would cause a reasonable person to inquire about the effect of such facts on the title being examined, the investigator should disclose such facts in their opinion so that the person for whom the opinion is written may determine whether to undertake an inquiry.

Inquiry notice is sometimes called implied notice, or imputed notice, and may sometimes be equivalent to constructive notice.

One of the more recent decisions is from the Missouri Supreme Court in the case of *Jo Levitt v. Merck & Company*, 17-2630 (8th Cir. 2019). This was a personal injury case, in which one of the issues had to do with the Missouri Statute of Limitations. In Missouri, the statute of limitations for personal injury claims is five years after the cause of action accrues. Mo. Ann. Stat. § 516.120.

"[T]he cause of action shall not be deemed to accrue when the wrong is done or the technical breach of contract or duty occurs, but when the damage resulting therefrom is sustained and is capable of ascertainment" (Mo. Ann. Stat. § 516.100).

The Missouri Supreme Court defined "capable of ascertainment" as when "the evidence [is] such to place a reasonably prudent person on notice of a potentially actionable injury." *Powel v. Chaminade Coll. Preparatory, Inc.*, 197 S.W.3d 576 582 (Mo. 2006) (emphasis removed). This "objective" test is from the standpoint of a "reasonable person in [plaintiff's] situation." Id. at 584, 586. Both the "character of the condition . . . and its cause" must be capable of ascertainment. *Elmore v. Owens-Illinois, Inc.*, 673 S.W.2d 434 (Mo. 1984).

While this is a Missouri decision, it may serve to emphasize a standard that should be followed by any investigator at any location.

Tiffany, *The Law of Real Property* contains a concise treatment of the effect of notice in a chain of title.[60] It states, supported by references to decisions in a variety of jurisdictions, "In so far as a purchaser has actual or constructive notice of a conveyance or other instrument executed by one previously owning or claiming to own the land, he is charged with notice of all matters stated or referred to in such conveyance, which may possibly affect the title, and he is bound to make any inquiries or researches suggested by such statements or references. For this purpose, a purchaser is charged with notice of any conveyance which occurs in the chain of title under which he claims, that is, he is charged with notice of all matters stated or referred to in any conveyance which is essential to support his claim, without reference to whether he has actual notice of such conveyance. And the fact that such conveyance in the chain of title is not of record, or is improperly recorded is ordinarily immaterial in this regard. And he is charged with notice of the contents, not only of instruments in his chain of title, but also of other instruments referred to in such instruments, although not of record, in so far, at least, as it is reasonably possible for him to acquire knowledge thereof. And it follows that notice of a prior conveyance thus acquired by reference thereto in the chain of title is sufficient to defeat any claim of priority based on the failure to record such conveyance. Being put upon inquiry by the recital or statement in a conveyance in the chain of title, the purchaser 'is bound to follow up this inquiry, step by step, from one discovery to another and from one instrument to another, until the whole series of title deeds is exhausted and a complete knowledge of all the matters referred to and affecting the estate is obtained.' Being thus put on inquiry, the purchaser is presumed to have prosecuted the inquiry until its final result and with ultimate success. Likewise, if a purchaser is charged with notice of an instrument, as being of record, or in his chain of title, and such instrument refers to a judicial proceeding, he is chargeable with notice of the character and validity of such proceeding, so far as the title is dependent thereon.

A purchaser has been regarded as charged with notice of a provision contained in a conveyance of neighboring land, made by one in his chain of title, when the purpose and effect of such provision was to create an easement or other servitude upon the land which he is purchasing."

The more extensive a research project is, the more a researcher is likely to be put on notice of certain purported facts. The basic rule is that when one is put on notice of some fact, or potential concern, one is chargeable by law to know about it and everything it would lead to.

---

[60] Third Edition by Basil Jones, § 1293.

Certain principles have been derived throughout the court system, which may be summarized as follows.

## PRINCIPLES REGARDING NOTICE

The following principles concerning notice may be derived from a study of the decisions discussing the issue. Additional information and verification may be obtained from Appendix B, which consists of a compilation of selected court decisions relative to notice, by state.

- When a person has sufficient information to lead him to a fact, he shall be deemed conversant of it.
- Reasonable diligence requires that a person make inquiry when it is reasonable and prudent to do so, and a person is charged with notice of all facts that a reasonable inquiry would reveal.
- Notice of facts and circumstances is equivalent to knowledge of all of the facts.
- The inquiry notice maxim is that the means of knowledge is equivalent to knowledge.
- Inquiry notice arises when a party becomes aware or should have become aware of a particular set of facts.

**Author's Note:** What chances are taken, or how much liability is incurred, when a professional attempts to shortcut the process by failing to acquire the original survey of every line in question?

# 5C Summary of Court Decisions Regarding Control of Original Survey

The function of county or other local surveyors begins when the surveyors undertake the identification of lands that have passed from the Government into private ownership, based upon the description derived from the original survey. Manual of Instructions, Section 5-5.

### *Sanders v. Webb,* **621 N.E.2d 420 (Ohio, 1993)**

Primary function of second surveyor is to find boundaries established by first surveyor.

### *Roth v. Halberstadt,* **258 Pa.Super. 401, 392 A.2d 855 (Pa.Super. 1978)**

The primary function of a court faced with a boundary dispute is to ascertain and effectuate the intent of the parties at the time of the original subdivision.

### *Rivers v. Lozeau,* **Fla. App. 5 Dist., 539 So.2d 1147 (Florida, 1989)**

When an original survey is used in a property description to transfer title to the property, monuments set by the original surveyor on the ground will control over discrepancies within the description of the total parcel and over all subsequent surveys attempting to locate the same line.

Concisely, the original survey is the controlling entity, regardless of conflicts due to inadequacy of later descriptions. Legitimate differences can occur, however, but should be explained and appropriate definitions only accepted according to the prevailing rules.

### *Van Blarcom et al. v. Kip,* **26 N.J. Law 351 (1857)**

Where boundaries of a tract of land are defined in the original patent, and the land afterwards passes through successive transfers without any description by metes and bounds, the boundary lines will be presumed to remain as fixed by the original title.

Where reliance is placed on original boundary description which has not been altered since its creation, and subsequent descriptions, of any type, conflict, the investigator or researcher should seek to identify why, and be able to explain the difference.

### *Hagerman v. Thompson,* **235 P.2d 750 (Wyo., 1951)**

The purpose of a resurvey is to ascertain lines of original survey and original boundaries and monuments as established and laid out by survey under which parties take title to land, and they cannot be bound by any resurvey not based on survey as originally made and monuments erected.

The rule seems to be well established that the purpose of a resurvey is to ascertain the lines of the original survey and the original boundaries and monuments as established and laid out by

DOI: 10.1201/9781003032557-9

the survey, under which the parties take title. Parties cannot be bound by any resurvey not based upon the survey as originally made and monuments erected. 11 C. J. S. 635. *Day vs. Stenger,* 47 Idaho 253, 274 P. 112; *Bayhouse vs. Urquides,* 17 Idaho 286, 105 P. 1066; *Wing vs. Wallace,* 42 Idaho 430, 246 P. 8, and numerous cases cited. All the three surveys in question here were resurveys, binding on no one, unless one of these perchance should ultimately in a proper proceeding be found to be correct. We have no way of knowing which one of them is correct. It is a well-known fact that surveyors are apt to differ from each other, and surveyors employed by the U.S. government are not immune from the frailties of their profession. Which one of these resurveys is correct is a question of fact. *United States vs. State Investment Company,* 264 U.S. 206, 44 S. Ct. 289, 68 L.Ed. 639.

## THE DOYLE CASE

Judge William J. Holloway, Jr., in the decision of *The United States of America v. Doyle,*[61] went to great length to distinguish an original survey from a retracement survey, and a lost corner from an obliterated corner, both significant in an attempt to location the controlling original survey line.

Precisely accurate resurvey cannot defeat ownership rights flowing from original grant and boundaries originally marked off.

Original survey as it was actually run on ground controls, even if boundary was incorrect as originally established.

Conclusiveness of inaccurate original survey is not affected by fact that it will set awry the shapes of sections and subdivisions.

This is an action brought by the Government alleging occupancy trespass by defendants of a portion of the Pike National Forest, and seeking injunctive relief against trespass and for removal of improvements from the property in dispute. The case was tried to the court. The trial court determined the boundary dispute in the Government's favor on the basis of a dependent resurvey. Injunctive relief was granted and this appeal followed.

The Government owns land in the Pike National Forest including the SW 1/4 SE 1/4, Section 1, Township 8 South, Range 71 West of the 6th Principal Meridian in Jefferson County, Colorado. The defendants are the owners of the north 250 feet of the east 700 feet of the NW 1/4 NE 1/4, Section 12, adjoining to the south. The dispute here concerns the north boundary line of the defendants' property which is formed by the section line between the described portions of Sections 1 and 12 as it runs along the north of their property. According to the Government the correct section line lies south of the property line claimed by the defendants. The defendants say that the true line is about 124 feet north on one end and 147 feet north on the other end of the section line that the Government claims to be correct.

The Government's position is that the correct section line and therefore the north property line of the defendants' property should follow a resurvey by the single proportionate measurement method made by a Government surveyor, Mr. Brinker, in 1965.

It says that a stone marker which was described in the original 1872 survey performed by a Mr. Oakes is lost and that the loss of this marker makes the quarter corner at the northwest corner of the NE 1/4 of Section 12 a lost corner. Therefore, the Government maintains that the resurvey made between the northeast and northwest corners of Section 12, and establishing a straight line between them, was the proper basis for locating the true section line and the quarter corner which was located at the midpoint of that line. The defendants, on the other hand, essentially argue that collateral evidence consisting of Forest Service signs, tree blazes, and testimony sufficiently established as correct the boundary relied on by them. They say that a determination that a corner

---

[61] 468 F.2d 633 (Colo., 1972).

is lost is disfavored and that the trial court applied the incorrect criteria and burden of proof in making its determination that the corner was lost and erred in accepting the boundary established by the 1965 resurvey.

The trial court found that a stone relied on by the defendants was not the actual quarter section marker, and that this corner was lost; that the tree blazes were too recent to be relied on; that the dependent resurvey was a proper basis for determining the boundary; and that, therefore, defendants were in trespass on the disputed strip of land. We are satisfied that the record supports findings by the trial court that the original quarter corner monument was lost and that a stone claimed by defendants to be the marker was not the original monument.

The guiding legal principles are not in dispute. Where there is no controlling federal legislation or rule of law, questions involving ownership of land are determined under state law, even where the Government is a party. *Mason v. United States,* 260 U.S. 545, 558, 43 S.Ct. 200, 67 L.Ed. 396; *United States v. Williams,* 441 F.2d 637, 643 (5th Cir.); *Standard Oil Co. of California v. United States,* 107 F.2d 402, 415 (9th Cir.). The rule is recognized implicitly by the federal statute permitting resurveys. See 43 U.S.C.A. § 772.

The original survey as it was actually run on the ground controls. *United States v. State Investment Co.,* 264 U.S. 206, 212, 44 S.Ct. 289, 68 L.Ed. 639; *Ashley v. Hill,* 150 Colo. 563, 375 P.2d 337, 339. It does not matter that the boundary was incorrect as originally established. A precisely accurate resurvey cannot defeat ownership rights flowing from the original grant and the boundaries originally marked off. *United States v. Lane,* 260 U.S. 662, 665, 666, 43 S.Ct. 236, 67 L.Ed. 448; *Everett v. Lantz,* 126 Colo. 504, 252 P.2d 103, 108. The conclusiveness of an inaccurate original survey is not affected by the fact that it will set awry the shapes of sections and subdivisions. See *Vaught v. McClymond,* 116 Mont. 542, 155 P.2d 612, 620; *Mason v. Braught,* 33 S.D. 559, 146 N.W. 687.

The actual location of a disputed boundary line is usually a question of fact. Gaines v. City of Sterling, 140 Colo. 63, 342 P.2d 651. "... [T]he generally accepted rule is that a subsequent resurvey is evidence, although not conclusive evidence, of the location of the original line." *United States v. Hudspeth,* 384 F.2d 683, 688 n. 7 (9th Cir.); accord, see *Ben Realty Co. v. Gothberg,* 56 Wyo. 294, 109 P.2d 455, 458, 459. And in its trespass action the burden of proving good title to the land rests on the *Government. Yakes v. Williams,* 129 Colo. 427, 270 P.2d 765; see also *Cone v. West Virginia Pulp & Paper Co.,* 330 U.S. 212, 67 S.Ct. 752, 91 L.Ed. 849.

The procedures for restoration of lost or obliterated corners are well established. They are stated by the cases cited below and by the supplemental manual on Restoration of Lost or Obliterated Corners and Subdivisions of Sections of the Bureau of Land Management (1963 ed.). The supplemental manual sets forth practices and contains explanatory and advisory comments.

Practice 1 of the supplemental manual recognizes that an existent corner is one whose position can be identified by verifying evidence of the monument, the accessories, by reference to the field notes, or "where the point can be located by an acceptable supplemental survey record, some physical evidence, or testimony." Practice 2 recognizes that an obliterated corner is one at whose point there are no remaining traces of the monument, or its accessories, but whose location has been perpetuated, or the point for which may be recovered beyond a reasonable doubt, by the acts and testimony of the interested land owners, competent surveyors, or other qualified local authorities, or witnesses, or by some acceptable record evidence. Practice 3 states that a lost corner is one whose position cannot be determined, beyond reasonable doubt, either from traces of the original marks or from acceptable evidence or testimony bearing on the original position, and whose location can be restored only by reference to one or more interdependent corners.

The authorities recognize that for corners to be lost "[t]hey must be so completely lost that they cannot be replaced by reference to any existing data or other sources of information." *Mason v.*

*Braught,* supra, 146 N.W. at 689, 690. Before courses and distances can determine the boundary, all means for ascertaining the location of the lost monuments must first be exhausted. *Buckley v. Laird.* 493 P.2d 1070, 1075 (Mont.); Clark, Surveying and Boundaries § 335, at 365 (Grimes ed. 1959); see advisory comments of the supplemental manual, supra at 10.

The means to be used include collateral evidence such as boundary fences that have been maintained, and they should not be disregarded by the surveyor. *Wilson v. Stork,* 171 Wis. 561, 177 N.W. 878, 880. Artificial monuments such as roads, poles, fences and improvements may not be ignored. *Buckley v. Laird,* supra, 493 P.2d at 1073; *Dittrich v. Ubl,* 216 Minn. 396, 13 N.W.2d 384, 390. And the surveyor should consider information from owners and former residents of property in the area. See *Buckley v. Laird,* supra, 493 P.2d at 1073-1076. "It is so much more satisfactory to so locate the corner than regard it as 'lost' and locate by 'proportionate' measurement" (Clark, supra § 335 at 365).

## THE LOST CORNER

Relevant to the Doyle case, but not referenced therein, is the Colorado case of *Lugon v. Crosier.*[62] In Lugon, the court concluded that a corner never set cannot be either lost or obliterated, since it never existed. The court called it a "myth," the rule being to set it where the surveyor would have set it had he done so. This is the same reasoning as applied to protracted lines – run the line where the surveyor would have run it had he done so.[63]

"This was a proceeding under Code 1921, c. 24, for the establishment of disputed boundaries of section 9, township 4 north, range 84 west of sixth p. m. in Routt county. The plaintiffs in error were defendants below. There were several reports by Richardson, the first commissioner, and one by Harkness, his successor. The latter was adopted, in toto, by the court, and a decree was entered accordingly. The case comes here on error. The dispute is fundamentally on the position of the northwest and northeast corners of the section, and, incidentally, upon the north, east, and west quarter corners thereof.

The township closes on the first correction line north, a standard parallel, and is somewhat more than one mile short, so that the north tier of sections, and say 150 feet of the next tier, is missing, and the survey of section 9 is otherwise irregular. These conditions do not all affect the questions before us, but they explain some things which otherwise would be confusing.

The correction line above mentioned was surveyed in 1873, and, according to the practice in government surveys, is marked by standard corner monuments upon the southeast and southwest corners of the southern tier of sections of township 5 north of said range 84.

Ordinarily, the north tier of sections being omitted, the northeast corner of section 9 would coincide with the southeast corner of 33, and the northwest corner of 9 with the southwest corner of 33, but, since these corners are on a correction line, and said corners of 33 were fixed first as standard corners, the east and west boundary lines of 9 were or should have been simply run to that line, the intersections forming the corners of section 9 and tied in the field notes to the standard corners.

1. The Harkness report in question fixes the northeast corner of section 9 by starting from the southeast corner of that section, which is undisputed, and running a line due north to its intersection with the correction line. Whether this method was right is one of the cardinal questions before us. The interior of said township 4 was surveyed for the government in 1881

---

[62] 78 Colo. 141, 240 P. 462 (Colo., 1925).

[63] Where the purpose is not to ascertain the position of lines and corners once actually run and established, but to construct a survey by making two lines never run, these lines should be fixed where the surveyor would have made them if he had run them out. *Mercer v. Bate,* 4 J.J. Marsh. 334 (Ky, 1830).

by one Smith. The commissioner started where Smith, in his notes, said he started, followed the course which he said he took to the line on which he said he stopped, and called that the corner. We cannot say that was wrong. Plaintiffs in error say that there was here a monument which was ignored, but the commissioner expressly finds that Smith never located or found the standard parallel (correction line), but calculated or guessed the ties from the field notes of the survey of 1873, and that this monument "is not the original government closing corner for the northeast corner of section 9." If so, it was properly ignored, and there was evidence to support the finding. The commissioner, an engineer and surveyor, who not only heard the witnesses, but went on the ground and re-ran the lines with the field notes of Smith's survey, and saw the topography therein mentioned and the alleged monuments themselves, is in a better position than we to determine whether the monument is genuine.

The plaintiffs in error say, however, that if the monument is rejected, the rule of the General Land Office as to restoring lost monuments must be followed. The rule invoked is Gen. L. O. Reg. 47:

> A lost or obliterated closing corner from which a standard parallel has been initiated or to which it has been directed will be reestablished in its original place by proportionate measurement from the corner used in the original survey to determine its position.

The corner in question is a closing corner, but not one from which a standard parallel has been initiated, nor one to which a standard parallel has been directed; we do not see, therefore, that the rule relates to this case, but if it did we doubt that the corner can be regarded as lost or obliterated. It never existed, and so cannot, strictly speaking, be said to be lost or obliterated. If the monuments were lost or obliterated, there would be some reason to attempt to relocate it, and perhaps the method prescribed in Rule 47 is as good a way as any other, but when it is a myth, never on the ground, the natural, straightforward, and sensible way is to establish the corner at the place where the original surveyor ought to have put it, and that is where the north course of the east line of the section meets the correction line at right angles, and that is where the report puts it. Everybody knows that that is where the section line ought to have closed, and where the original surveyor, honest or dishonest, meant to close it; that his duty required him to close it there, so that the inclosure of his lines might be a rectangle or nearly so. Why should courts be less reasonable than reasonable men?

2. The next question is the location of the northwest corner of section 9. The commissioner professed himself unable to find whether a certain stone located 655 feet east and 122 feet north of the standard southwest corner of said section 33 was the original government corner set as the closing corner of the west line of section 9, but leaves that to the court upon the evidence. The court, in effect, finds that it is such corner, and, in accordance with the commissioner's recommendation if such should be the finding, places the true northwest corner of 6 at the place where a true line from the southwest corner thereof to the said stone intersects the said correction line.

This was right. Being a true monument, it controlled the course (indeed it was very near it), but the correction line was also a monument, and, in view of the fact that the surveyor, Smith, had no right to cross it, it must be regarded as controlling the stone monument. Therefore the true corner is on the correction line, where the surveyor ought to have stopped, and where his notes say he stopped.

3. The plaintiffs in error complain of the costs, but the abstract does not show what the orders in that regard were; we cannot, therefore, say they were erroneous.

4. The court at one time directed Commissioner Richardson to fix the corners by said rule 47, and plaintiffs in error now claim that to be consequently the law of the case. We do not think so. The court may correct its own errors until final judgment and motion for new trial denied."

## RESURVEY CANNOT CHANGE LINES

No surveyor or court has the authority to alter or modify a boundary line once it has been created. It can be interpreted only from the evidence of where that boundary is located.[64]

Boundaries of original survey cannot be changed by resurvey to affect titles.[65]

In this case, the court stated, "Section 772, 43 U.S.C. A., permits resurveys and retracements of the old survey, but also provides that 'provided that no such resurvey or retracement shall be so executed as to impair the bona fide rights or claims of any claimant, entryman, or owner of lands affected by such resurvey or retracement' The evidence indicates that the lands of the defendants were located and patented according to the survey of 1883, and particularly the closing corner here in question, and the foregoing rule, if intended to be applicable to a case such as is before us, would be inconsistent with the statute just quoted. It is a general rule that the original corners as established by the government surveyors, if they can be found, or the places where they were originally established, if that can be definitely determined, are conclusive on all persons owning or claiming to hold with reference to such survey and the monuments placed by the original surveyor without regard to whether they were correctly located or not. 9 C. J. 164; 50 C. J. 912-914; 8 Am Jur. 788; *Henrie v. Hyer*, 92 Utah 530, 70 P.2d 154. See also *Bentley v. Jenne*, 33 Wyo. 1, 236 P. 509; *Porter v. Carstensen*, 40 Wyo. 156, 274 P. 1072. The United States Supreme Court stated the rule succinctly in *United States v. State Investment Co.*, 264 U.S. 206, 44 S.Ct. 289, 68 L.Ed. 639, thus:

"Although the power to correct surveys of the public land belongs to the political department of the Government and the Land Department has jurisdiction to decide as to such matters while the land is subject to its supervision and before it takes final action * * * this power of supervision and correction by the Department is 'subject to the necessary and decided limitation' that when it has once made and approved a governmental survey of public lands and has disposed of them, the courts may protect the private rights acquired against interference by corrective surveys subsequently made by the Department * * *. A resurvey by the United States after the issuance of a patent does not affect the rights of the patentee; the government after conveyance of the lands, having no 'jurisdiction to intermeddle with them in form of a second survey.' * * * And although the United States, so long as it has conveyed its land, may survey and resurvey what it owns, and establish and re-establish boundaries, what it thus does is 'for its own information,' and 'cannot affect the rights of owners on the other side of the line already existing.'"

In *Washington Rock Co. v. Young*, 29 Utah 108, 80 P. 382, it was held that where an original government survey of land was made before the township line was established, the fact that a retracing of such survey placed the corner of a section east of the township line as subsequently established, and in another township, could not injuriously affect the rights of a party holding under a government patent based on the original survey, and that the original survey is controlling. In *Harrington v. Boehmer*, 134 Cal. 196, 66 P. 214, it is stated that "a government township lies just where the government surveyor lines it out on the face of the earth." In *Galt v. Willingham*, 11 F.2d 757, the court stated: "The only thing on which appellants reasonably can rely is the failure of the government surveyor to run his range line due north and south. But in re-establishing the lines of the survey the footsteps of the original surveyor should be followed, and it is immaterial that the lines actually run by him are not correct." If that is true of a range line, it must be equally true of the lines of a standard parallel. The only distinctive feature in this case is that it is sought, under the provisions of rule 378 [56 Wyo. 309] supra, to deprive a true government corner, which is the only one in that particular place, or in the neighborhood thereof, which

---

[64] Principle 8, Section 2.7, *Brown's Boundary Control and Legal Principles*, 7th Edition.
[65] *Ben Realty Co. v. Gothbert*, 56 Wyo. 294, 109 P.2d 455 (Wyo. 1941).

marked the north boundary of the township and the closing corner of two sections, of the character as such corner. We find no justification therefor under the decisions, which are uniform to the effect that government corners mark the true boundaries. The trial court in *Galt v. Willingham*, 300 F. 761, said: "Granted that the rules governing surveyors of government land are required to run range lines due north and south, yet if the surveyor does not do this, as I understand the law, when it comes to re-establishing the lines, they are to be run as the surveyor ran them at the time of making his survey, and not what he ought to have done. And so strict is this rule that not even the government can change the lines to the detriment of private interests." So in this case, granted that the surveyors of 1883 should have established the closing corner here in question 218.01 feet north of where they did, yet since they did not do so, the rights of the patentees here in question and those of their successors in interest cannot be prejudiced thereby.

Resurvey to locate lines of original survey; if subsequent survey does not follow original, purchasers cannot be bound by it.[66]

1. The purpose of a resurvey subsequent to taking of title by purchasers and settlers is to ascertain the lines of the original survey and the original boundaries and monuments as established and laid out by the survey under which the parties originally took title.

### *Riley, administratrix, & c. v. Griffin, et al.*, 16 Ga. 141 (1854)

In ascertaining boundaries, the locations of the original surveyor, so far as they can be found, are to be resorted to; and where they vary from the proprietor's plan, the locations actually made, will control the plan.

Whenever, in a conveyance, the deed refers to monuments, actually erected as the boundaries of the land, it is well settled that these monuments must prevail, whatever mistakes the deed may contain, as to courses and distances.

Courses and distances are pointers and guides, rather to ascertain the natural objects of boundaries.

## MAGNETIC DECLINATION IN RETRACEMENT OF ORIGINAL LINES

In renewing lost corners, allowance should be made for variation in the magnetic needle since the date of the original survey, the courses and distances not departed from except where necessary, and, in making measurements, allowance should be made for unevenness of the ground.

### *Bryan, & c. v. Beckley,* 16 Ky (Litt Sel Case 91 (1809)

To restore lost lines and corners. Course and distance not to be departed from, but in cases of necessity. Distances must first yield. Allowances to be made for variation of the needle. Unevenness of ground to be allowed for. Mistake in distance originally committed in one line, could have affected only the opposite. A mistake in one course, not to be presumed to have affected any other course. Court bound to take notice that there is a magnetic variation from the true meridian. Surveyors generally took their courses from the magnetic meridian. Horizontal measure to be attained, being the basis of the art of surveying. The variation of the magnet since the original survey is the allowance to be made. Existing lines and corners to govern, however variant from the courses called for. When visible and actual landmarks fail, resort is to be had to courses and distances. Departure from distance not indulged, further than necessary.

### *Den on Demise of Joseph Norcom v. Thomas H. Leary,* 3 Iredell (25 N.C.) 49 (1842)

---

[66] *Bayhouse v. Urquides,* 17 Idaho 286, 105 P. 1066 (Idaho 1909).

When a course is resorted to for want of a better guide to find the terminus of boundary of a tract of land, it is the course as it existed at the time to which the description of the tract of land refers. If it appears that because of the magnetic variation, that course is not the same with that which the needle now points out, it is the duty of the jury to make allowance for such variation, in order to ascertain the true original line.

The work on the ground constitutes the original survey, NOT the plat or the resulting land description in a deed or other document.

### Morales v. CAMB, 160 P.3d 373 (Colo.App. Div. 2 2007)

Original monuments control plat. The court's rationale and review of existing case law presents a worthwhile study of how the system works and why others, with authority, should not speculate on how to fix a discrepancy.

Through various conveyances, defendant CAMB acquired title to lots 3, 4, and 5, and plaintiff obtained title to lot 6, which abuts lot 5 on its north. All of the pertinent conveyances referred only to the Vasquez Village subdivision plat for their legal descriptions.

In 2002, CAMB began planning to re-plat its three lots for development of a town home project. In re-surveying these lots, it was discovered that the monuments marking the boundary between lots 5 and 6 were inconsistent with at least one distance call shown on the Vasquez Village plat. While this distance was shown as 25 feet on the plat, the monument was placed some 38 feet from the pertinent prior point. Further, while the monument for the southeast corner of lot 6 was consistent with a distance call on the plat for that location, it is some 13 feet south of the location of the boundary line as depicted on the plat. Both monuments, therefore, exist some 13 feet south of the boundary between the two lots as shown on the plat.

As a consequence, if the monuments are determined to be the true points establishing the southern boundary of plaintiff's lot 6, that lot will have an additional strip of about 13 feet, containing about 1,197 square feet, added to the lot as shown by the line on the recorded plat. But if the boundary line on the plat is determined to represent the proper boundary, this strip would be a part of lot 5.

If there appears to be a misdescription in a deed, a court must ascertain the true intent of the parties. *Wallace v. Hirsch*, 142 Colo. 264, 350 P.2d 560 (1960); *see Lazy Dog Ranch v. Telluray Ranch Corp.*, 965 P.2d 1229 (Colo. 1998) (in construing a deed, it is paramount to ascertain intent of parties).

However, certain rules of construction are used to disclose that intent.

First, "[i]t is a well settled principle that when lands are granted according to an official plat of the survey of such lands, the plat itself, with all its notes, lines, descriptions and landmarks, becomes as much a part of the grant or deed by which they are conveyed, and controls so far as limits are concerned, as if such descriptive features were written out upon the face of the deed or grant itself." *Spar Consol. Mining & Dev. Co. v. Miller,* 193 Colo. 549, 568 P.2d 1159 (1977), citing *Cragin v. Powell,* 128 U.S. 691, 9 S.Ct. 203, 32 L.Ed. 566 (1888).

Here, then, the deeds conveying lots 5 and 6 to the parties incorporated all of the items of information on the plat, including the surveyor's certificate attesting that appropriate monuments had been placed on the ground, as required. *See Spar Consol. Mining & Dev. Co. v. Miller,* supra.

Further, it is a general rule that the monuments placed by the original surveyor are conclusive on all persons owning or claiming to hold with reference to such survey. *Everett v. Lantz,* 126 Colo. 504, 252 P.2d 103, 108 (1952). "Monuments control courses and distances, which are considered the least reliable of all calls." *Jackson v. Woods,* 876 P.2d 116 (Colo. App. 1994). "The courses and distances in a deed always give way to the boundaries found upon the ground, or supplied by the proof of their former existence, where the marks or monuments are gone." *Cullacott v. Cash Gold & Silver Mining Co.,* 8 Colo. 179, 6 P. 211 (1885) (citing *Lodge v. Barnett,* 46 Pa. St. 477 (Pa. 1864)); 12 Am.Jur.2d Boundaries § 74 ("Where land is disposed of by reference to an official plat, the boundary lines [as] shown on the plat control. In locating land upon the ground from the calls and descriptions in the map, plat, or field notes referred to, the same primary rules apply as exist in the locating of calls and descriptions in a deed containing no such reference, that is, the various calls are given the same order of preference. In case of conflict, monuments control plats or maps, and an actual survey controls over a plat or a map.")

In the trial court, CAMB presented an affidavit from a registered professional land surveyor who averred that, using the field notes for the Vasquez Village subdivision, the descriptions contained in those notes were consistent and allowed the exterior boundary lines of that subdivision to "close." However, CAMB's surveyor averred that, if the locations of the monuments were used as the boundary indicators, the resulting description of the subdivision's exterior boundary would not close. Hence, this expert concluded that the discrepancy between the monuments and at least one distance call on the plat resulted from the misplacement of the monuments, or a "field blunder," and that the distance calls and boundary line as reflected on the plat, rather than the monuments, should control the location of the pertinent boundary.

The trial court rejected this ultimate conclusion, and so do we.

Even if we assume that both monuments were misplaced, the rule that monuments control over distance and course calls on the plat is nevertheless applicable and the monuments still control the boundary location. *See Everett v. Lantz,* supra, citing *Ben Realty Co. v. Gothberg,* 56 Wyo. 294, 109 P.2d 460 (1941) (monument misplacing 8th standard parallel still controls description of land in grant).

### *Ripley v. Berry & al.,* 5 Me. 24 (1827)

Where land is conveyed by deed, referring to a plan, between which, and the original survey, there is a difference in the location of lines **anne**; the lines and, originally marked as such, are to govern, however they may differ from those represented on the plan.

Where lots have been granted, designated by number according to a plan referred to, which has resulted from an actual survey, the lines and corners made and fixed by that survey have been uniformly respected in this State, as determining the extent and bounds of the respective lots."

The court stated: "It is a well-settled principle, that whatever is included within the bounds of a lot, as it was actually located upon the face of the earth, is to be considered as a part of such lot, and, to use the language of the court in the case of Pike v. Dyke, 2 Greenl. 213," Where lots have been granted, designated by number according to a plan referred to, which has resulted from an actual survey, the lines and corners made and fixed by that survey have been uniformly respected in this State, as determining the extent and bounds of the respective lots."

### *Esmond v. Tarbox,* 7 Me. 61 (1830)

Where the plan and the monuments made by the original surveyor of a tract of land do not correspond, the monuments are to be resorted to, in order to ascertain the true location.

### *Home Owners' Loan Corporation v. Dudley et al.,* 141 P.2d 160 (1943)

The original location of a monument controls, and, if it is obliterated, the court is concerned in ascertaining where it was originally located.

### *Sellman v. Schaaf,* 269 N.E.2d 60 (Ohio, 1971)

Where original monuments as located by surveyor are still ascertainable, boundary lines determined by such monuments will determine boundaries of lots irrespective of deviation from course or distance as set forth in plat.

### *Diehl v. Zanger,* **39 Mich. 601 (1878)**

A re-survey, made after the monuments of the original survey have disappeared, is for the purpose of determining where they were, and not where they ought to have been.

Refer to case to explain why that is necessary.

### *Lawson v. Viola Tp.,* **210 N.W. 979 (S.D., 1926)**

Original monument, if found or established, marks true corner, though resurvey might indicate different location.

In boundary dispute where location of original monument can be found or can be established by evidence, such location is the true corner, regardless of the fact that resurveys may show that it should have been located elsewhere.

In boundary dispute, where location of original monument cannot be established by evidence, corner may be established by new survey made from points that can be determined and in accordance with field notes of original survey.

### *Froscher v. Fuchs,* **130 So. 2d 300 (1961)**

In locating disputed boundary, surveyors must follow original survey lines under which property in question and neighboring properties are held, notwithstanding inaccuracies or mistakes in original survey.

Purpose of resurvey is to locate as far as possible previously established lines.

The court: in cases deciding the boundary between two parcels of land, the law is settled that it is the duty of the surveyors to follow the original survey lines under which the property and neighboring properties are held notwithstanding inaccuracies or mistakes in the original survey. The purpose of this rule of law is that stability of boundary lines is more important than minor inaccuracies or mistakes. This rule was firmly established in Florida by *Akin v. Godwin,* Fla.1950, 49 So.2d 604. See also *Wildeboer v. Hack,* Fla.App.1957, 97 So.2d 29 and *Bishop v. Johnson,* Fla.App.1958, 100 So.2d 817, 820.

### *Watrous v. Morrison,* **33 Fla. 261, 14 So. 805 (Fla. 1894)**

Original survey controls; it is the survey as it was actually run on the ground that governs ".".

In the sale of lands in sections, or subdivisions thereof, including lots, according to the government survey, the survey as actually made controls. *Miller* **[33 Fla. 267]** *v. White,* 23 Fla. 301, 2 So. 614; *Liddon v. Hodnett,* 22 Fla. 442. It is the survey as it was actually run on the ground that governs, if the monuments, corners, or lines actually established can be located or proved. Courses and distances yield to such corners and lines so long as the latter can be located, and for the reason that the latter are the fact or truth of the survey as it was actually made, while the former are but descriptions of the act done, and, when inaccurate, they cannot change the fact. *McClintock v. Rogers,* 11 Ill. 279; *Yates v. Shaw,* 24 Ill. 367; *Bauer v. Gottmanhausen,* 65 Ill. 499; *Kincaid v. Dormey,* 47 Mo. 337; *Major's Heirs v. Rice,* 57 Mo. 384; *Willis v. Swartz,* 28 Pa. St. 413; *Riley v. Griffin,* 16 Ga. 141.

### *Overton v. Davis, No. E2006-01879-COA-R3-CV (TNCIV)*

Construction of deeds, uncalled for marker, fence a monument when called for; original survey, not where modern equipment and methods would place it

As noted in *Wood v. Starko*, 197 S.W.3d 255, 260 (Tenn. Ct. App. 2006):

> the question to be answered is not where new and modern survey methods will place the boundaries, but where did the original plat locate them. The main purpose of a resurvey is to rediscover the boundaries according to the plat upon the best evidence obtainable and to retrace the boundary lines laid down in the plat.... [T]he known monuments and boundaries of the original plat take precedence over other evidence and are of greater weight than other evidence of the boundaries not based on the original monuments and boundaries.

The following case describes in detail the analysis necessary when dealing with rectangular surveys and descriptions outside of the PLSS. Anyone surveying in eastern rectangular townships, or systems, should be familiar with this decision.

### Actual location upon ground of original lot lines will control, if capable of being ascertained. *Neill v. Ward*, 103 Vt. 117, 153 A. 219 (1930). Refer to Figure 5.2.

The controversy is over the location of the boundary line between lot 59 and lot 60 in the first division of lots which is owned by the defendant. Moretown, as shown by the original town plan, is bounded on the southeast by the town of Berlin, on the northeast by the Winooski River, on the northwest by Duxbury and on the southwest by Waitsfield, but now by Waitsfield and Northfield. There are three main divisions of lots extending northwesterly from the Berlin town line to the Duxbury town line. Starting at the town lines of Waitsfield and Northfield, and extending northeasterly, the divisions of lots are second, first, and third. The second division contains four tiers of lots, the first and third divisions, three tiers each. There are seventeen lots in each tier.

When lots are hereinafter referred to without stating the division they are in, first division lots are indicated. When other lots are referred to, the division they are in is given.

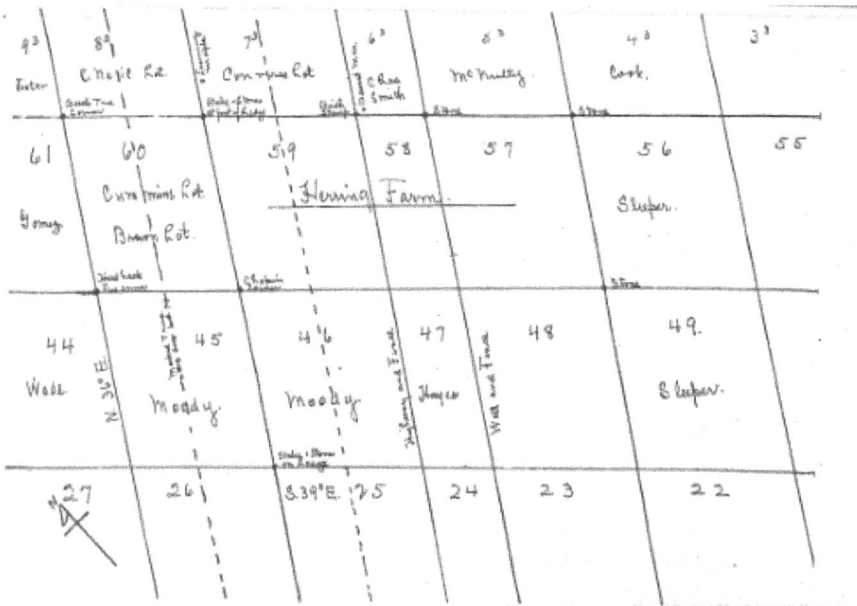

**FIGURE 5.2**  Diagram from *Neill v. Ward*.

The plaintiff owns what is called the Herring farm, which consists of lots 57, 58, and 59. Abijah Herring, the original owner of the Herring farm, purchased lot 58 October 16, 1847. The description in his deed is: 'Lot number fifty-eight (58) in the first division of lands in said town drawn to the original right of Ebenezer Brown.' He purchased lot 57 September 21, 1863. the description in his deed is: 'Lot number fifty-seven (57) in the first division of lands in s'd Moretown drawn to the right of James Wallace and commonly called the Spafford lot, and adjoining the farm of s'd Herring.' He purchased lot 59 October 23, 1865. The description in his deed is: 'Lot number 59 in the first division of lots in said town drawn to the original right of Lemuel Abbott, estimated to contain 116 acres of land more or less.' All of said deeds are warranty deeds. Certified copies of other deeds were received in evidence showing a title of record of lot 59 back to March 1, 1841.

Abijah Herring owned and occupied the Herring farm until January 9, 1907. On that day, by conditional deed, he conveyed his farm to Julius E. Martin, describing it as "Being my home farm where I now live in Moretown, and being lots 57, 58 and 59 in the first division lots of land in said Moretown." The several subsequent conveyances of this property in the plaintiff's chain of title describe it as "the Abijah Herring Farm," and refer to previous deeds and the records 'for a more particular description of said premises' The plaintiff purchased 'the Abijah Herring farm' January 13, 1921, and has since lived on and occupied the same. The marks indicating the lines between the lots forming the Herring farm are mostly obliterated now, as no attempt has been made to preserve them, but in some places they are still visible.

A. O. Cummins, by his warranty deed dated November 29, 1882, conveyed lot 60, commonly known, and referred to at the trial below, as the Cummins lot or the Brown lot, to Joseph M. Brown and Charles J. Brown. Said lot is described in said deed as 'Being all of the first division lot in said Moretown drawn to the original right of Nathaniel Barrel.' Reference is made in said deed to prior conveyances which show a title of record to said lot to 1830. Charles J. Brown was the son of Joseph M. Brown. Joseph M. Brown died in 1899, and Charles J. was his only heir. Charles J. Brown, by his warranty deed dated March 28, 1912, conveyed said lot 60 to his son, Leo F. Brown, and, on July 9, 1924, Leo F. Brown, by his warranty deed, conveyed the same to the defendant. The plaintiff conceded at the trial below that the defendant is the owner of lot 60.

The plaintiff claims that the dividing line between lots 59 and 60 is a little over 1,000 feet, or about a half a lot, northwest of where the defendant claims that such dividing line is located.

Before taking up the defendant's motion to set the verdicts aside, we will consider the defendant's exceptions to the admission in evidence of the field book, the town plan, and the original grant of the town of Moretown.

Lot 60 and the lots forming the Herring farm are in the third tier of lots in the first division. In laying out the lots, as shown by the town plan and the field book, this tier of lots starts at the Berlin town line with lot 53 and ends with lot 69 at the Duxbury town line. Lot 53 is described in the field book as follows. 'Lot 53, Begins at a maple in the E. line of the town being the N.E. corner of No. 52; thence N. 47/d W. 116[r] to a spruce; thence N. 28/d E. 144[r] to a beech; thence S. 47/d E. 116[r] to a hemlock in the Town line; thence S. 28/d W. 144[r] to the first bound.'

Lot 54 begins at the S.W. corner of lot 53, and each subsequent lot begins at the S.W. corner of the preceding lot. The courses and distances of the lines of the other lots are the same as those of lot 53, and a named tree is at each corner.

The plaintiff bases his case on the field book and town plan on the theory that the lines and dimensions of the lots of the third tier, as actually laid out on the ground, are substantially the same as described in the field book. In his opening statement to the jury, counsel for the plaintiff, after describing the layout of the lots, said: 'So that in getting at the location of the lot lines according to the land records (meaning the field book) we begin at the east on the Berlin town line and count off the lots until we get up through 60, the last one beginning where the previous one ended, according to the land records.' After stating that the defendant commenced cutting timber on lot 59 in the latter part of 1928, he further said: 'Mr. Neill immediately procured the services of an engineer to go up there and survey the lot for him. He went up there, this engineer did, and found lot lines substantially as I have told you already.'

The defendant claims that lot 60, as occupied by himself and his predecessors in title for more than fifteen years, has old corners and lines marked upon the ground, and that during that time they maintained exclusive possession up to the dividing line as claimed by him. He further claims that the lines and corners of the lots comprising the Herring farm, as marked upon the ground, are not substantially the same as described in the field book; that lot 58, as located on the ground, is not a full lot, as described in the field book, but substantially a half lot.

The plaintiff, after introducing in evidence certified copies of deeds of his title of record and a title of record of lot 59 back to 1841, offered in evidence the town plan and certain parts of a book labeled 'The Proprietors' First Book of Record and the Field Book of Moretown.' This book and the town plan came from the office of the town clerk of Moretown, and no question is raised as to their authenticity and genuineness.

It appears from the proprietors' records contained in said book that it was voted at the first meeting of the proprietors, held on June 13, 1798, to lay out the town into three divisions, 'the first to contain 104 acres, the second division to contain 114 acres, and the remainder in equal division for the third division as it shall(?) out.' It is apparent that the acreage specified in said vote referred to the lots of each division and not to the division itself, as the lots of the first division, as described in the field book, contain 104.4 acres each, and the lots of the second division contain 116 each. At the same meeting a committee, of which Abel Knapp was a member, was chosen to lay out the town. At a meeting of the proprietors held on September 28, 1798, it was voted to accept the survey and return of the committee, and a draft was had of the lots of the three divisions as recorded in said book beginning with page 15 and ending with page 20.

There were evidently errors in the proceedings of some of the earlier meetings of the proprietors, and no field book had been made, as the proprietors preferred their petition to the Legislature of 1803, stating that, through the neglect of their surveyors that surveyed the town, no field book had been made nor any minutes of said survey preserved whereby a regular field book could be made, and that the former proprietors' meeting was lost through mistake, and praying for a special act to call a meeting to regulate their affairs. The Legislature did pass such an act, and a meeting was held pursuant to it on September 11, 1804, at the dwelling house of Asa Sterne. The third article of the warning of the meeting was 'To see whether the proprietors will agree to survey & draft the town anew, or rectify any mistakes in their former survey or draft or any of their former proceedings and complete a plan and field book of said town.' It was voted at the meeting not to survey the town anew. It was also voted that the proprietors' committee be empowered to make an examination and to report 'what in their opinion ought to be done in order to rectify all former mistakes and also to make out a regular plan and field book of the survey of said township.'

At an adjourned meeting held on October 6, 1804, the committee reported that in their opinion all the doings of the proprietors after May 7, 1799, were illegal, and that there were mistakes in the draft and settling of lots in the first and second divisions that could not be rectified without a new draft of the lots of said divisions. It was voted to accept the report and to make a new draft of the first and second division lots. A new draft of the first division lots was then made.

At an adjourned meeting held October 8, 1804, a new draft of the second division lots was made, and it was voted 'to accept of the survey and draft of the lots in each division as now rectified & stand affixed to each proprietor's name on the new parchment plan of Moretown.' The meeting was then adjourned to November 26, 1804.

On November 7, 1804, the Legislature passed an act authorizing and empowering the proprietors 'at any legal meeting already named or hereafter warned for the purpose to ratify and confirm the division of said town into severalty, so far as the same has been made, in fact. And such division, so ratified and confirmed, shall be good and valid in law, to all intents and purposes, any law, usage, or custom to the contrary notwithstanding.'

At the adjourned meeting held November 26, 1804, it was voted to accept all the doings of the proprietors at their meeting held at the dwelling house of Asa Sterne on September 11, 1804, with all their doings at their several adjournments, and to ratify and confirm the draft and division of said town as it then stood on the new map of said town, 'agreeable' to the act of the Legislature passed in 1804. There were several adjournments of the proprietors' meeting until July 22, 1805, when, it having appeared that there were 'great errors and illegality' in the meeting of September 11, 1804, and at the several adjourned meetings, it was voted that all the votes and proceedings, except those relating to immaterial matters, had, done and passed at said meeting of September 11, 1804, and at any subsequent adjourned meeting, 'be and the same are reconsidered and declared to be entirely null and void.' The meeting then voted to dissolve.

A duly warned meeting of the proprietors was held at the house of Asa Sterne on September 20, 1805. One of the articles of the warning was to see if the proprietors would 'accept and confirm any former draft or division of said township into severalty in pursuance of an act passed by the Legislature in November 1804 and to quiet the settlers on their actual settlements in said town in lieu of their drafts so far as is agreeable to law and to complete the survey and field book of said town in lieu thereof.' Another article was to see if they would vote 'to quiet the settlers in the possession of the lots they are actually settled on in lieu of their drafts as far as the law admits and to make a draft and division of the remainder of said township into severalty and to complete a survey and field book of said town and to rectify all mistakes now existing in the present survey and field book or any other proceedings of said proprietors.'

At this meeting Wright Spalding was elected clerk. It was voted, among other things, to quiet settlers in the first division. The meeting adjourned to November 18, 1805. At the adjourned meeting on November 18, 1805, the previous vote to quiet settlers in the first division was reconsidered. It was then 'Voted to establish the draft of the proprietorship as recorded in this book beginning with page fifteen and ending with page twenty.' This is the draft of 1798. It was also 'Voted to accept of a Field Book of this town as exhibited to this meeting by Abel Knapp, Esqr. dated Octr. 1804'" After the transaction of some other business it was 'Voted to dissolve the foregoing meeting.'

On page 85 of the book is the heading: 'Field Book or Survey of Moretown in the County of Chittenden Completed October, 1804.' Then follows the descriptions of the lots of the town on pages 85 to 115 inclusive. At the conclusion of the description of the lots, on page 115, there is the following: 'Attest Abel Knapp, Surveyor. These may certify that the foregoing as recorded from page 85 to the present page is a true copy of a field book as exhibited to the proprietors of Moretown at their meeting held at the house of Asa Sterne on the 18th day of Novr. 1805. Attest Wright Spalding proprietors Clerk.'

The description of lots 60 to 53, inclusive, in the field book as recorded in 'The Proprietors' First Book of Record and the Field Book of Moretown' were received in evidence subject to the objection and exception of the defendant.

A general objection was that the field book had no legality under the act of the Legislature of 1804; that, when the proprietors ratified and confirmed the new draft made on October 6 and 8, 1804, at the adjourned meeting held on November 26, 1804, they exhausted their authority under the act of 1804;

that they had no authority to reconsider that vote at their adjourned meeting on July 22, 1805, and no authority to call the last meeting held on September 30, 1805, and no authority, at the adjourned meeting held on November 18, 1805, to establish the draft of 1798, and to accept the field book dated October, 1804, exhibited by Abel Knapp at that meeting. There is no merit in this objection.

The proprietors' meeting of September 11, 1804, was called according to the provisions of an act regulating proprietors' meetings. Two Tolman's Compilation, Laws of Vt. 315. This act provided that the warrant for such a meeting should set forth "the several matters and things to be transacted." The adjourned meetings up to and including the meeting of July 22, 1805, were but continuations of the same meeting without any loss or accumulation of powers. *Warner v. Mower*, 11 Vt. 385. Nothing could be transacted at any of the adjourned meetings unless it could have been transacted at the called meeting. 46 C. J. 1378. Since the statute provided that the warrant for a proprietors' meeting should specify the business to be transacted, any business transacted without an article in the warning therefor was void. *School District v. Smith*, 67 Vt. 566, 32 A. 484.

All deliberative or legislative bodies, during their session, have the power to do and undo, consider and reconsider, as often as they think proper, and it is the final result only which is to be regarded as the thing done. *State v. Foster*, 7 N.J.L. 101; *People v. Davis*, 284 Ill. 439, 120 N.E. 326, 2 A.L.R. 1650, 1655. And this Court has held that a town, like an individual, may change its purposes, and may express that change by its vote, and, unless some right in another has been acquired or has vested under its action, no one may complain of the change. *Stoddard v. Gilman*, 22 Vt. 568, 573; *Cox v. Mount Tabor*, 41 Vt. 28, 31; *Estey v. Starr*, 56 Vt. 690.

It appears from the records of the meeting of September 11, 1804, and its adjourned meetings that the proprietors acted upon various matters not specified in the warning, and which the law specifically provided should not be acted upon unless set forth in the warning. There may be some doubt in this respect as to some of the other business transacted. It does not appear, however, that any right in another had been acquired or had vested under their action, and, in support of the vote of reconsideration, it will be presumed that none had been acquired or had vested. The proprietors had the power and authority at the adjourned meeting of July 22, 1805, to reconsider and declare null and void the votes and proceedings 'done and passed' at the previous meetings, and, when such action was taken, they still had the authority and power to proceed under the act of November 7, 1804.

The proprietors' meeting held on September 30, 1805, was a legally called meeting. The business transacted at the adjourned meeting held on November 18, 1805, was set forth in the warning; and the votes establishing the draft of 1798, and accepting the field book exhibited by Abel Knapp were legal.

The defendant further says that the validity of the field book is open to serious doubt as it is apparent that, if it was completed in October, 1804, it had no relation to the old draft of 1798, but rather to the new draft authorized and made on October 6 and 8, 1804. We cannot agree with this contention. It appears from the records that only one survey was made, and that survey was the basis of the draft of 1798. It appears from the evidence that Abel Knapp was a surveyor, and from the records that he was a member of the committee chosen to make the survey. The field book itself had nothing to do with the drafts. It is simply a description of the courses and distances of the lines, and of the corners of the lots of the town as they were surveyed, and as they appear by number and division on the town plan.

The plaintiff offered in evidence the description of lot 59 as contained in the field book as showing its location and as giving color of title to the same, followed by possession by the plaintiff and his grantors since 1841. The defendant objected to the admission of such description of lot 59, and of the descriptions of any of the lots as contained in the field book, on the ground that such descriptions were not admissible for the purpose of showing the location of any of the lots in the absence of a survey showing where those lots were actually located on the ground. After considerable discussion, counsel for the plaintiff said: 'We know where the Abijah Herring farm is, that can be found, 56 can be found. We are going to have witnesses here to testify to it, all these lots can be found by the present owners, we will show where they are, beginning at the town line, and we offer this description by metes and bounds as I have indicated in connection with our possession by ourselves and our grantors since the first of March, 1841. * * * * * We offer certified copy of the description of lots 59 back to 53, inclusive, on the town line. We offer to show that town line and 53 can be identified and followed today.' The certified copy was admitted and the defendant was allowed an exception on the grounds stated.

In view of the theory on which the plaintiff tried the case and of the objection made by the defendant to the admission of the descriptions of the lots, we take it that the substance of the final offers of the plaintiff was that he would show that the lot lines of lots 59 to 53, inclusive, as actually surveyed and marked on the ground, were substantially the same as described in the field book. The actual location upon the ground of original lot lines will control, if capable of being ascertained. *Silsby v. Kinsley*, 89 Vt. 263, 95 A. 634. In connection with this offer, the town plan and the description of the lots in the field book were admissible. The fact that the plaintiff's evidence did not come up to the offer did not affect the correctness of the ruling of the Court. *Herrick v. Holland*, 83 Vt. 502, 511, 77 A. 6.

Percy G. Smith, a civil engineer, and a witness for the plaintiff, was employed by the plaintiff to survey his farm and locate the dividing line between lots 59 and 60. A plan which he made of the plaintiff's farm from his survey, and which shows the dividing line between these lots as claimed by the plaintiff, is Plff's Ex. 27.

He testified in direct examination that he surveyed on the lines of lots 53 to 59, inclusive, and of other lots, and that he followed the range line between the first and third divisions, which is the northerly boundary of lots 57, 58, 59, and 60, from the Berlin town line northwesterly.

It appeared that in the papers of the Surveyor General there was item of the year 1784; 'That in perambulating or running town lines throughout the State, one thirtieth part be allowed for swag of chain'; and Smith testified that the lots he measured figured that way.

He testified in cross-examination that he measured the range line from the Berlin town line across seven lots to the northwest corner of lot 59, as claimed by the plaintiff, with a steel tape, and the distance was 895 rods; that the actual distance from the Berlin town line to the northwest corner of lot 59 according to the town plan and field book, and with the allowance of one-thirtieth part added, was a little over 839 rods; that his actual measurements overran the distance given in the field book by about 56 rods.

The evidence showed that the northwest line of lot 59, according to the town plan and field book, is substantially where the defendant claims it is, and that the cutting complained of was within the overrun.

In the redirect examination of Smith, the plaintiff offered in evidence the original charter of the town of Moretown, having special reference to what it said about the way surveying should be done. The part of the charter referred to described the land granted as 'containing by Ad-measurement, 23,040 acres, which tract is to contain six miles square, and no more; out of which an allowance is to be made for highways and unimprovable lands by rocks, ponds, mountains and rivers,' etc. The court admitted the charter for what it was worth, subject to the objection and exception that it was irrelevant and immaterial to any issue, and vague as to its bearing on any issue in the case.

The plaintiff, in making his offer, declined to limit it or specify the particular purpose for which the charter was offered further than to say, in substance, that it was the foundation for all surveying in the town of Moretown; that it was a part of the field book and was offered from the field book.

To sustain an objection on the grounds of immateriality and irrelevancy alone it should be made to appear that the evidence is clearly immaterial and irrelevant. *Slayton v. Drown*, 93 Vt. 290, 107 A. 307; *In re Wells' Will*, 95 Vt. 16, 113 A. 822; *Gomez & Co. v. Hartwell*, 97 Vt. 147, 122 A. 461. We think the charter might be admissible for certain purposes, which it is not necessary to state, and that there was no error in admitting it against the objection made.

It appears, however, that the only use the plaintiff made of the charter was as a foundation for allowance for highways, ledges, etc., not mentioned in the field book, made by Smith in his surveying to account for the overrun of fifty-six rods. It is also apparent that this was the only purpose for which the charter was offered, although it was not revealed to the court. This was an improper use of the charter, and it would have been error for the court to have admitted it for that purpose alone. As the overrun of fifty-six rods is material in considering defendant's motion to set aside the verdict, we give the reasons why the use made of the charter by the plaintiff was improper.

The charter is not a part of the field book. It is a separate document recorded in the first pages of the proprietors' book of records as required by 'AN ACT, regulating proprietors' meetings,' passed March 9, 1787, and in force when the meetings of the proprietors of Moretown hereinbefore mentioned were held. 2 Tolman's Compilation, Laws of Vt., 315, 317. Nor is it the foundation, of the survey, town plan, and field book relied upon by the plaintiff. Section 2 of said Act of 1787 provides that the mode of division of land held in common in any township shall be as follows: 'When the proprietors are met, according to the warning, and have agreed upon the number of acres to be allotted or divided, to each proprietor, they shall chuse a committee to make survey thereof, which committee * * * shall lay out, and number one lot (or as many as the proprietors vote) to each right, and when such survey shall be made, they shall return a plan thereof to the proprietors, when met, describing the corner of each lot.' Then follows a description of the way in which the draft of the lots should be made. The records of the proprietors' meetings show that the survey, town plan, and field book of the lots in the town of Moretown were made according to this section of the act, aided by later enabling acts. The most that can be said of the quoted portion of the charter is that it gave the committee, chosen by the proprietors to lay out and survey the lots of the town, authority to make allowances called for by it in their layout and survey; and, unless they did so, there is no basis for anyone to make such allowances later. The presumption is that the committee, in making the survey, made all allowances permitted by law, and that the same are included in the lines of the lots as shown on the town plan and described in the field book. At the trial below the plaintiff claimed that the field book showed an actual survey, and, in locating his line between lots 59 and 60, he relied upon the location and boundaries of lots 53 to 59, inclusive, as shown by the town plan and field book. It necessarily follows that, in surveying the same, he is bound by the descriptions of the lots as given in the field book. To hold otherwise would make the town plan and field book meaningless as a basis for ascertaining the location and boundaries of any particular lot, and would permit a surveyor arbitrarily to manipulate lot lines to meet the exigencies of a particular case.

There are six grounds in the defendant's motion to set aside the general and special verdicts. The special verdict is that lot 58, as it now exists in the town of Moretown, is not substantially a half lot. We do not consider the third, fifth, and sixth grounds, as their substances is embodied in the other grounds of the motion. The grounds we consider are: (1) That the verdicts are wholly unsupported by the evidence; (2) that the jury, in returning said verdicts, disregarded the testimony; (4) that, assuming that the plaintiff's evidence made a case for the jury, it was so outweighed by the countervailing evidence that intelligent and fair-minded men could not reasonably reach said verdicts.

For a proper understanding of the questions raised by the motion, we first consider the actual location of certain lots upon the ground as indicated by old marked corners and lines.

H. H. Squires, a land surveyor of experience, was a witness for the defendant. He did surveying for the defendant for the purpose of locating the dividing line between lots 59 and 60. He ran the lot line dividing lots 59 and 60 its entire length from the Northfield town line to the Winooski river, and lines of other lots that have a bearing on the location of the disputed line. A plan he made from his survey is Dft's Ex. N. So far as it represents an actual survey, it shows the lines and corners of the lots as they are actually marked upon the ground, and not as they are located on the town plan and described in the field book. Other lines on his plan indicate the general relation of the lots shown, and the effect that the line claimed by the plaintiff would have upon other lots if extended through the town.

The situation can be better understood by referring to the accompanying diagram. The lines running S. 39/d E. are the range lines.

No question is made as to the location upon the ground of the range lines of the first division of lots. The lines running N. 36/d E. are the lot lines. The broken lines are the lines which the plaintiff claims are the northwesterly and southeasterly lines of lot 59. The solid lines are the lot lines as claimed by the defendant or about which there is no dispute. Lots 4 to 9, inclusive, are in the third division. The other lots are in the first division. The names on the lots are the names of present or former owners of such lots and by which the lots are commonly called by those who are familiar with them. Many of the witnesses did not know any of the lots, not even their own, by number, except as they located them on the plans in evidence, but knew and called them by the names of such owners.

The parties agreed at the trial below that, as located upon the ground, a stone on the northerly range line at the southeast corner of lot 5, third, marks the southeast corner of that lot, the southwest corner of 4, third, the northeast corner of 57, and the northwest corner of 56; that a stone on the southerly range line at the southwest corner of lot 56 marks the southwest corner of that lot, the southeast corner of 57, the northwest corner of 49 and the northeast corner of 48; and that a stone on the northerly range line at the northwest corner of 57 marks the northwest corner of that lot, the northeast corner of 58, the southwest corner of 5, third, and the southeast corner of 6, third. Mr. Sleeper, who owns lots 56 and 49, was a witness for the plaintiff. He testified that the southeasterly lines of 57 and 48 were surveyed forty-five years ago, and there were a lot of old marked trees on the lines at that time. The line between lots 57 and 58 runs through cleared land, and there are no marks upon the ground to show its location. It was agreed that a line, substantially parallel with the southeasterly line of 57, extended southwesterly from the northeast corner of 58 to the southerly range line, is the line between 57 and 58, and corresponds with a similar line on the plans received in evidence. The southeast corner of 58 is not marked except as it meets the line dividing lots 48 and 47, which is a continuation of the line between 57 and 58 and is marked by a fence and stone wall.

The evidence shows that a mound of decayed wood covered with grass where an old hollow birch stump with a stake stuck in it formerly stood, on the northerly range line, about sixty rods northwest of the northeast corner of 58, marks the northwest corner of 58, the northeast corner of 59, and the southeast corner of 7, third, called the Converse lot, and the southwest corner of 6, third, called the Charles Smith lot, as said lots are located on the ground. There is a stone monument, called the Dewart monument, about two rods northwest of this corner which was erected by Frank Dewart in 1925 when he surveyed the lines for the defendant. Mr. Dewart is dead.

Julius Converse, a man seventy years old, and who has always lived in Moretown, was a witness for the defendant. He bought the Converse lot from his father in 1887, and sold it five years ago. It had been in his family since 1854. He has known the lot since he was a boy, and was on it with his father. His father told him where the corners were. He told him that the hollows stump with the stake in it was at the southeast corner. He has seen the stump and stake a great many times, and it was a very old stump. He went to this corner the Saturday before he testified, but the stump was gone; there was a mound where the stump used to be, but all he found was 'rotten wood.' He was present at this old stump in 1925 when Mr. Dewart started his survey around some lots from it. He went with Dewart through two lots, but Dewart's compass must have varied because he was about two rods from the stump when he came back.

George Herring, a witness for the defendant, was a nephew of Abijah Herring. He and his wife owned and lived on the Herring farm from April, 1908, to May, 1914. Leo Brown then owned lot 60. The hollow birch stump with a stake in it marked the northwest corner of lot 58 while the witness lived on the farm. The stump was then rotted so that the stake leaned over. Every year he cut hay 'clear to the stump' and 'cut around the stump by hand.' It was in the fence between his land and the Charles Smith lot. The fence turned at the stump and went northerly along Smith's land. In November, 1913, he employed J. A. Chapin, a surveyor, to run the line between lots 59 and 60. Mr. Chapin started his survey at the hollow birch stump at the northwest corner of 58.

A plan made by him of his survey was received in evidence. It shows a stone at the northeast and southeast corners of 57, which it is agreed mark those corners. It shows a birch stump at the northwest corner of 58. It shows 57 and 59 as full-sized lots and 58 as a half lot.

The evidence shows that a marked line extending northeasterly from the old birch stump, on the same course as the other lot lines, is the dividing line between lots 6 and 7, third. There is old growth timber with some second growth all the way on the northwest side of this line. On the southeast side about a third of the way is open pasture, a third is brush and small trees, and a third is old growth timber with second growth mixed in with it. There are marked trees, marked five or six years ago, the whole distance of the line. Where there is old growth timber on both sides of the line there is a well-marked line of trees with old marks. The marks are of different ages; some trees have three sets of marks on them. The youngest of these old marks are not less than forty years old, and the oldest are at least seventy-five years old.

The fence described by the witness Herring as turning at the old birch stump and going northerly along Charles Smith's line was built about 1910. Smith had cleared some of his land next to the line and wanted to use it for a pasture. Julius Converse bought the wire and Smith built the fence. Converse told him to put it where he wanted to. The fence runs for some distance on lot 7, and then turns and runs easterly on lot 6 towards Smith's buildings. It was not built for a line fence.

The defendant's evidence tends to show that a pile of stones at the foot of a ledge on the northerly range line, about 126 rods northwesterly of the old birch stump, marks the northwest corner of 59, the northeast corner of 60, the southwest corner of 7, third, and the southeast corner of 8, third, as said lots are located on the ground.

It appears that when Mr. Chapin made his survey in 1913, he started with a wrong compass direction and ran the line from the old birch stump northwesterly in lot 7, third, to a point where there are now a stake and stones near a leaning maple tree, a short distance from the northwesterly line of 7 and about thirty rods northeasterly of the range line, and located the northwest corner of 59 at that point. At the trial below, this corner was referred to as the 'leaning maple corner.' That Mr. Chapin was skeptical of this corner is apparent from his plan. It does not appear who placed the stake and stones at that point or when they were placed there.

The defendant owned lot 7, third, in 1925. At that time there was a controversy between him and the plaintiff as to the location of the line between 59 and 7, third–the plaintiff claiming that the stake and stones near the leaning maple was the northwest corner of 59. At that time the plaintiff did not question the location of the dividing line between 59 and 60, as claimed by the defendant, but only its northwesterly terminus. He did not question its location until December, 1928, when Mr. Smith was surveying for him. Mr. Smith began his survey at the leaning maple corner but did not go far before he found that he was not on the right line, and he did not get onto the northerly range line until he came to the Dewart monument. Mr. Dewart was employed by the defendant in 1925 to locate the true dividing line between 59 and 7, third. He went to the leaning maple corner and soon found it was wrong. He located the pile of stones at the foot of the ledge as the true northwest corner of 59, and placed a stake in the pile of stones. The plaintiff admitted at the trial below that the leaning maple corner is wrong and that the true northwest corner of 59 is on the range line. He admits that the pile of stones at the foot of the ledge is on the range line, but claims now that the corner is 56 rods farther to the northwest.

Julius Converse testified that his father never showed him the southwest corner of 7, third, but told him when he bought the lot that that corner was at the foot of a ledge, that he would find a pile of stones there. He did not go to the corner at that time, but was there with Mr. Dewart in 1925, and at the foot of the ledge they found the stones 'laid around as you would lay a chimney.' George Herring testified that during the Chapin survey they found a stone pile at the foot of the ledge southerly of the leaning maple.

The plaintiff testified that he was present in 1925 and saw Mr. Dewart set the stake at the foot of the ledge. When he set the stake there was nothing to indicate a corner; no stones were piled there, and Mr. Dewart did not pile any; he 'just drove a stake in the ground.'

The evidence shows that the dividing line between lots 7 and 8, third, extends northeasterly from the stake and stones at the foot of the ledge, on the same course as the other lot lines, about 182 rods to the remains of a bridge which marked the northeast corner of said lot 8. The line is marked for some distance from the stake and stones by trees with line marks on them about forty years old.

The defendant's evidence shows that the dividing line between lots 59 and 60, which is a continuation of the line between lots 7 and 8, third, is a marked line extending southwesterly from the stake and stones at the foot of the ledge about 152 rods to a wooden stake in the southerly range line, which marks the southwest corner of 59, the southeast corner of 60, the northeast corner of 45, and the northwest corner of 46, as said lots are located on the ground. There are several large trees along the line marked on their northeast and southwest sides which indicates that they were marked as line trees. The marks on the trees are from forty to fifty years old. J. A. Chapin located this line as the dividing line between 59 and 60 when he surveyed the lines in 1913.

The stake on the southerly range line is set in wet, swampy ground. Three or four witness trees are near the stake. They are marked with a single blaze on the side facing the stake, which is the way corners are witnessed. The stump of another stake is in the ground about nine inches from this stake, the rest of the other stake having rotted off close to the ground. The stump in the ground is at least twenty-five years old and may be one hundred years old. When J. A. Chapin made his survey in 1913, he found at the southerly end of the line a stake lying on the ground by its stump so rotten that it could not be picked up. He set another stake there near the stump. There were old blazes on the witness trees at that time which were quite dim and he renewed them. The stake that he set at this place was referred to at the trial below as the 'Chapin stake' or 'corner.'

The evidence shows that a beech tree, now lying on the ground on the northerly range line about 126 rods northwest of the stake and stones at the foot of the ledge marks the northwest corner of 60, the northeast corner of 61, the southwest corner of 8, third, and the southeast corner of 9, third, as these lots are located on the ground. This tree is marked on four sides as a corner, the oldest marks being about fifty years old. When Mr. Dewart surveyed the lines in 1925, this tree was standing, and he marked it on four sides with the numbers of the lots cornering there.

The evidence shows that a stake and stones on the southerly range line near a hardhack tree with a blaze on one side, about 152 rods southwest of the beech tree corner, marks the southwest corner of 60, the southeast corner of 61, the northwest corner of 45, and the northeast corner of 44, as these lots are located on the ground. Mr. Squires testified that when he surveyed the lines in 1928, he found that the stake at this corner had been set recently, but the pile of stones was old, and an old stake was lying on the stones. It appeared that Mr. Dewart located the stake and stones as the southwest corner of 60 in his survey of 1925. At that time the old stake was standing in the pile of stones. He cut a new stake and put it in place of the old one and laid the old stake on the stones beside it.

It also appeared that about forty-six years ago one Fred Willey and his brother purchased a narrow strip of twenty-five acres of timber land lying northwesterly of lot 60. Their lower corner at which they started to measure out their land was the hardhack tree corner. The Brown lot cornered there.

At that time there was a stake with stones piled around it at the corner. There was a hardhack tree blazed on one side six or seven feet from the stake.

Albert Wade, a witness for the defendant, has owned lot 44, lying westerly of and cornering on lot 60, for twenty-eight years. Mr. Royce, the man from whom he bought, pointed out the hardhack tree and stake and stones to him as the corner of the lot before he bought it. At that time Royce told him that the original line ran right through there. He has been on the line running southwest from the corner (line between 44 and 45) and "it is marked on trees all the way along that original line." When he bought the lot, the marks on the trees were several years old, and the hardhack tree had marks on it that were several years old.

The evidence shows that a stake and stones on a ledge on the range line southerly of lots 45 and 46, at the northeast corner of lot 26, marks the northeast corner of 26, the northwest corner of 25, the south-west corner of 46 and the southeast corner of 45, as said lots are located on the ground. Henry Church, a witness for the defendant, and who owns a part of lot 25, lived on 45 or 46 when he was a boy.

Forty-five years ago, when he was twelve years old, a surveyor by the name of Holt surveyed the lines of lot 26. He was with Holt and his party during all of that survey, and knows where the lines and corners of 26 are located as marked upon the ground. At that time a pile of stones on a ledge marked the northeast corner of 26, and that corner has always been so marked.

Lot 60 and some of the lots in its immediate vicinity are timber lots. Where there is timber the range lines are well marked by line trees with old marks.

The evidence shows that the northwesterly lines of 8, third, 60, and 45 make one continuous straight line which is marked its whole length by line trees with marks of various ages ranging from five years to one hundred and thirty years, which shows that the line or parts of it have been surveyed more than once. Several of the marked trees were old trees, one hundred years or more old. Many of the marks were fifty or more years old. Mr. Squires found one very old marked tree on the north-westerly line of 8, third, sixteen or twenty feet from the beech tree corner. The tree was about thirty inches through and he thought it was about three hundred years old. He cut out one of the marks on it and it was approximately one hundred and thirty years old, which would indicate it was first marked about the time of the original survey. The evidence also shows that the continuous straight line intersects the northerly range line at the beech tree corner and the southerly range line at the hardhack tree corner.

There are the remains of an old camp on lot 60 southeast of the middle of the lot and on the land claimed by the plaintiff. C. J. Brown built the camp about forty-six years ago and cleared about half an acre of land around it. He and his family lived there two or three years and it was occupied by him for many years in his lumbering operations on the lot. In 1908, Leo Brown, son of C. J. Brown, took most of the remains of the camp and moved them to their home in East Moretown.

Herring brook flows southeasterly through the Herring farm. A branch of it flows down through lots 60 and 59. A log road formerly extended from the main highway near the buildings on 58 up through 58, 59, and 60 to the Brown camp. There were two bridges on the log road across the branch of the Herring brook near where the defendant claims the line between 59 and 60 is located. The first bridge was easterly of the line in 59, and the other bridge was westerly of the line in 60. The defendant's line is on the top of a rise between the bridges about a rod or so beyond the first bridge.

About 1895, C. J. Brown first pointed out the line at the top of the rise to Leo Brown as the line between his land and Herring's, and also pointed out marked trees along the line. Sometime before 1909, soon after the owner of lot 45 had his lines surveyed, C. J. Brown pointed out the stake at the Chapin corner and the stake and stones at the hardhack tree corner to Leo Brown as the southeast and southwest corners of his lot, and also pointed out some marked trees on the line between those corners.

James McNulty, a witness for the defendant, worked for C. J. Brown on lot 60 at different times begin-ning about thirty-five years ago. One time, when they were cutting timber near the Herring farm, Brown pointed out the line between his land and the Herring farm to them to limit their cutting. The line he pointed out was on the rise just beyond the first bridge. There were several trees with old marks on them on the line as far as they worked. They cut clear to the line. Another time he was chopping for Abijah Herring near the line, and Herring pointed out the same line to him as his line.

The foregoing evidence as to the marks upon the ground locating the corners and boundaries of 58, 59, 60 and adjacent lots is not contradicted except by the testimony of the plaintiff that there was no pile of stones at the foot of the ledge on the northerly range line when Mr. Dewart set a stake there, and by the testimony of Mr. Smith that when he went to that place in December, 1928, the stones piled around the stake were new stones and not old stones. But their testimony does not contradict the evidence that the stake and stones at that point in fact marks the northwest corner of 59 and the northeast corner of 60 as the boundaries of said lots are marked upon the ground.

Mr. Smith never surveyed the southeasterly and northwesterly lines of 60 as claimed by the defen-dant, and it does not appear that he ever saw the beech tree corner or the hardhack tree corner. He saw the Chapin corner on the southerly range line, but he did not consider that it and the stake and stones at the foot of the ledge on the northerly range line were corners because the stakes were new. He knew that the Dewart monument purported to be a mark at the northwest corner of 58, but he gave it no consideration as such in locating the lines of 58 and 59.

The plaintiff locates 58 and 59 by measuring said lots northwesterly from the northwest line of 57, as that lot is located on the ground, according to the dimensions of said lots as described in the field book. This, on paper, locates the line between 58 and 59 a half a lot northwesterly of the old birch stump, and the line between 59 and 60 where the plaintiff claims it is.

This method of locating said lots is based upon two assumptions. First: That the agreement of the parties as to the location of lot 57 on the ground is a concession by the defendant that its location on the ground is identical with its location as shown by the town plan and field book. Second: That lot 58 could not have been located on the ground as a lot smaller than it is shown to be by the town plan and field book.

The defendant denies that he ever conceded or agreed that the location of lot 57 on the ground is identical with its location as shown by the town plan and field book. The plaintiff's assumption of this claimed concession is unwarranted. When the parties agreed to the location of 57 on the ground nothing was said about that location being identical with its location as shown by the town plan and field book; and the plaintiff has failed to point out any evidence from which such concession can be inferred. In fact, there is no such evidence.

From the outset of the trial below, the defendant's claim has been based on the location of the lots in question as indicated by ancient marked lines and corners, and that lot 58, so located, is a half lot. Before the agreement was made, Mr. Smith testified that none of the lines of the lots surveyed by him conformed with their description in the field book, and that the dividing line between 59 and 60, as located by him, is 56 rods northwesterly from the same line when located by the field book measuring northwesterly from the Berlin town line; and it had appeared that, according to the field book, said dividing line is substantially where the defendant claims it is.

The claim of the plaintiff that lot 58, as located on the ground, could not be a small lot, requires a consideration of the evidence.

In support of the method by which he locates lots 58 and 59 on the ground, the plaintiff claims that, since the description of these lots in the deeds is by number, "it is the same as though the descriptions contained in the field book had been copied into the deeds, and the land embraced therein conveyed thereby"; that such deeds gave the grantees color of title to the lots as described in the field book, and that actual possession of a part of the lots by the grantees was extended by implication to the whole of the lots; that since a part of each lot, at least, has been occupied adversely by the plaintiff and his grantors for more than fifteen years under their deeds, he has a perfect title to the whole of the lots as shown on the town plan and described in the field book. This claim of the plaintiff is correct as a general statement of the law, and it applies with equal force to the limits of lot 60 and the title to, and possession of, said lot by the defendant and his grantors, as said lot is described in the defendant's chain of title as "Being all of the first Division lot * * * drawn to the original right of Nathaniel Barrel," which the town plan shows is lot 60.

It is conceded that the plaintiff owns lot 59 and that the defendant owns lot 60. The line between these lots is a common line. Each party and his grantors have had actual and constructive possession of his lot to this common line for more than sixty years. The burden is upon the plaintiff of showing that the location of this common line upon the ground is where he claims it is. *Downer v. Tarbell*, 61 Vt. 530, 533, 17 A. 482.

A precise statement of the effect of a description in a deed of a lot by reference to its number is given in *Spiller v. Scribner*, 36 Vt. 245, 247, and followed in *Silsby v. Kinsley*,[67] 89 Vt. 263, 95 A. 634, and in *D'Orazio v. Pashby*, 102 Vt. 480, 150 A. 70. It is there said that such a description "is a description in its legal effect according to the lines of such lot *as surveyed and established* in the original division of the town, and is just as definite, although not so particular, as it would be if the lines were given, and should receive the same construction and have the same legal effect in one case as the other." Described thus, the lot lines, if surveyed upon the ground, serve as monuments in fixing the boundaries. *Silsby v. Kinsley,* supra.

So far as it appears from the evidence, the lines and corners of lot 59, as claimed by the plaintiff, are not marked upon the ground. Mr. Smith testified that no southeast corner of 59, as laid out by him, can be found marked on the ground; and there is no evidence tending to show that there is a northeast corner of such lot marked on the ground or that there are any marks on the ground showing the location of the of the southeasterly line of such lot. Mr. Smith testified that there is no northwest corner or southwest corner of such lot marked on the ground; that he surveyed the northwesterly line of such lot, but found no marks along that line; that there were a number of "fairly old trees" along the line, but he found no marks on them. It does not appear from the evidence that he found any marks of any kind or age indicating the location of that line. The fact that there are no marks upon the ground to indicate the location of the lines of 59, as claimed by the plaintiff, is strong evidence, if not conclusive, that such lines were never surveyed, or if surveyed, they cannot be ascertained.

The actual location upon the ground of original lot lines will control, if capable of being ascertained. *Silsby v. Kinsley,* supra. When such lines have never been surveyed, or if surveyed, their location upon the ground cannot be ascertained, but the lines have been actually run and marked upon the ground, and have been recognized as correctly located for more than fifteen years by all parties in interest, such actual lines and monuments, marked upon the ground, constitute the survey, and, where found, will control the courses and distances named in the original layout. *Richardson v. Chickering*, 41 N.H. 380, 77 Am. Dec. 769; *Pyburn v. Campbell*, 158 Ark. 321, 250 S.W. 15; *Rowell v. Weinemann*, 119 Iowa 256, 97 Am. St. Rep. 310; *George V. Thomas*, 16 Tex. 74, 67 Am. Dec. 612; *Hall v. Davis*, 36 N.H. 569; *Warren v. Pierce*, 6 Me. 9, 19 Am. Dec. 189; *Le Compte v. Lueders*, 90 Mich. 495, 51 N.W. 542, 30 Am. St. Rep. 450; *Watrous v. Morrison*, 33 Fla. 261, 14 So. 805, 39 Am. St. Rep. 139, 142; *Martin v. Carlin*, 19 Wis. 454, 88 Am. Dec. 696.

In *Hall v. Davis,* supra, the court said: 'That, in the description of a line, what is most material and certain shall control that which is less material and uncertain; that boundaries marked on the land,

---

[67] This case is detailed in Appendix A.

as being most material and certain, are to govern courses and distances; that if the plan, or the line described in a deed or charter, and the monuments made by an original survey of a tract or township of land, do not correspond, the monuments are always to determine the true location and that the marks on the ground of an old survey, indicating the lines originally run, are the best evidence of the true location of that survey. * * * And it may be regarded as well settled, that where land is conveyed by a deed referring to a plan or to a charter line, between which and the actual original survey, as shown by fixed monuments upon the ground, there is a difference in the courses and distances, or in the location of lines and monuments, the lines and monuments, as originally located and marked on the ground, are to govern, however they may differ from those represented on the plan, or described in the charter.'

In *Martin v. Carlin,* supra, the court, quoting from *McClintock v. Rogers*, 11 Ill. 279, 296, a leading case, said: 'The original monuments, when ascertained, afford the most satisfactory and we may say conclusive evidence of the lines originally run, which are the true boundaries of the tract surveyed, whether they correspond with the plat and field notes of the survey or not. All agree that courses, distances and quantities must always yield to the monuments and marks erected and adopted by the original surveyor, as indicating the lines run by him. These monuments are facts; the field notes and plats indicating courses, distances and quantities, are but descriptions, which serve to assist in ascertaining those facts. Established monuments and marked trees not only serve to show with certainty the lines of their own tracts, but they are also resorted to, in connection with the field notes and other evidence, to fix the original location of a monument or line which has been lost or obliterated by time, accident or design.'

In *Hull v. Fuller*, 7 Vt. 100, 110, this court said: 'When the original monuments are found, no testimony can be received to show that the surveyor intended to locate the boundaries elsewhere. Were it otherwise, the boundaries of the whole State might be disturbed. A single error in the allotment of a town might lead to a new allotment throughout; and if ancient landmarks are to be disturbed, upon this principle, there would be no end to the consequences.'

The rule is similar where there is a conflict between courses and distances on the one hand, and known boundaries and monuments on the other, mentioned in the description in a deed. In such cases the courses and distances, as a general rule, must yield to the known boundaries or monuments, because it is more likely that there would be a mistake or misunderstanding about the course or distance than about the boundary or monument. *Keenan v. Cavanaugh*, 44 Vt. 268; *Fullam v. Foster*, 68 Vt. 590, 594, 35 A. 484, and cases cited; *Sowles v. Butler*, 71 Vt. 271, 276, 44 A. 355; *Vermont Marble Co. v. Eastman*, 91 Vt. 425, 448, 101 A. 151.

All lands are supposed to be actually surveyed, and where a deed describes a lot by its number according to a plan, the intent is to convey the land according to that actual survey. Consequently, if marked lines and marked corners are found, courses and distances must yield to them. *Riley v. Griffin*, 16 Ga. 141, 60 Am. Dec. 726, 729; *Bean v. Bachelder*, 78 Me. 184, 3 A. 279. In the latter case, where a plan had been made to delineate an actual survey, and a deed described a lot by number according to that plan, the court, in holding that an actual survey upon the ground, rather than the plan, fixed the location and boundaries of the lot, said: "The plan was merely a picture. The survey was the substance. The plan was not made to show where the lots were to be hereafter located, or how they were to be hereafter bounded. It was made as evidence of where they had before been located and bounded. The lot actually surveyed, bounded by the lines actually run, was the lot intended to be conveyed. The plan was named in the deed, rather as a picture, indicating the location and lines of the lot. Still the actual boundaries, rather than the pictured boundaries, were to be sought for. The picture might not be wholly accurate."

Another rule is that resort may be had to the lines of adjacent lots to determine the location of a lot when its location on the ground cannot be ascertained. *Silsby v. Kinsley,* supra; *Bristol Mfg. Co. v. Palmer*, 82 Vt. 438, 74 A. 76.

Since the plaintiff admits that there are no marks upon the ground locating the lines of lots 59 and 60 as claimed by him, it necessarily follows that the location of said lots must be determined by the actual lines and monuments as marked upon the ground, and, if necessary, resort may be had to the established lines of adjacent lots.

The northwesterly line of lot 58 is not marked upon the ground unless the old birch stump is a corner, and this is questioned by the plaintiff. Lot 6, third, lies directly northeast of 58 and lot 47 lies directly southwest. It is conceded that these lots are the same width and that their lot lines form two continuous straight lines parallel with each other. The plaintiff assumes these facts in locating his northwesterly lines of 58 and 59, and also assumes that 58, 47, and 6, third, are full-sized lots. No question is made as to the southeasterly lines of these three lots. The southeasterly line of 47 is marked by a fence and stone wall.

Aaron H. Martin bought 45 acres of land in the northeast corner of lot 5, third, in 1864. He purchased lot 6, third, in 1869. In 1890, the administrator of his estate conveyed said two parcels of land to the Northfield Savings Bank. After several mesne conveyances of said lands, in which they are described as being 115 acres, the bank again became the owner of the same. The bank conveyed the premises to Hiram Sanders October 5, 1895, the description in the deed being: "The farm now occupied by said Sanders, bounded as follows: North by land of James Kelley, east by the land of Jessie Willey and land occupied by the widow McNulty. South by the land of Abijah Herring and west by land of Julius Converse, containing 115 acres more or less." Sanders reconveyed the premises to the bank February, 1902, and the bank conveyed the same to Charles Smith, April 7, 1902; the description in the deed being: "The same premises deeded said Bank * * * by Hiram Sanders and wife. Reference to said deed being had and to other previous deeds for a more definite description. Said premises supposed to contain one hundred and fifteen acres more or less."

The description in the deed from the bank to Sanders of the land now owned by Charles Smith as bounded "west by land of Julius Converse" makes the Converse lot a monument; and this monument must control the location of the northwesterly line of 6, third, rather than the courses and distances described in the field book. *Park v. Pratt*, 38 Vt. 545, 552; *Church v. Stiles*, 59 Vt. 642, 10 A. 674; *Cutler v. Barber*, 93 Vt. 468, 474, 108 A. 400; *Viall v. Hurley*, 94 Vt. 410, 416, 111 A. 395; *Vermont Marble Co. v. Eastman*, 91 Vt. 425, 448, 101 A. 151; *Bryant v. Maine Cent, R. R. Co.*, 79 Me. 312, 9 A. 736; *Curtis v. Francis*, 63 Mass. 427, 435. The undisputed evidence shows that the southeasterly line of the Converse lot (7, third) abutting the westerly side of Smith's land is a line of marked trees running northeasterly from the old birch stump on the northerly range line. This definitely fixes that marked line as the northwesterly line of 6, third, and makes it a small lot. That it is a small lot, as marked on the ground, is further shown by its acreage. The deeds of the Smith farm describe it as containing 115 acres. As forty-five acres of this farm is in 5, third, the part of the farm comprising lot 6 is not more than seventy acres. It also appears that from 1856 to 1870, a time when it was shown as a distinct parcel of land, said lot 6 was in the grand lists and appraisals at sixty acres. Since 1870 the whole farm has been listed at 115 acres.

Lot 47 is owned by Cornelius Hayes. Thomas Winter and his wife conveyed to John Shanley 'the north half' of this lot together with eleven acres and forty-five rods off the south end. By successive conveyances, in which this piece of land is described as containing forty-two acres and twenty-five rods, the title to it came to Cornelius Hayes, December 3, 1895, and he has owned it since then. This piece of land, from 1852 to 1866, a period during which it can be identified on the grand lists and appraisals, was set to the owners at forty-two acres. If eleven acres and forty-five rods of this piece of land is in the south end of 47, it follows that the north half of the lot is about thirty-one acres, and the whole lot is sixty-one and three-fourths acres, which makes it a small lot. There is a highway along its northwesterly side. Cornelius Hayes was a witness called by the plaintiff. He testified in direct examination that he was on one side of the highway and Moody, who owns lots 45 and 46, was on the other side, that there is a fence on each side of the highway. He testified in cross-examination that the fence on the northwesterly side of the highway is the dividing line between his land and Moody's.

The town plan and field book show that 46 abuts 47 on the northwest. It appears from Smith's plan, Plff's Ex. 27, that the northeast line of Hayes' land is 1,000 feet long and the southwest line is 963 feet long, which makes 47 a half lot.

It appears from the plans in evidence that if the northwesterly line of 47, as marked on the ground by the fence on the northwesterly side of the highway, is extended northeasterly on the course of the lot lines, it is substantially parallel with the southeasterly lines of 47, 58, and 6, third, and meets the northwesterly line of 6, third, and intersects the northerly range line at, or substantially at, the old birch stump. This makes 58 a small lot; but it is not the only evidence that it is a small lot. Thirty years ago Abijah Herring wanted George Herring to take a deed of 58 and care for him the rest of his life. At that time he told George Herring that lot 58 was a small lot; that it was not a full lot. It has already appeared that more than thirty years ago he pointed out the marked line between 59 and 60 to James McNulty, who was then chopping for him, as the northwesterly line of 59.

There is not space enough between the marked northwesterly line of 59 and the northwesterly line of 57 for two full-sized lots, but only for one full-sized lot and a half lot. It is conceded that 59 is a full-sized lot. Herring's deed of it describes it as 'estimated to contain 116 acres of land more or less.' Lot 58, then, as located on the ground, must be a half lot. *Warren v. Pierce*, 6 Me. 9, 19 Am. Dec. 189. In that case the proprietors voted to lay out their town in one hundred-acre lots. The question was as to the location of lot 8, which was owned by the plaintiff. At the trial in the lower court, the plaintiff located lots 7 and 9 on the ground beyond controversy. The area between these lots was 200 acres. The trial judge instructed the jury that irrespective of acreage, 8 must be presumed to extend from 7 to 9 unless a different location was shown. The supreme court said: 'He would have been justified in using stronger language, and in stating that 8 did and must extend from 7 to 9, unless a different original location could be shown. * * * The proprietors voted, it seems, to lay out their town in one hundred-acre lots. But it is of no consequence what they proposed or intended to do; and the question is, What have they done by their surveyors or other agents duly authorized? Their intention as manifested by their vote was very inaccurately executed. * * * If the defendant could have shown original corners, or a line dividing the space between 7 and 9, the case would have been differently presented.'

Mr. Smith testified that the acreage of the Herring farm, as claimed by the plaintiff, is 334 acres divided as follows: Lot. 57, 107 acres; lot 58, 110 acres; lot 59, 117 acres. It appears, however, that from 1856 to 1863, during which time Abijah Herring owned 58, all the land set to him in the grand lists and appraisals was 60 acres; that after he bought 57 in 1863, his farm was listed in 1864 and 1865 at 171 acres; that after he bought 59 in 1865, his farm was listed from 1866 to 1874 at 277 acres, and from 1874 to the present time it has been listed at 280 acres. Mr. Smith testified that the difference of 56 acres between the acreage of the Herring farm, as claimed by the plaintiff, and its acreage, as shown by the grand lists and appraisals, is the acreage of the land in dispute.

It also appears that for a few years prior to 1856, lot 6, third, was listed at 112 acres, and lot 58 at 110 acres. It does not appear why the acreage of these lots was changed in 1856, unless it can be inferred that it was changed to conform to the fact that the lots are small lots as located on the ground.

The plaintiff, in support of his northwesterly line of lot 59, relies upon a line of marked trees with old marks on them extending southwesterly from the southerly range line about half the length of lot 45. He claims that these marked trees are on the dividing line between 45 and 46. The true dividing line between 45 and 46 is a continuation of the true northwesterly line of 59. The plaintiff, to have this marked line a continuation of the northwesterly line of 59, assumes that 47 is a full-sized lot, and Cornelius Hayes owns only the southeast half of it. He arbitrarily locates his southeasterly line of 46 a half a lot northwesterly of Hayes' northwesterly line. He then measures 2,180 feet to the line of marked trees, and calls that line the dividing line between 45 and 46. This makes that northwesterly line of 46 a continuation of plaintiff's northwesterly line of 59, but it places the line a half a lot northwest of the Chapin corner and the stake and stones on the range line southerly of 45 and 46, which are respectively the northwest and southwest corners of 46 as marked on the ground, and leaves a half of a lot between 46 and Hayes' land that is owned by no one.

The defendant claims that this line of marked trees is nothing but an old marked line dividing the northeast and northwest quarters of lot 45. Lot 47, as we have already said, is a small lot, and the southeasterly line of 46 is the fence on the northwesterly side of the highway, as said lots are located on the ground. With the lots so located, the line of marked trees is in the middle of the northerly half of 45. The trees are marked as line trees and it appears that lines between divisions of lots are sometimes marked that way. It also appears that for many years before 1892, Patrick Keough owned the northeast quarter of 45 and John Carrigan owned the northwest quarter. In 1892 these quarters were conveyed to a common grantee, and 45 and 46 are now owned by Mr. Moody. There is nothing to indicate that the line of marked trees was intended as a lot line. There is no marked corner at either end of it, and, if extended either way, it does not intersect any range line at a marked corner nor coincide with any marked lot line; but it is consistent with the theory that it marked the dividing line between the two quarters of the northerly half of 45 when they were owned separately.

There are certain definite conclusions that can be drawn from the fact hereinbefore set forth and the evidence: (1) That there was an original survey of the lots in question; (2) that there are no marks upon the ground from which it can be inferred that lots 58, 59, and 60 were surveyed and located on the ground, as claimed by the plaintiff; (3) that there are old marked lines and corners on the ground, existent when first seen by the several witnesses, which have been recognized and acquiesced in as the true boundaries of the lots in question by their owners and those interested for more than thirty years, and that such marked lines and corners are evidence of the lines originally run, and determine the boundaries of the lots in question as they are located on the ground; (4) that the line of marked trees between lots 59 and 60, as claimed by the defendant, is the true dividing line between said lots as located on the ground.

As hereinbefore stated, it appears from Mr. Smith's testimony that, if the theory on which the plaintiff started to try the case is followed by starting at the Berlin town line with lot 53 and measuring off the lots according to their descriptions in the field book, the distance to the northwest corner of lot 59 is 839 rods, which locates the northwesterly line of 59 where, or very close to where, the defendant says it is. The distance from the Berlin town line to the northwest corner of lot 59 as claimed by the plaintiff, and measuring 58 as a full-sized lot, is 895 rods, an overrun of 56 rods. At the trial below, the plaintiff attempted to account for this overrun by making allowances not contained in the field book, but we have already held that he could not do that; that, in measuring lots according to their descriptions in the field book, he is confined to the courses and distances of the lot lines as actually given in the book, plus an allowance of one unit in thirty for swag in chain. Mr. Smith started at the northeast corner of lot 57 when he measured the distance of 895 rods, and measured both ways, 525 rods through four lots to the Berlin town line, and 370 rods through 57, 58, and 59 to his northwest corner of 59.

The total width of the four lots from the Berlin town line to the northeast corner of 57 according to the town plan and field book is 479 rods, which shows an overrun of forty-six rods in the width of said lots as located on the ground. The total width of lots 57, 58, and 59, according to the town plan and field book, is 360 rods, an overrun of ten rods in the width of said lots as located on the ground by the plaintiff. These measurements show that lot 57, as located on the ground, is forty-six rods farther to the northwest than is shown by the town plan and field book.

It is apparent that there was a departure from the town plan and field book in the actual surveying and marking of the boundaries of the lots on the ground; in fact, it appears from the records of the proprietors' meetings that there were mistakes in the original survey, and trouble in the early settlement of the lots in the first division. It is clear that there was an expansion of lots between 57, as located on the ground, and the Berlin town line; and it is a reasonable explanation to say that 58 was located on the ground as a small lot to compensate for the same. Mr. Smith, in locating the plaintiff's northwesterly line of 59, according to the town plan and field book, ignored the forty-six rods overrun in the lots northeast of 57 and the ten rods overrun in lots 57, 58, and 59, and errone-ously assumed that the location of 57 and 58, as marked on the ground, was the same as their location as shown by the town plan and field book. If, in surveying the Herring farm, he had followed the boundaries of the lots as marked upon the ground, he would have located his northwesterly line of 59 where the line of marked trees is.

There remains the question of adverse possession as affecting the location of the dividing line between 59 and 60 on the ground. As we have hereinbefore said, where a deed describes a lot by its number, according to a plan, the intent is to convey the land as actually surveyed and marked upon the ground. Since it appears that the location of the dividing line as marked upon the ground and as shown by the town plan and field book are substantially the same, the actual possession of a part of 59 by the plaintiff and his grantors would not extend by implication beyond that line. *Shedd v. Powers*, 28 Vt. 652; *Fullam v. Foster*, 68 Vt. 590, 596, 35 A. 484; *Rice v. Chase*, 74 Vt. 362, 368, 52 A. 967; *Silsby v. Kinsley*, 89 Vt. 263, 273, 95 A. 634. There is no evidence that Abijah Herring or any subsequent owner of 59 claimed to own any land northwest of the dividing line as marked upon the ground until the plaintiff made such claim a short time before this suit was brought. Abijah Herring claimed to own only to this marked line; and the undisputed evidence shows that C. J. Brown, Leo Brown and the defendant have had actual and constructive possession of lot 60 to this marked line under a claim of title since 1882.

We have considered all of the evidence in the case and are satisfied that on the undisputed facts lot 58 is a small lot, and that the plaintiff has failed to show title to any land northwest of the dividing line between lots 59 and 60 as said line is marked upon the ground. The court below erred in denying the motion of the defendant to set aside the verdicts.

The defendant asks in his motion that the verdicts be set aside and judgment rendered for the defendant.

It has long been the established rule of practice in cases brought into this Court upon exceptions to finally dispose of the case here, particularly, when the question raised was as to the sufficiency of the undisputed evidence to support a verdict. If the exception was to the overruling of a motion for a verdict, and the exception was sustained, this Court rendered such judgment as the trial court should have rendered, unless a jury trial became necessary or the decision of this Court placed the case in such a state that either party had a right to a trial by jury. *Bass v. Rublee*, 76 Vt. 395, 400, 57 A. 965; *Riggie v. Grand Trunk Ry. Co.*, 93 Vt. 282, 107 A. 126.

When the question as to the sufficiency of the evidence to support a verdict was raised in some way other than by a motion for a directed verdict, and the exception was sustained, the judgment was usually reversed and the cause remanded. But under our present practice the rule is that when it is clearly apparent that on another trial the party against whom reversal is made cannot strengthen his case by a proper amendment of his pleadings, or by the introduction of new and additional evidence, we render such judgment as the trial court should have rendered, thereby saving the parties the need-less trouble and expense of a new trial. *Wetherby's Admr. v. Twin State Gas Co.*, 83 Vt. 189, 201, 75 A. 8, 25 L.R.A. (N.S.) 1220, 21 Ann. Cas. 1092; *Derosia v. Ferland*, 83 Vt. 372, 385, 76 A. 153, 28 L.R.A. (N.S.) 577, 138 Am. St. Rep. 1092; *Davis v. B. & M. R. R.*, 86 Vt. 205, 210, 84 A. 818; *Johnson v. Bennington, etc., Ry. Co.*, 87 Vt. 519, 523, 90 A. 507; *Globe Granite Co. v. Clements*, 92 Vt. 383, 104 A. 104; *Riggie v. Grand Trunk Ry. Co.*, 93 Vt. 282, 287, 107 A. 126; *Booth v. N.Y. C. R. R. Co.*, 95 Vt. 9, 16, 112 A. 894. But if it appears, or is made to appear, that it is probable that the party against whom reversal is made will be able to make a stronger case on another trial than he made at the former trial, or that an injustice will be done by rendering final judgment in this Court, the cause will be remanded. *Kennett v. Tudor*, 85 Vt. 190, 81 A. 633; *Lapoint v. Sage*, 90 Vt. 560, 99 A. 233; *Manley v. B. & M. R. R.*, 90 Vt. 218, 222, 97 A. 674; *Hebard v. Cutler*, 91 Vt. 218, 99 A. 879; *Rice v. Bennington County Bank*, 93 Vt. 493, 512, 108 A. 708; *Gaines v. Baldwin*, 92 Vt. 451, 104 A. 825;

*Bradley v. Blandin*, 94 Vt. 243, 256, 110 A. 309; *O'Boyle v. Parker-Young Co.*, 95 Vt. 58, 63, 112 A. 385; *Peters v. Estate of Poro*, 96 Vt. 95, 106, 117 A. 244, 25 A.L.R. 615; *Parker v. Bowen*, 98 Vt. 115, 120, 126 A. 522; *Weinberg v. Roberts*, 99 Vt. 249, 131 A. 14.

Our disposition of the case depends upon the conclusiveness upon the plaintiff of the dividing line between lots 59 and 60 as marked upon the ground.

Since upon the plaintiff's theory of the case the dividing line, according to the town plan and field book, is substantially where the line of marked trees is, we are unable to see how the plaintiff can make a better case on a new trial.

The undisputed evidence of the age of the marked lines and corners of the lots in question and of adjoining lots, and the facts that such lines have been recognized and acquiesced in as the true lot lines by all parties in interest for so many years, and that other surveys of these lots have followed these same lines, are, in the absence of some proof of other lines located and fixed in the original survey, conclusive that they are the true lines and corners located and marked in the original survey. *Hanlon v. Ten Hove*, 235 Mich. 227, 209 N.W. 169, 46 A.L.R. 788; *Dupont v. Starring*, 42 Mich. 492, 4 N.W. 190; *Haring v. Van Houten*, 22 N.J.L. 61.

In *Dupont v. Starring*, Justice Cooley said: "It has been repeatedly held by this court that a boundary line long treated and acquiesced in as the true line, ought not to be disturbed on new surveys. * * * Fifteen years' recognition and acquiescence are ample for this purpose, * * * and in view of the great difficulties which often attend the effort to ascertain where the original monuments were planted, the peace of the community requires that all attempts to disturb lines with which the parties concerned have long been satisfied should not be encouraged." Diehl v. Zanger, 39 Mich. 601.

In *Hanlon v. Ten Hove*, the court said: 'The original survey may have been inaccurate, the lines may not correctly fix the boundary, but if they have been acquiesced in for a sufficient length of time they fix the true line' as matter of fact and as matter of law.'

In *Haring v. Van Houten*, the court said: 'A wrong acquiescence, even in an erroneous location, will conclude parties. Considerations of public policy forbid that such errors should be corrected after long acquiescence, where, as in the present instance, the consequence would be a corresponding change in the possession of a whole neighborhood.'

Furthermore, the marked line between lots 59 and 60 has been recognized and acquiesced in as the true dividing line between the lots by their owners for more than thirty years, accompanied by continued possession with reference thereto. The rule is well settled that adjoining owners may thus establish the division line of their lands, which will be binding upon them and their privies. Soulia v. Stratton, 99 Vt. 304, 131 A. 610, and cases cited.

In *Clark v. Tabor*, 28 Vt. 222, it is held that the recognition by the owners of adjoining lots of a particular line as their division line, and their acquiescence in it for a period of fifteen years, will establish it and make it thereafter binding, if, during that time, the owners had a continued, although it was only a constructive, possession of their lots.

In *Burton v. Lazell*, 16 Vt. 158, 161, this Court said: 'And when a line is marked between the lots to which both parties claim as the division line for more than fifteen years, this is considered as decisive evidence that the line thus acquiesced in is the true line; and it is not to be disturbed by any survey or new line made after that period has elapsed.'

Moreover, although the defendant fully briefed this question that on the uncontradicted evidence, the verdicts should be set aside and judgment rendered for him in this Court, the plaintiff has failed to make it appear that it is probable that he could make a stronger case on another trial, and has presented no reason why the case should be remanded.

Other questions are raised by the exceptions in respect to the admission and exclusion of evidence and the court's charge to the jury; but, in view of the disposition we make of the case, it is not necessary to consider the same.

The defendant has brought a petition for a new trial, based upon newly discovered evidence. In view of the disposition we make of the case, it is not necessary for us to consider it, and it is dismissed, without costs.

ON MOTION FOR REARGUMENT AND REMAND.

After the foregoing opinion was handed down, the plaintiff moved for leave to reargue the case, and requests the Court, in case of reversal, to remand the same, pending which the entry of judgment has been withheld.

Many of the reasons assigned in support of the motion are a reiteration of the claims made in the plaintiff's brief and in oral argument, and have already been fully examined by the Court. The motion does not challenge the correctness and applicability of the law of the case as stated in the opinion, and plaintiff's counsel, when arguing the motion, said the Court had stated it correctly. The burden of the plaintiff's complaint is that the Court, in treating certain evidence as conclusive, has infringed upon the province of the jury to his prejudice.

A similar question was raised on motion for a reargument in Spaulding v. Mut. Life Ins. Co., 94 Vt. 42, 56, 109 A. 22, 29, and is considered fully in the opinion. It is said there that it is well-settled that the court may withdraw the case from the jury altogether, and direct a verdict for the plaintiff or the defendant, as the one or the other may be proper, where the evidence is undisputed or is of such conclusive character that the court in the exercise of a sound judicial discretion would be compelled to set aside a verdict returned in opposition to it; and that it would be an idle proceeding to submit the evidence to the jury when they could justly find only in one way. The Court further said: "It is essential to the due administration of justice that the jury should reason correctly. It is the unquestioned right of the party to insist that this reasoning be exercised, and it becomes the duty of the court so to supervise the trial as to assure this result. Among other things, upon occasion, it is the duty to rule as to what is or is not rationally possible for the jury to do. In exercising this function the court does not decide questions of fact but is ruling on a matter of law. * * * * When the facts are such that reasonable men can fairly draw but one conclusion, the court may, and on motion should, withdraw the case from the jury." It is with these principles in mind that we consider the plaintiff's motion.

The plaintiff's chief criticism is to the holding that on the evidence in the case, lot 58, as marked on the ground is a small lot. This question was briefed fully by the defendant and argued by both parties, so it required our consideration. We are inclined to think that it has been given more importance by the parties than it is entitled to, as, in the view we take of the case, the size of this lot is not determinative of the main question, and is not necessarily involved as an independent fact.

The plaintiff also complains because the Court referred to the Chapin plan and the Squires plan in the opinion. These plans and the plan made by Mr. Smith for the plaintiff were not received as independent evidence, and were not considered as such by us. They were received merely to aid the jury to a proper understanding and application of the testimony of the witnesses as to the location of the physical objects involved in the controversy. Hassam v. Safford Lumber Co., 82 Vt. 444, 74 A. 197.

In determining whether the location of the line between lots 59 and 60 is a jury question, so that the case must be remanded, it is not necessary to consider the plans nor the size of lot 58, as marked on the ground, so we exclude the plans and the evidence relating to the size of 58, which includes the grand lists and all evidence relating to the size of lot 47 and lot 6, third, from our consideration of this question.

The plaintiff criticizes the conclusiveness given in the opinion to the testimony of Fred Willey and Albert Wade that the hardhack tree corner is the southwest corner of lot 60, the testimony of Henry Church that a pile of stones on a ledge marks the northeast corner of lot 26, and the testimony of James McNulty that Charles Brown pointed out the marked line between 59 and 60 to him as the dividing line between his land and Herring's land, on the ground that the credibility of these witnesses is for the jury, and the jury might disregard their testimony if they so desired. But this is not the law. Those witnesses were credible witnesses, and had no interest in the outcome of the case. Their testimony was direct and positive, and it is not contradicted by cross-examination or other testimony, facts or circumstances. The general rule is that where a credible witness testifies distinctly and positively to a fact and is not contradicted, and there is no circumstance shown from which an inference against the fact testified to can be drawn, the fact may be taken as established, and a verdict directed on such evidence. *Howard Nat. Bank v. Wilson*, 96 Vt. 438, 453, 120 A. 889; *Crichton v. Barrows Coal Co.*, 100 Vt. 460, 139 A. 252; *Potts v. Pardee*, 220 N.Y. 431, 116 N.E. 78, 8 A.L.R. 785. And a claim by counsel that uncontradicted testimony in a case is not true does not require a submission of the case to the jury. *Boudeman v. Arnold*, 200 Mich. 162, 166 N.W. 985, 8 A.L.R. 789.

The witness Willey testified that about forty-six years ago he and his brother purchased some land lying northwest of lot 60, and their lower corner at which they started to measure out their land was the hardhack tree corner; that the Brown lot cornered there. Counsel say that the witness did not remember any marks on the hardhack tree, so the jury had a right to consider his lack of memory or other weaknesses and find that the tree where he started to measure in fact had no marks on it. The testimony of the witness relative to the tree being marked is as follows: 'Q. Were there any marks on it when you bought? A. I don't really remember that, it was just hewed a little on one side. Q. Blazed? A. Yes, blazed they call it.' There are other perversions of the testimony in the motion, and we call attention to this one that it may be known that they have not passed unnoticed.

The line we hold to be the true dividing line between lots 59 and 60, hereinafter referred to as the marked line, is an old marked line extending northwesterly from the Chapin stake on the southerly range line to the stake and stones at the foot of the ledge on the northerly range line. There are no marks upon the ground to indicate the location of the line claimed by the plaintiff, but if there were such marks, that line would be approximately fifty-six rods northwesterly of the marked line.

It appears from the proprietors' records that they voted to divide the land in the town into three divisions, the first division lots to contain 104 acres each; that the committee appointed to make the survey made a report of such survey, and their survey and return were accepted; that a plan of the town was made and a field book of such survey was accepted and recorded. It appears from the town plan and the field book that all of the lots in the town, except the river lots, are 116 rods wide. The legal presumption arising from this evidence is that such a survey, as it is shown by the town plan and field book, was made; and the burden of proof is upon the party who claims that a different survey and division was made, to establish that fact. Beach v. Fay, 46 Vt. 337.

The plaintiff began his case on the theory that the lines and dimensions of the lots, as shown on the town plan and described in the field book, control the location of the line in question, as his counsel said in his opening statement to the jury: 'So that in getting at the location of the lot lines according to the land records (meaning the town plan and the field book as recorded in the proprietors' records) we begin at the east on the Berlin town line and count off the lots until we get up through 60, the last one beginning where the previous one ended.' It was on this theory that he offered the descriptions of lots 53 to 59, inclusive, and that the court received them in evidence.

It appears from the testimony of the plaintiff's surveyor, who measured the same on the ground, that if a start is made at the Berlin town line with lot 53, and the width of that lot and each succeeding lot up to and including lot 59, is measured as it is described in the field book, plus one-thirtieth for swag of chain, the northwesterly line of 59 is located very close to where the marked line is. It may not actually coincide with the marked line, but it is close enough to be the same line for all practical purposes.

Counsel say in the motion that it is true that there is no question as to where seven lots of 116 rods each will measure to from the Berlin line, but it is in dispute as to just what lot measured on the ground this distance will reach to; that there appears to be an overrun of some fifty-six rods which the Court hold material in considering the motion to set aside the verdict. The answer to this is, that so long as the plaintiff sticks to his original theory that all of the lots are each 116 rods wide, there can be no dispute as to what lot seven lots measured from the Berlin line will reach to; they will reach to the marked line between lots 59 and 60. Nor is there then any question of overrun in the lots, because there is no overrun in seven lots each 116 rods wide from the Berlin town line to the marked line. The question of overrun comes into the case only when the plaintiff attempts to establish the dividing line between lots 59 and 60 fifty-six rods northwesterly of the marked line.

In the descriptions of the lots in the field book, each corner is marked by a certain tree. The plaintiff says that the description of the lots is to and from given objects, and that he is entitled to show that lot 59 runs to such given objects or to the corner where such objects once stood even though the distances between them overrun when measured on the ground; that the same can be shown regarding lots 58 to 53, inclusive, to the Berlin town line.

The corner trees named in the field book or, if destroyed, the places on the ground where they stood were monuments, and, if they could be identified, they would control over courses and distances. No evidence was introduced on the trial below identifying any of the corner trees or their location; and the plaintiff admits that this cannot be done, as he says further in his motion: 'The fact that there is a survey recorded wherein the lines are run from tree to tree (all of which were in the ground) is some evidence of a survey. The fact that the course of nature has now destroyed these trees does not make this evidence of a survey any less.' The rule is that, if the existence or location of monuments is not proved, courses and distances will govern the location of lots on the ground. Bagley v. Morrill, 46 Vt. 94. It is true that the destruction of the corner trees does not lessen the evidence of an original survey, but it does confine the plaintiff to the courses and distances given in the field book in locating the lots upon the ground, and, as we have said before, when located thus, the dividing line between lots 59 and 60 comes where the marked line is, and there is no overrun or expansion of lots.

What we have said so far has to do with the inevitable result reached when the line between lots 59 and 60 is located according to the theory upon which the plaintiff began the trial of the case.

The trouble with the plaintiff's case is that, while he still adheres to his original theory in argument, yet in locating his line between 59 and 60 he departs from it in part and follows it in part. Instead of beginning at the Berlin town line and measuring off seven lots according to their courses and distances as given in the field book, he starts at the northwesterly line of lot 57, the actual location of which as marked upon the ground, is at least forty-six rods farther to the northwest than its location as shown by the town plan and field book, and then measures off two full lots according to the courses and distances of the field book, in utter disregard of corners and lines marked upon the ground, and calls the point reached by said measurement the location of the dividing line, although there is nothing marked upon the ground to indicate the location of a line at that place. It is only when the plaintiff does this that the overrun or expansion of lots comes into the case. But, as we say in the opinion, the line cannot be established in this way, and, in the absence of evidence of the location upon the ground of the lot lines as originally surveyed, the marked line must control.

The plaintiff says that the Court gives too conclusive a character in the opinion to the question of adverse possession. This question is treated fully in the opinion, and we have nothing to add to what is said there.

All that has been presented to us in support of the motion for reargument has been carefully considered, and no ground for the motion is found.

We rendered judgment for the defendant because on the undisputed evidence the court below should have rendered such a judgment, and we were unable to see how the plaintiff could make a better case on another trial. On the motion for a remand the plaintiff has not satisfied us that it is probable that he will be able to make a stronger case on another trial than he made on the trial below. He has not brought to our attention any new and additional evidence that might strengthen his case. His principal ground for a remand is that he may have further opportunity to do more surveying northwesterly from lot 59 to the Duxbury town line in the hope that he may discover something that will aid him. But this is not a ground for a remand. Counsel when arguing the motion stated that he did not expect to find anything that would substantially change the location of the line claimed by the plaintiff.

We have already held that his present line cannot be maintained, and nothing has been presented that makes it appear probable that another line near it could be maintained. We are not dealing now with lines on maps or plans, but with lines marked on the ground. The marked line on the ground dividing lots 59 and 60 and the distance between it and the Berlin town line cannot be affected or changed by any surveying done northwesterly of it.

This decision is a magnificent example of working through the evidence to determine the location of the original survey – the actual acts of the original surveyor – by applying the appropriate rules and following appropriate procedures.

## GOVERNING RULES AND LAWS AT THE TIME OF CREATION, TITLE, OR SURVEY

There must be familiarity with the laws and rules governing the survey process at the time that the original survey was performed. Only then, can the follower understand terminology, procedures, and the aim and goal of the original surveyor, or deed compiler, whichever the case.

Examples include state statutes requiring that surveys be related to true north, standard allowances to measurements based on regulations or customary practice, terminology, and procedures employed, depending on the nature of the conveyance, creation, or transfer of rights, and whether required follow-up procedures have been properly adhered to. Depending on the jurisdiction, and the time frame, there are others that must be accounted for.

# 6 Protracted Surveys

[t]his is not a question of tracing an actual boundary, or of discovering a lost one, or one which may be presumed to have been completed; but of constructing a survey by adding two lines which were never actually run. And the cardinal object is to ascertain what the surveyor would have done if he had gone on to complete the work.

*Beckley v. Bryan & c.* **2 Bibb 493 (1801)**

There are legitimate protracted surveys,[1] within and without the Public Land Survey System (PLSS). Others are constructed for mere convenience for conveyancing purposes, generally avoiding the time and expense of a proper survey. Nevertheless, they exist and are governed by different rules than those accepted as standard for survey discrepancies and ambiguities.

Technically, protracted lines are not surveys, although they are commonly called such. Because they are not surveyed lines, rules and procedures for surveyed lines do not apply, but, in some cases, may be used as guidance. It is important to make the distinction, and necessary to apply the appropriate rules. The cases are replete with discussion of a survey being that which takes place on the ground, whereas protracted lines are not placed on the ground, and only appear on paper as a diagram, or as words within a document.

## PROTRACTION WITHIN THE PLSS

Presently, the Manual of Surveying Instructions published by the United States Department of the Interior, Bureau of Land Management (2009) is the most comprehensive of a series of manuals, and Special Instructions for selected areas, since 1804.[2] Since not everything within the Public Land System has been surveyed and located on the surface of the earth, some alternative measures have been used from time to time, one of which is protraction. Section 3-138 begins the section on Protraction Diagrams, and presents the following:

Official protraction diagrams are intended to provide a basis for the administration and management of unsurveyed Federal lands for all purposes short of conveying title. Such protractions can become the basis of land location for leasing purposes and for various administrative boundaries, including wilderness, National Recreation Areas, special use areas, withdrawals and selections.

"Protraction diagrams should not be treated as 'protracted subdivision township surveys.' The latter typically have run and marked exterior township lines and protracted section lines. The protracted section lines are represented as dashed lines indicating that they were not run and marked and the distances given are parenthetical distances.

The process of surveying a protracted tract or legal subdivision while protecting its location based upon the protraction diagram can involve extensive work. First, all the corners on the exterior of the unsurveyed area controlling the corners to be established must be found or reestablished by dependent resurvey. Second, using the protraction diagram as the record, the protracted

---

[1] This term is actually an oxymoron, and misleading, since protracted lines are not part of a survey, as they have not been located on the ground. However, the term is, and has been, used so frequently as it pertains to location, it is retained here.

[2] Manual of Surveying Instructions, 2009.

DOI: 10.1201/9781003032557-10

township corners must be located. Only then can the location and establishment of the needed township subdivision lines take place, followed by the needed monumentation.[3]

"Where protraction diagrams show plans of extension of the rectangular system that front potentially measurable water bodies, the meander lines may be indicated by an irregular line traced from existing map or photographic information. The irregular line is in lieu of a plot of a traverse.

"Protraction diagrams are preparatory for a plan of survey and to be able to describe and lease unsurveyed Federal lands. The protraction diagrams are of unsurveyed lands, and all lines, including the meander lines, are calculated from existing records or photography. Where water is segregated, the water area is taken out of the protracted upland area. The plan of survey protracts but does not determine the meander lines or riparian lots.

A lot shown as riparian by protraction may not be riparian upon survey. Therefore, when it becomes time to survey a township with protracted segregated water bodies the surveyor will not survey the water bodies where protracted but where the water bodies physically are. All surveys of protractions diagrams shall protect bona fide rights as to location.[4]

## PLATS OF PROTRACTION DIAGRAMS

Protraction diagrams have been developed in two forms. Prior to 1998, corner positions were defined by bearing and distance with reference to the exterior boundary of the protraction. Subsequently, the process was amended and corner positions are now defined by coordinates, often called an amended protraction diagram.[5]

The authority for the preparation of protraction diagrams issues only from the BLM Director. The instructions for contacting the Director, as well as the instructions of the preparation of protraction diagrams may be found in Sections 9-114 through 9-118 in the *Manual*.

## PROTRACTION WITHIN EARLY GRANTS PRIOR TO, OR SEPARATE FROM, THE PLSS

Chapter Four in *Boundary Retracement: Processes and Procedures*, is devoted to Protracted Surveys, and contains an extensive review of court decisions resolving them. They are in a category of their own, where standard rules do not apply, and as noted in several decisions protracted lines are not part of the survey.

The major caution with a protracted survey, similar to the instructions in the *Manual*, is that protracted lines are not surveyed lines. Therefore, when dealing with an early grant in the colonial states where the grant is based on a few surveyed lines with the rest of the lines protracted, two sets of rules apply, since the *protracted lines are not part of the survey*.

Clayton Orn's *Texas Law Review* article[6] on following the footsteps begins with its opening paragraph stating "When the public domain of Texas was surveyed in the early days two kinds of surveys were made—one, a survey on the ground, and the other a survey in an office. Even when surveys were made on the ground, not all lines were actually run and monumented, but many were projected. The projected lines thus, in effect, became office calls."

Protraction is the method whereby boundary lines are created without being run on the ground. They may be drawn on a plat, or merely written out in a land description.

---

[3] Manual, Section 3-140.
[4] Manual, Section 8-196.
[5] Manual, Section 9-114.
[6] Vanishing Footsteps of the Original Surveyor, *Baylor Law Review*, Vol. 17, No. 3, Spring 1952.

The basic problem with protracted surveys was reported in a West Virginia case,[7] wherein the court stated:

"The result of this loose, cheap, and unguarded system of disposing of public lands was that in less than 20 years after the adoption of the system nearly all of them were granted, the most part to mere adventurers, in large tracts or bodies, containing not only thousands, but in many cases hundreds of thousands, of acres in one tract. Often the grantees were nonresidents, and few of them ever saw their lands or expected to improve or use them for purposes other than speculation. The entries and surveys under warrants so cheaply and easily obtained were often made without reference to prior grants, thus creating interlocks, thereby covering land previously granted, so that in many instances the same land was granted to two or more different persons. Sometimes upon one survey actually located others were laid down by protraction—constructed on paper by the surveyors, without ever going upon the lands, thus creating on paper blocks of surveys containing thousands of acres, none of which were ever surveyed or identified by any marks or natural monuments. Thus, while the state was rapidly disposing of her large domain at 2 cents per acre, it was not attaining the main objects in view—the settlement of the country and revenue from the owners of lands. It was found that the grantees not only failed to settle upon and improve their lands, but in most instances they wholly neglected to pay the taxes due thereon, whereby revenue failed and the improvement of the territory was embarrassed and retarded. Nonoccupation of the lands and nonresidence of the owners made a resort to the lands themselves the only fund for delinquent taxes."

Protracted surveys are the subject of great detail in the case of *Rowe v. Kidd*.[8] It states that a protracted line is one that is only run mentally, and therefore not marked.

**Note:** Since following the boundary lines of the original creators, whether a surveyor or a scrivener, is the basic rule, it is necessary to know by what method(s) the title, and ultimately its boundaries, was created.

Lines of rectangular lots, both inside and outside the PLSS, may have been created by protraction, that is drawn on paper, but not marked on the ground. Or, in the alternative, as has been noted in a number of decisions, once surveyed and marked, but now unable to be identified on the ground. Either way, the procedure for resolution is the same. However, a word of caution is in order. Before a conclusion is drawn that such lines have not been surveyed (and therefore no actual "footsteps"), all possibilities must be explored, including original town layouts, original subdivision layouts, proprietors' records and the like, otherwise there is no proof that there was no original survey.

Protraction often leads to multiple markers set for the same corner, known widely as "pincushions," due to different surveyors' opinions, procedures, or legitimate measurement error.

As stated by Justice Bibb and reported at the beginning of the chapter, [for protracted, or projected, lines], the cardinal object is to ascertain what the surveyor would have done if he had gone on to complete the work.[9]

## MISCELLANEOUS PROTRACTIONS

This category contains examples of extended lines and parcels based on the locations of surveyed lines and parcels as well as protracted lots within a prior surveyed perimeter. Conveyances are made based on the protracted lines.

---

[7] *Fay v. Crozer*, 156 F. 486, S.D.W.Va., 1907, citing Hutchinson on Land Titles, pp. 1, 2.

[8] 249 F. 882, E.D. Ky, 1916.

[9] *Beckley v. Bryan & c.* 2 Bibb 493 (1801).

There are protracted original grant surveys done to move the process along, all of which have presented unique problems and challenges.[10]

The case of *Dicus v. Allen*[11] presented an interesting twist, in that there were no original locations in the vicinity, therefore nothing to base a retracement survey on. The court wrestled with this situation, and presented the following:

"The guides to locating boundaries are set forth in *McKinley v. Hilliard,* 248 Ark. 627, 454 S.W.2d 67 (1970), wherein the court, quoting from *Ewart v. Squire,* 239 F. 34 (4th Cir. 1916), stated: In ascertaining location the guides in the order of importance are: (1) Natural objects; (2) artificial objects; (3) adjacent boundaries; (4) courses; (5) distances; (6) quantity. But the rule is flexible, and it does not control against the intention of the parties as shown by the description taken as a whole …. The order of the importance of the guides is manifestly the more flexible when the description of subdivisions of a tract is ascertained by protraction and not by actual survey. (Emphasis supplied.)

Consistent with the foregoing guidelines, the record reflects that the tracts adjoining appellants' property on the west, north and east sides have corresponding boundaries and descriptions. There are no overlaps or discrepancies in the legal descriptions of any of the four adjoining tracts. Moreover, under the Whitfield surveys, all of the adjoining owners have at least the same quantity of acreage called for in their respective deeds. When appellees' surveyor, set out to check [appellant surveyor's] surveys, it was his duty solely to locate the lines of the original survey where title to land has been established under a previous survey. He cannot establish a new corner, nor can he even correct erroneous surveys of earlier surveyors. *McKinley v. Hilliard.*

## COURT PROCEDURE IN ASCERTAINING LOCATION; PROTRACTION VS. ACTUAL SURVEY

Two basic rules summarize the procedures for the difference between the two. "In determining the location of an actual survey, the fundamental principle is that it is to be located where the surveyor ran it. As it has been put, the thing to be done is to track the surveyor."[12] And with a protracted line, "the cardinal object is to ascertain what the surveyor would have done if he had gone on to complete the work."[13]

## ORIGINAL SURVEY IS WITHOUT ERROR

*The established rule of property is that the original United States Government survey is prima facie correct* and surveys must conform as nearly as possible with the original government survey. *Carroll v. Reed,* 253 Ark. 1152, 491 S.W.2d 58 (1973). Emphasis supplied by author.

In the case of *Little v. Williams,* 88 Ark. 37, 113 S.W. 340, this court approved the following statement by Mr. Justice Brewer, speaking for the court, in *Russell v. Maxwell Land Grant Co.,* 158 U.S. 253, 15 S.Ct. 827, 39 L.Ed. 971: "In the nature of things, a survey made by the Government must be held conclusive against collateral attack in controversies between individuals. There must be some tribunal to which final jurisdiction is given in respect to the matter of surveys, and no other tribunal is so competent to deal with the matter as the land department.

---

[10] See Chapter 4, Wilson, D.A., *Boundary Retracement,* for an extensive treatment of protracted surveys (which, by the way, are not really surveys, nor are they treated as such).

[11] 2 Ark.App. 204, 619 S.W.2d 306 (Ark.App. 1981).

[12] *Rowe v. Kidd,* 249 F. 882 (E.D.Ky., 1916).

[13] *Beckley v. Bryan & c.* 2 Bibb 493 (1801).

None other is named in the statutes. If in every controversy between neighbors the accuracy of a survey made by the Government were open to question, interminable confusion would ensue."

The court in *Russell v. Maxwell Land Grant Company*[14] stated, "These authorities are decisive upon this question, and in the nature of things, a survey made by the government must be held conclusive against any collateral attack in controversies between individuals. There must be some tribunal to which final jurisdiction is given in respect to the matter of surveys, and no other tribunal is so competent to deal with the matter as the Land Department. None other is named in the statutes. If in every controversy between neighbors the accuracy of a survey made by the government was open to question, interminable confusion would ensue. Take the particular case at bar. If the survey is not conclusive in favor of the plaintiff, it is not conclusive against it. So we might have the land grant company bringing suit against parties all along its borders, claiming that, the survey being inaccurate, it was entitled to a portion of their lands, and as in every case the question of fact would rest upon the testimony therein presented, we should doubtless have a series of contradictory verdicts, and out of those verdicts, and the judgments based thereon, a multitude of claims against the United States for return of money erroneously paid for land not obtained, or for a readjustment of boundaries so as to secure to the patentees in some other way the amounts of land they had purchased.

> It may be said that the defendants have the same right to rely upon the regular surveys that the plaintiff has upon the survey of this special land grant. This is undoubtedly true, but the survey is one thing and the title another. If sectional lines had been run through the entire limits of the Maxwell Grant, it would not thereby have defeated the grant, or avoided the effect of the confirmatory act. A survey does not create title; it only defines boundaries. Conceding the accuracy of a survey is not an admission of title. So the boundaries of the tract claimed by defendants may not be open to dispute, but their title depends on the question whether the United States owned the land when their ancestor filed his homestead claim thereon. If at that time the government had no title, it could convey none.

**Author's Note:** Two very important statements appear in this decision that demand emphasis. "The survey is one thing and the title another," and "a survey does not create title; it only defines boundaries." These principles should be kept in mind at all times, as each has its own rules and procedures, and each governed by separate authority. Title matters are decided by legal authorities, while boundary location matters are strictly a matter for a registered or licensed land surveyor.[15] The often-quoted principle in many decisions applies here: *What are boundaries is a matter of law, where boundaries are is a matter of fact.*[16]

## MORE THAN ONE ORIGINAL SURVEY – THE PERIMETER PLUS THE INTERIOR LOTS

If an interior lot abuts a perimeter line, two retracements may be necessary, and they may conflict with one another. Then the question becomes which line, or survey, is the senior, and therefore controlling. Refer to Figures 6.1 and 6.2.

---

[14] 15 S.Ct. 827, 158 U.S. 253 (1895).

[15] Although title attorneys and others who regularly work with them develop expertise as to land descriptions, the only professional authorized to locate land lines on the ground is a registered land surveyor. In fact, the definition of a legally sufficient real property description is one that can be located on the ground by a surveyor. However, in the absence of statute, a surveyor is not an official and has no authority to establish boundaries; like an attorney speaking on a legal question, he can only state or express his professional opinion as to surveying questions. *Rivers v. Lozeau*, Fla.App. 5 Dist., 539 So.2d 1147 (Fla., 1989).

[16] "the rule of law is that the court adjudge what are the boundaries of a conveyed tract of land, and the jury ascertain where they are." *Redmond v. Stepp*, 100 N.C. 212, 6 S.E. 727 (N.C. 1888).

**FIGURE 6.1**  Perimeter survey of Washington, NH, April 1753.

This situation is commonplace in the colonial states where rectangular lotting has taken place. Often there exists two (sometimes even more) separate surveys, one of the perimeter of the township, and another of the interior lotting. The latter is sometimes done in phases where there are several divisions within the township. In other cases, the interior lotting lines are protracted, resulting in an even more difficult situation with potential conflicts and ambiguities.

This plan Described the tract of Land Called Manadnock Number Eight (or New Concord) Granted unto Cap^t Peter Prescott & Others, Lying in y^e Province of New Hamp^r—part of Masons Patent Containing 33280 Acres the Bounds Course & Length of Lane as Delineated On the plan herewith Layed in a Scale of One mile to an Inch.

Acceptance of survey by proprietors, January, 1753: "… Grant the Contents of Eight milles Long, & Six miles & an half wide part of said Lands Bounded as follows viz^t Beginning at a Stake, & Heap of Stones, the Norwesterly Corner of Monadnock Number Seven, (So Called,) Granted to Reuben Kidder, & others, which Stake & Stones is in the Patent Line, lately run by Joseph Blanchard jun^r Esq^r from thence runs by said N^o Seven South Eighty Degrees, East Six miles & an half to a Stake & Heap of Stones, from thence North Twenty Eight Deg^r East Eight miles on Ungranted Lands, to a Beach tree from thence North Eighty Degrees West, six miles & an Half, to the patent Line aforesaid, from thence Southerly by said patent Line to y^e first Bounds mentioned…"

"The Above plan Describes the Township Called Manadnock N^o 8 or New Concord which Ly's in the Province of New Hamp^r within Mason's Grant As the Same is Lotted out And Divided, &

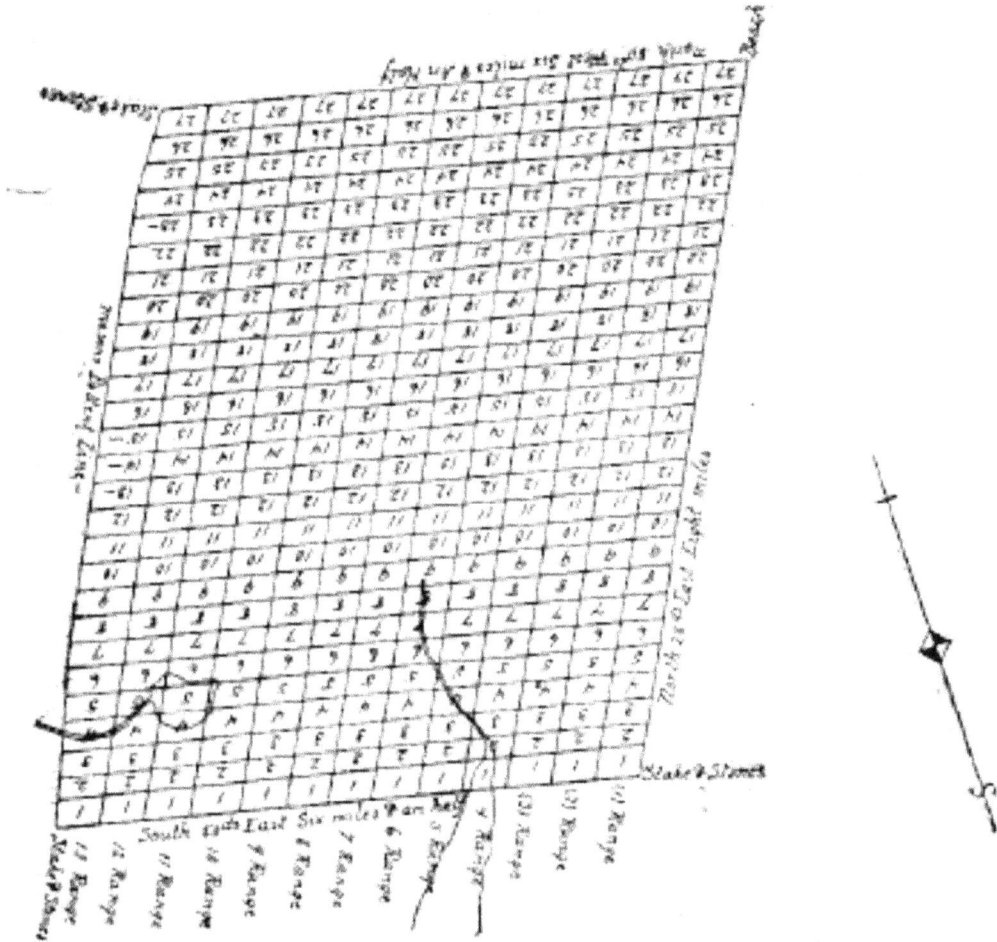

**FIGURE 6.2**    Lotting plan of Washington, NH, December 1753.

Nombered on the Severall Lotts in the plan which lotts Contain One Hundred Acres Each And Are One Hundred & Sixty rods in length And One Hundred rods in Breadth All exclusive of the Westernmost Range, the length of Which are as lay'd Down in the plan December, 1753. Lay'd Down in a Scale of One mile in An Inch."

**Author's Note:** This plan appears to be a protracted lotting designating to the lot drawers which lot(s) they are entitled to as there is no return of survey, nor an attestation that the lots were marked in any fashion, merely depicted on a plan. Obviously, over time, as lots were marked on the ground, the results are a conglomeration of locations, which do not tie together well. This is all the more reason to understand the creating documents, and how they were derived. Unfortunately, reality is that many locations are found to be incorrect according to original creations and original titles.

The problem described is also of concern to any retracement surveyor in the PLSS when the survey of the township and the subsequent layout of sections are done independently and at different times. To further complicate the process, a later subdivision of a section adds another layer, with one or more subsequent surveys, some of which are original, but not all, as some are existing surveys created before. Each survey contains its own characteristics and its own set of conditions, as well as its own set of errors. The case of *Rivers v. Lozeau*[17] presented in Chapter 1 illustrates the inherent problem associated with this, where a small lot was subdivided from a larger tract. The surveyor was faced with his original (subdivision) survey while retracing the previous survey of the parent tract. He failed to distinguish the two, treating the created lot as a single survey, thereby doing it incorrectly, interfering with the rights of the parties involved and creating a subsequent lawsuit.

As in the South Dakota case of *Titus v. Chapman*,[18] the original monuments, or the places where they stood, govern over unknown markers with no known basis and official maps with protracted, unsurveyed lines. The map is nothing more than a picture by a draftsperson, and the protracted lines are not part of the survey.

This decision underscores the definition that a *corner* is a point not a physical entity, by stating the monument, or the place where it stood (the point, or the position). Several cases have stressed this before (Maine) by stating, two different ways, that so long as the corner (the *point*) can be determined, or located, it remains controlling over lesser, conflicting calls.

The Vermont case of *Brooks v. Tyler*[19] presents an interesting dilemma relating to ancient bearings when no other information is available. The case had to do with conveying a lot by number according to a plan without any monumentation except at the starting point and reciting courses and distances for the lines. The original plan of the town showed the lot in question with a maple tree and a bearing for the east line of the lot. The original question for the court was the timing of using the directions presented in the description. As the court recited, "the question is whether the line in the declaration is to be run according to the present direction of the compass, or according to the direction when the lot was surveyed, and as the lot *was actually surveyed*" [emphasis by the court]. It should be obvious what the correct answer and application should be, and ultimately was, but the court's discussion is noteworthy:

"It is not always the case that lots and small tracts of land can be described by natural monuments. If artificial ones are erected, they may be soon prostrated, or decay by time; and we are, therefore, obliged to rely for the direction of boundary lines upon the compass, as the best means in our power. But a given line fifty years ago is not the true line now. A variation of from three to fifteen degrees has taken place within the last half century. 1 *Williams' History*, p. 473. What then is to be done? There may be no natural boundaries, one corner only remains, and the ancient survey shows the direction from that corner. It cannot be doubted that the parties are to hold according to this ancient survey; but the question returns, how is the line to be described in the declaration? If we declare for a piece of land giving the present direction of the compass, then our proof will not support the declaration: there will be a variance between the declaration and the proof offered. The simple way seems to be to declare for the land just as it was surveyed, and then, after ascertaining what the variation is, to run the line with reference to the variation. Practical surveyors have no difficulty in calculating this variation. In this way courts of justice can at all times give to parties the identical land which was originally surveyed out to them. This is the most feasible mode, and we do not perceive that it is open to any objection.

"We are aware that parties may declare for their lands by the number of the lot. But unless the lines have been marked and known, the same difficulty will arise here. The lines must be ascertained by the compass, and no one will pretend that the present direction is to be taken, but such a one as will give the ancient line. But parties are not obliged to declare for their lands by number. This may not always be practicable. It may often happen that no other description can

---

[17] 539 So.2d 1147 (Fla., 1989).
[18] 2004 SD 106, 687 N.W.2d 918 (S.D. 2004).
[19] 2 Vt. 348 (Vt. 1829).

be given than that of the ancient survey; and unless the mode we have attempted to pursue can be sustained, parties may be compelled to lose their lines."

## LOCATING BLOCKS OF LAND

Blocks of land were also an issue in the Pennsylvania case of *Clement v. Packer*.[20] Reciting an earlier decision, *Coal Co. v. Clement*,[21] "When original surveys have been made and returned as a block into the land-office, the location of each tract therein may be proved by proving the location of the block. In ascertain the location of a tract, the inquiry is not where a tract should or might have been located, but where it actually was located. Every mark on the ground tending to show the location of any tract in the block is some evidence of the location of the whole block, and therefore of each tract therein." The court went on to emphasize "that the block of 1793, as returned to the land-office, must be located by its own marks, and not by calls *of later surveys, or by marks found upon the ground younger than 1793*" [emphasis from the court].

Another early decision is that of *Ralston v. M'Clurg*.[22] Strictly dealing with protracted lines, the court addressed the issue as follows, distinguishing the problem from a straight retracement situation:

"[t]his is not a question of tracing an actual boundary, or of discovering a lost one, or one which may be presumed to have been completed; but of constructing a survey by adding two lines which were never actually run. And the cardinal object is to ascertain what the surveyor would have done if he had gone on to complete the work. *Beckley v. Bryan &c.* 2 Bibb 493, I Ky (Ky. Dec.) 91 (1801). This is to be ascertained, not by vague conjecture, but by rational deductions from his report, as compared with the existing facts."

## WORKING WITH AN ERRONEOUS SURVEY

The Florida decision in the case of *Hardee v. Horton*[23] addressed protracted lines based on an erroneous official survey.

This case had to do with deed descriptions referencing a map of rectangular (PLSS) layout in the Everglades region of Florida. "The 'map adopted as official' according to which the state trustees made conveyances of unsurveyed lands in the Everglades area, was manifestly intended to indicate the approximate location of the section, township, and range lines protracted on the map, so that purchasers of the unsurveyed lands in any part of the vast unsurveyed area would know the approximate location of the sections and townships that were to be surveyed and numbered according to the federal system. This import of the map appearing in the public records approving the map as 'official' is binding on all parties. Conveyances of unsurveyed lands in the Everglades are of a designated acreage in stated numbered sections, townships, and ranges that are to be surveyed and the exact location and boundaries thereof ascertained and established according to the United States system of surveys, so that the townships will contain 36 sections of 640 acres each; the 'map' of merely protracted lines being referred to as indicating the approximate location of the sections, townships, and ranges stated in the conveyances. The contemplated survey has reference to a comprehensive legal survey of the Everglades area. Separate surveys to locate and establish the exact boundaries of unsurveyed lands described by legal subdivisions in deeds of conveyance would lead to intolerable confusion and uncertainty

---

[20] 125 U.S. 309, 8 S.Ct. 907, 31 L.Ed. 721 (1888).
[21] 95 Pa, St, 126.
[22] 39 Ky. 338 (Ky.App. 1840).
[23] 90 Fla. 452, 108 So. 1889 (Fla. 1925).

in land titles and boundaries; and the governmental power of the state may be exerted to prevent such a general chaotic condition. Such surveys are not contemplated by parties to conveyances of unsurveyed public lands. In this case the parties are bound by the public records relating to conveyances of unsurveyed public lands, and such records contemplate a legal survey of the Everglades area according to a comprehensive system that is regulated by law and in general use in surveying public lands. The state has made the contemplated survey, and the lines established by the state are binding on all parties; no fraud being involved. The statute approving the survey as established by the state does not change boundaries or impair vested rights, for the reason that in conveyances of unsurveyed public lands, rights in particular boundaries do not exist until the boundaries are established by public authority. *Everglades Sugar & Land Co. v. Bryan,* 87 So. 68, 81 Fla. 75.

"The conveyance here considered is not of surveyed lands described by metes and bounds, or by sections, townships, and ranges, according to a map referred to which purports to represent survey lines that had been actually located and established on the ground. Neither does the description refer to established objects on the ground. In such cases the survey lines established on the ground or the objects on the land that are referred to may control. See *East Coast Lumber Co. v. Ellis-Young Co.,* 45 So. 826, 55 Fla. 256; *Stonewall Phosphate Co. v. Peyton,* 23 So. 440, 39 Fla. 726; *Alden v. Pinney,* 12 Fla. 348; *Andreu v. Watkins, 7 So.* 876, 26 Fla. 390; *Davis v. Rainsford,* 17 Mass. 207; *Cragin v. Powell,* 9 S.Ct. 203, 128 U.S. 691, 32 L.Ed. 566; *Higuera v. United States, 5 Wall.* 827, 18 L.Ed. 469; *Kirch v. Persinger,* 100 So. 166, 87 Fla. 364.

"Nor was the conveyance made according to a map whose lines with practical accuracy correspond with the surface area of unsurveyed lands represented by the map, a portion of which unsurveyed lands is intended to be conveyed. In such cases slight excesses of surface area over that indicated by the map might be contemplated by the parties and might not affect a legal survey in locating the lands described.

"In construing a deed conveying lands and the maps referred to in describing the lands, the nature, origin, and purpose of the map, the position of the contracting parties and the circumstances under which they acted, should be considered, and the language used should be interpreted in the light of all the pertinent circumstances so as to give effect to the intent of the parties, even if an erroneous part of the description has to be disregarded in effectuating the general intent of the conveyance. 18 C.J. 285; 9 C.J. 228, 229; *Campbell v. Carruth,* 13 So. 432, 32 Fla. 264.

In 1908, Walter R. Comfort, for $2 an acre, purchased a stated number of acres of the unsurveyed lands of the Everglades from the state trustees. The intent of the parties to the deed from the state trustees to Comfort covering stated sections, townships, and ranges "according to map adopted as official," and containing a specific number of acres, was the conveyance of the stated number of acres in the specified sections as indicated by the map, which sections were to be thereafter surveyed and their exact location and boundaries determined and established by an official survey having relation to the unsurveyed area, that would give the sections an area of 640 acres each with appropriate numbering according to the U.S. system, the township and range numbers being indicated by figures on the margins of the map.

The complainant alleges that the lands "had not been surveyed," and that the "map was made" "without actual survey in the field." There is uncontradicted evidence "that the lands, when conveyed, were not surveyed, but it was understood that a subsequent survey was to be made in describing the lands."

As the lines appear on the map referred to in the conveyance, the north line of township 54 south, range 40 east, and south line of township 53 south, range 40 east, coincide; but this is not controlling because the map being only a protracted drawing of uncertain accuracy, representing a large area of unsurveyed public lands, and showing townships of uniform size, a legal survey of the Everglades area according to the U.S. system of surveys was necessary and contemplated to locate on the ground the boundaries of lands conveyed, and the system of U.S. surveys expressly provides that material errors in adjoining surveys should not be extended in making surveys in adjacent unsurveyed territory. Section 246 of Manual of Instructions referred to in Section 2399, Rev. Gen. Stats. of U.S. (U. S. Comp. St. § 4807). And the federal statute and authorized rules of survey thereunder provide for special instructions by administrative officers to control surveys where errors and irregularities appear. See, also, *Ruffner v. Hill, 7 S. E.* 13, 31 W.Va. 428.

The conveyance to Comfort was not of lands beginning at the north line of township 54 as theretofore surveyed, but it was of unsurveyed lands described as being in stated sections and subdivisions of sections in township 53 south, range 40 east. Township 53 had not been surveyed as a legal subdivision, and township 54 south, range 40 east, had been only partially surveyed. The fact that the east two miles of the north line of township 54 south, range 40 east, and the west line of township 53 south, range 41 east, had been run, does not give to township 53 south, range 40 east, its definite location and exact boundaries, particularly where the federal system of surveys expressly provides that errors discovered in one survey are not to be extended into an adjoining survey and special instructions are authorized to control the making of a survey where errors appear in an adjacent survey. See *Ainsa v. United States,* 16 S.Ct. 544, 161 U.S. 208, 40 L.Ed. 673.

The U.S. survey lines on the east and west sides of the Everglades were run from the north, and in the survey of township 53 south, range 40 east, the state surveyor ran the line south from the northwest corner of township 51 south, range 41 east, a verified established corner several townships to the north of township 53; and by running south 18 miles found that the south line of township 53 south, range 40 east, as correctly surveyed, did not reach the north line of township 54 south, range 40 east, as previously surveyed by the United States authorities, and that there were 1,635 acres of surplus land between the correct south line of township 53 south, range 40 east, and the north line of township 54 south, range 40 east, as previously run by the federal government. The result was the formation of six lots between the south line of township 53 south, range 40 east, and the north line of township 54 south, range 40 east, such lot being numbered according to the federal system. The lot between Section 35, township 53, and the north line of township 54, is No. 2, and contains 237.4 acres.

Horton's subsequent private survey in 1920 was not authorized by law. It began at the northeast corner of township 54 south, range 40 east, and running west and north surveyed the southwest quarter of southwest quarter of southwest quarter, Section 35, township 53 south, range 40 east, as being on the north line of township 54 south, range 40 east, in the southwest portion of lot 2 that had been established by the previous state survey. Horton's surveyor testified that no survey had been made by the United States of township 53 south, range 40 east, and that he did not locate the southeast corner of township 53 south, range 40 east. The survey made for Horton shows that it is of an abnormal and not of a normal Section 35, township 53 south, range 40 east, containing 640 acres, as is contemplated by the map referred to in Horton's deed. When Horton's private survey was made, township 53, including Section 35 thereof, had been surveyed by the state, and apparently the survey had for some years been acquiesced in as a proper location of the boundaries of Section 35, township 53 south, range 40 east. The records of the state land office show that township 53 south, range 40 east, was fully surveyed in 1912, though the hiatus lots south of township 53 were not surveyed till 1918.

As the title to the unsurveyed lands embraced in patent No. 137 had vested in the state without the lands being surveyed, any legal survey of the lands had to be made under state authority; and as the map referred to in Comfort's deed was by the record of its adoption as "official," shown to be only as near a correct map of the land lines as could be furnished "without an actual survey of the same," and as a proper legal sectional survey of the unsurveyed lands was contemplated so that the specific number of acres purchased in the designated sections could be legally located and their boundaries definitely fixed, it necessarily follows that the parties to the Comfort deed intended that the contemplated legal survey should be made and that the survey should be according to the federal system, which requires a township to be 6 miles square, containing 36 sections of 640 acres each. In making surveys under the federal system, relatively small excesses of land as they appear are usually added to some section or sections in a township; but where relatively large areas of excess lands appear, some other appropriate and permissible disposition of the excess land should be made; and this is contemplated and provided for by the federal rules.

The evidence refers to a "Plat of Hiatus," otherwise known as "township 44 1/2 south, range 42 east (between townships 44 and 45 south, range 42 east) Tallahassee Meridian, Florida, surveyed by William H. Richards Jr., U.S. surveyor, from April 21 to May 1, 1915, incl., under special instructions from the Commissioner of the General Land Office bearing date February 5, 1915, approved by the United States Commissioner of the General Land Office September 26, 1916," which plat shows lots formed by survey lines between the stated townships covering different areas from 29.73 to 48.50 acres, totaling 731.19 acres, the same being quite similar to the "hiatus" lots of surplus lands formed by the state survey in 1918 between the south line of township 53 and the north line of township 54 south, range 40 east. See defendant's Exhibit A; volume 13, p. 278, Minutes Trustees, I. I. Fund.

Section 2399, U.S. Revised Statutes as amended April 26, 1902, authorizes the Commissioner of the General Land Office to give "instructions" for surveying public lands, to meet peculiar conditions encountered in making surveys, therefore it appears that the survey made by the state in forming the hiatus lots south of township 53 was in accordance with the system and practice in making U.S. surveys, and that such a survey was contemplated by the conveyance to Comfort in 1908.

Section 2399, U.S. Revised Statutes as amended in 1902, is as follows:

"The printed manual of surveying instructions for the survey of the public lands of the United States and private land claims, prepared at the General Land Office, and bearing date Jan. 1, 1902, the instructions of the Commissioner of the General Land Office, and the special instructions of the surveyor-general, when not in conflict with said printed manual or the instructions of said Commissioner, shall be taken and deemed to be a part of every contract for surveying the public lands of the United States and private land claims."

See, also, Section 453, U.S. Rev. Stats. (U. S. Comp. St. § 699).

The printed Manual of Surveying Instructions, dated January, 1902, referred to in Section 2399, contains the following and also provides for special instructions to meet particular cases:

246. When new surveys are to be initiated or closed upon the lines of old surveys, which although reported to have been executed correctly, are found to be actually defective in alinement, measurement, or position, it is manifest that the employment of the regular methods prescribed for surveying normal township exteriors and subdivisions would result in extending the imperfections of the old surveys into the new, thereby producing irregular townships bounded by exterior lines not in conformity with true meridians or parallels of latitude, and containing trapezium-shaped sections which may or may not contain 640 acres each, as required by law.

247. Therefore, in order to extend such new surveys without incorporating therein the defects of prior erroneous work, special methods, in harmony as far as practicable, with the following requirements, should be employed, viz.:

The establishment of township boundaries conformable to true meridian and latitude lines.

The establishment of section boundaries by running two sets of parallel lines governed respectively by true meridians and parallels of latitude, and intersecting each other approximately at right angles at such intervals as to produce tracts of square form containing 640 acres each.

The reduction to a minimum of the number of fractional sections in a township, and consequently of the amount of field and office work.

In Sections 283 et seq. of the Manual, provisions are made for surveying "Hiatuses and Overlaps."

The contemplated legal survey had reference to a comprehensive system applicable to the unsurveyed area, and not to individual surveys of particular portions conveyed. It is shown that the state survey was made according to a comprehensive authorized system; and there is nothing to indicate that any portion of the state survey had referenced to the inclusion or exclusion of any particular land in making a subdivisional survey.

The conveyance to Comfort and the map referred to contemplated a system of legal surveys embracing many townships, including township 53 south, range 40 east, and not a particular survey of a subdivision of a section; therefore the contention that after the state survey had been made, the quarter-quarter, quarter-section in township 53 that was conveyed to Horton could and should be separately surveyed, and its location and boundaries established without reference to the comprehensive legal survey that was contemplated by the parties to the conveyance, is untenable. The federal "system is such that a township is surveyed as a unit." *Santa Fe Pac. R. Co. v. Lane*, 37 S.Ct. 714, 715, 244 U.S. 492, text 495 (61 L.Ed. 1275). In *State ex rel. Kittel v. Jennings*, 35 So. 986, 47 Fla. 307, the fractional township had been surveyed and the boundaries of fractional Section 16 definitely located as school land under the grant of 1845.

When the survey lines of township 53 south, range 40 east, were established by the state, the lines were not changed, but they remained as surveyed and established.

The complainant does not allege that Comfort purchased unsurveyed land in township 53 south, range 40 east, with reference to peculiar qualities of particular parts of the land, even if that would affect the rights acquired by the conveyances. Horton's conveyance was executed long after the state survey was made.

The court concluded, "a map or plat which represents no survey, but is prepared by projecting lines of a prior erroneous government survey on paper over a space representing a large area of unsurveyed lands, purporting to represent section, township, and range lines according to the rectangular method of surveying, although adopted and referred to in deeds of conveyance as the official map of the grantor, when shown by competent testimony to be inaccurate and unreliable as an aid to locate the unsurveyed lands which are conveyed by description according to the rectangular method of describing lands, is insufficient as a survey of said lands.

A complete title to unsurveyed public lands does not vest in the grantee until the lands conveyed have been identified by an authorized survey; land where unsurveyed lands are conveyed by description according to the rectangular method of describing lands, although the deed be a grant *in praesenti*, the title vests in the grantee upon delivery of the deed subject to the right and duty of the political authorities of the state to identify and separate by a survey the lands conveyed from the unsurveyed lands within which they are included.

According to Orn,[24] ground surveys were required by the Mexican government and by the Republic of Texas before public lands could be granted or patented. Since 1879, Texas statutes have required that field notes of public lands contain a certificate that the survey was actually made on the ground. Consequently the law presumes that a survey of public lands was actually made in the field, and that boundaries were established on the ground, but, according to *Stafford v. King,*[25] and others, this presumption is not conclusive.

Even with the statutory requirements in place, the courts have held that if a survey was not made in the field, but in an office, or the lines were projected, the grant or patent is not void if the field notes described the land sufficient certainty and definiteness that the boundaries can be located on the ground."

---

[24] Vanishing Footsteps of the Original Surveyor, *Baylor Law Review*, Vol. TV, No. 3, Spring 1952.
[25] 30 Tex. 257 (1867).

# 7A Special Cases
## *In General*

*The real world is a special case.*

**Proverbs**

## LEGISLATIVE ACTS

### DEFINITION

Legislation (or "statutory law") is a law which has been promulgated (or "enacted") by a legislature or other governing body or the process of making it. Before an item of legislation becomes law it may be known as a bill, and may be broadly referred to as "legislation," while it remains under consideration to distinguish it from other business. Legislation can have many purposes: to regulate, to authorize, to outlaw, to provide (funds), to sanction, to grant, to declare or to restrict. It may be contrasted with a non-legislative act which is adopted by an executive or administrative body under the authority of a legislative act or for implementing a legislative act.
*Wikipedia*

### SIGNIFICANCE TO THE LAND SURVEYOR

It is very important, when answers are needed, or the created paperwork is not found in traditional locations, past legislation should be searched. The appropriate answers may be found there, or at least perhaps a clue appears as to what else might be searched.

The following are categories relevant to boundaries and titles, with examples of each. There are many more, and there may be other categories of interest, depending on the situation at hand.

Legislative acts usually create a new title, therefore often creating one or more new boundaries. Such boundaries would be considered original since they appear for the first time, and therefore fall into the category of an original survey. Some descriptive elements are based on measurements, others are not. In the case of *Bartlett Land and Lumber Company vs. Saunders*,[1] it was claimed by the demandant, and proof was offered to show, that the western boundary of [a town], being in a wild and mountainous region, had never been located on the ground in 1830, and could not be located from the description contained in the grant, because it was too vague and uncertain to admit of a fixed and definite survey. But the plat annexed to the grant, and referred to by the grant for greater certainty, did show a boundary line, laid down to a scale. The court ruled, "if there was no other evidence on the subject, this would be sufficient to show that [the town] had a boundary, and a definite one, whether it was ever actually run out on the ground or not."

Examples of legislative acts creating titles, boundaries and other rights include:

- Turnpikes and others, e.g., plank roads
- Highways
- Toll roads and bridges

---

[1] 103 U.S. 316, 26 L.Ed. 546 (1881).

DOI: 10.1201/9781003032557-11

- Railroads
- Trolleys and street railways
- Canals
- Ferry landings (with access)
- Municipal boundaries
- Cession of lands

With many of these categories, jurisdiction and consolidation may have changed their characteristics and use. Special care must be exercised to understand the history of the subject matter, to ensure finding the original creation of title and the original survey location. For example, there has been an extensive consolidation of railroads through the years, along with an abundance of changes in names. Many which began as small short line railroads have since become part of an extensive system carrying the name of the parent company.

## MILITARY BOUNTY LANDS

### DEFINITION

The military bounty system has traditionally been a method of stimulating military enlistments with inducements of land or money. For Indian and French campaigns, colonies offered cash inducements for both enlistments and supplies. During the Revolution, the states and Congress

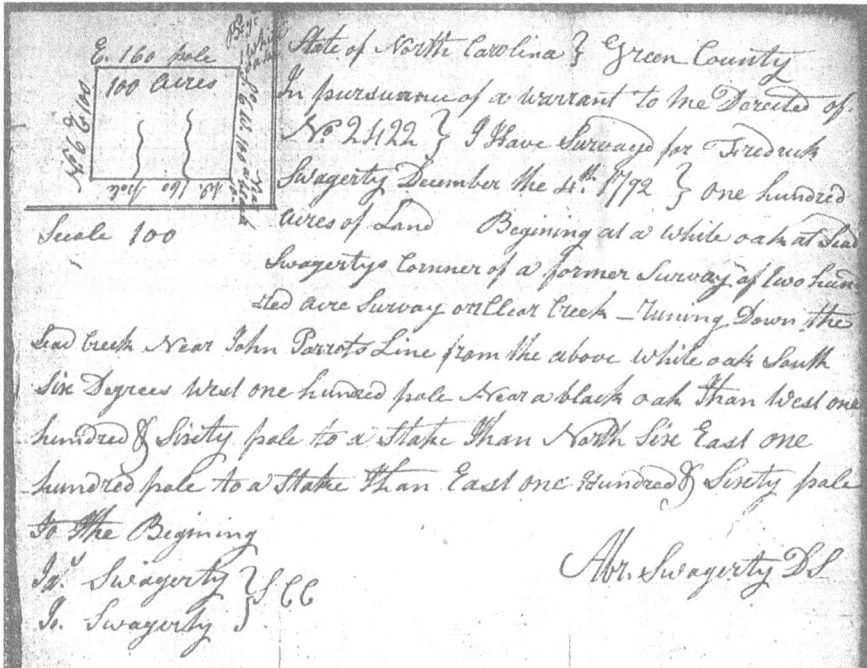

**FIGURE 7.1** Survey of Frederick's 100-acre tract along clear creek. (Land entered February 5, 1780; surveyed December 4, 1792; North Carolina State Grant, February 23, 1793; Tennessee State Library and Archives, Nashville, Tennessee – Microfilm Collection #1177 [Earliest Tennessee Land Records & Earliest Tennessee Land History, Irene M. Griffey, p384, Clearfield Co., 2003, Baltimore].)

often bid against one another for recruits. Sums mounted and bounty jumping and reenlisting were prevalent.[2]

## Bounty Land Served as Both an Incentive and a Reward for Military Service

Bounty Land is land awarded to soldiers for one of two reasons. It is either given as an inducement to serve like a recruiting bonus or as part or sum of a pension package in compensation for services rendered. In America, the practice began in Colonial Times and continued through the Revolutionary War and all conflicts up to but not including the Civil War.

The federal government provided bounty land for those who served in the Revolutionary War, the War of 1812, the Mexican War, and Indian wars between 1775 and 1855. It was first offered as an incentive to serve in the military and later as a reward for service. Prior to the Revolution, bounty lands were awarded by the King of England to British subjects who either served in the French and Indian War(s), or suffered as a result of it. This latter category mostly applied to captives by Indians or the French.

An act of March 3, 1855 (10 Stat. L. 701) extended military bounty land laws to Indians, entitling veterans from the Revolutionary War and the Indian Wars of 1818 and 1836 to warrants that could be exchanged for public lands. A few earlier acts had specified bounty lands for Indians, but this act marked the first time land was made available on a large scale. For example in the form of a survey and description, refer to Figure 7.1.

## Applications for Indian Bounty Lands

Applications were taken by Indian agents in the Indian Territory west of Arkansas in the years immediately following the act.

## French and Indian War

At the conclusion of the French and Indian War, England's King George III issued a proclamation on October 7, 1763, which authorized the governors of the British colonies in North American to issue grants for bounty land for officers and soldiers who had participated in the conflict. Field officers were to receive 5,000 acres, captains 3,000 acres, subalterns (officers below the rank of captain, especially second lieutenants) 2,000 acres, noncommissioned officers 200 acres, and privates 50 acres.

In addition, lands were awarded for hardships suffered by families, and captives. One such captive was Hannah Duston of Haverhill, Massachusetts, who was abducted by Indians in 1697. She and her neighbor, also taken captive, traveled for about two weeks when they were left with a Native American family, men, women, and children and another young English captive. At Duston's request, he asked, and was shown the proper way to kill someone with a tomahawk. One night when the Indian family was sleeping, the three armed themselves and killed and scalped ten of the Indians. They then left in a canoe, traveling down the Merrimack River to Massachusetts where they presented the scalps to the Massachusetts General Assembly, receiving an award of 50 pounds. Though Hannah Duston is all but forgotten today, she was probably the first American woman to be memorialized with a public monument, and this statue is one of three built in her honor between 1861 and 1879.

The land west of the Appalachian Mountains to the Mississippi River was not open for bounty land. However, land in Nova Scotia, New York, and West Florida (recently won from the Spanish) was offered for bounty lands. Some colonies, such as Massachusetts, had no more available land

---

[2] Encyclopedia.com.

within their borders to serve as bounty land, so their grants were located in other colonies or the recently opened new land in West Florida.

Wars included in at this time were the Pequot War (1637), Narraganset War (1675), King Phillip's War (1675–1676), King William's War (1689–1697), Queen Anne's War (1702–1713), War of Jenkins' Ear (1739–1748), King George's War (1740–1748), and the French and Indian War (1756–1763). British colonies involved were Connecticut, Georgia, Maryland, Massachusetts, New Hampshire, New York, North Carolina, Nova Scotia (Canada), Pennsylvania, Rhode Island, and the Providence Plantations, South Carolina, Virginia, and West Florida.

The award of military bounty lands was a long-established tradition of the British crown. As early as 1646, the colonial government of Virginia gave lands surrounding frontier forts to the soldiers defending them. Colonial Connecticut, Georgia, Maryland, Massachusetts, New Hampshire, New York, North Carolina, Pennsylvania, Rhode Island, South Carolina, and West Florida also awarded military bounty lands.

Between 1788 and 1855, the newly organized U.S. government continued the tradition, offering lands in exchange for service during the Revolutionary War, War of 1812, Mexican War, or ongoing conflicts with Native Americans. Many of these bounties were awarded in the 1850s, as the government retroactively broadened the scope of eligible applicants. Earlier awards were made within the U.S. Military Tract in Ohio. Later, federal awards could be used for just about any unclaimed public domain lands.

States with disposable lands also awarded property for military service, though not always within their modern boundaries. Georgia, Maryland, New York, Pennsylvania, and South Carolina did award lands inside the state. Virginia awarded lands in modern-day Kentucky, Ohio, and Indiana. Massachusetts gave away parts of Maine until Maine took over the process when it became a state in 1820. North Carolina did the same in Tennessee until 1797, when the new state of Tennessee took the reins. (If land was awarded in a state not already mentioned (like Arkansas), look to federal sources.) Connecticut didn't award military service per se, but gave lots in northern Ohio to those whose property was burned by the British during the Revolution. Within many of these regions, there were defined "military districts."

The Bounty System was a program of cash bonuses paid to entice enlistees into the army. Since it was frequently abused, it was outlawed in 1917 by the Selective Service Act. Land grants were prevalent during the French and Indian Wars, the Revolutionary War, the War of 1812, the Mexican War, and the Civil War. Land grants were made for enticements and service in all of the wars except the Civil War, which was by cash only. In addition to service, grants and payments were made for relief to widows, heirs and hardships.

## Where to Find Records Which Include the Original Surveys

One can expect to find two types of records: smaller tracts, which were completely surveyed, and larger tracts measured in part and protracted in part.

Search for original records in historical societies or archives of the states that awarded them. For example:

- *Connecticut* recorded deeds to recipients in the Connecticut town where the recipient lost property to the British during the Revolution. These have been compiled and indexed in Bockstruck's *Revolutionary War Bounty Land Grants.*
- *Georgia.* The Georgia Archives has both bounty records and headright grants, which were issued to heads of household. The Family History Library has some transcribed Georgia records, too.

**TABLE 7.1**

**Partial Listing of Land Bounty Payments by Rank during the Revolutionary War — There Were Many Variations Depending on State, and the Particular War**

## Revolutionary War Bounty Land Acreage

|  | Private | Officer | Ensign | Lieut. | Captain |
|---|---|---|---|---|---|
| United States | 100 | 100 | 150 | 200 | 300 |
| Georgia | 230–287 | 345 | 460 | 460 | 575–690 |
| Maryland | 50 | 200 | 200 | 200 | 200 |
| Massachusetts | 100 | 100 | 100 | 100 | 100 |
| New York | 500 | 500 | 1,000 | 1,000 | 1,500 |
| North Carolina | 640 | 1,000 | 2,560 | 2,560 | 3,840 |
| Pennsylvania | 200 | 250 | 300 | 400 | 500 |
| South Carolina | 100 | 100 | 100 | 100 | 100 |
| Virginia | 100–300 | 200–400 | 2,666 | 2,666 | 4,000 |

- *Maryland.* Plat maps and ledgers of Revolutionary War era-awards are at the Maryland State Archives. Various published indexes of Maryland's bounty land awards may be found at major genealogical libraries.
- *Massachusetts.* Search Archives.com's collection "Maine, Revolutionary War Pension Applications" for awards made by the state of Massachusetts, but within present-day Maine.
- *New York.* The State Archives holds military and other land grant records dating from 1642. *The Balloting Book*, an 1825 index to New York bounty records, is searchable at siteWeb. An index to the New York Military tract on microfilm can be rented through a Family History Center near you.
- *North Carolina* warrants date before the Revolutionary War, too. Post-war bounties were distributed northeast of modern-day Nashville, Tennessee. The State Archives has all kinds of land warrants (see the Adjutant General's Office Records for military ones). These have also been published in several indexed volumes of *Tennessee Land Entries Military Bounty Land* (1783–1841) by AB Pruitt, available at major research libraries.
- *Pennsylvania* Land Office records are now at the Pennsylvania State Archives (click on "Land Records"; look for military bounty lands in both "Donation Land" and "Depreciation Land" records).
- The South Carolina Department of Archives and History has four volumes of bounty grant. Some of these are duplicated within the state's main grant books.

- *Virginia* offers searching of both Revolutionary War-era warrants and rejected applications at VirginiaMemory. Nearly 5000 Revolutionary War warrants awarded in modern-day Kentucky are digitized and searchable at the Kentucky Land Office website.

## FEDERAL LAND PATENTS

The Bureau of Land Management undertook a project to scan the patents so as to provide images online. The project involved scanning 1135 volumes of patents of military bound land, in addition to many more non-military patents.

## PROCEDURE

Christine Rose's book, *Military Bounty Land, 1776–1855*,[3] is an extremely valuable resource for the researcher in locating military land grants, both state and federal.

## COURT DECISIONS

Lands were frequently granted without having been surveyed and monumented, resulting in overlaps and gaps as well as disputed claims. Numerous early court decisions are found, particularly in the states of West Virginia, Tennessee, Kentucky, as well as surrounding states. Even worse, this problem was exacerbated as a result of later surveyors not extending the appropriate amount of effort to accurately locate such grants, leading to many later problems. A researcher, or a surveyor, would be well advised to be conscious of this fact and be on the alert for potential title issues.

The 1829 Kentucky decision of *Mercer v. Bate*[4] is one of the earliest decisions, and noteworthy since it involved conflicts between two military grants both founded on the proclamation of 1763. Reference to this decision is recommended. It is too lengthy and without proper illustration(s) to be included here.[5]

---

[3] San Jose: CR Publications. 2011.
[4] 27 Ky. 334 (Ky.App., 1829).
[5] The case is highlighted in *Boundary Retracement: Principles and Processes*.

# 7B Special Cases
## *Land-Based Situations*

## ROADS, STREETS, AND HIGHWAYS

### HIGHWAYS

#### DEFINITION

The term "highway" is the generic term for all kinds of public ways, whether carriage-ways, bridle-ways, foot-ways, bridges, turnpike roads, railroads, canals, ferries, or navigable rivers.[6]

"Highway" is defined as a main road or thoroughfare; hence a road or way open to use of public, including in broadest sense the term ways upon water as well as upon land; "highway" is generic term for all kinds of public ways, whether it be carriage-ways, bridle-ways, foot-ways, bridges, turnpike roads, railroads, canals, ferries, or navigable rivers, and includes county and township roads, railroads and tramways, bridges and ferries, canals and navigable rivers. In short, every thoroughfare is a "highway."[7]

A highway is a way open to the public at large, for travel or transportation, without distinction, discrimination, or restriction, except such as is incident to regulations calculated to secure to the general public the largest practical benefit therefrom and enjoyment thereof.[8]

It is a generic term and includes all public roads and ways. In its broad or general sense, it covers every common way for travel in any ordinary mode or by any ordinary means, which the public has the right to use either conditionally or unconditionally, and thus may include turnpikes and toll roads, bridges, canals, ferries, navigable waters, lanes, pent roads, and crossroads. In a limited sense, however, the term means a way for general travel which is wholly public.[9]

#### CASE LAW DEFINITION

Since the argument in the case of *Sproul v. Foye*[10] had to do with the location of whether a layout or an actual route of travel constituted a highway in order to satisfy a call in a description, it was necessary for the court to define the word "highway." Refer to Figure 7.2.

The dispute was whether a call for a road in a description could be satisfied by an actual route of travel, or by a layout for a contemplated route. The court stated, "when a road is referred to in a deed as one of the boundaries of the land conveyed, we should ordinarily suppose that something more than a mere location was meant. A road is a way actually used in passing from one place to another. A mere survey or location of a route for a road is not a road. A mere location for a road falls short of a road as much as a house lot falls short of a house. 'Bounded by the new county road leading from Wiscasset to Dresden.' Can the proposition be maintained that an invisible and unwrought location answers such a call better than a visible wrought road over which the public

---

[6] Black's Law Dictionary.
[7] *City of Long Beach v. Payne*, 44 P.2d 305 (1935).
[8] *39 Am Jur 2d, Highways, Streets, and Bridges, § 1.*
[9] Ibid.
[10] 55 Me. 162 (1867).

DOI: 10.1201/9781003032557-12

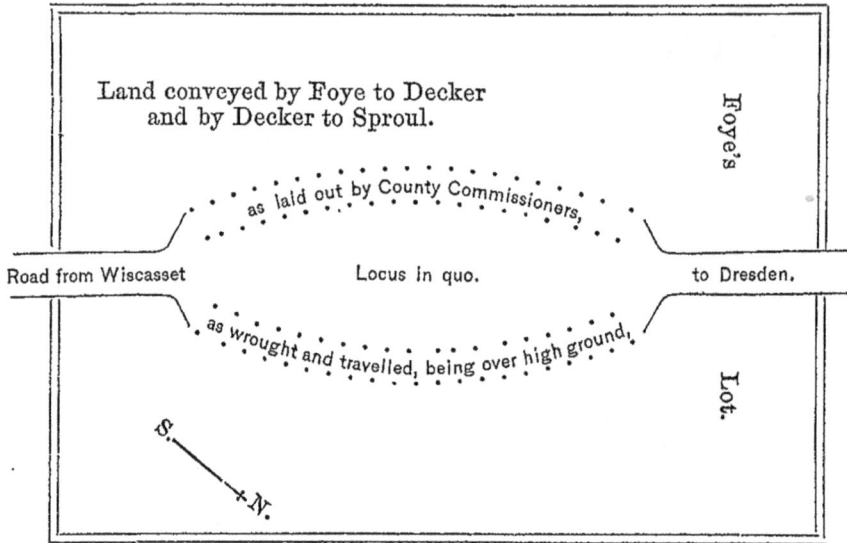

**FIGURE 7.2** Diagram from *Sproul v. Foye*.

travel is passing daily? We think not. We think such a call in a deed requires something more than a mere location to answer it."

Further definition may be found in Black's Law Dictionary: "Highway: a free and public road, way, or street; one which every person has the right to use. The generic name for all kinds of public ways, whether carriage-ways, bridle-ways, foot-ways, bridges, turnpike roads, railroads, canals, ferries or navigable rivers."

## ROAD LINES (HIGHWAY RIGHT-OF-WAY)

"In order to lay the basis for determining whether a fence belonging to the defendant constitutes an encroachment, it is necessary for the People to establish as nearly as may be, the lines of the Douglass survey according to which the rights of all the parties were fixed and still remain fixed, and these lines and these rights cannot be changed by subsequent surveys which ignored the original survey in accordance with which the highway was established.

"Where county surveyor surveyed and laid out proposed public highway, and his report, showing lines followed and government corners found, was adopted by county board of supervisors, highway so surveyed became fixed, and no change could thereafter be made, so far as county's rights were concerned and rights of landowners adjacent thereto were affected.

"In making resurvey to determine boundaries of lands which have been granted, tracks of original survey should be followed, so far as it is possible to discover and locate lines of lands patented, as nearly as may be, according to original survey thereof, and in reference to which lands have been sold."[11]

## HIGHWAY CREATION

Some highways are created according to centerline description, some by sideline description, and some more general than specific, especially the older ones.

---

[11] *People v. Covell*, 62 P.2d 602, 17 Cal.App.2d 627 (1936).

Some are ancient, some private, some public, any or all may be difficult and demand not only concentrated effort by themselves, but also the addition of specialized insight.

## What If Not Created by Survey?

Some highways have very general parameters, without any survey data. To determine the characteristics of such a way demands considerable investigation for consideration of location and width. Numerous court decisions at all levels and from a variety of jurisdictions serve to illustrate procedures when confronted with less than desirable, or adequate, definition.[12]

## Highway Creation

Highways may be created by a variety of methods. It must be kept in mind that any highway creation may be for public use, private use, restricted use, seasonal use, and conditional use.

## Creation by Grant

The most common method of this type is by some form of deed, will, or similar process. Generally, the biggest problem is in the interpretation of the grant, whether fee title or easement.

## Creation According to the Prevailing Statute at the Time (Statutory Layout)

Ordinarily, a designated authority may lay out a road according to the terms of the statute. Depending on the particular state, such layouts are frequently an easement of use and a responsibility for maintenance.

## Creation through Dedication (and Acceptance)

Acceptance may be formal or informal (such as a use, maintenance, or acknowledgement).

## Creation through the Eminent Domain Process, by the Appropriate Authority

Generally, absent a special statute, such takings result in an easement. Even though there may be a statute allowing a fee taking, there is no guarantee that fee will result.

## Through Prescriptive Use

This implies some form of adversity, or encumbrance in conflict with the underlying fee ownership.

## By Layout over Public Land

This is just an added improvement to the underlying fee ownership.

---

[12] Also refer to several ALR's dealing with poorly or inadequately described highways. For Example, 110 ALR 174, *Locating Easement of Way Created by a Grant Which Does Not Definitely Describe Its Location*; 130 ALR 768, *Extent and Reasonableness of Use of Private Way in Exercise of Easement Granted in General Terms*; 28 ALR2d 253, *Width of Way Created by Express Grant, Reservation, or Exception Not Specifying Width*; 5 ALR3d 439, *Extent of, and permissible Variations in, Use of Prescriptive Easements of Way*; 24 ALR4th 1053, *Location of Easement of Way Created by Grant Which Does Not Specify Location*; 36 ALR4th 769, *Locating Easement of Way Created by Necessity*.

## According to Legislative Act

Acts of the state or federal legislative body may create separate title entities. What results depends on the circumstances of the creation and the interpretation of the wording in the specific act, or amendment thereof. See foregoing discussion relative to legislative acts.

## Through Custom

(Customary use by a relatively small group of people, which may be a neighborhood, a municipality, but not the public at large.) It may result in a road of some characters or the use of an area for a specific purpose.

Subdivision roads and streets are created by their surveys. Certain specified procedures are required to result in their being in one category or another. The important consideration, however, is that when shown on a plat (or overall conceptual design which may or may not include a plat) all lot owners on the plat acquire certain rights in the roads, whether constructed, useable, or not. One needs to pay careful attention to the fact that issues with subdivision roads and streets (particularly if never constructed or accepted by the appropriate authorities), are commonplace. Traveled portions exist outside of platted (and sometimes accepted) definitions, differences of opinion regarding right-of-way determination are common, and pertinent records, especially older ones, are often not in proper custody, properly protected nor made readily available.

A road layout, often according to an applicable statute, is created by a survey. However, since roads are built without exactly conforming to their surveyed location, courts have tended to adopt the traveled location.[13]

## RIGHTS OF WAY

There are rights of way that are not highways per se. They may be private, they may be of limited duration, and they may be limited in their use. Winter roads, scenic trails, and byways, along with other more obscure definitions generally fall in this category.

## Relocating a Private Right-of-Way

Courts have dealt with the process of moving the location of a right-of-way.[14] Formal processes, particularly for public ways, involved new creation documents and, sometimes, votes for vacating rights and conveyancing. In the past, there were sometimes formal exchanges of parcels of land.

When two parties, the dominant and the servient owners, wish to relocate a right-of-way, generally the process is as follows: both parties must agree to a relocation and agree as to the new location. When the parties cannot agree, depending on the circumstances, other procedures may be available.

Private rights of way may be relocated by a variety of means: inadvertently, deliberately (formal process), weather, or by convenient use.

## Right-of-Way Line

A right-of-way line, whether public or private, is a line of separation between two estates, usually either fee or easement, or sometimes a combination of both for various parts of the right-of-way.

---

[13] Reference *Sproul v. Foye*, 55 Me. 162 (1867), and related decisions from a variety of jurisdictions.
[14] See 80 ALR2d 743, *Relocation of Easements (other than those originally arising by necessity); Rights as Between Private Parties.*

Its original creation arises when the title is established. Sometimes this is done by means of a survey, other times by words only without benefit of a survey. If the way is fee, then it exists as a separate parcel with a separate title. If it is an easement, it still has a separate title, but exists as a burden on its servient estate, or the property(ies) it exists over, on, over or under.

## CHARACTER OF RIGHT-OF-WAY LINE

To begin, the Supreme Court of Ohio stated in 1992 in the case of *Manufacturer's National Bank of Detroit v. Erie County Road Commission*,[15] "border of right-of-way is "boundary line" like any other." As an argument in the case, the private landowner claimed that requiring a land user to determine exactly where the right-of-way lines are located places too great a burden on the land user. The court replied, "a landowner or occupier is under an obligation to know the boundaries of the property. The border of the right-of-way is a boundary line like any other." Consequently, when easements are involved in a survey, they should be approached like any other survey, or boundary location. That is, they are either equivalent to the survey of a subject parcel, or of an abutter, depending upon their character or relationship to the land in question.

An easement is a separate estate in land, specifically an interest. Some states have ruled that easement surveys are land surveys, and fall under the purview of the land surveyor. Since an easement has a chain of title with a description of its function and its boundaries, it has a location on the surface of earth, and should be treated the same as any tract of land. The same rules apply, and its creation establishes one or more original boundaries. The following decisions focus on that principle:

One of the earlier decisions concerning survey location is *The City of Racine v. The J.I. Case Plow Company*,[16] which discusses proper procedure for locating streets according to a plat.

In ascertaining the true location of the streets, lots, and blocks in a city, according to the plat and survey thereof, regard is to be had (1) to the natural monuments referred to therein, and (2) to the artificial monuments placed by the surveyor to mark lines or boundaries, before resorting to the courses and distances marked on the plat or survey.

The complaint is for an injunction to restrain the defendant company from encroaching upon Water Street, in Harbor addition to the village (now city) of *Racine*, by constructing buildings upon block fifteen projecting into said street, and for the removal and abatement of any such projections already constructed within such street, and for the establishment of said street according to the survey and plat of said addition, and as marked thereon. The answer denies the encroachment, and raises the only question in the case, and that is, what is the true location of Water Street according to the plat and survey? By the courses and distances appearing on the plat and survey, there appeared to be no encroachment, and that the north side of the street was far south of the constructions complained of. These courses and distances, if relied on in determining the true lines and boundaries of the streets, blocks, and lots in this addition, will very materially change all the north and south lines of all of the streets, blocks, and lots from those which have been generally and uniformly recognized and acquiesced in, and according to which they have been occupied and used, and to which the buildings and improvements on the lots, and contiguous to the streets, have been adjusted for over 30 years after the same was platted, and would cast all the lines into confusion, and create conflict and litigation in respect to the whole plat; and this would probably be the result, in at least a majority of cities and villages, by adhering to such a criterion in determining the location of their streets and lots. The original plats, maps, and surveys of western cities and villages, in respect to figures of measurement, and courses and distances marked thereon, in a large majority of cases have been found notably imperfect, incorrect, and unreliable.

---

[15] 63 Ohio St.3d 318; 587 N.E.2d 819.
[16] 56 Wis 539, 14 NW 599 (1883).

The early settlers, who first buy and build upon the lots, do not attempt to ascertain their lines by a computation of measurements of all the other lots and blocks by the figures on the plat, or stated in the certificate of survey, or the courses and distances marked thereon, or by a resurvey from the starting point of the first one. But they consult the stakes, and other monuments and land-marks, either natural or artificial, fixed and placed at the time of the original survey, if any, and such is generally the case, and such is the method adopted by those who buy and build afterwards, if such land-marks still exist; and afterwards, and after such monuments or land-marks have been destroyed or removed, such lines are ascertained by constructions of a permanent character which were built according to such original monuments, and finally, as time goes on, long usage, prescription, antiquity, and reputation may be the only means of determining the true lines and boundaries, and these methods in this order are to be preferred to courses and distances and figures marked on the original plat and survey, as the higher degrees of evidence. At almost any time in the course of municipal history, to rely upon the figures, courses, and distances of the original plat and survey, or upon a resurvey upon the *data* thereof, would be utterly subversive of the rights of real property, and of public and private interests. So far as the figures, courses, and distances on the plat and certificate of survey of this addition is concerned, it is candidly admitted by the learned counsel of the respondent that the 304 feet mentioned in such certificate as the north and south extension of the block directly south of and contiguous to the Water Street in question is obviously a mistake, and does not agree by some 95 feet with the figures designating the width of the lots in said block, and it is argued that it is a mistake because they do not agree. This shows the utter unreliability of such figures, courses, and distances marked on the plat and survey, for who can say what figures shall be taken as correct when they do not agree. And yet the circuit court predicated its findings for the defendant in this case upon such evidence alone.

But, without further argument, the rules of evidence in such cases have become so cardinal and elementary that the citation of many authorities in which they are recognized and repeated, is not necessary. "The principle on which these rules are founded is that effect should be given to those things about which men are least liable to make a mistake." *Davis v. Rainsford,* 17 Mass. 207; *M'Iver's Lessee v. Walker,* 9 Cranch 173; 1 Greenl. on Ev., § 301, and note. These rules are founded upon that principle, and are ranked in degree as follows: *first,* the highest regard is had to natural boundaries; *second,* the lines actually run and corners actually marked on the ground at the time of the making of the plat and survey; *third,* the lines and courses of an adjoining lot or block, if called for or ascertained; and, *fourth,* the courses and distances marked on the plat or survey. (See above note.) There is yet another criterion not mentioned in the note, which is to be preferred to courses and distances. If no monuments are mentioned or in existence, evidence of long-continued occupation, though beyond the given distances, is admissible. *Owen v. Bartholomew,* 9 Pick. 520. And yet another. If the description is ambiguous or doubtful, parol evidence of the practical construction given by the parties by acts of occupation, *recognition of monuments* or boundaries, is admissible. *Stone v. Clark,* 1 Met. 378. These rules in respect to the comparative weight to be given to evidence in such cases, especially as between monuments, natural or artificial, designated or fixed at the time of the survey, and courses and distances marked on the plat or survey, have been frequently recognized by this court. *Vroman v. Dewey,* 23 Wis. 530; *Marsh v. Mitchell,* 25 Wis. 706; *Fleischfresser v. Schmidt,* 41 Wis. 223; *Nys v. Biemeret,* 44 Wis. 104; *Lampe v. Kennedy,* 49 Wis. 601, 6 N.W. 311; *S. C., ante,* p. 249.

The testimony was overwhelming that at the time the plat and survey of this addition were made stakes were placed along both sides of Water Street, between this block 15 and block 14, south of it, to designate these lines, and that soon thereafter fences were built according to these stakes on both of these lines, and remained there over 30 years. In building the fence along the south line of block 15, posts were set in the exact place of the stakes at each corner, and this fence was built within a few inches north of two trees which are still standing, and a natural sidewalk was left along the fence towards the street and a ditch dug outside of that, and such ditch remained until the defendant purchased block 15 in 1877. It was further shown that on the southeast corner of block 17, west of block 15, and across a street, there was also placed a stake at the line of the survey on the exact line of the stakes above mentioned, and that afterwards a shanty was built on that corner according to such stake, which still remains; and that there were other buildings constructed on the line of these stakes, which still remain.

Nearly all of these material facts were proved by the testimony of witnesses who were personally cognizant of the original survey, and two of them respectively first owned blocks 14 and 15, and continued to own them until quite recently, and their testimony was abundantly corroborated by other witnesses.

According to the south line of block 15 and the north line of Water Street, established by this evidence, the defendant company had encroached and threatened to encroach upon said street with its buildings about 7-1/2 feet. All the testimony in relation to these stakes, fences, buildings, and other monuments along these lines was first admitted under objection, and afterwards ruled out by the court. It is claimed, however, by the learned counsel of the respondent, that all of this evidence was not ruled out, but only that relating to the fences, and not that relating to the stakes or other monuments set at the time of the survey. But we think it is clear that the court finally rejected all of it, or that which was admitted under objection. It would have been idle to admit that in relation to the stakes without also admitting evidence where the stakes were placed. But if the circuit court rejected any of it, as is admitted, it was clearly error, and it was equally erroneous to find where the line of Water Street was in utter disregard of this evidence. It is claimed also, by the learned counsel of the respondent, that all of this evidence was outside of the case as made by the pleadings, and irrelevant, as the complaint avers the location of Water Street and the encroachment thereon, according to the recorded plat. The plaintiff is therefore bound by the figures, courses, and distances appearing thereon, and the certificate of survey, and it may be that the court adopted the same view in rejecting and disregarding such evidence, and in predicating its findings upon what were deemed the courses and distances marked upon the plat and survey.

This position is, certainly, very critical, if not captious. By the rules of evidence above referred to the only proper way of ascertaining the true location of the streets, lots, and blocks, according to the plat and survey, is to consult – *first,* the natural monuments referred to therein; and, *second,* the artificial monuments placed by the surveyor to mark the lines or boundaries thereof, before resorting to the uncertain courses and distances on the plat and survey. In any case, the lines are determined according to the plat and survey, and as much in one case as another. The recorded plat and survey must be resorted to in order to show the existence of the street or block as such. There is no exception taken or objection made to the plat and survey, and they are both, probably, correct, and the monuments are only resorted to in order to establish the lines and corners according to the same. The courses and distances marked thereon may be, and in this case they are, incorrect, and therefore will not and do not establish the lines and corners according to the plat and survey. The only difference in the two methods is that the monuments are the more certain one, and better evidence to ascertain the same fact rather than the courses and distances, and that fact is where Water Street really is according to the plat and survey. In our opinion the plaintiff conclusively proved the encroachment complained of, and should have had judgment.

*By the Court.* The judgment of the circuit court is reversed, and the cause remanded, with direction to render judgment according to the prayer of the complaint.

In *Andreu v. Watkins,*[17] the court reported, "Where a deed states that the land conveyed is bounded on the west by a specified street, according to a map referred to, the meaning of the deed

---

[17] 26 Fla 390 (1890).

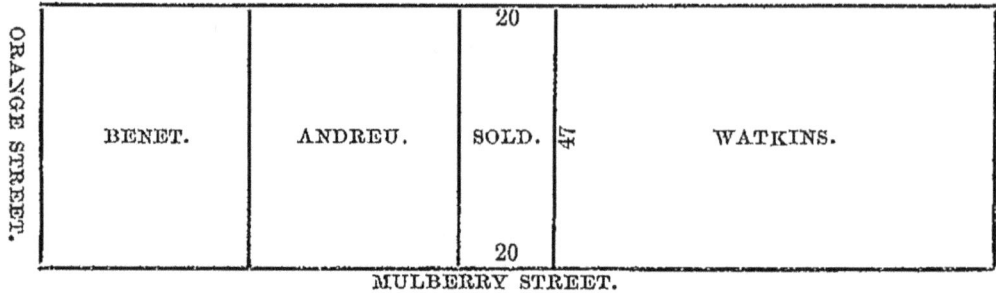

**FIGURE 7.3**  Diagram from *Andreu v. Watkins*.

is that wherever the eastern line of the street as it was laid out, or actually surveyed, is, there is also the western boundary of the land conveyed." Refer to Figure 7.3.

Appellee recovered judgment in an action of ejectment brought against appellants (husband and wife) for a piece of land in the city of St. Augustine, and described in the declaration as follows: 'A strip of land being the east end or portion of a certain lot conveyed by Lizzie Andreu and Michael Andreu, her husband, to Mildred Watkins, by deed dated April 5, 1884, recorded in Book CC, on page 435, etc., of the public records of St. Johns county, commencing at the south west corner of the said lot on the north side of Mulberry street, being 20 feet west of the place where the old line of fence stood, which, up to the time of the said conveyance, marked the dividing line between the land of said Mildred Watkins and said Lizzie Andreu; thence northerly 47 feet; thence east 5 feet; thence southerly parallel to the first line 47 feet to Mulberry street; thence west along Mulberry street to the place of beginning; and containing 235 square feet. The plea was the general issue.'

Syllabus by the Court

1. When there is no ambiguity in the description of land in a deed of conveyance, parol testimony is inadmissible to vary or qualify the description.

2. When a deed conveys to W. 'a strip of land, being the east end or portion of a certain lot deeded' to the grantor 'by Abbott, by deed' of a specified date, and recorded in a designated book, stating the page, 'the west half having been sold to D. B.; the piece now intended to be conveyed by this deed measures as follows: Beginning at the south-east corner of said lot, which is now the dividing line between this lot and the lot of' the grantee, 'thence northerly forty-seven feet to the north-east corner of said lot,' thence west, and thence south, and thence 'eastwardly along Mulberry street to the place of beginning,' giving the length of the lines, the meaning of the deed to W. is that the true east line of the 'lot conveyed to her grantor is the east line of the 'strip' intended to be conveyed to W., and parol evidence is inadmissible to show that a fence (not mentioned in the deed) west of such line, which fence had under a mistake been recognized by the grantor and W. as the true line for a period less than that required by the statute to bar a recovery by the grantor of her land lying east of the fence, was intended by the deed as the eastern boundary of the strip conveyed.

3. Where a deed states that the land conveyed is bounded on the west by a specified street, according to a map referred to, the meaning of the deed is that wherever the eastern line of the street as it was laid out, or actually surveyed, is, there also is the western boundary of the land conveyed.

4. In description of lands, monuments control distances where there is an inconsistency between the two as given in a deed of conveyance.

5. What the boundaries of land as actually given by a deed are, is a matter of law to be ascertained from the meaning of the deed; where they are, is a question of fact, and to show this, or apply the description to its subject-matter, parol testimony is admissible.
6. A deed which fails to describe a part of the land intended to be conveyed does not convey the land thus omitted, and no recovery of the part so omitted from the description can be had under the deed.
7. Where one deed refers to another or to a map or plan of a survey for a description, the deed, map, or plan referred to becomes as much a part of the instrument making the reference as if actually copied into it.

To maintain her action the plaintiff introduced in evidence the following instruments:

A deed from Lucy B. Abbott, dated May 17, 1883, conveying to plaintiff all that certain lot of land in the city of St. Augustine described as follows: Beginning at the intersection of Mulberry and Water streets, from the point northerly along said Water street 158 feet; thence south 47 feet to Mulberry street; thence eastwardly along north side of Mulberry street 158 feet to beginning corner; and is bounded on north side by land of said Lucy B. Abbott; south by Mulberry street; east by Water street; and west by lot of Andreu, and is in that part of the city of St. Augustine known as the 'North City,' and conveyed to said Lucy B. Abbott by I. W. Starke on the 28th day of August, 1872, and recorded in Book I, p. 505; 'also reference to map on file in county clerk's office of the Mary Ann Davis tract, to which reference may be had.'

A deed dated April 5, 1884, from Mrs. Andreu and her husband, conveying to the plaintiff 'a strip of land being the east end or portion of a certain lot deeded to party of first part by Lucy B. Abbott, by deed dated June 23, A. D. 1882, recorded in Book AAA, p. 123, the west half having been sold to Dora Benet. The piece now intended to be conveyed by this deed measures as follows: Beginning at the south-east corner of said lot which is now the dividing line between this lot and the lot of Mildred Watkins; thence northerly 47 feet to the north-east corner of said lot; thence west 20 feet; thence south, and parallel with the east side, 47 feet to Mulberry street; thence eastwardly along Mulberry street 20 feet to the place of beginning; the front and rear being 20 feet each, and the east and west sides 47 feet deep.' Then follows a diagram thus:

Miss Watkins, the plaintiff, testified: That the dividing line between herself and the Andreus was, at the time of their conveyance to her, marked by a fence. Andreu's lot was inclosed by a fence on the east. That this fence was understood between them as dividing their lots. That she had not received possession of the 20 feet conveyed to her by the Andreus. That they had given her only 15 feet and one inch. That the fence was moved after their conveyance to her. That Andreu said he had only given witness this quantity. That the fence is now on the same line where Andreu moved it, 15 feet and 1 inch from where it stood, so far as witness knows. It was moved several months after she bought. That they agreed to let the fence remain for a while, as she had a crop on the land. She could not swear that any one measured the distance between where the fence now stands and the line where it used to stand. It was 158 feet from the line of Water street to where the line of Andreu's fence stood. That Andreu was, at the time of the deed from Miss Abbott to witness, in possession of his lot, and his fence was there, the fence that was understood to make the east line of this lot.

On cross-examination she testified that she had put up a fence on the east side of her lot on the line of Water street, but had never fenced it on Mulberry street.

James L. Colee, a witness in behalf of plaintiff, testified that some six weeks ago he measured the Watkins lot, and the adjacent lots as they stand, and made a plat 'showing the lots as they now stand, and as they stood several years ago.' The red lines show the lots as they stood at the time Miss Watkins and Mr. Benet bought. The black lines as they now are, are shown by the fences. The present fence stands 15 feet and 6 inches from where the old fence stood. The present fence is now 173 feet and 6 inches from Water street. In making my survey I began at the south-east corner of Miss Watkins' lot. The plat was introduced, but is not necessary to an understanding of the case.

On cross-examination he said that he being his survey at the south-west corner of Water and Mulberry streets, and measured along Mulberry street 158 feet, and drove a stake; did not measure from Orange street; that he accepted, as the east line of Miss Watkins' lot, her fence built on the line of Water street; that he 'went' according to her deed; cannot say when she bought; that he did not have any data as to where the fence then stood; took Mrs. Benet's statement and that of Mrs. Andreu as to where the old fence stood.

Upon redirect examination he said that Mrs. Andreu could not show him where the old fence stood. They said their fence on Orange street was moved back five feet, and that it was done to straighten the street, and that they intended to have the same amount of land they bought from Miss Abott. Could not say if Andreu said the fence was moved in from Orange street before or after the sale by him to Miss Watkins.

The testimony in behalf of the defendant is as follows:

A deed from Miss Lucy B. Abbott, dated June 23, 1882, recorded in Book AAA, p. 123, of St. Johns county records, and conveying to Mrs. Andreu 'all that certain piece or parcel of land in the North city lying and being in the city of St. Augustine, * * * described as follows: Bounded on the north by lands of the party of the first part, on the south by Mulberry street, on the west by Orange street, and on the east by lands of the party of the first part; and measuring on the north and south sides 150 feet, and on the east and west ends 47 feet, as by reference to the map of the North city of the Mary Ann Davis tract filed in the office of the clerk of the circuit court of St. Johns county, Fla., May 14, 1877, and is part of the same land that was conveyed to party of first part by John W. Starke, by deed dated August 28, 1872, recorded in Book T, p. 505.' The map mentioned in this deed was also introduced in evidence.

Miss Lucy B. Abbott testified that she was grantor in the deeds to Miss Watkins and Mrs. Andreu; that she had owned the property since 1873; Van Ness and witness owned the whole Mary Ann Davis tract; that upon a survey made of that part of the tract covered by the property in controversy a few years since she was present, and the old stake set on the line of Water street was found; that Miss Watkins' fence on the east end of her lot is on the line of Water street as platted, and as found by said survey; that the distance by the original survey of the tract from Water street to Orange street is 300 feet; that at the time Andreu built his fence witness could not get a surveyor to locate the lines of his lot, and he built his fence five feet into Orange street.

William Mickler deposed that he knew the location of the streets in North city, and the location of the land in litigation; that he had lately made a survey and measurements of the blocks lying between Water and Orange streets, and found the distance between these streets, measured along Mulberry street, to be 302 feet; that Orange and Water streets are each 30 feet wide at Mulberry street, and Mrs. Watkins' fence on Water street is on a line with the other fences on that street; that he did not measure the lots, but only the blocks; that the fence of the Benet lot on Orange street is on a line with the other fences on that side of the street. If the fence on west end of Benet's lot was moved five feet to the west, it would reduce the width of Orange street to 25 feet.

The other facts are stated in the opinion.

**OPINION**

RANEY, C.J. (after stating the facts as above.)

It is certain that the land intended to be conveyed by Miss Abbott to Mrs. Andreu by the deed of June 23, 1882, was bounded on the west by Orange street; or, in other words, that the eastern line of Orange street, as that street was actually located by the survey which the map of the North city of St. Augustine, filed in the office of the clerk of the circuit court of St. Johns county on the

14th day of May, 1877, was intended to represent, was made the western boundary, and the land conveyed was that extending east from Orange street 150 feet, and north from Mulberry street 47 feet, the land north and east being at the time of the conveyance the property of the grantor. The meaning and effect of this deed was that wherever the eastern line of Orange street, as it was laid out, was, there also was the western boundary of the land conveyed; and wherever the northern line of Mulberry street was, there was the southern boundary of the grant; and that the eastern boundary line was 150 feet east of the former street, and the northern line 47 feet north of the latter street. These are the boundaries which, as a matter of law, are given by the deed. *Abbott v. Abbott,* 51 Me. 575, 581. To apply this description, or identify the line described, or, in other words, locate the boundaries, it is necessary to find the described lines of the two streets, as they were actually laid out, and measure the distances given in the deed. Where the boundaries of land described in any deed really are, is always a question of fact; and parol testimony is admissible to show where they are, or apply the description to its subject-matter. Abbott v. Abbott, supra; *Reamer v. Nesmith,* 34 Cal. 624; *Turnbull v. Schroeder,* 29 Minn. 49, 11 N.W. 147.

The deed from Miss Abbott to Miss Watkins, the plaintiff, was executed about 11 months after the above conveyance, and makes the intersection of Mulberry and Water Street on the north side of the former street the south-east and initial point of the description of the ground, and the former street the southern boundary, the latter street for the distance of 47 feet the eastern boundary, and a line parallel with and 47 feet north of Mulberry street the northern boundary, and one parallel with Water Street the western boundary. It is true that it gives the distance along Mulberry street, and the length of the northern parallel line, as 158 feet; but it also says that the land granted is bounded on the west by the "lot of Andreu," ( *Flagg v. Thurston,* 13 Pick. 145,) and it makes reference to the same map to which the deed to Andreu refers.

The proof is that only one map answering the designation of the two deeds is on file in the clerk's office in St. John's county; and it should be remarked or the map that [26 Fla. 402] the block of land out of which the above conveyances were made is not subdivided into lots, but is, according to the map, one solid piece, measuring 300 feet east and west, by 340 feet north and south, and bounded east, south, and west by the streets mentioned above, and north by Locust street.

These are the conveyances from the common grantor, existing at the time of execution of the deed of April 5, 1884, from Mrs. Andreu and her husband to Miss Watkins, upon which deed the latter relies for a recovery of the land sued for. The question, of which a correct answer, when made, will afford a solution of this controversy, is, what land does this deed convey? By its terms, it conveys "a strip of land being the east end or portion of a certain lot deeded to party of the first part by Lucy B. Abbott, by deed dated June 23, A. D. 1882," (stating, as supra, the book and page in and upon which it is recorded, and showing it to be the deed to Mrs. Andreu,) "the west half having been sold to Dora Benet. The piece now intended to be conveyed by this deed measures as follows: Beginning at the southeast corner of said lot, which is now the dividing line between this lot and the lot of Mildred Watkins; thence northerly 47 feet to the north-east corner of said lot; thence west 20 feet; thence south and parallel with the east side 47 feet to Mulberry street; thence eastwardly along Mulberry street 20 feet to the place of beginning."

There is no ambiguity in this description. Its purpose is to convey the east end of the lot conveyed by Miss Abbott to Mrs. Andreu, and it makes the south-east and north-east corners of the land so conveyed the north and south-east corners of the piece intended to be conveyed by this deed to Miss Watkins. The eastern boundary of the lot conveyed to Mrs. Andreu is made the eastern boundary of the piece intended to be conveyed by her and her husband to Miss Watkins, and no other piece of land than one whose eastern boundary is located 150 feet east of the eastern line of Orange Street will answer the calls of this deed. No other eastern boundary will effectuate a conveyance of the eastern end of the lot. This deed, in giving the measurement as "beginning at the southeast east corner" of the lot conveyed to Mrs. Andreu, and stating that such corner "is now the dividing line between" such lot and the lot of Miss Watkins, and following with the words "thence northerly 47 feet to the north-east corner of said lot," means that the south-east corner on Mulberry street, 150 feet east of Orange street, was the commencement of the dividing line between the Andreu and the Watkins lot, and that the line run northerly to the north-east corner was the dividing line, and that this line should be the eastern boundary of the piece of land conveyed. The intention of the parties, as shown by this deed, was that it should convey the east-end piece of the lot, the same to measure 47 feet north and south, by 20 feet east and west, and be bounded on the south by Mulberry street, and having its south-east corner 150 feet east of Orange street; and to locate the boundaries of the land it is necessary to find the eastern boundary of Orange street as it had been located by the survey represented by said map, and from these locate the initial point of description 150 feet east, on the north side of Mulberry street, and trace the boundaries according to the directions and distances given in the deed.

There can be no doubt from the testimony in this case that the eastern boundary of Orange Street was, at the time Miss Abbott conveyed to Andreu, and when Andreu built the fences, at least five feet east of where he put the western fence; and the only conclusion that can be drawn, particularly in view of Mass. Abbott's testimony, is that he located it where he did through a mistake as to where the eastern line of the street was, and this mistake naturally resulted in and accounts for his putting the eastern fence five feet west of where, by the calls of his deed, it should have been. Admitting as a matter of fact that both of the Andreus and Miss Watkins, when the deed from the former to the latter was made, understood the eastern fence to be on the eastern line of the lot conveyed to Mrs. Andreu to Miss Abbott, this misunderstanding cannot change the meaning and effect of the deed to Miss Watkins, which deed shows that it was the intent to convey the eastern 20 feet off the land actually conveyed to Mrs. Andreu by the deed of June 23, 1882. Whether or not the western fence had been moved in before the conveyance to Miss Watkins is not so certain. According to this deed, Mrs. Andreu owned at least five feet east of her fence, and the record of her deed was constructive notice to Miss Watkins, even if Miss Watkins' deed from the common grantor did not put her on notice by its reference to Mrs. Andreu's lot, and Mrs. Andreu could have recovered against Miss Watkins, even assuming the latter's possession to have been adverse, as the period of the statutory bar had not run.

There can be no doubt but that Miss Watkins' fence is on the line of Water Street as it was laid out by the original survey. Miss Abbott testifies to the finding of the old stake at the south-east corner, and the evidence is otherwise entirely satisfactory on the point. It is true that Mickler's measurement of the distance from Water Street to Orange Street, as the fences now stand, was 302 feet, and Colee's survey makes it 303 feet, whereas the above plan of the original survey states it to be only 300 feet. These discrepancies can be of no benefit to the plaintiff, nor any harm to the defendant, in their controversy as it appears now, for it is not pretended that Mrs. Andreu claims further east than 130 feet from the fence, or east line of Orange Street; or, in other words, that she denies Miss Watkins' right to the east 20 feet of the 150 feet of land conveyed to her by the original deed from Miss Abbott. Assuming that the east line of Orange Street and the west line of Water Street were, as actually laid out on the ground by the surveyor, more than 300 feet apart, the mistake as to their distance from each other would not affect their actual location as boundaries. *Turnbull v. Suhhroeder,* 29 Minn. 49, 11 N.W. 147. The description of the street in the deed from Miss Abbott to Mrs. Andreu would nevertheless mean its east line as it was actually run out. Whether or not, in the absence of any other proof than that as to finding the old stake on Water street, the eastern line of Orange street would not be assumed to be 300 feet west of that stake, is immaterial to this case.

The contention of Miss Watkins that the fence, which was 5 feet west of the true line, or only 145 feet east of Orange street, is the eastern boundary of the land conveyed, is not supported by the deed. It is not mentioned in the deed as the eastern boundary. If it had been made the eastern boundary by proper words, then of course the 5 feet east of it would not have passed by the deed; but as it was not, the fact that either or both of the parties to the deed may have understood it to be will not include it in the grant. A deed which fails to describe any part of the land intended to be conveyed does not convey the part omitted from the description.

The parol evidence as to where the fence stood was inadmissible to prove that any other than the east true line of the lot conveyed to Mrs. Andreu was the east line to the piece conveyed by her and her husband to Miss Watkins; and no other land passed by the latter deed than a piece having the same eastern boundary, and extending west 20 feet, and otherwise answering the calls of the deed, and hence the land sued for is not included in that deed, and cannot be recovered under it. *Cornell v. Jackson, 9 Metc.* 150; *Northrop v. Sumney,* 27 Barb. 196; *Tymason v. Bates,* 14 Wend. 671; *Crosby v. Parker,* 4 Mass. 110; *Armstrong v. Du Bois,* 90 N.Y. 95; *Cornell v. Todd, 2 Denio,* 130; *Clark v. Baird, 9 N. Y.* 183, 199 et seq.; *Drew v. Swift,* 46 N.Y. 204; *Linscott v. Fernald, 5 Greenl.* 496; *Bell v. Morse, 6 N. H.* 205; *Van Wyck v. Wright,* 18 Wend. 157; *Clark v. Wethey,* 19 Wend. 320; Sedg. & W. Tr. Title Land, § 798a.

There is in the authorities presented in behalf of the appellee nothing in conflict with the views given, or authorities we have cited. The cases relied upon to sustain the introduction of parol evidence to show that the fence was the dividing line referred to by the deed are such as present a latent ambiguity in the description, as in *Abbott v. Abbott,* supra, and *Hedge v. Sims,* 29 Ind. 574, or where stakes or other monuments were held to control distances, as in *Turnbull v. Schroeder,* supra; or where, as in *Reamer v. Nesmith,* supra, *Waterman v. Johnson,* 13 Pick. 261, and *Claremont v. Carlton, 2 N. H.* 369, parol testimony was admitted to explain particular expressions which did not of themselves convey a definite meaning.

Where one deed refers to another, or to a map or plan of a survey, for a description, the deed, map, or plan referred to becomes as much a part of the instrument making the reference as if actually copied into it. *Chaffin v. Chaffin, 4 Gray,* 280; *Allen v. Bates, 6 Pick,* 460; *Foss v. Crisp,* 20 Pick. 121; *Vance v. Fore,* 24 Cal. 436; 3 Washb. Real Prop. (4th Ed.) 427, 428, 430; Gould, Waters, § 194.

As the plaintiff has not been in possession of the land sued for the period and under the circumstances necessary to create in her a statutory title by adverse possession, nor in fact in possession of it at any time in so far as this record discloses, and as such land is not covered by her deed from Mrs. Andreu, and parol testimony is not admissible for the purpose for which it is attempted to be used, a new trial should be granted.

One of the most significant decisions concerning original location is that of *Gaspar Orena v. City of Santa Barbara*[18] where the court stated, "Where a deed from a city bounded the land granted by a street which had been previously located and surveyed by a city surveyor, and the grantee had for many years held possession of and fenced the lot nearly according to such location and survey, and the street was in use by the public accordingly, the fact that a new survey of the street is afterwards made, changing its line so as to exclude therefrom a strip of land adjoining the lot granted, cannot entitle the grantee to remove his fence to the line of the street as fixed by the new survey, so as to include such strip in his lot, but he is restricted to the lot as bounded by the line of the street as originally surveyed."

**Gaspar Orena, Respondent, v. City of Santa Barbara, Appellant No. 14118,l 91 Cal. 621, 28 P. 268.**
**Supreme Court of California, 1891.**

---

[18] 91 Cal. 621 (1891).

The plaintiff's title must be based on his deed of July, 1867, obtained from the town trustees in lieu of his old grant, or confirming that grant, which was not definitely located, and by this deed plaintiff's land is bounded on two sides by State and Gutierrez streets; and he is estopped by his petition for this deed, and the deed itself, from claiming otherwise. (*Stanley v. Green*, 12 Cal. 163; *Sneed v. Woodward*, 30 Cal. 434; *Bollo v. Navarro*, 33 Cal. 466; *Phelps v. McGloan*, 42 Cal. 298; *McFadden v. Ellmaker*, 52 Cal. 348; *People v. Blake*, 60 Cal. 497, 503, 510; *Moore v. Jones*, 63 Cal. 12.) If the land in controversy was part of Gutierrez Street, no adverse claim, use, or possession could give plaintiff title to it, and the town or city could not convey it to him, or grant him permission to privately use it, as a municipality has no right to sell its streets or to grant such permission. (*City of Visalia v. Jacob*, 65 Cal. 436; 52 Am. Rep. 303; *Hoadley v. San Francisco*, 50 Cal. 265; *Cohn v. Parcels*, 72 Cal. 367, 370; Dillon on *Municipal Corporations*, 3d ed., secs. 669-671, 675, 683; *Barter v. Commonwealth*, 3 Pa. 253; *Burbank v. Fay*, 65 N.Y. 57.) The city was not estopped by either ordinance No. 28 or ordinance No. 39 from asserting that Gutierrez Street is where it was originally located by the Haley survey, and the objections of defendant to those ordinances, as to the competency of the municipal authorities to pass them, should have been sustained. Each of these ordinances was passed after plaintiff obtained his deed from the town, and each of them is void. (Stats. 1863-64, p. 502; Stats. 1856, pp. 198-203; Stats. 1873-74, pp. 330-332; Stats. 1875-76, p. 285; Pol. Code, sec. 4408, subd. 19; *Polack v. San Francisco Orphan Asylum*, 48 Cal. 491; *McCracken v. San Francisco*, 16 Cal. 591; *Jersey City v. State*, 30 N. J. L. 521; *Cross v. Morristown*, 18 N. J. Eq. 305.) Neither ordinance 28 nor ordinance 39 alters Gutierrez Street, or vacates any portion of it, and no power to vacate a street existed at the time these ordinances were passed. (*Wilbur v. Washburn*, 47 Cal. 67; *Breed v. Cunningham*, 2 Cal. 361; *Pearson v. Baker*, 4 Dana, 323.) Parol evidence is admissible to show conflict between the map and the survey, and this can be done, notwithstanding the deed to plaintiff refers to the official map of the town; and as a discrepancy has been shown to exist between the map and the survey, therefore the survey must prevail. (*Whiting v. Gardner*, 80 Cal. 78; *O'Farrel v. Harney*, 51 Cal. 125; *Penry v. Richards*, 52 Cal. 673.) The recital in Orena's deed, in regard to the land being "known and described on the official map of the said town," included, first, as a matter of law, a reference to an actual survey; and second, it included such reference as a matter of fact by reason of the statement on the map itself; and thirdly, an actual survey was proven, not simply of an initial point and base-line, but of all streets, etc., in the town. (*O'Farrel v. Harney*, 51 Cal. 125; *Penry v. Richards*, 52 Cal. 672. See also *Jackson v. Cole*, 16 Johns. 257; *Ferris v. Coover*, 10 Cal. 630; *Colton v. Seavey*, 22 Cal. 503; *Kittle v. Pfeiffer*, 22 Cal. 491; *Franklin v. Doland*, 28 Cal. 178; 87 Am. Dec. 111; *Piercy v. Crandall*, 34 Cal. 334; *Jackson v. Freer*, 17 Johns. 29; *Chinoweth v. Haskell's Lessee*, 3 Pet. 93; *McIvers's Lessee v. Walker*, 9 Cranch, 178; *Wendell v. People*, 8 Wend. 190.) Parol evidence is admissible to show location of lost stakes, and the places where monuments once stood may be ascertained and identified by evidence. (*Bradford v. Hill*, 1 Hayw. 22; 1 Am. Dec. 548, note, *Robinson v. Kime*, 70 N.Y. 154; *Colton v. Seavey*, 22 Cal. 497; *Dale v. Smith*, 1 Del. Ch. 1; 12 Am. Dec. 64, note; *Coate v. Speer*, 3 McCord, 227; 15 Am. Dec. 628; *George v. Thomas*, 16 Tex. 74; 67 Am. Dec. 620; *Riley v. Griffin*, 16 Ga. 141; 60 Am. Dec. 726.) In the scale of evidence as to the identity of land, courses and distances occupy the lowest grade. (*Riley v. Griffin*, 16 Ga. 141; 60 Am. Dec. 726. See note to *Heaton v. Hodges*, 30 Am. Dec. 731.) Courses and distances will be controlled by reference to a fixed line of a street. (*Faris v. Phelan*, 39 Cal. 613.) The initial point in the survey or the beginning corner in the plat is of no higher dignity or importance than any other corner in the survey, and the actual survey governs. (*Pearson v. Baker*, 4 Dana, 323. See Code Civ. Proc., secs. 1854, 1859, 2077; *Riley v. Griffin*, 16 Ga. 141; 60 Am. Dec. 729; *George v. Thomas*, 16 Tex. 74; 67 Am. Dec. 615; *Williams v. Spaulding*, 29 Me. 112; *Hall v. Davis*, 36 N.H. 569; *Heaton v. Hodges*, 14 Me. 66; 30 Am. Dec. 742; *Cheney v. Slade*,

3 Murph. 86; *Esmond v. Tarbox*, 7 Greenl. 61; 20 Am. Dec. 346; *Pike v. Dyke*, 2 Greenl. 213.) A line which is actually marked part of the way, though not a right line from corner to corner, will control the boundary, and such boundary for the residue of the distance will be a direct line to the corner or point of intersection called for, -- in this case, State Street and Gutierrez Street. (*Heaton v. Hodges*, 14 Me. 66; 30 Am. Dec. 738, note; *Thornbury v. Churchill*, 4 T. B. Mon. 29; 16 Am. Dec. 125, 129; *Bryan v. Beckley*, Litt. Sel. Cas. 91; 12 Am. Dec. 276; *Lyon v. Ross*, 1 Bibb, 466; *Vance v. Marshall*, 3 Bibb, 151, 152; *George v. Thomas*, 16 Tex. 74; 67 Am. Dec. 615, 616, 621, and note; *Saltonstall v. Riley*, 28 Ala. 164; 65 Am. Dec. 335; *Wynne v. Alexander*, 7 Ired. 237; 47 Am. Dec. 326; 2 Washburn on Real Property, 675; Tyler on Boundaries, 132.) Ancient reputation and possession are entitled to infinitely more respect in deciding upon the boundaries of lots than any experimental surveys. (*Ralston v. Miller*, 3 Rand. 44; 15 Am. Dec. 704.) Common repute and hearsay evidence are admissible to prove boundaries. (Code Civ. Proc., sec. 1870, subd. 11; *Mortin v. Folger*, 15 Cal. 277-282; *Cornwall v. Culver*, 16 Cal. 424, 428; *People v. Velarde*, 59 Cal. 459; Greenl. Ev., sec. 145, and notes; Wharton on Evidence, sec. 187; *McCoy v. Galloway*, 3 Ohio, 282; 17 Am. Dec. 593; *Coate v. Speer*, 3 McCord, 227; 15 Am. Dec. 628; *George v. Thomas*, 16 Tex. 74; 67 Am. Dec. 621, and note; *Riley v. Griffin*, 16 Ga. 141; 60 Am. Dec. 726, 730.) Acts and declarations of a deceased surveyor, while surveying adjoining lots, are admissible as evidence. (*Adams v. Blodgett*, 47 N.H. 219; 90 Am. Dec. 569-571, and note.)

*W. C. Stratton*, for Respondent.

What the boundaries of a street are is a matter of law, but where they are is a matter of fact. (*White v. Spreckels*, 75 Cal. 610.) In the absence of monuments, the field-notes of the survey will govern and determine the true line. (*Hubbard v. Dusy*, 80 Cal. 281; *Piercy v. Crandall*, 34 Cal. 344; *Jackson v. Wendell*, 5 Wend. 146, 147; *Walsh v. Hill*, 38 Cal. 486; *McIver's Lessee v. Walker*, 9 Cranch, 173; 2 Am. & Eng. Ency. of Law, 503, 508; note to *Heaton v. Hodges*, 30 Am. Dec. 740.) The map will control as to the boundaries, as the defendant failed to find the original monuments at the corner of State and Gutierrez streets, and has not satisfactorily established the place where they stood, nor does it appear from the evidence that the true line cannot be shown by courses and distances. (*Whiting v. Gardner*, 80 Cal. 78.)

**OPINION**

Temple, Judge

This is an action to quiet title to a strip of land about sixteen feet wide, claimed by plaintiff to be a part of block 268 in the city of Santa Barbara, and by the defendant to constitute a portion of Gutierrez Street.

In 1851 the city of Santa Barbara, being the proprietor of the ungranted lands within its limits, contracted with Haley for a map of the city, and for a survey and division into blocks and streets of a certain portion of the lands. Haley agreed to divide the city into squares of 150 yards by streets sixty feet wide, and "to mark each angle of said squares with substantial redwood stakes eighteen inches long, not less than two inches in diameter, and **[91 Cal. 626]** to sink said stakes into the ground sixteen inches; to make an accurate map of said streets and squares," etc.

March 1, 1851, Haley reported "that the city was now surveyed in the manner pointed out in the contract." The report did not contain field-notes, or state the initial point or base-line, or that any had been established.

In November, 1852, the city employed Wackenreuder to make certain surveys and maps, which were reported April 22, 1853, with two maps marked 1 and 2. These maps were, by ordinance passed August 8, 1855, declared official maps of the city of Santa Barbara.

They were maps showing, or purporting to show, the Haley survey.

April 29, 1853, an ordinance was approved declaring certain streets, as surveyed by Haley and mapped by Wackenreuder, open; Gutierrez Street was one of these streets.

Haley made his survey according to contract, and actually located all the streets and blocks, marking each block by a stake driven at each corner. These stakes remained for several years, or at least many of them did, and were known and recognized as Haley's stakes. Lots were sold by the city, and conveyed according to the survey; or at least such is our inference from the evidence, although the fact is not made very clear.

The plaintiff claims under a deed from the town of Santa Barbara, dated July 9, 1867, which describes his property as "one hundred varas lot situated in the southwest corner of block 268, in the town of Santa Barbara, containing 278.1 square feet, bounded on State and Gutierrez streets, and at right angles to said streets 278.1 feet."

It does not appear that there was, at the date of grant, any official map of the town of Santa Barbara, except the Haley, sometimes called the Brady, map, and the Wackenreuder maps.

It is plain, therefore, that plaintiff's lot is bounded by Gutierrez Street as actually located by Haley, and the question in the case is as to this location.

The court apparently failed to find evidence of the existence of any of Haley's stacks upon block 268, or of the precise point where they had actually been placed.

Haley was a witness at the trial, and testified, substantially, that the intersection of Carrillo and State streets was the initial point of his survey; that the streets, as he surveyed them, were exactly sixty feet wide, with the exception of Carrillo and State streets, which were eighty feet each; and that the blocks were 450 feet square, as required by his contract with the city.

The plaintiff had been in possession of his lot for twenty-five years, and had, through tenants, fenced it. Up to about 1876, his fence along Gutierrez Street corresponded very nearly with the line of Gutierrez Street, as claimed by the defendant. At that time, a survey was made from the intersection of Carrillo and State streets, which, allowing sixty feet for the street and 450 feet for blocks, according to the projected plan of the city and Haley's report of his survey, located his line as he now claims it. Thereupon he moved his fence out, as he testified, some fourteen feet; as defendant claims, the whole sixteen feet in dispute.

If the survey of Haley had been exact, the plaintiff would be justified in his conclusion, whether the point measured from were the initial point or any other ascertained point in the Haley survey, for all would correspond; but so soon as it was determined that Haley's survey was inaccurate, as admittedly it was, the value of such measurements from the initial point, or any other, as evidence, was materially impaired. For instance, a point in Haley's survey was pretty well established, at the corner of State and Montecito streets, distant the width of one block and one street easterly from the disputed boundary; and another at the corner of Haley and State streets, the width of one block and two streets westerly. Between these two points, assuming them to have been established, there is an excess of over sixteen feet. According to the testimony of a former city surveyor, who ran the line of Gutierrez Street, by the line of improvements and reputed monuments in the Haley survey, as he determined them, fourteen feet of this excess is between Haley and Gutierrez streets. This is in the block immediately opposite plaintiff's premises on Gutierrez Street, and is between plaintiff's premises and the assumed initial point. On plaintiff's theory, he would get this entire excess. In other words, he gets the benefit of the mistakes in the survey, which, in order to make out his case, he assumes to have been mathematically correct.

But the initial point and base-line, if they had been marked on the map, and returned in the notes of the survey, instead of existing only in the memory of the surveyor, as they did in this case, would not be necessarily more controlling than other ascertained points in the survey in ascertaining the actual location of streets and blocks; whether the initial point be of greater or less importance than other ascertained points, would depend on circumstances, their proximity and relation to the point to be located.

In determining the line of the street, measurements on that street would naturally be of more value than elsewhere, and if they, or the places where they were, cannot be located, it would be important to ascertain the boundaries of the street as actually opened and used; and if such location has been generally acquiesced in by the public, by lot-owners and the municipality, in the absence of more certain evidence, it will be conclusive.

In this case, the very fact that the plaintiff has inclosed his full quantity of land within the block, as claimed by the defendant, and that, after acquiescing in this line for a number of years, he put out his fence, intruding upon the street then actually used by the public, is a strong circumstance against him. That the fences were built by his tenants makes no difference. He let and relet the premises as inclosed, and still occupies and claims to the fences. He cannot deny his inclosure.

If the entire survey could be revised and made to conform to the plan of the city as projected, and the plaintiff could be awarded his lot under the revised survey, he would gain nothing. There is no reason in his contention except upon the idea that he will retain the full quantity which he now holds, and so much more.

But there is no more reason for adjusting the lines of Gutierrez Street to the supposed base and initial point than the other streets and blocks where no monuments of Haley's survey are found. This is obviously impracticable, for it affects the oldest and best-improved portion of Santa Barbara, and the evidence shows, what should have been expected, after the lapse of nearly forty years, that it is difficult to locate such monuments.

The learned judge who tried the case adopted the theory of the plaintiff, and he was probably induced to do this by certain ordinances of the city of Santa Barbara, which were admitted against the objections of the defendant, and which rulings are assigned as error in the statement on motion for a new trial, and on this appeal.

One was ordinance No. 28, passed November 3, 1870. It was here ordained that the iron stakes at the intersection of State and Carillo streets, and at the easterly corner of block 142, "be from now henceforth the initial points of the town survey and of all locations of lots and streets. Satisfactory evidence shows that the aforesaid corners of blocks have been the initial points of the Salisbury Haley surveys,. ... therefore. ... all surveys made thereafter that in any manner deviate from the said initial points, and vary with the courses and distances as set forth in said S. Haley's maps, are hereby declared null and void," etc.

This ordinance is void, and should not have been received in evidence. The city council could not change the location of streets and highways in this manner, nor could it affect the rights of lot-holders under grants previously made. Much less could it lay down rules of evidence by which the courts should determine the location of points, nor determine the fact as to the initial point.

Even if it had been made in advance it would have been utterly impracticable, and would not have controlled the actual survey, although they were shown to be incorrect. No absolutely accurate survey ever was or ever will be made. In order to build cities and towns there must be some finality as to the location of blocks and streets. Under such an ordinance, if valid, there could be none. All lines would be forever subject to be revised and corrected, or changed by new surveys not really more accurate, but only for the time deemed so.

The ruling admitting ordinance 39 may be disposed of in the same way.

The respondent claims that these ordinances, under the circumstances, constitute an estoppel against the city; but independently of the fact that so far as locating these lines are concerned they are absolutely void, we see no element of estoppel in them, and the plaintiff has not expended money or done anything to his injury in reliance upon them.

All these ordinances, to which objection was made, were enacted after the grant to plaintiff and after the street had been opened. In no event could they affect the location of plaintiff's lot. We cannot say that they did not affect the judgment of the court below in its determination of the cause.

In his complaint, the plaintiff avers that for more than sixteen years before the commencement of the action, he had been in the actual, open, notorious, quiet, peaceable, and exclusive possession of the premises in controversy, holding and claiming the same adversely to the whole world, and that the defendant was not seised or possessed of the land or any part thereof for five years before the commencement of this action.

The second finding of the court is in nearly the same language. It is also found that the plaintiff is the owner. This may have been a conclusion from the fact of the adverse possession found. The fourth finding is to the effect that the premises in dispute do not constitute a portion of Gutierrez Street. This also may have been a conclusion from the same finding of adverse possession. The fifth finding is to the effect that the plaintiff did not declare that the premises in dispute were a part of Gutierrez Street, or consent to its use, and the same has never been dedicated to public use.

This is all that is found as to the title of plaintiff, and all may have been inferred from the supposed adverse possession of plaintiff. But the question of adverse possession is a false quantity in the problem to be solved.

The burden was upon the plaintiff to show title in himself. He can do that only by proof that the premises are within his grant. Proof of his supposed adverse possession does not make a *prima facie* case. This claim assumes the very question in controversy, for if the premises are a part of the public street, there could be no adverse possession. Individuals may intrude upon or obstruct a public thoroughfare, but the public cannot be disseised of such lands, and such intruder acquires no rights.

We think the judgment and order should be reversed, and a new trial ordered.

The Court. — For the reasons given in the foregoing opinion, the judgment and order are reversed and a new trial order.

Another early decision from the Wisconsin court is *The City of Racine v. Emerson*,[19] wherein the court commented on the location of a street line according to an 1842 plat.

The ruling question in this case is: Where is the east line of Wisconsin street in front of the lot in question, according to the Vilas plat of 1842? *Miner v. Brader*, 65 Wis. 537, 27 N.W. 313. It is not, Where is such line according to any subsequent survey or plat? All resurveys or subsequent surveys are of no effect except to determine that question. A resurvey that changes lines and distances and purports to correct inaccuracies or mistakes in the old plat is not competent evidence in the case. There are only two questions: (1) Where is the true line fixed by the original plat? (2) Is the fence in question on that line? A resurvey that changes or corrects the old survey and plat can never determine the first question. A resurvey must agree with the old survey and plat to be of any use in determining it. The survey made on the arbitrary plan established by the common council in 1881 does not agree with the old plat in courses or distances, in the dimensions of blocks and lots, or in the lines of the streets. It seems to have been made to correct the old plat, to straighten the streets, and make a better plat than the old one. Resurveys for the lawful purpose of determining the lines of an old survey and plat are generally very unreliable as evidence of the true lines. The fact, generally known and quite apparent in the records of courts, is that two consecutive surveys by different surveyors seldom, if ever, agree; and the greater number of surveys, the greater number of differences and disagreements will occur. When two surveys disagree, the correct one cannot be determined by still another survey. It follows that resurveys are of very little use in such a case as this, except to confuse it. In *Miner v. Brader*, 65 Wis. 537, 27 N.W. 313, there were two surveys, and they disagreed; and the court had to resort to the evidence of a practical location of the lines by monuments. Monuments set by the original survey in the ground, and named or referred to in the plat, are the highest and best evidence. If there are none such, then stakes set by the surveyor to indicate corners of lots or blocks or the lines of streets, at the time

[19] 85 Wis 80 (1893).

or soon thereafter, are the next best evidence. The building of a fence or building according to such stakes, while they were present, become monuments after such stakes have been removed or disappeared, and the next best evidence of the true line.

This case does not differ materially from *Racine v. J. I. Case Plow Co.* 56 Wis. 539, 14 N.W. 599; *State v. Schwin*, 65 Wis. 207, 26 N.W. 568; *Miner v. Brader*, 65 Wis. 537, 27 N.W. 313; *Koenigs v. Jung*, 73 Wis. 178, 40 N.W. 801, and some other cases in this court, and is ruled by them. The fence in front of this lot was evidently built according to stakes still standing, which were set by the surveyor Vilas; and this fence is on the line with 200 feet of fence built according to the stakes then standing in the blocks on the south side of Ninth street, and fences and buildings in the next block south built according to stakes then standing by Mr. Dunrand, only three or four years after the plat was made. This fence also agrees with buildings on the north side of said lot, set according to the original survey. This testimony is almost as conclusive that this fence was built on the line of Wisconsin street as if the original stakes of the survey were still standing there to indicate it. When the testimony is undisputed that this defendant and several witnesses have been there present with these fences forty-five years, and that they have not been materially changed in their location during all that time, the above facts would seem to be the most conclusive evidence that those fences were built on the true line according to the original plat.

All the defendant needs to show is that the fence in question is on the line of Wisconsin street according to the plat of 1842. That plat became a part of the deeds executed under and in reference to it. *Shufeldt v. Spaulding*, 37 Wis. 662. The defendant and others in the vicinity obtained their titles and went into possession and made their improvements, set out shade trees, and built their buildings with reference to that plat, soon after it was made, and according to the stakes set out by the surveyor to mark the lines of that street then existing in many places. We are satisfied that the defendant resorted to the best evidence in existence of the true line of Wisconsin street in front of his lots. It is fortunate that this evidence is yet in existence. The time will soon come when it will have been lost by the destruction of all monuments, natural or artificial, and by the death of the old inhabitants. Then resort must be had to evidence of lesser degree to establish ancient boundaries, and long-continued occupation with respect to unchanged lines, and reputation, even, may be the best evidence available. In any case of disputed boundary the testimony, or even the acts, of the surveyor who originally established it, and who pointed out the stakes set by himself to mark the line so many years ago, accompanied by continued use and occupation in recognition of such line, is not only proper, but strong, evidence that such was the true line, and better evidence than a new survey made more than forty years afterwards, which changes such line. *Koenigs v. Jung*, 73 Wis. 178, 40 N.W. 801. That case would seem to justify fully the testimony of Vilas's statements which were objected to by the respondent in relation to his recognition of the stakes and pointing out the true line in accordance with them. This line so practically located has become an ancient boundary, in favor of which the rules of evidence are and should be liberal.

There have been so many cases similar to this in this court, and all the various questions here involved have been so repeatedly settled, that it is supererogation to again repeat them. In addition to the above cases are the following: *Hrouska v. Janke*, 66 Wis. 252, 28 N.W. 166; *Vroman v. Dewey*, 23 Wis. 530; *Marsh v. Mitchell*, 25 Wis. 706; and *Nys v. Biemeret*, 44 Wis. 104; and many cases in other states are cited in the excellent brief of the learned counsel of the appellant which sustain the same principles. Such cases of the disturbance of the ancient lines and boundaries of streets, lots, and blocks in our cities and villages by arbitrary resurveys under the authority of their officers ought not to be encouraged. The public and private owners have acquiesced in the lines established by the first and original survey and plat, and by practical location and undisturbed possession for a great many years, and there does not seem to have been any necessity to disturb them at this late day.

Some selected decisions from various jurisdictions follow.

### *Pereles v. Gross, 126 Wis. 122 (1905).*

In resurveying a tract of land according to a former plat or survey, the surveyor's only function or right is to relocate, upon the best evidence obtainable, the corners and lines at the same places where originally located by the first surveyor on the ground. Any departure from such purpose and effort is unprofessional, and, so far as any effect is claimed for it, unlawful.

This case had to do with the location of a street line by the city engineer.

### *Johnson v. Westrick, 200 Wis. 405 (1930).*

The corner of a city block involved in the determination of boundaries is where the original surveyor of the plat fixed it by the setting of a stake or other monument, regardless of where present surveys may fix the centerline of a street, and regardless of inaccuracies in his measurements judged by such center line.

"It was held that the east line of the street was where the original surveyor placed it, not where it should be according to resurveys or subsequent surveys; that subsequent surveys are worse than useless; they only serve to confuse, unless they agree with the original survey."

### *People v. Covell, 62 P.2d 602; 17 Cal.App.2d 627 (1936).*

Original survey granting and establishing certain rights, whether accurate or inaccurate, fixes rights not only of government, but of landowners, and government, after establishing such a line and granting and conveying certain rights, possesses no power thereafter to change course of that line.

Where county surveyor surveyed and laid out proposed public highway, and his report, showing lines followed and government corners found, was adopted by county board of supervisors, highway so surveyed became fixed, and no change could thereafter be made, so far as county's rights were concerned and rights of landowners adjacent thereto were affected.

In making resurvey to determine boundaries of lands which have been granted, tracks of original survey should be followed, so far as it is possible to discover and locate lines of lands patented, as nearly as may be, according to original survey thereof, and in reference to which lands have been sold.

State's and county's allegations that fence built by defendant encroached on public highway set forth no cause of action, in absence of showing that fence encroached on highway according to lines established by county surveyor in laying out highway in 1872, and adopted by county board of supervisors, since according to such lines, rights of all parties were fixed and remain fixed.

## ROAD LAYOUT DESCRIPTIONS

Based on survey and location information, descriptions of road layouts have the same inherent faults as any similar descriptions. Modern retracements are continually frustrated by inaccurate measurements and loss of monumentation. The older the layout and its description, the more likely it is to be troublesome. When early roads were laid out, their centerline was often marked, which abruptly disappeared as soon as construction began. The early descriptions were very sketchy, little monumentation, sometimes directions and distances though not always precise, and coordinated with current landowners, which must be tracked forward to determine present day conditions. The older the description, the more difficult it is to work with.

## LACK OF DOCUMENTATION

Like the creation of unwritten rights affecting property ownership and related interests, there may be no documentation in existence. Such is the case with prescriptive ways, those resulting from implied dedication and those created by customary use (custom). An inherent pitfall when no documentation is available is that prescription, implied dedication and custom all "look alike" in their appearances, yet each has a different basis in creation by law. The tendency is to treat them as prescriptive ways, yet they may not be that at all, which may affect how they may be treated.

Locating poorly described roads and highways is an *art* in itself. Locating any surveyed line, when according to specific measurements (with error) or by general description, is difficult at best, whether property sideline description, general location, or specific entity such as a highway or other class of right-of-way.

This can also be called a *science* in itself.[20] There are variations among the states, sometimes statutory, as well in interpretations from various courts.

A few courts have expressed a comparatively strict standard regarding the designation of an easement's location, requiring that the easement description meet the general conveyancing standard for identifying a parcel of land.

Some courts apply the "surveyor sufficiency" test: certainty such that a surveyor can go upon the land and locate the easement from the description. This is the same standard applied to whether a deed description or other land description is sufficient to satisfy the requirements for a valid contract. See *Rivers v. Lozeau*.[21] After Judge Cowart stated, "Although title attorneys and others who regularly work with them develop expertise as to land descriptions, the only professional authorized to locate land lines on the ground is a registered land surveyor," he went on to add, "In fact, the definition of a legally sufficient real property description is one that can be located on the ground by a surveyor." This is because a requirement for the validity of a contract is that it must identify the subject matter of the contract. If it does not do that, the contract may be void for insufficiency as a matter of law.

Outside of Washington, the statute of frauds property description requirements are more forgiving. In general, a property that is described with "reasonable certainty" such that it "may be identified by a competent surveyor" will satisfy the statute of frauds. *McKevitt v. Sacramento*, 55 Cal App 117, 203 P 132 (1921). This is true even if the parties must resort to extrinsic evidence. "When [identifying the property with reasonably certainty] is possible, either with or without the aid of extrinsic evidence, the description is sufficient. *Id.*

For more on the issues of the sufficiency of legal descriptions, see *Vinson v. Brown*, 80 S.W.3d 221 (Tex. App. Austin 2002), *Douglas v. Vorpahl*, 167 Wis 244, 166 NW 833 (1918), and *R. H. Lindsay Co. v. Greager*, 204 F.2d 129 (10th Cir. 1953). Thus, when executing real estate documents, it would be wise to include the most detailed or "full" legal description available. Given that many real estate documents are digitized and made available on the internet, your local county recorder's office would be the first place to look.

(*Note*: Express easements are treated differently than regular deed grants when it comes to specific descriptions. Most jurisdictions require the legal description of the servient estate (property burdened by the easement) to contain the legal description, while the description of the easement itself can be more general.)

The Idaho Supreme Court issued a decision that reaffirms an Idaho rule concerning property descriptions. The Court held that a conveyance deed is not enforceable unless the property description in the deed describes the property so that it is possible for someone to identify exactly what property is being conveyed. This description must be written in so that quantity, identity, or boundaries can be determined. The description must stand alone, without support from outside evidence.

---

[20] Cross-reference this with later footnote. "Few subjects in real estate law are as difficult to decipher as the nature of an ancient way supposedly extant since the reign of George III." [1760–1820].

Complaint made by a Massachusetts Land Court judge in a 1991 decision (JOHNSON, 1991).

[21] 539 So.2d 1147 (Fla. App. 5 Dist. 1989).

In this recent decision, *The David and Marvel Benton Trust v. McCarty*,[22] McCarty asserted that the following legal description in a quitclaim deed was sufficient:

The property at 550 Linden Drive and the building known as Benton Engineering building located upon the property and all adjacent parking lots to the South of the Building and to the West of the Building and right of access into the parking lot located at 550 Linden Drive, Idaho Falls, Idaho located in Bonneville County and more commonly known as the Benton Engineering Office Building.

After reviewing prior case law addressing proper legal descriptions, the Court disagreed, explaining that the quitclaim deed's description was legally deficient and unenforceable. Among other things, it explained that a physical street address – like that one contained in the description – is insufficient. Also, the landmarks referenced in the description did not adequate describe the property because McCarty claimed the quitclaim deed conveyed additional land beyond what these landmarks encompassed. Finally, the Court explained that the most damning evidence against McCarty was the fact that she submitted various explanations of the property – a revised quitclaim deed and a survey – that both conflicted with the quitclaim deed's description. Thus, even resorting to outside evidence (or extrinsic evidence), it was unclear what property the quitclaim deed conveyed.

Interestingly, the Court explained that extrinsic evidence cannot be used to clarify a legal description but used extrinsic evidence to show that there was ambiguity as to what exact property was allegedly transferred. Perhaps the Court should have disregarded the revised quitclaim deed and survey and instead focused solely on whether the quitclaim deed – standing alone – described a definite portion of property. In any event, the takeaway from this case is that parties to a land transaction should take care to clearly and accurately describe the transferred property. Any ambiguity could render the transaction void and create headaches for subsequent owners. At times, a survey may be necessary to avoid there problems.

The description of land from a deed must identify the land, or it must refer to something that will identify it with certainty; otherwise, the description is void for uncertainty. Parol evidence is admissible to fit the description to the land. N.C. Gen. Stat. § 8–39. Such evidence cannot, however, be used to enlarge the scope of the descriptive words. The deed itself must point to the source from which evidence aliunde (from another source) to make the description complete is to be sought (clarification added). *Baldwin v. Hinton*, 243 N.C. 113, 117, 90 S.E.2d 316, 319,

When considering sufficiency of description, it is always important to keep in mind the earlier discussion citing 1 Meigs' Digest, p. 540, as follows:

"Title to land cannot exist without boundary; the plaintiff must show a marked boundary, or some proof from which boundary can be ascertained, before he can say to the defendant, even though he is a naked possessor: 'I have a better right to possess this particular piece of land than you have.' But then, in order to establish boundary, it is not indispensably necessary that some particular corner or marked line should be proven to exist. If it be proven to have existed, or any monument, corner, or marks from which the boundaries called for in a grant or deed can be satisfactorily ascertained, according to any easy and natural interpretation, it is sufficient. Nevertheless a grant can be lost for uncertainty in its boundary, there being no means by which its boundary can be defined. But not if by any reasonable means the intention of the contracting parties can be ascertained."

The North Carolina courts, however, have been known to take a significantly more restrictive approach to the description issue, holding that an easement grant or reservation is void for uncertainty if it does not adequately describe the location of the easement. See *Allen v. Duvall*, 311 N.C. 245, 316 S.E.2d 267 (1984) and a line of additional cases.

---

[22] 161 Idaho 145, 384 P.3d 392 (Idaho 2016).

## THE COURT, IN QUOTING OUTSIDE AUTHORITY, INCLUDED THE FOLLOWING IN ITS DECISION

It is stated in 110 A.L.R., Annotation ... "where the grant of an easement of way does not definitely locate it, it has been consistently held that a reasonable and convenient way for all parties is thereby implied, in view of all the circumstances" ... "It is a settled rule that where there is no express agreement with respect to the location of a way granted but not located, the practical location and user of a reasonable way by the grantee, acquiesced in by the grantor or owner of the servient estate, sufficiently locates the way, which will be deemed to be that which was intended by the grant." 237 N.C. at 542, 75 S.E.2d at 543.

Continuing, "When an easement is created by deed, either by express grant or by reservation, the description thereof 'must either be certain in itself or capable of being reduced to a certainty by a recurrence to something extrinsic to which it refers.... There must be language in the deed sufficient to serve as a pointer or a guide to the ascertainment of the location of the land.'" *Thompson v. Umberger,* 221 N.C. 178, 180, 19 S.E.2d 484, 485 (1942) (and cases cited therein) (emphasis ours). See *Oliver v. Ernul,* supra, 277 N.C. 591, 178 S.E.2d 393.

"It is to be stressed that an alleged grant or reservation of an easement will be void and ineffectual only when there is such an uncertainty appearing on the face of the instrument itself that the court--reading the language in the light of all the facts and circumstances referred to in the instrument --is yet unable to derive therefrom the intention of the parties as to what land was to be conveyed. *Thompson v. Umberger,* supra."

The Supreme Court of Appeals of West Virginia also has adopted the approach that an easement is void if it is not adequately described in the instrument seeking to create it.

*Highway Properties v. Dollar Sav. Bank,* 189 W.Va. 301, 431 S.E.2d 95 (1993).

Quoting *Allen v. Duvall* with approval.

Even when poorly described, and possibly even falling short of the sufficiency test, often times an experienced forensic surveyor can determine the location of an easement. As the Mississippi court stated in *Peacher v. Strauss,*[23] "No deed or conveyance of land was ever made, however minute and specific the description, that did not require extrinsic evidence to ascertain its location; and this is so whether the description be by metes and bounds, reference to other deeds, to adjoining owners, watercourses, or other description of whatever character."

Every contract or deed for the conveyance of land must define its identity and fix its locality, or there must be such a description of the land as, by the aid of parol evidence, will readily point to its locality and boundaries.

Looking through the authorities, it will be seen that, under every conceivable state of facts, and in every imaginable circumstance, the cases may be counted by hundreds, if not by thousands, where contracts, wills and deeds are made effective by the identification, by extrinsic evidence, of the person or subject intended, yet in no wise violating the rule, that such evidence "cannot be admitted to contradict, add to, subtract from, or vary the terms of a written instrument. A simpler rule, perhaps, in most cases, is this: that evidence may explain but cannot contradict written language."

Available tools include both aerial and terrestrial photographs, ancient maps and sketches, and parol evidence from historians and local residents, as well as former owners. See forestry map at Figure 7.4.

Without some form of creating entity, it is not possible to identify the creation of a line, so it is vitally important to take advantage of whatever information may be available, no matter how obscure or how difficult to obtain.

---

[23] 47 Miss. 353 (1872).

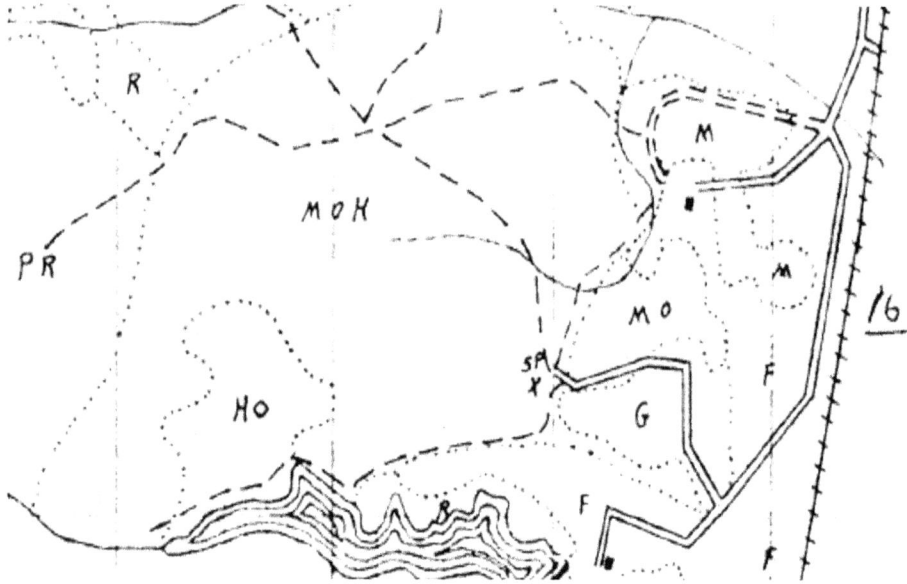

**FIGURE 7.4**  Section of forestry map showing secondary roads, both solid and dashed double lines, and trails, single dashed lines.

## STATUTORY GUIDANCE

In Maine, for example, there is a statute[24] which addresses unknown, or unspecified right-of-way widths.

### § 2102. Lost or disregarded locations; stay

When a highway is laid out through a town and an agent appointed by the county commissioners to open and make it, and the record location thereof cannot be found on the face of the earth or consistently applied thereto or said agent is not making said highway according to the record location, the municipal officers or town agent may file a complaint in the Superior Court setting forth the facts and praying an injunction to stay the proceedings of said road agent. The court shall issue a summary notice to said road agent to appear before it to answer said complaint and on a hearing of the parties may issue a temporary injunction upon such terms and conditions as it deems reasonable. Subsequent proceedings on the complaint shall be similar to proceedings in other civil actions in which equitable relief is sought.

### § 2103. Lost or unrecorded boundaries

When a highway survey has not been properly recorded or preserved or the termination and boundaries cannot be ascertained, the board of selectmen or municipal officers of any municipality may use and control for highway purposes 1 1/2 rods on each side of the center of the traveled portion of such way.

When any real estate is damaged by the use and control for highway purposes of such land outside the existing improved portion and within the limits of 1 1/2 rods on each side of the center of the traveled portion, they shall award damages to the owner as provided in section 3029.

---

[24] Title 23. Transportation; Part 1. State Highway Law; Chapter 11. Laying Out, Altering and Discontinuing Highways; § 653. Highway boundaries. Part 2. County Highway LAW; Chapter 205. Fixing of Boundaries or Locations.

1. **Authority.** The department may establish the boundary lines, limits or locations of any or all state highways and state aid highways and cause durable monuments to be erected at the angles thereof.
2. **Reestablishment of lost or doubtful boundaries.** Whenever in the opinion of the department the boundary lines, limits or location of any state highway or state aid highway or any part thereof are lost, uncertain or doubtful, the department may reestablish those lines, limits or location; land lying within those lines is a part of the highway right-of-way. The department shall file with the town clerk of the town in which the highway is located and with the registry of deeds in the county in which the highway is located maps showing the boundary lines, limits or location of such a reestablished highway, and those lines, boundaries, limits and location are those of the reestablished highway. The department shall post descriptions of those parts of such highways that lie within towns in one conspicuous place in those towns and at 2 points along the highway, and it shall publish a description of those parts of highways that lie within any county in a newspaper, if any, in that county.

In the absence of record, plan, or layout sufficient to reestablish the boundary lines, limits, or location of a state highway or state aid highway, the width of a state highway or state aid highway is deemed to extend to and include the area lying outside the shoulders and ditch lines and within any landmarks or historic features such as fences, fence posts, tree rows, stone walls, corner stones, or other monuments indicating the boundary line.

In the absence of record, plan, or layout or any landmarks or historic features that evidence the location of the boundaries of the right-of-way, the width of a state highway or state aid highway is deemed to extend to and include the sidewalks, shoulders, and ditch lines adjacent to that highway and to the top of cuts or toe of fills where they exist.

**Author's Note:** It is strongly suggested that one be particularly mindful of wording such as this. While offering an alternative solution in cases of doubt, often it comes with a caveat, or has specific requirements. Too often an unskilled person will come to an improper conclusion based on their opinion, yet the basic requirement has not been met, thereby the resulting procedure is improper, incorrect, and perhaps contrary to law. This can only lead to making problems worse, rather than solving them. In the above statute, the wording, "has not been properly recorded or preserved or the termination and boundaries cannot be ascertained," should be understood to mean that ordinary procedure cannot be implemented. Frequently that is not the case, although it may appear to be to the novice, or less than experienced in such matters.

In Vermont, the following statute[25] exists:

The right-of-way for each highway and trail shall be three rods wide unless otherwise properly recorded. Any highway which had been designated as a trail prior to July 1, 1967 and later becomes a trail shall retain the same width of right-of-way as a trail as it had as a highway, but not exceeding three rods.

According to the Vermont court,[26] "When the survey of a highway has not been properly recorded or the records preserved, or, if termination and boundaries cannot be ascertained, the board of selectmen may use and control for highway purposes one and one-half rods each side of the travelled portion thereof."

"The evidence is clear that the land in dispute on Nelson Pond is within one and a half rods of the center line of the present travelled portion of the highway."

---

[25] Title 19. Highways; Chapter 7. Laying Out, Discontinuing, and Reclassifying Highways; § 702. Width of highways and trails.
[26] *Savard v. George*, 125 Vt. 250, 214 A.2d 76 (Vt. 1965).

New York has a provision for minimum width on sections of abandoned highways:[27]

Whenever, pursuant to this chapter or under the provisions of any statute, any town road shall have been widened, straightened, extended, drained, paved, or otherwise improved and in the process thereof a town shall have acquired from an adjacent owner certain lands necessary for said right-of-way by purchase, condemnation, or as a gift and where under such circumstances either the grantor of said new right-of-way shall own the property on both sides thereof for the full length of the new taking or the consent in writing of any and all other owners within such area be given, and there shall be sections of the old road as it existed before said improvement which are of no further use for highway purposes to said town, the town board of said town in which said land is situated, upon the recommendation of the town superintendent of highways, may adopt a resolution, with the consent of the county superintendent of highways, to abandon to the abutting owner or owners such sections or parts of the old road as it existed before said improvement which are of no further use for highway purposes, providing the road after such abandonment shall be not less than three rods in width, and the supervisor of said town is hereby authorized to execute and acknowledge in the name of the town and affix the seal of the town to a quit-claim deed or deeds of the land so abandoned and to deliver the same to the abutting owner or owners for such consideration and upon such terms and conditions as the town board of the said town shall deem proper.

In the New York decision of *Clark v. The State of New York*,[28] the court stated as follows: "In the absence of evidence to the contrary the earliest known center line of the traveled highway is considered the center line of the highway right-of-way and when the original survey of such road runs in a single line such a line is presumed to be the center line of the highway. The failure of the State to open the highway to its full width does not operate to extinguish the rights of the public to the parts unopened." *Mullen v. State*, 17 Misc.2d 63, 185 N.Y.S.2d 732.

It may be said that it is very unusual to not have appropriate information, unless the only pertinent records have been destroyed, or actually lost.[29] Usually, such records are not at hand because the investigator did not search long and hard enough. This may result from limiting parameters, such as insufficient time, funds, or resources. Most times, failure to have the necessary information results from not searching in the right place(s). When a decision is made without proper creating information, and the correct information is later discovered, there is automatically now a two-fold problem: the original problem still exists, and the subsequent situation based on an incorrect premise. The worst-case scenario is when parties actions have relied on the latter, perhaps even executed title documents, or erected improvements.[30]

In Iowa, for example, the law under which the resurvey was made is found in section 964 of the Code, and is as follows: "When by reason of the loss or destruction of the field notes of the original survey, or in cases of defective surveys or record, or in cases of such numerous alterations of any highway since the original survey that its location cannot be accurately defined by the papers on file in the proper office, the board of supervisors of the proper county may, if they deem it necessary, cause such highway to be re-surveyed, platted and recorded as hereinafter provided." The question to be determined, is, whether the re-survey complained of was authorized by this section of the statute. The facts as they appear in the record are, that a public road was established in the year 1856, and the survey and field notes thereof were filed and recorded. By these field notes it appears that the initial point of the road was at the south-east corner of the S.W. 1/4 of the S. E. 1/4 of section 20, township 72, range 2 west, and the calls are by course and

[27] Article 8. Town Highways; § 212-A. Abandoning of parts of town highways.
[28] 246 N.Y.S.2d 53 (1963).
[29] There are numerous examples where "lost" records mysteriously appear at a later date. Usually people give up too quickly.
[30] See *Klose v. Mende*, following.

distance, and do not seem to correspond with the lines of the government surveys, and do not mention stakes or monuments, excepting the terminus of the survey is named as being at "the south-east corner of the old college lot at Kossuth." It does not appear that these recorded field notes are so obscure, repugnant, or otherwise defective, that they cannot be traced upon the face of the earth. On the contrary, the bill of exceptions shows that it was conceded upon the trial "that a line of road as originally surveyed and established could be traced by the field notes on the ground." It is claimed, however, that the original survey as shown by the field notes does not correspond with the route as actually surveyed and marked upon the ground by the surveyor, and upon the hearing before the board of supervisors, evidence was introduced which established that fact to the satisfaction of the board.

It is claimed that as the record showed the road was originally established upon the line designated by the field notes and survey, no change could be made therein by parol evidence showing that the survey was actually made upon a different line. It is a general rule in all questions of disputed lines of surveys that course and distance must yield to actual monuments fixed and recognized by the survey. And this is upon the principle that the establishment of the actual line as surveyed and measured upon the ground is the object always to be attained in controversies of this character. We think, then, it was competent for the board to hear evidence as to where the survey was actually made, and, upon being satisfied from the evidence that the re-survey was upon the line as originally surveyed, to approve and confirm the re-survey. The very object of a resurvey is to ascertain the location of a road already established. *Carey v. Weitgenant*, 52 Iowa 660, 3 N.W. 709. *Blair v. Boesch*, 59 Iowa 554, 13 N.W. 662 (Iowa 1882).

## COURT INTERPRETATIONS

Courts have historically stated in their decisions that roadway width necessarily is interpreted to be more than just a traveled way. Obviously, roadway construction being what it is, is at the minimum, necessary to have slopes and ditches, extending width beyond the traveled portion. With some rights of way, particularly railroads, a cleared swath constitutes the entire width of the right-of-way. Extended width is a critical consideration when dealing with rights by prescription, dedication, and custom, where no width has been specified.

## PERTINENT COURT DECISIONS

Of great assistance are a fair number of decisions from various courts which provide guidelines for answering questions about ways. A selected few follow:

Inclusion of road on maps or the existence of cellar holes may be evidence of a road, as would authenticated photographs or the testimony of long-time residents of the community. *Town of Weare v, Paquette*, 121 N.H. 653, 434 A.2d 591 (1981).

Evidence of wheel tracks is competent to show the limits of a highway established by use. *Plummer v. Ossipee*, 59 N.H. 55 (1879).

Walls built on either side of highway established by prescription may constitute evidence tending to show location of side lines in part if course of highway where no walls have been erected. *Hoban v. Bucklin*, 184 A. 362, 88 N.H. 73.

Where the location of a highway established by dedication or prescription is indefinite and uncertain, its limits and boundaries will ordinarily be determined by the common and ordinary use of the highway. The easement is, however, not necessarily limited to the traveled path, but may include such width as is necessary for the convenience of the traveling public. *Levine v. Town of West Haven*, 120 Conn. 207 (1935).

Where a road is established solely by an implied dedication or by prescription, its width is not extended by statute beyond fences on each side of the way constantly maintained by the owner of the land throughout the period of user. *Reip v. County Court of Colhoun County*, 110 W.Va. 7 (1931).

In *Wilson v. DeGenaro*,[31] the court dealt with a 1927 conveyance whereby defendant's predecessor in title conveyed property by deed and reserved a right-of-way "as now laid out." The quoted words, given their ordinary meaning, clearly referred to the "location" of the right-of-way when it was reserved in the 1927 deed; thus it would be illogical to assume that the grantor intended to specify the right-of-way's location (boundaries) – but without so stating, not specify its width – which was inevitably part of those boundaries. The location of the right-of-way being shown by facts, at question was its width.

Evidence indicated it was 8 feet, but 12 feet had been used, (because of the nature of the material – oyster shells) – the evidence "convinced the court that the traveled portion was 8 feet" and "there was evidence that the road consisted of more than the traveled portion. The evidence supports the reasonable conclusion that it could not rise perpendicularly from its base.......... but was somewhat sloped. It is logical to conclude that the two sides of the road accounted for some additional footage." It was 12 feet at the time of trial, the court decided the right-of-way shall not exceed the width of 12 feet.

Interestingly, the court did not specify a particular width for the right-of-way, but since evidence indicated that 12 feet had been used in the past, logically stated that the width of the right-of-way *shall not exceed* 12 feet.

The Pennsylvania case of *Bartholomew v. Baker*[32] addresses a road, Valley Forge Road, a public highway, as to its status and its location. The argument had to do with the placing of a hedge and monuments allegedly within the right-of-way, to which the lower court ordered removal.

The appeals court stated at the outset that there was no dispute as to the facts, but the problems raised were pure questions of law, the principal one being whether the western line of Valley Forge Road, a county line highway, was to be located, under the peculiar circumstances present, by measuring from the centerline of the road as it had been traveled for a number of years or by some other method.

According to the testimony, the road had "become filled with small crooks and bends" and had become narrowed to an average width of 17 feet. The establishment of the highway was to be 565 perches in length and 33 feet in width, and the center was to be straight line dividing the two counties on a course south twenty-five degrees east. The road had never been opened to its full width, but use by the public had established a well-defined beaten track with some bends but substantially on the general course called for in the report of viewers.

The lower court found, among others, the important facts that the eastern line of defendant's property, the center line of Valley Forge Road as laid out and ordered to be realigned in 1870, and the division line of the counties of Chester and Delaware are coincident; that at no point along defendant's land was the western edge of the road as traveled more than 16-1/2 feet west from the center line of Valley Forge Road as laid out in 1870 (that is, that the beaten track for its entire length along the defendant's property was wholly within the limits of the road as then laid out); that the hedge and monuments lately placed by the defendant are within 16-1/2 feet of the centerline of the beaten track of the road as it had been used for a number of years; and that the western line of the road north of Exeter Avenue and defendant's property is for some distance located from 18 inches to two feet outside and west of the west line of the road as it was laid out by the viewers in 1870. The traveled portion of the road in front of defendant's property is from

---

[31] 415 A.2d 1334; 36 Conn. Sup. 200 (1979), adopted 181 Conn. 480 (1980).
[32] 143 Pa.Super. 149, 17 A.2d 724 (Pa.Super. 1941).

12 to 15 feet in width and is for the most part in Chester County. There are other important facts which were referred to in the course of the opinion.

The court below reached the conclusion, which would support the decree, that the western line of the road was to be determined by measuring 16-1/2 feet from the center of the beaten track. In arriving at that result, the court relied upon the principles announced in the leading case of *Furniss v. Furniss,* 29 Pa. 15, and followed in many subsequent decisions. It was there held that "'where a road has been opened by supervisors, its location cannot afterwards be altered by another supervisor, for the purpose of placing it on what he may suppose to be its proper site;' and this is in accordance with practical experience. It is certainly the duty of supervisors, in executing an order for the opening of a road, to follow the report of viewers as nearly as possible. But we all know that in practice it is rarely carried out with fidelity. Marshy ground, an increased elevation, a deep ravine, and many other things, may induce a deviation; yet when once laid, there it must remain. The authority under the order to open is exhausted by the action of those to whom it is directed, and cannot be resumed. It can only be altered by a new proceeding under the road law. Where a public road has been opened, and worked on ground different from that marked out and reported by viewers (I speak not of roads through unimproved lands where the original marks are plainly visible), it will be a safe rule to extend its breadth, when necessary, from the middle of the road so worked and used, to the limit allowed by law."

By joint action of the courts of Chester and Delaware Counties, this road was not only realigned and straightened, but was made a *county line* road with all the characteristics incident to a public highway on a county line. Such a road differs in many respects from an ordinary township road and is specifically authorized by the statutory road law. The only appellate court case which we have been able to find that deals with such a road is *Roaring Creek Road,* 11 Pa. 356 (1849).

Roaring Creek marked the dividing line between two counties and the viewers attempted to lay out a road on the banks of the creek, at some places in one county and at other places in another county, crossing and recrossing the stream. The Supreme Court set aside the proceedings in both counties, saying "The roads contemplated to be laid out, as this road was, under the 26th section before recited, are roads laid out on the division line between the counties, the half of the breadth of the road in each county. The line of the counties is the centre of the road."

There was no direct evidence that when this road was opened it was placed in the precise location where it is now found. Plaintiffs rely solely on the inference that is to be drawn from its present location. The evidence on this subject was not satisfactory, for the bald testimony that the road has been in the same location for forty years is not very definite evidence from which to determine the centerline of the beaten track of an unimproved road within two or three feet and, in turn, find the precise location as made by those whose duty it was to realign the road after the court order in 1870. It is a matter of common knowledge that beaten tracks on unimproved roads shift with the weather, with the seasons, and after storms. That portion of the principles announced in the Furniss case dealing with the fixing of boundaries of public roads by reference to the center of the beaten track was a rule of necessity and, in part, a rule of convenience. It is a wise rule if properly applied. It was never intended to supplant direct evidence as to where the road was actually located when opened. If when the road was opened its course was marked either by marking the boundaries or the center line of the road, then the limits of the road would not shift if the center line of travel deviated from time to time: *Com. v. Shoemaker,* 14 Pa.Super. 194. A public road once laid out cannot be changed by the supervisors; it can only be changed by a new proceeding under the road law: *Holden v. Cole,* 1 Pa. 303; *McMurtrie v. Stewart,* 21 Pa. 322.

"This road was a public highway for more than a century prior to 1870. At that time the courts of the two counties ordered that it should be realigned so as to form a straight line for its entire course and that the county line should mark the centerline of the road. There is here no difficulty in fixing the location of the county line and that line therefore determines the location.

In fixing the boundaries of a public highway, the property rights of abutting owners are always affected. When such an owner is adversely affected by the opening of or changing the location of a road, he is entitled to damages. If the boundaries are to correspond always with the center of the traveled highway, the owner would have great difficulty in protecting his rights in case of a deviation in the traveled route, which we have seen is bound to occur on an unimproved road. The disadvantages that had resulted from a strict application of the principles announced in the Furniss case were such that the subject was given consideration by the Legislature. If there be any doubt remaining as to the correctness of the result reached by us, it is removed by the Act of June 19, 1901, P. L. 573 (36 PS § 1905), which provides: "In all cases where public roads in this commonwealth have been or may hereafter be laid out by viewers, and the report of such viewers has been or may hereafter be confirmed by the court according to law, the width thereof fixed and ordered to be opened, and the same has been or may hereafter be opened, with the roadbed or track thereof traveled by the public located within the lines of such road as originally laid out, such lines shall be and remain the boundary lines of such road, unless the location of such road has been or may be changed by due course of law."

Even if this road had been an ordinary township road or if its centerline had not been marked in any way, this statute would require a reversal of the decree, for the traveled portion of the highway in front of defendant's land was wholly within the boundaries fixed by the court order of 1870. The court below was of the opinion that this act was not applicable because the traveled portion of the road at a point north of Exeter Avenue and plaintiff's property extended, for a short distance, from eighteen inches to two feet beyond the western line of the road as laid out in 1870. The effect of such a construction is to hold that if the traveled portion of a road at any point is outside the boundaries fixed by the order opening or aligning the road, then the act is not applicable. We might have a road many miles in length and if at one point the traveled portion of the road was found to be even in part outside the lines as fixed by the court order, then one would be compelled to depend upon inference to fix the location of that road for its entire length. We cannot believe that such was the intention of the Legislature.

When we consider the mischief to be remedied and the objects to be attained, we cannot give the law such a narrow construction as that placed upon it by the court below. To so interpret the statute would be to defeat its prime purpose. It seems clear to us that it was the intention of the Legislature that when supervisors should undertake to widen the traveled portion of a highway, they be limited to the boundaries fixed by the court order, except at such places on the road as the track traveled by the public was in whole or in part outside the lines of the road as originally laid out. The Legislature did not intend to arbitrarily divide roads affected by that act into two classes, one where the traveled portion of the road was for its entire course within the original lines as ordered, and the other where the traveled portion of the road was at some one or more places without those lines, and make such classification the basis of the application of the relief granted. Inasmuch as instances are rare where county roads do not at some point depart from the court order, the act would afford little relief if given the limited meaning contended for. The intent was to protect the rights of abutting property owners by providing in substance that where the traveled portion of the road is "within the lines of such road as originally laid out," such lines shall remain the boundary lines until changed by due course of law. The adverb "where" refers to a specific location. The traveled road in front of defendant's land at no point extended west of the original line as ordered in 1870. Therefore the supervisors have no right to extend the road west of the original line at any point in front of defendant's land.

In construing the Act of 1901, the court below relied upon *Waldschmidt v. Glenfield Boro.,* 60 Pa.Super. 538, and *Gray v. North Versailles Twp.,* 208 Pa. 77, 57 A. 190, but we find nothing in those cases contrary to the conclusion reached here. It appears very clearly from the statements of facts in those cases that the place of alleged departure was at the precise location where the line was in controversy.

In determining the rights of the parties to this controversy, we are not concerned with rights acquired by prescription for the public has not occupied or exercised any rights in the land in controversy. A different question may arise where the traveled portion of the road is in part beyond the limits fixed by the order of 1870. Since such questions will involve persons who are not parties to this action, we are not concerned with them here and they will have to be met if they arise."

**Author's Note:** This case provides guidance for the surveyor concerning unimproved roads and puts in perspective the difference between a layout and an existing route of travel, when the two do not agree. Note particularly that a layout governs, until changed according to proper procedure. Note as well that the constructed layout governs over the contemplated layout.

The court in the Maine case of *Sproul v. Foye* introducing this section, came to the same conclusion (see). *A layout for a highway is not a highway.*

## POLICY GUIDELINES

In some states, the department of transportation has a policy where, in the absence of definitive information, reliance for location is the historical traveled way. A width is determined, if possible, by one of several methods, and sideline location is based on evidence, such as fences or other improvements. One must use extreme caution with this approach however, as it is only a valid procedure when better methods fail, or better information is absent.

In this regard, the Illinois decision of *Klose v. Mende*[33] is an important study. This case found its way into the court system several times. Beginning in 2000, the township sent notice to abutting property owners that it intended to improve two roads, and included a copy of pages from the township ledger demonstrating that there was a statutory dedication of a 66-foot right-of-way. No other documents were produced, including the survey made at the time of the dedication. The Kloses, owning property that adjoined both roads, filed a complaint, and the court stated there was a valid dedication so the Kloses appealed. The appeals court subsequently found that the township's claim to the roads failed because the documents offered by the township were incomplete and thus insufficient to establish a valid dedication of the roads. Specifically, this court determined that the ledger entries presented by the township failed to satisfy the requirements for a valid dedication as the statutory requirements include the petition requesting permission to build the roads, a surveyor's report, survey and plat, as well as the road commission's order granting the dedication. This court also found that the township had acquired an easement by prescription for the portion of the roads in use at the time of the action. The court vacated the trial court's dismissal of the Kloses' complaint and ordered the trial court to enter an order establishing the Kloses' fee simple title and the township's prescriptive easement rights in the roads.

The Kloses filed a motion to reinstate, to enter a second amended complaint, to set a date for the township's answer, and to order the township to produce right-of-way maps. In March of 2003, the township discovered documents in a previously unavailable closet which included the original order dedicating the roads at issue, the original surveyor's report and plat, and receipts from adjoining landowners indicating the compensation that was paid for the taking of their land for the roads. Pursuant to a motion by the Kloses, the township's original petition was dismissed in May 2003 based on the petition's failure to adequately assert due diligence by the township. An amended petition was filed in June of 2003, which set forth the township's diligence efforts and included affidavits from the commissioner and the township clerk attesting to their efforts. A second amended petition followed.

---

[33] 378 Ill.App.3d 942, 882 N.E.2d 703 (Ill.App. 3 Dist. 2008).

Following the hearing, the trial court stated that the newly discovered evidence presented by the commissioner constituted the records that the appellate court determined were necessary to establish a valid dedication. The court also determined that the township was not negligent and was not lacking in diligence in searching for the original records and that the evidence established that a reasonable search was conducted of all the logical places in which the records would be stored. The trial court further found that the town clerk's record-keeping was reasonable, not negligent; that the existence of the original records constituted newly discovered evidence; The Kloses again appealed.

The decision was upheld, however a dissenting opinion stated as follows: "I would hold, that the township had a duty to maintain the records in the first place and cannot escape the consequences of its failure to properly maintain the records, whether intentionally or in good faith. A township must maintain the records and failed to do so, I would hold that the township's failure to maintain the records showed a lack of due diligence in presenting a fact or claim to the trial court at the time of the original action. Thus, I would hold it was error for the trial court to grant the township's petition."

Whether agreeing or disagreeing with the outcome of this case, a valuable lesson can be learned. A question often arises as to the extent of a search or an investigation, which may ultimately come down to reasonableness. As the Maryland court stated in the case of *Maryland Coal and Realty Co. v. Eckhart*,[34] "an albatross in the wake of every title searcher is the ominous question of whether he has gone back far enough in the chain of title." The same can be advised with any search. After all, notice exists, resources are available, and what are the ultimate consequences of failing to review something that is required. The facts don't change automatically, and should a question be reopened in the future, as in the *Klose* case, results could be consequential, possibly disastrous.

Even though there may not be a definition, there is still a location, which includes beginning and ending point, and there are still boundaries. The principle being, every title has a location, and every title, of necessity, has one or more boundaries.

All this requires a search of the title and the records relating to any right-of-way whether private, public, highway, railroad, trail or whatever, of whatever character. The date of creation is important as the laws and regulations in place at the time likely have a bearing on the details at issue. There never is any substitute for a complete history, title and otherwise, concerning any right-of-way. When dealing with any easement, there is, at the least, a dominant and a servient tenement (owner), sometimes more than one. That means that there are, at a minimum, two chains of title.

Once an easement has been created, difficulties often arise with respect to its location and dimensions.[35] The mode of creation bears on the resolution of these issues.[36]

### Bartholomew et al., Supervisors of Easttown Tp. v. Baker, 17 A.2d 724 (1941)

It is common knowledge that beaten tracks on unimproved roads shift with the weather, with the seasons, and after storms. The rule authorizing boundaries of public roads to be fixed by reference to center of the beaten track is a "rule of necessity" and in part a "rule of convenience," and does not supplant direct evidence of where the road was actually located when opened.

If when road was opened its course was marked by marking either the boundaries or the center line, limits of the road would not shift if center line of travel deviated from time to time.

---

[34] 25 Md.App. 605, 337 A.2d 150 (1975).

[35] One court went so far as to state the following: "Few subjects in real estate law are as difficult to decipher as the nature of an ancient way supposedly extant since the reign of George III." [1760–1820].
    Complaint made by a Massachusetts Land Court judge in a 1991 decision (Johnson, 1991).

[36] See *Wright v. Horse Creek Ranches*, 697 P.2d 384 (Colo., 1985) and Powell on Real Property, § 34.12 (1994).

**Author's Note:** A word of caution is in order. Many of these fallback procedures rely on a basic caveat, that something is unknown, missing, or impossible. Many times what seems out of the question, is actually a mistaken belief by someone of limited experience, or not as diligent as an experienced investigator. As one author stated, *impossible is not a fact, it is an opinion.*[37] If a decision is made and a later determination finds otherwise, it may cause one or more documents to be void in the final analysis. A common example is the misuse of a boundary line agreement, whereby the true, or existing, line originally created must be unknown or unascertainable. If that requirement is later found to be untrue, it may serve to void not only the proceeding itself, but also the later titles dependent on it. The California courts have a long history of such cases.[38]

## RAILROADS

There are several types of railroads, e.g., common carriers, street railways, private short line railroads, as well as others. Nearly all are created by charter, through legislative act, except for minor, local and personal lines.

The permission to the appellee to enter the forest reserve and construct its road having been granted on the 19th of April, 1909, it was in the months of August and September, when it constructed its road, acting within its legal rights in definitely locating its right-of-way "by the actual construction of the road." When the question as to whether the actual construction of the road caused the title of the right-of-way to be vested in the railroad company was before the Supreme Court in *Jamestown & N. R. Co. v. Jones,* supra, Justice McKENNA used this language:

"The ruling gives a practical operation to the statute, and we think is correct. It enables the railroad company to secure the grant by an actual construction of its road, or in advance of construction by filing a map as provided in section 4 (Act of March 3, 1875). Actual construction of the road is certainly unmistakable evidence and notice of appropriation."

In this opinion is quoted with approval what Secretary Vilas said in *Dakota Central R. R. Co. v. Downey,* 8 L.D. 115, as follows:

"'As to the roadway the construction of the road fixes the boundaries of the grant, and fixes it by the exact rule of the statute.'"[39]

The biggest problem the researcher is likely to face is whether a section of railroad is an easement or a fee title (or some lesser estate). The basic suggestion for this type of investigation is to begin with the Washington decision of *Brown v. State,*[40] which lists guidelines followed by the court system to decide this question.

In addition, there are many more aspects of railroad title than merely the area of the track. Maintenance areas, roundhouses, stations, gravel pits for material, all are necessary in the establishment and maintenance of a railroad. Many of these other uses may have been acquired as a fee title, even if the main route of travel was acquired as an easement.

## CEMETERIES AND BURIAL SITES

Areas for burial of the dead may be public, owned and maintained by local government or in the case of federal burial sites by the federal government, or may be private. This latter category is where the researcher, whether lawyer or surveyor (or other, such as historian) may encounter difficulty.

---

[37] Muhammad Ali.
[38] See also, Chapter 9 in *Boundary Retracement: Processes and Procedures*, for numerous examples.
[39] *Van Dyke v. Arizona Eastern R. Co.*, 18 Ariz. 220, 157 P. 1019 (Ariz. 1916).
[40] 130 Wn.2d 430, 924 P.2d 908 (Wash. 1996).

Family burial sites. Many small family plots remain in all states, whether they are actively used today or not. Fortunately, many of these are documented, in family records, public records, on maps and within historical collections. Extensive research should be undertaken to insure the nature and extent of any burial site. Not all sites are marked, or fenced, and in some cases burials are within and without a designated area. More than likely certain persons are guaranteed access for visitation and maintenance, and in many states, construction must not take place within a specified distance of the burial ground. On occasion, graves are unmarked and therefore not visible or obvious.

Why should we care about cemeteries and burial sites?

- It is often a different title – must research and/or acknowledge abutters
- If the same title, may be a different estate and have access attached
- Minimum standards for survey
- It has boundaries like any other parcel
- May be the subject of a survey
- Identify boundaries for proper setback
- Access may be a concern
- May affect or create a right-of-way
- May be an encumbrance on the subject property
- May need names and relationships for genealogy/title/research purposes
- May be protected under the law

From a title perspective, a cemetery, including a family burial ground, may fall into one of four categories: created by deed, by reservation or exception, through dedication, prescription or by custom. They may be fee title, lease, trust, dedicated to public use, or customary or prescriptive.

The title acquired by the establishment of a cemetery on the property of another, or by exception of an established cemetery from a conveyance to a stranger has been variously described as an easement, a trust, a license, and a fee. In some cases the nature of the title is determined by the manner in which the cemetery was established.

Somewhat akin to a subdivision, a burial site may be governed by statute, surveys and sketches of location may be in form of written description, or may be shown on a plan, often not of public record but in the hands of the Sexton or overseer of the cemetery.

Some are well-marked, others not well marked. Still others have a defined perimeter, with additional graves outside of the marked definition.

While it is often assumed that a cemetery is public, and owned in fee simple and a deed to it should exist, such is not always the case. Family burial grounds are an obvious exception, and may be part of the land that is occupied for burial purposes. It therefore, from the standpoint of outstanding rights, may exist as a separate entity or may be part of the parent tract. In addition, the extent to which heirs and other family members have rights, with access as well, may be far-reaching. Any burial site, obvious, marked, documented or otherwise, is a separate title, and must be explored accordingly.

A special category is the native American burial ground, which is likely governed and protected by federal laws. They may be considered historical sites, or, in particular, religious sites. Great care must be exercised not to interfere with such burial sites.

## CONDOMINIUMS, PARTY WALLS

A **condominium**, often shortened to **condo**, in the United States and in most Canadian provinces, is a type of living space similar to an apartment but independently sellable and therefore regarded as real estate. The condominium building structure is divided into several units that are each separately owned, surrounded by common areas that are jointly owned. Similar concepts in other English-speaking countries include **strata title** in Australia, Malaysia, New Zealand, and the Canadian province of British Columbia; **commonhold** in the United Kingdom; and **sectional title** in South Africa.

Residential condominiums are frequently constructed as apartment buildings, but there has been an increase in the number of "detached condominiums," which look like single-family homes but in which the yards, building exteriors, and streets are jointly owned and jointly maintained by a community association.

Unlike apartments, which are leased by their tenants, condominium units are owned outright. Additionally, the owners of the individual units also collectively own the common areas of the property, such as hallways, walkways, laundry rooms, etc., as well as common utilities and amenities, such as the HVAC system, elevators, and so on. Many shopping malls are industrial condominiums in which the individual retail and office spaces are owned by the businesses that occupy them while the common areas of the mall are collectively owned by all the business entities that own the individual spaces.

The common areas, amenities, and utilities are managed collectively by the owners through their association, such as a homeowner association.

Scholars have traced the earliest known use of the condominium form of tenure to a document from first-century Babylon. The word *condominium* originated in Latin.[41]

There are many forms of real estate ownership that are similar to condominiums but not identical.

Classic privately owned detached houses on privately owned lots may be part of a community that has a homeowner's association. Such an association may administer a common park area, for example, or an access road, or architectural standards for the houses.

This may give rise to what is known as the **common scheme doctrine.**

A classic case is the Massachusetts Land Court Decision of *Campbell et al. v. Nickerson, et al.*[42]

In a case such as this, it may be necessary to assemble all of the conveyances involved in the development. That requires researching all of the titles involved back to their origin. In the *Campbell* case, that date was 1711. This demonstrates once again that there is no substitute for a complete history of land titles, including related easements. This is not an isolated case, nor is the concept confined to colonial states.

In a townhouse complex, multiple physical houses are combined into a single architectural building. Each unit owner owns an identified plot of land and the building affixed to it, but that building is physically part of a larger building that spans lots. There is a continuous roof and foundation and a single wall divides adjacent townhouses. If there is an apartment below not owned by owner of townhouse, then it is not a townhouse, just a bi-level apartment/condominium. Legally, this is very similar to detached houses, but because of the intertwining of interests in the single architectural building, a homeowner's association is required. It would be impractical, for example to replace the roof of just one townhouse. But unlike the condominium, the townhouse complex's HOA owns none of the building or the land under it. It is essentially under contract to the townhouse owners to maintain the parts of the building that are hard to divide. Even the walls between townhouses are usually outside the purview of the HOA, being jointly owned and maintained by the owners of the townhouses on either side. Like the condominium, the townhouse complex often has common areas for roads, parking, clubhouses, and such.

---

[41] Wikipedia.
[42] 73 Mass.App.Ct. 20 (2008). See Wilson, *Easements Relating to Land Surveying and Title Examination*. Wiley, 2013. See also, Kline, Kristopher, *Unmistakable Marks: Common Scheme Doctrine*, POB, March 2016.

A rowhouse is like a townhouse except that the houses are not physically connected. They are independent structures that simply have no space between them. Technically, they are detached.

A building with multiple residential units may simply be owned in common by multiple people, with each having specific rights to a particular unit and undivided interest in the rest. This is like a condominium, but there is no HOA with legal powers. It is much harder to govern, as the individual unit owners often have to agree unanimously or court intervention is required.

States must be examined individually for the requirements at the time the development was established. Examination of Statutes governing condominiums and planned unit developments (PUD'S) should provide information as to the type of development. Then it is a matter of studying the condominium documents, title papers and survey plats for details of the project. There is widespread variation, although basic concepts remain somewhat uniform.

For Example, California statutes recognize three kinds of "common interest developments": condominium, townhouse, and community apartment, with the latter being the owned-in-common concept described above.

The Maine court stated in *Olson v. Albert*[43], the following:

"Although we have never adopted the doctrine of implied restrictive covenants, we have noted that many jurisdictions apply the doctrine when all of the following conditions are met:

1. a common owner subdivides property into a number of lots for sale;
2. the common owner has a "general scheme of development" for the property as a whole, in which the use of the property will be restricted,
3. the vast majority of subdivided lots contain restrictive covenants which reflect the general scheme;
4. the property against which application of an implied covenant is sought is part of the general scheme of development; and
5. the purchaser of the lot in question has notice, actual or constructive, of the restriction.

This court also stated in *Chase v. Burrell*[44], "Many jurisdictions have adopted the doctrine of 'reciprocal negative servitudes, or 'implied equitable servitudes,' in which a restrictive covenant will be implied in a silent deed when certain conditions are met. Although the criteria for application of the doctrine vary somewhat among different jurisdictions, generally, the doctrine is applied when" [reciting the above restrictions]. Citing several previous authorities, including 20 Am Jur. 2d Covenants, Conditions, and Restrictions, §§173, 175, 299 (1965); 7 G. Thompson in Real Property, §§ 3163, 3171 (1962).

## Party Walls

Apart from special statutory definitions, the term "Party Wall" may be used in four different legal senses.

It may mean:

1. a wall of which the adjoining owners are tenants in common;
2. a wall divided longitudinally into two strips, one belonging to each of the neighboring owners;

[43] 523 A.2d 585 (Me. 1987).
[44] 474 A.2d 180 (Me. 1984).

3. a wall which belongs entirely to one of the adjoining owners, but is subject to an easement or right in the other to have it maintained as a dividing wall between the two tenements;

4. a wall divided longitudinally into two moieties, each moiety being subject to a cross easement, in favor of the owner of the other moiety.[45]

## Definition: Party Wall

n. A wall shared by two adjoining premises which is on the property line, such as townhouses, condominiums, row houses, or two units in a duplex. Both owners are responsible for maintaining structural integrity of the wall, even if the wall is entirely on the property of one of the parties.

Copyright © 1981–2005 by Gerald N. Hill and Kathleen T. Hill. All Right reserved.

PARTY WALL. A wall erected on the line between two adjoining estates, belonging to different persons, for the use of bothestates. 2 Bouv. Inst. n. 1615.

2. Party walls are generally regulated by acts of the local legislatures. The principles of these acts generally are, that the wall shall be built equally on the lands of the adjoining owners, at their joint expense, but when only one owner wishes to use such wall, it is built at his expense, and when the other wishes to make use of it, he pays one half of its value; each owner has a right to place his joists in it, and use it for the support of his roof. When the party wall has been built, and the adjoining owner is desirous of having a deeper foundation, he has a right to undermine such wall, using due care and diligence to prevent any injury to his neighbor, and having done so, he is not answerable for any consequential damages which may ensue. 17 John. R. 92; 12 Mass. 220; 2 N. H. Rep. 534. Vide 1 Dall. 346; 5 S. & R. 1.

3. When such wall exists between two buildings, belonging to different persons, and one of them takes it down with his buildings, he is required to erect another in its place in a reasonable time, and with the least inconvenience; the other owner must contribute to the expense, if the wall required repairs, but such expense will be limited to the costs of the old wall. 3 Kent, Com. 436. When the wall is taken down, it must be done with care; but it is not the duty of the person taking it down to shore up or prop the house of his neighbor, to prevent it from falling; if, however, the work be done with negligence, by which injury accrues to the neighboring house, an action will lie. 1 Moody & M. 362. Vide 4 C. & P. 161; 9 B. & C. 725; 12 Mass. R.220; 4 Paige's R. 169; 1 C. & J. 20; 1 Pick. 434; 12 Mass. 220; 2 Roll., Ab. 564; 3 B. & Ad. 874; 2 Ad. & Ell. 493 Crabb on R.P. Sec. 500. In the excellent treatise of M. Lepage, entitled "Lois des Batimens," part 1, c. 3, s. 2, art. l, will be found a very minute examination of the subject of party walls, with many cases well calculated to illustrate our law. See also Poth. Contr.de Societe, prem. app. n. 207; 2 Hill.: Ab. 119; Toull. liv. 2, t. 2, c. 3.

Party walls are not a new concept, although their common use has become a standard procedure with condominium and condominium-type ownership. For example, early 19th-century deeds in Portsmouth, New Hampshire describe the use of party walls.

Party walls as an established boundary between titles may be ancient, making yet another case for thorough research. A major case[46] in Portsmouth, New Hampshire, a very old city (incorporated 1653), demonstrates property descriptions based on party walls. See Figure 7.5.

---

[45] Wikipedia.

[46] This case was tried in both federal and state (county) courts, based on two different jurisdictional issues. The two basic issues in the case were title and boundary.

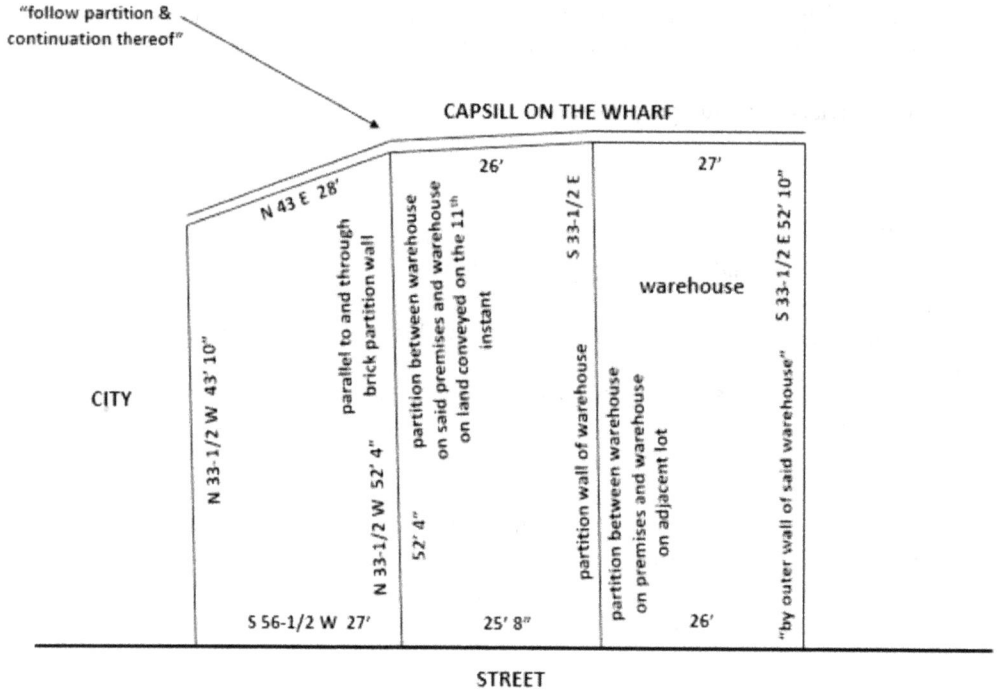

FIGURE 7.5   Three adjacent lots described in 1837 using partition (party) walls.

## OWNERSHIP AND USE OF AIRSPACE

This is more a creation of rights than a survey, although there may be descriptive definitions around airports, airfields, and landing strips.

### EASEMENTS THROUGH AIR

Communication, solar, avigation, view, pedestrian walks, pole lines

As new "green ideas" expand, use of airspace will become more of an issue. Wind farms are constructed on ridgelines and on the tops of mountains. Cell towers, for maximum coverage, are either built on high ground or extended upward high into the airspace above it.

Use of airspace, along with claims, becomes an issue in not only the above examples, but also solar energy collection, especially on an individual basis whereby panels are installed on roofs or in back yards. The erection of a tall building on an adjoining lot could easily interfere with the access to the sun, particularly at certain times of the year when the angle of sunlight is critical.

Scenic views enhance the value of real estate, sometimes significantly, to the point where, under certain circumstances, parties endeavor to protect such rights by acquiring, or designating, an easement of view. Owners with property in the mountains value their views of the landscape. Owners on water bodies value not only their access rights to the water, but the view across the same.

Access in and out of airports whereby planes take off and land on fixed runways, has long been an issue, particularly in neighborhood of airports. Three-dimensional easements, known as avigation easements, are acquired at the ends of runways where other ownerships are concerned. They are wedge-shaped with the taller end of the wedge outward from the end of the runway.

An avigation easement is an easement or right of overflight in the airspace above or in the vicinity of a particular property. It also includes the right to create such noise or other effects as may result from the lawful operation of aircraft in such airspace and the right to remove any obstructions to such overflight. Hence, avigation easement permits aircraft approaching an airport to fly at low elevations above private property. This in effect prevents the landowner's near airports from building above a set height or requires the trimming of trees. There is a real and important difference between a clearance easement and an avigation easement and that the prior existence of one does not as a matter of law preclude the possibility of inverse condemnation of the other. *Adams v. United States,* 230 Ct. Cl. 628 (Ct. Cl. 1982).

## GLIDE PATH

An instrument landing system glide path, commonly referred to as a glide path (G/P) or glide slope (G/S), is "a system of vertical guidance embodied in the instrument landing system which indicates the vertical deviation of the aircraft from its optimum path of descent," according to *Article 1.106* of the ITU Radio Regulations (ITU RR).[47]

## UNDEFINED RIGHTS IN AIRSPACE

Solar and wind access are becoming increasingly more important. Although rules, regulations, and practices are not new, they generally lack extensive guidelines and definitions. Some insight may be gained from the ancient custom known as the Doctrine of Ancient Lights. Generally defined as a doctrine of English common law giving a landowner an easement or right by prescription to the unobstructed passage of light and air from adjoining land if the landowner has had uninterrupted use of the lights for twenty years. Once a person has the right to ancient lights, the owner of the adjoining land cannot obscure them, such as by erecting a building. If the neighbor does so, he or she can be sued under a theory of nuisance, and damages could be awarded.

In 1984, voters in San Francisco passed Proposition K, which prevents construction of any building over 40 feet (12.2 m) that casts a shadow on a public park, unless the Planning Commission decides the shadow is insignificant. This proposition causes problems for a proposed 75-story building south of Market Street, which would cast a shadow on a public park ten blocks away, for one hour of the day in the fall, as well as St. Mary's Square, Justin Herman Plaza, and Union Square for significant parts of the year. Massachusetts has similar laws against the casting of shadows on Boston Common, the Public Garden, and other important public open spaces.[48]

As solar access becomes more critical, air rights in regards to sunlight will become of interest. An added consideration is the increased interest in home gardening, whereby shade from surrounding trees, sometimes on another's property, may affect the amount sunlight received, to the point where successful raising of plants may not be possible.

While state and federal laws in the United States have not supported the English doctrine, local municipal regulations and zoning laws may offer the prevention of some interferences.

## AIR RIGHTS

Air rights are the property interest in the "space" above the earth's surface. Generally speaking, owning, or renting, land or a building includes the right to use and develop the space above the land without interference by others.

This legal concept is encoded in the Latin phrase *Cuius est solum, eius est usque ad coelum et ad inferos* ("*Whoever owns the soil, it is theirs up to Heaven and down to Hell.*"), which appears

---

[47] Wikipedia.
[48] Wikipedia.

in medieval Roman law and is credited to 13th-century glossator Accursius; it was notably popu-
larized in common law in *Commentaries on the Laws of England* (1766) by William Blackstone;
see origins of phrase for details.

## NATURAL OCCURRENCES AFFECTING RESULTS

Changes in air and water currents can alter the normal results of normal and expected progres-
sion. An interesting case in point is the Fairbanks, Alaska litigation over changes in the Chena
River. Pancratz gained land attached by accretion to his upland, despite claims by the State that
the alluvion was deposited on the bed of the river, owned by the public. A subsequent bridge
constructed at the site and erosion were contemporaneous, and Pancrantz blamed the state for
altering the river current because of the placement of one of the supporting piers of the bridge.
The court in the second case found otherwise.

As tree grow and age, they naturally cast more shade, blocking more sunlight as time passes.
Natural processes, if left unchecked or not taken into consideration, can easily affect the outcome
on anticipate results.

## SUBTERRANEAN RIGHTS

Since some rights extend underground, or independently underground, they have a title and a
location. Natural features, such as caves, and artificial features such as mines and tunnels, can
be measured and mapped. Locating them is essentially a survey of an existing set of conditions,
resulting, likely, in a first survey and, perhaps, subsequent surveys.

Aquifers relied on for domestic uses and commercial uses such as mill operation are a complex
concern for title and sometimes location (see subsequent section on mill rights). Highway tunnels,
subway lines, sewer lines, underground transmission lines, and water lines all have different aspects
of title considerations. Locations, although sometimes complex, are generally straightforward.

## MINERAL RIGHTS AND INTERESTS

### MINERAL RIGHTS

Mineral rights are property rights to exploit an area for the minerals it harbors. Mineral rights can be
separate from property ownership. Mineral rights can refer to sedentary minerals that do not move
below the Earth's surface or fluid minerals such as oil or natural gas. There are three major types of
mineral property; unified estate, severed or split estate, and fractional ownership of minerals.

Ordinarily, when one thinks about mineral rights and interests, gold and silver, and perhaps
other precious metals, come to mind. Nothing could be further from the truth. All types of minerals, including sand and gravel, have value, may be mined, and sold. Oil and gas fall into this category. Minerals may be surface, underground, or under water. In the recent past, oil and gas wells
under the ocean have become numerous, such that sales and leases have been extended from the
land to the offshore waters, very much in a global sense. The very reason for the recent controversy
and boundary resolution in the Caspian Sea was because of the abundance of oil and gas resources
in the seabed, and which of five nations is entitled to what proportionate share of the bottom.

### MINERAL SURVEYS

Mineral surveys, by their very nature are three-dimensional. Excavations below the surface, may
result in materials and excess spoil deposited in piles above ground.

# 7C Special Cases
## *Water-Related Situations*

## WATER BOUNDARIES AND RIPARIAN ISSUES

There have been several texts published dealing with water boundaries and title issues concerning various categories of water. Such a complex topic and subject to many variations such that only a superficial treatment is included here as relating to original surveys.

Water is continually moving, due to natural processes of gravity and wind, as well as artificial influences of dikes, dams, levees, dredging, and other manmade structures.

Running water was characterized by Blackstone as *transient property*.[49]

## MILL SITES

### MILLS, MILL RIGHTS, AND MILL PRIVILEGES

Mills and mill sites, which number in the thousands in the United States alone, occupy a special niche in the world of surveys, boundaries, and titles. Not well understood by the average person, the law is very old, very complex, and exceedingly far-reaching.

> The application and use of flowing water to work machinery is as old as the law. Corn mills have existed from time immemorial, and it appears, from old legal authorities, that fulling and other mills worked by water for the purpose of manufacture are of a very ancient date. And it is not too much to say, that the value of actual or supposed water rights of this character throughout England may be estimated by hundreds of thousands, perhaps millions.

Getzler, *A History of Water Rights at Common Law,* quoting *Nuttall v. Bracewell,* L.R. 2 Ex. 1, 9, 1866.

### The theory of water rights in Blackstone's *Commentaries on the Law of England*[50]

Blackstone's analysis of water law is scattered throughout his commentaries, and a reassembling follows. The essence of his water doctrine can be compassed in five points:

1. Water is a type of corporeal right, a transient element to the public but subject to a qualified individual property or title during use.
2. The first appropriator of water wins title under a natural principle defending occupation.
3. Title subsists only during time of use, as water cannot be possessed or appropriated in the manner of land.

---

[49] Blackstone, William. Commentaries on the Laws of England.
[50] Sir William Blackstone, 1765–1769. Blackstone (1723–1780) was Professor of Common Law, Oxford University and an eminent, prolific English authority on common law. Common law derives from a long line of rights developed over centuries on behalf of the people. The common law is part of the underlying law used in England and the United States.

DOI: 10.1201/9781003032557-13

4. The prior appropriation theory is distinct from theories of acquisition of incorporeal rights by prescriptive long user.
5. Actions for wrongs, such as nuisance and trespass, are used to defend such water rights; and this raises the issue of identifying actionable damage.

## SELECTED COURT DECISIONS

### Definition of Mill Privilege

The right of a riparian proprietor to erect a mill on his land and to use the power furnished by the stream for the purpose of operating the mill, with due regard to the rights of other owners above and below him on the stream. **Black's Law Dictionary**, citing *Gould v. Boston Duck Co.*, 13 Gray (Mass.) 452; *Hutchinson v. Chase*, 39 Me. 511, 63 Am. Dec. 645; *Moore v. Fletcher*, 16 Me. 65, 33 Am. Dec. 633; *Whitney v. Wheeler Cotton Mills*, 151 Mass. 396, 24 N.E. 774, 7 L.R.A. 613; *Rome Ry. & Light Co. v. Loeb*, 141 Ga. 202, 80 S.E. 785, 787, Ann. Cas. 1915C, 1023.

The term "mill privilege" means the land and water used with the mill, and on which it and its appendages stand. *Moore v. Fletcher*, 16 Me.(4 Shep.) 63, 65, 33 Am. Dec. 633.

The raising of the head of water to drive a mill constitutes a mill privilege. *Pettee v. Hawes*, 30 Mass. (13 Pick.) 323–326.

The conveyance of a mill privilege will operate to convey the land occupied for the purpose, unless there be in the conveyance something indicating a different intention. *Farrar v. Cooper*, 34 Me. 394, 397.

The grant of "mill privileges" without special mention of water rights, gives a right to the actual flow of the stream, subject to its reasonable use by upper owners. *Whitney v. Wheeler Cotton Mills*, 24 N.E. 774, 775, 151 Mass. 396, 7 L.R.A. 613.

The grantee of a "mill privilege" without special mention of water takes a right to the actual flow of the stream, subject to its reasonable use by upper riparian owners. *Rome Ry. & Light Co. v. Loeb*, 80 S.E. 785, 787, 141 Ga. 202, Ann.Cas. 1915C, 1023.

A "mill privilege" is the right to the use of a water power in its existing state, and therefore where the owner of a parcel of land to which mill privileges are appurtenant, is so seised of such land, and he conveys the same, he cannot convey therewith the right of flowing lands above the lands conveyed, though such land is held by the grantor in common with another. *Hutchinson v. Chase*, 39 Me. 508, 511, 63 Am.Dec. 645.

Riparian owner in exercise of mill privilege has right to erect mill and dam on stream, where owners above and below are not injured, but must so operate dam as to let natural volume of stream pass through and to permit floating of logs. Terms "mill seat," "mill site," and "mill privilege" are synonymous, and are used interchangeably to name a location on a stream where by means of a dam a head and fall may be created to operate water wheels. *Bean v. Central Maine Power Co.*, 173 A. 498, 503, 133 Me. 9.

It is held that the term "mill privilege" in a conveyance or grant embraces the right which the law gives the owner to erect a mill thereon, and to hold up or let out the water, at the will of the occupant, for the purpose of operating the same in a reasonable or beneficial manner. It is said that it must of necessity include a reasonable amount of fall below the mill for the purpose of letting the water flow off without obstruction, and that it is not reasonable to confine the right of the mill owner to the exact amount of the fall, to the fractional part of an inch, which will enable the water to flow away from his wheels without obstruction. *Gould v. Boston Duck Co.*, 79 Mass. (13 Gray) 442, 452.

## MILL SITE

The grant of a "mill site" to one exclusively, entitled him to the exclusive use of the water power. *Mandeville v. Comstock*, 9 Mich. 536, 537.

The grant of a "mill site," etc., should be construed to include a water power, together with the right to maintain a dam wherever such dam would be suitable for the convenient and beneficial appropriation of the water power. *Stackpole v. Curtis*, 32 Me. 383, 385.

An exception of a "mill site" in a grant or lease should be construed to operate as an exception of the soil of the mill site, and so much land as is necessary for the mill pond and erecting and carrying on the business of the mill. *Hasbrouch v. Vermilyea*, N.Y., 6 Cow. 677, 681; Burr v. Mills, N.Y., 21 Wend. 290, 294.

A conveyance of a certain "mill site," with the sawmill, machinery, etc., thereon standing, "meaning to convey all the premises which A.B. purchased of C.D. by deed, dated, etc., with all the privileges and subject to all the restrictions therein expressed, reference thereto being had for a more particular description of the premises," should be construed to include the whole land under the mill, notwithstanding the land acquired by the deed to which reference was had, covered by a part of the premises on which the mill was erected. The term "mill site" embraces all the land the mill covers. *Crosby v. Bradbury*, 20 Me. (7 Shep.) 61, 65.

It is held that the term "mill site," in a conveyance or grant, embraces the right which the law gives the owner to erect a mill thereon, and to hold up or let out the water, at the will of the occupant, for the purpose of operating the same in a reasonable or beneficial manner. It is said that it must of necessity include a reasonable amount of fall below the mill for the purpose of letting the water flow off without obstruction, and that it is not reasonable to confine the right of the mill owner to the exact amount of the fall, to the fractional part of an inch, which will enable the water to flow away from his wheels without obstruction. *Occum Co. v. A. & W. Sprague Mfg. Co.*, 35 Conn. 496, 512.

## MEANDER LINES

Where the call is a meanderable object, such as a river, *mesa*, mountain, or arroyo, it should be meandered, to the extent that it constitutes such boundary. *Stoneroad v. Stoneroad,* 4 N.M. 181, 12 P. 736 (N.M. 1887).

*Quoting from the New Mexico court,* "It is conceded by both parties that, where lands are to be surveyed by our government, the rules adopted by the interior department in surveys of grants like the one in question, where boundary calls are given, is to draw a straight line north and south, or east and west, through such point, on the side of the tract of which it constitutes such boundary, to the intersection of the boundaries on the other sides; and where the call is a meanderable object, such as a river, *mesa*, mountain, or arroyo, it should be meandered, to the extent that it constitutes such boundary, within the projected lines of the other sides. See *Ortez Mine Grant*, 2 Copp, Pub. Land Laws, 1276; *U.S. v. Soto*, 1 Hoff. L. Cas. 68; Tyler, L. Bound. 29, 187, 188."

While the word *meander* brings readily to mind meandering water boundaries in the PLSS, the word may also have other meanings. For example, in the Kentucky case of *Rowe v. Kidd*,[51] the court spoke of the surveyor meandering a riverbank, a cliff, and the river itself. While he did not follow a set of instructions as if in the PLSS, he did not extend his survey beyond these features, running his traverse lines not along, but back from, the features mentioned. This was not an unusual practice with certain features, some of which were boundaries, and likely such practice gave rise to the accepted rules for riparian lands in the PLSS.

Strictly speaking, a *meander* is defined as a turn or winding, as of as stream; hence, a winding path or course; a labyrinth.[52]

---

[51] 249 F. 882 (E.D.Ky. 1916)
[52] Webster's New Collegiate Dictionary.

To meander means to follow a winding or flexuous course, and when it is said, in a description of land, "thence with the meander of the river," it must mean a meandered line – a line which follows the sinuosities of the river – or, in other words, that the river is the boundary between the points indicated.[53] This term is used in some jurisdictions with the meaning of surveying and mapping a stream according to its meanderings, or windings and turnings.[54]

## MEANDER LINES IN THE PLSS

### DEFINITION

A meander line is the traverse run at the line of mean highwater of a permanent natural body of water. In original surveys, meander lines are not run as boundary lines. They are run to generally define the sinuosities of the bank or shoreline and for determining the quantity of land in the fractional sections remaining after segregation of the water area. *Glossaries of BLM Surveying and Mapping Terms.*

While a meander line is not a boundary, it may tie into or be somehow connected to a boundary, and therefore lend some insight to the original location(s) of the boundary(ies). It is part of the survey, although not intended to define a boundary, or the boundary of any designated tract of land. Being part of a survey, it is therefore original, and should be retraced, as nearly as possible, according to the same rules for identification and retracement of original surveys.

Being a meander line and near to the water, it is subject to the forces of nature, and therefore a subsequent locator may find some, if not all, of the original evidence now missing, or disturbed. Partial recovery, or a failure of recovery, will result in additional problems that likely will need to be addressed on a case by case basis. In extreme cases, meander lines once near water but on dry line may now be completely submerged, presently further difficulty.

### MEANDER CORNER

A corner marking the intersection of a township or section boundary and the mean high-water line of a body of water. Also a corner on a meander line. (ASCE Definitions.)

### HOW SURVEYED, AND WHERE CORNERS WERE PLACED

There is considerable variation within meander lines, depending on the individual surveyor running the line, and the conditions under which he was working (e.g., water level at the time and the character of the topography). Interpretations may have to be made on a case-by-case basis, but some court decisions may provide guidance.

### MEANDER POSTS

Often posts were set along meander lines, and noted and called out by number.

Meander corner monuments are to consist of the regulation posts used for the monumentation of the public land surveys.[55] Meander corners are marked "MC" on the half toward the meanderable body of water.[56]

---

[53] Black's Law Dictionary, citing *Turner v. Parker*, 14 Or. 341, 12 P. 495; *Schurmeier v. St. Paul & P. R. Co.*, 10 Minn. 100, 88 Am. Dec. 59.
[54] Ibid. See *Jones v. Pettibone*, 2 Wis. 317.
[55] Manual of Surveying Instructions, Sec. 8-11.
[56] Manual of Surveying Instructions, Sec. 4-39.

## VARIATION AMONG STATES

In the Michigan decision of *Pigorsch v. Fahner*,[57] the court had these words:

First. A meander line is simply a surveyor's line. When used in connection with inland lakes, it refers to a line run by the government land office surveys, upon which titles to Michigan real estate were originally patented to individual owners by the United States.

Original surveys contained meander lines around most of the larger inland lakes, especially where section lines would have been under the water. The meander line was largely a convenience for the surveyor; it did not represent a classification of the lakes. Some larger lakes were not meandered, especially where they did not touch upon the section lines. Some smaller lakes were meandered because section lines could not be run otherwise.

The meander line generally represented the shoreline. It was not necessarily exact. While some states have held that title to the lands within the meander line remained in the United States, and upon statehood vested in the state, Michigan does not regard a meander line as a boundary. We have held that meander lines were merely a means of computing the acreage in the uplands – no charge being made by the government for subaqueous lands – and under our cases, the littoral owner has been held to take title to the land under the lake. The logic of the Michigan rule leaves something to be desired. If the patentee did not pay for the subaqueous lands, there is no reason to assume he took title to them.

If the meander line was merely a means of computing useable upland acreage, it would follow that the purchaser of government lands upon which a non-meandered lake stood, was paying for unusable acreage.

Whatever the weakness of our rule, it is too well settled at this point in time to question that littoral or riparian owners take title to subaqueous lands; the only distinction between meandered and non-meandered lakes being that in the former, the riparian owners take title to the center of the lake along pie-shaped extensions of their shoreline boundaries – while in the latter case, the riparian owner takes title only to the bottom land lying within the description conveyed to him.

Lots patented under the public land laws according to a plat showing them bordering on a lake extend to the water as a boundary and embrace pieces of land found between it and the meander line of the survey where the failure to include such pieces within the meander was not due to fraud or mistake, but was consistent with a reasonably accurate survey, considering the areas included and excluded, the difficulty of surveying them when the survey was made, and their value at that time. P. 664.

A Nebraska decision, *Bissell v. Fletcher*,[58] was an action of ejectment, brought by the plaintiff against the defendant to recover possession of certain land which the plaintiff claimed belonged to lot 3, sec. 31, T. 2 N., R. 18 W., in Harlan county. On the trial of the cause, the court below found the issues in favor of the defendant and dismissed the action.

There was no dispute about the essential facts in the case, and they were as follows: the plaintiff is the owner of lot 3 in sec. 31, T. 2 N., R. 18 W., and is entitled to and is in possession of the same except as hereinafter stated. This lot is shown by the plat and patent introduced in evidence; contains 52 60/100 acres. The plat shows that the south-west corner of the lot extends to the north channel of the Republican river, and there is some testimony tending to show a meander line between the river and the south line of said tract. The defendant is the owner of lots 6 and 7 in said section and was the owner thereof at the commencement of this action. These lots lie south of lot 3, and in fact between it and the river as it flows at present. The contention of the plaintiff is that lot 3 extends to the river; and notwithstanding the fact that lot 3 contains all the land the

[57] 386 Mich. 508; 194 N.W.2d 393 (1972).
[58] 19 Neb. 725, 28 N.W. 303 (Neb. 1886).

plaintiff purchased and paid for, and the effect of the extension of the line would be to give him about 117 acres of land to which he seems to have no equitable right, still he contends the law declares the land to be his. It is also shown that lots 4 and 5 in said section, if the plaintiff is entitled to recover, would also be extended to the river and absorb a considerable portion of the tract that the plaintiff claims. There is a paucity of testimony as to the character of the channel of the river along which the meander lines were run; whether in fact the river flowed there at the time the original surveys were made in 1865 and 1869 is not proved. The testimony tends to show that the river at the present time flows about three-quarters of a mile south of lot 3. As to when this change of the channel took place, there is no proof. It is not claimed, however, nor is there any proof that lots 6 and 7 are an accretion to lot 3. The case therefore, is similar in most respects to that of *Lammers v. Nissen*, 4 Neb. 245. In that case this court, GANTT, J. delivering the opinion of the court, said (page 251): "The mere fact that it is run and is designated upon the plats as a meandered line, certainly cannot be conclusive in the matter. To establish the doctrine that such meander line is conclusive, would estop the government from disposing of lands left unsurveyed between such line and the bank of the stream, and would prevent the correction of mistakes made by surveyors in such case, and would be in direct conflict with the well settled rule of law defining what is an accretion to land." This, we think, states the law correctly. The rule seems to be that if, when the entry of public land is made, the bank of the river at an ordinary stage of water was in fact where the meander line was represented by the survey, and land has since been formed by accretion, it will become the land of the person who has title to the land immediately behind it. *New Orleans v. United States*, 10 Peters 717. *Granger v. Swarts*, 1 Woolw. C.C.R. 91. But the plaintiff does not claim the land in controversy as an accretion, and has no right or title to lots 6 and 7 as a part of lot 3. The judgment of the district court is clearly right and is affirmed.

An interesting federal decision offering guidance is as follows:

**274 F. 290 (5th Cir. 1921)**
**LANE et al.**
**v.**
**UNITED STATES and four other cases.**
**Nos. 3536-3540.**
**United States Court of Appeals, Fifth Circuit.**
**June 22, 1921**
BRYAN, Circuit Judge.

"These suits were brought by the United States to recover several small tracts of land in Caddo parish, La., and also to compel an accounting for oil and gas produced therefrom by appellants. The lands lie in separate tracts along the edge of Ferry Lake, in township 20 north of range 16 west, Louisiana meridian. This township was first surveyed in 1839 by United States Deputy Surveyor A. W. Warren, and the plat filed by him represents several fractional sections bordering on the lake. According to that plat the lands in litigation were included within these fractional sections.

Because of the discovery of oil thereon the township was resurveyed in 1917, under the direction of Arthur D. Kidder, Supervisor of Surveys for the General Land Office. By this later survey the lands in litigation are represented as lying between the meander lines of Warren's survey and Ferry Lake, and are designated as new subdivisions omitted from Warren's survey. The fractional sections surrounding the lake had been sold and patented from time to time, and acquired by appellants long before the resurvey of 1917 was made.

There are no disputed questions of fact. It was admitted by stipulation that Kidder's survey was accurately made, and that Kidder correctly re-established and reproduced Warren's survey; that the contour line of Ferry Lake, at mean high water, in 1812, when Louisiana was admitted into the Union, in 1839, when Warren made his survey, and in 1917, when Kidder made his survey, was 173.09 feet above mean Gulf level; that all the tracts of land in litigation lie above that elevation, and are upland in character; and that Ferry Lake is, and was in 1812 and 1839, a navigable body of water. It was also shown beyond question that Warren in his survey ran the section lines to, and accurately established corners and set meander posts on, the contour line of Ferry Lake.

In running out his meander lines, Warren did not always follow the exact contour of the lake, but he approximated it very closely in so far as the lands involved in these suits are concerned. In some instances, irregular areas of land which jut out into the water are outside the meander lines, and in other instances, where the land sharply recedes, water is inside the meander lines. Very little difference in acreage between the land so excluded and the water so included is shown by the testimony. In at least one of the cases the area of water included exceeds the area of land excluded.

In the first mentioned of these suits Warren's survey represents the land in that fractional quarter quarter section as containing 26.80 acres; Kidder's survey adds as a new subdivision 5.67 acres. In the second suit, Warren's survey represents 114.80 acres; Kidder's survey adds 11.49 acres. In the third suit, Warren's survey represents 155 acres; Kidder's survey adds 27.87 acres. In the fourth suit, Warren's survey represents 23 acres; Kidder's survey adds 12.72 acres. In the fifth and last suit, Warren's survey represents 83.75 acres; Kidder's survey adds 22.71 acres. It is for these additional acres the government sues.

The cases were consolidated, and the court entered decrees sustaining the right of the government to recover the lands, and also holding appellants liable to account for various amounts realized from the sale of oil and gas, but allowing credit for the costs of production. The government by cross-appeal in the first, second, third, and fourth suits contends that appellants should not be allowed to deduct these expenses, but should be held liable as willful trespassers. In the fifth suit, that of *United States v. La Robadierre and the Gulf Refining Company, No.* 3540, the original defendants have not appealed, but the government by direct appeal assigns as error the deduction of expenses of operation and the failure to hold the Gulf Refining Company liable in solido with the other defendant.

It is the contention of the United States that the lands in litigation, lying between the meander lines of Warren's survey and Ferry Lake, are public lands which were never surveyed until 1917. Appellants, on the other hand, assert title as owners of the fractional sections.

Section 2396 of the Revised Statutes (Comp. St. Sec. 4804) deals with the boundaries and contents of public lands, and provides that boundary lines shall be ascertained by running straight lines from established corners to opposite corresponding corners, except that, in those portions of a fractional township where corresponding corners cannot be fixed, the boundary lines shall be ascertained by running from established corners 'to the water course, Indian boundary line, or other external boundary of such fractional township.' There is no other provision of law for the running of meander lines along a water course, but they are usually run by surveyors for the purpose of computing acreage. It is proper that they should show the sinuosities of the banks of the water course; for the water course, and not the meander line, is the true boundary. *St. Paul & P. Railroad Co. v. Schurmeier, 7 Wall.* 272, 19 L.Ed. 74; *Hardin v. Jordan,* 140 U.S. 371, 11 Sup.Ct. 808, 838, 35 L.Ed. 428; *Mitchell v. Smale,* 140 U.S. 406, 11 Sup.Ct. 819, 840, 35 L.Ed. 442; *Producers' Oil Co. v. Hanzen,* 238 U.S. 325, 35 Sup.Ct. 755, 59 L.Ed. 1330.

It was Warren's duty to bound the fractional sections on the lake, and it is apparent from his survey and field notes that it was his intention to do so. The survey is not invalidated by the failure to include within the meander lines small, irregular areas of land. In Mitchell v. Smale, supra, the survey called for 4.53 acres, but it was held that title to 25 additional acres also passed by the patent. In that case the Supreme Court said:

"The difficulty of following the edge or margin of such projections, and all the various sinuosities of the water line, is the very occasion and cause of running the meander line, which by its exclusions and inclusions of such irregularities of contour produces an average result closely approximating to the truth as to the quantity of upland contained in the fractional lots bordering on the lake or stream. The official plat made from such survey does not show the meander line, but shows the general form of the lake deduced therefrom, and the surrounding fractional lots adjoining and bordering on the same. The patents when issued refer to this plat for identification of the lots conveyed, and are equivalent to and have the legal effect of a declaration that they extend to and are bounded by the lake or stream. Such lake or stream itself, as a natural object or monument, is virtually and truly one of the calls of the description or boundary of the premises conveyed, and all the legal consequences of such a boundary, in the matter of riparian rights and title to land under water, regularly follow."

Meander lines, and not water courses, are held to be boundaries, where there is no body of water within a reasonable distance therefrom, as in *Horne v. Smith,* 159 U.S. 40, 15 Sup. Ct. 988, 40 L.Ed. 68; Producers' Oil Co. v. Hanzen, supra; or where there is no body of water at all, as in *French-Glenn Live Stock Co. v. Springer,* 185 U.S. 47, 22 Sup.Ct. 563, 46 L.Ed. 800; *Gauthier v. Morrison,* 232 U.S. 452, 34 Sup.Ct. 384, 58 L.Ed. 680; *Chapman & Dewey Lumber Co. v. St. Francis Levee District,* 232 U.S. 186, 34 Sup.Ct. 297, 58 L.Ed. 564; and *Lee Wilson & Co. v. United States,* 245 U.S. 24, 38 Sup.Ct. 21, 62 L.Ed. 128; or where there is gross fraud, as in *Security Land & Exploration Co. v. Burns,* 193 U.S. 167, 24 Sup.Ct. 425, 48 L.Ed. 662. These cases are relied upon by the government, but in our opinion they are inapplicable to the facts under consideration. In none of the cases cited on the government's briefs is the rule stated in Mitchell v. Smale overruled or even modified; but, on the contrary, in several of them that case is cited with approval.

Here the water course was in existence and was actually meandered. There is no suspicion or suggestion that Warren's survey of the lands in litigation was fraudulent. That part of his survey which is in question in these cases was made with a fair degree of accuracy, and for the purposes of computing the acreage and of showing the sinuosities of Ferry Lake as a boundary. It is not impeached by the fact that a subsequent survey, made with great minuteness because of the discovery of oil, disclosed comparatively slight variations from the established boundary.

There being no appeal by the original defendants in *United States v. La Robadierre et al., No.* 3540, the decree in that case is affirmed. In the other cases, the decrees are reversed on the original appeals, with directions to dismiss the bills of complaint, and the cross-appeals are dismissed.

### LINE FIXED BY REFERENCE TO MEANDER CALL

The Texas decision of *Maddox v. Dayton Lumber Co.*[59] demonstrates that meandering bodies of water, survey locations and meander calls are not restricted to the PLSS. This case dealt with meander calls, rather than meander lines or meander corners. See Figure 7.6.

While not an example of a meander line, it is nevertheless related since it deals with meanders, per se. Mrs. Maddox instituted an action against the Dayton Lumber Company for recovery of title and possession to, and remove cloud from, 1,778.7 acres of land described as Shattuc surveys 62, 63, and 64 in Liberty County, Texas; and also to establish the true boundary lines existing between the said Shattuc surveys and the Martinez leagues 6 and 9 on the east, and between the Shattuc surveys and I & G.N. surveys 38 and 39 on the west.

The Miller league referred to in the Martinez field notes was surveyed prior to the location of the Martinez grants.

---

[59] 188 S.W. 958 (Tex.Civ.App., 1916).

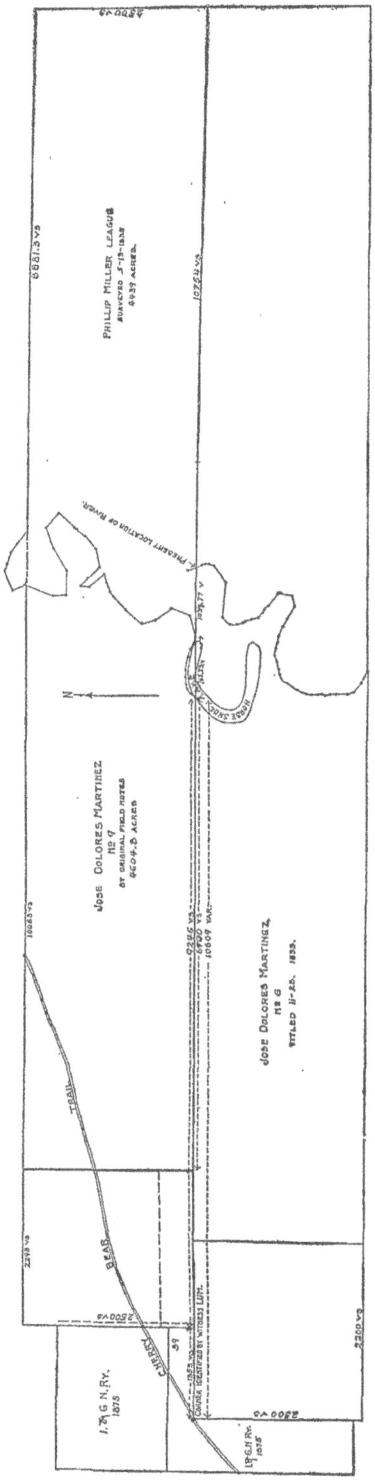

**FIGURE 7.6(a)** Diagram from *Maddox v. Dayton Lumber Co.*

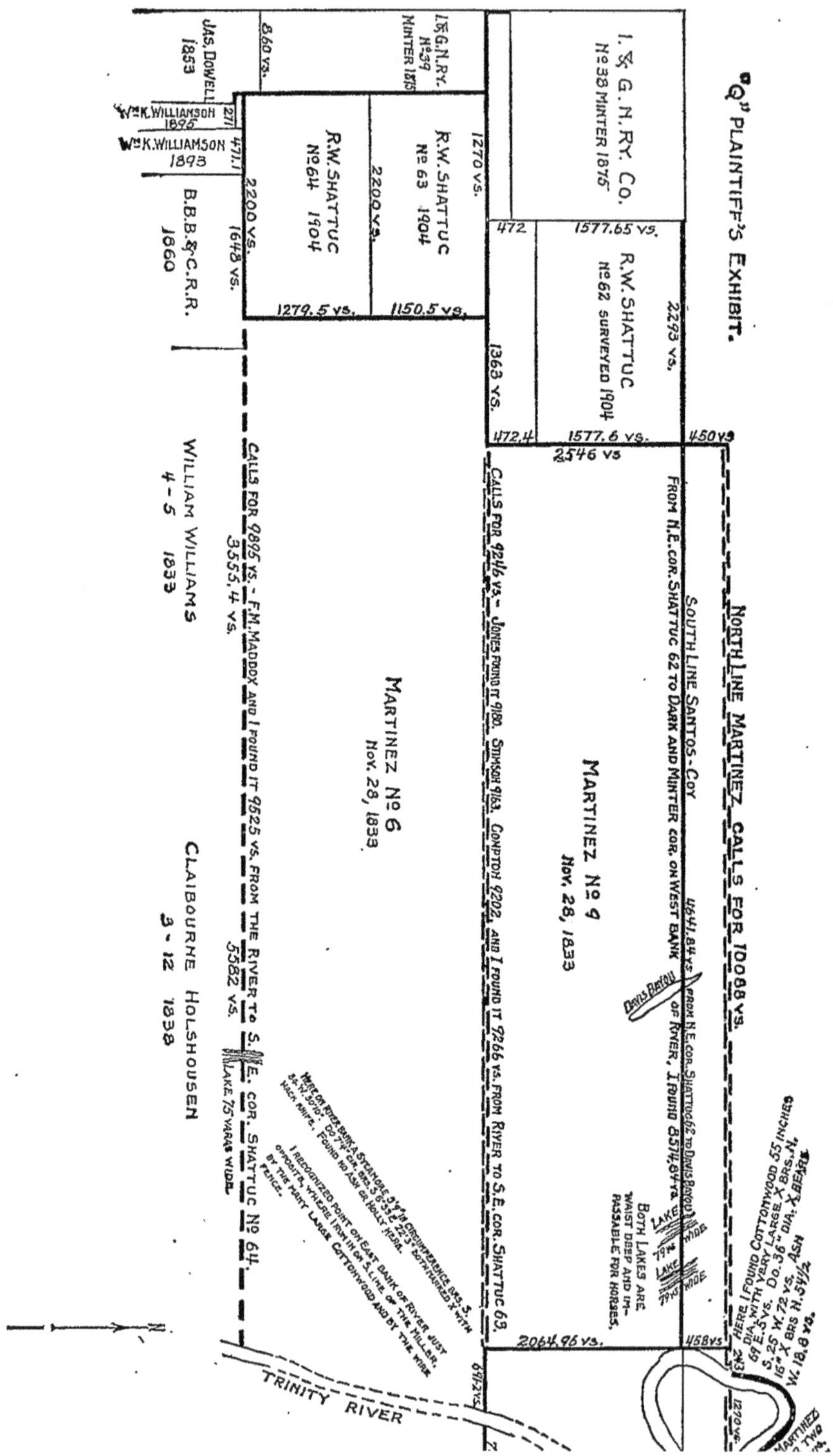

**FIGURE 7.6(b)** Diagram from *Maddox v. Dayton Lumber Co.*

Martinez leagues 6 and 9 were titled in 1833 to Jose Dolores Martinez. Both parcels were surveyed and their field notes filed in the General Land Office. I. & G.N. surveys 38 and 39 were located in 1875, and there is no controversy as to their location on the ground. The east line of the two surveys and the west line of the two Martinez surveys 6 and 9 are called for in the field notes of the two I. & G.N. surveys as their common boundary.

In 1904 a deputy survey for Liberty County, acting on the assumption that he had discovered a vacancy between the west lines of I. & G.N. surveys 38 ad 39 surveyed and located the Shattuc surveys 62, 63, and 64, placing them in the alleged vacant area between said surveys, and plats and field notes of said Shattuc surveys were filed, and the land included there was held to be "scrap land" by the state and put on the market for sale. Both land and the timber were then sold by the state.

After hearing the evidence, the lower court fixed the boundaries of the two Martinez grants, so as to include all the land constituting the Shattuc surveys, and the appellants appealed.

The field notes indicate that the western boundary line of the Miller grant calls for the Trinity River and gives the meanders of that river in variations, degrees, and minutes, in its extension from the inner northwest corner to the southwest corner of said league. The present location of the river (1916), with its meanders, as it flows across the north and south lines of the miller, does not now fit the meander calls for the river made in 1833, when the Miller west line was surveyed. The Trinity river is now at a point on the south line of the Miller, 1,039.77 varas east from where it calls to be in the original field notes at the southwest corner; in other words, if the field notes for the west line of the Miller were ignored, and the southwest corner located according to its call on the Trinity river where it now is, it will be 1,039.77 varas east of where it was when the survey was made in 1833.

The appellants contend that the southwest corner of the Miller must be placed at the intersection of the south line of said league with the river as it is now located, and in projecting the south line of Martinez grant No. 9, the distance of 9,246 varas must be measured, they said, from that point. On the other hand, it is contended by appellees that the course of the river has changed since the original survey of the Miller in 1833, and that what is now known as "Horseshoe Lake" on the south line of Martinez No. 9, was then part of the Trinity river, and that in measuring the distance from the south line call of said league, it must be projected from a point on the west side or bank of said lake; that point, at the time survey was made, being, as they assert, the west side of the river as called for in the field notes.

There was much evidence introduced upon this subject. Some of the surveyors testified that they found the bed of an old river, that it is now a slough, that it is plainly visible on the ground, that its meanders practically fit the meander calls for the river made in the Miller field notes in 1833, and that said old river bed or channel was contiguous with and included Horseshoe Lake. We have carefully examined the evidence in the record on this issue, and find that there is just and ample warrant for the conclusion that the channel of the river has changed since 1833, and that the Miller field notes, giving the meanders of the river, correctly located the channel of the river at that time, and that Horseshoe Lake was therefore then a part of the channel of said river. The following maps are made a part of this opinion, and reference is hereby made to the same for elucidation of the points at issue and the conclusions of the court:

It is to be observed that the west line calls of the Miller field notes places all of the land *east* of the river in the Miller grant, and that the river forms a part of the north, as well as the west boundary line of said survey-winding around the west and a part of the north sides of said grant; also that a line projected due west from the most easterly intersection of the Trinity river with the well-marked north line of the Miller grant, as found on the ground, extends a distance of 1,270 varas before it would again meet the furthermost west bank of said river, and all of the land adjacent to said line until it strikes the bank of said river is included, according to the field notes, in the Miller grant.

It is contended by appellants that the northwest corner of the Miller is at a point where the north line of the Miller, as now found on the ground, first intersects the Trinity river from the east. While on the other hand, appellees contend, and the trial court impliedly found in fixing the northeast corner of Martinez No. 9 as it did, that the northwest corner of the Miller, as it affects the beginning, or the northeast corner of Martinez No. 9, was at a point 1,270 varas further west from the point contended for by appellants, as such northwest corner of the Miller.

It is a cardinal principle of law that in determining the boundaries of conflicting surveys, that the footsteps of the surveyor are the determining factor, and where such footsteps as called for in the field notes are found and identified they must control. *Taft v. Ward,* 58 Tex.Civ.App. 259, 124 S.W. 437; *Byrd v. Langbein,* 135 S.W. 206.

The footsteps of the surveyor can be traced in fixing the boundaries of the Philip Miller, and the west line of that survey must therefore be constructed according to its field notes, and in accordance with the meander calls of the Trinity river. To accept the contention of appellants in placing the northwest corner of the Miller at the point they claim it should be would ignore the principle herein announced, and do actual violence, besides, to the field notes of that survey, as it would exclude territory which, beyond all doubt, is included in the Miller field notes.

The fact that the meander calls of the Philip Miller do not now balance-that is, that they do not close by an excess of westings of 114 varas in reaching the south line in running the same from the northwest to the southwest corner-is of no material importance. Such discrepancies are not unusual in making long-distance surveys. According to the testimony of one experienced surveyor, it was his observation that when an old survey has more than four sides, it cannot be made to close, especially along a river bank. The discrepancy is due, according to the testimony of such experienced surveyor, to the difference in accuracy in measuring with a chain or a tapeline. The chain always runs excessive to the tapeline measurement. These meanders, together with the old north and south well-marked lines of the Miller, are sufficient to support the conclusion of the court in placing the western and a part of the northern boundaries of the Miller as herein determined.

The field notes of the Martinez No. 9 do not give the meanders of the river on its east line. But the east line of Martinez No. 9 and the west line of the Philip Miller being common lines, and the Miller being the older survey, the meanders of the Trinity river, as specifically given in the latter survey, will control the contour of the Martinez east line.

The record is practically conclusive that none of the original witness or bearing trees called for in the field notes of Martinez No. 6 or 9 can now be located on the ground. The witness Jones, who located the Shattuc surveys, reported in his field notes, and also testified on the trial of the cause, that he located the witness trees at the northwest corner of Martinez No. 9, and also at the southwest corner of Martinez No. 9 *at the places he contends said corners should be.*

Every other witness, including surveyors who checked the work of the witness Jones, witnesses both for the appellants and appellees say that they could find no such original bearing trees as Jones said he found, and again, on cross-examination, the witness Jones admitted that he made the Shattuc surveys and located the vacancy for himself. He says:

> When I found out there was a vacancy in there, I got a man to settle on it, and apply for it. I did not get the land myself. *** I was to get my part of it, of course. Jett was to get all of the land; he and I together and Shattuc was to get 160 acres. I was acting as deputy surveyor at the time. *** I know that under the law a surveyor cannot become interested in any lands that is located, but I don't know any in the state that hasn't got land in that way. I know that it is against the law, but we do lots of things against the law.

If the witness trees be located at the points claimed by the surveyor, Jones, and the north and south lines of Martinez No. 9 be allowed their full distance calls, there will necessarily be a conflict between the boundaries of the Philip Miller and the said Martinez. This evidence is so inconsistent and totally at odds with all of the other testimony in the record that we feel justified in ignoring it, as was done by the trial court and jury.

We do not believe that the lines of Martinez 6 and 9 were ever run on the ground by the original surveyors; and the presumption that they were actually run, in our opinion, is overcome by the following evidence: (1) the field notes of neither one of said grants designates any sort of an identification mark on the witness trees placed at the corners of said grant. (2) No object called for in the field notes has ever been identified, except the Trinity river, in front of the Miller league. (3) The actual meanders of the Philip Miller places the river in a radically different position from that shown by the original plat of Martinez No. 9 filed in the General Land Office by the surveyor. (4) No original corners, bearing trees, line, or line trees of 6 or 9 have ever been identified, or located anywhere at any time. This is a most singular situation, since all of the lines of the two Martinez grants 6 and 9 run through a deeply forested country, and the north and south lines are from 4 to 5 miles in length. (5) The field notes of 6 and 9 both show that the surveyor did not meander the river. The calls therein for the river are: "Following the turns of the river upward." (6) Except the Trinity river, no natural object is called for, although the south line of No. 6 crosses Gaylor's creek within 86 varas of the "L" corner of the Dowell, crosses Gaylor's creek seven times on the north line of the William Williams, and on the north boundary line of the Claiborne Holshousen crosses a deep lake 75 to 160 varas wide, and also crosses two creeks 9 varas wide before reaching the river. (7) The common lines of 6 and 9 cross Davis bayou 30 varas wide three times. (8) The north line of Martinez No. 9 crosses a pond 150 varas wide, two boggy lakes 70 and 79 varas wide, waist deep, and impassable for horses, and then crosses Davis bayou 30 varas wide, and still no natural object except the Trinity river is anywhere mentioned in either of the two Martinez grants, or shown on the original plat filed by the surveyor in the General Land Office.

A landowner witness testified to the location of corners and the common reputation of the lands in the community. The court stated, inasmuch as appellants' assignments of error 1 to 6 attack the admissibility of such evidence, it may be appropriate to here consider the rules governing the admissibility of evidence of this character.

In the evolution of practice and procedure through the long grind of the centuries, the character of the jury changed from one which had personal knowledge before the trial of the matters in controversy, or which acquired information through its own initiative by conversing with the neighbors who were thought to have knowledge of such matters, to one which knew nothing about the issues in controversy, or whose mind was free from opinions on questions about to be submitted for its determination, and whose source of information must be confined exclusively to the narration of facts by witnesses called to testify in their presence in open court. Under the latter conditions the rules of evidence, as we understand them today, had their origin. Cross-examination was soon recognized as such a vital test of the accuracy of statements that the courts eventually came to the adoption of the hearsay rule. However, one of the several exceptions to that doctrine allows the introduction of general reputation, which in its nature is still a traditional resort in a modified sense, to common repute as a source of knowledge.

This exception to the hearsay rule is based upon the principle of *necessity* and the principle of *circumstantial guaranty of trustworthiness.*

As applied to boundary cases the *necessity* is to be found in the general dearth of other satisfactory evidence of the desired facts, the matter usually being an ancient one, and no living witnesses found; and the *circumstances creating a guaranty of trustworthiness* are found when the topic is such that the facts are likely to have been generally inquired about, and of sufficient importance to have become the subject of neighborhood discussion, and discussed by persons having personal knowledge of the facts involved, and thus, it is presumed, that the community's conclusion, if it has been so founded, is likely to be worthy of belief and trustworthy in character.

In discussing this kind of evidence, Justice Coke, in the case of *Stroud v. Springfield,* 28 Tex. 668, says:

> The admission of evidence of common reputation as to old boundaries, which frequently cannot possibly be proved by direct and positive testimony, is based on the extreme probability of the truth of a fact received, assented to, and acted on as true by the common consent of a community having peculiar means for correct information, and no interest to warp their judgment in forming a conclusion. In the absence of direct and positive testimony which, when they are ancient, cannot usually be had to establish boundaries, common reputation is perhaps as little liable to error as any other species of evidence that can be resorted to for the purpose, and, indeed, is frequently the only resort. The general rule is undoubted, that common reputation is admissible as evidence in questions of boundary, but there is much diversity of opinion as to its proper application. The unrestricted admission of this species of evidence would be fraught with the most dangerous tendencies, and violative of the best dictates of experience. The admissibility, as well as the value and weight, of general reputation must, from its nature, depend very much upon the circumstances of the case in which it is offered. It cannot, of course, be received as to title. It is admissible only as to the locus in quo of the boundary; a fact of which the community or neighborhood around it is supposed to be peculiarly well informed. The boundary must be an ancient one, and its supposed locality must be of sufficient interest and note in the neighborhood or community to have been the subject of observation and conversation among the people. *** There, weight of opinion or neighborhood report is not common reputation. The reputation or understanding must have been formed and in existence before the controversy commenced in which it is used as evidence. Men are not presumed to be indifferent in regard to matters in actual controversy, for when the contest has begun, people generally take one side or the other, and, if they are disposed to speak the truth, facts are or may be seen by them through a false medium. *** For this reason it is necessary that proof of common reputation must have reference to a time ante litem motam. The question propounded to the witness Copps ('whether or not the location of the Powell league as represented on the map had been uniformly and generally regarded and accredited as the true location by the community around it'), is general, and not limited even to the filing of the suit. It is not limited to any particular time. *** The court ruled upon the question as propounded, and the ruling is correct. Hearsay evidence is generally inadmissible. Desiring to introduce evidence under an exception to that general rule, it devolved upon the appellant to bring himself within the limits of the exception, which he failed to do.

Evidence of this character must be general, concurrent, and certain as to the subject-matter. *Matthews v. Thatcher,* 33 Tex.Civ.App. 133, 76 S.W. 64. It must be reputation, and not individual assertion. Wigmore on Evidence, § 1584; *Russell v. Hunnicutt,* 70 Tex. 660, 8 S.W. 500.

An analysis of the witness Lumm's testimony shows that it does not, at least, comply with two of the requisites hereinabove stated, viz.: first: it was not general in its character. The cross-examination shows that the conversation with his father-in-law in the woods was the first and only time he ever heard the matter mentioned, and that he never heard anybody else say anything about it. That "people didn't talk land matters then, as the land was not worth anything, and they didn't pay any attention to it." Second: it is not certain as to the subject-matter, and therefore lacks the element of trustworthiness required by the rule, in that he did not know where the corner was to which his father-in-law referred, and he frankly admits that he cannot say whether the reputation was before or after 1875. (It being conceded by all of the parties that the corners and lines now found on the ground of the Martinez grants were put there in that year.) In our opinion, this evidence was not admissible, or, if admitted, it is of such a flimsy character as to wholly lack sufficient force to overcome the other controverting facts on this issue, and we hardly feel justified in establishing it as a precedent for the location of boundary lines, or corners, so as to overcome the dignity of a call for distance.

The field notes of the junior surveys introduced in evidence by appellees, wherein calls are made for the lines and corners of the Martinez surveys, cannot be used to establish general reputation that the lines and corners of the latter surveys were at the places appellees contend for, for the reason that they, too, lack the element of trustworthiness essential to that kind of evidence, and are not concurrent with the date of the Martinez grants. It is not shown that such surveyors making the junior surveys had knowledge of the actual location of the Martinez surveys. The evidence in question does not come within the rule announced in the case of Reeves v. Roberts, 62 Tex. 552, where the surveyor making the declaration was shown to have actually seen the old marked corner, and to have found the monument at the disputed point, and have used such point on numerous occasions to make other surveys.

The Dowell survey was made 20 years after the Martinez surveys, and calls for the bearing trees as being at the southwest corner of Martinez No. 6, which do not correspond to either the size, mark, or direction of any of the trees contained in the Martinez field notes and designated by such field notes as being in that corner. The decisions of our courts have settled the proposition that lines and boundaries cannot be construed with reference to objects that may be found upon the ground, as indicating the footsteps of the surveyor, when there are no calls in the grant for such objects. *Railway Co. v. Anderson,* 36 Tex.Civ.App. 121, 81 S.W. 781.

The I. & G. N. surveys were not made until 42 years after the location of the Martinez grants, and concededly at a time when the west line of the two Martinez surveys were marked on the ground. The calls contained in such surveys for the Martinez lines and corners cannot meet the requisites of the rule of general reputation as hereinabove stated. Such surveys are certainly not concurrent with the establishment of the boundaries of the Martinez grants. This can also be said of the other junior surveys introduced in evidence. *Linney v. Wood,* 66 Tex. 22, 17 S.W. 244.

The admission of the junior surveys generally in evidence in boundary suits is fully discussed and the correct rules announced in the former opinion in this cause. 159 S.W. 391.

In justice to the trial court, it may be said that the junior surveys were admissible for the purpose of showing the general location of such surveys on the ground, and we think the instructions which the court gave to the jury when such evidence was admitted, to the effect that the junior surveys calling for the Martinez lines "were to be considered by the jury for all purposes, except that you will not consider the bald declaration as to the location of the line," was sufficient to limit the legitimate purpose of such evidence, and was so understood by the jury. After giving the jury such verbal instructions, it was not necessary to again cover the same matter in the court's written charge.

This disposes of assignments of error 1 to 6, and No. 19. The other several assignments of error have been considered, and, finding no merit in them, they are overruled.

It follows from what has been said that the Martinez grants were not surveyed on the ground. There is no marked west line or west line corners or other marked lines or corners on the ground which have been found or identified as having been made in 1833 by which appellees can overcome the field note calls for course and distance, so as to extend the lines and corners of said grants as claimed by them. These grants were "paper" or "office" surveys only, and their lines must be constructed by the calls for course and distance from the Trinity river, by the calls for the Trinity river, and the west line calls of the Philip Miller, and by so doing we think they can be fully identified and located on the ground.

Article 29 of decree 190 of the Congress of the state of Coahuila and Texas, dated in the city of Leona Viscaria April 28, 1832, among other things, provided:

> The surveys of vacant lands that shall be made upon the borders of any river, running rivulet, or creek, shall not exceed one-fourth of the depth of the land granted, should the land permit. Gammell's Laws of Texas, p. 302.

This decree was in force when the two Martinez grants were made, and it is evident from the shape of the surveys and the length of their north and south lines that the surveyor was endeavoring to conform to this degree.

Looking to the field notes of survey No. 9, we find that it is described as being "on the west bank of the Trinity river and in front of the league occupied by Philip Miller."

Construing this language, the Court of Civil Appeals for the Third Supreme Judicial District, in the former opinion, said:

> Looking to the field notes of said surveys, we find that survey No. 9 is described as being in front of the league occupied by Philip Miller. Its lower corner is the same as the upper corner of No. 6, which is described as being 'adjoining in the middle of the river with the southwest corner of the league occupied by a colonist named Philip Miller.' As both the Philip Miller survey and Martinez No. 9 were leagues that were required by law to be of the same width fronting on the river, it would appear from this description that by the words 'in front of the league occupied by Philip Miller' would mean that the lines of the Philip Miller, if extended across the river the distance called for in the field notes of No. 9, would trace the upper and lower lines of said survey No. 9.

We are of the opinion that this is the proper construction to be placed upon the language of said field notes, and we adopt the same as our conclusion.

Extending the north line of the Philip Miller across the river to the west bank of the Trinity river where such line intersects the river at its most westerly point, we have the northeast or beginning corner of Martinez No. 9, and thus do not disturb any of the land included in the field notes of the Philip Miller; from there to the west 10,088 varas the second corner is reached; from there to the south 2,500 varas the third corner is reached; and from that point east 9,246 varas to the west bank of Horseshoe Lake, which we have heretofore determined was at the time of making such survey the west bank of the Trinity river, the fourth corner is reached, and from there following the turns of the old river channel upward (in accordance with the meander calls of the Trinity river, given in the Miller field notes), until arriving at the point of commencing, the league is constructed, as we believe it was the purpose and intent of the original surveyor to construct it, and without making any portion of its east and south lines conflict with the Philip Miller grant.

There is no dispute about the boundary lines of the two I. & G. N. surveys as they are located on the ground. Shattuc survey No. 62 will therefore only include such territory as lies between the east line of I. & G. N. survey No. 38 and the west line of Martinez No. 9, as that line has been fixed by this court.

Martinez league No. 6, as described in the original field notes, "is situated on the western side of Trinity river adjoining in the middle of the river with the southwest corner of a league occupied by a colonist named Philip Miller, commencing from the laurel 12' in diameter upon the margin of the river," etc.

As we have heretofore determined, the true Philip Miller southwest corner is at a point on the east side of Horseshoe Lake, which at the time of the survey of the Philip Miller was a part of the Trinity river. Therefore, the beginning corner of the Martinez survey No. 6 is at a point on the west side of the bank of Horseshoe Lake. This survey should then be constructed by running course and distance as follows: thence west 10,609 varas second corner; from there to the south 2,500 varas, third corner; from thence to the east parallel with the north line of said survey to the east bank of the Trinity river; thence following the turns of said river, as the same is now found on the ground, upward, until a point is reached on said river opposite a deep gully which is a continuation of Horseshoe Lake; thence following the meanders of said gully and lake in westerly, north, and northeast directions to the place of beginning.

The boundaries of Shattuc 63 and 64 should therefore be placed between the east lines of I. & G. N. surveys 38 and 39 and the west lines of Martinez grants 6 and 9, as fixed and determined by this court, and will only include such territory as lies in said area.

The judgment of the lower court is reformed and affirmed in accordance with this opinion.

Orn,[60] in his treatise, makes a point of stating that in Texas, the rule that the footsteps of the surveyor are to be followed is not without an exception. The footsteps will not be followed in the full extent where the monuments are established along a meander line, and the field notes call for the river. In such cases, the calls for the river control, and the lines will be extended beyond the monuments in order to intersect the river.[61]

This is because, as stated numerous times in many cases, meander lines are not run as boundaries of the land surveyed, but for the purpose of defining the sinuosity of the river, and also for determining the quantity of land included in the survey. The river, rather than the meander line as monumented, is the boundary.[62]

## ROLLING EASEMENTS

Rolling easements[63] – while not strictly a protracted easement by definition, still a non-surveyed, or non-located, line of separation between two estates. Its difficulty is that it is continually shifting, in some jurisdictions the effect of which is to affect the title.

Based on previous discussions concerning boundary and title establishment, boundaries, and properties that change with varying natural conditions, such as storms and other natural processes, may pose a unique dilemma. While original title lines are fixed, under some conditions that may change, and locations become a definite challenge. Timing of natural events is critical in identifying location at the moment of a change in title. Following recent, and ancient, court determinations tends to acknowledge physical land movement, additions and loss of property and property rights, depending on how they were created and the nature of the process(es) affecting them. Some publications and available information provide insight to problem solving and identification, as well as a number of court decisions in coastal states available for guidance.

Rising sea levels have a profound effect on coastal regions, beaches, wetlands, and estuaries. As pointed out in recent review, the Environmental Protection Agency recently produced a technical primer summarizing possibly approaches to allow wetlands, beaches, and barrier islands to migrate inland.[64] It also discusses what rolling easements can accomplish, with legal possibilities for creation, advantages and disadvantages, where they may be applied and subsequently managed.

This manual defines rolling easements as:

1. "Regulation or an interest in land in which a property owner's interest in preventing real estate from eroding or being submerged yields to the public or environmental interest in allowing wetlands, beaches, or access along the shore to migrate inland.
2. An interest in land along the shore whose inland boundary migrates inland as the shore erodes.
3. In Texas, an easement along the shore whose inland boundary migrates inland or seaward as the shore erodes or accretes.

Additionally, "there is generally a rolling design boundary seaward of which the restrictions apply, such as the dune vegetation line. At a minimum, a rolling easement prohibits hard shore protection and other structures that prevent the landward edge of wetlands or beaches from migrating inland or block public access along the shore. A rolling easement may also require

---

[60] *Vanishing Footsteps of the Original Surveyor* (1952).
[61] *Dutton v. Vierling*, 152 S.W. 450 (Tex.Civ.App., 1912).
[62] *Koepsel v. Allen*, 68 Tex. 446 (1887).
[63] "Ambulatory boundaries" are boundaries that migrate with a shifting shore.
[64] James G. Titus. *Rolling Easements*, 2011. Washington, DC: U.S. Environmental Protection Agency. 176 pp.

removal of preexisting buildings as they become nonconforming structures seaward of the rolling design boundary. Along estuaries, a rolling easement may also prohibit grade elevation of dry land, which would tend to squeeze wetlands."[65]

The potential impact of rising sea level on public access depends on how the public obtained access.

If the public-trust doctrine is the source of public access, then the impact of sea level rise on access is similar to the impact on wetlands and beaches. Where there is no shoreline armoring or other obstruction, shoreline erosion causes the landward boundary of public access to move inland. Any seaward boundaries for specific types of access move inland as well: for example, if driving on the beach is prohibited within 50 feet inland of the high-water mark, then as the shore erodes, that boundary will migrate inland. Similarly, pedestrian access is generally impractical seaward of the mean high tide line in areas of wave runup: as the shore erodes, the mean high tide line retreats as well. Wherever the shore is armored, pedestrian and vehicular access can be eliminated as the access ways are squeezed between the retreating shore and the shoreline armoring.

Wherever the public has access for reasons other than the public trust doctrine, shore erosion can eliminate access whether or not the shore is armored.

*Public Trust Lands.* Where property lines follow a shoreline, the rule for several centuries has been that the property lines advance or retreat whenever shores gradually advance or retreat. The principal is generally known as the "law of accretion and reliction (sea level drop)" because the law originally evolved as courts decided cases between the King of England and waterfront landowners regarding the ownership of newly created lands. But the same rule applies

when the shore erodes, which is part of the rule's justification.

When the shoreline migrates suddenly, by contrast, the property line does not move, under the "law of avulsion." Although somewhat counterintuitive, courts treat avulsion and accretion differently for several reasons. Originally, all lands had fixed boundaries, so when large areas of land suddenly appeared over what had been water, early courts had little reason to change the rule that what had been the King's water was now the King's land. When the state fills a body of water to create land, the state owns that land under the law of avulsion, although there may be provisions to ensure that the littoral landowner continues to have access to the water. The courts in some states, however, view the new land as an artificial accretion and award it to the waterfront landowner. Another example of avulsion would be a river changing course or the sudden creation of an inlet through a barrier island. If one's home is originally west of a channel, and a storm causes the channel to switch to a point west of the home, then under the law of avulsion the same person still owns the home (see Figure 7.4).

The law of avulsion has a clear rationale when land is created or a channel switches, but the logic for the rule is not as clear in the case of a sudden retreat of the shoreline. Most ocean beaches have had at least one storm that caused substantial erosion since the land was originally transferred from the government to a private landowner. If courts follow the doctrine of avulsion, then boundaries remain out in the ocean at the location where they had been before the avulsive storm. Finding such boundaries would be difficult. Moreover, if the original intent of a land grant from a state (or the King) was for the public to own the wet beach below the high water mark, it seems unlikely that the state would want continued public ownership of the wet beach to depend on whether shore erosion was caused by severe storms or more gradual processes. For this reason, Texas has decided not to follow the rule of avulsion for the impact of shore erosion on the seaward boundary of privately owned land.

Many states that observe the law of avulsion provide the waterfront land owner with the right to fill and thereby recover the lost dry land, but eventually move the boundary inland if the owner

---

[65] Beachapedia.

fails to do so. The right to recover lost land has limited utility: federal and state laws require a landowner to obtain a permit before filling open water or wetlands with soils to create or reclaim land from the sea, and obtaining such a permit may be difficult. Nevertheless, the landowner's right to reclaim land implies that when a governmental beach nourishment project reclaims the land shortly after it is lost, the reclaimed land belongs to the private landowner, though otherwise land created by beach nourishment would be an avulsion that belongs to the state.

*Access along Privately Owned Lands.* As we discuss in the previous subsection, the public has access to many privately owned beaches, for one of two reasons: (a) under the public trust doctrine of a few states, the public retained access to the beach when the state (or King) transferred the land to a private owner; or (b) the public re-acquired access from a private landowner. The impact of sea level rise on access along the shore is different for those two situations:

- The public access way reserved by the public trust doctrine migrates inland as shores erode.
- A public access way acquired from a private landowner does not migrate if that landowner's parcel is submerged; so access along a beach can become impractical.
- The impact on access is ambiguous (depends on state-specific law and site-specific facts) if public access is acquired from a private landowner and only a portion of her parcel is submerged.

Under the public trust doctrine, the inland boundaries of public access are based on environmental features of the shore. Therefore, when the shoreline moves gradually, the inland boundary of public access also moves. In New Jersey (and possibly Oregon), as the dune vegetation line retreats, the public has access to the new area of beach that was formerly part of the dune. In the five states where private land extends to mean low water, the public continues to have access up to mean high water (for fishing, hunting, navigation) as the ordinary high-water mark advances inland. The impact of avulsive shore erosion on public access is less clear. If avulsion does not change a property boundary, one might assume that it would not change the inland boundary of public access. Yet, the practical need for access along a beach depends on where the shoreline is now, while the need for established property lines for mineral royalties or port facilities would not require boundaries to move instantaneously to be effective. Few, if any, cases have addressed the distinction between access and ownership as defined by the public trust doctrine in the context of an avulsive loss of land.

Public access usually does not migrate inland where it has been obtained by means other than the public trust doctrine. As a general rule, a landowner can grant someone else the right to cross her own land. (Such a right is generally called an "easement.") But a waterfront owner cannot sell what she does not own, such as the right to cross a neighbor's land. Therefore, the dry beach easement conveyed by the owner of one parcel cannot migrate to an inland parcel. Consider the many communities where government agencies have purchased or otherwise acquired public access along privately owned beaches whose title extends to mean high water. The public access is along beaches over parcels that are waterfront today, but not across parcels that are not even along the water. Suppose the shore erodes so that today's beaches become water and the beach migrates onto land that currently is the second row of lots back from the ocean. The public will not have access along the new dry beach. It will still have access across land that was previously the dry beach; but pedestrian access will not be feasible if the mean high tide line is regularly flooded by the runup from large waves.

There is no clear rule about whether existing public easements migrate inland within a given parcel of land. If the normal rule for easements applies, then the inland boundaries probably do not move inland. Some state courts have explicitly declared that easements do not roll. In Texas, the public access boundary within a given parcel moves if the shore erodes gradually, but does not move if the shore retreats suddenly during a hurricane. If avoiding such ambiguities is important, deeds that provide public access should specifically say whether the access migrates with the changing shore.

*Shoreline Structures.* Homes standing on the beach can impair access along the shore, by blocking vehicles and creating a hazard to anyone on the beach. Where the shore is armored, pedestrian and vehicular access along an eroding shore is generally lost because the beach is eliminated.


## FERRY LANDINGS

Ferry landings have generally been created by act of the state legislature. Therefore, legislative documents must be consulted to determine what was granted and its extent. Early grants may still exist as an outstanding title from its source, or as a stand-alone parcel with its own title. Many are defined as to location and extent, and often access was created at the same time, which may result in the source document for a road in existence today.

Provisions for ferry landings may sometimes be found in Statute law, and may today be governed by such regulations. It is not unusual to find that they have been surveyed, monumented and platted.


## SHORELINE DIVISION

Methods of Division of Shore

Extending upland side lines without change of direction

- Projection at right angles to thread of stream
- Projection at right angles to course of river (thread)
- Projection at right angles to line of channel
- Projection at right angles to pierhead, bulkhead, or harbor line

If done this way, and subsequently changed, it is a matter of reconstruction

- Projection at right angles to general line of shore
- Projection perpendicular to old shore
- Projection perpendicular to new shore
- Perpendicular to nearest point on the bank
- Projection at the shortest distance to the water
- Proportionate shoreline method (distance)
- Proportionate shoreline method (area)
- Personal agreement

—*Apportionment and division of area of river as between riparian tracts fronting on the same bank, in absence of agreement or specification, 65 ALR2d 143.*


## WHARVES AND HARBOR RIGHTS

Wharves and harbor rights exist both on tidal waters and inland waters. Different states have different regulations, while tidewater mostly comes under federal jurisdiction. Some tidewater rules, because of international commerce, are global in scope.

Colonial states on the east coast, states on the west coast, and those on the Gulf of Mexico all have some unique court decisions, both state and federal in nature, due to a variety of issues and underlying conditions. The port of Boston, Massachusetts, for instance, has been chanllenged for many years over wharfs attached to both natural shorelines and filled land, which is in abundance. New York harbor, Chesapeake Bay, the ports of Charleston and Savannah, ports along the Gulf of Mexico, California and the pacific northwest all have produced decisions deciding a variety of aspects of shore and harbor use.

Inland water body concerns may be found on many of the larger lakes in most states where they exist. The Great Lakes and a few other border waters are subject to federal jurisdiction, and many are considered to be "inland seas."

## HARBORS AND ROADSTEADS

Roadsteads are areas seaward of the coast or a harbor that are used for loading or anchoring ships. Article 9 of the 1958 *Convention on the Territorial Sea and the Contiguous Zone* provides that "Roadsteads which are normally used for the loading, unloading and anchoring of ships, and which would otherwise be situated wholly or partly outside the outer limit of the territorial sea, are included in the territorial sea.[66] The coastal state must clearly demarcate such roadsteads and indicate them on charts together with their boundaries, to which due publicity must be given."[67,68]

## BULKHEAD LINE

Bulkhead line is an officially set line along a shoreline, usually beyond the dry land, to demark a territory allowable to be treated as dry land, to separate the jurisdictions of dry land and water authorities, for construction and riparian activities, to establish limits to the allowable obstructions to navigation and other waterfront uses.

In particular, it may limit the construction of piers in the absence of an official pier line (pier-head line).

Various jurisdictions may define it in different ways. A formal definition may read as follows: *a geographic line along a reach of navigable water that has been adopted by a municipal ordinance and approved by the Department of Natural Resources, and which allows limited filling between this bulkhead line and the original ordinary high water mark, except where such filling is prohibited by the floodway provisions.* (Several municipalities in Wisconsin use wording closely approximating this sample.)

1. An officially set line along a shoreline, usually outside of the dry land, to demark a territory allowable to be treated as dry land, to separate the jurisdictions of dry land and water authorities, for construction and riparian activities, to establish limits to the allowable obstructions to navigation, etc.
2. (U.S.) A line in a navigable waterway established by the U.S. Army Corps of Engineers, beyond which solid fill is not permitted.

---

[66] Territorial Sea, aka Marginal Sea, is defined as the water area bordering a nation over which it has exclusive jurisdiction, except for the right of innocent passage of foreign vessels. It is a creation of international law, although no agreement has thus far been reached by the international community regarding its width. It extends seaward from the low-water mark along a straight coast and from the seaward limits of inland waters where there are embayments. The United States has traditionally claimed three nautical miles as its width and has not recognized the claims of other countries to a wider belt. *Shore and Sea Boundaries*, Volume 1.

[67] The requirement for charting roadsteads is found in Article 16 of the 1982 Convention.

[68] Reed, Michael W. *Shore and Sea Boundaries*. Volume 3.

# BULKHEAD LINES ON INLAND WATERS

Example:

CHAPTER 482 ESTABLISHING BULKHEAD LINES ALONG THE PINE RIVER BETWEEN THE SIXTH STREET BRIDGE AND THE MAIN STREET BRIDGE AND NORTHERLY AND EASTERLY FROM THE NORTH STREET BRIDGE 482.01 TITLE. This Chapter shall be known, cited and referred to as the Richland Center Krouskop Park-Allison Park Bulkhead Line Ordinance, except as referred to herein where it shall be known as "this Chapter". 482.02 INTENT AND PURPOSES. The intent and purpose of this Chapter is to establish bulkhead lines in the public interest pursuant to Section 30.11, Wisconsin Statutes, along the portions of the Pine River hereinafter described so as to provide regulations for the deposition of materials or construction of any structures upon the bed of any navigable water that lies on the landward side of the established bulkhead line. The further intent and purpose of this Chapter is to establish the basis for t determination of the maximum elevation for the deposition of said materials or the construction of said structures. The establishment of said bulkhead lines and elevations is made for the purpose of protecting the public rights and the navigable waters of said Pine River and conforms as nearly as possible to existing shorelines of said River and to the requirements of the floodway as determined by flow studies of the Pine River made jointly between the Department of Natural Resources and said City and to establish elevation of said deposits or construction so as not to affect the elevation of the regional flood on this portion of the Pine River in excess of the limits set forth by the Department of Natural Resources. 482.03 RULES AND DEFINITIONS. (1) Rules: (a) Words used in the present tense shall include the future; and words used in the singular number shall include the plural number, and the plural the singular. (b) The word "shall" is mandatory and not discretionary. (c) The word "may" is permissive. (d) The word "lot" shall include the words "piece" and "parcel"; the word "building" includes all other structures of every kind regardless of similarity to buildings; and the phrase "used for" shall include the phrase "arranged for", "designed for", "maintained for", and "occupied for". (e) The word "typical section" shall be the typical section appended to and made a part of this Chapter and which pictorially defines the stream beds, shorelines and bulkhead lines and their relationship to each other as referred to herein. (2) Definitions: 482-1 (a) Bulkhead Line: A geographic line along a reach of a navigable stream that has been adopted by a municipal ordinance and approved by the Department of Natural Resources pursuant to Section 30.11, Wisconsin Statutes, and has been established by a description prepared by a registered Land Surveyor and which limits the encroachment upon the stream bed for the deposition of materials or construction of any structure and which allows the deposition of materials on the landward side except where Flood way and Flood Fringe areas regulations would prohibit or limit such filling. (b) Navigable Waters: All streams, sloughs, bayous and marsh outlets which are navigable in fact for any purpose whatsoever are navigable. (c) Flood: A temporary rise in stream flow or stage that results in inundation of the areas adjacent to the channel. (d) Flood Frequency: A means of expressing the probability of flood occurrences as determined from a statistical analysis of representative stream flow records. It is customary to estimate the frequency with which specific flood stages or discharges may be equalled or exceeded, rather than the frequency of an exact stage or discharge. Such estimates by strict definition are designated "exceedence frequency" but in practice the term "frequency" is used. The frequency of a particular stage of discharge is usually expressed as occurring once in a specified number of years. Also see: Recurrence interval. (e) Flood Plain: the land adjacent to a body of water which has been or may be hereafter covered by flood water including but not limited to the regional flood. (f) Flood Profile: A graph showing the relationship of the water surface elevation of a flood event to a location that generally is expressed as a distance upstream from a designated point on a stream or river. (g) Floodway: The channel of a stream and those portions of the flood plain adjoining the channel that are required to carry and discharge the flood water or flood flows of any stream or river including but not limited to flood flows associated with the regional flood. (h) Regional Flood: A flood determined or approved by the Department which is representative of large floods known to have generally occurred in Wisconsin and which may be expected to occur on a particular stream because of like physical characteristics. The regional flood generally has an average frequency of the one hundred (100) year recurrence interval flood. (i) Reach: A hydraulic engineering term to describe longitudinal segments of a stream or river. A reach will generally include the segment of the flood plain where flood heights are primarily controlled by manmade or natural flood plain obstruction or restrictions. In an urban area, the segment of a stream or river between two consecutive bridge crossings would most likely be a reach. 482.04 STATUTORY AUTHORIZATION. This Chapter for the establishment of these bulkhead lines is adopted pursuant to the authorization contained in Sections 62.23 and 30.11 and Chapter 87, all of the Wisconsin Statutes. 482.05 ESTABLISHMENT OF BULKHEAD LINES. (1) The following described bulkhead line is hereby established along the south side of the Pine 482-2 River from the Sixth Street Bridge to the Main Street Bridge and then to the East and that the maximum elevation of materials deposited on the landward side of said bulkhead line within the flood plain limits shall not exceed 734.0 feet and is shown on the Flood Plain Map which is attached hereto and is a part of this Chapter, all subject to the approval of the Department of Natural Resources, to-wit: Commencing at the Southwest Corner of Lot 6, Block 2 of House- holder's Addition, Schoolcraft, City of Richland Center, Richland County, Wisconsin; thence N 43 35' 40" E, 337.56 feet to the point of beginning of the bulkhead line on the existing shoreline of the Pine River; thence N 84 55' 51" W, 425.50 feet; thence S 81 36' 37" W, 165.32 feet; thence S 88 01' W, 29.95 feet; thence N 85 31' 13" W, 56.45 feet; thence N 87 03' 59" W, 117.46 feet; thence N 89 25' 14" W, 113.63 feet; thence N 86 14' 51" W, 105.19 feet; thence S 54 52' 18" W, 1,103.79 feet; thence S 34 53' 17" W, 134.99 feet; thence S 42 06' 25" W, 85.68 feet; thence S 53 32' 12" W, 174.44 feet; thence S 48 57' 50" W, 193.08 feet to end of said bulkhead line on the existing shoreline of the Pine River; thence S 61 35' 36" E, 870.96 feet to the South east Corner of Lot 1, Block 1, Haseltine's Heirs Addition to the City of Richland Center, Wisconsin. The above described bulkhead line being located partly in the S 1/2 of the NW 1/4 of Section 16, and partly in the SE 1/4 of the NE 1/4 of Section 17, and partly in the NE 1/4 of the SE 1/4 of Section 17, all in T. 10 N., R. 1 E., City of Richland Center, Richland County, Wisconsin. (2) The following described bulkhead line is hereby established along the north side of the Pine River from a location north of the Sixth Street bridge to the Main Street bridge and then northerly along the West side of the Pine River and that the maximum elevation of the materials deposited on the landward side of said bulkhead line between the Sixth Street bridge and the Main Street bridge within the flood plain limits shall not exceed 733.0 and that the maximum elevation of the materials deposited on the landward side of the bulkhead line northerly from the Main Street bridge and between that portion of the bulkhead line and Main Street (Wisconsin State Trunk Highways 56 and 80) shall not exceed 731.0 and said bulkhead line is shown on the Flood Plain Map which is attached hereto and is a part of this Chapter, all subject to the approval of the Department of Natural Resources, to-wit: Commencing at the Northwest Corner of the NW 1/4 of Section 16, T. 10 N., R. 1 E., City of Richland Center, Richland County, Wisconsin; thence South 918.01 feet along the West line of said Section 16; thence N 83 09' 43" W, 370.25 feet; thence S 88 31' 40" W, 469.15 feet; thence S 4 05' 00" E, 593.00 feet; thence S 0 23' 40" W, 483.3 feet; thence S 1 43' 40" W, 141.9 feet; thence S 9 02' 00" E, 482-3 49.83 feet; thence S 64 31' 0" E, 64.95 feet to the point of beginning of the bulkhead line on the existing shoreline of the Pine River; thence N 45 05' 03" E, 114.95 feet; thence N 55 30' 00" E, 150.00 feet; thence N 71 00' 00" E, 150.00 feet; thence N 88 15' 27" E, 655.65 feet; thence N 80 30' 08" E, 95.57 feet; thence S 88 09' 14" E, 147.16 feet; thence S 89 58' 52" E, 114.24 feet; thence S 88 44' 59" E, 104.73 feet; thence S 88 22' 28" E, 146.57 feet; thence S 60 38' 00" E, 75.58 feet; thence S 73 41' 01" E, 42.03 feet; thence S 73 57' 05" E, 103.02 feet; thence N 60 21' 09" E, 34.44 feet; thence N 61 49' 41" E, 141.65 feet; thence N 47 40' 58" E, 60.86 feet; thence N 29 03' 53" E, 105.44 feet; thence N 12 37' 26" E, 151.68 feet; thence N 01 46' 46" W, 293.48 feet; thence N 11 41' 19" W, 87.30 feet; thence N 35 50' 52" W, 72.35 feet; thence N 65 58' 15" W, 60.98 feet; thence N 36 47' 12" W, 198.05 feet; thence N 07 35' 36" E, 81.25 feet to the end of the bulkhead line on the existing shoreline of the Pine River; thence N 43 16' 15" W, 319.46 feet; thence N 18 52' E, 69.67 feet to the Southwest Corner of Block 1, W.H. Collins Addition to the City of Richland Center, Wisconsin; The above described bulkhead line being located partly in the W 1/2 of the NW 1/4 of Section 16 and partly in the E 1/2 of the NE 1/4 of Section 17, all in T. 10 N., R. 1 E., City of Richland Center, Richland County, Wisconsin. 482.06 APPROVAL BY STATE OF WISCONSIN AND FILING. This Chapter shall not be in full force and effect until three certified copies of the Chapter together with three true and correct copies of the Flood Plain Map are submitted to and approved by the Department of Natural Resources and one approved copy is on file in the office of the City Clerk and one approved copy is recorded in the office of the Register of Deeds of the County of Richland, State of Wisconsin. 482.07 VALIDITY AND SEVERABILITY. It is hereby declared to be the intention of the Council that the several provisions of this Chapter are separable in accordance with the following: (1) If any court of competent jurisdiction shall adjudge any provisions of this Chapter to be invalid, such judgment shall not affect any other provisions of this Chapter not specifically included in said judgment. 482-4 (2) If any court of competent jurisdiction shall adjudge invalid the application of any provision of this Chapter to a particular property, building, or other structure, such judgment shall not affect the application of said provisions to any other property, building, or structure not specifically included in said judgment. 482.08 PENALTIES. Any person, firm or corporation who violates, disobeys, neglects, omits or refuses to comply with, or who resists the enforcement of any of the provisions of this Chapter, shall forfeit not less than $25 nor more than $400, together with the costs of prosecution, and in case of failure to pay such forfeiture and costs may be imprisoned in the county jail of Richland County until such forfeiture and costs are paid but not to exceed 90 days. Every day of violation shall constitute a separate offense. 482.09 EFFECTIVE DATE. This Chapter shall be in full force from and after September 1, 1970.

## PIERHEAD LINES

A pierhead line is, in essence, a legal boundary beyond which artificial structures (such as piers) may not be built into navigable waters. *New Jersey v. Delaware* (2008) 552 U.S. 597, 642 [128 S.Ct. 1410, 1438, 170 L.Ed.2d 315, 345]. As part of the Rivers and Harbors Act of 1899, the United States may fix pierhead and bulkhead lines as part of its supreme powers to regulate navigable waters. 33 U.S.C.S. § 400, et. seq. Many states have also established their own pierhead or bulkhead lines at various locations.

## OIL AND GAS RIGHTS (MINERALS)

Offshore drilling is a mechanical process where a wellbore is drilled below the seabed. It is typically carried out in order to explore for and subsequently extract petroleum which lies in rock formations beneath the seabed. Most commonly, the term is used to describe drilling activities on the continental shelf, though the term can also be applied to drilling in lakes, inshore waters, and inland seas.

Offshore drilling presents environmental challenges, both offshore and onshore from the produced hydrocarbons and the materials used during the drilling operation. Controversies include the ongoing U.S. offshore drilling debate.

There are many different types of facilities from which offshore drilling operations take place. These include bottom founded drilling rigs (jackup barges and swamp barges), combined drilling and production facilities either bottom founded or floating platforms, and deepwater mobile offshore drilling units (MODU) including semi-submersibles and drillships. These are capable of operating in water depths up to 3,000 meters (9,800 ft.). In shallower waters, the mobile units are anchored to the seabed; however, in deeper water (more than 1,500 meters (4,900 ft.), the semi-submersibles or drillships are maintained at the required drilling location using dynamic positioning.

## WIND FARMS

Offshore wind power or offshore wind energy is the use of wind farms constructed in bodies of water, usually in the ocean on the continental shelf, to harvest wind energy to generate electricity. Higher wind speeds are available offshore compared to on land, so offshore wind power's electricity generation is higher per amount of capacity installed, and NIMBY opposition to construction is usually much weaker. Unlike the typical use of the term "offshore" in the marine industry, offshore wind power includes inshore water areas such as lakes, fjords, and sheltered coastal areas, using traditional fixed-bottom wind turbine technologies, as well as deeper-water areas using floating wind turbines.

At the end of 2017, the total worldwide offshore wind power capacity was 18.8 gigawatt (GW). All the largest offshore wind farms are currently in northern Europe, especially in the United Kingdom and Germany, which together account for over two-thirds of the total offshore wind power installed worldwide. As of September 2018, the 659 MW Walney Extension in the United Kingdom is the largest offshore wind farm in the world. The Hornsea Wind Farm under construction in the United Kingdom will become the largest when completed, at 1,200 MW. Other projects are in the planning stage, including Dogger Bank in the United Kingdom at 4.8 GW, and Greater Changhua in Taiwan at 2.4 GW.

# 7D Special Cases
## *Land and Water Uses*

## AQUACULTURE

Aquaculture, on land and in the sea, involves, beyond the necessary regulatory permits, definition of ownership, titles, and boundaries. It comes in many forms, and encompasses a variety of mostly animals, and some plants. Along our coasts, and even offshore, we find fishing and farming for many types of fin fish and shellfish, along with seaweed and other fauna such as sea worms. Some areas are public, while others are private; some are owned, while others are leased, and all are protected by law. Ocean areas are no different than on-land areas, each has a title element with accompanying boundarie. Many of these boundaries are defined, marked and mapped. The following court decisions illustrate their attributes, characteristics, and accompanying problems. Aquaculture is generally considered to be an agricultural endeavor, and has been legislated as such in some states.

### LOBSTER WARS/OYSTER WARS

Claims by nation-states have sometimes led to conflicts involving fishing rights, for both fin fish and, often, shellfish. While they are often called wars, they are generally non-violent, but occasionally lead to increased hostilities. Ocean resources are usually jealously guarded, and some disputes continue to arise on an irregular basis. Others lay idle until a time where value increases, a hostile trespass action occurs, or an arrest or other confrontation takes place. In relatively recent times, a New Hampshire lobsterman was arrested by Maine Sea and Shore Fisheries wardens for possessing undersized lobsters based on a length limit between the two states differing by 1/16th of an inch. The lobsterman insisted he was fishing in New Hampshire as entitled, while the Maine wardens insisted he set his traps in Maine waters. Since there was a dispute over the location of the state line, the case went to the Supreme Court of the United States. The dispute is not entirely satisfactorily settled today.

Nation-states can be very protective of territorial waters, in the interests of defense, ocean resources such as minerals, and fishing rights of a variety of types, including fin fish and shellfish.

### CLAM FLATS

Clam flats may be natural, artificial (seeded), public land/waters, or privately owned or leased. While public areas i.e., flats open to the public, or on public land, there may be licensing requirements, either state, municipal, or both, and there may be areas designated for shellfish harvesting. Therefore, boundaries are involved, which may be surveyed, platted or otherwise defined, any of which has an origin of its definition. As flats are opened and closed, due to regulated seasons or health concerns such as red tide, boundaries may change from time to time, depending on varying conditions.

DOI: 10.1201/9781003032557-14

This case compares the difference between marine resources, for the purpose of regulation. In *State v. LeMarch,*[69] the court began with the regulation enacted by the Town of Woolwich:

> No person, firm or corporation shall, within the limits of the town of Woolwich in the county of Sagadahoc, dig or take any clams, clam-worms, sand-worms or blood-worms, without having first obtained a license from the municipal officers of said town of Woolwich, who are authorized to grant and issue such licenses and fix the fee there for. No license shall be granted or issued to any person, firm or corporation unless such person, firm, or corporation is a resident of said town of Woolwich. Nothing herein shall prohibit a riparian owner of shores or flats in said town of Woolwich from digging and taking claims there from for food for himself and family without license. For the purposes of sections 85 to 87, inclusive, the term 'a resident' shall mean a person, firm or corporation who has resided in this state for a term of at least 6 consecutive months and in the town of Woolwich for at least 3 consecutive months prior to making application for license.

The particular offense with which the respondent is charged is that, being a resident of Dresden in the County of Lincoln, he did at Woolwich, in the County of Sagadahoc, on June 9, 1951, unlawfully and without right 'dig bloodworms[70] without first having obtained a license from the Municipal Officers of said Town of Woolwich.' The complaint was brought in the Municipal Court at Bath in said County of Sagadahoc; the respondent, after a plea of not guilty, was found guilty and fined $15 and costs; and appealed to the Superior Court; that the only issue was 'the question of the constitutionality of Chapter 34, Section 77, Revised Statutes of Maine, as revised by Chapter 332, Section 85, of the Public Laws of 1947, and incorporated in the Second Biennial Revision of the Sea and Shore Fisheries Laws of the Public Laws of 1949, which the respondent contends deprives him of his property without due process of law; abridges his privileges and immunities; denies him the equal protection of the laws in contravention of the Fourteenth Amendment to the United States Constitution, and precludes him from acquiring and possessing property, and of pursuing and obtaining safety and happiness in violation of Article I, Section 1, of the Constitution of the State of Maine'.

The facts are not in dispute. The agreed statement reads in part as follows:

"On the ninth day of June, A. D. 1951, Robert Lemar, of Dresden, Maine, was arrested by a Warden of the Maine Sea and Shore Fisheries Department for unlawfully digging blood-worms from the flats within the limits of the Town of Woolwich, he, the said Robert Lemar, not then having a written permit of the Municipal Officers of said Town of Woolwich and paying the fee there for, and he not being a riparian owner of shore or flats in said Town.

"It is admitted that said Robert Lemar dug said worms as above set forth; that he was a resident of Dresden at that time; that he had no license from said Town, but that he did have a valid commercial fishing license issued by the Commissioner of Sea and Shore Fisheries of the State of Maine permitting him to dig and sell worms taken from the flats and coastal waters of the State; that said respondent was not a riparian owner of the shore or flats within said Town of Woolwich; also that said respondent is a legal resident of the State of Maine within the terms and provisions of the Sea and Shore Fisheries Laws of said State, to wit: Chapter 34, Revised Statutes of 1944, as revised."

The only issue is the constitutionality under the state and federal constitutions of the statute here involved. That issue seems to have been settled by the case of *State v. Leavitt,* 105 Me. 76, 72

---

[69] 147 Me. 405, 87 A.2d 886 (1952).

[70] A seaworm which lives in mud flats and is highly prized as a fishing bait. As such, they are sold commercially to sporting anglers, similar to angleworms dug on land and used for freshwater angling. Collecting for fishing bait is a multimillion-dollar business on the east coast of North America.

A. 875, 26 L.R.A., N.S., 799, except that that case applies to clams and this to blood worms. The principles of law involved in each case are exactly the same.

In the Leavitt case, the opinion of the court discusses exhaustively the very complex question of the rights of riparian owners in navigable rivers to the shore front where it is bounded by the sea and where the tide ebbs and flows. It is unnecessary to discuss the question as to what extent the rule of law laid down in that case over forty years ago may have been modified by recent decisions of the Supreme Court of the United States. It is sufficient to say that that case still remains the law in this state in so far as it involves the right of this state to control by legislation public fishing rights in the waters along our shores. In other words, this state, unless it has parted with title, owns the bed of all tidal waters within its jurisdiction as well as such waters themselves so far as they are capable of ownership, and has full power to regulate and control fishing therein for the benefit of all the people. *Commonwealth v. Hilton,* 174 Mass. 29, 54 N.E. 362, 45 L.R.A. 475. The legislature has the right to authorize the selectmen of each town within the state to make a regulation forbidding the taking of clams without a permit and to provide that permits shall be granted only to the inhabitants of the town. Commonwealth v. Hilton, supra. Such a regulation is not in violation of the Fourteenth Amendment of the constitution of the United States.

As was said in *State v. Leavitt,* supra, 105 Me. at page 85, et seq., 72 A. at page 879:

> Since it must be assumed that the public interest required some limitation upon the right of clam fishing, it does not seem to us that it is unreasonable or arbitrary for the state having a proprietary interest as well as a governmental power all for the public benefit to give the preference to those whom the law for more than two hundred and fifty years has given a preference, and who were enjoying a preference when the fourteenth amendment was adopted, namely, the inhabitants of the town within which the fisheries are located. The discrimination between them and the inhabitants of other towns seem to us to 'bear a just and proper relation' to the difference in situation, in locality, and in the actual enjoyment of prior legal rights or privileges. It is not unreasonable that they to whose doors nature has brought these 'succulent bivalves' shall be entitled to them before those who are less favorably situated whenever there must be restriction. And we do not think that the legislative recognition of this existing superiority in situation and privilege denies to others the equal protection of the law.

> And it may be said further that if the state may, under the circumstances, prefer some, it may, so far as the fourteenth amendment is concerned, entirely exclude others. A preference violates equality as certainly as exclusion does.

> The reasons suggested by us for holding that this discriminating legislation is not inimical to the equal protection clause of the fourteenth amendment apply alike to the inhabitant of the town who takes clams for his own use and to the hotel keeper in the town who takes them for use in his hotel.

As we said earlier, the same rule which applies to clams applies also to blood worms. The respondent would have us overrule *State v. Leavitt*, supra, which has been the law in this state for many years. This we are not prepared to do.

**Author's Note:** This case serves to illustrate the complexity of public, particularly ocean, resources. A simple violation of a jurisdictional regulation resulted in a constitutional issue, and involved much more than a simple local court proceeding.

On all coasts, particularly on the Pacific Coast, clams are an important resource. The case of *Washington State Geoduck Harvest Association v. Washington State Department of Natural Resources*[71] had to do with State process in leasing certain areas of public lands for harvesting clams.

---

[71] 124 Wn.App. 441, 101 P.3d 891 (2004).

"Washington State Geoduck Harvest Association (Association) appeals the trial court's grant of summary judgment in favor of the Department of Natural Resources (DNR) and Douglass Sutherland, Commissioner of Public Lands. The Association argues that DNR has no authority to regulate geoduck harvesting and that DNR regulations for auctioning geoduck harvesting rights (1) violate the public trust doctrine; (2) violate equal protection safeguards; and (3) are inconsistent with RCW 77.04.012. Finding no error, we affirm.

"The Association represents a group of commercial geoduck harvesters. Geoducks are within the class of animals known as mollusks. They are large clams that average two pounds each, that occur naturally along the Pacific Northwest coast. It takes them four to five years to reach an average harvestable size of one-and-a-half pounds. They burrow into the ocean floor to a depth of as much as three feet and commercial divers harvest them from the beds of the state's navigable waters.

"Unlike other clam resources, the state does not lease lands containing geoducks for long-term cultivation. Instead, it leases tracts of aquatic land yearly until the geoduck resource is harvested down to a set density. The tract is then placed into recovery status, a period that may take from 11 to 42 years. DNR's management plan thus relies on natural replenishment of geoducks to provide sustained harvest opportunities rather than cultivation and enhancement.

The legislature has authorized DNR to regulate commercial geoduck harvesting in Washington State waters. RCW 79.96.080. DNR implements an established auction process under which geoduck harvesters must comply with numerous requirements prior to and after successfully bidding on harvesting rights.

RCW 79.96.080 requires that geoducks "be sold as valuable materials under the provisions of chapter 79.90 RCW." Chapter 79.90 RCW outlines DNR procedures and responsibilities for auctioning valuable materials. Interested parties must present a $50,000 deposit to bid on a tract of harvestable ocean bed. The highest bidder at the public auction must then prove itself to be a "responsible bidder" as defined by the statute. RCW 79.90.215. If it satisfies the enumerated criteria, then DNR permits the successful bidder to harvest geoducks from the relevant tract.

Asserting that DNR's auction process has unlawfully restricted their right to harvest geoducks in Washington, the Association filed a complaint for a declaratory judgment in the superior court. The complaint sought to have RCW 79.96.080 declared invalid under the public trust doctrine and equal protection principles. It also generally challenged DNR's right to manage resources on aquatic lands and argued that DNR's auction process was inconsistent with RCW 77.04.012.

The Association and DNR brought cross motions for summary judgment. The court granted DNR's motion and denied the Association's motion for summary judgment.

The Association then filed this timely appeal, challenging (1) the right of DNR to regulate geoduck harvesting; (2) DNR's methods for regulating geoduck harvesting under the public trust doctrine and equal protection principles; and (3) DNR's use of the procedures under RCW 79.96.080 for commercial geoduck harvesting instead of Department of Fish and Wildlife (DFW)'s general mandate under RCW 77.04.012.

ANALYSIS

I. STANDARD OF REVIEW

Although the Association articulates three challenges to DNR's regulation of geoduck harvesting, resolution of the narrow issue of the facial constitutionality of RCW 79.96.080 under the public trust doctrine and equal protection principles answers all of the Association's claims. Statutes are "presumed constitutional and the burden [falls] on the party challenging the statute to prove its unconstitutionality beyond a reasonable doubt." *Tunstall ex rel. Tunstall v. Bergeson,* 141 Wash.2d 201, 220, 5 P.3d 691 (2000).

II. DNR'S AUTHORITY TO REGULATE GEODUCK HARVESTING ON PUBLIC LAND

The Association first argues that because geoducks are not on "state lands" as defined by the Public Lands Act (PLA), state agencies cannot regulate them as "valuable materials." Br. of

Appellant at 14. But the Association confuses DNR's authority to regulate resources on state-owned "public lands" with the Act's more limited definition of "state lands."

Under the PLA, "public lands" include both "state lands" and "aquatic lands." RCW 79.02.010(a). "Aquatic lands" are specifically defined as "all *state-owned* tidelands, shorelands, harbor areas, and the beds of navigable waters as defined in chapter 79.90 RCW that are administered by [DNR]." RCW 79.02.010(1). [2] DNR has express statutory authority to administer state-owned aquatic lands. RCW 79.90.465(12).

The PLA acknowledges state ownership of aquatic lands, and directs DNR to manage the resources on that land in the public interest. And the public trust doctrine ensures state management of public lands, in part, through our constitution's express reservation of "the beds and shores of all navigable waters in the state" for state ownership. Wash. CONST. art. 17, § 1.

DNR has the authority to sell "valuable materials" on "tide or shore lands belonging to the state." RCW 79.90.090. And geoducks are to be "sold as valuable materials" under the requirements of chapter 79.90 RCW and the PLA. RCW 79.96.080. Thus, DNR may manage and sell geoducks on public lands, which includes state-owned aquatic lands, under chapters 79.02 and 79.90 RCW.

The Association also contends that geoducks cannot be "valuable materials" under RCW 79.96.080 because they are living creatures and not "materials" in the same sense as the enumerated items in RCW 79.90.060.

This argument is flawed because RCW 79.96.080 does not state that geoducks "are" valuable materials. The statute only states that they shall be "sold as" valuable materials, with sales regulated under chapter 79.90 RCW. Thus, RCW 79.96.080 references RCW 79.90.060 to invoke the sale requirements of chapter 79.90 RCW, not to equate geoducks with the materials listed in the statute.

### III. DNR'S AUCTION PROCESS AND THE PUBLIC TRUST DOCTRINE

The Association argues that DNR's procedures for auctioning harvesting rights under RCW 79.96.080 violate the public trust doctrine. It contends that the doctrine provides a "right to fish" that DNR's sale of geoduck harvesting rights precludes.

Under the public trust doctrine, DNR must protect various public interests in state-owned tidelands, shore lands and navigable water beds. The traditionally protected interests include commerce, navigation, and commercial fishing. *Orion Corp. v. State,* 109 Wash.2d 621, 641, 747 P.2d 1062 (1987). But our Supreme Court has expanded this list to include "incidental rights of fishing, boating, swimming, water skiing, and other related recreational purposes generally regarded as corollary to the right of navigation and the use of public waters." *Caminiti v. Boyle,* 107 Wash.2d 662, 669, 732 P.2d 989 (1987) (quoting *Wilbour v. Gallagher,* 77 Wash.2d 306, 316, 462 P.2d 232 (1969)). This necessarily obligates the state to balance the protection of the public's right to use resources on public land with the protection of the resources that enable these activities.

The principal question is whether the public trust doctrine applies to DNR's sale of commercial geoduck harvesting rights on public lands. We hold that it does.

The Association argues that DNR's regulation of commercial geoduck harvesting substantially impairs the public's right to use the state land for that purpose. DNR responds that under *State v. Longshore,* the state has the right to sell shellfish embedded in public aquatic lands free from any claims under the public trust doctrine. 141 Wash.2d 414, 5 P.3d 1256 (2000).

Typically, animals found in the wild are *"ferae naturae,"* indicating that no person owns the animal until it is reduced to possession. *Longshore,* 141 Wash.2d at 421-22, 5 P.3d 1256. But the sedentary nature of clams and oysters has resulted in statutory and case law that departs from this general rule and distinguishes shellfish because they are "so closely related to the soil." *Longshore,* 141 Wash.2d at 423, 5 P.3d 1256 (quoting *Sequim Bay Canning Co. v. Bugge,* 49 Wash. 127, 131, 94 P. 922 (1908)). This divergence is rooted in the perception that the "fixed habitation [of clams] when imbedded in the soil" makes them, "in a very material sense, belong with the land." *State v. Van Vlack,* 101 Wash. 503, 505-06, 172 P. 563 (1918); *Sequim Bay,* 49 Wash. at 131, 94 P. 922. "In this respect clams differ from fish, game birds and game animals in their wild or natural state." *Van Vlack,* 101 Wash. at 506, 172 P. 563.

In *Longshore,* the court considered whether, under Washington law, the public trust doctrine gives the general public a right to take naturally occurring or cultivated clams from property that the state had sold into private ownership. *Longshore,* 141 Wash.2d at 428, 5 P.3d 1256. It concluded that there is no such right because the clams were bundled with the property owners' land rights. *Longshore,* 141 Wash.2d at 422, 428, 5 P.3d 1256.

DNR asserts that *Longshore* is determinative of whether the doctrine provides a public right to harvest geoducks from *state-owned* land. But case law concluding that embedded shellfish "belong with" the land indicates that whether the state or a private entity owns the land is of critical importance in assessing whether the public trust doctrine applies. *Longshore,* 141 Wash.2d at 422, 5 P.3d 1256 (quoting *Sequim Bay,* 49 Wash. at 131, 94 P. 922).

The public trust doctrine "prohibits the State from disposing of its interest in the waters of the state in such a way that the public's right of access is substantially impaired, unless the action promotes the overall interests of the public." *Rettkowski v. Dep't of Ecology,* 122 Wash.2d 219, 232, 858 P.2d 232 (1993) (citing *Caminiti,* 107 Wash.2d at 670, 732 P.2d 989).

Here, the state, not a private party, owns the navigable beds. Thus, DNR has a continuing obligation under the public trust doctrine to manage the use of the resources on the land for the public interest. And *Longshore* is consistent with the conclusion that shellfish embedded on public property are resources that invoke a public right under the public trust doctrine. 141 Wash.2d at 422, 5 P.3d 1256.

The United States Supreme Court has indicated that shellfish harvesting is an interest protected by the public trust doctrine. *Phillips Petroleum Co. v. Mississippi,* 484 U.S. 469, 476, 108 S.Ct. 791, 98 L.Ed.2d 877 (1988) (public lands may be used for shellfish harvesting); *McCready v. Virginia,* 94 U.S. 391, 395-397, 24 L.Ed. 248 (1876) (restrictions on oyster harvesting permitted under concept of state authority to regulate public waters); *Smith v. Maryland,* 59 U.S. (18 How.) 71, 75, 15 L.Ed. 269 (1855).

But each state individually determines the public trust doctrine's limitations within the boundaries of the state. And state regulation of shellfish harvesting on public land is consistent with our state's protection of commerce, navigation, commercial fishing, and incidental recreational activities. *See Caminiti,* 107 Wash.2d at 670, 732 P.2d 989. As a result, the doctrine applies to protect the public interest in harvesting geoducks embedded in the navigable water beds of state-owned lands.

Because the public-trust doctrine applies, we must determine whether DNR has violated the doctrine through its management regime. We ask: "(1) whether the State, by the questioned legislation, has given up its right of control over the jus publicum; and (2) if so, whether by so doing the State (a) has promoted the interests of the public in the jus publicum; or (b) has not substantially impaired it." *Caminiti,* 107 Wash.2d at 670, 732 P.2d 989. We apply heightened scrutiny, as the statutes are essentially being measured against constitutional protections for public access to unique resources. *Weden v. San Juan County,* 135 Wash.2d 678, 698, 958 P.2d 273 (1998).

In *Caminiti,* the court concluded that the state had not impermissibly given up control of the jus publicum by allowing private landowners to build recreational docks in public waters. 107 Wash.2d at 675, 732 P.2d 989. There was no significant loss of control because (1) the state had not conveyed title to the land; (2) DNR was authorized to regulate access rights and could revoke dock rights; (3) local regulations governed construction, size, and length of the docks; and (4) the docks had to comply with the Shoreline Management Act, the hydraulics act and state flood control laws. *Caminiti,* 107 Wash.2d at 672-73, 732 P.2d 989.

Here, the state allows commercial geoduck harvesting only through specific procedures and requirements that DNR implements and enforces. Under these procedures, no title to state land is conveyed, DNR is responsible for appraising the resources in the water beds, the resource bidders must provide an estimate of resources to be removed, and the state may apply "such terms and conditions deemed necessary ... to protect the interests of the state." RCW 79.90.310; *see generally* RCW 79.90.180-310, and Washington State Administrative Code 220-52-019.

Additionally, DNR has a right to revoke or suspend a commercial harvesting agreement, and the harvester must comply with applicable commercial diving safety standards and federal occupational safety and health administration regulations. RCW 79.96.080. Thus, under the relevant statutory framework, the state has not given up its right of control over the state's geoduck resources.

Further, the state's action is only improper where it does not promote, or substantially impairs, the public interest. Here, the opposite is true.

The public trust doctrine, as applied to DNR's regulation of commercial geoduck harvesting, protects the public right to recreation, commerce, and commercial fishing, all of which are bolstered by the state's system of facilitating sustainable geoduck harvesting and natural regeneration of the resource. And the proceeds from the sale of harvesting rights go to support aquatic resource management and enhancement of aquatic lands for all uses by the public. RCW 79.90.245; RCW 79.96.080. Thus, DNR's procedures and regulation of commercial geoduck harvesting serves the public, satisfies the public trust doctrine's requirements, and is not an unconstitutional infringement on the public's rights.

IV. EQUAL PROTECTION PRINCIPLES

The Association next argues that RCW 79.96.080 is unconstitutional because it violates equal protection principles. But these protections do not apply where the challenged law does not create a suspect class of persons. *Neah Bay Chamber of Commerce v. Dep't of Fisheries,* 119 Wash.2d 464, 476, 832 P.2d 1310 (1992).

Article I, section 12 of the Washington Constitution requires that no law grant "any citizen, class of citizens, or corporation other than municipal, privileges or immunities which upon the same terms shall not equally belong to all citizens, or corporations." And neither RCW 79.96.080 nor any statute it references singles out any group of persons for distinct treatment. RCW 79.96.080.

The Association in essence asserts that because commercial geoduck harvesting is regulated differently from other fish and shellfish harvesting, the equal protection clause is implicated. But geoducks are not citizens and are not protected by our constitution's equal protection clause. *See Cetacean Community v. Bush,* 386 F.3d 1169, 1179 (9th Cir.2004).

Even expanding the Association's argument to assert differential treatment of geoduck harvesters from other shellfish harvesters, its claim is unsuccessful. The challenged statute applies only to geoduck harvesting, does not create classes of harvesters, and, thus, does not trigger equal protection principles. RCW 79.96.080.

V. DNR'S GEODUCK HARVESTING PROCEDURES AND RCW 77.04.012

The Association asserts that DNR's geoduck management does not comply with RCW 77.04.012 because it fails to "maintain the economic well-being and stability of the fishing industry." RCW 77.04.012. This statute provides a set of general management objectives for the Fish and Wildlife Commission, its director, and DFW to guide their management of wildlife, fish "and shellfish in state waters and offshore waters." RCW 77.04.012.

But the statutes clearly separate the obligations of DFW and DNR. Chapter 77.04 RCW establishes the responsibilities of DFW. RCW 77.04.012 does not control

DNR's regulation of commercial geoduck harvesting under RCW 79.96.080. Under RCW 77.60.070, DFW "may not authorize a person to take geoduck clams for commercial purposes outside the harvest area designated in a current [DNR] geoduck harvesting agreement issued under RCW 79.96.080." RCW 79.96.080 is a specific directive that requires DNR to regulate the sale of geoducks, and allows DNR to "place terms and conditions in the harvesting agreements as [it] deems necessary."

The two previous decisions, one on each coast, relate to jurisdiction and regulatory processes within state offshore boundaries. While complex, they serve to illustrate a fact of aquaculture that can be extremely important and involving a number of issues. While each state abutting tidewater, as well as each country with ocean boundaries, has similar issues and disputes, the following decisions within the United States are noteworthy, and summarized as follows:

The next case, *Payne & Butler v. Providence Gas Co.*,[72] is an early case having to do with injury to private property in the form of marine resources, oysters, and clams.

It was an action of trespass on the case brought by the plaintiffs against the defendant to recover damages for injury to their oysters and quahaugs and to their grounds, and for the expense of cleaning the same caused and made necessary by the deposit in the Providence river of tar, oils, and other deleterious substances manufactured by the defendant.

The plaintiffs' declaration is in five counts, whereof the first count reads as follows: "For that on, to wit, the 1st day of October, 1903, and on divers other days and times between that day and the commencement of this suit, the defendant corporation was located and engaged in the business of the manufacture of illuminating gas and kindred products in the city of Providence having two gas works or plants located on the west bank of the Providence river. That in the process of the manufacture of said gas and kindred products large quantities of coal, petroleum, oils, and other materials were used; that from said manufacture of gas and kindred products large quantities of coal tar, water gas tar, oils, and other substances which are the products of and the refuse from the manufacture of illuminating gas and kindred products, were produced. That on, to wit, the days and times aforesaid, said defendant discharged and suffered and allowed to escape from its premises to and into said Providence river large quantities of said coal tar, water gas tar, oils, and other substances. That said coal tar, water gas tar, oils, and other substances floated in and upon the waters of said Providence river and were carried by the currents and tides in said Providence river down said river and were deposited in large quantities in the waters and on the bottom of said river and of the upper part of Narragansett Bay into which said river empties. That said coal tar, water gas tar, oils, and other substances aforesaid are of a nature very injurious and poisonous to the healthy growth of all kinds of shellfish which grow in said waters and with which they come in contact. That the plaintiffs on, to wit, the 1st day of October, 1903, and from that time to the time of bringing the suit, were the holders and occupants of a large acreage of ground located under the waters of said Providence river and the upper part of Narragansett Bay. That said ground consisted of, to wit, three lots or parcels of land, the first of which is located on, to wit, the easterly portion of Great Bed, to wit, westerly and northerly of the Beacon, so called, the second lying northerly of Sabin's Point and easterly of the channel, the third southerly of Sabin's Point and northerly of Crescent Park Wharf and easterly of the channel, being in area in the aggregate amount of, to wit, 28 acres. That said land was held by virtue of valid leases and granted by the state of Rhode Island to them for the purpose of growing shell-fish, and which were in full force and effect throughout all said times. That said ground during the time alleged was used and occupied by the plaintiffs, being properly and sufficiently staked, marked, designated, and identified by said plaintiffs. That upon said ground the plaintiffs at the time aforesaid had a large quantity of oysters, quahaugs, and other shellfish upon said ground occupied, as aforesaid, which were the property of the plaintiff, being in the amount of, to wit, 40,000 bushels. That they were engaged in the business of selling shellfish and had a large trade

---

[72] 31 R.I. 295, 77 A. 145 (1910).

and business which they supplied from the shellfish upon said grounds. That said shellfish were ready for the market and were being grown for the market, and said shellfish and trade were of great value. That thereafter on said days and on the times heretofore alleged said coal tar, water gas tar, oils, and other substances discharged, suffered, and allowed to escape into the Providence river by the defendant corporation, as aforesaid, were carried down said river by said currents and tides and deposited in and upon said grounds used and occupied by said plaintiffs as aforesaid and in and upon said oysters, quahaugs, and other shellfish deposited on the grounds described as aforesaid, and greatly injured and destroyed the same, impairing their growth and rendering them unfit for market. That said plaintiffs were at said several times engaged in the business of selling shellfish and were dependent upon the shellfish upon those grounds to supply their trade and customers --whereby and by means thereof said oysters, quahaugs, and other shellfish of the said plaintiffs were killed and destroyed, and covered by the said coal tar, water gas tar, oils, and other substances and the like, and greatly hindered in growth and value, and said plaintiffs were greatly hindered and prevented from taking, selling, and marketing the same and were deprived of all the profits and benefits which would have arisen and accrued to them therefor, whereby and by means whereof said plaintiffs have suffered great loss and been greatly injured."

The fourth count differs in one material respect from the foregoing counts in not declaring upon the leases of oyster beds as the basis of recovery. The allegation of property in the shellfish is made to depend on possession and is set out as follows: "That the plaintiffs on, to wit, the 1st day of October, 1903, and throughout the period from that time until the commencement of this suit, were the sole holders and occupants of a large acreage of ground of said Providence river and the upper part of Narragansett Bay covered by the tide waters of said river and bay. That said ground consisted of, to wit, three lots or parcels. (Here follows specific designation.) That said ground when taken and accepted by them did not have other shellfish growing upon it, but was free from the same. That said ground at the times alleged was used and occupied by the plaintiffs and was properly and sufficiently staked, marked, designated, and identified by said plaintiff, so as to clearly indicate the ground to which they claimed the use and possession for the planting and growing of oysters, quahaugs, and other shellfish and the gathering of the sets of oysters, quahaugs, and other shellfish. That upon said ground occupied, as aforesaid, the plaintiffs at the said several times had a large quantity of oysters, quahaugs, and other shellfish, which were the property of the plaintiff, being in the aggregate amount of, to wit, 40,000 bushels. ***"

The defendant pleaded the general issue to each of the five counts, and also pleaded specially that the plaintiffs were not the owners of said oysters, quahaugs, and other shellfish, thereby raising the question of the validity of the leases declared upon in the first, second, and third counts. The plea to the fourth count denies that the plaintiffs were the sole holders and occupants of the ground described in said count, and also denies that said ground was free from all shellfish. The plaintiffs' replications were formal, and on the issue thus joined the case proceeded to trial. The verdict of the jury was in favor of the plaintiffs for $17,280. During the course of the trial numerous exceptions to the introduction of testimony, to the charge of the court to the jury, and to his refusal to charge as requested, were taken by the defendant, and two constitutional questions were raised. After the rendition of the verdict, the defendant filed its motion for a new trial which was denied. The defendant has brought the constitutional questions to this court by certification from the judge presiding at the trial as well as by bill of exceptions.

The bill of exceptions filed in this court sets forth 22 grounds of exception, of which only the first, second, third, fourth, fifteenth, sixteenth, eighteenth, and nineteenth are now pressed. These exceptions are as follows:

"Second. During the trial of said cause the plaintiffs offered in evidence a lease of oyster beds numbered 299 from the state of Rhode Island to William L. Sunderland, dated April 13, 1901; also lease of oyster beds numbered 454 from the state of Rhode Island to William L. Sunderland, dated August 1, 1902; also lease of oyster beds numbered 466 from the state of Rhode Island to Payne & Butler, dated January 24, 1903; also lease of oyster beds numbered 320 from the state of Rhode Island to Walter H. Butler, dated June 17, 1901--which said leases are marked respectively plaintiffs' Exhibits 1, 2, 3, and 4. The defendant objected to the introduction of leases numbered 299, 454, and 320 on the ground that they were not leases made to the plaintiffs in this case, and there were no assignments thereof in writing, and to the introduction of all the leases on the ground that it did not appear that in making said leases, or in agreeing to let the property covered by the leases, the shellfish commissioners proceeded in accordance with the requirements of the statutes of the state, or that they acted within the jurisdiction conferred upon them by the statutes of the state.

"Third. The presiding justice allowed one of plaintiffs' witnesses to testify in regard to the amount of quahaugs and little-necks that were on the land known as the beds south of Sabin's Point, which William L. Sunderland had leased from the state, but which Payne & Butler claim to hold under a verbal arrangement with Sunderland; the purpose of such testimony being to lay a claim for damages by reason of injury to such quahaugs and little-necks. Objection was made to this testimony on the ground that the land in question was shown by the evidence introduced by the plaintiffs to be under a lease, not to the plaintiffs, but to William L. Sunderland, and also on the further ground that it appears by the leases and by the statutes of the state that these lots were not leased for the cultivation of little-necks and quahaugs, but that they are and must be leased for the cultivation of oysters only.

"Sixteenth. This ground of exception is based upon the refusal of the trial judge to give the following instructions as requested by defendant's counsel: '(1) The plaintiffs cannot recover on the ground that they are lessees of any of the grounds described in the declaration, because the laws under which the alleged leases purport to be made are in contravention of section 17 of article 1 of the Constitution of the state of Rhode Island, and are therefore void. (2) The plaintiffs cannot recover on the ground that they are lessees of any of the grounds described in the declaration and in the leases produced in evidence because the shellfish commissioners, in making said leases, did not proceed in accordance with the requirements provided by the state authorizing them to make such leases, and the leases are therefore void. (3) The plaintiffs cannot recover on the ground that they are lessees of any of the grounds described in the declaration and in the leases produced in evidence because the shellfish commissioners had no jurisdiction or authority to make said leases. (4) The injury suffered by the plaintiffs, if any, is an injury suffered by them in common with the general public; and they have no right of action for such injury. (5) The plaintiffs cannot recover for injury to quahaugs located on any of the premises described in their declaration. *** (8) The plaintiffs cannot recover for injury to quahaugs on land leased to W. L. Sunderland. (9) The plaintiffs cannot recover for injury to quahaugs on land leased to W. H. Butler. (10) The plaintiffs cannot recover on the ground that the leases, although originally invalid, were made valid by the act passed at the January session, A. D. 1908, because:

In the consideration of the first exception, that the verdict is against the law and the evidence, and that the damages awarded are excessive, a brief statement of the facts may be of service:

It appears that, some time before 1901, the plaintiffs had obtained possession of some small pieces of oyster land and were engaged in the business of oyster culture separately in a small way. About 1901 they began to carry on business jointly and to secure more ground. Walter H. Butler then held two acres of oyster ground south of Sabin's Point, under lease No. 320. In 1902 the plaintiffs sublet four acres of ground adjacent to this ground from W. L. Sunderland who held it under lease No. 454. Previously John S. Payne had sublet of said W. L. Sunderland two acres of ground adjacent to this land located south of Sabin's Point. In consequence of this subletting the plaintiffs were permitted by Sunderland to have the use of said ground for planting oysters and other shellfish for a consideration paid to him, and thereafter they entered upon the land and enjoyed this use. The plaintiffs planted all this ground--eight acres in all-- lying south of Sabin's Point, with oysters and little-necks or quahaugs, and used it generally as a private and several oyster fishery. It was buoyed, staked, and defined as such grounds commonly are, so as to clearly define the ground so exclusively occupied and claimed by them. It was further shown by the plaintiffs that all the eight acres of ground which lay south of Sabin's Point and opposite Camp White had a hard mud bottom; the result of the ground having been used a few years before by the city of Providence as a dumping ground for the mud taken from the river in connection with certain excavations. The ground was not "a natural oyster ground," and it did not have oysters and other shellfish growing upon it naturally when it was occupied and planted by the plaintiffs, or either of them, or by W. L. Sunderland. The other piece of land occupied and planted by the plaintiffs at the time of the injuries complained of was a tract of 20 acres situated to the west and north, and near to the Beacon, so called, south of Starve Goat Island and west of the channel. This land was leased by the plaintiffs under lease No. 466. This ground was occupied by the plaintiffs in the fall of 1902 and was planted by them with a quantity of oysters of varying ages and sizes. This ground was buoyed off, marked, and designated by the plaintiffs in the customary manner adopted by all holders of oyster grounds in Narragansett Bay. It was occupied exclusively by the plaintiffs, and no claim was ever made of a right to occupy the same by any other person.

The defendant corporation was engaged in the manufacture of illuminating gas and its by-products in the winter of 1903-04, and up to the present time, for use in the city of Providence and adjacent territory. It was the only corporation there engaged in the manufacture of gas in any quantities and the only one which had a gas plant located upon or near to the shore of any of the waters flowing through the city of Providence. The defendant had two plants, one located upon the west bank of the Providence river a little to the south of Point Street Bridge, known as the North Station, or the West Station, either name being used, and the other located on the west bank of the Providence river but farther south at the foot of Public street, and was known as the South Station. Prior to March 28, 1904, coal gas and water gas were manufactured at each of these two plants, though water gas was manufactured much more extensively at the South Station. The ground in the vicinity of the South Station and especially immediately to the south of it was "made" land composed chiefly of cinders, ashes, and general refuse matter. This is largely true of the ground upon which the South Station stands.

It is the injury caused by the escape of water gas tar, and other injurious by-products and refuse from the manufacture of gas from the plants of the defendant, into the Providence river, and thence by means of the currents and tides to and upon the oyster beds and tide-flowed land of the plaintiffs, and the damages consequently following, of which the plaintiffs complain. The oyster beds and tide-flowed land of the plaintiffs are situate about two miles below the defendant's gas plant.

It appears that the winter of 1903-04 was a particularly severe one, and that the upper part of the bay and the river were frozen over in January, 1904, and remained so for some eight weeks or so, during which time the plaintiffs as well as other oyster men were prevented from taking oysters from the oyster beds on account of the freezing of the river. At the time they left off work in January, 1904, the oysters and little-necks were apparently in good marketable condition; but immediately upon resuming the work of catching oysters and little-necks in March and April, 1904, after the ice had broken up, some of the shellfish were found to be dying, and the flavor of those which were not dying was so affected as to render them unmarketable. They had the peculiar flavor of gas which is pungent, disagreeable, and so repellant to the taste as to make them uneatable. The escape of the water gas tar and refuse matter into the river continued throughout the years of 1904 and 1905 and caused more or less damage to the shellfish during both years.

The plaintiffs claimed damages for injuries caused to them both by acts of commission and omission on the part of the defendant corporation to the following property: (1) To the oysters upon their several beds; (2) To the little-necks or quahaugs upon the beds; and (3) to the ground itself by rendering the same unfit for the planting of oysters and other shellfish because of the settling upon it of water gas tar and other by-products from the defendant's gas plant.

Water gas tar is a peculiar and distinct chemical substance which is a by-product of the manufacture of water gas. Water gas, which of itself is nonluminous, requires the injection into it of carbons, and this is done by the use of a low grade of petroleum which is passed into a chamber containing the gas at a high temperature, where it is decomposed into a gas and mixed with the water gas, and thereby supplies the necessary particles of carbon to cause light when the gas is burned. From this process there is a refuse which is collected when the gas is cleansed and cooled which is known as "water gas tar." This substance has practically the same density which salt water of the upper waters of Narragansett Bay has. It will, therefore, float in suspense in the water for long distances before it settles. It was easily carried by the tides and currents to the plaintiffs' beds and even farther south. Water gas tar is of a viscous or sticky nature. It is repellent to the nature of the oyster which will not admit it, and while this is present in the water in any quantity the oyster will not feed, and will therefore eventually starve itself to death. It also kills the vegetable organism in the water which naturally exists in large amounts, and which furnishes the chief food upon which the oyster and other shellfish live. By this lack of food the oyster is starved and destroyed.

The defendant claimed: (1) That the oysters, etc, were killed by anchor ice; (2) that they were "drowned" by the heavy freshets of fresh water coming down the three rivers which flow into Providence river; (3) that the plaintiffs cannot recover for shellfish which the plaintiffs had placed upon the ground sublet from W. L. Sunderland; and (4) that all the leases were void; and, therefore, the plaintiffs could not recover.

There was no dispute as to the market value of the oysters or little-necks. There was a dispute as to the amount of shellfish which the plaintiffs claimed to have. The fact that the ratio of deaths among oysters and little-necks was from 50 per cent. to 75 per cent. in 1904 was not disputed. The testimony further shows that all the ground lying south of Sabin's Point was not a natural oyster or quahaug ground. The ground was free from either when it was buoyed, staked in, and occupied by the plaintiffs.

The defendant's chief defense is a claim that all the leases in issue are void through a failure on the part of the commissioners of shellfisheries to issue them in accordance with the terms of the law.

It is sufficiently shown that the defendant is responsible for the damage done, and there is ample evidence as to the amount of the damages. In these respects the verdict is supported by the evidence. But by the pleadings the title of the plaintiffs is put in issue, and it is necessary for them to maintain the same by a fair preponderance of the evidence.

As to the property of the plaintiffs upon the eight acres of tide-flowed ground, lying south of Sabin's Point, sublet by W. L. Sunderland to the plaintiffs, none of which was a natural oyster or quahaug bed, and upon which the plaintiffs deposited oysters and quahaugs, otherwise known as little-neck clams, or little-necks, belonging to them for the purposes of growth and culture of the same for the use and benefit of the plaintiffs, which land was buoyed, staked, and defined by them, so as to give public notice, in the customary manner, of the fact that they had exclusive possession of the same for the purposes aforesaid, we are of the opinion that the plaintiffs have shown sufficient title by proof that they were in possession, under a claim of right, not disputed by any one claiming to have a better or other or any title to the same, and that by so depositing the shellfish upon said land in the public waters of the state the plaintiffs did not abandon the same, but continued to maintain their possession of said shellfish in a manner fully recognized and held to be sufficient from time immemorial. See *Fleet v. Hegeman,* 14 Wend. (N.Y.) 42; *Brinckenhoff v. Starkins,* 11 Barb. (N.Y.) 248; *Sutter v. Van Derveer,* 47 Hun (N.Y.) 366; *Id.,* 122 N.Y. 652, 25

N.E. 907; *Cook v. Raymond,* 66 Conn. 285, 33 A. 1006; *Grace v. Willets,* 50 N. J. Law, 414, 14 A. 559; *Decker v. Fisher, 4 Barb. (N.Y.)* 592; *Wooley v. Campbell,* 37 N. J. Law, 163.

In Sutter v. Van Derveer, supra, the plaintiff deposited small oysters in Jamaica Bay in a bed defined by stakes and buoys and left them there to grow. Later when he was raising them for the purpose of sending them to market, the defendant drove off the plaintiff's men and took possession of the oysters. Defendant attempted to justify this action under a lease or license obtained from the board of auditors granted under a statute which authorized any inhabitant of Jamaica to use a portion of the land under the public waters of the town on which there was no planted bed of oysters for the purpose of planting oysters therein, and, provided the land was not a planted bed of oysters, such board was authorized to give the applicant a certificate of license. It was held that the certificate of the defendant did not cover the oyster bed of the plaintiff, because that was already planted and occupied. The court said, at page 367 of 47 Hun: "Jamaica Bay is an arm of the sea, and its waters are tidal and public, and all the inhabitants of the state have a common right to the use of the same for navigation, fishing, and all other purposes of a public character, and the plantation of oysters in such waters is a legitimate exercise of such common right, and they remain the property of the person so planting them, and their conversion by another furnishes a cause of action to the owner. *Post v. Kreischer,* 103 N.Y. 110 [8 N.E. 365]. Oysters planted in tidal waters on a well marked and clearly defined bed, where there were no natural oysters before, are the personal property of the planter, and he may maintain an action for their conversion. *Fleet v. Hegeman,* 14 Wend. [N.Y.] 42; *Decker v. Fisher, 4 Barb.* [N.Y.] 592; *Lowndes v. Dickerson,* 34 Barb. [N.Y.] 586. Oysters are classified in law with animals *ferae naturae* in which a qualified property may be acquired, by possession, both by the civil and common law, and such property continues so long as the dominion of the owner is maintained. It is protected by the law to the same extent as an absolute property, and every invasion of the same is redressed in the same manner. When they are reclaimed or taken from their natural beds and collected together for the purposes of growth or multiplication, they cannot be retained in actual manual possession, and they must be deposited in their native element, and so long as they remain within such control as will enable the owner to resume actual possession of them at his pleasure his ownership will continue. *** Oyster beds are marked and defined by stakes and buoys to give notice to the world of the exercise of the common right to use the public waters of the state by the occupant, and that the oysters have not been abandoned. They are, when so planted, entirely within control; and the purpose is to take them up at pleasure when they become suitable for the market. The traffic in oysters has become very expanded, and their cultivation has come to be a very extensive business. The oysterman plants his oysters in their natural soil and element with regularity, and expects to gather their growth with as much certainty as the farmer plants his grain in the soil and expects to harvest and enjoy the crop; and, for all practical purposes, the title to the oysters is as valuable and certain as the title to grain or any other article of commerce."

As to the other 20 acres of oyster ground whereon oysters grew naturally, other and different questions are presented, and among them arises the question of the constitutionality of the legislation relative to the leasing of oyster grounds in this state. A consideration of the subject involves an inquiry into the nature of the right thereby granted. It has been decided that the privilege of locating oyster beds on public lands, and of planting and taking oysters therefrom, is merely a license which may be revoked at the pleasure of the Legislature, and which ceases with the use of the land for that purpose. It is also subject to the public's right of navigation and of fishery, and, if it interferes therewith, the oysters or clams, etc., may be removed as a nuisance.19 Cyc. 998, and cases cited. Even if it should be regarded in this state as a license for a definite term, it is still but a license, for there is no right of renewal in the lessees at the expiration of their term. The defendant claims that "Pub. Laws, c. 853 (passed March 23, 1901), as interpreted by the shellfish commissioners in leasing the state oyster lands to the plaintiffs, is unconstitutional because it grants an exclusive fishery." This is a peculiar method of raising a constitutional question. There is only one body that is authorized to interpret the statutes of this state with the view of determining their constitutionality. Under article 12 of the amendments to the Constitution of the state, adopted in November, 1903, section 1, "The Supreme Court shall have final revisory and appellate jurisdiction upon all questions of law and equity." Furthermore, a statute is constitutional or not as the case may be without reference to the interpretation of any board or boards of commissioners. It is deemed to be constitutional until this court shall have decided that it is not. In answering the constitutional question sought to be raised, we will consider the question to be whether said chapter 853 is in derogation of article 1, § 17, of the Constitution: "The people shall continue to enjoy and freely exercise all the rights of fishery, and the privileges of the shore, to which they have been heretofore entitled under the charter and usages of this state. But no new right is intended to be granted, nor any existing right impaired, by this declaration." What rights, then, had the people under the charter and usages of this state? By the common law all persons have a common and general right of fishing in the sea, and in all other navigable or tidal waters; and no one can maintain an exclusive privilege to any part of such waters unless he has acquired it by grant or prescription. This right is subject to the paramount right of navigation. 19 Cyc. 992, and cases cited. According to some authorities, the crown has no power since the passage of Magna Charta to grant a several right of fishery in navigable tidewaters, except where the right had already been enjoyed for some time prior to that statute. By other authorities, however, it is held that Magna Charta has no effect and in no way restricts the power of the crown to grant such a several right of fishery. 19 Cyc. 994, note, and cases cited. There is no question made but that since Magna Charta, the crown and Parliament together could grant such a right. If we concede that under the charter from the crown this state received only such rights as the crown could grant, and even that under Magna Charta the crown could not grant an exclusive or several right of fishery, then the state indeed lacked the authority that remained in the Parliament of Great Britain.

In *Lakeman v. Burnham, 7 Gray (Mass.)* 440, Chief Justice Shaw, referring to the cases of *Weston v. Sampson, 8 Cush. (Mass.)* 347, 54 Am.Dec. 764, Com. v. Alger, 7 Cush. (Mass.) 53, and *Dunham v. Lamphere, 3 Gray (Mass.)* 268, said: "By these cases we think the following propositions are well established: That by the Revolution, and by the acknowledgment of our independence by the British government, the state of Massachusetts succeeded to all public rights of British subjects, whether originally belonging by prerogative to the crown, or exercised and administered by Parliament in due course of law; that by the charter of the colony of Massachusetts the people and settlers of the territory acquired not only the right of soil, but a right to the shores and arms of the sea, for all useful purposes of navigation and fishery; that since that charter no exclusive right of navigation or fishery, on the seashores or in the bays or arms of the sea, could be acquired, except under the authority of the colonial, provincial, or constitutional government, administered by the Legislature by act or resolve." This reasoning is equally appropriate to the state of Rhode Island. As "by the Revolution and by the acknowledgment of our independence by the British government," this state "succeeded to all public rights of British subjects, whether originally belonging by prerogative to the crown, or exercised and administered by Parliament in due course of law," the public rights to which we so succeeded were thereby added to the rights and powers derived from the charter, and the public rights and powers formerly exercised and administered by Parliament naturally and necessarily came under the control of the Legislature. So at that time we had the authority of both crown and Parliament. And we continued to have that unrestricted until the adoption of the federal Constitution. Thus, the Legislature of this colony had from the date of the granting of the charter, July 8, 1663, to the time of the Revolution, 1776, or, according to Chief Justice Shaw, later to the date of the acknowledgment of our independence by the British government, the powers conferred by the charter; from the Revolution or the time of the acknowledgment of our independence to May 29, 1790, when we adopted the federal Constitution, the Legislature had the combined powers of crown and Parliament and thereafter until May, 1843, the date of the adoption of our Constitution, the General Assembly had the full powers of crown and Parliament less whatever power had been granted to the federal government. After the adoption of the state Constitution, the Legislature had the powers of crown and Parliament aforesaid, less the power taken therefrom for the federal government, and also minus whatever powers were taken from it by the Constitution of the state.

"In every sovereign state there resides an absolute and uncontrolled power of legislation.

In Great Britain this complete power rests in the Parliament; in the American states it resides in the people themselves as an organized body politic. But the people, by creating the Constitution of the United States, have delegated this power as to certain subjects, and under certain restrictions, to the Congress of the Union; and that portion they cannot resume, except as it may be done through amendment of the national Constitution. For the exercise of the legislative power, subject to this limitation, they create, by their state Constitution, a legislative department upon which they confer it; and, granting it in general terms, they must be understood to grant the whole legislative power which they possessed, except so far as at the same time they saw fit to impose restrictions. While, therefore, the Parliament of Britain possesses completely the absolute and uncontrolled power of legislation, the legislative bodies of the American states possess the same power, except: First, as it may have been limited by the Constitution of the United States; and, second, as it may have been limited by the Constitution of the state. A legislative act cannot, therefore, be declared void, unless its conflict with one of these two instruments can be pointed out. It is to be borne in mind, however, that there is a broad difference between the Constitution of the United States and the Constitutions of the states as regards the powers which may be exercised under them. The government of the United States is one of enumerated powers; the governments of the states are possessed of all the general powers of legislation. When a law of Congress is assailed as void, we look in the national Constitution to see if the grant of specified powers is broad enough to embrace it; but, when a state law is attacked on the same ground, it is presumably valid in any case, and this presumption is a conclusive one, unless in the Constitution of the United States or of the state we are able to discover that it is prohibited. We look in the Constitution of the United States for grants of legislative power, but in the Constitution of the state to ascertain if any limitations have been imposed upon the complete power with which the legislative department of the state was vested in its creation. Congress can pass no laws but such as the Constitution authorizes either expressly or by clear implication; while the state Legislature has jurisdiction of all subjects on which its legislation is not prohibited. 'The law-making power of the state,' it is said in one case, 'recognizes no restraints, and is bound by none, except such as are imposed by the Constitution.' That instrument has been aptly termed a legislative act by the people themselves in their sovereign capacity; and is therefore the paramount law. Its object is not to grant legislative power, but to confine and restrain it. Without the constitutional limitations, the power to make laws would be absolute. These limitations are created and imposed by express words, or arise by necessary implication. The leading feature of the Constitution is the separation and distribution of the powers of the government. It takes care to separate the executive, legislative, and judicial powers, and to define their limits. The executive can do no legislative act, nor the Legislature any executive act, and neither can exercise judicial authority." Cooley, Const. Lim. (7th Ed.) p. 241.

Moreover, the question is no longer an open one, for it has been finally determined in this state. The case of *State v. Cozzens,* 2 R.I. 561, is decisive upon this point:

"This was an indictment under the 'act for the preservation of oysters and other shellfish within this state,' and charged the defendant with stealing oysters of the value of $40, from a private oyster bed in Narragansett Bay. The indictment was tried in the court of common pleas, before Staples, J., and the defendant found guilty. The cause was brought before this court upon the following exceptions to the rulings in the court below, to wit:

"(1) Because the defendant requested the court, by his counsel, to charge the jury that the eighth, ninth, tenth, eleventh, twelfth, and thirteenth sections of the act, entitled 'An act for the preservation of oysters and other shellfish in this state,' and the acts in amendment and explanatory of said act, were unconstitutional and void, and in violation of and repugnant to the seventeenth section of the declaration of certain constitutional rights and privileges in the first article of the Constitution of this state, which section provides that 'the people shall continue to enjoy and freely exercise all the rights of fishery and the privileges of the shore to which they have been heretofore entitled under the charter and usages of this state,' which charge the court refused to give, but did charge the jury that said several acts and sections were constitutional and binding upon the citizens of this state.

"(2) Because he offered evidence to show that the place where the said offense was alleged to have been committed is and has been freely enjoyed and used from the earliest settlement of this state to the time of the offense alleged in the indictment, as a common and public fishery, where the people of the state have been accustomed, under the charter and usages of this state, to fish for oysters and other shellfish, which said evidence the court refused to allow to pass to the jury.

"(3) Because the defendant offered to prove and did prove that the said places in the indictment mentioned had been freely used and enjoyed by the people of this state down to the commission of the said alleged offense, as a public and common quahaug fishery, and that said fishery for quahaugs has been interrupted, interfered with, and destroyed by the planting of the oyster beds in question, and requested the court to charge the jury that, if they believed said beds did interfere with the public use and enjoyment of said quahaug fishery, then the act of the commissioners in leasing said beds was in violation of and repugnant to the said seventeenth section of said Bill of Rights, which charge the court refused to give.

"(4) Because the defendant's counsel requested the court to charge the jury that no power to lease a natural oyster bed exists in the Legislature of this state or in the commissioners by them appointed, and that any lease so executed is unconstitutional and void, which charges the court refused to give.

"(5) Because the court refused to charge the jury that said acts and parts of acts and the doings of the commissioners under the same are in violation of the principles of the common law.

"(6) Because the court refused to charge the jury that no exclusive right could exist in the proprietors of any private oyster bed to take from said bed oysters, the natural growth of said bed, and that the defendants had a perfect right to take such oysters, so naturally growing on any natural oyster bed, without molestation, and that the crime alleged in said indictment could not be committed by taking any such natural oysters; but did charge that the proprietors of said bed had, under the lease granted by the said commissioners, the exclusive right to take oysters planted or naturally growing thereon.

"Greene, C. J.,

"We are of opinion that the declaration of rights and privileges, contained in the seventeenth section of the first article of the Constitution, was intended to be carried into effect by legislative regulations; such regulation having for its object to secure to the whole people the benefit of the constitutional declaration, and being necessary for that purpose. In pursuance of this object, the first section of the act, under which this indictment was preferred, provides that no oysters shall be taken from the common fisheries between the 15th day of May and the 15th day of September. The second section prohibits the taking of quahaugs or clams from Long Bed, West Bed, or from Great Bed, during the same time. The third section provides that no person shall take from any public oyster bed more than three bushels in twenty-four hours, nor plant on a private bed oysters taken from a public bed. And section 4 prohibits the use of dredges in taking oysters. These sections restrict the right of fishery, and it will not be contended that they are unconstitutional. We refer to them to show that legislative restriction is indispensable to secure to the public the benefit of the oyster fishery.

In the case of *Clark v. Providence,* 16 R.I. 337, 340, 15 A. 763, 765, 1 L.R.A. 725, which was a suit in equity for an injunction to restrain the defendant city from filling in part of the cove basin and from alienating or diverting the land surrounding said basin, now used as a public park, but formerly tide-flowed and reclaimed by filling, to other uses, this court said: "Second. The complainants ask an injunction, because they are citizens and residents of the city of Providence, and as such entitled to enjoy and freely exercise all the rights of fishery and privileges of the shore to which the people of the state have been heretofore entitled under the charter and usages of the state, and because, the cove basin being covered by tide water, an act of the General Assembly authorizing the filling or partial filling of it is in derogation of these rights and privileges and in conflict with Const. art. 1, § 17. This section reads as follows: 'The people shall continue to enjoy and freely exercise all the rights of fishery, and the privileges of the shore to which they have been heretofore entitled under the charter and usages of this state. But no new right is intended to be granted, nor any existing right impaired, by this declaration.' It is clear that this section leaves the rights of the people as they existed previously to the Constitution. It neither diminishes nor adds to them, as was decided by this court in *State v. Medbury,* 3 R.I. 138. It is necessary, therefore, to know what the rights and privileges of the people were under the charter and usages of this state before the Constitution was adopted, in order to decide whether they are infringed by the act of the General Assembly. The act must be taken to be constitutional unless the contrary appears, for in our opinion the General Assembly has in this matter the authority, not simply of the English crown, but of both crown and Parliament, except so far as it has been limited by the Constitution of the state, or by the Constitution and laws of the *United States. Gould v. Hudson R. R. Co.,* 6 N.Y. 522; *King v. Montague, 4 B. & C.* 598. There is little in the Rhode Island reports to enlighten us on this point. In *State v. Medbury,* 3 R.I. 138, the court held that the provision of the colonial charter, in regard to fisheries, does not relate to the shellfisheries of the state, but was designed to continue to the people of New England the right more especially to prosecute the cod fishery in Rhode Island waters, as they had been accustomed to prosecute it before the charter was granted. In *State v. Cozzens,* 2 R.I. 561, the court held that the General Assembly has power to lease portions of the tide waters of the state for private oyster fisheries to the complete exclusion, for the exercise of the shellfisheries thereon, of all but the lessees, even though the leases include portions of natural oyster and quahaug grounds, notwithstanding said section 17. Neither of these decisions gives any support to the claim of the complainants, but, so far as they bear, are adverse to it. It is common knowledge that the citizens of the state have always been accustomed to dig clams freely along the shores of the bay and river wherever they could be found, and, subject to some legislative regulations, to fish in the deeper waters; and because this has been so, and because formerly citizens were accustomed to clam and fish in the cove, the complainants take it for granted that they are entitled, under section 17, to clam and fish there in the same manner forever, and that therefore any act of the General Assembly which authorizes the filling or partial filling of the cove, thus lessening their ability to do so, must be unconstitutional and void. But it is also common knowledge that there are many places where fish and clams were formerly taken which are now solid land made by filling out the shores, either with the tacit acquiescence or with the implied or express consent of the General Assembly. The acts of the General Assembly establishing, or authorizing the establishing, of harbor lines, some of which existed before the Constitution (Digest of 1822, pp. 484, 485), go to show that the assumption of the complainants is too broad, and that these rights of clamming and fishing are enjoyed in subordination to the paramount authority of the General Assembly to regulate and modify, and, to some extent at least, to extinguish them.

Prior to the adoption of the state Constitution, the Legislature of the state had full authority to grant exclusive rights of fishery in the public waters of the state. And in at least two instances they exercised this right, as appears from the following "Charter to L. Wilcox and E. Carpenter," granted January, 1822:

"Whereas Leonard Wilcox and Earl Carpenter have preferred their petition to this assembly, setting forth that they have made arrangements for importing and keeping on hand a regular supply of oysters for the use of the inhabitants of the town of Providence and its vicinity, and prayed that an act may be passed securing to them the exclusive right to take oysters on a piece of ground in Providence river, to be by them cultivated, and improved by them for that purpose:

"Section 1. Therefore, be it enacted by the General Assembly, and by the authority thereof it is enacted, that the said Leonard Wilcox and Earl Carpenter, their heirs, executors, administrators and assigns, may and they are hereby authorized and empowered to, import from without this state, and deposit for growth and increase, on the piece of ground, situated in Providence river, to be hereafter more particularly described, such quantities of oysters as they may think proper, not less, however, than three hundred bushels per year; and to have and enjoy the entire and exclusive right to fish and take oysters from the same; that the piece of land or flat on which the said oysters shall be deposited and taken as aforesaid shall be located by George Field and Philip Allen, Esquires, who are hereby appointed a committee for that purpose, and are hereby authorized to set off and locate two acres of land or flat in such part of Providence river as they may think proper to be improved by them, the said Wilcox and Carpenter, for the purposes aforesaid; and to make their report into the secretary's office when and in what manner they shall have located the same.

"Sec. 2. And be it further enacted, that any person or persons who shall fish for and take oysters on such piece of land or flat as said committee shall locate as aforesaid, shall forfeit and pay, for every such offense, to and for the use of the said Wilcox and Carpenter, their heirs, executors, administrators or assigns, a sum not less than ten nor more than fifty dollars, to be recovered by action of debt in any court proper to try the same; and the boat or boats, with the other fishing tackle employed in taking such oysters, shall be forfeit: Provided however, that before any oysters shall be deposited on said piece of ground, to be located by said committee as aforesaid, the said Wilcox and Carpenter, their heirs, executors, administrators or assigns, shall erect and cause at all times during the continuance of this act to be kept up, at each of the two corners of said piece of land or flat nearest the shore, a permanent post, with painted boards on each post, with hands pointing towards the bounds of the said piece of land or flat, and with the penalty for fishing thereon, painted in legible letters on each board; and whenever the said posts and boards shall not be kept up in manner as aforesaid, the said Wilcox and Carpenter shall not be entitled to recover the forfeitures aforesaid; and any person or persons who shall pull down or damage such posts or boards, or shall make fast any vessel or boat to them, shall be liable to pay a sum not less than two nor more than ten dollars, to and for the use of the said Wilcox and Carpenter as aforesaid, to be recovered as aforesaid: And provided also, that the said Wilcox and Carpenter shall first agree with the owner or owners of the adjacent land or shore: And provided further, that nothing in this act shall prevent the owner or owners of said shore from erecting any wharf or wharves thereon.

"Sec. 3. And be it further enacted, that this act shall continue in force for seven years, and no longer.

"Sec. 4. And be it further enacted, that if the said Leonard Wilcox or Earl Carpenter, or either of them, shall take, or cause to be taken, any oysters, in any part of this state, without the limits of said two acres, after such limits shall have been defined by the committee aforesaid, and for the purpose of depositing said oysters, so taken, within said limits, and shall so deposit the same within said limits, the said Wilcox and Carpenter, or either of them so taking and depositing as aforesaid, shall forfeit and pay a sum not less than ten nor more than fifty dollars; to be recovered by action of debt in any court proper to try the same; one moiety of said sum to and for the use of any person who shall sue for and recover the same, and the other moiety thereof to and for the use of the state.

"Sec. 5. And be it further enacted, that this act, together with the report of the committee locating and defining the limits of the two acres of land or flat, as hereinbefore required, shall be published in one of the newspapers printed in Providence, three weeks successively."

And also in "An act to authorize Ephraim Gifford to plant a bed of oysters in Mount Hope Bay," passed at the October session, 1827, as follows:

"Section 1. Be it enacted by the General Assembly, and by the authority thereof it is enacted, that Ephraim Gifford, of Bristol, in this state, and his heirs, executors, administrators and assigns, shall be and hereby are authorized to import from without the same, and deposit, for growth and increase, on a certain piece of ground situated in Mount Hope Bay, and contiguous to common fence point, near the mouth of town pond creek, not exceeding forty rods along the shore of said point, nor twenty rods from low water mark, towards the channel of w water mark, towards the channel of said bay, such quantities of oysters as he or they may think fit, not less, however, than fifty bushels; and the said Gifford, and his said representatives shall have and enjoy the exclusive right to take oysters from the said piece of ground, for the period of ten years from the passage of this act.

"Sec. 2. And be it further enacted, that any person or persons who shall fish for and take any oysters on said piece of land, until the expiration of said ten years, shall forfeit and pay for every such offence, to and for the use of said Gifford, or his said representatives, the sum of ten dollars, to be recovered in an action of debt, before any justice of the peace in the county of Newport or Bristol; and the boat or boats, so employed in taking any such oysters, shall be forfeited: Provided, however, that before any oysters shall be deposited on said piece of ground as aforesaid, the said Gifford shall erect and cause to be kept up, at each of the corners of said piece of ground, a buoy or a permanent post, with a painted board on each post or buoy, with a hand thereon pointing inward, to designate the spot, with the penalty for fishing for oysters thereupon; and whenever said posts or buoys shall cease to be kept up, neither of said penalties shall be incurred; and any person or persons who shall remove, pull down or destroy, or in any way damage either of said posts, buoys or boards, shall forfeit and pay the sum of ten dollars, to be recovered and appropriated as aforesaid: Provided, however, that before any part of this act shall take effect, said Gifford shall obtain the consent of the owner or owners of the land on the shore adjoining the said piece of land, for the appropriation thereof as aforesaid.

"Sec. 3. And be it further enacted, that if the said Gifford, or any one or more of his said representatives, shall take or cause to be taken any oysters in any part of this state, without the limits of said piece of land, for the purpose of depositing the same thereon, he or they, as the case may be, shall forfeit and pay, to and for the use of this state, the sum of twenty dollars for every such offence, to be recovered in an action of debt, in any court proper to try the same.

"Sec. 4. And be it further enacted, that this act shall be published in the newspaper printed at Warren, and in one of the newspapers printed at Newport, three weeks successively before the same shall take effect.

These are instances of the manner in which the Legislature exercised its powers over the rights of fishery, to which the people were entitled under the charter and usages of the state, and illustrate the fact that they were subject to the control of the General Assembly. The rights which the General Assembly had in 1822 and 1827 were not abridged by the foregoing provisions of article 1, § 17, of the Constitution, for it is expressly stated therein "no new right is intended to be granted, nor any existing right impaired, by this declaration." In other words, no change was made. No greater privileges were reserved to the people than they already had, and no powers or rights of the General Assembly were thereby abridged. Therefore the whole subject of fisheries, floating and shellfish, and all kinds of shellfish whether oysters, clams, quahaugs, mussels, scallops, lobsters, crabs, or fiddlers, or however they may be known and designated and wherever situate within the public domain of the state of Rhode Island, are under the fostering care of the

General Assembly. It is for the Legislature to make such laws regulating and governing the subject of lobster culture, oyster culture, clam culture, or any other kind of pisciculture, as they may deem expedient. They may regulate the public or private fisheries. They may even prohibit free fishing for a time and for such times as in their judgment it is for the best interest of the state so to do. They may withhold from the public use such natural oyster beds, clam beds, scallop beds, or other fish beds as they may deem desirable. They may make a close time within which no person may take shellfish, or other fish, and generally they have complete dominion over fisheries and fish as well as all kinds of game. We find no limitation, in the Constitution, of the power of the General Assembly to legislate in this regard, and they may delegate the administration of their regulations to such officers or boards as they may see fit. Considered in this light, chapter 853 aforesaid is not obnoxious to the Constitution. The defendant's second exception must be overruled. The leases therein mentioned were not subject to attack in a collateral proceeding. The leases are valid on their face. The defendant cannot, therefore, attack their validity collaterally. The validity of such grants can only be attacked by a proceeding in equity or in proceedings between those who are claiming title to the same property but through different grants. If good upon its face, then the defendant who makes no claim to the title is bound by it. It is not harmed in its own property rights, and it cannot go behind the lease.

The presumption is that the lease is valid if good upon its face, and it is presumed that the acts of the judicial body which granted it are all valid. *"Omnia prcesumuntur rite et solemniter esse acta."* They are presumed to know the law and to have followed it in case of each necessary step. This presumption is necessary for the security of titles. It is easy to understand how very uncertain all titles would become if they were subject to collateral attack. Such a rule would practically allow new and distinct issues to be raised and tried within a case. This is not the policy of the law. It would tend to unsettle titles and lay them open to frequent and unexpected attacks. Grants are evidences of title of a higher order than the ordinary transfer of property rights and should not be thus subject to collateral attack. The law has wisely placed this safeguard about this form of conveyance in order that titles may have stability and security. The justice of the superior court was right when he admitted those leases, as they were valid on their face, and the defendant did not raise a question as to anything appearing upon the face of the instruments It attempts to attack the leases by going behind the face of the instrument and attacking the actual transactions. As a third party and a stranger to the title, and one who makes no claim to the title, it cannot raise this issue.

The principle that a grant or patent from the state cannot be collaterally attacked and can be questioned only in a direct proceeding in equity, unless the grant is void upon its face, is a rule of law which has been recognized throughout the history of jurisprudence in this country, and has received general recognition by the various states in the Union. As early as 1815, Marshall, C. J., in *Polk's Lessee v. Wendal et al., 9 Cranch,* 87, 98, 3 L.Ed. 665, approved the principle in the following words: "The laws for the sale of public lands provide many guards to secure the regularity of grants, to protect the incipient rights of individuals, and also to protect the state from imposition. Officers are appointed to superintend the business; and rules are framed prescribing their duty. These rules are in general directory and when all the proceedings are completed by a patent issued by the authority of the state, a compliance with these rules is presupposed. That every prerequisite has been performed is an inference properly deducible and which every man has a right to draw from the existence of the grant itself. It would therefore be extremely unreasonable to avoid a grant in any court for irregularities in the conduct of those appointed by the government to supervise the progressive course of a title from its commencement to its consummation in a patent." And concluded by stating that any review of such a grant is a matter of equitable jurisdiction unless the grant is absolutely void upon its face. This has continued to be the rule recognized by the federal courts. *Chandler v. Calumet, etc., Min. Co.,* 149 U.S. 79, 13 S.Ct. 798, 37 L.Ed. 657;

*Patterson v. Winn,* 11 Wheat. 380, 6 L.Ed. 500; *Bealmear v. Hutchins,* 148 F. 545, 78 C.C.A. 231; *Oliver v. Pullam (C. C.)* 24 F. 127.

In *New York Central & Hudson River Ry. Co. v. Aldridge,* 135 N.Y. 83, 91, 32 N.E. 50, 52, 17 L.R.A. 516, which was an action to recover land, both parties to the suit claiming title through patents from the state, but defendants' patent having been granted prior to plaintiff's in point of time, the plaintiff contends that, although the defendant's patent was prior in time, it was invalid because the defendant was not the owner of the adjoining upland, and that consequently it could not obtain a grant to the particular lands in question, being lands under tide water, and the statute making it necessary to own the upland in order to have a valid title to the tide-flowed lands. The court says: "The patent to the defendant, being prior in point of time to that granted to plaintiff, would seem to make the title of the defendant the better of the two. The plaintiff answers this by stating that the defendant was not the owner of the adjoining upland, and consequently could not legally obtain a grant of these lands under water. *** The simple claim that the patent of the defendant is void because he was not the upland proprietor (even if well founded in fact) could not be urged in this action, which is a purely legal one, to obtain possession of real estate on the ground that plaintiff is the owner thereof. There is no fact set up in the complaint calling for the exercise of the equitable powers of the court, and the plaintiff must succeed by showing the better legal title. It has been frequently held that a patent such as this, which is not void on its face, and which required evidence dehors the instrument to show its invalidity, can only be assailed in a direct proceeding to review the action of the commissioners, or by an action in equity to set aside the patent. *Blakslee Co. v. Blakslee's Sons Iron Works,* 129 N.Y. 155 [29 N.E. 2], and cases cited. The claim that defendant could not legally obtain a grant of these lands because not the owner of the upland is thus disposed of, so far as this action is concerned."

The rule is reasserted in *Archibald v. N.Y.C. & H. R. Ry. Co.,* 157 N.Y. 574, 581, 52 N.E. 567, 568, where plaintiff's title from the state was attacked because it was granted upon certain conditions, and it was contended that these conditions had not been complied with. The court says: "It is said that the grantees did not apply the property to any such uses within the time specified, and we will assume that that is so. But this does not enable the defendant to treat the prior patent as invalid or void, or permit an attack upon it in a collateral way. If there had been a breach of the conditions subsequent in the grant, the defendant is not at liberty to take advantage of that omission. The state only can claim the right to vacate the patent for breach of conditions subsequent, and then only in a direct action or proceeding for that purpose. The defendant can raise no such question in this court." *Brady v. Begun,* 36 Barb. (N.Y.) 533; *Morgan v. Turner,* 35 Misc. 399, 71 N.Y.S. 996; *Blakslee Mfg. Co. v. Blakslee's Sons Iron Wks.,* 129 N.Y. 155, 29 N.E. 2; *Jackson v. Marsh,* 6 Cow. (N.Y.) 281; *Jackson v. Hart,* 12 Johns. (N.Y.) 77, 7 Am.Dec. 280; *Jackson v. Lawton,* 10 Johns. (N.Y.) 23, 6 Am.Dec. 311.

In *Houston et al. v. State* (1905) 124 Ga. 417, 419, 52 S.E. 757, 758, which was an indictment charging the taking of oysters from the private bed of the complainant, the defendant attempted to justify on the ground of no title in the complainant, by stating that the state could grant only inhabitable or cultivable land. The deed under which the complainant held had attached to it a plat which was marked "marsh land." The court, after stating that such a patent could not be collaterally attacked unless it was apparent on its face that it was void, and stating further that the mere fact that the words "marsh land" appearing on the face of the document was not sufficient to say that it was not issued in accordance with law, says: "When the application for the grant was before the ordinary, his decision that the land could be lawfully granted was judicial in its nature, and it is entitled to all the presumptions which usually attach to judgments rendered by a court of competent jurisdiction." And in replying to the contention that the grant was procured by fraud the court says: "The fraudulent procurement of the grant would not authorize a collateral attack upon it. Only in a direct proceeding to which the state was a party could it be set aside on the ground that it was improvidently granted or was procured by fraud. *Calhoun v. Cawley,* 104 Ga. 335 [30 S.E. 773]. Even if the grant was not authorized by law, it is not open to collateral attack: its illegality not appearing on its face. *Vickery v. Scott,* 20 Ga. 795. It follows that the grant was properly admitted in evidence." *Patterson v. Buchanan,* 37 Ga. 560; *Tison v. Yawn,* 15 Ga. 491, 60 Am.Dec. 708; *Sykes v. McRory,* 10 Ga. 465, 54 Am.Dec. 402; Winter v. Jones; 10 Ga. 190, 54 Am.Dec. 379.

In *Chauvin et al. v. Louisiana Oyster Commission et al.* (1907) 121 La. 10, 15, 46 So. 38, 39, the opinion shows that the oyster commission of Louisiana was given jurisdiction by a broad statute over all beds of rivers, lakes, bays, sounds, etc., and the shellfish growing thereon, with power to lease the same. The commission attempted to lease land which the plaintiff held under a title from the state. The commission in its official capacity attacked the plaintiff's title on the ground that a portion of the tract conveyed to the plaintiff consisted of tide-flowed land, that such land could not legally have been conveyed to the plaintiff by the state, and that therefore the commission had jurisdiction over this land and might lease it. The court says: "The answers to this contention are: (1) That neither the state nor any of its agents can attack its solemn patent, valid on its face, in a collateral proceeding." If the commission, a legally constituted agent of the state, could not in this capacity attack the validity of such a title, certainly an entire stranger to the title, and particularly a wrongdoer, could not come in and make such an attack.

In *American Ass'n, Limited, v. Innis, etc.* (1901) 109 Ky. 595, 60 S.W. 388, which was an action to enforce a vendor's lien, it was alleged that the patents issued to the defendant were fraudulent, etc., and that the conditions precedent to granting the title required by the statute had not been conformed to. Among others was a condition that not more than 200 acres should be granted to any person. A former opinion of the court had held, in the face of this statute, that any person might file any number of orders or requests for surveys for 200-acre strips. The court says that, while such an interpretation does not seem to it to be altogether reasonable, yet the fact that many persons have acted upon it and thereby acquired more than 200 acres apiece, and that consequently vast vested interests having been built up, and so many innocent persons would be injured by upsetting the opinion, that they will not disturb it. See page 606, 109 Ky. 60 S.W. 388. The court further says, at page 607 of 109 Ky., at page 391 of 60 S. W.: "It is settled law that the validity of a patent cannot be inquired into in a collateral proceeding, unless the patent is void upon its face, or has been issued under such circumstances as the statute declares to be fraudulent. *** Another point relied on is that appellees could not acquire title by merely surveying a base line and then platting the entries upon this line. *** The identical question was raised and construed in the case of *Cain v. Flynn, 4 Dana (Ky.)* 499. It was held that a patent could not be questioned collaterally on this ground or parol evidence admitted to show the fact." *Frazier v. Frazier,* 81 Ky. 137; *Little v. Bishop, 9 B. Mon. (Ky.)* 240; *Taylor v. Fletcher, 7 B. Mon. (Ky.)* 80; *Hardin v. Cain, 2 B. Mon. (Ky.)* 56; *Underwood v. Crutcher, 7 J. J. Marsh. (Ky.)* 529; *Jennings v. Whitaker, 4 T. B. Mon. (Ky.)* 50; *Allen v. Pulliam,* 66 S.W. 722, 23 Ky. Law Rep. 2129; *Uhl v. Reynolds,* 64 S.W. 498, 23 Ky. Law Rep. 759.

In *American Dock & Improvement Co. v. Trustees of Public Schools,* 39 N.J.Eq. 409, the court adopts for that state the law as laid down by Marshall, C. J., in *Polk's Lessees v. Wendal, 9 Cranch,* 87, 3 L.Ed. 665.

In the case of *Harrington v. Goldsmith,* 136 Cal. 168, 169, 68 P. 594, which was a partition suit, the defendant denying that either the plaintiff or himself was owner of any part of the land, questioning accuracy of the description in the patent, etc., the court says: "The patent from the state of California, which included the lands in question, was evidence of the title of the grantees therein, and was admissible in evidence in support of the plaintiff's claim. It was not void upon its face, and could not be attacked collaterally for any irregularity of proceedings upon which it was issued. See *Doll v. Meader,* 16 Cal. 295." *Peabody v. Prince,* 78 Cal. 511, 21 P. 123; *Worcester v. Kitts,* 8 Cal.App. 181, 96 P. 335.

In *Martin v. Brown,* 62 Tex. 485, 488, the court held: "Generally if a patent has been issued by an officer authorized to grant it, and whose duty it was to examine and pass upon the evidence upon which it was issued, and some irregularity or illegality has supervened in the preliminary proceedings, the patent could only be impeached therefore by the state or some individuals having an equity in the land at the time the patent issued. Todd v. Fisher, 26 Tex. 239." *Carter v. Clifton,* 44 Tex.Civ.App. 132, 98 S.W. 209; *Heil v. Martin* (Tex.Civ.App. 1902) 70 S.W. 430; *Greenwood v. McLeary* (Tex.Civ.App. 1894) 25 S.W. 708. The presumption is in favor of the regularity of the grant. There is also a strong presumption that duly constituted officials will perform their official duty and that all proceedings by them are regular. Wigmore states the rule in section 2534, where he says: "The general experience that a rule of official duty, or a requirement of legal conditions, is fulfilled by those upon whom it is incumbent, has given rise occasionally to a presumption of due performance. *** It may be said that most of the instances of its application are found attended by several conditions: First, that the matter is more or less in the past, and incapable of easily procured evidence; secondly, that it involves a mere formality, or detail of required procedure, in the routine of a litigation or of a public officer's action; next, that it involves to some extent the security of apparently vested rights, so that the presumption will serve to prevent an unwholesome uncertainty; and, finally, that the circumstances of the particular case add some element of probability."

A similar principle was laid down in the case of *State v. Board of Aldermen,* 18 R.I. 381, 382, 28 A. 347, with regard to official acts, when the court said: "The rule is well established that, if an inferior court has jurisdiction, every intendment is to be made to support its judgment. *** The principle applies equally to statutory tribunals as to courts, and on certiorari to review the proceedings of commissioners, boards, or other officers, the presumptions are all in favor of their rightful action, and of their proceeding in a manner authorized by law."

The courts are nearly all of one opinion upon this question. The patent or lease from the state is treated as of a higher order than an ordinary grant of an individual. This rule is a necessary and essential rule to quiet titles, and prevent such attacks as are attempted to be made in this case. For the same reason every presumption is made in favor of the acts of all state bodies having judicial powers. There would never be a certainty in any property resting upon a grant if every stranger could attack its validity.

Furthermore, if the proceedings of the shellfish commissioners could be inquired into in this proceeding, it could only be with reference to the 20 acres of ground on Great Bed. We have examined the record regarding the action of the shellfish commissioners in respect of the letting and find no illegality or irregularity in the same.

Summarized, the objections of the defendant may be said to be the lease is void because: (1) the commissioners agreed to lease more than one acre at a time to one firm; (2) the commissioners violated the statute in making the lease by receiving, advertising, and granting an application for a lease of lots in bulk, instead of requiring, advertising, and granting leases upon separate applications for each lot of one acre; (3) the shellfish commissioners violated the law in agreeing to lease and in leasing the 21 acres to Payne & Butler on one and the same day; (4) the commissioners failed to give notice, in the advertisement, of the day, hour, and place where the land will be let.

The claim of the defendants is based upon an entire misapprehension of the object and scope of the statute. Pub. Laws, c. 853, passed March 29, 1901, which reads as follows:

"Section 1. There shall be elected by the general assembly, in grand committee, five commissioners of shellfisheries, one from each county, who shall hold office for the term of five years and until their successors, respectively, shall be elected and qualified to act. Any vacancy that may occur in said offices while the General Assembly is not in session may be filed by the Governor until such time as some person elected by the General Assembly, in grand committee, to fill such vacancy, shall be qualified to act. Any person elected by the General Assembly to fill such vacancy shall hold office for the unexpired term of the person whose place he is elected to fill. They shall have power and authority to elect a clerk and prescribe his duties: Provided, that nothing in this act shall be so construed as to affect the tenure of office of the present commissioners of shellfisheries, who shall continue to hold their offices for the terms for which they were elected.

"Sec. 2. The said commissioners, previous to entering upon the duties of their office, shall severally give a bond, with sureties satisfactory to the general treasurer, in the sum of one thousand dollars, with condition to faithfully perform the duties of the office according to law.

"Sec. 3. The said commissioners shall make annual report to the General Assembly at its January session of their doings and the condition of this department of the public service, including a detailed statement of all moneys received and expended on account thereof.

"Sec. 4. The said commissioners shall have an office in the State House in the city of Providence, where the maps, charts, books, leases, and other property connected with said commission shall be kept.

"Sec. 5. Each of said commissioners shall, by virtue of his office, make complaints for any violation of the laws of this state relating to shellfisheries, and of any subsequent amendments thereof, without giving recognizance or surety for costs.

"Sec. 6. The said commissioners may appoint such deputies as they shall deem necessary for the detection and prosecution of any violation of the laws of this state relating to shellfisheries. Each of said deputies appointed as aforesaid shall be, by virtue of his office, a special constable, and as such deputy may, without warrant, arrest any person found violating any of said laws, and detain him for prosecution not exceeding twenty-four hours, and may seize any boat or vessel used in such violation, together with her tackle, apparel, and furniture, and all implements belonging thereto. Said commissioners may make all necessary regulations for the enforcing of said laws, and they shall be allowed their actual disbursements made in carrying the same into effect.

"Sec. 7. Said commissioners may, unless otherwise by statute prohibited, agree to lease in the name of the state, by public auction or otherwise, to any suitable person, being an inhabitant of this state, any piece of land within the state, covered by four feet of tide water at mean low tide as delineated upon the plats in the office of commissioners of shellfisheries, and not within any harbor line, to be used as a private and several oyster fishery for the planting and cultivation of oysters thereon, upon such terms and conditions as they may deem proper, but not for a longer term than ten years or for a shorter term than five years, nor for a rent of less than ten dollars per annum for every acre to be leased, where the water is of the depth of less than twelve feet at mean low water, as shown on the plats in the office of the commissioners of shellfisheries, and not agreeing to lease more than one acre at a time in one lot or parcel to one person or firm; but in drawing such leases said commissioners may include in the instrument of lease one or more acres of land so leased by them, and all such leases shall be made and executed free of expense to the lessees; and neither of such commissioners shall at any time be interested in any lease of ground for planting oysters, or in the cultivation or product thereof: Provided, however, that in Little Narragansett Bay, and in Pawcatuck river below 'Pawcatuck rock,' so called, the said commissioners may let such land on terms as to time and rentals as may seem to them best.

"Sec. 8. The said commissioners may let and lease any lands within the state covered by tide water where the said water is of the depth of at least twelve feet according to the plats in the office of the commissioners of shellfisheries at the average low water, for the purpose of having the said land used in planting and cultivating oysters in the deep waters of Narragansett bay and tributaries, at an annual rental of not less than five dollars per acre, for a term not exceeding ten years from such letting.

"Sec. 9. Any person who shall wrongfully make claim to any public oyster ground, of which he has no lease or title from the state, by erecting bounds or monuments thereon of any description, or otherwise claiming title to such land, shall for the first offense pay a fine of twenty dollars, and costs, and for every subsequent offense pay a fine of fifty dollars and costs, one-half thereof to the use of the state and the other half to the complainant.

"Sec. 10. The said commissioners shall cause the original surveying and platting of all lands for planting and cultivating oysters under provisions of this chapter to be done at the expense of the state and without charge to the lessees; and the state auditor shall draw his order for the payment of said surveys and platting upon the general treasurer, upon properly presented vouchers approved by said commissioners, and the general treasurer shall pay said orders, out of any moneys that may be in his hands not otherwise appropriated.

"Sec. 11. The said commissioners may at the request of the lessee, for cause shown, cancel or modify any lease, or they may remit or abate the rent reserved therein if it shall be made to appear to the satisfaction of the commissioners that it would be equitable so to do.

"Sec. 12. The said commissioners shall not let any land north of a line extending across Providence river from Field's Point to Kettle Point; or let any lands west of a line drawn from Warwick Neck light to Pojack Point, at Potowomut Neck; or let any land between Pomham Light and Nayatt Light or between Pawtuxet Neck and Rocky Point in-shore from land already leased; or let any of the ponds in Little Compton, Charlestown, South Kingstown, New Shoreham, Tiverton, or Westerly; or let the channel between Long Neck and Marsh Island flats, from the channel in Providence river to the bridge in Pawtuxet: Provided, however, that nothing in this act shall be so construed as to affect any of the lands which have been leased or the releasing thereof.

"Sec. 13. The said commissioners shall give notice of every application for a lease of land for the planting of oysters by publication twice a week for two successive weeks in some daily newspaper published in the city of Providence, and also once a week for two successive weeks in some newspaper published in the county nearest to which the ground is located, describing the land therein applied for and giving the name and residence of the applicant and the day, hour, and place where the land will be let; which day shall in all cases where the first hearing upon such an application is to be had upon the first or third Friday of the month, and the commissioners may give such further notice of such application as they may deem to be necessary to inform persons interested of the pendency of such application, and the actual costs of publishing said notices shall be paid by the applicants.

"Sec. 14. Said commissioners may adjourn such hearing from time to time, and may issue process to compel the attendance of witnesses for either party, and shall give notice to all parties who have appeared before them upon any application of the time and place when their decision will be given; and such decision shall be final, unless appellate proceedings are taken and prosecuted as hereinafter provided.

"Sec. 15. Any person aggrieved by the decision of the commissioners upon any application for a private or several oyster ground or oyster fishery may petition the common pleas division of the supreme court within and for the county nearest to which said land so applied for lies, for a reversal or modification of such decision, in like manner and with the same procedure, excepting where a different procedure is provided in this act, as prescribed in sections fifteen and sixteen of chapter forty-six in the case of petition for relief for over-assessment for taxes.

"Sec. 16. Application for citation in such case shall be made to the clerk of said common pleas division within five days from the day such decision shall have been made, and the petitioner shall, at or before the time for filing his petition, file with said clerk a copy of the proceedings before the commissioners, and a bond, signed by him or by someone in his behalf, with sufficient surety, in the sum of fifty dollars, payable to said clerk for the use of the state, with condition to prosecute such petition to final judgment and to pay such witness fees and the costs of summons incurred by any party opposing such petition as the court shall award, in case the decision of the commissioners shall not be reversed.

"Sec. 17. Such case shall be heard and tried in the same manner as other cases entered upon the docket of said court, and the judgment of the court (which shall be entered immediately upon the rendition of decision or verdict) shall be conclusive upon the question whether said land shall or shall not be leased, and the commissioners shall grant or refuse a lease accordingly.

"Sec. 18. Such leases shall be executed by such lessee, as well as by said commissioners, in two parts, one part thereof to be delivered to such lessee and the other part thereof to be retained by said commissioners and recorded in a book kept for that purpose, and shall contain proper covenants for the payment of rent and the performance of the conditions and observance of the restrictions therein set forth, with proper clauses reserving to said commissioners a right to re-enter on behalf of the state and to terminate said lease for breach of any of said covenants.

"Sec. 19. Said commissioners shall before granting any such lease cause the land to be leased as aforesaid to be surveyed and platted, and shall in all cases cause proper bounds with marks thereon to be set up either on the shore opposite and nearest to such land to be leased as afore-said, in order to define the limits thereof, or shall cause such land to be leased as aforesaid to be marked with stakes or buoys at the corners of the ground leased, with such marks thereon as they may direct. Such bounds, stakes, or buoys, with the marks thereon, shall be renewed whenever the commissioners shall direct.

"Sec. 20. The drawing and executing of such leases, the original surveying and plating, shall be done by said commissioners without expense to the lessees. The setting up of the bounds, stakes, or buoys shall in all cases be done by the lessee under the direction of the commissioners.

"Sec. 21. Every person who shall willfully injure, deface, destroy, or remove such marks or bounds, or deface any mark thereon, or shall tie or fasten any boat or vessel to any such stake or buoy, shall be fined twenty dollars for each offence, one-half thereof to the use of the state and one-half thereof to the use of the complainant. Every such person shall, in addition thereto, be liable in an action of the case to pay double damages and costs to the person who shall be injured by having the marks and bounds, stakes, or buoys of their said lots injured, defaced, removed, or destroyed as aforesaid.

"Sec. 22. The oysters planted or growing in any private oyster ground leased as aforesaid shall, during the continuance of the lease, be the personal property of the lessee of such oyster ground.

"Sec. 23. Every person who shall work a dredge, pair of oyster tongs or rakes, or any other implement for the taking of shellfish of any description, upon any private and several oyster ground or bed without the consent of the lessee or owner thereof, or who shall, while upon or sailing over any such ground or bed, cast, haul, or have overboard any such dredge, tongs, rake, or other implement for the taking of shellfish of any description, under any pretence or for any purpose whatever, without the consent of such lessee or owner, shall for the first offence be fined not exceeding twenty dollars or be imprisoned not exceeding thirty days, and for every subse-quent offence shall be fined not exceeding one hundred dollars or be imprisoned not exceeding six months.

"Sec. 24. Said commissioners shall from time to time diligently inspect and ascertain whether or not the terms and restrictions of the leases are kept and performed in a just and proper manner, and whether or not the rents are punctually paid; and in case said terms and restrictions are not kept and performed, or said rents are not punctually paid, the commissioners shall forthwith enter upon the land so leased and terminate the lease.

"Sec. 25. The commissioners may, in the name of the state, institute any legal proceedings that may be necessary for the collection of such rent. The commissioners may take possession of any lot leased, upon which the rent shall not have been paid, and may dispose of said lot with all the oysters thereon at public auction to the highest bidder, first giving notice of the time and place of sale by publishing the same at least once each week for two successive weeks in some newspaper published in the city of Providence, with power to adjourn such sale from time to time, giving like notice of such adjournment; to make and execute to the purchaser at such sale a good and sufficient conveyance of all the right, title and interest of said lessee in and to the lot leased, together with the oysters thereon; and to receive the proceeds of such sale, and from said proceeds to retain all sums due and owing the state for rent as aforesaid, together with all expenses incident to such sale, rendering and paying the surplus of said proceeds of sale, if any there be over and above the amounts so to be retained as aforesaid, to said lessee, his heirs, executors, administrators, or assigns.

"Sec. 26. Every person who shall take oysters from any private and several oyster bed, except between the hours of sunrise and sunset, shall be fined twenty dollars for each offence, one-half thereof to the use of the state and one-half thereof to the use of the complainant; and every boat or vessel used or in any way employed in so doing shall, together with its tackle, apparel, furniture, and implements on board, be forfeited.

"Sec. 27. Every person who shall wrongfully take and carry away oysters from a private oyster bed shall for the first offence be fined fifty dollars and be imprisoned for thirty days, and for every subsequent offence shall be fined one hundred dollars and be imprisoned for six months.

"Sec. 28. Any police constable may in view of the commission of any offence against the provisions of this act upon any of the public waters of the state arrest the offender without warrant and detain him for prosecution not exceeding twenty-four hours.

"Sec. 29. Every person who shall willfully break up, damage, or injure any bed of oysters, or any tract of land leased from the state for an oyster bed, by depositing thereon earth, stones, or dredgings or scoopings from the river or docks, or in any other manner, shall be fined not exceeding five hundred dollars, one-half thereof to the use of the state and one-half thereof to the use of the complainant; and shall forfeit his boat or vessel, with her tackle, apparel, and furniture, and all the implements by him used in injuring such oyster bed.

"Sec. 30. Every person convicted a second time of a violation of any of the provisions of this act shall, in addition to the penalties herein before mentioned, be deprived of the privilege of fishing for oysters in the waters of the state for the space of three years thereafter, under penalty of thirty days imprisonment for each offence.

"Sec. 31. Every person who shall take more than two bushels of oysters during any one day from Trustan pond, in South Kingstown, shall be fined not less than five dollars nor more than twenty dollars for every bushel so taken above two bushels.

"Sec. 32. Each of said commissioners shall be by virtue of his office a special constable, and, as such commissioner, may arrest any person found violating any of the provisions of this act, and may seize any boat or vessel, with her tackle, apparel, and furniture, and all implements belonging thereto, when employed in taking oysters or in injuring any oyster bed in violation of the provisions of this act, and shall make complaint when called upon to do so for all such violations, and in any such complaint he shall not be required at the time of complaint or thereafter to enter into recognizance or in any way to become liable for the costs that may accrue thereon; and the Attorney General shall, when notified to do so by the complainant, prosecute all such complaints in the court where the same shall be made or be pending; and all cases of appeal thereof from the sentence of a district court, and all questions arising under the same, or under any complaint and warrant made under the provisions of this act, in either division of the Supreme Court, shall be conducted by said Attorney General.

"Sec. 33. A surveyor may be employed to fix the place or otherwise to designate the locality of any violation of the provisions of this act, and reasonable charges of such surveyor for such service shall be allowed by the court, if said employment shall be by said court deemed to have been necessary; and such charges when allowed as aforesaid shall be taxed in the bill of costs.

"Sec. 34. All leases of oyster grounds heretofore granted by the commissioners of shellfisheries to any party or parties residents of this state are hereby validated and confirmed.

"Sec. 35. Chapter 170 of the General Laws and all acts and parts of acts inconsistent herewith are hereby repealed, and this act shall take effect from and after its passage."

It is evident that by the provisions of section 7 the commissioners are authorized to agree to lease to certain suitable persons any piece of land within the state covered by four feet of tide water at mean low tide, without reference to the size or extent of the same, for certain prices to be determined by them within certain limits, and not agreeing to lease more than one acre at a time in one lot or parcel to one person or firm; but in drawing the leases said commissioners may include one or more acres of land so leased by them. The meaning of the section with reference to the agreements to lease is perfectly plain, whatever may be the reason for the requirement, and that is that the minds of the proposed lessee and those of the commissioners shall meet upon separate propositions to lease each acre. And as this may be done by auction, for example, acre lots could be put up for sale separately even on the same day, and as often as the bidding ceased and the hammer fell and the lot, offered for sale, was struck off to the highest bidder, that would constitute an agreement to lease, or, if, as in the present case, blank leases had been prepared with the numbers, designating each acre lot proposed to be leased, filled in, and before the execution of the same by the parties the clerk of the commissioners should read to the proposing lessee, the numbers of each of the lots in the proposed lease, seriatim, while the expectant lessee followed the same on a plat, whereon the land to be leased was delineated, and which contained the corresponding numbers to those set out in the lease, and if the clerk should pause after each number so read by him to give the applicant an opportunity to signify his acceptance or rejection of the offer of the commissioners to lease said land thus tendered to him by their clerk, an affirmative answer either orally expressed or by nods or signs, or by acquiescence however indicated, would constitute an agreement to lease which would be effective as to each and every lot upon which the minds of the contracting parties so met. It matters not how short the interval of time, between the reading of the numbers, if it was long enough to enable the tender to be accepted, and, as a matter of fact, it was accepted. The fact that the agreements of lease were all made in one day, or one afternoon, or one hour or less, is not of the slightest importance. The rule that the law knows no fraction of a day is inapplicable in a case of this kind. The agreements were not agreements "to lease more than one acre at a time" if they were made at different instants of time.

The provisions of sections 13, 14, 15, 16, and 17 relate to applications for a lease of land for the planting of oysters, and the notice to be given thereon in order that the public may be heard upon the question whether said land shall or shall not be leased. The question is not to whom, if any one, it shall be leased. That question is not before the commissioners or before the court upon appeal. The question is one of public interest, to wit: shall certain land be taken from the free and common oyster fisheries for the purposes of a private and several oyster fishery? This is the only question that can be heard by the commissioners at the first hearing upon the application, or by the court upon appeal. The proceedings had before the commissioners and upon appeal before the court are analogous to the proceedings had upon the "writ of ad quod damnum," which is defined to be: "A writ issuing out of and returnable into chancery, directed to the sheriff, commanding him to inquire by a jury what damage it will be to the king, or any other, to grant a liberty, fair, market, highway or the like. The name is derived from the characteristic words denoting the nature of the writ, to inquire how great an injury it will be to the king to grant the favor asked. Whishaw, Fitzherbert, Nat. Brev. 221; Termes de la Ley." Bouv. Law Dict. The commissioners, in the first instance, and upon appeal the court or jury, are called upon to inquire whether the granting of such application will be injurious to the public. In the "oyster act" of January, 1844 (Dig. 1844, p. 535) § 10, commissioners of the shellfisheries, or any two of them, were required, before granting any leases of private or several oyster grounds, or oyster fisheries, to personally inspect the land asked to be leased, and to decide upon the propriety of leasing the same, taking special care not to include in the land so leased any old oyster bed, or any part of any old oyster bed which in their opinion can for the greater advantage of the public be used as a free and common oyster fishery; but their decision in the premises proved by the execution of the lease shall in all cases and for all purposes be final and conclusive thereupon. This legislation, passed less than a year after the adoption of the Constitution by the people of the state, is a pretty fair contemporaneous construction of the powers of the General Assembly in the matter of shellfisheries under that instrument. The fact that in said act no appeal was provided to be taken from the decision of the commissioners is evidence that the General Assembly deemed that the public rights were amply conserved without further provision.

Chapter 853, aforesaid, section 19, provides that the commissioners shall before granting any such lease cause the land to be leased to be surveyed and platted. It is manifest that before the land is surveyed and platted it cannot be definitely located in acre lots, and, therefore, before such surveying and platting is accomplished that it would be impossible for the parties to enter into an agreement of lease concerning acre lots thereafter to be ascertained, and, as the commissioners are required to survey and plat the land before granting the lease, they are forbidden to grant or even to agree to lease before such survey and plat has been made. Furthermore, the question of to whom the land shall be let after said surveying and platting is a matter entirely within the discretion of the commissioners from which no appeal is provided. In case of two or more applicants for the same land, the decision of the commissioners must prevail, for even if a person deeming himself aggrieved by the decision of the commissioners should attempt to take an appeal under the provisions aforesaid, and if he should succeed in getting his case heard by the court or jury, the only judgment that could be rendered in the case is, as we have already seen, that the "land shall or shall not be leased." And so if the jury should find that the land shall not be leased – a finding that would be nugatory if the commissioners had previously decided that it shall be leased and no appeal had been taken from their finding in that regard – the verdict would not assist the appellant who wished the land to be leased and leased to himself. And if the jury should find that the land shall be leased, that would not assist him because he was not aggrieved in that part of the finding by the commissioners, and the jury are powerless to decide to whom it shall be leased, and such a verdict would therefore be but a barren victory, if, indeed, an appellant could ever hope to get as far as that on such a claim.

There is nothing irregular, or illegal, in the application, advertisement, or proceedings before the commissioners.

The transcript shows that on June 17, 1902, John S. Payne and Walter H. Butler of East Providence made application for 31 (21?) acres of oyster ground on Great Bed southerly of Starve Goat Island and near the same. It also shows that on June 21, 1902, the application was advertised in the Providence Telegram, as follows: "John S. Payne and W. H. Butler both of East Providence make application for twenty-one acres of oyster grounds in said river on Great Bed all southerly of Starve Goat Island and near the same. Friday, July 4, A. D. 1902, at ten o'clock, at the office of the commissioners of shellfisheries, State House, is hereby appointed the time and place for consideration of the same. The above hearing to be held at the same time and place on the following day, July 5th, to which time and place the hearing will be adjourned." On July 5, 1902, the records of the shellfish office contained the following: "Office of the Commissioners of Shellfisheries, Providence, July 5, A. D. 1902. Meeting called for the purpose of considering applications of Robert Pettis for land near Bullock's Point; John S. Payne and W. H. Butler, George C. Bell of Warwick, Charles C. Miller of Providence, Sidney A. Balcom of East Providence, for land in Providence river, being a part of Great Bed. Present, members of board. To the granting of land applied for by Mr. Pettis there was no objection, and the same was granted. To all the land applied for on Great Bed a strong protest was presented signed by eighty-four persons claiming the said land to be a source of great revenue to the public fishers. To support the claims of the protestants appeared Abner Hart of Pawtuxet, a shellfish dealer, but not a fisherman; Mr. Charles A. Aldrich of East Providence, who was a small bed holder; and a gentleman by the name of Woodward who had followed a number of callings and who now claims to be a fisherman. The claims of the applicants were ably presented by Mr. Payne, Mr. Bell, and Mr. Butler, basing their claims upon a desire of shellfishermen of small means who are desirous of becoming oyster producers upon their own holding, upon such lands as they with their means would make them--enable them to hold. After a very full hearing the commission took the matter into consideration and examination, and voted to continue said application until the second day of December, A. D. 1902, at ten o'clock a.m." On December 11, 1902, the following entries are made: "Meeting this day called to order. A quorum present, Mr. Wilbour, the president, in the chair. Reading of the records of the last meeting approved. The question before the commission being the lease of land applied for by Messrs. Payne and Butler, George C. Bell, Charles C. Miller, Sidney A. Balcom was continued until this meeting. After brief consultation of the same it was unanimously voted to grant the said application as applied for." Upon this action being taken, the word "granted" was written opposite the above advertisement referring to the Payne and Butler application.

The defendant thereupon makes the following claim: "Whatever may be the legal effect of these recorded acts, it is entirely clear that on December 11, 1902, the application of Payne & Butler for a private oyster fishery was granted, and the clerk of the shellfish commission himself so testified. In response to the questions: 'What was done after the granting--after the commission decided to grant the application, what was done? What were the other steps?' He answered: 'The engineer was informed of the granting of the application, the hearing had been held, and directed to go down there and make a survey.'" And argues as follows:

"Counsel for the defense contend that, both by the intention of the parties as disclosed by the record, and because of the provisions of the statute, the legal effect of the granting of this application was the acceptance by the commission of the offer contained in the application consummating a contract to lease, and that, at that moment, the agreement to lease contemplated by section 7 was made. The record of the shellfish commissioners shows that on July 5th at a meeting called 'for the purpose of considering applications filed, the land was granted. (To the granting of land applied for by Mr. Pettis there was no objection and the same was granted.)' The record also shows that the granting of the application to Payne & Butler and others on Great Bed was opposed, that a hearing at which different interests were represented was held, and that the issue discussed was whether the land should be leased. It also appears that, after consideration of the issue thus framed, the application was continued, and that at an adjourned meeting where 'the question before the commission [was] the lease of land applied for by Messrs. Payne & Butler,' etc., it was said, 'After brief consultation of the same it was unanimously voted to grant the said application as applied for.' In the face of this testimony, it is inconceivable that it was understood by the parties that the granting of the application meant merely that 'they [the state engineer] might survey.' If the legal effect of granting the application was merely an authorization of a survey by the state engineer, as is stated by Mr. Collins, it can readily be seen that the granting of the application confers absolutely no rights upon the applicant. And if this was the intention of the parties, it is inconceivable not only that they should confine their discussion before the commissioners to the question as to whether the land should be leased, but that there should be any opposition at all to the granting. No one could or would object to a survey of the public waters by the state engineer, nor is that official required to be authorized for that purpose by the granting of an application for an exclusive fishery. It is also inconceivable that the commissioners would have held several meetings at which apparently the only question considered was the advisability of leasing the land before granting the application, if the only effect of granting the application would be an authorization of a survey giving no rights to the applicant. The disclosed intention of the parties, aside from any necessary legal consequences attached to the act by the statute, shows that the granting of the application was something more than permission to survey, and that it was in fact, in the contemplation of the parties, a contract, an agreement to lease when the subsequent statutory steps had been performed.

"It is clear, however, from the statute that the granting of an application does confer substantial rights on the applicant. Pub. Laws, c. 853, § 15, as amended by Court and Practice Act, 1905, § 1209, provides that 'any person aggrieved by the decision of the commissioners upon any application for a private or several oyster fishery, may petition the Supreme Court for a reversal or modification of such decision.'

"Since the granting of an application by the commissioners after a full and extended hearing must be a decision upon the application, it necessarily follows that the granting of the application, in the contemplation of the statute, confers substantial rights on the applicant, since, otherwise, no one interested in the granting or withholding of the land from lease could be aggrieved, and the provision for appeal would be useless. The granting of the application cannot, as contended by the clerk of the shellfish commission, spend its force in directing a survey of the land to be made without affecting in any manner the applicant's interest in the land applied for.

"It should also be noted that if the clerk of the shellfish commission is correct in his views as to the legal effect of granting the application, and if the granting of the application confers no rights on the applicant, then the advertisement published is clearly defective. Pub. Laws, c., 853, § 13, is peremptory that the notice of the application shall contain, not only a description of the land, and the residence of the applicant, but also 'the day, hour and place where the land will be let.' The only day, hour, and place advertised by the commissioners, as is shown by the record, is a time and place for a consideration of the application. If, therefore, the consideration of the application culminating in its being granted means merely permission to the engineer to survey, without giving any contract rights to the applicant, the day set out in the advertisement can under no circumstances be the day, hour, and place where the land will be let, and the advertisement is for that reason defective.

"The clerk of the shellfish commission evidently believed that the granting of the application could not confer rights upon the applicant because a survey and plat of the land had not yet been made. Pub. Laws, c. 853, § 19, however, simply prohibits the granting of the lease before survey and platting, and in no way affects the question as to whether the granting of the application is or is not an agreement to lease. It is believed that effect cannot be given to the several portions of this statute unless the granting of the application is an agreement to lease followed, provided an appeal is not taken within five days, and provided the land is platted, staked, and buoyed, by the granting of the lease itself. It is submitted, therefore, that both the intention of the parties, and the provisions of the statute, show that the legal effect of granting an application is the consummation of a contract to lease, and is the agreement to lease under section 7.

"If the contract to lease was made at the time the application was granted it is manifest, since 21 acres were applied for and granted, that an agreement to lease more than one acre at a time in one lot or parcel was made. It is also clear under the decision of *State v. Burdick* (1886) 15 R.I. 239, 2 A. 764, that because of this defect the lease (plaintiffs' Exhibit 3) is void.

"For the reasons given above, counsel for the defense believe that the agreement or agreements to lease the 21 acres to Payne & Butler on Great Bed were made on December 11, 1902, the date the application was granted. If, however, the court should be of the opinion that this was not the legal effect of the granting of the application under the facts disclosed in this case, and that the testimony of the clerk of the shellfish commission was sufficient to negative the consummation of the agreement to lease, as of that date, then the undisputed testimony of the same witness and of Mr. Payne, the plaintiff executing the instrument of lease for Payne & Butler, shows affirmatively that no agreement or agreements to lease were made until after the instrument of lease (plaintiffs' Exhibit 3) was drawn, and read to Mr. Payne on the day of, and immediately prior to, its execution. Mr. Collins, the clerk of the shellfish commission, says: 'No lease can contain more than one acre at a time, and some of them less than an acre--fraction of an acre--are read over to the applicant, and he assents, he is required to assent to the desire to take those lots. If we read over, five, six, eight, and he says, "Nine, I don't want," as we let them lot by lot, I suppose we are compelled to give him a lease up to that point; and then, if it is all satisfactory, all looked over carefully lot by lot, then he consents to that, the agreement is made, it is put in one lease, or is in one lease, and then it is signed by the parties.' Payne's description of the procedure followed in connection with the granting of the lease (plaintiffs' Exhibit 3) is to the same effect. Plaintiffs' counsel said: 'You are testifying in regard to the date of the lease. What was done at that time?' Mr. Payne: 'Why, the clerk of the commission read the lease over--had two leases, and he read one of them over, what was said on it, the numbers of acres, and so forth, all down the whole of the list, passed me the other one, and I read that, and then he handed me the one he had and asked me to sign it, and I am pretty sure I signed for the firm. I am pretty sure I did.' Mr. Collins also says that it was customary for the commissioners to hear objections to running out the leases even after the numbers of the lots were put on the leases from the charts and plats, long after the application had been granted. So long as either party was at perfect liberty to withdraw, and so long as the commissioners were in a position to consider the advisability of letting the land, it is clear that there could be no agreement to lease. Payne, however, and indeed any applicant, as is shown by this testimony, had the right to refuse to take lots even when the duplicate instrument of lease was being read on the day of, and immediately prior to, signing. So, also, Mr. Collins says positively that the agreement to lease was made immediately prior to the signing, and necessarily, therefore, on the date the instrument was executed. If, therefore, the legal effect of granting the application is not an agreement to lease as contended by counsel for the defense, then the undisputed testimony conclusively shows that the agreement or agreements to lease the lots embraced in one instrument of lease were made on a single day; that day being the day of the execution of the instrument of lease.

"The testimony also shows that, according to the customary practice of the shellfish commission, the only attempt at a separate agreement to lease each acre in a group of lots containing more than one acre and included in the same instrument of lease consisted in reading the lot numbers seriatim from the previously drawn instrument of lease. Thus, Mr. Collins in describing the customary procedure of commission in agreeing to lease, and in leasing land, says, 'If we read over 5, 6, 8, and he says, "Nine, I don't want," as we let them lot by lot, I suppose we are compelled to give him a lease up to that point.' Assuming for the moment that the procedure outlined by Mr. Collins does affect a separate agreement to lease each separate acre, and assuming also that this customary procedure was followed in agreeing to lease the 21 acres on Great Bed to Payne & Butler, it is apparent that these separate agreements made on the same day, are part of one and the same transaction in point of time. Counsel, therefore, contend, even though these assumptions were found to be true, that under the meaning naturally attributed to the words, 'at a time,' and under the meaning given them by judicial construction, such agreements would violate that portion of section 7 of the statute which prohibits the commissioners from 'agreeing to lease more than one acre at a time in one lot or parcel,' since they are agreements to lease more than one acre in one lot or parcel at a time.

"Thus in Hunter et al. v. Wetsell (1881) 84 N.Y. 549 [38 Am.Rep. 544], where the plaintiff sued for the purchase price under an oral contract covered by that portion of the statute of frauds which provides that 'every contract for the sale of any goods, etc., shall be void *** unless the buyer shall at the time pay some part of the purchase money,' and the evidence showed that contemporaneously with the oral statement of the contract plaintiff delivered the check, but there was no evidence as to time of its payment, Finch, Justice, after holding that the giving of the check was not payment, said (page 554, 84 N.Y. [38 Am.Rep. 544]): 'It would be an entirely reasonable and just construction to say that the delivery of the check and its presentment and payment constituted one continuous transaction, and should be taken as such without reference to the ordinary delay attendant upon turning the check into money. The statute does not mean rigorously, eo instanti. It does contemplate that the contract and the payment shall be at the same time, in the sense that they constitute part of one and the same continuous transaction. We think, therefore, that there was payment "at the time" within the meaning of the statute.'

"In *United States v. Buchanan* [D.C.] 9 F. 689 (1881), it was held, in an indictment under section 3324 of the Revised Statutes of the United States, providing a penalty for failure to obliterate the revenue stamp at the time of emptying such cask, that the words 'at the time' did not require an obliteration at the very instant the cask is emptied, but the act ought to be done in a convenient time considering the surrounding circumstances affording evidence of reasonable delay. The same construction of these words is adopted in *Rice v. K. P. Ry. Co.* (1876) 63 Mo. 314, and in *Goggin v. K. P. Ry. Co.* (1874) 12 Kan. 416.

"It is clear from these authorities that under the customary and usual construction of the words 'at a time' they include more than an instantaneous fraction of a second. It necessarily follows, therefore, that the procedure adopted by the shellfish commission in the granting of this lease, in which the agreement or agreements to lease 21 acres were separated only by the time it took the clerk to read the number of the lots, made all the agreements part of one transaction, a transaction which included also the signing of the instrument of lease, and is not in compliance with the provision of the statute against leasing more than one acre 'at a time.'"

The defendant fails to take into consideration the successive steps by which the leasing is to be accomplished: first, the application for a lease of oyster ground, upon which notice is to be given by the commissioners to the public of the time and place where the land will be let, that is, where the question, shall the land be leased or not, will be determined, which hearing may be adjourned from time to time. If the commissioners shall at the hearing grant the application, that is, if they shall conclude that the land shall be leased, that is, shall be taken from the public and let for private purposes in the future, the public question and the only public question in the matter will be thereby determined, and from such determination an appeal is given to any person aggrieved. If no appeal is taken, the next step will be to have the land surveyed and platted. After the land has been surveyed and platted into acre lots, then, but not before then, the question, to whom shall this land be let, can arise. This is not a public question and is of interest only to applicants and in the case of competing applicants, or in case, the commissioners should refuse to grant in its entirety the application of a sole applicant, the decision of the commissioners in respect to the same will be final and conclusive, and no appeal therefrom is provided. As the public has no interest in the question, to whom shall the land be leased, no notice of the time and place of the hearing upon that question, by publication or otherwise, is necessary. When the matter does come up for determination, after the land has been surveyed and platted into acre lots, then the applicant and the commissioners meet and for the first time the question of leasing the lots is to be considered lot by lot, and, whenever the minds of the parties concur in regard to the letting of a particular lot, an agreement to lease that lot is thereby accomplished. There is no limit to the number of acres that may be included in any one lease under the statute, so long as the agreements to lease the lots were separately made. That is, the statute requires the commissioners to consider the question of leasing acre by acre, so that the advisability of letting the same shall be scrutinized in particular and not in general. They are forbidden to determine the question, in this manner: we will let you this tract of 21 acres if you will lease the same. But they may dispose of the matter in this manner: Of the 21 acres delineated on this plat as lots numbered 1, 2, 3, 4, 5, 6, 7, 8, 9, 10, 11, 12, 13, 14, 15, 16, 17, 18, 19, 20, and 21, each containing one acre of land, do you want to lease lot No. 1? If you do, you may have it. If the applicant signifies his acceptance of the offer so made, an agreement to lease that lot is reached. And so, if the same method is pursued as to each of said lots with the same result, an agreement to lease each and every lot has thereby been made. And the length of time consumed in making the agreements is of no consequence so long as they were made separately. In this manner the statute would be complied with literally. It was not necessary for the plaintiff to prove that this method was followed in the case of the lease in question, although he has done so, for the lease recites the fact, and all presumptions are in favor of the legality of the acts of the commissioners.

Each lease is regular and valid on its face.

The lease in question reads as follows: "No. 466. This indenture, of two parts, made and entered into on this 24th day of January in the year of our Lord one thousand nine hundred and three by and between the state of Rhode Island and Providence Plantations, on the one part, and John S. Payne and W. H. Butler both of East Providence in the county of Providence in said state, of the other part witnesseth, that the said state doth hereby lease, devise, and let unto the said Payne & Butler a certain piece of land in Providence river lying and being, and covered with tide water, containing about 21 acres and bounded and described as follows, to wit: Being lots numbered 6 to 8 incl. 14 to 16 incl. 22 to 24 incl. 47 to 49 incl. Section 3. 189 to 191 incl. 168 to 173 incl. Section 4-- in the book of 'Leased Oyster Beds Surveyed and Platted under the direction of the Shellfish Commissioners, 1880,' by Shedd & Sawyer, Civil Engineers. Said lots were leased separately, but are included in one lease for convenience. To have and to hold to them the said Payne & Butler executors, administrators and assigns, to their use as a private or several oyster fishery, for the planting and producing of oysters, for and during the term of ten years from the day of the date hereof, on the terms and conditions (among others) that the said lessee, their executors or administrators, shall pay therefor to the General Treasurer of said State, during the said term, the yearly rent per acre, of 10 dollars in manner hereinafter provided. And the said state both hereby covenant with the said lessee, their executors, administrators and assigns, that they may and shall occupy the premises hereby leased during the term aforesaid peaceably and quietly, and free from all lawful claim and demand of all persons whomsoever, other than as herein before or hereinafter set forth: the said lessee for themselves their executors, administrators and assigns (with a reservation of his rights to claim remission or abatement, as by law provided), doth covenant with said state, that they will pay to the General Treasurer, for the use of said state, the sum of two hundred and ten dollars on the first day of January in each year during the term aforesaid. Furthermore: This lease is made and accepted--subject to the provisions of existing laws relating to oyster fisheries, and to a reserved right of the state to amend said laws as it shall deem expedient (reference to the same being here made); and also subject to the further conditions following, to wit: First. That he shall at all times erect, place or renew the bounds, stakes or buoys, with marks thereon for defining the premises, as and when required by the Commissioners. Second. That he shall pay all expenses of renewing stakes or bounds, and rent, to the General Treasurer, as aforesaid. Third. That he shall not underlet or assign the premises to any person whomsoever, without the assent, in writing, of the commissioners. Fourth. That he will not knowingly or wilfully violate any provision of the laws at any time in force relating to the oyster grounds or oyster fisheries within the state. And, fifth, that in event he shall refuse or neglect to comply with or conform to these conditions, or any or either of them, the said Commissioners may, on the part of said state, re-enter upon said leased premises and terminate the lease, and declare the same forfeited, and dispose of the lessee's interest in the said land, together with all the oysters thereon, at public auction, to the highest bidder, upon giving one week's notice of such sale in some newspaper printed in the city of Providence; and the lessee, their executors, administrators or assigns, shall be holden to pay all damage that shall thereby be sustained by said state. In witness whereof, the commissioners of shellfisheries hereunto subscribe the name of said state, and set their names as commissioners, and the said lessees hereunto set their hands the day and year aforewritten. The State of Rhode Island and Providence Plantations, by [Signed, sealed and delivered in presence of:] P. H. Wilbour,

James M. Wright, Samuel B. Hoxie, Jr., John H. Northup, Wm. T. Lewis Jr., Commissioners of Shellfisheries. Payne & Butler, Lesse."

As the leases in the case are valid, they need no validating. Therefore it becomes unnecessary to consider the constitutionality of the validating act (Pub. Laws 1908, c. 1574), which may have been passed out of abundant caution.

There is no merit in any of the defendant's exceptions, and the constitutional question must be answered in the negative. The defendant's exceptions are therefore overruled, and the case is remitted to the superior court, with direction to enter judgment on the verdict.

Jamaica Bay is an arm of the sea, and its waters are tidal and public, and all the inhabitants of the state have a common right to the use of the same for navigation, fishing, and all other purposes of a public character, and the plantation of oysters in such waters is a legitimate exercise of such common right, and they remain the property of the person so planting them, and their conversion by another furnishes a cause of action to the owner. *Post v. Kreischer,* 103 N.Y. 110 [8 N.E. 365]. Oysters planted in tidal waters on a well marked and clearly defined bed, where there were no natural oysters before, are the personal property of the planter, and he may maintain an action for their conversion. *Fleet v. Hegeman,* 14 Wend. [N.Y.] 42; *Decker v. Fisher,* 4 Barb. [N.Y.] 592; *Lowndes v. Dickerson,* 34 Barb. [N.Y.] 586. Oysters are classified in law with animals *ferae naturae* in which a qualified property may be acquired, by possession, both by the civil and common law, and such property continues so long as the dominion of the owner is maintained. It is protected by the law to the same extent as an absolute property, and every invasion of the same is redressed in the same manner. When they are reclaimed or taken from their natural beds and collected together for the purposes of growth or multiplication, they cannot be retained in actual manual possession, and they must be deposited in their native element, and so long as they remain within such control as will enable the owner to resume actual possession of them at his pleasure his ownership will continue. \*\*\* Oyster beds are marked and defined by stakes and buoys to give notice to the world of the exercise of the common right to use the public waters of the state by the occupant, and that the oysters have not been abandoned. They are, when so planted, entirely within control; and the purpose is to take them up at pleasure when they become suitable for the market. The traffic in oysters has become very expanded, and their cultivation has come to be a very extensive business. The oysterman plants his oysters in their natural soil and element with regularity, and expects to gather their growth with as much certainty as the farmer plants his grain in the soil and expects to harvest and enjoy the crop; and, for all practical purposes, the title to the oysters is as valuable and certain as the title to grain or any other article of commerce.

While seemingly strictly a trespass case, the review of related cases and the emphasis on surveying and location issues makes this a noteworthy study. A number of issues were raised during the trial of the case, which fell under existing statutes and common law principles. The basic fact stressed here was that title to both personal and real property are protected, measurable and definable by ordinary means.

The following case is similar, as it involves the disturbance of clam and oyster beds of several holders, which are protected under the statute and are superior title to the later activity of a construction project. It also discussed constructive notice, since the governing statute requires recordation of the leases from the State of New Jersey. While it is illustrative of similar rights and legal protection, it is located in a different state.[73]

Defendant Ocean City Automobile Bridge Company appeals from judgments against it in favor of plaintiffs in eight separate suits and defendant Hill Dredging Company appeals from judgments against it, as well as the Bridge company, in favor of two of the plaintiffs. The cases were tried together.

---

[73] *Jesse Thomas et al. v. Ocean City Automobile Bridge Company et al.,* 108 N.J.L. 143, 156 A. 493 (1931).

These eight suits were begun by the lessees of certain clam and oyster grounds in Great Egg Harbor River, in Atlantic county, demised by the State of New Jersey, acting through the board of shell fisheries, for the period of one year beginning January 1st, 1927. These plaintiffs had been tenants of the same grounds for some years prior to 1927. During the year 1927, the bridge company acquired a sixty feet wide strip of lands under water from the State of New Jersey, acting through the board of commerce and navigation, for the purpose of constructing a toll bridge across Great Egg Harbor river from Somers Point, in Atlantic county, to Beasley's Point, in Cape May county. The bridge company contracted with the defendant Hill company to make the fill. The work was done in 1927.

Plans and specifications called for a fill of the width of at least three hundred and forty feet, as follows: A sixty feet strip coinciding with the riparian grant, to be bulkheaded on each side and filled to a point about six feet above ordinary high water; a forty-feet strip, likewise bulkheaded, and to be filled to about three feet above ordinary high tide, on each side of the sixty-feet wide strip granted by the state; and beyond the forty-feet strips a further fill one hundred feet wide on each side, with no bulkhead or other means of retaining it. There were two sections to be filled; one extending from the Atlantic county shore line southwesterly about seven hundred and thirty-five feet; the other extending from Drag channel southwesterly about one thousand two hundred and twenty feet to the main channel.

The suits were all brought against the bridge company and the dredging company upon two counts, one for negligence and the other for trespass. See Figure 7.8.

**FIGURE 7.8** 1886 USGS topographic map of Great Egg Harbor.

The lands held by only two plaintiffs, Lloyd and Tallman, were actually traversed by the fill. The complaint was that the manner of doing the work under the contract, plans and specifications resulted in sand being distributed for great distances by the action of the tide, and of the dredges and other apparatus used, and spreading upon the lands of all of the plaintiffs upon which oysters and clams were planted, killing such oysters and clams. The trial resulted in verdicts in favor of all of the plaintiffs, as to Lloyd and Tallman against both defendants for trespass, and as to the six other plaintiffs against the bridge company only for negligence, a verdict having been entered in favor of the dredging company as to all plaintiffs excepting Lloyd and Tallman.

Several grounds for reversal are argued. The first and second grounds are that the trial judge should have nonsuited at the end of the plaintiffs' case and that he should have directed a verdict for defendants upon the whole case. The principal ground urged is that the bridge company by its grant had a superior right to the use of the land so granted to it, notwithstanding the prior grant to the plaintiffs Lloyd and Tallman. We are unable to agree with this contention. The leases to plaintiffs were made upon statutory authority (Pamph. L. 1917, p. 116; *Cum. Supp. Comp. Stat.*, p. 585), and gave plaintiffs the exclusive privilege of planting and cultivating oysters and clams upon said lands. The subsequent grant of the riparian right to the bridge company could not and did not wipe out the rights of the plaintiffs-lessees to the use of the land. Such subsequent grant was subject to any prior rights acquired from the state. Both defendants were bound to know of such leases. The statute provides for the recording of such leases, so that all persons have notice of the rights of lessees of such lands from the state. In addition to the constructive notice to defendants by the recording of the leases, it is in evidence that the defendants had actual knowledge of the use and occupation of the lands by the plaintiffs. In this situation, to construct a fill upon the lands of Lloyd and Tallman was a clear trespass. Counsel for both defendants admitted that the proof was that both defendants actually went on the lands of Tallman and Lloyd. Further, the fill was made many feet wider than any alleged grant warranted. As to the other plaintiffs, the proofs are clear that the manner of constructing the fill, and a failure to provide against the tides carrying sand loosened and put in motion by the acts of defendants resulted in injury and damage to such plaintiffs, for which plaintiffs are entitled to recover. *Paul v. Hazleton*, 37 N.J.L. 106; *Wooley v. Campbell, Ibid.* 163; *Grace v. Willets*, 50 N.J.L. 414.

We think the motions to nonsuit and to direct verdicts in favor of defendants were properly denied.

The third ground for reversal is that the court erred in refusing to strike out testimony of the plaintiffs as to the quantity of oysters and clams plaintiffs had on the leased lands. In each instance complained of, the testimony came from witnesses who were familiar with the planting, development and gathering of shell fish. The weight of such testimony was for the jury. There may be doubt whether or not the motions to strike this testimony came too late. We think, however, the testimony was competent and that the motion to strike it out was properly denied.

Grounds four and five relate to requests to charge (a) that a verdict must be returned for the bridge company if it used diligence in selecting the Hill company to do the pumping, and that the sand going on plaintiffs' property was a direct result of the contract. The difficulty with this request is that the contract provided for the work to be done in such manner that it did damage to plaintiffs' property. The bridge company owed a duty to plaintiffs not to be a party to creating a nuisance. *Jenne v. Sutton*, 43 N.J.L. 257. In addition, this request, as framed, also would apply to Lloyd and Tallman and as to them it was clearly improper.

We think the court properly refused to charge (b) if the "roadway could have been filled in without permitting any sand to go on the property of plaintiffs," there should be a verdict for the bridge company. It is enough to say that the request, as framed, likewise applied to all plaintiffs, and it would clearly not apply to Tallman and Lloyd, whose lands by the terms of the contract were to be traversed by the fill.

The sixth ground for reversal is directed to the charge of the trial judge as follows: "In performing the work under the contract in this case, defendant was bound to know the effect of the currents of the river upon the sand pumped and agitated by the pumping machine." When a person undertakes work, he is obliged to know and anticipate those things that will reasonably result from his action. We see no error in the statement complained of.

The seventh ground is that the court charged: "The stakes marking the oyster and clam grounds at the time defendant commenced the pumping and agitating of sand in question were notice to defendant to some right in plaintiff in said land." We think this is so. The statute provides for marking the leased beds and such marking is notice to the public of some claim of right to such beds. In addition, it was in evidence that the person in charge for defendants saw the stakes and knew they marked oyster and clam lands. There was no error in this charge.

We have examined the other grounds for reversal urged by both of the defendants and find no merit in them. Upon the whole case, we conclude that the case was properly submitted to the jury both upon the liability of defendants and the amount of damages.

## Mussel Farms

Mussels are Canada's top shellfish aquaculture product, produced in Newfoundland and Labrador, Nova Scotia, Prince Edward Island, and Quebec. Mussels are available year round, and is the second most important species in terms of volume and farm-gate value produced in Canada.

Mussel farming is also found immediately south of the above described area, in Maine, and a recent decision is that of F. Austin Harding v. Commissioner of Marine Resources.[74]

The commissioner of the Department of Marine Resources appeals from a judgment of the Superior Court, Knox County, vacating a series of aquaculture leases granted to Mike & Joe's Sea Farm, Inc. (Sea Farm) and Great Eastern Mussel Farms, Inc. (Great Eastern) pursuant to 12 M.R.S.A. § 6072 (1981 & Supp. 1985). The Superior Court vacated the leases due to the Department's failure to consider evidence of the impact of the proposed aquaculture project on the property values of F. Austin Harding (Harding), an intervenor in the proceedings. Because we agree with the Department that 12 M.R.S.A. § 6072 does not require consideration of the effect of the proposed aquaculture project on Harding's property values, we vacate the judgment of the Superior Court and determine that the leases should be issued to Sea Farm and Great Eastern consistent with the commissioner's original order.

In June of 1983, Sea Farm and Great Eastern filed a joint application to the Department of Marine Resources seeking a series of aquaculture leases in submerged land directly off the coast of Vinalhaven for the purpose of growing mussels. The tracts of land sought by the applicants totaled 50 acres and included 15 acres in Old Harbor and 35 acres in Crockett Cove. See Figure 7.9.

In August of 1983, Harding requested intervenor status for himself as property owner and trustee of 140 acres of land adjacent to Crockett Cove alleging that the aquaculture leases, if granted, would adversely affect his property values. Although intervenor status was granted to Harding, the Department made it clear that he would not be permitted to submit testimony regarding alleged diminution of his property values as a result of the proposed aquaculture operations.

At the public hearing on the leases, the applicants testified that the project involved harvesting seed mussels from areas outside the proposed lease site and transferring these mussels to submerged land within the lease area at lower densities to encourage rapid growth for marketing. Both harvesting and seeding operations were to be conducted by means of a lobster boat. A 4-1/2 foot chain sweep drag would be used to harvest the mussels from the seabed. No permanent surface or subsurface structures were to be employed in the operation, except four buoys to mark the lease site.

---

[74] 510 A.2d 533 (Me., 1986).

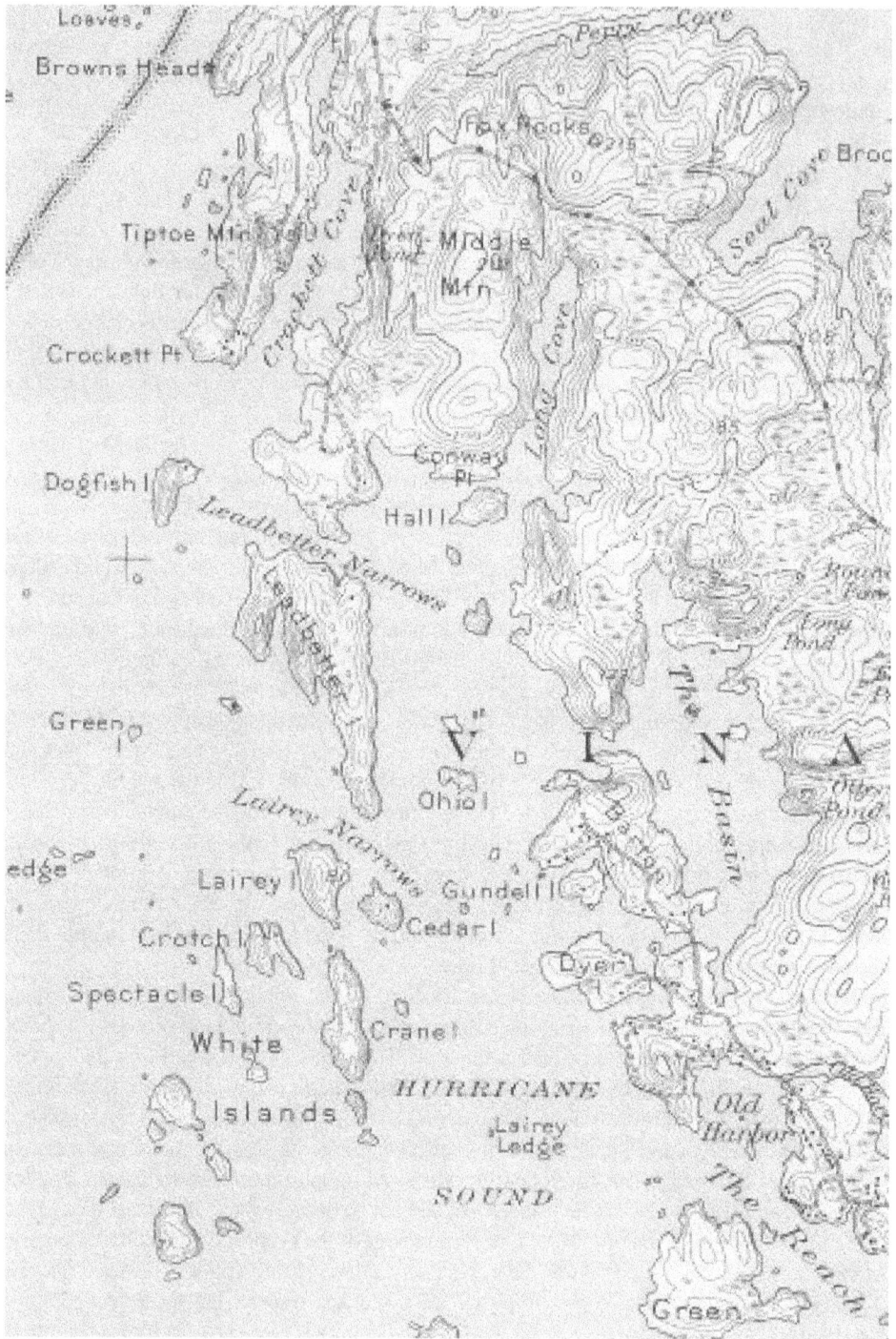

**FIGURE 7.9**   Section of Vinalhaven showing Crockett Cove and Old Harbor. (1944 USGS topographic map.)

At the hearing Harding sought to present testimony concerning the effect of the proposed aquaculture operation on his property values. The deputy commissioner, however, refused to receive such evidence consistent with the Department's initial determination that Harding would be precluded from presenting such testimony.

Following the public hearing, the commissioner granted the leases and found that the proposed project fulfilled all the criteria set out in 12 M.R.S.A. § 6072(7) in that it (1) would not unreasonably interfere with the ingress or egress of any riparian owners, navigation, fishing, or other uses of the area; and (2) would not conflict with shoreline zoning. The commissioner also reiterated in his decision that he did not consider the potential impact of the leases on individual upland property values to be a relevant decisional criterion under the aquaculture leasing statute.

Harding sought judicial review by the Superior Court pursuant to M.R.Civ.P. 80C alleging, inter alia, that the Department erred as a matter of law by denying him the opportunity to present evidence concerning the impact of the leases on the value of his property. Harding did not then, and does not now allege that the impact amounted to a taking of his property. Although it affirmed the Department's decision in all other respects, the Superior Court agreed with Harding and vacated the leases in light of the commissioner's exclusion of evidence relating to the effect of the project on Harding's property values. Although the remand was limited to the presentation of additional evidence rather than an entirely new hearing, the Superior Court did not retain jurisdiction. The commissioner appeals from that Superior Court judgment.

Because of the procedural posture of the case at bar, we first address the question whether to entertain this appeal. In ordinary circumstances we would decline to exercise our appellate jurisdiction because the order of remand is considered interlocutory and not final. See *Harris Baking Co. v. Maine Employment Security Commission,* 457 A.2d 427, 428 (Me.1983). Contrary to the loose language of some of our opinions, see, e.g., *Town of Kittery v. White,* 415 A.2d 1087, 1089 (Me.1980) (appeal dismissed for lack of jurisdiction because no final judgment), our final judgment rule is not jurisdictional but merely a prudential rule to avoid piecemeal review and promote judicial economy. See *Maine Central Railroad v. Bangor & Aroostook Railroad,* 395 A.2d 1107, 1113 n. 7 (Me.1978). See generally 2 Field, McKusick & Wroth, Maine Civil Practice, § 73.1 (2d ed. 1970 & Supp.1981). In *Bar Harbor Banking & Trust Co. v. Alexander,* 411 A.2d 74, 77 (Me.1980), we recognized the propriety of fashioning an exception when necessary to avoid undue disruption of administrative process. In the case before us, the commissioner performs more than a simple adjudicative function; he is charged with administrative enforcement as well. Because of the impact of the Superior Court decision upon aquaculture leasing procedures and because that decision may otherwise escape appellate review, we determine that the case before us presents an appropriate occasion to apply the Bar Harbor Banking exception to the final judgment rule.

We recognize that by expanding the Bar Harbor Banking exception, we aggravate the already difficult task of determining when a Superior Court decision relating to governmental action is ripe for appellate review. Compare, e.g., Harris Baking, 457 A.2d 427, and *Town of Pittsfield v. Chandler,* 457 A.2d 1122 (Me.1983) (Mem.Dec.) with *Chandler v. Town of Pittsfield,* 496 A.2d 1058 (Me.1985) and *Sanborn v. Town of Eliot,* 425 A.2d 629 (Me.1981). In an appropriate case, however, the Superior Court can avoid the uncertainty by expressly retaining jurisdiction of the case pending further administrative proceedings. See Sanborn, 425 A.2d at 631. Retaining jurisdiction may be particularly appropriate when the further administrative proceedings directly affect the substantive issues presented to the Superior Court and when the result will be a more comprehensible record for review by the Law Court. In the case before us, an administrative determination as to whether the leases adversely affect the plaintiff's property values will not resolve the substantive question.

Maine's aquaculture leasing statute provides a rather narrow list of factors that a commissioner must consider before granting a lease in submerged lands. This list does not include property value diminution among the criteria to be examined. Instead, before an aquaculture lease can be approved, the commissioner must ensure only that the proposed project does not unreasonably interfere with the ingress and egress of riparian owners, navigation, fishing, or other uses of the areas and is not in conflict with applicable coastal zoning laws. 12 M.R.S.A. § 6072(7).

Despite the fact that consideration of upland property values is not required by the plain language of section 6072(7), Harding argues that the stated condition of section 6072(7) that leases not "unreasonably interfere with ... other uses of the area" requires a commissioner to allow adjacent property owners to submit property devaluation evidence. We disagree.

Although an examination of the legislative history of the phrase "other uses of the area" provides little guidance, it is unlikely that the Legislature intended to infuse this phrase with property value concepts when such factors are already given separate consideration in the same sentence within the limitation that leases may not conflict with applicable coastal zoning laws. In fact, the prohibition that leases may not interfere with "other uses of the area" more likely is meant to protect lobstering, clamming, scalloping, swimming, mooring of boats, and those other activities that traditionally take place in the areas where aquaculture is to transpire. As a result, we find that the department was correct in concluding that section 6072(7) on its face does not require consideration of the effect of aquaculture operations on upland property values.

In spite of the absence of express statutory language requiring the commissioner to consider property value diminution, Harding argues alternatively that the so-called public-trust doctrine compels consideration of this factor by implication. We also disagree with Harding on this score.

Maine law has long recognized that public rights inherent in submerged lands owned by the State and that the State must manage these lands for the benefit of the public. *Moor v. Veazie*, 32 Me. 343, 356 (1850); *Moulton v. Libbey*, 37 Me. 472, 485-87 (1854). In fact, individual Justices of the Supreme Judicial Court and the Legislature itself have acknowledged the existence of the so-called "public trust doctrine." See Opinion of the Justices, 437 A.2d 597 (Me.1981); 12 M.R.S.A. § 559(1) (1981). Despite the apparent presence of the doctrine in Maine law, however, there appears to be no authority for the proposition that private land values must be considered by the State before it can grant aquaculture leases in the State's submerged lands.

Historically, the public rights to be protected in management of submerged lands included navigation, fishing and fowling – common public uses. The Justices opined that the needs of a growing society may lead to a wider variety of public uses to be protected by the doctrine. Opinion of the Justices, 437 A.2d at 607. We note the familiar rule that an advisory opinion binds neither the Court nor the individual Justices who gave the opinion. *Martin v. Maine Savings Bank*, 154 Me. 259, 269, 147 A.2d 131, 137 (1958). We need not decide, however, the precise scope of the public trust doctrine nor whether the doctrine achieves constitutional status under Me. Const. art. IV, pt. 3, § 1 as intimated by the Justices. Opinion of the Justices, 437 A.2d at 606, 607. Even if we were to adopt a more expansive view of the doctrine, we are unpersuaded that public trust considerations require inclusion of individual private property values in the legislative formulation. Moreover, to the extent that individual property values have an effect upon the "public interest," they are protected by compliance with applicable zoning restrictions.

As a result, the Department was correct in refusing to consider individual property values in its determination. Consequently, the judgment of the Superior Court must be vacated, and the decision of the commissioner to grant the leases should be affirmed.

## Oyster Lands

The term oyster lands implies areas where oysters thrive, while the term oyster farm implies cultivation, either of natural areas or areas planted and maintained.

## OYSTER BEDS

Off-bottom oyster farming occurs on state-owned submerged lands and, in some cases, private lands. As such, authorization to use and occupy those lands is required from the landowner, which is often the respective state, depending on who has title to the seabed. In general, authorization takes the form of a lease. Through the leasing process, the landowner grants the oyster farmer the right to use the submerged lands and the water column for a period of time in exchange for some sort of payment.

Arguably, oysters are the most popular shellfish, and therefore have the most activity, in culture and in the court system. Several states lead the production of harvested oysters, Connecticut, New York, and Maryland being among the highest in volume, along with the Gulf Coast states.

The case of *White v. Petty et al.*[75] is an early Connecticut decision having to do with trespass on a private ownership of oyster lots.

The complainant alleges that she is the owner of various oyster-lots in the town of Darien, upon which she has, in good faith, planted large quantities of oysters, and prays for an injunction restraining the defendants from entering upon them, removing the stakes, and taking up the oysters. The defendants deny that the plaintiff has any title to the ground, alleging that it is a natural oyster-bed. The court decided that none of the ground in question was a natural oyster-bed. The court finds that a part of the ground was, pursuant to the statutes, (Rev. St. 1875, p. 214, § 3) in the summer of 1881, designated to the plaintiff and others, who afterwards assigned their rights to her, and the other part of the ground was designated to the plaintiff's assigns in the months of February and March, 1881. The court also finds that the ground so designated in 1881 was first inclosed by buoys and stakes on the 29th day of September, 1882, and that until then the public had no knowledge of such designation.

Upon the trial the defendants introduced evidence to prove that all the ground in question was, and for 10 years before had been, a natural oyster-bed. To rebut that evidence, the plaintiff offered in evidence the report of a committee appointed by the superior court, pursuant to the act of 1881, (p. 104, § 12, Sess. Laws 1881) to locate and describe, by proper boundaries, all the natural oyster-beds in the town of Darien, and to make report, etc. The report was dated March 26, 1883, and was recorded March 27, 1883, and did not describe any part of the oyster ground in question as a natural oyster-bed. The defendants introduced evidence that in May, 1883, certain other parties brought a petition to the superior court, under the provisions of the act, (Rev. St. 1875, p. 215, § 11,) praying the court to declare the designations of all the land in question void, and to order the removal of the stakes inclosing it, upon the alleged ground that it was a natural oyster-bed, which action was still pending. The defendants claimed that the committee appointed under the act of 1881 had no jurisdiction over ground which had been designated before that act went into effect, and hence that their report, not describing that part of the ground in question which was designated in March and September, 1881, as natural oyster ground, was no evidence that it was not such. They assign the refusal of the court so to rule as their first specific reason of appeal.

The defendants rely in support of their claim upon the case of *In re Oyster Ground Committee*, 52 Conn. 5. That case was a remonstrance against the report of a committee appointed under the act of 1881 to designate the natural oyster-beds in the town of Clinton. "Several persons," in the language of the court, "who had oyster grounds set out to them in severalty prior to the passage of the act, and who had acquired valuable interests therein, and whose interests might be seriously affected if the report of the committee should be accepted, appeared and remonstrated against its acceptance." The court accepted the report of the committee, saying "that the act of 1881, notwithstanding its general language, does not apply to oyster ground previously designated."

---

[75] 57 Conn. 576, 18 A. 253 (1889).

The case is undoubtedly a decisive authority for the defendants if the word "designated," as there used, is to be taken as referring merely to the act of the committee authorized to determine and inform applicants what ground they may occupy. Before the applicant can acquire any right to the exclusive occupation of the ground, he must "mark and stake out the place." This he may or may not do after the committee has acted. Until he does it, the rights of the public in the ground assigned to him are not affected.

It is apparent, from the statement of the case referred to, that the remonstrants had perfected rights to the ground in question in that case, and the word "designated" was used to describe such rights. In the present case, the ground was designated to the plaintiff, or her assigns, by the committee some months before the act of 1881 went into effect; but no steps were taken by her to appropriate it to her use until September 29, 1882, more than a year after the act took effect, and when the committee would have been justified in presuming that she had abandoned any claim to it, if another application for it had been made. The *status* of the ground in question was thus fixed on March 27, 1883, the date when the report of the committee is found to have been accepted and recorded; the provision of the statute being that "such report, when accepted by said court, and recorded in the records thereof, shall be a final and conclusive determination of the extent, boundaries, and location of such natural beds at the date of such report."

The defendants claimed in the court below that the report of the committee did not tend to show the condition of the ground in 1881, and assign as a reason of appeal that the court did not so decide. Natural oyster-beds are not usually found so suddenly that the fact that the ground was not such in 1883 would not tend to show its condition in that respect two years before. But the finding is that the report was offered merely in rebuttal of the defendant's evidence that the ground was, at the time of the trial in 1886, and for 10 years before (covering the date when the report was recorded) had been, a natural oyster-bed. In either view it was proper evidence.

Another reason of appeal is that the court erred "in holding that the finding of the committee could be treated as *res adjudicata* against the defendants." We do not understand from the finding that the question was made upon the trial. The defendants' claim there was in these words, "no finding of said committee could be treated as *res adjudicata* against a private right." The defendants claimed no private right, but such right only as they had in common with the rest of the public. We understand that by the expression "private right" they intended the right which they claimed the plaintiff had acquired by the action of the committee in February and March, 1881. But if the claim now made had been made in the court below, it should have been overruled. There is no ground for a distinction between the defendants and the rest of the public in respect to the conclusiveness of the committee's report.

The defendants assign two other reasons of appeal, predicated upon the finding that there was an action pending against the plaintiff, brought under the statute of 1875, to procure the removal of the stakes inclosing the ground in question, etc. That statute provides, if it is found in such suit that the stakes have been improperly set, the defendant therein shall be entitled to remove the planted oysters within a limited time. The court finds that the plaintiff planted a certain quantity of oysters and gravel upon the ground in question after that suit was brought. The two reasons are as follows: (1) That the court erred in holding that a permanent injunction could be granted while the title was in dispute, and the question of the title was then in the superior court upon a complaint under a statute specifically providing a way to try said title; and (2) In holding that there could be any good faith in law as to the 600 bushels of seed and the 3,000 bushels of gravel planted while a petition was pending to set aside the designation.

It is enough to say that the suit brought under the statute of 1875 has no relation to this one, and evidence regarding it was not properly in the case. The purpose of the statute of 1875 is not, as the defendants claim, to provide a way for the trial of questions of title, but to effect the removal of stakes improperly set up, by means of which the public are deterred from exercising their rights upon ground belonging to them.

Two other reasons of appeal are stated, but the facts which they assume are not found, and they were properly abandoned upon the argument. There is no error.

## OYSTER FARMS

One of the issues concerning an oyster farm is whether it is natural, artificial, or a combination of both. Natural oyster beds, or farms if being farmed (and harvested), are property of the owner of the land on which they reside, but may be leased to others, or used by others with permission. Artificial farms are planted by seed on (mostly) land barren of shellfish, and cultivated over time, including being reseeded. Combination beds are those that occur naturally but are maintained or expanded by seeding in more oysters. Being natural on public (state) land, beds are open to the public at large, following enacted regulations. Those within someone's riparian zone, being personal and part of the realty, are off limits to others without permission or right.

In the New Jersey case of *The State, Robert H. McCarter, Attorney-General, Plaintiff in error, v. The Sooy Oyster Company, Defendant in error*,[76] the single question was raised on the refusal of the court to receive the oral testimony offered by the plaintiff to impeach the validity of the defendant's riparian grant for the lands in dispute, the purpose of the plaintiff being to show by such testimony that such land was in point of fact natural oyster beds within the meaning of the statute of 1888. *Pamph. L., p.* 140.

If this testimony was properly rejected it is not contended that the direction of a verdict was otherwise erroneous.

The bill of exceptions shows that the land was covered by tidal waters which made a *prima facie* case for the plaintiff, to meet which the defendant introduced in evidence a grant of said lands made by the riparian commissioners on November 25, 1903, to E.T. and W.T. Sooy under the great seal of the state pursuant to an act of the legislature approved March 21, 1871. It was admitted that the defendant had a regular chain of title under this grant. At this stage of the trial, the plaintiff offered to prove, in rebuttal, by witnesses, who knew the characteristics of the land in dispute, that such lands had on them at the time of the making of said grant natural oyster beds within the meaning of the amendment to the Riparian act of 1888, which reads as follows:

"That no grant or lease of lands under tidewater whereon there are natural oyster beds shall hereafter be made by the riparian commissioners of this state, except for the purpose of building wharfs, bulkheads or piers."

Objection to the introduction of this testimony being made by the defendant, the court ruled that the grant could not be attacked by such proof in an action of ejectment, and, no other testimony being offered, directed a verdict for the defendant.

The exception allowed the plaintiff to each of these rulings presents the same question which, for convenience, may be divided as follows: *first,* whether or not the proffered proof was a collateral attack upon the riparian grant? *Second,* whether the validity of such grant could be thus impeached in an action of ejectment? And *third,* whether the fact that the plaintiff was the attorney-general acting in his official capacity made any difference?

1. That the testimony offered by the plaintiff was a collateral attack upon the defendant's riparian grant is too clear to justify extended discussion. The plaintiff by his action of ejectment sought to obtain a judgment awarding possession of the premises; when his course to this end was barred by the introduction of the defendant's riparian grant, the plaintiff sought to get rid of

---

[76] 78 N.J.L. 394, 75 A. 211 (1910).

such grant in such action by impeaching its validity by oral testimony as to a matter of fact not suggested on the face of the grant. This was of the very essence of collateral attack.

2. The more debatable question is whether the proffered testimony tended to show that the grant was *dehors* the jurisdiction conferred upon the riparian commissioners, for, if it was, the invalidity of the grant arising from such unwarranted assumption of jurisdiction could be set up anywhere by anyone who was injured by it.

Counsel have therefore correctly apprehended that the question on which their controversy turns is one of jurisdiction, but counsel for the plaintiff in error, in contending, as he does, that the rejected testimony would have shown that the riparian commissioners acted outside of their jurisdiction, loses sight, as it seems to me, of the distinction between jurisdiction in its comprehensive sense and the adequate authority to make a particular grant which is loosely called the jurisdiction to make such grant. This distinction between jurisdiction over a subject-matter generally and the possession of the adequate authority to make a particular disposition respecting it goes to the very heart of the present controversy, for it is the absence of the former alone that exposes such particular act to collateral attack, transgressions of the latter being reviewable only by a direct proceeding.

The distinction between these two conceptions or sorts of jurisdiction is pointed out with great clearness by Mr. Justice Woodhull in his opinion in *Ritter v. Kunkle,* 10 *Vroom* 259, construing that section of our Small Cause act, which provides that where the justice has jurisdiction no judgment shall be removed by *certiorari.* The opinion points out that in that section jurisdiction is used in the sense of the adequate authority to render the particular judgment that was rendered, but that, in the first section of the act, jurisdiction is used in its comprehensive sense as defining the class of cases cognizable by justices of the peace.

It is evident that in a large number of cases the vital question must be whether a given evidential fact affects generic jurisdiction or whether it appertains only to that sort of jurisdiction that Mr. Justice Brown calls *quasi* and that we have agreed to call specific. At all events that is the precise point in controversy in the present case, for if the fact that the lands under tidewater were also natural oyster beds affected the generic jurisdiction of the riparian commissioners, proof of such fact ought to have been received in the action of ejectment, whereas if such fact, if proved, went only to the specific jurisdiction of the commissioners to make the particular grant, such proof was properly rejected.

It is quite aside from this legal proposition to say, however truly, that the riparian commissioners were without authority to grant land on which there were natural oyster beds, for that is only to say that they lacked specific jurisdiction to make the grant, whereas, in order to justify the introduction of the testimony that was rejected at the trial, such proof must have tended to show that the commissioners were without generic jurisdiction in the premises. It is not a question of the lack of authority to make the grant, but of how such lack of authority may be taken advantage of, *i.e.,* whether by direct or by collateral attack. In other words in order to convict the court below of legal error it is necessary for the plaintiff in error to maintain the proposition that lands under tidewater, if they be also natural oyster beds, are not within the generic jurisdiction of the riparian commissioners. Looking more closely at this proposition, two things are to be noted – *first,* that the more inclusive term, viz., lands under tidewater describes *eo nomine* an integral portion of the public domain, and *second,* that the less inclusive term, viz., natural oyster beds indicates merely certain characteristics or qualities that may or may not be possessed by any given portion of such tidal lands, viz., rigid points of fixation to which embryo oyster spats have attached themselves. This being so, and the riparian commissioners being empowered to grant lands under tidewater, and being forbidden to grant such parts of such lands as possessed the foregoing characteristic, it is entirely obvious that in the performance of their duties such commissioners must of necessity determine whether a given piece of tidal land does or does not possess the characteristics that constitute natural oyster beds, and that such determination must enter into and be evidenced by the act done in performance of this official duty. If this be so, then upon entirely familiar principles, the making of a grant by the riparian commissioners imperatively imports and evidences that such determination, whether right or wrong in point of fact, had been made by them. A riparian grant therefore, in the present state of our statutory law, bears within its own bosom the necessary implication that the commissioners have determined that the lands thus granted did not possess the statutory characteristics that would have prohibited their being granted. If the actual fact be otherwise, the riparian commissioners were, none the less, acting within their jurisdiction in determining the question of fact with respect to which, upon the hypothesis, they reached an erroneous conclusion. For the rule already announced has this corollary, viz., that the existence of generic jurisdiction includes the right to determine questions of fact on which the existence of a specific jurisdiction depends; which determination, being judicial in character, falls within the established rule respecting the conclusive effect to be given to the findings of special tribunals acting within the scope of their authority.

Assuming, therefore, that, in the case in hand, the riparian commissioners, by reason of their erroneous determination of a question of fact, were without authority to make the grant under which the defendant claimed title, it is none the less true that such error did not go to the generic jurisdiction of the commissioners, but, on the contrary, presupposes such jurisdiction as the basis of the exercise of the very judgment which, upon the hypothesis, turned out to be erroneous. The conclusion therefore is irresistible that testimony whose only tendency was to show that the riparian commissioners committed error in the determination of a question of fact which they had the requisite jurisdiction to determine appertained only to the specific jurisdiction of the commissioners to make the particular grant, and hence, under the rule already laid down, was not available for the purposes of a collateral attack upon the validity of such grant.

It will serve, I think, a useful purpose to say, at this point, that if the legislature had itself specified what lands should not be granted, as, for instance, if it had forbidden the granting of lands beneath the Mullica river or under the Kill von Kull, a different question might be presented. For, in such case, the essential feature on which the present case turns would be wholly lacking, viz., the necessity of a determination by the riparian commissioners of a question of fact on which their specific jurisdiction to make the particular grant depended. In such supposed case the lands so specified by the legislature would, in the language of the cases, "be withdrawn from sale." See cases cited, *post*.

Precisely what we are deciding therefore is that in the exercise of their general jurisdiction to grant lands under tidal waters it was in the present case necessary for the riparian commissioners to determine whether or not the lands in dispute were natural oyster beds and that the subsequent making of a grant thereof was in legal intendment tantamount to a provisional adjudication that such lands did not possess such characteristics and that such adjudication, however erroneous in point of fact, and whether induced by error of judgment or by active or passive misrepresentation, cannot be called in question upon the trial of an action of ejectment by the production of testimony as to the actual character of said land.

The result thus reached by the consideration of the nature and incidents of grants made by our riparian commissioners accords with judicial decisions elsewhere regarding the grants or patents of the land departments of both the state and federal governments.

"A patent," said Mr. Justice Grier, in *United States v. Stone,* 2 Wall. 525, "is the highest evidence of title and is conclusive as against the government and all claiming under junior patents or titles until it is set aside or annulled by some judicial tribunal."

In *Johnson v. Towsley,* 13 Wall. 72, Mr. Justice Miller, speaking of the general land office, stated the general doctrine to be "that when the law has confided to a special tribunal the authority to hear and determine certain matters arising in the course of its duties the decision of that tribunal within the scope of its authority is conclusive upon all others."

To the same effect is the language of the same judge in *Moore v. Robbins,* 6 *Otto* 530.

In *St. Louis Smelting Co. v. Kemp,* 14 *Otto* 636, Mr. Justice Field, speaking of a patent of the land office, said, touching a decision of the officers of that department:

In that respect they exercise a judicial function, and therefore it has been held in various instances by this court that their judgment as to matters of fact, properly determinable by them is conclusive when brought to notice in a collateral proceeding. Their judgment in such cases is like that of other special tribunals upon matters within their exclusive jurisdiction, unassailable except by a direct proceeding for its correction or annullment."

This language is quoted with approval by Mr. Justice Brown in the case of Noble *v.* Union River Logging Railway Co. already referred to. *Ard v. Brandon, 39 U.S. (L. ed.)* 525.

It is needless to continue citations to this same effect that might be multiplied to an almost indefinite extent.

There is, however, a class of cases that presents so strict an analogy to the one under consideration as to justify one further citation. The cases referred to arose under the act of congress granting swamp lands to the states, and the following quotations from the opinion of Mr. Justice Miller in *French v. Fyan,* 3 *Otto* 169, make it perfectly clear that the question in that case was the same as the one we are considering, viz., the legal effect of a determination by the proper public officer as to the characteristics of certain lands upon which the authority of such officer to make a grant of such lands depends. The similarity of the two cases both as to the legal question involved, and the manner in which it was raised in the collateral proceeding, appears from the opinion, the pertinent parts of which are as follows:

"An action of ejectment in this case was tried by the court below without a jury by agreement of the parties and the only finding made by the court was a general one in favor of the defendant on which judgment was rendered in bar of the action. The single question in this case is raised on the refusal of the court to receive oral testimony to impeach the validity of a patent issued by the United States to the State of Missouri for the land in question under the act of 1850, known as the swamp-land grant, the purpose being to show by such testimony that it was not in point of fact swamp land within the meaning of that act. The bill of exceptions shows that the land was certified in 1854 (under another act to a mesne grantor), and the plaintiff, by purchase, became vested with such title as this certificate gave. To overcome this *prima facie* case, defendant gave in evidence the patent issued to Missouri in 1857 under the Swamp Land act, and it was admitted that defendant had a regular chain of title under this patent. It was at this stage of the proceeding that the plaintiff offered to prove, in rebuttal, by witnesses, who had known the character of the land in dispute since 1849 till the time of the trial, that the land in dispute was not swamp and overflowed land made unfit thereby for cultivation, and that the greater part thereof is not, and never has been, since 1849, wet and unfit for cultivation. But the court ruled that since the defendant had introduced a patent from the United States to the state for said land under the act of September 28th, 1850, as swamp land, this concluded the question, and the court rejected the said parol testimony to which ruling of the court the plaintiff then and there excepted."

It is evident that the legal question thus presented is precisely the same as the one we are called upon to decide.

In affirming the judgment of the court below and approving its ruling in rejecting the proffered proof, Mr. Justice Miller said that the only question was "whether in an action at law in which these evidences of title came in conflict, parol testimony can be received to show that the land in controversy was never swamp land, and therefore the patent issued to the state under that act is void."

The analogous question before us is whether parol testimony can be received to show that the land in dispute was natural oyster beds, and hence the grant thereof made under our Riparian act was void.

In disposing of this question Justice Miller said: "We have so often commented in this court on the conclusive nature and effect of such a decision when made and evidenced by the issuance of a patent that we can do no better than to repeat what was said in the case of *Johnson v. Towsley,* 13 Wall. 72." The opinion then quotes liberally from the earlier decision and concludes: "We see nothing in the case before us to take it out of the operation of that rule, and we are of opinion that, in this action at law, it would be a departure from sound principle and contrary to well-considered judgments in this court and in others of high authority to permit the validity of the patent to the state to be submitted to the test of the verdict of a jury on such oral testimony as might be brought before it. * * * The principle we have laid down is in harmony with the system which governs the relations of the courts to the officers of the executive departments, especially those having charge of the public lands, as we have repeatedly decided, and we must abide by them."

There is no conflict between the views expressed by Mr. Justice Miller in this case and those expressed by him in *Reynolds v. Iron Silver Mining Co.,* 116 U.S. 687; *Doolan v. Carr,* 125 U.S. 618; *Iron Silver Mining Co. v. Campbell,* 135 U.S. 286.

All confusion between cases like the swamp-land grants and those in which placer mines and mineral lodes are involved is avoided by bearing in mind the obvious distinction between two things that are totally different from each other and two different states or conditions of the same thing. The distinction is fundamental. A mineral lode is a totally different thing from a placer mine and not a mere state or condition of such mine. A placer is "earth, sand or gravel containing valuable mineral in particles" (Webster), whereas a lode is "a solid vein of metal." They are two distinct things, not two different states of the same thing. Hence, a reservation from sale of one of these distinct things leaves nothing to the determination of the state's sale agent. The same is true of saline or other discrete mineral or chemical deposits. Swamp land, on the contrary, is simply land that is too wet for cultivation, a characteristic by which it is distinguished from arable land, but it is none the less land; it is a mere state or condition of land the existence of which must of necessity be passed upon by the state's agency as an incident of its office.

*So natural oyster beds are not a distinct and different thing from tidal lands; they are simply tidal lands possessed of certain qualities* [words from the court, emphasis supplied by author].

In view of this obvious distinction the opinions of Mr. Justice Miller do not conflict or leave the law as laid down by the Supreme Court of the United States in hopeless confusion.

Assuming, however, for the argument's sake, that such confusion exists, I am clear that we should adopt that rule of procedure (and it is a mere question of procedure) that will give to our riparian grants stability and value rather than that which will detract from, if it will not destroy, such value by exposing them to constant collateral attack and invalidation by the verdict of juries of the vicinage. It is one thing to have the validity of a riparian grant judicially decided by the Court of Chancery and quite a different thing to have it successfully assailed before a jury of the vicinage. The value of such a grant depends upon the fact that it is not subject to such assault. "It is this unassailable character," said Mr. Justice Field in St. Louis Smelting Co. v. Kemp, "which gives it its chief, indeed its only value, as a means of quieting its possessor in the enjoyment of the lands it embraces. If intruders upon them could compel him, in every suit for possession, to establish the validity of the action of the land department and the correctness of its ruling upon matters submitted to it, the patent, instead of being a means of peace and security, would subject his rights to constant and ruinous litigation. He would recover one portion of his land if the jury were satisfied that the evidence produced justified the action of that department and lose another portion, the title whereto rests upon the same facts, because another jury came to a different conclusion. So his rights in different suits upon the same patent would be determined, not by its efficacy as a conveyance of the government, but according to the fluctuating prejudices of different jurymen or their varying capacities to weigh evidence."

Is this the procedure we should deliberately adopt? If our riparian grants are to be subject to such indirect attacks it is difficult to believe that any prudent man would purchase so insecure a tenure. Under such rule of interminable collateral attack, tidal lands, improved at no matter what expense, would revert to the state or pass to other hands whenever the allegation of a plaintiff in ejectment that there were natural oyster beds on such land obtained the concurrence of a jury of the vicinage.

The case moreover is not one which involves the annulling of the grant. The riparian commission certainly has the power to grant lands under tidewater which are not natural oyster beds. This grant, like any other, may be entirely valid as to a portion of the land granted, but may not operate to convey such portion as is natural oyster beds. In such a case, the grant could not be annulled, and it is not the function of a suit in Chancery to determine the boundaries of a grant. The boundaries are to be determined as any other question of boundaries is determined, ordinarily by the verdict of a jury. The grant, if by mistake it conveys land which was not meant to be conveyed, is to be rectified by a bill in Chancery for reformation. Such a proceeding is quite inconsistent with the theory that the question of fact whether any of the land was natural oyster beds had been adjudicated by the riparian commission adversely to the claim of the state; for if that question had been before them and decided, it could not be contended that by their subsequent grant they did not mean to include the land, the character of which had been thus determined. It is incorrect to say that in this proceeding the attorney-general sought to have a grant declared void. He claimed the whole of the land, not because the grant was invalid, but because it was inoperative, just as if it had conveyed land which the state had previously granted to others.

I am aware of the importance of sustaining the validity of riparian grants when legally made, but it is equally important to maintain the rights of the state as against individuals who seek to acquire lands contrary to law. We pursue the safer course when we decide such cases in accordance with the logical consequences of established legal principles without regard to those questions of public policy which may well affect legislative action, but ought not to control the action of the courts. If I entertained the fear of the result of jury trials which is sometimes expressed, a fear which hardly comports with the long-continued use of the jury in cases involving the title to land, I should hesitate to say that the court was justified in adopting a rule which would substitute another tribunal. It has always been held that the citizen has a constitutional right to submit the title to his land to the verdict of a jury, and the doubt which has been felt as to the power of the legislature to confer upon the Court of Chancery jurisdiction to remove a cloud upon title, the limitations upon the power of the court in those cases and the express reservation of the right to take the verdict of a jury, combine to warn us how careful the law has been to preserve the prerogatives of the jury in cases where the title to land is involved. The care thus evinced is persuasive evidence that upon the whole substantial justice has been done by juries. The occasional miscarriages of justice have been redressed by the corrective power of the courts, and it may be a debatable question whether such miscarriages have been more frequent than they would have been if the decision had been as in some countries committed to the court alone. The intent of the legislature is plain. At the beginning of our legislation as to riparian grants by section 12 of the act of March 31st, 1869 (*Gen. Stat., p.* 2789, *pl.* 19), the commissioners were authorized to proceed in the name of the state by ejectment or otherwise against trespassers upon or occupants of lands of the state under tidewater. No question has been raised by plea or in the argument as to the right of the state to maintain ejectment or to the form in which this action is brought. It is irregular to entitle the cause in the name of the attorney-general at the instance of nominal plaintiffs, but the declaration avers that the action is brought for the state, and concludes by the averment that the defendant wrongfully deprives the state of the possession; the plea is that the defendant is not guilty of the injury whereof the state hath complained. In substance, it is an action of eject-ment by the state, and counsel were right in so treating it; the mere formal introduction of the dec-laration is amendable. What, then, is the situation which the case presents? It is recognized that even after a grant under the great seal, the state may assert rights not only by bill in Chancery to annul the grant, but by entry upon the land. It was so held by one of our most learned judges, the late Chief Justice Depue, in the great case of *American Dock and Improvement Co. v. Trustees of Public Schools,* 12 Stew. Eq. 409 (at *p.* 418). What he said was in form only a charge to a jury, but in fact it was an opinion disposing of the case, and prepared with a care and thoroughness commensurate with the magnitude of the interests involved, and the pains that great judge was accustomed to take. At common law the sovereign could not make an entry in person, but as was said by Mr. Justice Nelson in *United States v. Repentigny,* 5 Wall. 211, 268: "The mode of assert-ing or of assuming the forfeited grant is subject to the legislative authority of the government. It may be after judicial investigation, or by taking possession directly under the authority of the government without these preliminary proceedings."

At common law the king could not maintain ejectment because by reason of his legal ubiquity he could not be dispossessed. 3 Bl. Com. 257. The difficulties in the way of entry and of eject-ment were similar. When our legislature by section 12 of the act of 1869 authorized proceedings "by ejectment, or otherwise, against persons and corporations trespassing upon or occupying the lands of the state under water," it evidently meant to do away with the technical difficulties inter-posed by the common law. The words "or otherwise" can hardly refer to any proceeding against occupants as distinguished from trespassers unless by way of entry. Occupants, if not trespassers, are rightfully in possession, and the proceeding by the state must be in the nature of an entry to terminate that right. The actions provided for by this section were evidently common-law actions, since the act provides that the charges and counsel fees shall be taxed by the Chief Justice, not by the Chancellor. This statute itself is a sufficient answer to the suggestion that the only remedy of the state is by bill in Chancery. We may, in the absence of plea or suggestions to the contrary, assume that an action of ejectment instituted by the attorney-general for the protection of the rights of the state has been duly authorized; if not, that is a question between the riparian com-mission and the attorney-general, and want of authority to sue would afford no justification to the defendant in an action by the state if he is trespassing upon or occupying its lands. The real ques-tion is, therefore, whether it was competent for the attorney-general in the present case to show that any part of the lands belonged to the state, for if section 12 was meant to provide a method of testing the title, in addition to the method by bill in Chancery, and an equivalent for the right of entry by a private individual, it would be no answer to produce the grant under the great seal; that is the very thing to be questioned.

In the determination of this question of title, the state made its *prima facie* case by proving that the lands were under tidewater. Such lands belong to the state unless they have been con-veyed away. In answer to this, the defendant made a *prima facie* case for itself by producing the grant from the riparian commission. Such a grant is valid as far as affects lands under tidewater that are not natural oyster beds. The application of that grant to the *locus in quo* depends, how-ever, upon the authority of the riparian commissioners. It is not a question of attacking a grant under the great seal, but of locating that grant upon the ground. If in so locating it, part of the land is found to be natural oyster beds, that part cannot pass to the grantee because the legislature has expressly forbidden the commission to make such a grant, and, so far as they exceed their powers, their act is not the act of the state at all. No one can contend that if the riparian commis-sion made two grants which overlapped, either one on its face would be conclusive; it would be a question for a jury which grant was the earlier. It is true that a controversy of that kind would arise between private individuals, but can there be any difference in legal principle between the case of land previously conveyed which the state is therefore without power to grant, and the case of land which the legislature has forbidden the riparian commission to convey? As Mr. Justice Van Syckel said in *Polhemus v. Bateman,* 31 *Vroom* 163:

"It is well settled that officers entrusted with power of sale exercise a naked power, and no title passes unless the conditions exist upon which the exercise of the power depends."

He adds:

"A grant which, by the terms of the Riparian act, the riparian commission were disabled to make, and which, by the express language of the conveyance executed by them, they declared they did not intend to make, passed no title to the grantee, and could not have the effect to make the act of Polhemus a trespass."

It is true that in that case the grant itself embodied the limitation of the statute, but Mr. Justice Depue, also speaking for this court, had previously said, in Fitzgerald v. Faunce, 17 Vroom 536 (at p. 594):

"Under the act of 1871, no one but a riparian owner can apply, and the grant by the commissioners to anyone else would be *ultra vires.*"

If a grant which the commissioners were empowered to make to a riparian owner is *ultra vires,* when the grantee is not in fact the riparian owner, much more must their act be *ultra vires* when it attempts to convey land which the commissioners were expressly forbidden to convey. No machinery is provided by which the commission can determine either the fact of the ownership of the *ripa* or the fact of the existence and location of natural oyster beds. In truth, as I shall show, the latter fact has been left to the determination of another tribunal. The riparian commission may well make a general grant, and for the state's protection, rely upon the express language of the act of 1888, limiting its power, of which everyone must take notice. The fact that the great seal of the state is physically affixed cannot enlarge the grant. It is of no effect because affixed by the riparian commission, so far as they exceed the power given them by the legislature.

The grantee taking the title from the riparian commissioners is, of course, chargeable with notice of the powers and the limitations upon the powers of the commissioners imposed by the public acts of the legislature, and if the act is unauthorized, he is conclusively presumed to know the fact, and acts at his peril, just as anyone who chooses to deal with an agent, with knowledge that the agent is transcending his powers.

The error assigned is the exclusion of all evidence as to the character of the land conveyed. We must assume that the state could have proved that all the land was natural oyster beds and within the statutory prohibition. If this is so, there was no jurisdiction in the commissioners. If it is not so, it was permissible for the state to show the character of the land in order to ascertain the extent of the conveyance, precisely as in the cases of the mining grants at Leadville from the United States government, to be hereafter cited, it was permissible to show that the grantee under a placer patent knew of the existence of lodes and veins within its boundaries so that his conveyance from the government was less extensive than it purported to be on its face. I have likened the case of a grant of natural oyster beds, forbidden to be conveyed, to the case of lands already conveyed to another, where the state has no title. The comparison is not original with me. In 1815, Chief Justice Marshall, in delivering the opinion in *Polk v. Wendal,* 9 *Cranch* 87, said that the question was: "Is it in any, and if in any, in what cases, allowable in an ejectment to impeach a grant from the state, for causes anterior to its being issued?" He then held that it would be unreasonable to avoid a grant for irregularities in the conduct of public officers appointed to supervise the matter, and commented upon the greater advantage of a court of equity where the question was of essentials rather than of mere irregularities, and added: "But there are cases in which a grant is absolutely void; as where the state has no title to the thing granted; or where the officer had no authority to issue the grant. In such cases the validity of the grant is *necessarily examinable at law.*"

Numerous cases have followed Polk v. Wendal. As recently as 1896, in Burfenning v. Chicago, St. Paul, &c., Railway, Mr. Justice Brewer said: "It has undoubtedly been affirmed over and over again that in the administration of the public land system of the United States, questions of fact are for the consideration and judgment of the land department, and that its judgment thereon is final. Whether, for instance, a certain tract is swamp land or not, saline land or not, mineral land or not, presents a question of fact not resting on record, dependent on oral testimony; and it cannot be doubted that the decision of the land department, one way or the other, in reference to these questions, is conclusive and not open to relitigation in the courts, except in those cases of fraud, &c., which permit any determination to be re-examined. But it is also equally true that when by act of congress a tract of land has been reserved from homestead and pre-emption, or dedicated to any special purpose, proceedings in the land department in defiance of such reservation or dedication, although culminating in a patent, transfer no title and may be challenged in an action at law. *In other words, the action of the land department cannot override the expressed will of congress or convey away public lands in disregard or defiance thereof.*"

So, in *Noble v. Union River Logging Co.,* 147 U.S. 165, Mr. Justice Brown said (at *p.* 173):

"It is true that in every proceeding of a judicial nature there are one or more facts which are strictly jurisdictional, the existence of which is necessary to the validity of the proceedings, and without which the act of the court is a mere nullity."

He gives as an example of such a jurisdictional fact, without which the act is a mere nullity, the case where "the land department issues a patent for land which has already been reserved or granted to another person," in which case he adds, "The act is not voidable merely, but void." After distinguishing between such jurisdictional facts and those which he calls *quasi*-jurisdictional, he continues:

"This distinction has been taken in a large number of cases in this court in which the validity of land patents has been attacked collaterally, and it has always been held that the existence of lands *subject to be patented* was the only necessary prerequisite to a valid patent. In the one class of cases it is held that if the land attempted to be patented had been reserved, or was at the time no part of the public domain, the land department had no jurisdiction over it and no power or authority to dispose of it. In such cases its action in certifying the lands under a railroad grant or in issuing a patent is not merely irregular, but absolutely void, and may be shown to be so in any collateral proceeding."

There is a class of cases of which *French v. Fyan,* 93 U.S. 169, is a good example, where the determination of the land office as to the character of the land—in that case swamp land or not -- was held final. In all these cases, as far as I know, it will be found that the land covered by the patent was land which the land office had power to grant, and the only difference was a difference of procedure, of the *method* by which the title of the government was to be conveyed, not of the *right* of the land commissioner to convey. In other cases, where the land was not subject to disposition, the result was different. I need refer only to cases which have a more or less close analogy to the case at bar.

In *Morton v. Nebraska,* 21 Wall. 660, Morton and others, in an action of ejectment, claimed lands by virtue of patents under the Military Bounty Land act. The State of Nebraska claimed the lands under acts of congress reserving saline lands from sale. It was held that the title of the state must prevail. Mr. Justice Davis said:

"It does not strengthen the case of the plaintiffs that they obtained the certificates of entry, and that patents were subsequently issued on these certificates. It has been repeatedly decided by this court that patents for lands which have been previously granted, reserved from sale, or appropriated, are void. The executive officers had no authority to issue a patent for the lands in controversy, because they were not subject to entry, having been previously reserved, and this want of power may be proved by a defendant in an action at law."

In that case the lands in question had been noted on the field books as saline, but these notes had not been transmitted to the register's general plats. The case turned upon the fact that the land was not only saline but known to be such, and the court reserved the question what effect the statute reserving such lands might have on salines hidden in the earth.

In *Reynolds v. Iron Silver Mining Co.,* 116 U.S. 687, the mining company claimed under a patent for placer mines. The defendant asserted a right under what are called lode claims. The court held that the act of congress provided for three classes of cases:

1. When the applicant for a placer patent is at the time in possession of a vein or lode included within the boundaries of his placer claim he shall state that fact, and on payment of the sum required for a vein claim and twenty-five feet on each side of it at $5 per acre and $2.50 for the remainder of the placer claim, his patent shall cover both.

2. Where no such vein or lode is known to exist at the time the patent is applied for, the patent for a placer claim shall carry all valuable mineral and other deposits which may be found within the boundaries thereof.

3. But in cases where the applicant for the placer patent is not in possession of such lode or vein within the boundaries of his claim, but such a vein is known to exist, and it is not referred to or mentioned in the claim or patent, then the application shall be construed as a conclusive declaration that the claimant of the placer mine has no right to the possession of the vein or lode claim.

"It is this latter class of cases," says Mr. Justice Miller, "to which the one before us belongs," and it was held that the placer patent conferred no title, and that the mining company, which sought to recover possession in an action at law, could not succeed against the defendants who were working the lode, even though they were mere trespassers. The importance of this decision in the present case is that the effect of the placer patent was made to depend upon a matter resting purely in parol, as to whether or not at the time the application was made for the placer patent a lode or vein was known to exist within its bounds. The case is, therefore, authority for determining by the verdict of a jury in an action at law a fact resting purely upon parol evidence, and operating to limit the general language of the patent and to define the extent of the property conveyed.

In *Doolan v. Carr,* 125 U.S. 618, Carr brought ejectment claiming title under a patent from the United States to the Central Pacific Railroad Company and a subsequent deed from the railroad company to himself. The defendants offered to show that this patent was void, to which the plaintiff objected upon the ground that the patent could not be collaterally attacked, that it could be attacked by bill in equity only, that it was conclusive evidence in the pending action that the legal title of the lands therein described was granted and transferred by the United States to the grantee named in the patent. The court sustained the objection, and on writ of error the judgment was reversed for this error. Mr. Justice Miller said:

"There is no question as to the principle that where the officers of the government have issued a patent in due form of law which, on its face, is sufficient to convey the title to the land described in it, such patent is to be treated as valid in actions at law as distinguished from suits in equity, subject, however, at all times, to the inquiry whether such officers had the lawful authority to make a conveyance of the title. But if those officers acted without authority, if the land which they purported to convey had never been within their control, *or had been withdrawn from that control at the time they undertook to exercise such authority,* then their act was void -- void for want of power in them to act on the subject-matter of the patent, not merely voidable, in which latter case, if the circumstances justified such a decree, a direct proceeding with proper averments and evidence would be required to establish it was voidable, and should, therefore, be avoided. The distinction is a manifest one, although the circumstances that enter into it are not always easily defined. It is, nevertheless, a clear distinction established by law, and it has been often asserted in this court that even a patent from the government of the United States, issued with all the forms of law, may be shown to be void by extrinsic evidence, if it be such evidence as by its nature is capable of showing a want of authority for its issue."

Mr. Justice Miller's opinion is valuable for the reason that it contains a review of prior decisions in point. It is true that he refused to decide how far parol evidence could be received in an action at law for the purpose of impeaching the patent, holding that the evidence in the case was entirely documentary. This could be true only in a qualified sense, since a part of the evidence was necessarily parol testimony to identify the land and the boundaries. Chief Justice Waite, in dissenting, expressly said that the ground of his dissent was not that in a proper case the validity of a patent of the United States for the conveyance of land might not be attacked in a suit at law by proving it was issued without the requisite authority, but that this is not a proper case for the application of that rule. He thought that such proof could only be made by one who held a right at law or in equity, which is prior in time to that of the patentee, or by one who claimed under the United States by a subsequent grant or some authorized recognition of title; but even on the view which he took, the State of New Jersey, in the present case, would have a right to prove the want of authority of the riparian commissioners to issue the patent in question.

In *Iron Silver Mining Co. v. Campbell,* 135 U.S. 286, the controversy arose in an action in the nature of ejectment between the owner of a patent for a placer mine and the owner of a patent for a lode within the boundaries of the land covered by the placer patent. The placer patent was prior in point of time, but the trial court held that it must be conclusively presumed from the fact that the lode patent had been subsequently issued that the facts necessary to make it a valid grant had been ascertained by the land office and could not be questioned in an action of ejectment. The essential fact was whether the existence of the lode was known to the patentees of the placer patent at the time they acquired the same, a fact resting purely in parol; but the Supreme Court of the United States held that the trial court was in error, and that, in order to sustain the validity of the lode patent, it was necessary for the claimant thereunder to establish the existence of the lode and the knowledge thereof by the patentee under the elder patent at the time he acquired the same. It would be difficult to find a case where the validity of a patent rested more thoroughly upon the verdict of a jury rendered upon parol evidence as to a fact of which there could be no record. The case is also interesting because it distinguishes *French v. Fyan*, relied upon in the present case, upon the ground that in that case the secretary had certified that the lands in controversy were swamp lands. In the present case there is nothing to show whether or not the riparian commission considered the question whether the land in question was natural oyster beds.

The litigation in these cases of the Iron Silver Mining Company was important and prolonged, since it involved grants at Leadville. The matter came before the court again in *Iron Silver Co. v. Mike & Starr Gold and Silver Mining Co.,* 143 U.S. 394, which was twice argued and held for mature consideration, the contention being that a known vein must be a located vein or lode, but it was held that this was not correct, and the matter was treated as a question of fact to be determined upon parol evidence by a jury. The importance of this case to the present lies in the fact that it shows clearly that lands may be reserved from sale by a general description, and that a specific description by metes and bounds is not necessary. Just as known lodes or veins were reserved from grant under a placer patent, leaving the fact that they were known as well as their exact boundaries to be determined by the verdict of a jury, so natural oyster beds are reserved from the riparian lands which the commissioners are authorized to convey.

In *Davis v. Weibbold,* 139 U.S. 507, which was an action for the possession of mining land, the plaintiff Weibbold claimed title under a patent for mineral land. The defense justified under a patent prior in date for a town site which, pursuant to law, exempted from the grant any mine of gold, silver, cinnabar or copper, and any valid mining claim or possession held under existing laws of congress. If the mineral rights claimed by the plaintiff were within this exception, his title was good, and it was sought to sustain it on the ground that the land office in issuing the patent must have determined that the facts existed which would make the patent valid. The Supreme Court held, reversing the trial court, that by a proper construction of the defendant's patent, only such lands were excepted as were at the time of the grant known to be so valuable for their minerals as to justify expenditure for their extraction. It said that the question was not whether there were valuable minerals at the time that the patent was issued, but whether such minerals were known to exist within the premises at the date of the town site patent to the probate judge.

"The plaintiff not having offered any proof upon this point, but having relied upon the fact as a matter of presumption merely, the defendant should have been permitted to establish the negative of it. The absence of any proceedings required by law or the custom of the mining district to initiate a right to a mining claim which he might perhaps have shown would have been very persuasive that no mine was known to exist."

The case, therefore, made the validity of a patent issued by the land office depend upon a matter resting entirely upon parol evidence as to whether minerals were known to exist at the time the town site patent was granted. The court treated the mining patent as affording a presumption that the existence of such minerals was known, subject to be rebutted by proof on the part of the claimant under the town site patent. The case is important, because the patent which was pronounced inoperative contained nothing on its face to indicate its defect. The defect arose out of the fact that the minerals in question were held not to be within the reservation of the prior town site patent, and what that reservation was, depended upon no record, but upon parol testimony entirely. At first blush it might appear as if the ruling of the court in this case upon the effect of the town site patent conflicted with the rule that lands reserved from sale do not pass by the patent. An examination of the opinion, however, shows clearly that such is not the case, for the court held that while the language of the exception in the town site patent would seem on first impression to constitute a reservation of such mines in the land sold and all mining claims on them to the United States, that such was not the necessary meaning, since, in strictness, the words imported only that the provisions by which title to the land in such town sites is transferred shall not be the means of passing a title also to mines of gold, silver, & c. And this explanation by the court distinguishes also its subsequent decision in *Barden v. Northern Pacific Railroad Co.,* 154 U.S. 288 (at *pp.* 323, 324). In that case all that the court decided was that the congressional grant of lands to the Northern Pacific Railroad, which reserved mineral lands, reserved not only lands which were known to be mineral at the time, but those which were subsequently discovered to be such. In order to meet the argument that that construction of the grant made the title uncertain, the court said, *arguendo,* that the title would be made certain when the patent was issued, but this conclusion of the court rested upon the view suggested in Davis v. Weibbold, that there was power in the land office to issue a patent even for mineral lands, and that it was only the procedure which was different. I do not, of course, question that when a patent is once issued, or a riparian grant is once made, defects in procedure cannot be questioned collaterally.

All the cases treat lands reserved from sale and lands previously conveyed as coming within the same class. The law is thus summed up in *St. Louis Smelting and Refining Co. v. Kemp,* 104 U.S. 636. Mr. Justice Field, while maintaining the conclusive presumption attending a patent for lands, used the following language:

"Of course, when we speak of the conclusive presumptions attending a patent for lands we assume it was issued in a case where the department had jurisdiction to act and execute it; that is to say, in a case where the lands belonged to the United States and provision had been made by law for their sale. If they never were public property, or had previously been disposed of, or if congress had made no provision for their sale, or had reserved them, the department would have no jurisdiction to transfer them, and its attempted conveyance of them would be inoperative and void, no matter with what seeming regularity the forms of law may have been observed. The action of the department would, in that event, be like that of any other special tribunal not having jurisdiction of a case which it had assumed to decide. Matters of this kind disclosing a want of jurisdiction may be considered by a court of law. In such cases the objection to the patent reaches beyond the action of the special tribunal, and goes to the existence of a subject upon which it was competent to act."

And again: "The doctrine declared in these cases as to the presumptions attending a patent has been uniformly followed by this court. The exceptions mentioned have also been regarded as sound, although from the general language used, some of them may require explanation to understand fully their import. If the patent, according to the doctrine, be absolutely void on its face it may be collaterally impeached in a court of law. *It is seldom, however, that the recitals of a patent will nullify its granting clause, as, for instance, that the land which it purports to convey is reserved from sale.* Of course, should such inconsistency appear, the grant would fail. Something more, however, than an apparent contradiction in its terms is meant when we speak of a patent being void on its face. It is meant that the patent is seen to be invalid, *either when read in the light of existing law, or by reason of what the court must take judicial notice of, as, for instance, that the land is reserved by statute from sale or otherwise appropriated, or that the patent is for an unauthorized amount, or is executed by officers who are not entrusted by law with the power to issue grants of portions of the public domain.*"

For the same view see *Lake Superior Ship Canal, & c., Co. v. Cunningham,* 155 U.S. 354 (at *p.* 373), and *Burfenning v. Chicago, St. Paul, & c., Railway Co.,* 163 U.S. 321.

In *Mahn v. Harwood,* 112 U.S. 354, which was a suit to recover damages alleged to have resulted from the infringement of letters patent, and therefore a collateral attack on the patent, Mr. Justice Bradley said: "Where it is evident that the commissioner, under a misconception of the law, has exceeded his authority in granting or reissuing a patent, there is no sound principle to prevent a party sued for this infringement from availing himself of the illegality, independently of any statutory permission so to do. This is constantly done in land cases, where patents have been issued which the land officers had no authority to issue, as where the lands have been previously granted, reserved from sale, or appropriated to other uses."

These cases justify the statement that a grant by the riparian commission of lands which they have been expressly forbidden by the legislature to sell conveys no title, and is subject even to collateral attack, and that in a proper case this attack may be made by parol proof of the actual character of the lands.

It does not help us much to say that the action of the riparian commission cannot be questioned as to matters of fact within their jurisdiction. The important question is what fact was essential in order that the commission might have jurisdiction. No better description of such jurisdictional fact can be had than has been given in the cases above cited and all agree that if the legislative department has reserved the land from sale, the administrative department of the government has no power over the land so reserved. None of the cases hold that this reservation must be of a particular tract described by metes and bounds or identified by specific name. Morton *v.* Nebraska, as far as it goes, seems to indicate the contrary, and clearly the Leadville mining cases are authority for the assertion that such definite description is unnecessary. A lode of mineral in the earth may be marked when it is finally explored or worked with all its spurs, dips and angles, but until that is done, it is at least as hard to locate definitely as a natural oyster bed. *Natural oyster beds, as the term is used in our legislation, seems to mean something different from land under tidewater suitable for growing oysters. From the beginning of legislation on the subject of oysters in 1846 the legislature has constantly made reference to natural oyster beds as something well known and distinct from tidal lands suitable for the planting of oysters. The original act, by section 14 (Gen. Stat., p. 808), authorized the owners of flats and coves along the shore to stake out land and plant oysters, provided they did not include any natural oyster beds always covered with water beyond low-water mark, and section 20 prohibited, under a penalty, the removal of old shells from any natural oyster banks or beds. The amendment of 1890 (Gen. Stat., p. 813, pl. 43) authorized the staking out of ground and the planting of oysters on flats or in coves upon which*

*there had not been theretofore any natural oyster beds. The supplement of 1891 (Gen. Stat.,
p. 813, pl. 44a) confirmed the rights of persons using or occupying grounds under tidewater for
the planting or cultivation of oysters, "said grounds not being natural clam grounds or natural
oyster seed beds;" it further declared (Gen. Stat., p. 814, pl. 44c) that any person who should
plant oysters upon any of the natural beds should be deemed a trespasser and such planted oysters should be forfeited to the public. Surely the legislature would not have decreed a forfeiture
unless the natural beds were of such a character that the planter could not be mistaken, and
trespass innocently* [emphasis supplied by author].

Section 8 of the act of 1882 (*Gen. Stat., p. 823, pl.* 95), as to Maurice river cove and Delaware
bay, makes it a misdemeanor, punishable by imprisonment, to take oysters from natural oyster
beds for the purpose of planting in the waters of any other state. It would be most unreasonable
to punish such an act as a crime unless the natural oyster beds were of so defined a character that
the offender must know that he was violating the statute.

The act of 1883 as to Maurice river cove and Delaware bay (*Gen. Stat., p. 824, pl.* 104), the
supplements of March 16th, 1893 (*Gen. Stat., p.* 830, *pl.* 129), of April 5th, 1893 (*Gen. Stat.,
p.* 830, *pl.* 131), contain similar recognition of natural oyster beds. The Ocean County act of
1886 for the first time makes a distinction between natural oyster beds in general and those "now
known and recognized," a distinction which suggests the provisions of the United States statutes
as to known mineral lodes in the cases already cited.

The act of April 4th, 1893 (*Gen. Stat., p.* 833, *pl.* 148, *ff.*), for the first time provides for an
oyster commission. By its first section the natural oyster beds of the state are divided into seven
districts, of which District No. 5 is "the bays and waters of Atlantic county." It may be contended
with some reason that the effect of this act was to declare all the bays and waters of Atlantic
county to be natural oyster beds, and if the only meaning of the expression was to designate a
quality or characteristic of the land, I should think this the proper construction; it is, of course,
not the proper construction, if my view is correct, that natural oyster beds are a definite portion of the land under water sufficiently recognizable to justify the legislature in making it a
crime to remove oysters therefrom. My view is supported by section 3 of the act (*Gen. Stat.,
p.* 833, *pl.* 150), which makes it the duty of the commissioners to make a careful inspection of
the natural oyster grounds in their respective districts and to cause a supply of shells to be spread
on the ground. The commissioners could hardly make this careful inspection unless the natural
oyster beds were defined with some approach to accuracy, perhaps as close an approach as would
be possible in the case of a mineral lode with its veins and all the spurs, dips, and angles. If such
natural oyster beds were sufficiently defined to be recognized by the commissioners, they could
be recognized with equal facility by the purchaser of a riparian grant.

Similar provisions are found in the act of May 17th, 1894. *Gen. Stat., p.* 835, *pl.* 161,
and *p.* 836, *pl.* 166.

New districts were created by the act of 1896. *Pamph. L., p.* 186. The waters of Atlantic
county were still included but divided between two districts. With the act of 1899, relating to
Delaware bay and Maurice river cove, a new policy began and leases by the state oyster commission were authorized, but certain named oyster beds and any other commonly known natural oyster bed in Delaware bay or Maurice river cove, or the creeks and rivers emptying therein,
were reserved from the power to lease. The policy of leasing was extended to Ocean county in
1902. *Pamph. L., p.* 170. The act was not extended to Atlantic county until 1905, which was
after the grant in the present case was made. Under such legislation the oyster grounds subject
to lease by the oyster commission must necessarily be withdrawn from the jurisdiction of the
riparian commission. The importance of the legislation in its bearing on the present grant lies
in the fact that it shows a consistent purpose on the part of the legislature to separate lands
under tidewater into two classes, those which are natural oyster beds and those which are not

natural oyster beds. Since 1888 the first class have been withdrawn from the power of the riparian commission.

No one would assert that if the legislature had forbidden the commission to convey lands under the waters of the Mullica river, the commissioners could convey any such lands and have their conveyance free from collateral attack in an action of ejectment. But such a description would be only a little, if at all, more definite than the exception of natural oyster beds. It would certainly be necessary to determine where the Mullica river is, and while, no doubt, a natural boundary of that kind generally has a definite location at any particular point of time, not only may the river shift its bed (some rivers shift very far and very often) so that the determination of its location at the time of the grant would rest in parol evidence, but the bounds of the river itself may be indefinite. We have only to recall the controversy lasting for so many years between the colony of New Jersey and the colony of New York as to the proper location of the river Delaware at forty-one degrees forty minutes north latitude, to realize that the geographical location of a river may be open to dispute, and, of course, the determination of the question must depend on parol evidence.

I have assumed that the act of 1888 is still in force. The act of March 20, 1891, evinces no intention to repeal the act of 1888. It was evidently meant to apply only to such lands as were still within the control of the riparian commission, and to change the procedure for acquiring grants, not to extend the power of the commission to lands which only three years before had been withdrawn and which were the object of the sedulous care of the legislature for the common benefit, as shown by the series of statutes above referred to.

**CONCUR BY:** MINTURN

MINTURN, J. Concurring in the views expressed by Mr. Justice Swayze, I am led to oppose this grant by the riparian commissioners upon the further ground that the only tenable view which its supporters advance to uphold it in a court of law against collateral attack is based upon the assumption that somewhere between the lines of the legislation affecting this commission there exists in view of the general power conferred to sell riparian lands, a power "tantamount to an adjudication" to determine whether or not the lands so conveyed are part of the natural oyster grounds of the state. But manifestly if it can be demonstrated that the power to make such an adjudication has been expressly vested by the legislature, not in the riparian commissioners, but in an entirely distinct, independent and co-ordinate commission, it must be perceived that the syllogism upon which such a claim is constructed is entirely fallacious.

In 1888 the legislature, for the protection of the great oyster industry of the state, and as a sequence to the consistent protective legislation which had its genesis in the year 1846 (Act for the protection of clams and oysters, approved April 14th, 1846), and which has been supplemented by legislation almost annually since, enacted that "No grant or lease of lands under tidewaters wherein there are natural oyster beds shall hereafter be made by the riparian commissioners of this state except for the purpose of building wharves, bulkheads and piers." *Pamph. L. 1888, p.* 140.

This act declared a legislative public policy relative to these lands, and was equivalent to a declaration, that a conveyance of the natural oyster beds of the state, excepting as indicated in the excepting clause of the act, should be invalid, and *ipso facto* void.

It is significant that in no instance where the legislature has undertaken to deal with legislation enlarging or supplementing the powers of the riparian commissioners, do we find any intent, express or implied, that their duties shall extend to a determination tantamount to an adjudication, regarding the location of natural oyster beds; and it is a logical and legal inference that if the legislature intended to repose that power in these commissioners it would have so expressed itself. But *per contra* we find that in 1893, in an act entitled "An act to promote the propagation and growth of seed oysters, and to protect the natural oyster beds of this state"

(Pamph. L. 1893, *p.* 503), oyster districts were created, beginning with District No. 1, and terminating with District No. 7, and that in one of the districts thus enumerated were included the lands in controversy. The first section of the act provides "that the natural oyster beds shall be and they are hereby divided into seven districts, as follows: District No. 3, from Gunning river south to Rose's Point; District No. 4, from Rose's Point south to the division line between Atlantic county and Ocean county." The second section of the act provides for the appointment of commissioners for the respective districts, who were required to take an oath for the faithful performance of their duties; which by the third section were prescribed *inter alia,* to be the supplying of shells upon these grounds, for the purchase of which the state appropriated annually $5,000. From this appropriation the act appropriated the sum of $500 "for the mouth of Mullica river and adjacent waters, known as graveling oyster beds."

The same act provides that "In the event that it may become necessary that any one particular district shall require a greater expenditure than above provided, the said commissioners, in meeting assembled, may determine the proportion to be allotted to such district."

In 1899, by an act under a similar title, the legislature divided these lands under water into six districts, denominating them all "The natural oyster seed grounds of this state." *Section* 1. The centre line of Mullica river was made the division line between Districts Nos. 3 and 4, and three commissioners were appointed for the former districts and two for the latter. The act confers upon the commissioners powers substantially similar to those conferred by the act of 1893, and in its sixth section provides that "it shall be the duty of the said oyster commissioners to strictly enforce all existing laws regulating the natural oyster seed grounds of this state." And concludes by appropriating the sum of $12,000 *annually* to meet the requirements of the act.

This legislation, manifesting in every section the dominant purpose of the legislature to protect this great natural industry, contains within its provisions a specific designation of the *locus in quo* as natural oyster beds, and vests in the oyster commissioners the power and duty to protect the lands and regulate their use. If the power be not vested in these commissioners by this legislation to select and designate the natural oyster beds of the state, how can it be said, by mere constructive inference, that this power resides in a commission created and designated for a purpose entirely distinct from and foreign to such a duty?

Supplementing, in effect, the act of 1888, prohibiting sales of these lands as a declared public policy of the state, we thus have the distinct enumeration by the legislature and a description by natural boundaries of the natural oyster beds of the state, and the power is thereby vested in a distinct commission to perpetuate and protect the use of such lands. These acts, of course, are public acts and must be judicially noticed in any determination of this question, and it must also be presumed that the oyster commissioners did their duty, and expended under the requirements of this legislation in seeding the lands the large appropriation annually made for that purpose by the state. These lands were thus segregated by legislative fiat from the great body of riparian lands, and the protecting arm of the legislature was thrown about them.

The United States Supreme Court under similar conditions, in dealing with government grants by federal agencies of timber, mineral and swamp lands that had been "reserved from sale," by acts of congress, has held "that a patent may be collaterally impeached *in any action,* and its operation and conveyance defeated by showing that the department had no jurisdiction to dispose of the lands; that the laws did not provide for selling them, or *that they had been reserved from sale or dedicated to special purposes.*" *Wright v. Roseberry,* 121 U.S. 488; *Noble v. Union River Logging Co.,* 147 U.S. 165; *Rice v. Minneapolis Northwestern Railway Co.,* 1 *Black (U.S.)* 358; *United States v. Stone,* 2 Wall. 525; *Polk v. Wendal,* 9 Cranch 87.

The record of this case shows that, upon the trial at the Circuit, Mr. Cole, the counsel for the state representing the attorney-general, made this offer: "We propose to show by this witness and by many others that the lands comprised within the grant from the riparian commissioners to the

Sooys is natural oyster beds within the meaning of the amendment to the Riparian act of 1888," which testimony, upon objection by defendant's counsel, was excluded, and an exception was entered to the ruling. The court thereafter, without the introduction of further testimony, directed the jury to find for the defendant, upon the ground that the deed of the riparian commissioners could not be attacked collaterally, to which ruling also exception was taken. These exceptions present the questions here discussed, and upon both questions the rulings were erroneous for the reasons here advanced. Had the proffered testimony been allowed it might have been shown that for generations a hardy race of seafaring men had made an independent existence from these beds, and that the existence of the beds was as notorious as was the beacon light at Barnegat, of which no state board dealing with the subject-matter could plead ignorance. It might have been shown that for many years the state had expended thousands of dollars developing this great industry under the public policy indicated in this legislation, and these facts might have been established by the oyster commissioners as well as by the members of the vicinage. But it may be asked here how could the riparian commissioners, when making a conveyance, distinguish oyster lands from other lands under tidal waters? The answer must be obvious. They as commissioners were supposed to know of the existence of the acts of the legislature upon this question. Side by side with them in legislative and legal contemplation was a co-ordinate commission constituted for the purpose of dealing with that very subject-matter, and entrusted, as these commissioners must have known, with the guardianship of this vast industry. Had they but knocked the door of this storehouse of information would have been opened to them. Had they but asked and confessed their ignorance they would have received the necessary information. But, in any event, neither their ignorance nor their negligence in the face of positive prohibitive legislation can breathe vitality into a grant of lands that had been both reserved from sale, and dedicated to special use by express legislative enactment. Nor can I acquiesce in the scholastic refinement and subtlety of distinction found necessary to differentiate the federal decisions in order to vindicate this grant, for the reason that, conceding the existence of this oyster legislation and its applicability to the case at bar, the question of construction cannot enter. The legislative intent in these acts is plain and leaves no room for construction, and in any case the rule is fundamental that he who seeks to maintain a claim in derogation of a public right can leave nothing to inference or conjecture. *Black Int. L.* 300; *Sewall v. Jones,* 91 Pick. 99; *Water Commissioners v. City of Hudson,* 2 Beas. 420; *Black v. Delaware and Raritan Canal Co.,* 9 C.E. Gr. 455.

But *argumentum ab inconvenienti* is invoked, and we are asked to support such grants upon a rule of constructive procedure which will accord them as is said: "Stability and value rather than that which will detract from, if not destroy such value by exposing them to constant collateral attack and invalidation by the verdict of juries of the vicinage." I am not of those who share disquieting fears regarding the results of the jury system; for with Blackstone it can be said that "it is a system of trial that hath been used time out of mind in this nation, and seems to have been coeval with the first civil government thereof." An institution, says De Lolme, that is the "noblest invention for the support of justice ever produced." An institution, remarks Hume, "admirable in itself and the best calculated for the preservation of liberty and the administration of justice that ever was devised by the wit of man." Before the Twelve Tables of the Roman Law or the Pandects of Justinian were conceived, it found its genesis upon the banks of the Rhine and within the shadow of the round towers of Scotia Major, where the legions of the imperial Caesars beheld it flourishing. *Tacitus Germanicus XII., Mooney's Gaelic Laws, &c.* Against it the conspiring forces of feudal despotism hurled themselves in vain. Against it was pitted through centuries of scholastic controversy a theocratic order which inherited the jurisprudence of classic Greece and Rome, combined with an aristocracy of class prejudice, which labored steadfastly for its downfall. It has withstood the ruthless hand of the invader and the bloody scenes of internecine strife. It has survived the destruction of dynasties, and the wreck and ruin of empires, and stands to-day unique and indestructible in the modern jurisprudence of the world, the sheet anchor of every nation, wherever popular government can claim an advocate, and the hope of every citizen, wherever humanity can command a champion.

It may well be asked, therefore, why such an institution, entrusted as has been this immemorially, not only with the disposition of the property and sacred honor of a people, but with their very lives, should be challenged in this day as a fit medium to determine conflicting claims in an action in ejectment — that their verdicts may vary must be expected where the facts are variant. That divers minds may vary upon a similar state of facts, is as demonstrable when applied to courts as when applied to jurors, as the decisions from the days of the year books will attest, except where the doctrine of *stare decisis* has compelled judicial consensus. But in any view of the jury system, as applied to civil procedure, the corrective power of the court in cases of passion, prejudice or unfair influence has invariably been extended to induce a result consistent with legal principles. And if an impartial jury of the vicinage cannot be procured, both common law and statute provide for a change of venue. 1 Tidd. Pr. 654; *Practice Act,* § 203.

A misconception as to the function of a jury under this legislation, seems to prevail as a basis for the argument, for we find it urged that the *locus in quo* must revert to the state "whenever the allegation of a plaintiff in ejectment, that such land was a natural oyster bed obtained the concurrence of a jury of the locality." It is not the plaintiff's view, nor yet the jury's view, that under this legislation stamps the character of oyster lands upon the *locus in quo*. It is nothing less than the legislative fiat which existed at the execution of the grant, that such lands were and should remain inalienable that furnishes *ratio decidendi* for either — court or jury. This legislation, by metes and bounds, describes the lands; segregates them from the other lands of the state; and commands that they shall not be sold, but shall be denominated "natural oyster seed grounds of this state." *Pamph. L.* 1899, *p.* 160, § 1.

Under this legislation no claim could be supported for the existence of oyster lands in the Hudson, the Passaic or the Hackensack rivers, and the court would properly prevent speculation in the matter, by controlling the verdict because these streams are not within the language of the act. But where, as in the case at bar, the legislation designates "the mouth of the Mullica river and adjacent waters known as graveling oyster beds," as natural oyster beds, uncertainty vanishes, and the state was clearly entitled to the direction of a verdict upon the completion of the proffered proof.

But *argumentum ab inconvenienti* is further pursued, and it is urged that the state's revenue might be depleted; and that the revenue produced from sales of these lands is irrevocably applied to the support of free public schools. Logically considered, this contention might receive condemnation as a species of *petitio principii;* and legally it may be criticized as an argument of public policy for submission to the legislative branch of government, and therefore clearly *coram non judice* upon this argument. But it presents a misconception of the effect of the riparian legislation in this state.

The act of 1869 creating the riparian commission appropriated such income, first to the administrative purposes of the commission, then to the payment of the state debt, and finally to the trustees for the maintenance of free schools. *Pamph. L.* 1869, *p.* 1017. The act of 1871 devoted such receipts to the support of free schools. *Pamph. L.* 1871, *p.* 98. The act of 1890 devoted them to the necessary expenses of the state (Pamph. L. 1890, *p.* 92), and the act of 1894 finally appropriated them to the support of free public schools. If the maxim of construction, *contemporanea expositio est optima et fortissima in lege,* is to apply, we have here presented to us a legislative declaration antagonistic to the present contention. The only authoritative deliverence in support of this notion is an opinion of the late Attorney-General Grey, but neither his language nor any inference that can reasonably be derived from his language will support the contention. Of course, under the constitutional provision upon this subject, the revenue now in hand, whether derived from this or any other source, cannot be diverted to other uses; but there is no authority for the assumption that either the revenue *in futuro* or the riparian lands *eo nomine* are irrevocably pledged to the support of any department of the state government. Against the contention for a public policy, which is construed to require a sale of these lands for revenue, rests the accusation that in so doing the state is parting with the birthright of its people for a mess of pottage. Buckle, in his "History of Civilization in England," tells us that parliamentary agitation and legislation during fifty years prior to the time he wrote, was almost a continuous effort of the house of commons to regain and retrieve from class ownership not only the lost privileges of the people, but the lost highways, roads, greens and commons that form such a picturesque background to the history of early English rural life. And need we look beyond our own day in this land where the efforts of two administrations of the federal government have been directed to recovering lost forest, mineral and swamp lands, the common heritage of the nation? In the light of history it requires no exuberance of fancy to picture the dire results to the state if the public policy inspired by this grant be adopted, as a result of which this great natural industry, the prolific toiling-place of generations of independent self-supporting citizens shall be aliened forever, and they themselves evicted as completely and as effectually, by the mere presence and potency of the great seal of the state, as were the thrifty Highland crofters of Scotland under the great "Sutherland clearances" at Lochaber, when a whole people were swept into exile, to make way for sheep walks and pasture lands. As against a so-called public policy, that would ignore the independent existence and self-supporting happiness of a people, and set it in the balance as *quid pro quo* for what Tennyson calls the "Jingling of the guinea," we may well invoke the condemnation of John Stuart Mill; "that when the inhabitants of a country quit it, because the government does not leave them room to live in it, that government is already judged and condemned." From whatever aspect, therefore, we may view this grant, it is defenceless in law and insupportable in policy.

The court being equally divided in opinion, the judgment under review must be affirmed.

**Author's Comment:** This is a lengthy and significant decision with extensive analysis of legal mechanisms, and interesting comparison to mineral interests and characteristics. Drawing on numerous similar cases from a variety of jurisdictions, instructive in courts thinking and reliance on precedence. It contains some good points on oyster lands.

The Rhode Island decision in the case of *State v. Burdick and Others*,[77] relates to an original survey and maps of one-acre lots made in 1882, being the source document, along with related deed to the various lots, of the original survey.

This case comes up from the court of common pleas on exceptions for errors in ruling at the trial in said court. The case is on an indictment against the defendants for taking and carrying away oysters from the private oyster bed of George H. Ward, in Powaget or Charlestown pond, in the town of Charlestown. The statute (Pub. St. R.I. c. 146, § 3) provides that the commissioners of shell-fisheries may lease, by public auction or otherwise, to any suitable person, being an inhabitant of the state, "any piece of land in Powaget or Charlestown pond, covered by tide-water, * * * not leasing more than one acre in one lot or parcel to any one person or firm." Under this section the commissioners, on September 18, 1882, executed a writing purporting to be a lease to George H. Ward, the granting part of which is in the words following: "The state doth hereby lease, demise, and let unto the said George H. Ward a certain piece of land, in Charlestown pond lying and being, and covered with tide-water, containing about (10) ten acres, and bounded and described as follows, to-wit: being lots numbered 24, 25, 26, 27, 23, 22, 21, 20, 16, and 17 on a plat of oyster ground, in one-acre lots, Charlestown pond, R.I., surveyed by John L. Kenyon, September 18, 1882; said plat being on file in the office of the commissioners of shell-fisheries. Said lots were leased separately, but are included in one lease for convenience." At the trial the state introduced evidence to show that oysters were taken by the defendants

[77] 15 R.I. 239, 2 A. 764 (R.I. 1886).

from lot 26, October 1, 1883; Ward then being still lessee under the lease, if it be valid. Two of the exceptions reserved by the defendant were for the refusal of the court below to instruct the jury in compliance with the requests of the defendants as follows, to-wit: "(1) The lease in the case is inoperative and void for the reason that by it there is an attempt made to lease more than one acre in one lot or parcel to the said George H. Ward, to-wit, ten acres. (2) The following language contained in said lease, viz., 'said lots were leased separately, but are included in one lease for convenience,' are ineffectual, and cannot operate to take this case out of the limitation of the statute, which provided that not more than one acre, in one lot or parcel, shall be leased to any one person or firm, and said language is used merely for the purpose of evading the statute."

The statute is clear that no more than one acre shall be leased, in any one lot or parcel, to any one person or firm. It is not disputed that the meaning is that no single letting can include more than one acre in any one lot or parcel. The lease described the premises let as "a certain piece of land" – words which prima facie a signify a single parcel or lot – and the rest of the description is to the same effect. There is no dispute but that the ten acres lie together in a single parcel. The lease is therefore ineffectual, unless effect can be given to it by virtue of the phrase, "said lots were leased separately, but are included in one lease for convenience." It is not stated how the lots were leased separately, whether orally or in writing. The statute (chapter 146, § 12) requires, by necessary implication, that they shall be in writing, with certain covenants and reservations in favor of the state. Now, if the letting was by written leases, as required by section 12, then those leases passed to George H. Ward his title, and were the leases which should have been put in evidence to prove it; the paper adduced being inoperative as a lease, whatever value it might have as a memorandum of the terms on which the actual leases were given. There is no pretense, however, that such leases were given; a lease including all the lots being given to avoid the necessity of the several leases. Undoubtedly, the so-called separate letting was oral, and, being oral, it was null and void because it did not meet the requirements of the statute. The exceptions to the two refusals must therefore be sustained. Exceptions sustained.

Another case from the same jurisdiction stressing the importance of maps as required by law is *State v. Nash*,[78] decided in 1892, relying on the wording of the current statute.

The defendant, having been convicted in the criminal court of common pleas for Fairfield county of a violation of Gen. St. § 2358, namely, of taking oysters from a place in Bridgeport, in said county, designated for their cultivation, appealed to this court. The taking was admitted; also the designation, on October 20, 1890, of the territory named by the board of shell fish commissioners of this state, pursuant to Gen. St. § 2317; but the defendant claimed that the ground was a natural clam and oyster bed, and therefore not duly designated within the meaning of the law, and that the grant of the franchise, and a subsequent assignment thereof, were, for that reason, void. And the broad question presented by the various reasons of appeal in this case, confessedly brought to this court as a test one, in the interest of a class known as "natural growth oyster-men," who earn their livelihood by procuring oysters from the public natural beds of the state, is whether it is competent to defend against a charge of this character by evidence that the territory from which the oysters were taken, not included in the locations and descriptions embraced in Gen. St. § 2328, is in fact, and notwithstanding its designation and grant subsequent to the enactment of that statute, a natural oyster bed under state jurisdiction. Accordingly as this question shall be answered by this court in the affirmative or negative, the judgment of the court below must be vacated or sustained, and we will therefore consider it by itself, without separate reference to the various rulings and to the reasons of appeal.

The counsel for the defendant say in their brief that it has been the policy of the state to encourage the planting and cultivation of the large area of territory in Long Island sound by

---

[78] 62 Conn. 47. 25 A. 451 (Conn. 1892).

private enterprise, by every legitimate and reasonable method of protection; but that the legislature has shown an equal solicitude to preserve the natural beds of the state for the public use, and that "it cannot be claimed that it is not wise and beneficial to protect the industry of oyster planting, which in the past ten years has become second to none in this state, but that that industry is not to be advanced to the prejudice of those persons who are not so fortunate as to control capital, and who rely upon the natural beds to furnish them occupation and support." In this we fully concur, and these balanced considerations, if kept in mind, will aid materially in arriving at the intention of the legislature in those enactments, which the defendant's counsel further declare to be "a compilation of divergent views and inconsistent provisions, resulting from the perpetual conflict between the two classes of oystermen."

This court held in *Averill v. Hull,* 37 Conn. 320, that the defendant, in a proceeding brought for the confiscation of a vessel used in taking oysters from grounds designated by a town committee, might show that such place was in fact a natural oyster bed, and that such designation would be invalidated by such proof. It is claimed that the reasoning and authority of that decision apply, and demonstrate, that the evidence rejected in this case was admissible; and, unless the cases are distinguishable by virtue of subsequent legislation, it may be freely conceded that this contention of the defendant is correct. As the law stood when Averill v. Hull was decided, and prior to 1881, since the state had made no effort to designate and point out its natural oyster beds, nor had empowered the town committees to designate those in town jurisdiction, this was of necessity a proper defense in each case; since otherwise, as the court (SEYMOUR, J.,) said in that case, if such evidence was "not admissible, the natural oyster beds of the state are subjected to the control of a committee of the town, without a hearing and without appeal." But it is manifest that, although necessity made such defense relevant, the right in every case to collaterally attack, by parol evidence, in trials to the jury, the title of those to whom designations were made, must have been disquieting in the extreme to persons desiring to embark in the business of oyster culture; and it was, as we believe, largely for that reason that the legislature, by an act passed in 1881, amended in 1882, now Gen. St. § 2315, enacted that the state should exercise exclusive jurisdiction and control over all shell fisheries located in a certain designated area of the state, and further provided, in what is now Gen. St. § 2316, that a survey and map should be made of all the grounds within that area which had been or might be designated for the planting or cultivation of shell fish, and also a survey of all the natural oyster beds in the area, which should be located and delineated on the map. The object of this survey seems manifest, and, it having been accomplished, the legislature in 1885, by public act, (Acts 1885, c. 118,) expressly provided that "the locations and descriptions of the natural oyster beds respectively under state jurisdiction are as follows, to wit," enumerating them with the utmost particularity and precision. And the legislature in 1887 re-enacted that act by approving it as section 2328 of the General Statutes, to take effect January 1, 1888; thus declaring, not alone what the locations and descriptions of such beds were in 1885, but that they remained unaltered in 1888, and were intended, as they must have been, unless we impute folly to the legislature, as authoritative, permanent, and exclusive designations of such grounds under state jurisdiction, subject only to such alterations, if any, as the legislature might, from time to time, prescribe, or as might be made in accordance with existing statutory provisions. And the effect of these enactments is that in a proceeding like the present, and in a case where the grant of the franchise to private parties has been made since such enactment, while the fact that such ground is a natural oyster bed would render the grant invalid, the only proof of such fact which is admissible by way of collateral attack is not by parol evidence, but by showing that such ground is embraced in the locations and descriptions contained in the statute of the natural oyster beds under state jurisdiction. Any seeming difficulties in the way of our present conclusion resulting from the provisions of certain of the sections of the act of 1881, which are continued in the present revision of the statutes, will disappear if we remember, as we should, that such provisions, having been enacted prior to the authoritative determination by the legislature in the act of 1885, are now, and so far forth as they continue existent provisions to be, construed in the light of the changed condition which that statute has produced.

It was suggested that these grounds, if not natural oyster beds in 1885, might have been such prior to that time, and within ten years previous to their designation in 1890. The answer to this suggestion may be found in the language of this court in *White v. Petty,* 57 Conn. 579, 580. 18 A.Rep. 253. There is no error in the judgment appealed from.

The next case has to do with an accusation of trespass and, again, differentiating between natural and cultivated oyster beds. *White v. Petty, et al.*[79] involves a complainant alleging that she is the owner of various oyster-lots in the town of Darien, upon which she has, in good faith, planted large quantities of oysters, and prays for an injunction restraining the defendants from entering upon them, removing the stakes, and taking up the oysters. The defendants deny that the plaintiff has any title to the ground, alleging that it is a natural oyster-bed. The court decided that none of the ground in question was a natural oyster-bed. The court finds that a part of the ground was, pursuant to the statutes, (Rev. St. 1875, p. 214, § 3) in the summer of 1881, designated to the plaintiff and others, who afterwards assigned their rights to her, and the other part of the ground was designated to the plaintiff's assigns in the months of February and March, 1881. The court also finds that the ground so designated in 1881 was first inclosed by buoys and stakes on the 29th day of September, 1882, and that until then the public had no knowledge of such designation.

Upon the trial the defendants introduced evidence to prove that all the ground in question was, and for 10 years before had been, a natural oyster-bed. To rebut that evidence the plaintiff offered in evidence the report of a committee appointed by the superior court, pursuant to the act of 1881, (p. 104, § 12, Sess. Laws 1881) to locate and describe, by proper boundaries, all the natural oyster-beds in the town of Darien, and to make report, etc. The report was dated March 26, 1883, and was recorded March 27, 1883, and did not describe any part of the oyster ground in question as a natural oyster-bed. The defendants introduced evidence that in May, 1883, certain other parties brought a petition to the superior court, under the provisions of the act, (Rev. St. 1875, p. 215, § 11) praying the court to declare the designations of all the land in question void, and to order the removal of the stakes inclosing it, upon the alleged ground that it was a natural oyster-bed, which action was still pending. The defendants claimed that the committee appointed under the act of 1881 had no jurisdiction over ground which had been designated before that act went into effect, and hence that their report, not describing that part of the ground in question which was designated in March and September, 1881, as natural oyster ground, was no evidence that it was not such. They assign the refusal of the court so to rule as their first specific reason of appeal.

The defendants rely in support of their claim upon the case of *In re Oyster Ground Committee,* 52 Conn. 5. That case was a remonstrance against the report of a committee appointed under the act of 1881 to designate the natural oyster-beds in the town of Clinton. "Several persons," in the language of the court, "who had oyster grounds set out to them in severalty prior to the passage of the act, and who had acquired valuable interests therein, and whose interests might be seriously affected if the report of the committee should be accepted, appeared and remonstrated against its acceptance." The court accepted the report of the committee, saying "that the act of 1881, notwithstanding its general language, does not apply to oyster ground previously designated."

The case is undoubtedly a decisive authority for the defendants if the word "designated," as there used, is to be taken as referring merely to the act of the committee authorized to determine and inform applicants what ground they may occupy. Before the applicant can acquire any right to the exclusive occupation of the ground, he must "mark and stake out the place." This he may or

---

[79] 57 Conn. 576, 18 A. 253 (Conn. 1889).

may not do after the committee has acted. Until he does it, the rights of the public in the ground assigned to him are not affected.

It is apparent, from the statement of the case referred to, that the remonstrants had perfected rights to the ground in question in that case, and the word "designated" was used to describe such rights. In the present case the ground was designated to the plaintiff, or her assigns, by the committee some months before the act of 1881 went into effect; but no steps were taken by her to appropriate it to her use until September 29, 1882, more than a year after the act took effect, and when the committee would have been justified in presuming that she had abandoned any claim to it, if another application for it had been made. The *status* of the ground in question was thus fixed on the 27th of March, 1883, the date when the report of the committee is found to have been accepted and recorded; the provision of the statute being that "such report, when accepted by said court, and recorded in the records thereof, shall be a final and conclusive determination of the extent, boundaries, and location of such natural beds at the date of such report."

The defendants claimed in the court below that the report of the committee did not tend to show the condition of the ground in 1881, and assign as a reason of appeal that the court did not so decide. Natural oyster-beds are not usually found so suddenly that the fact that the ground was not such in 1883 would not tend to show its condition in that respect two years before. But the finding is that the report was offered merely in rebuttal of the defendant's evidence that the ground was, at the time of the trial in 1886, and for 10 years before (covering the date when the report was recorded) had been, a natural oyster-bed. In either view it was proper evidence.

Another reason of appeal is that the court erred "in holding that the finding of the committee could be treated as *res adjudicata* against the defendants." We do not understand from the finding that the question was made upon the trial. The defendants' claim there was in these words, "no finding of said committee could be treated as *res adjudicata* against a private right." The defendants claimed no private right, but such right only as they had in common with the rest of the public. We understand that by the expression "private right" they intended the right which they claimed the plaintiff had acquired by the action of the committee in February and March, 1881. But if the claim now made had been made in the court below, it should have been overruled. There is no ground for a distinction between the defendants and the rest of the public in respect to the conclusiveness of the committee's report.

The defendants assign two other reasons of appeal, predicated upon the finding that there was an action pending against the plaintiff, brought under the statute of 1875, to procure the removal of the stakes inclosing the ground in question, etc. That statute provides, if it is found in such suit that the stakes have been improperly set, the defendant therein shall be entitled to remove the planted oysters within a limited time. The court finds that the plaintiff planted a certain quantity of oysters and gravel upon the ground in question after that suit was brought. The two reasons are as follows: (1) That the court erred in holding that a permanent injunction could be granted while the title was in dispute, and the question of the title was then in the superior court upon a complaint under a statute specifically providing a way to try said title; and (2) In holding that there could be any good faith in law as to the 600 bushels of seed and the 3,000 bushels of gravel planted while a petition was pending to set aside the designation.

It is enough to say that the suit brought under the statute of 1875 has no relation to this one, and evidence regarding it was not properly in the case. The purpose of the statute of 1875 is not, as the defendants claim, to provide a way for the trial of questions of title, but to effect the removal of stakes improperly set up, by means of which the public are deterred from exercising their rights upon ground belonging to them.

Two other reasons of appeal are stated, but the facts which they assume are not found, and they were properly abandoned upon the argument. There is no error.

The following decision points out clearly, in the dissenting view, the value of oyster beds, and their place in the law as real estate. It also discusses prescriptive rights, and analyzes the facts in light of previous relevant decisions. Town of *Clinton v. Bacon*[80] begins with a petition for the removal of stakes set up by the defendant, inclosing ground for planting oysters, brought under Gen. St. § 2356, the material parts of which are as follows: "When any natural oyster-bed, or any part thereof, is designated, inclosed, or staked out contrary to the provisions of this chapter, the superior court, as a court of equity, in any county in which said oyster-bed is situated, upon the petition of any individual aggrieved, or by the town in which said oyster-bed is situated, against the person claiming the same, * * * shall appoint a committee, who, having been sworn, and having given notice to the parties, shall hear said petition, and report the facts thereon to such court; and, if it shall appear that such oyster-bed has been improperly staked out, the court may order said committee to remove the stakes inclosing the same; the costs to be paid at the discretion of the court." The case was reserved by the superior court for the advice of this court.

The committee to whom the case was referred finds that the defendant in March, 1863, applied to the committee of the town of Clinton appointed for that purpose, to designate the lot in question to be staked out to him for planting oysters, and that thereupon the lot was so designated by the committee, and the defendant staked it out, and has kept it inclosed with stakes continuously to this time, and has used the lot each year for laying down or planting oysters upon it. The committee also finds that the lot in question was, and for more than 30 years had been, a natural oyster-bed, when it was designated to the defendant in 1863. Upon the trial the defendant filed a written claim that the occupation of the ground for such a length of time, without disturbance from any source, barred the right of the town to claim that the designation should be set aside; and the same claim is made before this court. As bearing upon this claim, the committee further finds that since 1863 the defendant has all the time claimed that the designation of the lot to him gave him the right to the use of it for laying down and planting oysters upon it, and adds that he has claimed to have on said ground just such rights, no more and no less, as such designation gave him. The meaning of this finding seems to be that he claimed no right except under the designation, but has claimed that he thereby acquired the right to so use the lot. A question was made in the argument of the case as to what were the nature and extent of the right conferred by a valid designation of a lot which was not a natural oyster-bed; the plaintiff claiming that the person to whom it was made took only a revocable license to use it, and the defendant claiming that he acquired the title to the lot upon which the oysters were planted. The statute regarding the powers of the state commissioners of shell-fisheries (Gen. St. § 2317) provides that they may grant "perpetual franchises" in the land subject to their jurisdiction, which perhaps suggests the nature of the right conferred by designations by town committees. It is unnecessary, however, to decide this question; because, if we assume that a valid designation of the lot would have given the defendant a title to it, and that, therefore, upon the facts found, he has claimed such title since 1863, he has gained nothing by reason of the use of the lot under such claim. The defendant had originally no right to the lot in question, or to the exclusive use of it, unless he derived it from the state, the owner of it, and has no right to it by adverse possession, except such as he has acquired against the state. The town of Clinton never had any legal interest in it, nor any power except to appoint a committee which, acting for the state, might designate for individual use lots which were not natural oyster-beds. The lot in question is conclusively found to be a natural oyster-bed in this proceeding, which, although the state is not a nominal party to it, is brought by its authority and on its behalf. It is elementary law that a statute of limitations does not run against the state, the sovereign power.

---

[80] 56 Conn. 508, 16 A. 548 (Conn. 1888).

But the defendant claims that by his occupancy of the lot since the designation in 1863 he has gained the rights which a valid designation of it would have given to him. In *Town of Derby v. Alling,* 40 Conn. 436, in which it was contended that the town had lost the right to build a highway over land formerly dedicated to their use for that purpose, by delay in laying out a street across it, Judge SEYMOUR, in giving the OPINION, says: "The public could not be technically disseised, but public as well as private rights may be lost by unreasonable delay in asserting them. They may also be lost by an abandonment of them by those interested in their enforcement. Such abandonment may be inferred from circumstances, or may be presumed from long-continued neglect. The question whether the state has abandoned its rights, or forfeited them by neglect, must be largely dependent upon the circumstances of each particular case. It seems to us that neither should be inferred in this case." The action of the committee of the town of Clinton and of the defendant himself informed the public that the ground in question was not a natural oyster-bed, and the people of that town and the public acquiesced in the decision of the committee by leaving the defendant in the undisturbed possession of the lot until this proceeding was commenced. The committee finds that the people of Clinton had been accustomed every year to resort to the ground in question while it was a natural oyster-bed, and before it was designated to the defendant to gather native oysters; but the knowledge of the persons who went there at that time can hardly be imputed to the state or the public at large, who have the beneficial interest in the land.

But the defendant claims that if the designation was originally void by reason of the lot being a natural oyster-bed, it has been legalized by the following statute, enacted in 1877: "All designations of places for planting or cultivating oysters, within the navigable waters of any town, which have heretofore been made by authority of such town, through its selectmen or oyster-ground committee, are hereby validated and confirmed." Sess. Laws 1877, c. 94, § 2. This claim cannot be sustained. The oyster-ground committee had no jurisdiction over the lot in question. The pretended designation of it by them was a nugatory and void act, not performed "by authority of the town," as the town could not authorize it, and so it is not within the language of the healing act. Besides, it is clear that it was not within its intent. The rational construction of the act is that it was designed to cure mere irregularities in the action of the committee, or the omission by the persons to whom lots had been designated of the steps required to be taken by them to make the designation effectual. The legislature has jealously guarded the interest of the public in the natural oyster-beds of the state against appropriation by individuals. It is not to be presumed that it has reversed this policy in behalf of persons who are attempting to monopolize their use under a pretense of title.

The defendant also claims that the statute under which this proceeding is brought is in violation of the constitution, because it does not provide for a trial by jury. By it the petition is to be brought to the superior court as a court of equity, either by the town or a party aggrieved. The facts are to be determined by a committee to be appointed by the court, and reported to the court, which, if the allegations of the petition are found to be true, may administer specific relief by causing the stakes unlawfully set up to be removed. We fail to see that the constitutional provision referred to is violated by this statute. The defendant by collusion or mistake has procured the form of a designation to himself of natural oyster-ground. The pretended designation is void, and the title of the public to the ground is not affected by it; but it, and the stakes set up by the defendant as evidence of it, constitute a cloud upon the title to the ground, and tend to deter the public from the enjoyment of it. It would seem that, independently of the statute, the case would be a proper one for equitable interference, and that the state might upon general principles maintain a petition for the removal of the cloud thus cast upon its title by the cancellation of the designation, and the destruction of the evidence of it. The defendant claims that the provision in the statute that the petition may be brought in the name of the town or person aggrieved, and that the hearing shall be had before a committee, makes the proceeding one before a statutory tribunal, instead of one before a court of equity. This claim is answered by the statute, which expressly confers upon the superior court as a court of equity jurisdiction of the proceeding. Although the state is not a nominal plaintiff, it is a party in interest, represented by the town, or person who brings the suit by its authority, and to enforce its rights. The provision that the court should appoint a committee to hear the case was doubtless made in view of the peculiar nature of the question involved. The legislature has required the court to do what in other equity suits it has left to its discretion. No reason appears to question its power to make the enactment. The superior court is advised to render judgment for the plaintiff.

PARK, C.J., and PARDEE and LOOMIS, JJ., concur.

CARPENTER, J., ( *dissenting.*)

I cannot concur in the result to which the court has come. There are cases in which persons may acquire prescriptive rights, and even titles by adverse possession, against the public. In *Tracy v. Railroad Co.,* 39 Conn. 382, SEYMOUR, J., speaking for this court, says: "It is settled law in Connecticut that the title to an island, emerging, as this did, in navigable waters, vests in the state; and is also settled law that a grant from the state may be presumed in favor of long-continued exclusive and adverse possession." In *Town of Derby v. Alling,* 40 Conn. 410, this court, through the same learned judge, he then being chief justice, says "There is no statute of limitations which as such is applicable to the case. The public could not be technically disseised, but public as well as private rights may be lost by unreasonable delay in asserting them." As I view this case, it is not true that the state is the party beneficially interested. The title to the land covered by the sea is in the state, but it is a mere naked title. The state, as such, has never put it to any use. It builds no wharves or other structures, makes no use of it for purposes of navigation, and has never engaged in cultivating oysters. Within the last 50 years it has been discovered that Long Island sound is a field in which a large and important industry may be developed; and it has been the policy of the state to foster it, not by engaging directly in the business, but by allotting the ground in suitable lots to individuals, thereby encouraging and stimulating private enterprises, by which means a business of immense proportions has been built up. This has not been done merely by giving a license, which may be revoked at pleasure, but each proprietor over the area allotted to him has absolute dominion; as much so, for all practical purposes, as a proprietor in fee of the upland. That must be so, for in no other way could there be such an exclusive right as the statute calls for. And there must be permanency in the interest granted; otherwise the business would be too uncertain and hazardous. I cannot believe that the state intended that the grantee should take anything less than a fee so long as he, his descendants or assigns, should continue to cultivate oysters. To that extent, for the time being, at least, the state parted with all its interest. I think this will not be disputed as to any ground not a natural oyster or clam bed.

But in this case it is found that the ground allotted to the defendant was a natural oyster-bed; and the statute expressly prohibits the designation of such grounds, and therefore it is claimed that the designation is void. Even if this is so, it does not follow that the designation for all purposes is wholly inoperative. It gave to him a *prima facie* title, which must be regarded as good until determined by a judicial decree to be void; and after such a judgment it still would remain true that until then he had a colorable title, and occupied under a claim of right. Now, the exception in the statute was not for the benefit of the state at large, but for the benefit of the people who had been accustomed or desired to take natural oysters. Neither is this suit prosecuted in behalf of the state. The action is special, under a special statute authorizing it, in the name of the town, and in behalf of the local public. The state needed no enabling act to vindicate its rights. It was competent to sue, and had its agents under existing laws. It had no occasion to bring a suit in the name of the town. Nor do I believe that the state of its own motion originated the act under which this suit was brought. We can hardly close our eyes to the fact that this act was passed, as such acts usually are, at the instigation of parties specially interested. While the suit relates to a public matter, in this, that a portion of the public is interested in its object, yet it is in law, in name, and in reality a suit between private parties. If, therefore, the subject-matter had been upland of which any one had been disseised, the statute of limitations would have applied. *But it is oyster-ground. Nevertheless it is property; it is owned, bought, sold, and taxed. It is permanent and fixed, and therefore in the nature of real estate, and the laws governing real estate are to a great extent applicable to it* [emphasis supplied by author]. In this case it is true that the plaintiff was not strictly disseised; yet it, and those it represents, were deprived of a right which they previously had, under a claim of title, accompanied with exclusive possession for 22 years before this suit was brought. The case seems to me to be directly within the principle recognized as "well settled law" in *Tracy v. Railroad Co.* and *Town of Derby v. Alling,* supra. I cannot recognize as wise or just a decision which deprives the defendant of his property under such circumstances.

## OWNERSHIP OF OYSTER BED

The fascinating elements of the case of *Seth R. Robins v. Andrew Ackerly*[81] are the references to early historical title creations based on original patents, and to the applicable law at the time. The oyster bed is located within the Town of Huntington, created from patents in the 17th century. The basic law stems back to Blackstone's Commentaries. This case underscores the origin and preservation of the array of property rights contained within a title. See Figure 7.10.

The main question of title to these lands under water in Northport harbor is "res adjudicata." (*Lowndes v. Dickerson,* 34 Barb. 586.) One cannot acquire an exclusive right of property in oysters planted in a bed where oysters grow naturally. (1 Wend. 237; Osgood's Case, 6 City H. Rec. 4; 14 Wend. 42; N.Y. Statute, 1866, chap. 753; 4 Barb. 592; 11 *Id.* 298; 34 *Id.* 586; 8 N.Y. 475.) The grant of the "haven or harbor," assuming the same to be valid, may have carried with it the soil and still left the fishery subject to the public right. (2 Blackst. Comm. 40; *Brink v. Richmyer,* 14 Johns. 265.) The general form and configuration of the coastline indicate the reason and the propriety of a different rule applying to the case of the South Bay and Northport harbor, even if the granting clauses of the respective patents were the same. (60 N.Y. 57.)

The town of Huntington owns these lands and through their trustees (in whom is the legal title) can grant an exclusive use thereof. (Nicoll Patent, dated 1666, pp. 5–7; Dongan Patent, dated 1688, pp. 7–17; Fletcher Patent, dated 1694, pp. 17–27; *The Trustees of Brookhaven v. Strong,* 60 N.Y. 56; People v. Van Rensselaer, 9 *Id.* 291, 346, 347, 348; *McCready v. State of Virginia, 4 Otto,* 391; First Constitution of State of New York, § 36.) The words "All harbors, havens, waters," used in the grants specifically identify the premises. (*Rogers v. Jones, 1* Wend. 237; *People v. Vanderbilt,* 26 N.Y. 293.) The legal title to these premises is in the trustees of the town; they hold in trust for the benefit of the people of the town. (*Denton v. Jackson, 2 Johns. Ch.* 320; *Jackson v. Louw,* 12 Johns. 252; Foster v. Rhoads, 19 *Id.* 191; Jackson v. Lawton, 10 *Id.* 23.) Prescription supposes and must be based on a supposed grant; in the case of a user by the public there is no grantee and no prescriptive right can be obtained. (*Munson v. Hungerford,* 16 Barb. 265; Curtis v. Keesbi, 14 *Id.* 511; *Clements v. Village of West Troy,* 10 How. 199; *Post v. Pearsall,* 22 Wend. 425, 440; Pearsall v. Post et al., 20 *Id.* 121-125; *Bland v. Lipscombe,* 30 Eng. L. and Eq. 189, note.) The right to take and carry away fish such as is claimed here is in the nature of a profit, and is what was called in the early books "profit a prendre." (2 Washburn on Real Property [3d ed.], 276, subd. 3; Washburn on Easements [3d ed.], 126, § 21; *Id.* [3d ed.], 6 and 7, § 6.) A right of "profit a prendre" cannot be gained by custom. (Washburn on Easements [3d ed.], 125, § 20.) It must be gained, if at all (when not claimed by grant), by prescription. (Washburn on Easements [3d ed.],

---

[81] 91 N.Y. 98 (1883).

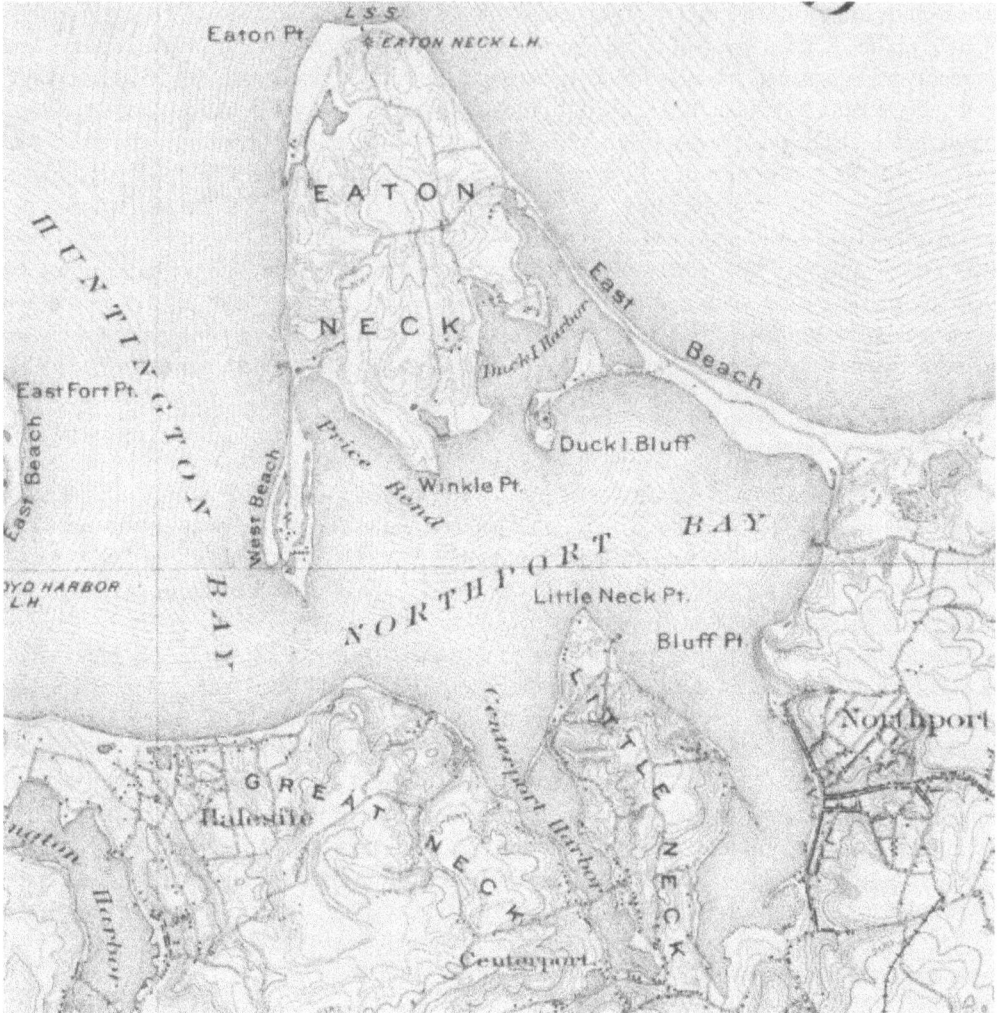

**FIGURE 7.10**   Northport Harbor, New York. (1901 USGS topographic map.)

15, § 14; *Id.* 125, § 20.) A town meeting has power to direct use of corporate property, etc. (1 R. S. [5th ed.], 87, § 9, subd. 5, 9, 11.)

The court stated, this action involves the right of the plaintiff to the use of land under water in Northport harbor, in the town of Huntington, Suffolk county, for the purposes of an oyster bed. The plaintiff's title is derived from a lease executed and delivered to him upon the 1st day of January, 1879, from the trustees of the town of Huntington, and the most important question involved is in regard to the legal title of said trustees to the land in question and their right to grant the same for the purposes named in the lease.

The title of the town of Huntington is derived from several patents issued at different times by the colonial governors of the colony of New York, the first of which bears date on the 30th day of November, 1666, and the last upon the 5th day of October, 1694.

Upon the trial no question was made as to the eastern and western boundaries of the patents, and there was testimony establishing the other boundaries, and that the town had claimed title to the land covered by water in the harbor and the fishing privileges and from time immemorial treated the same as the property of, and as belonging to the town, and also claimed an exclusive right to the same. The evidence shows that it regulated and exercised control over the fishing and shooting in the waters of the harbor, had passed resolutions to prevent strangers, who were not inhabitants, from fishing therein, and had made leases for marine railways and docks, and had executed the present lease for the use of the land under water, therein described, for oyster fishing.

The judge upon the trial found that the title of the lands covered by water in Northport harbor, with the shell-fish growing thereon, was in the trustees of said town under ancient patents, and the testimony sufficiently sustains such finding.

The question arising as to the rights acquired and the effect to be given to grants of the character of those herein referred to were the subjects of consideration, and substantially passed upon by this court in the case of *The Trustees of Brookhaven v. Strong* (60 N.Y. 56), and it was there decided that by the common law the king had the right to grant the soil under water and with it the exclusive right of fishery, and that a grant by the colonial government confirmed by subsequent legislation conveyed an exclusive right to the oyster fishery.

The patents under which the claim was made in the case cited were issued about the same time and were of a similar import as those relied upon in the case at bar. A part of the same South bay granted by the Brookhaven patents is also covered by the patents introduced in evidence upon the trial of this action.

The learned counsel for the defendant claims that a distinction exists between the two cases; that the locus in quo is different; that the charters and the surroundings were not the same, and that a continuous possession and use by the town, in the Brookhaven case, was relied upon to supply defects. It is true that the patents embraced different territories and the charters of the towns are not perhaps precisely the same in all respects, but a continuous possession and use of the land under water was an important part of the proof in the case at bar and greatly relied upon. Nor is there any serious question that the town exercised a control over the fisheries for a number of years so as to ripen into and strengthen its right thereto. Although the town did not lease any of the oyster beds until 1879, it did execute leases of other portions of the land covered by water and thus indicated its right to execute leases. It certainly as owner and as being in possession had a right to lease which is sufficient to uphold its claim to the land. The evidence that persons caught oysters there without paying for the privilege does not necessarily show that the town had no right to the oyster-fisheries and only furnishes some evidence which was to be weighed and considered by the court in determining the rights of the parties. The leases made show a more absolute title to the lands under water than a lease for fishing purposes, and in connection with the proceedings of town meetings and of grants for railroad purposes show that a claim has always been made of title, as well as to the right of fishery.

The fact that the trustees have allowed the lands to be enjoyed in common does not destroy their claim of title. Although the court say in the Brookhaven case that the elements of title derived from the patents were very much strengthened by possession and user, it disposes of the question of title mainly upon the authority of the patents themselves.

The counsel for the appellant claims that there is no recognition by the legislature of the title of the town of Huntington to the land under water of Northport bay or harbor. We think that the act of 1691, passed for the purpose of quieting and confirming titles, confirmed all royalties and other franchises which had been previously granted, and among these were those included in the charter of the town of Huntington. This is expressly held in the case of *Brookhaven v. Strong* (supra), and also in the case of *The People v. Van Rensselaer* (9 N.Y. 291). If the land in question is covered by the charters to which reference has been had, the cases cited dispose of the question at issue. We think it is established by sufficient evidence that the boundaries of the patents included the oyster bed which is the subject of this controversy. As already stated there is no question as to the western and eastern boundaries. The northern boundary of the town is the sound. This includes, we think, Northport harbor where the oyster beds in question are located. The language of the grant includes "all havens, harbors, creeks" as well as "fishing, hawking, hunting and fowling." In the Brookhaven case, the south boundary was the ocean, and there was a sandy flat or beach between the ocean and the bay, and the question was raised that the South bay was not within the grant. This court held that this objection could not be sustained and that the southern boundary which was the ocean, included the beach and of course the bay, etc. The patent under which the plaintiff claims is bounded on the north by the sound, adjacent to the sound is Eaton's neck and Eaton's neck beach, and south of this is Northport harbor. By analogy, both Eaton's neck and Eaton's neck beach are within the patents and necessarily the harbor also. The boundary by the sound includes all the land south of the sound. That this was intended is indicated by the use of the words in the grant, "harbor, havens," etc. That Northport harbor was included within the limits of the boundaries was proved by the undisputed evidence of the surveyor and others. It was also proved that Eaton's neck beach was leased by the town. It should also be noticed that Northport harbor is land-locked and has always been used and distinguished as a harbor. It is very evident that there was testimony showing that Northport harbor was within the limits of the grant of the patents and, as a question of fact, which was determined by the findings of the court upon the trial and sustained by the General Term upon appeal, the subject is not now open for review.

The case of *Lowndes v. Dickerson* (34 Barb. 586) is relied upon by the counsel for the appellant. It was there held that the harbor of Northport was not within the limits of the town of Huntington, and the inhabitants thereof had not the exclusive right to take fish therein; that the right of fishing is a common right inherent in the people by the common law, and that nothing passed by the grant of the colonial government to the town by implication. It was there stated in the opinion that the weight of authority was adverse to the existence of any power in the British crown to grant to an individual the right to take fish in the sea and in an arm thereof in exclusion of the common liberty. Since that case was decided this court has held, as we have seen in the Brookhaven case, that the crown had authority to grant the town, as such, a right of fishery within its borders, thus overthrowing the doctrine enunciated in the case of *Lowndes v. Dickerson* (supra), and this case has been also overruled by the decision of the same General Term in the case at bar. The claim, therefore, that that case is conclusive as to this, and that the question is res adjudicata is, we think, not well founded. The defendant there claimed the right to take oysters as a citizen of the town of Huntington. Neither party had any title or lease from the town, and it may be assumed that in that case the proof did not entirely establish that the bay and harbor in question were within the limits of the town of Huntington, as is the fact here, and as the court found, and that proof was not produced to establish, as is the case here, that from time immemorial the trustees of Huntington treated the harbor in question and the land covered by water therein as under the control and belonging to the town, regulated the fishing and shooting, prescribed penalties for any infringement upon the right of fishery, made leases and performed other acts of ownership which evinced that they held title to the same. Nor does it appear that there was any proof in that case that Northport harbor was a land-locked harbor; that Eaton's neck and Eaton's Neck beach lay north thereof, and that the sound was north of this; in fact there was no proof that Northport harbor was actually south of the north boundary of the patent. These defects and deficiencies show that the case cited was different in these material and important features from the case now considered. They have been supplied by proof upon the trial in this case. The doctrine laid down in the Brookhaven case brings Northport harbor directly within the boundaries of the town of Huntington as established by the grants made to them, and the lease executed by the town was legal and valid. (See, also, *Rogers v. Jones, 1 Wend.* 237.)

We have examined the other questions raised by the counsel for the appellant and in none of them do we find any reason which would lead us to a conclusion different from that which has already been expressed, or to the reversal of the judgment, which was clearly right and should be affirmed.

## SURVEY OF OYSTER BED

The Maryland decision of *Cox v. Revelle*,[82] is a proceeding by Samuel Revelle and others against George A. Cox, under Acts of 1914, c. 265, to have declared natural oyster bottom, and so subject to condemnation, beds previously leased to Cox. There was a judgment that the leased land was a natural bed, and Cox appeals.

The board of shellfish commissioners, acting under authority conferred by the act of 1906, chapter 711, leased to the appellant, on May 20, 1912, for the term of 20 years, a lot of ground in the bed of Manokin river, in Somerset county, to be used for the purposes of oyster culture. It was the intent and policy of the law that natural oyster beds or bars should not be subject to lease for private use, but should be reserved as public oyster fisheries for use in common by the people of the state under suitable regulation and license. The land demised to the appellant had originally been classed as ineligible for leasing, as the investigation and survey of oyster grounds for which the act made provision had officially shown it to be included in one of the areas designated as natural beds or bars. This ascertainment, however, was later reviewed by the circuit court for Somerset county, upon a petition by residents, as authorized by the statute, and the lot in question, as part of a larger tract, was determined to be "barren bottom," and therefore available for leasing, and the report and plat of the survey were accordingly amended. By the terms of the law, as it then existed, such an adjudication was final. It was about four years after the survey was thus revised that the appellant's lease was executed. In August, 1914, the appellees filed a petition in the court below, alleging that the ground leased to the appellant was natural oyster bar, and praying that it be so declared. This action was taken by virtue of the act of 1914, chapter 265, which amended the act of 1906 by providing, in part, by section 94B, that:

"Three or more residents of this state may at any time before January 1, 1915, file a petition in writing, alleging that five or more adjacent acres of natural beds or bars situated in the Chesapeake Bay outside county waters, or one or more acres of natural beds or bars within the territorial limits of any county of this state, to be described in said petition have been excluded from the surveys or resurveys of natural beds or bars of this state, such petition to be attested by the several oaths of the petitioners, and to be filed in the circuit court for the county in which or nearest to which the area in question is located."

The shellfish commissioners and the lessees, if any, of the disputed ground are required to be made defendants, and after summons and due opportunity to answer, the court is authorized and directed to "proceed promptly to hear all evidence adduced by the parties" and to "decide whether the area described in said petition is or is not a natural bed or bar as defined in section 83, and judgment shall be entered accordingly." Section 94B continues:

"The hearing in said circuit court shall be before a jury, unless jury trial be waived by all parties, in which event the hearing shall be before any judge or judges of said court. An appeal to the Court of Appeals of Maryland may be taken by either party to said case from the judgment of said circuit court within thirty days thereafter, and the Court of Appeals shall have power to review all questions of fact or law involved. If the final decision shall be that the area in question is a natural bed or bar, amended plats shall be made and copies filed as provided in section 94A."

A further amendment, contained in section 94C, enacts that:

The rights and interests of lessees under leases outstanding and in force at the time of the passage of this act (April 3, 1914), covering areas within the limits of natural beds or bars which may be established by the resurveys provided for by section 94A, or by proceedings taken under section 94B, and the oysters belonging to such lessees, located on such areas, shall be condemned by the state of Maryland for the use of the public.

---

[82] 125 Md. 579, 94 A. 203 (Md. 1915).

To the petition of the appellees the board of shellfish commissioners filed an answer neither admitting nor denying the allegations as to the character of the ground leased to the appellant, but demanding strict proof. The appellant, by his answer, denied that the lot embraced in his lease was natural bar, and relied upon the previous finding of the court on that subject as the basis of a plea of res adjudicata. In support of the latter theory of defense, the answer alleged that in November, 1912, a bill of complaint was filed in the circuit court for Somerset county, in which the present petitioners, with others, were plaintiffs, and the present respondent was one of the defendants, and in which the allegation was made that the area found by the court in the former proceeding to be barren bottom was in fact natural bar, and the judgment to the contrary had been obtained by fraud, and, together with the lease in controversy, should, for that reason, be set aside, and a decree passed to that end was reversed by the Court of Appeals and the bill dismissed, in *Cox v. Bennett,* 123 Md. 356, 91 A. 141, whereby the prior determination as to the leasable nature of the area was left in full force and effect. The answer also averred that the act of 1914, which permitted the action now pending, is in conflict with section 10 of article 1 of the Constitution of the United States, which prohibits state legislation impairing the obligation of contracts, and that it likewise violates section 40 of article 3 of the Constitution of Maryland which forbids the enactment of any law authorizing private property to be taken for public use without just compensation being first paid or tendered. The petitioners demurred to the lessee's answer in so far as it set up the defense of former adjudication and the constitutional objections we have noted. The demurrer was sustained, and, the case having proceeded to trial upon the question of fact raised by the pleadings as to whether the leased land was a natural bed or bar, a verdict was rendered in the affirmative upon that issue, and a declaratory judgment to the same effect was duly entered. In the course of the trial two exceptions were reserved by the lessee, one relating to the admission of evidence and the other to the action of the court on the prayers. The objections presented in this form were identical with those which had been overruled on demurrer.

The defense on constitutional grounds will be first considered.

It is entirely clear that in merely providing for a reopening of the investigation as to the nature of the ground leased by the state to the appellant, the act of 1914 does not impair the obligation of their contract. The judgment entered in the proceeding which the act allowed for the purposes of such an inquiry is simply a formal declaration as to an ascertained condition. It makes no reference whatever to the existing lease. Notwithstanding such a determination, the law recognizes the contractual rights of the lessee to their full extent. The interests created by the lease are given the same consideration that any other property is entitled to receive at the hands of the state. The lot demised having been found to be a natural oyster bar, provision is made for its condemnation for public use upon the distinct theory that the adjudication as to its real nature has not impaired or affected the vested estate of the lessee. In directing the acquisition of such a leased lot the act expressly declares, in section 94C, that: "No right, interest or property" of the lessee "shall be divested until the compensation awarded in such condemnation proceedings *** has first been paid." The appellant's contractual rights are thus duly regarded and protected by the law, and are required to yield only to the sovereign power of eminent domain.

It is, of course, not disputed that a leasehold interest acquired from the state is as completely subject to condemnation for public use as a title derived from any other source, or that the exercise of the power with respect to such an estate does not cause an impairment of contract within the meaning of the federal *Constitution. Long Island Water Co. v. Brooklyn,* 166 U.S. 685, 17 S.Ct. 718, 41 L.Ed. 1165; *West River Bridge Co. v. Dix,* 6 How. 507, 12 L.Ed. 535; *New Orleans Waterworks Co. v. Rivers,* 115 U.S. 674, 6 S.Ct. 273, 29 L.Ed. 525; *St. James Church v. B. & O. R. Co.,* 114 Md. 442, 79 A. 35; *Turnpike Road v. Railroad Co.,* 81 Md. 248, 31 A. 854; *B. & H. Turnpike Co. v. Union R. Co.,* 35 Md. 224, 6 Am. Rep. 397. But it is urged that the act of 1914, in authorizing proceedings like the one before us, directed the ascertainment of natural beds and bars to be made according to a specified standard which is different from the one applied in the original survey. This contention has reference to the fact that the act of 1906 contained no definition of such areas, while the act of 1914, by an amendment of section 83, provides that the term "natural beds or bars" shall be construed to mean "all oyster beds or bars under any of the waters of this state whereon the natural growth of oysters is of such abundance that the public have successfully resorted to such beds or bars for a livelihood, whether continuously or at intervals, during any oyster season within five years prior to" the new inquiry under the terms of the statute. The annual report of the shellfish commissioners submitted in 1912 shows that they adopted, for the purposes of their survey and location of natural oyster bars under the act of 1906, a definition having practically the same meaning and effect as the one later approved by the Legislature. If, however, it be assumed that the act of 1914 prescribed an entirely different rule for determining what areas were to be devoted to the use of the public, we are unable to find in that fact any support for the appellant's claim that his contractual rights under his lease are being impaired. It would have been competent for the state to provide for the appropriation for public use of all the leased oyster grounds, regardless of the question as to whether they were natural beds or barren bottoms. If this course had been pursued in the exercise of the right of eminent domain, and with due provision for just compensation, the lessees could not complain that the constitutional prohibition against legislative impairment of contracts was violated.

The policy of the existing law with respect to the use and disposition of oyster grounds is a question with which this court has no right to be concerned. The sole authority to deal with that subject is vested in the Legislature. It is within the discretion of that department of the state government to decide whether the whole or any part of either the natural bars or the barren bottoms shall be open to free or to qualified public use, or shall be reserved and controlled for conservation, cultural, or revenue purposes. In the exercise of its ample power, the Legislature has determined that oyster areas having certain characteristics shall be devoted to the use of the public. It has provided for a judicial ascertainment of the facts in reference to any location to which leasehold rights have attached, and, in a proper case, for the acquisition of such interests for the public by condemnation. The lessee is summoned and heard upon the preliminary question as to the character of the leased ground, and, by express enactment, he is protected in his title and possession until the state shall have paid him the just compensation which may be awarded him by due process of law. In view of the safeguards thus placed by the statute around the rights and interests of the appellant, there is clearly no impairment of contract to which the constitutional provision relied upon can be applied.

The contention that the act of 1914 is in conflict with section 40 of article 3 of the Maryland Constitution, prohibiting the taking of private property for public use without just compensation, requires no extended discussion. The provision of the statute against any divestiture of the property rights to be condemned until the compensation awarded is paid has already been emphasized. As the act, by its own terms, gives expression and effect to the constitutional rule to which the appellant refers, there is no room for the theory that the limitation it imposes has, in this instance, been disregarded.

It was urged that the act does not state for what public use the natural bars under lease are to be condemned, and that the court is consequently not in a position to decide whether the intended use is in fact of a public nature. The further point is made that if the lots no acquired by the state are designed to be used for the purposes of a public oyster fishery under existing law, regard must then be had to the statutory provision that the right to take oysters for sale from grounds within the limits of any county shall be confined to its own residents, and it is said that this is not to be considered a public use within the intent of the Constitution. As to the question thus proposed, we can have no doubt or difficulty. The plain purpose of the act is to secure all natural beds or bars for the public use to which oyster areas owned by the state may be subjected under laws now in force or hereafter enacted. There can be no doubt as to the public nature of the use to which such grounds are susceptible. The taking of oysters by the public, under license of the state, from lands and waters subject to its ownership and control, is undeniably a public use. *Smith v. Maryland,* 59 U.S. (18 How.) 71, 15 L.Ed. 269; *State v. Applegarth,* 81 Md. 293, 31 A. 961, 28 L. R. A. 812; *Hess v. Muir,* 65 Md. 599, 5 A. 540, 6 A. 673. In order to answer that description it is not necessary that the use, as to every natural oyster bar, shall be open to the public generally, but the right may be validly restricted to the citizens of the county within whose territory the fishery is located. The principle was stated and applied in *Webster v. Pole Line Co.,* 112 Md. 429, 76 A. 254, 21 Ann. Cas. 357, that a public use need not be available to the whole public, and that it may be confined to the inhabitants of a designated locality, provided it is exercisable in common and is not limited to particular individuals.

The remaining question to be considered relates to the defense of former adjudication. In our opinion the principle upon which such a defense must depend is not applicable to the case presented. This is a statutory proceeding to determine the desirability of property for public use, according to a prescribed standard, as a preliminary to its acquisition for the public by condemnation. An adjudication as to the leasable character of certain oyster grounds under the act of 1906 could not justify a denial by the court of the state's power to condemn such property for public purposes under a later enactment. The legislative judgment as to the expediency of such an appropriation cannot be thus controlled by judicial action. The courts have undoubtedly authority to decide whether the use for which private property is proposed to be taken under the power of eminent domain is public in its nature, but when this inquiry is answered in the affirmative, the question as to the propriety of exercising the power is committed exclusively to the discretion of the *Legislature. Shoemaker v. United States,* 147 U.S. 282, 13 S.Ct. 361, 37 L.Ed. 170; *Arnsperger v. Crawford,* 101 Md. 252, 61 A. 413, 70 L. R. A. 497; *New Cent. Coal Co, v. George's Creek C. & I. Co.,* 37 Md. 560; 15 Cyc. 629. By the judgment of the circuit court for Somerset county rendered in 1908, it was simply declared that the ground under consideration was barren bottom, and the only effect of the decree of this court in the equity proceeding to invalidate that finding was to dismiss the bill of complaint on the theory that the charge of fraud had not been sustained. *Cox v. Bennett,* 123 Md. 361, 91 A. 141. It would be giving the doctrine of res adjudicata a very anomalous application to hold that it can prevent the exercise by a judicial tribunal of a special statutory jurisdiction to reopen an inquiry previously made, under authority similarly conferred, although ample security is afforded to every vested interest. It has already been observed that the Legislature could have directed the condemnation of the leased oyster lots for public use without a prior inquiry by a court proceeding as to the special character of the ground, and notwithstanding a previous adjudication that they were leasable. There is hence no possible prejudice to the lessee in the fact that a preliminary investigation, in which he has an opportunity to be heard, has been provided for the ascertainment of the actual condition of the property to be acquired. The theory of estoppel by former judgment was therefore properly rejected.

The various questions raised by the record have now been considered, and we agree with the court below as to all of the rulings.

Judgment affirmed, with costs.

The follow-up decision of *Board of Shellfish Com'rs of Maryland v. Mansfield,*[83] continues the issues from the previous case. This is a proceeding by Joseph Mansfield and others, under Acts of 1914, c. 265, to condemn as natural oyster bottom a bed previously leased as barren. The lessee failed to appear, and the Board of Shellfish Commissioners answered to the petition. From a judgment declaring the leased ground to be a natural bed, the Shellfish Commissioners appeal. Reversed, and new trial awarded.

---

[83] 125 Md. 630, 94 A. 207 (Md. 1915).

The principal questions involved in this appeal have been decided in the case of *Cox v. Revelle,* 94 A. 203. This proceeding originated in a petition filed under the act of 1914, chapter 265, in the circuit court for Talbot county, praying that certain oyster ground in Miles river, which had been previously leased to a private individual upon the theory that it was barren bottom, might be judicially determined to be a natural bed or bar, as the result of which finding it would be subject to acquisition for public use by condemnation as provided in the statute. The lessee, though summoned, did not appear and make defense, but the board of shellfish commissioners filed an answer to the petition, disputing the allegations of fact and denying the constitutionality of the act upon which the proceeding was based. A demurrer to the averments of the answer which questioned the validity of the statute was sustained, and upon the issue of fact a judgment on verdict was rendered, declaring the leased ground to be a natural oyster bed. The shellfish commissioners have appealed. In addition to the questions raised on demurrer, there are several exceptions to the exclusion of evidence to be reviewed.

The first objection to the act, as stated in the answer, is that its provisions are so defective and contradictory as to make it incapable of enforcement. In support of this position, the appellants refer to the provision for the ascertainment of the outlines of natural oyster bars by testimony taken in court. It is said that while such evidence may prove whether a certain location represented on a chart is included within a natural oyster area, it is impracticable by this means to determine the exact limits of the bar. The act, however, does not depend upon the theory that the outlines of the submerged oyster beds are capable of being ascertained with precision. By section 90 it was provided that in the original survey to be made by the shellfish commissioners, they should use their judgment "liberally in favor of the natural beds and bars and allow a reasonable margin of the barren bottoms rather than encroach on a natural bed or bar," and in the same section there is a further provision that "all natural beds or bars shall be surrounded by neutral zones 200 yards wide in the Chesapeake Bay and Tangier Sound, and 50 yards wide elsewhere," and that these zones shall not be leased under the act or appropriated to private use. The Legislature evidently recognized the difficulty of defining with exactness the boundaries of the bars or beds which were intended to be excluded from the leasing system established by the state. We see no reason to hold that a substantially accurate location of oyster fisheries in the manner authorized by the act is not feasible, and we are certainly unwilling to declare the statute invalid on that ground. There are other provisions which are said to be open to the objection we have just noticed, but they have no relation to the case before us; and, even if we discovered any force in the contention, it could not now be appropriately made the subject of decision.

It is next asserted that the act contravenes article 3, § 40, of the state Constitution in that it contemplates the taking of private property for a use not public, and without just compensation being first paid or tendered. This question has been considered, and determined adversely to the view here urged, in the case of *Cox v. Revelle*, supra.

Another contention is that the statute authorizes the taking of property without due process of law, and is therefore in conflict with the 14th amendment of the Constitution of the United States and article 23 of the Maryland Declaration of Rights. This objection is not sought to be applied to the section upon which this proceeding is founded, but to separate any distinct provisions of the law which are not involved in the present case, and hence need not be discussed.

The further point is made that the right of trial by jury as to the issue of fact in cases of this nature is not protected by the act, although the Constitution of the state (article 15, § 6) declares that the right to such a trial in civil proceedings in the several courts of law, where the amount in controversy exceeds the sum of $5, shall be inviolably preserved. The statute provides that upon the filing of such a petition as the one here pending, the clerk shall docket a suit at law in which the issue as to the character of the disputed ground is to be tried before a jury, or before one or more of the judges, if a jury trial is waived. An appeal may be taken by any of the parties to the Court of Appeals, which is empowered "to review all questions of fact or law involved." In committing to this court the ultimate determination of the issue of fact joined in the case, it is urged that the act nullifies the right of jury trial secured by the Constitution. Assuming that the provision relied upon is applicable to a judicial inquiry under the statute as to the real character of areas alleged to be natural oyster bars and as such designed by the law to be available for public use, and assuming further that the language quoted is to be construed as meaning that the Court of Appeals should have authority to pass upon the issue of fact in such a proceeding, it does not follow that the entire act or section, or even the whole of the clause relating to the appeal, must be held invalid. The objectionable part of the provision is contained in the single word "fact," and that term may be readily eliminated, and the scope of appellate review thus unmistakably limited to questions of law, without any impairment of the general effect of the statute. The court is not justified in declaring the whole of an act unconstitutional merely because one of its provisions may be open to that objection, unless the void and valid portions are so dependent upon each other as to raise the presumption that the law would not have been passed if the invalid provision had been omitted. This is a sound and familiar principle, and it is clearly applicable to the present case. *Painter v. Mattfeldt,* 119 Md. 474, 87 A. 413; *Somerset County v. Pocomoke Bridge Co.,* 109 Md. 1, 71 A. 462, 16 Ann. Cas. 874; *Steenken v. State,* 88 Md. 708, 42 A. 212.

The two remaining objections on constitutional grounds refer to proceedings which the act permits the shellfish commissioners to take on their own initiative for the re-survey or re-examination of oyster bottoms with a view to the more complete reservation of natural bars for the use of the public. The powers thus conferred have not been exercised with respect to the area here in dispute. This is not a voluntary reinvestigation undertaken by the commissioners for the purposes indicated, but it is an adverse proceeding by private parties under a separate provision of the law. The two methods of inquiry are entirely different and disconnected, and it is only the involuntary mode of procedure with which we are now concerned.

The rulings on exceptions to testimony are yet to be reviewed. It was proposed by the shell-fish commissioners, in the course of the trial, to prove that portions of the ground in controversy had been held for the purposes of private oyster culture under licenses which were issued in 1905, by virtue of a pre-existing act, but which have since become forfeited. The issue of fact, as defined by the act of 1914, was whether, during the five years preceding the filing of the petition in this case, the natural growth of oysters had been so abundant, on the area in question, that the public had successfully resorted to it, continuously or at intervals, during any oyster season in that period, for the taking of oysters as a means of livelihood. The theory upon which proof was offered as to the prior use of the ground for oyster culture was that this fact might account for the presence of oysters found there subsequently, and would at least tend to rebut the inference that oysters taken during the five-year period were the product of "natural growth" within the statutory meaning, especially as only barren bottoms could have been legally appropriated for oyster culture under the licenses mentioned. In our opinion the proffered evidence on this subject was relevant to the question as to whether the tract under inquiry was a natural bed or bar as defined by the act, and the record does not enable us to say that the exclusion of such proof was not prejudicial to the interest which the appellants have in the correct determination of that issue.

While we fully concur with the trial court in its disposition of the demurrer, we must reverse the judgment for the error we have noted in the rulings upon the exceptions.

Another original colony on the East Coast, North Carolina, decided a case in 1890, that reviewed the long history of oyster farming in various parts of the world, and the ancient laws governing the practice. This decision is noteworthy for that reason alone.

*State v. Willis*[84] is a case wherein a warrant was issued against the defendant charging him with a violation of the provisions of the last clause of section 3393 of the Code, in taking oysters from the oyster garden of one Chadwick, without permission. He pleaded "not guilty," and contended that the bed staked off and inclosed was a natural oyster-bed, and that any citizen had the right to take oysters from any natural oyster-bed, under the last clause of section 3390 of the Code. The jury found the facts, in a special verdict, to be as follows: "That the prosecutor, Chadwick, has had his oyster-bed staked off according to the requirement of the statute, and obtained a license or grant from the clerk of the superior court of Carteret county. That the defendant took oysters therefrom, within the stakes where the garden was so laid off, without permission of said Chadwick. That there were no oyster-rocks within the stakes where the garden was so laid off, but there were some oysters within the same; but they were scattering, and oysters naturally grew there. That the area embraced within the stakes, which was 2 3/4 acres, was not such as would, within itself, induce the public to resort to it and get oysters; but, in connection with the oyster-rocks and oysters nearby, the public were in the habit of taking oysters from the said area and adjacent territory for livelihood, and for market. If, upon the foregoing statement of facts, the court be of opinion that the defendant is guilty, then the jury say he is guilty; but, if upon said statement, the court be of opinion that defendant is not guilty, then the jury find he is not guilty." Whereupon the court adjudged that the defendant is not guilty, and from this judgment the solicitor for the state appealed.

As early as 1822 the attention of our legislators was directed to the protection of the oyster interests in the waters of this state, and a statute was enacted inflicting a penalty upon "masters and skippers" for transporting oysters out of the state, and also prohibiting the use of any instrument except tongs in their taking. In those days the natural oyster-beds were considered amply sufficient to meet the demands of the public, and it was only deemed necessary to extend to them the protection of the law as above stated. In 1858 the law-makers, appreciating the importance of increasing the quantity, as well as improving the quality, of oysters, passed a statute very similar in its provisions to those of sections 3390–3393 of the present Code. These sections provide that any inhabitant "may make a bed in any of the waters of this state, and lay down or plant oysters or clams therein, having first obtained a license" from the clerk of the superior court of the proper county. It is further provided that in his petition he shall describe particularly the place where he desires to make such bed, and that he may stake out such grounds, not exceeding 1 acre, and that he shall hold the same in fee. If, however, he has included within his stakes any natural oyster or clam bed, or a space containing more than 10 acres, or if he shall fail, for the period of two years, either to use such bed, or to keep it properly designated by stakes, he shall forfeit such license. The act also provides that if any one shall injure such bed or stakes, or shall gather or take away any oyster within the lines of the same, without the permission of the owner, he shall be subject to a penalty, and also to indictment. There is a proviso, however, "that nothing herein shall be construed to authorize any person to stake off and inclose any natural oyster or clam bed, or in any wise to infringe the common right of the citizens of the state to any such natural bed." Much uncertainty existing as to what parts of the waters were subject to entry and grant under this law, a commissioner was appointed under chapter 119, Acts 1887, to make a survey, and finally determine and locate the natural oyster-beds, etc., in order that such entries and grants could be intelligently and safely made. This was done; and it seems that the purpose of the law has been accomplished, so that no such question as is here presented can generally arise except as to grants made before that time.

For some reason best known to the legislature, the above act was confined to the waters north of Cove sound, and has no application to this case. Cove sound, and all the waters south of it,

---

[84] 104 N.C. 764, 10 S.E. 764 (N.C. 1890).

therefore, are governed by the sections of the Code referred to; thus leaving the location of the natural oyster and clam beds in said sound to the courts and juries as the cases arise. We are therefore directly confronted with the difficult duty of defining a "natural oyster-bed." Although the cultivation of oysters obtained among the Romans in the days of Pliny, who speaks of it in one of his writings, and has since been carried on extensively in Italy, France, England, and other countries, we have been unable to obtain from them any light upon the particular question before us. In England, as with us, great care has always been taken to preserve to the public the common right of fishery, so that in granting privileges for oyster culture this public right has not been impaired to any material extent. Hence, as we have seen, the legislature, while encouraging the cultivation of oysters, has provided that natural oyster or clam beds shall not be subject to entry. In 1885, Lieut. Francis Winslow, of the United States Navy, was detailed, at the request of the legislature, to make a "survey of the natural oyster-beds and private oyster gardens, together with an examination of the waters of the state, with reference to the possibilities for the culture of shell-fish," etc. His report presents with much intelligence the different theories advanced as to what is a natural oyster-bed. He says: "In carrying out the first special requirement of the resolution, a difficulty experienced in all oyster localities was at once encountered. The question arose here [as elsewhere] as to what was properly a 'natural oyster-bed.' Naturally, that question had to be answered before the 'natural beds' could be surveyed and located. Very few people know what is, or what constitutes, a natural oyster-bed. Indeed, it is only a matter of opinion, at the best; and opinions are likely to be influenced largely by self-interest. A large number of persons make a distinction between oyster-beds that ebb dry and those that are covered at all states of the tide,--a distinction which, it is needless to say, does not have any sound foundation to rest upon. Many also appear to think that a natural bed is not a 'natural bed,' in the meaning of the law, because it is a little one. On the other hand, there are some whose definition of 'natural bed' is so liberal that it not only covers all places where oysters were in the past, or are in the present, but includes any area where they might, could, would, or should grow in the future. Arguments have been made to the effect that, as the drifting spat was evidently a product of nature, wherever the spat attached, or oysters grew, that spot became a natural oyster-bed. Evidently such a view would preclude any and every system of oyster culture. On the other hand, it has been argued that small groups and bunches of oysters, separated and distinct from any considerable area, were not 'natural beds,' within the meaning of the law. A legal decision [by Judge GOLDSBOROUGH, of Maryland,] defines a 'natural bed' as one not made by man, and of sufficient area to have been profitably worked by the general public, as common property, within some recent period of time. This decision has been practically adopted by the shell-fish commission of Connecticut in defining the natural beds of that state; and their course has been approved by legislative enactment. Useful as a guide, however, it would not be proper to be strictly governed by the GOLDSBOROUGH decision in defining the natural beds of North Carolina."

In this state the oyster industry is yet in its infancy. The population is too sparse, and the present demand too slight, to have caused any continuous fishery, or even any general knowledge of the positions or areas of the natural beds. Mere testimony as to previous fishery or non-fishery would not, therefore, in all places, be conclusive. It is not easy, from these conflicting theories, to deduce a satisfactory definition. We think that it is the capacity of the bottom to attach what fishermen call the drifting "spawn" or "spat" which distinguishes a natural from an artificial oyster-bed; but it will not do to confine this capacity to the inherent character of the soil, since many beds which are now considered natural may owe their origin to accidental causes, such as the deposit of brush, shells, drift-wood, and other objects to which the young oysters have adhered, and thus, after many years, resulted in the formation of a stratum which fulfills in every way the common idea of a natural oyster-bed. Neither can it with reason be said that every part of the bottom to which oysters may adhere, or to which they do adhere, and grow, will constitute such a bed, as they may be found scattered here and there in such small quantities as to be of but little value to the public; and such a theory would prevent entries, and thus defeats the purpose of the law in encouraging their cultivation. Something more permanent and valuable is meant by the word "bed." Webster's (1st Ed.) and the Century Dictionary give, as one of the definitions of "bed:" "A layer; a stratum; an extended mass of anything, whether upon the earth or within it, as a bed of sulphur, a bed of sand or clay;" and so the verb "bed:" "To lay in a stratum; to stratify; to lay in order or flat; as, bedded clay," etc. This view is well-illustrated by the stratum of marl to be seen in the banks of many of our eastern rivers, and in the marl pits in the eastern part of this state; the same being composed mainly, and in many cases, entirely of oyster shells, with alluvial deposits above. These considerations would exclude, therefore, the scattering growth of oysters which is to be found in many parts of the waters, and which is too small in quantity to be of value to the public. We think that a natural, as distinguished from an artificial, oyster-bed is one not planted by man, and in any shoal, reef, or bottom where oysters are to be found growing, not sparsely, or at intervals, but in a mass or stratum, and in sufficient quantities to be valuable to the public. This definition, we think, is more in accord with the spirit of our legislation than that of Judge GOLDSBOROUGH. The latter, in our opinion, lays too much stress upon the area, and involves an inquiry into the methods of taking oysters, and remuneration for the labor and capital employed. Too many elements of uncertainty enter into it to be of practical use in this state, where the cultivation of oysters is in its infancy, and their taking by the public is not exclusively for the purpose of sale and profit. While it seems impossible to give a more particular definition, it is believed that the one which we have adopted reflects the true spirit of the law, and may be of some practical use in ascertaining where grants may be made.

The application of the principles we have laid down to the case before us is easy. The special verdict finds that, although oysters naturally grew within the stakes, they were scattering, and insufficient in quantity to induce the public to resort to them alone as a means of livelihood, or "for market." The verdict also negatives the existence of "oyster-rocks," a species of oyster-beds. It is very clear that the verdict does not bring this case within the definition which we have formulated, and we are therefore of the opinion that his honor committed an error when he held that the area in question was a "natural oyster-bed." Error.

The decision in the case of *Charles H. Vroom v. John Tilly and Jacob I. Housman*[85] stresses the state's role and subsequent oyster farmer's role in mapping oyster beds and marking their extent. Such mapping and marking, including some surveying, defines the property limits of rights and claims.

This case also presents an interesting analysis of the differences between real property and personal property, as applies to a living animal, which is subject to the movement of the currents.

The plaintiff cultivated oysters on forty acres of land under the waters of Long Island sound, upon the supposition that the tract was within the bounds of lands granted to his licensors for such pursuit by the State. For six years he took out a large quantity of oysters. The defendants had acquired from the State a similar franchise in lands adjacent to those of the plaintiff. In fact the said 40 acres were within the bounds of the defendants' lands, although there was no indication thereof. This fact was ascertained by an official survey obtained by the defendants six years after the plaintiff had begun cultivation. Thereupon, in the face of the plaintiff's protest and explanation, the defendants took up for their own use the oysters from this tract. This action is for a conversion, and the plaintiff has gained the judgment.

The court found that the oysters taken up by the defendants were the result of the plaintiff's cultivation, and I think that the evidence justifies this finding.

The learned counsel for the appellants concedes that "the rule governing ownership of oysters, wrongfully planted or placed on the lands under water belonging to another, is that a

---

[85] 91 N.Y.S. 51, 99 A.D. 516 (1904).

man does not lose title to personal property, which can be identified, by the fact that he is a trespasser," but he insists that there is a distinction between the case of a man planting or placing oysters on such lands and a man who, like the plaintiff, simply prepares such land and, as a result of such preparation, gathers the germs floating in the water, which, under his care and culture, develop into oysters. I think, however, that these oysters were the property of the plaintiff. (*Grace v. Willets,* 50 N. J. Law, 414; *McCarty v. Holman,* 22 Hun, 53.) In *Grace v. Willets* (supra), Willets deposited a boatload of oyster shells and marked the deposits by stakes, and the germs of oysters floating in the water attached themselves to the shells. The court, per VAN SYCKEL, J., says: "Assuming, as we must, from the case as presented, that it was necessary to deposit the natural shell in order to attract the germ or sprout, and thereby in the order of natural growth produce the oyster, it seems as incontrovertibly to follow that the full-grown oyster is the property of him who planted the shell, as that the oyster when of marketable size belongs to him who planted it in its infant state, or as that the title to the colt is not lost by its growth and development into the horse." In *McCarty v. Holman* (supra), the plaintiffs planted both seed oysters and many scallop shells. GILBERT, J., speaking of the spat, said: "They are wafted away by currents, and would be lost unless they found an object to which they could adhere. The plaintiffs provided means within the bed, which they planted, to save the spat of oysters, and we are of opinion that their property in the oysters grown from the spat so preserved is quite as good as that in the parent oysters, whether the spat proceeded from oysters which they planted, or from other oysters." Ownership of the land whereon the oysters are deposited is not a prerequisite to ownership in the oysters. (*Davis v. Davis,* 72 A.D. 593; *McCarty v. Holman,* supra; *Fleet v. Hegeman,* 14 Wend. 42; *State v. Taylor,* 27 N. J. Law, 117; *Post v. Kreischer,* 103 N.Y. 110.) If the plaintiff were guilty of trespass in planting or cultivating oysters on the lands of another, such fact does not authorize the owner to take those oysters to his own use (*Davis v. Davis,* supra), although the owner might compel him to take them up or might remove them as a nuisance. (*State v. Taylor,* supra; *Sutter v. Van Derveer,* 47 Hun, 366.) Oysters reproduce by eggs, from which there hatches out a small free-swimming larva, becoming in a few days spat. These seek and attach themselves to some solid support, where they remain. (Ency. Am.) The property right in such oysters has been said to be akin to that gained over animals ferae naturae (*McCarty v. Holman,* supra; *Fleet v. Hegeman,* supra), though this principle has been sharply criticized by the Supreme Court of New Jersey in *State v. Taylor* (supra). If we regard them as ferae naturae, then the plaintiff has reclaimed them so far as is possible to such animals, in that he has caught them and confined them, and, therefore, the defendants cannot thereafter appropriate them, even though they are upon their lands. This question is well discussed by NELSON, J., in *Fleet v. Hegeman* (supra).

The judgment should be affirmed, with costs.

All concurred, except WOODWARD, J., who read for reversal.

WOODWARD, J. (dissenting):

I am unwilling to concur in the opinion about to be handed down by this court, because, in my judgment, it is fraught with great mischief, and is intended to defeat the policy of the State in respect to its oyster fisheries. It is undoubtedly true that at common law oysters planted in tidal waters on a well-marked and clearly defined bed, where there were no natural oysters before, were the personal property of the planter and he might maintain an action for their conversion (*Sutter v. Van Derveer,* 47 Hun, 366, 368, and authorities there cited), but in the year 1887 the State of New York undertook to 'promote and protect the cultivation of shell-fish within the waters of this State' (Laws of 1887, chap. 584), and by the provisions of this act the common law, in so far as it relates to this subject, was abrogated, and the rights of oystermen became regulated by this act. If I am right in this proposition, all of the cases decided before the modification of the common law, and which have been relied upon to support the subsequent decisions, are important only as affording a knowledge of what the law was, and not what it is to-day. It is important, therefore, to consider the provisions of the statute.

Chapter 584 of the Laws of 1887 provided in its 1st section that the commissioner of fisheries theretofore appointed, and his successor in office, shall be 'known as the Shell-fish Commissioner,' and it was his duty to 'finish and complete the survey now being made under his direction of all the lands under the waters of the State suitable for use for the planting and cultivation of shell-fish, and shall make a map thereof as heretofore provided. He shall finish and complete the survey now being made of all the beds of oysters of natural growth located in the waters of the State, and such beds of oysters of natural growth shall be set apart and preserved, and shall not be deemed to be included in the lands for which franchises are to be sold under the provisions of this act. Said commissioner shall ascertain the occupants of all lands claimed to be in the possession or occupation of any person or persons, and no grant of lands so occupied or possessed shall be made, except to the actual occupant or possessor thereof; provided said occupant or possessor, within one year from the passage of this act, shall make application for, and purchase the same.' That is, the State assumed dominion over all of the waters of this State suitable for oyster culture, undertook to preserve to the public the natural oyster beds, and to guarantee to those who had staked out claims under the common law the right to purchase a franchise for the same at a nominal figure, provided application was made within one year. This act, by necessary implication, denied the right of any individual to mark out and appropriate to his own use any of the lands under the waters of this State suitable for the planting and cultivation of shell-fish, at least in so far as such waters were embraced within the boundaries of the map which was authorized and directed to be made of such waters. In other words, the common-law right to plant and cultivate oysters was denied to individuals generally, and in its stead a special privilege was granted to such as should comply with the conditions of the statute. The act, after providing in section 2 for the appointment of an additional commissioner, who should be an expert oysterman, and in section 3 for a meeting of such commissioners for the purpose of formulating such rules and regulations as shall be deemed necessary as preliminary to hearing and granting applications for perpetual franchises for the purpose of shell-fish cultivation on the lands under the waters of this State, declared in section 3 as follows: 'After such rules and regulations shall have been agreed upon and formulated, the said Commissioners of Fisheries shall proceed to grant franchises for the purpose of shell-fish cultivation, as hereinafter provided. But no such franchise shall be granted until one month's notice of the application for a franchise or franchises shall have been given by posting in a conspicuous place, in the office of the Shell-fish Commissioner, and in the office of the town clerk of the town nearest to the lands applied for.' It was also provided in section 4 that these franchises were to be granted only to those who had resided in this State at least one year preceding the date of application for a franchise, and the amount of land allowed to be held by any person, firm or corporation was limited to 250 acres at any one time, so that every individual, firm or corporation engaged in shell-fish culture on the lands under water within this State since the year 1887 must, if obedient to the law, have operated under a franchise granted upon notice and under rules and regulations which, it is fair to presume, required that such franchises should be determined with reference to the map directed to be made, and presumably on file in the office of the Shell-fish Commissioner. This view seems irresistible, for it is provided in sections 5 and 6 as follows: 'When the conditions precedent to the granting of franchises, mentioned in the foregoing sections, have been complied with, the Commissioners of Fisheries are hereby empowered, in the name and behalf of the People of the State of New York, to grant, by written instruments under their hands and seals, perpetual franchises for the purposes of shell-fish cultivation in the lands applied for under the waters of the State, for the consideration of not less than one dollar per acre, if the lands are unoccupied or unused, and not less than twenty-five cents per acre if the lands are in present use and occupation, and the right to use and occupy said grounds for said purposes shall be and remain in the said grantee, his legal representatives or successors forever; provided only that the said grantee, his legal representatives or successors shall actually use and occupy the same for the purposes of shell-fish cultivation, and for no other purpose whatever. * * * *Immediately after the receipt of the aforesaid instruments of convey-ance, the grantee shall at once cause the grounds therein conveyed to be plainly marked out by stakes, buoys or monuments, which stakes, buoys or monuments shall be continued by said grantee, his legal representatives or successors.'* [Emphasis supplied by author. This statement in the law requires monumentation of the described parcel to be used, providing notice to others.]

The plaintiff in this action acquired all of his right to plant and cultivate oysters under the provisions of this statute, by reason of the original grant made to Elizabeth V. Merrill on the 8th day of December, 1891, which right was assigned to John H. Post in 1896, and the plaintiff and his uncle, Joseph Vroom, under a license from Post, undertook to locate the Merrill tract, and subsequently deposited a quantity of shells upon the Housman tract, which had been granted by the State to Nicholas P. Housman at the same time that the Merrill grant was made, upon the mistaken theory that they were making the deposit upon the latter tract. Later the plaintiff took an assignment of the Merrill tract from Post, as well as an assignment of his uncle's interest, and brought this action for the conversion of 1,600 bushels of oysters by the defendants, who claim to be the owners of the Housman grant. The plaintiff had absolutely no right to plant or cultivate oysters anywhere within the public waters of this State, except upon the lands under water which were granted to Elizabeth V. Merrill; he was her successor, and the statute made it his duty to maintain the 'stakes, buoys or monuments' which the grantee under the statute was bound to place 'immediately after the receipt of the aforesaid instruments of conveyance.' (Laws of 1887, chap. 584, § 6.) This was one of the conditions of his right to undertake oyster culture, and if he failed to stake out and indicate the lines of his franchise, as provided by law, and thus carried his operations over his line, he has only himself to blame, and he cannot plead good faith as a justification, for he was bound to know the law. The fact that the owners of the Housman tract had failed to erect stakes, buoys or monuments defining their boundaries can give the plaintiff no rights, where he has himself failed to comply with the requirements of the law. The Housman tract is bounded on the north by the south line of the Merrill tract, as well as by a township line, and if we assume that the Housman tract had been abandoned or forfeited to the State by reason of a failure to define it and to use it for the purposes provided by the statute, the plaintiff cannot, by a violation of the statute, and through ignorance of a fact which the law makes it his duty to ascertain and point out by stakes, buoys or monuments, gain any rights outside of the boundaries of his own franchise. Since the abolition of the rule of the common law in respect to these waters, the plaintiff is confined strictly to his franchise, and if he goes outside of its limits and makes improvements on the franchise of another, or upon lands which are in the custody of the State, he gains no right of property which gives him a right to maintain an action for conversion. It is not necessary to determine whether the defendants had any special rights by reason of their alleged ownership of the Housman tract, or whether the defendants were guilty of conversion as against the State of New York, for, in the view which I take of this question, the plaintiff, by proceeding unlawfully outside of the limits of his franchise, could not gain any property rights in the oysters which were developed upon the Housman tract, and that disposes of this case. The plaintiff in law could not make a mistake which would give him any right of property, because the law made it his duty to know and to mark out the limits of his franchise, so that if it be conceded that the evidence would warrant the conclusion that the plaintiff had, in good faith, made a mistake in his boundaries, this would not be a good excuse, for the law does not permit of the mistake. If it should be conceded that the defendants, by reason of their failure to stake out and make use of the Housman tract, had forfeited their rights, neither the plaintiff nor any one else would have the right to stake out the premises and occupy the same for oyster culture without a grant from the

State of New York, because in the exercise of its high sovereign powers the State has taken away the common-law right, and has taken this common lying under water under its police regulation, parcelling it out under such conditions as in its judgment is most likely to conserve the public welfare. To permit individuals to stake out claims outside of their franchises, or to enlarge them by overreaching and disregarding the plain provisions of the law, is to pave the way for endless friction and constitute a grave menace to the peace and good order of the State.

I find no evidence in this case to warrant the conclusion that the plaintiff has ever planted any seed oysters upon this tract; all that appears to have been done was to scrape the surface, to distribute over the same a quantity of shells, and to catch out star fish, which, I assume, are inimical to young oysters. The authorities seem to be agreed that the oyster is hatched from eggs laid by the female and fertilized by chance contact with the emissions of the male, and that for a period of a few days after being hatched the young oyster is gifted with a certain power of locomotion. If, during this period, he comes in contact with any firm substance in the water he attaches himself to such substance, and then goes through the process of development until he reaches maturity. The shells deposited by the plaintiff outside of his franchise did not produce the oysters which the defendants have taken; their purpose was that of a trap to catch and hold the young oysters during their swimming period of existence as they were carried through the waters by the tides or currents, and if we consider the oyster as ferae naturae he cannot be said to have been reduced to the possession of the plaintiff by being caught in a trap upon the premises of the defendants, or at least outside of the lawful franchise of the plaintiff. The act of reducing an animal ferae naturae into possession, where title thereby is created, must not be wrongful, and if such an act is effected by one who at the moment is a trespasser, no title to the property is created. (2 Am. & Eng. Ency. of Law [2d ed.], 345, and authorities cited in notes.) For the purposes of this action the defendants, as assignees of the interests of Nicholas P. Housman, must be deemed to be the owners of the franchise known as the Housman tract, and upon which it is conceded the plaintiff has deposited these shells, operating, as we have seen, as a trap to catch the infant oyster, and under the rule last above cited I am unable to see how the plaintiff has any title whatever to the catch made in such traps. In Rhode Island, where a trespasser placed a box in a tree on another's land for bees to hive in, it was held that he could not maintain an action for trover against a third person for taking bees and honey from the box (*Rexroth v. Coon,* 15 R. I. 35), and I am unable to distinguish this case in principle from the one now before us. The plaintiff is a mere trespasser, setting traps upon the premises of the defendants, and he gains no property right in the catch of these traps, and the plaintiff, having operated outside of his franchise rights, is not in a position to complain because the defendants, rather than some other person, have benefited by reason of his aid to the producing capacity of the lands in question. It has been held by high authority in England that game found and killed by a trespasser under such circumstances that it would be the absolute property of the owner of the soil if it had been found and killed by such owner instead of by the trespasser, does in law become the absolute property of the proprietor of the soil or privilege immediately on its being so caught and killed by the trespasser (*Blades v. Higgs,* 11 H. L. Cas. 621; 34 L. J. C. P. 286; 11 Jur. [ N. S.] pt. 1, 701; 12 L. T. [ N. S.] 615; *Sutton v. Moody, 1 Ld. Raym.* 250), and it is difficult to suggest a distinction in principle between oysters caught in a trap upon the premises of the defendant, and game killed upon the premises of the owner of the soil or of the privilege.

It should be the purpose of the court, not so much to do abstract justice in a particular case, as to establish and make certain the law and the public policy of the State, and in the present instance I am of opinion that both purposes will be served by a reversal of the judgment. The plaintiff, by a disregard of duty in ascertaining and marking the limits of his own franchise, has trespassed upon the franchise of another, and the mere fact that this trespass has resulted in improving the value of the defendants' franchise does not justify the trespass, nor give the plaintiff any right of property in the catch of oysters which has resulted. The plaintiff had rights within the limits of his own franchise which it was the duty of the State to protect, but when he disregarded his duty and became a mere trespasser the law owed him no further obligation, and the courts should now refuse to grant him a remedy against those who have merely exercised their rights under their franchise and have gathered the oysters which have been developed within the limits fixed by the grant. To do otherwise is to defeat the policy of the State, which places upon such holder of a franchise the duty of knowing and designating the limits of his franchise, and makes each case depend upon the evidence of good faith on the part of the trespassers, when, as I have pointed out, there can be no such good faith, because it constitutes a violation of a duty imposed by law

The judgment appealed from should be reversed.

**Author's Note:** The dissenting opinion in this case presents an interesting position. It is contrary to the basic decision, for different reasons, both of which are based on the fundamental concepts of property ownership, separated by the differences between real and personal property. Because of this discussion and reasoning, it is a case having value of study.

Further down the Atlantic coast, in South Carolina, the court in 1901 issued a decision in the case of *Alston et al. v. Limehouse, et al.*[86] As in several cases from other jurisdictions, the history of the title back to its origin was presented to examine the rights emanating from the original title holder, in this case King George II. It is a proper demonstration of how to understand the extent of the rights inherent, or attached to, the creating title document. *Because the case has more to do with procedure and less to do with the subject matter of this section, only the relevant portion of the decision is included here.*

The appeals herein are from orders of his honor, Judge Gage, granting a temporary injunction, and from and order of his honor, Judge Gary, referring it to the master to take the testimony in the above entitled cause, and report the same to the court. As the questions presented by the exceptions are largely dependent upon the pleadings, it is necessary to set out the complaint, which is as follows: "First. The plaintiffs are seised in fee simple, and are in actual possession of a tract of land in Georgetown county, in the state of South Carolina, on Waccamaw Neck, in what was the old parish of All Saints, Waccamaw, containing about 2,000 acres, butting and bounding north on land of Ward, east on the Atlantic Ocean, south on lands of Donaldson, and west on the Waccamaw river. Second. That the said tract of land includes a large areas (about 600 acres) of salt marsh, more or less subject to the daily flux and reflux of the tide, and lying between the eastern part of said tract, commonly called 'Dubordieu Island,' and the highland of the rest of the tract, which said marsh area is interspersed and intersected by runnels or small creeks and natural drainways, wherethrough the water daily brought in by the tide returns to the ocean. Third. That the said salt marsh is the breeding place and habitat of clams and other shellfish, and is also resorted to by wild ducks and other birds, and said creeks and runnels are in places the site of oyster beds and banks, and also the resort of fish. That in and under the original grant from which the title of these plaintiffs is derived, all and singular the entire marsh lands and beds of the creeks were granted and included by direct metes and bounds of the grant; the same running back from the Atlantic Ocean to the Waccamaw river, and including all the marshes and creeks between the two. That, in addition thereto, the grant gave in express terms to the grantees the exclusive and sole right and privilege of hunting, fowling, and fishing within the limits of the said grant. Fourth. That the said land has been owned and in the exclusive possession of the plaintiffs and their ancestors for more than a century, and of the parties through whom they claim since the date of the grant, in 1733. That for more than a century the plaintiffs or their ancestors have exercised and maintained the exclusive possession and dominion over all the said marshes,

---

[86] 60 S.C. 559, 39 S.E. 188 (S.C. 1901).

creeks, and drainways, according to the nature of the property, and their possession has never been before contested. Fifth. That now so it is that certain parties have undertaken to trespass and invade upon plaintiffs' said marsh and creeks, and to take and remove the clams and other shellfish from the beds of the creeks, and also fish and seine and remove the fish from said creeks, and, in addition, to habitually trespass upon, shoot, frighten, and scare off the game upon the said described property. Sixth. That the said marshes are dependent for their value in great measure for their use for the purpose of maintaining and preserving the game, and for maintaining and preserving the oyster beds and banks and claims and other shellfish therein and by fishing in said creeks, and of the continuous trespass thereon by parties, destroys the value thereof. Seventh. That certain parties, to wit, the parties above named as defendants herein, J. F. Limehouse, Jonas Happy, Abner Leonard, Sim Leonard, Oliver Sellers, and A. M. Hills, whose names are as above given, have, as plaintiffs are informed and believe, been the parties who have been engaged in the said trespass and invasion and depredation. That the said parties are, as plaintiffs are informed and believe, and so allege and charge, without financial ability to meet any judgment or execution at law, and the same, if even brought for damages against them, would be valueless, and that proceedings at law would necessitate continuous and incessant and a numerous multiplicity of suits against each successive trespasser for each successive trespass. Wherefore plaintiff's pray judgment that the said parties be permanently enjoined from in any wise trespassing upon said property of plaintiffs, and that in the meantime, until the hearing of the case on the merits, a temporary injunction do issue from this honorable court, restraining and enjoining them from such trespass." The answer of the defendants to the foregoing complaint in substance denies all the material allegations thereof.

On the hearing of the case in this court, a preliminary question was raised that the court was without jurisdiction, as the said orders were not appealable. We will first consider whether the orders of his honor, Judge Gage, were appealable. Those orders are as follows:

"These are two actions for injunctions. They were heard together, are dependent on the same fact, and I will make one order, to stand as the order for each case, just as if separately entitled therefor. The motions before me are for a continuance of the temporary injunctions heretofore granted until the issues raised by the pleadings have been tried. The motions were heard on complaint and answer, and affidavits submitted by both sides. The arguments was elaborate and helpful. The cause is very interesting. The plaintiffs claim title to several thousand acres of land in Georgetown county, stretching from Waccamaw river on the west to the Atlantic Ocean on the east. Within the description of the land are certain water ways, the habitat of clams, oysters, fish, ducks, and birds. The defendants are fishermen, follow that craft for a livelihood, and have been accustomed to catch shellfish in the said water ways, and they do so under claim of right, to wit, because the streams are navigable. The plaintiffs vest their title in the beginning on letters patent from Charles II. to the eight lords proprietors, 24th March, 1663; a grant from the rest (seven) of the lords to one of them, John Lord Cateret, 5th December, 1718; a deed of lease and release from Cateret to John Roberts, 18th and 19th February, 1730; a grant from George II. to John Roberts, 13th September, 1736. In the grant last named there is this language, to wit: 'Together with all woods, underwoods, timbers, timber trees, lightwood pitchings, lakes, ponds, fishing waters, water courses, pastures, feedings, marshes, swamps, ways, easements, profits, commodities, advantages, emoluments, hereditaments, and appurtenances, *** together with the privileges of hunting, hawking, fishing, and fowling in and upon the same, etc.' The contention of the plaintiffs is this: That the absolute title to the entire area embraced within the boundaries of their grant is in them, except the beds of such streams as are shown to possess the capability of floating useful commerce, and in those streams they own to low-water mark absolutely, and below that they have the exclusive right of fishing. The contention of the defendants is this: That the streams in which they have taken oysters are navigable, that in such streams the plaintiffs have no exclusive title below highwater mark, and the plaintiffs are not entitled under the terms of their paper title to the exclusive right to make oysters. I think there is not question but that the alleged place of trespass lies with the boundaries of the grant, and there is not question but that plaintiff's have proved a good paper title to the soil, above water, within that area. I do not understand the defendants to contend, by answer, proof, or argument, that they have acquired a right by prescription to take oysters from the lands and waters in question. The real contest before me was about these issues, to wit: (1) At what particular place did defendants take oysters? (2) Were the waters at those places navigable? (3) Do the plaintiffs' exclusive rights stop at low or high water on navigable streams? (4) What exclusive right of fishing did the words of the aforementioned grant convey to John Roberts, and were the same rights transferred to the plaintiffs?

"The first two are questions of fact; the last two are questions of law. The verified complaints do not specify in what water way the oysters were taken, or at what spot on the plaintiffs' lands the trespasses were committed, but alleges generally a trespass thereon. At what place the trespasses was done is matter of proof. The verified answers deny entry at every place except in navigable tide water streams and creeks, and the defendants' affidavits specify the creeks to be those called Jones, Town, Old Man's, and Dubordieu. Therefore, the only issue now is the right of defendants to take oysters in these four creeks. The testimony tending to prove these four water ways navigable is scant. Only two of the witnesses swear to facts. They are Cain and Munnerlyn. If the issues were contested, I should not feel warranted in finding that they were water ways of sufficient depth and width to float useful commerce. But the plaintiffs offer no testimony contra, so I assume the streams above named are navigable. It is not contended by the plaintiffs that the aforementioned grant undertook to convey to the grantee the soil underneath a navigable stream. But the plaintiffs do contend that the grantees took exclusive title to the soil down to low-water mark; and that the defendants deny. That issue is hardly relevant now, for the proof shows the defendants gathered oysters in the natural growth lands in the beds of the four navigable streams. It does not show that they gathered oysters on that area which lies betwixt high and low tide. Nevertheless, the question has been made, and I shall not shun it. The fact is, on the Georgetown coast flood tide covers a vast area of land, which at low tide is exposed to view. For convenience, I shall term that area 'marsh lands.' In this country the tides have no relevancy to navigability. It was otherwise in England, whence the common law and its terminology came. There tide waters and navigable waters were convertible terms. Here, if a water course is navigable, it is so because the depth and width of it are sufficient to float useful commerce. If the depth and width of a stream are augmented by a periodical increase of water, called 'tide,' that fact may make the stream navigable at those points in it where it is so in fact, to wit, in its channel, but not navigable where it is not so in fact, to wit, out of the channel in the marshes. The state owns (because it has refused to sell) the beds of navigable streams, not because they are covered with water, tide, or otherwise, and not because inhabited by fish, but because they are ways, convenient to float useful commerce. There is no greater reason why it should preserve for the public the fish in navigable than in nonnavigable streams, or the fish in water than game on the land. 'By the common law, the doctrine of the dominion over and ownership by the crown of lands within the realm under tide waters is not founded upon the existence of the tide over the lands, but upon the fact that the waters are navigable; "tide waters" and "navigable waters," as already stated, being used as synonymous terms in *England*.' *Illinois Cent. R. Co. v. Illinois*, 146 U.S. 387, 13 S.Ct. 110, 36 L.Ed. 1018. The reasonable conclusion, therefore, is that the marsh lands within the plaintiffs' boundaries belong absolutely to the plaintiffs.

"This brings us to the next inquiry, and that is the construction of the words in the grant from George II. to John Roberts, hereinbefore quoted. The language purports to convey, in brief, fishing waters, water courses, feedings, marshes, swamps, easements, profits, advantages, emoluments; together with the privileges of hunting and hawking and fishing. If the king had the powers to make the grant, the plaintiffs' predecessors took the rights for which the plaintiffs now contend. It makes little difference that the subsequent conveyors did not, in ipsissima verba, convey the aforementioned rights when they made deeds to the lands. The rights were first conveyed with the lands, and a subsequent conveyance of the land carried the rights are well. I have not examined if the king had the power to convey the rights and privileges he undertook to convey to John Roberts on 15th September, 1736. I assume that he did. Of late years the general assembly has undertaken to regulate oyster fishing in the 'public waters' of the state. 20 St. at Large, p. 1097. That statute devolves on the sinking-fund commission the power and duty to make a survey of oyster lands in public waters, and to lease the same in perpetuity. There is a provision in the statute which exempts from its operation those lands occupied by persons under existing laws, and held under grants issued under the laws of the state. The agent of the sinking-fund commission swears that the waters in dispute are, in effect, such waters. If it is true, the terms of the statute and the affidavit of Mr. Gibbes are not conclusive of the question here, but they are relevant to the issue.

"Finally, are the plaintiffs entitled to injunctions until legal issues made by the pleadings have been tried before the proper tribunals? Issue has been joined, and the action is ripe for trial. The rule is, the party asking for an interlocutory injunction must show (1) a clear legal right, and (2) Well grounded apprehension of immediate danger thereto. I have found the legal right. The affidavits of the defendants show likelihood of immediate injury thereto, if not restrained. It is therefore ordered that the injunctions heretofore granted to be continued with like force and effect until the further order of the court. Geo. W. Gage, Circuit Judge. Chester, S. C., 13th July, 1900."

"It having been brought to the attention of the court that no restraining order has heretofore been issued by this court, but the same was heard upon a rule to show cause why a temporary injunction should not be issued, and, therefore, that there was no injunction pending at the date of the filing of the order hereinbefore made to be continued: Now, in order to carry out the decision herein filed, it is ordered that the defendants, Mitchell Nesbit, Faith Johnson, Max Sindab, Cain Rutledge, James Greer, Saul Car, Sam Car, and D. H. Smith, and each of them, the servants, agents, employees and attorneys of them, and each of them, be, and they are hereby, enjoined and restrained from in any wise hunting, fishing, fowling, or otherwise trespassing upon the lands and premises of the plaintiffs described in the complaint herein, until this case be heard and decided on the merits, or until further order of the court. Geo. W. Gage, Circuit Judge. 28th July, 1900."

On a motion for a temporary order of injunction, the circuit judge, in considering the issues raised by the pleadings, should indicate that their consideration is solely for the purpose of determining whether the plaintiff has a prima facie right to an order of injunction. His order should not purport to dispose of the issues upon the merits, as was done in this case. The language of Judge Gage cannot be construed as a finding upon the facts in such a manner as to affect the merits of the case. It must be regarded as used for the purpose of showing that he was justified in granting the temporary order of injunction, and not as in any manner affecting the other question in issue. No fact decided upon such motion is concluded thereby, and when the other issues are brought to trial they are to be determined without reference to said orders. In the case of *South Carolina & G. R. Co. v. East Shore Terminal Co.,* 48 S.C. 315, 26 S.E. 613, the court says: "The order was necessarily made without prejudice to the rights of the parties upon the final hearing of the case; as much so, as if the words 'without prejudice,' etc., had been inserted in the orders. The circuit judge did not have the power, on the hearing of said motion, even if he had so desired, to decide the case upon its merits. The effect of said order was the same as if the circuit judge had stated in the order that it was only to remain in force until a decision could be made upon the merits" – citing *Garlington v. Copeland,* 25 S.C. 41, and *Sease v. Dobson,* 34 S.C. 345, 13 S.E. 530. Ordinarily, an order granting a temporary injunction is not appealable, and under the foregoing construction of said orders we see no reason why this case should not fall within the operation of the general rule. Having reached the conclusion that the said orders are not appealable, the other exceptions to these orders will not be considered.

We will next consider whether the order of Judge Gary was appealable. That order is as follows: "A motion has been made in both of these causes for an order of reference; and, after hearing counsel, and it appearing that the applications in both cases depend on the same questions, and have been heard together, and it being agreed that one order shall be made to be entered in both causes: Now, therefore, it is ordered that it be referred to F. L. Wilcox, Esq., to take the testimony in both of the above-entitled causes, and report the same to the court." Both equitable and legal issues were raised by the pleadings. The equitable issues were triable by the court in the exercise of its chancery powers, while the legal issues were triable by jury. The right to a perpetual injunction was dependent upon the result of the legal issues. The parties, therefore, had the rights to demand that the legal issues first be tried. One of the incidents of this mode of trial is the right to have the testimony taken before the jury, and not by a referee. It has been suggested that the circuit judge had a right to order a referee to take the testimony preparatory to a trial of the equitable issues. The answer to this is that there is no such limitations in the order. On the contrary, it is in general terms to take all the testimony in the case; not only that pertaining to the equitable issues, but also that pertaining to the legal issues. This latter could not be done without depriving the defendants of a substantial right, and this rendered the order appealable. Besides, the order is premature, and may become wholly unnecessary; for, as has been said, if the determination of the legal issues be in favor of the plaintiffs, then, in all probability, no further testimony will be needed to dispose of the equitable issues. If, on the other hand, the determination of the legal issues be in favor of the defendants, that will effectually dispose of the equitable issues. This order was, therefore, not only appealable, but, for the foregoing reasons, erroneous. It is the judgment of this court that the appeals from the orders of Judge Gage be dismissed, and that the order of Judge Gary be reversed.

Another case of whether an oyster bed was natural or planted, the case of *Jones v. Oemler*[87] stresses not only the governing statute, but also the use of the government chart depicting locations. Once again, survey and location data play an important role identifying boundaries, the extent and definition of property rights.

1. The state of Georgia, as owner of the beds of all tide waters within its jurisdiction, has absolute power to sell or lease such beds, or any portion thereof, to any of its citizens, upon any terms or conditions which its legislature may prescribe.

2. An attack upon the constitutionality of an act of the general assembly, based on the grounds that "it contains matter different from what is expressed in the title thereof, and refers to more than one subject-matter," without further specification, is too vague and indefinite for consideration by a court. The same is true of an attack upon an act alleging in general terms that the same "is invalid and unconstitutional."

3. Where an act which, among other things, prescribed "the method of lease of public domain within the state of Georgia for oyster planting, propagation, and cultivation," provided that a lessee should have "no authority to sublet or to assign his lease until after the expiration of five years from the date of his entry thereunder," and the act in question was subsequently amended by striking out the provision just referred to, it would then be the right of any lessee to assign his lease as soon as he acquired the same; and this is true although a provision in the original act was left of force, which declared "that, in the event the said lessee shall fail to comply with the requirements of [the law] as to the cultivation of said territory, he shall forfeit so much of said territory as has not been cultivated as hereinbefore required." In case of an assignment, the duty of complying with the provisions of the law would devolve upon the assignee, and in case of his

---

[87] 110 Ga. 202, 35 S.E. 375 (Ga. 1900).

failure to meet the legal requirements his right to the lease would be forfeited. But the right to proceed in courts to have such forfeiture declared vests in the state alone.

4. When a statute expressly declares that a specified chart made and published by the United States geodetic survey shall be conclusive evidence of the location of natural oyster beds upon the coast of Georgia, and the state, in pursuance of such a statute, makes contracts with its citizens, whereby it leases to them territory for oyster propagation, which, according to such chart, embraces no natural oyster-beds, such contracts are absolutely binding, and their validity cannot be affected by the passage of a subsequent statute amending the former by striking therefrom the word "conclusive," or by parol evidence showing or tending to show that the territory in question did in fact embrace natural oyster beds.

5. That unlawfully taking oysters from private beds held under laws from the state may be indictable and punishable under the criminal laws does not prevent the owners of such beds from enjoining insolvent persons from committing depredations thereon, when it is apparent that without such remedy the damage will be irreparable.

6. Under the evidence disclosed by the record, and in view of the law applicable thereto, there was no error in granting the injunction.

Augustus Oemler brought suit in Chatham superior court against William Jones and others, about 40 in number, for damages growing out of an alleged trespass by defendants on certain territory covered by Oyster Creek and Shad river, in the county of Chatham, which petitioner alleges he holds by virtue of certain leases made by the state of Georgia; and also for the purpose of enjoining defendants from further trespass upon the property. The petition alleged that these leases were made under several acts of the legislature, including the act approved August 22, 1891, which acts were made a part of said leases, and conveyed to petitioner the exclusive privilege of bedding or planting oysters on the territory in question; that these leases were duly accepted, and large sums of money expended in planting oysters and cultivating the same, amounting to $10,000. The lease contracts were made in the year 1892, and petitioner alleges that the lessees from the state, including petitioner, have for over seven years had possession, and planted and cultivated these lands without molestation by any one, except occasional parties who had infringed the criminal laws in order to procure oysters at these points. The effect of the cultivation and planting was to increase very largely the number of oysters in such territory; and that the parties who failed to take out leases, or, having taken out leases, failed to cultivate them had procured oysters wherever they could find them, and have so depleted the oyster beds along the coast that the sources of supply have been reduced and limited practically to private beds planted by private parties. This condition of affairs made it necessary for parties who had no such territory to take desperate measures in order to procure oysters for consumption and sale, and hence it was they would trespass from time to time on private beds; but, finding the number secured in this way too small for their purpose, a combination was formed, under color of the laws of the state leaving certain territory open to the citizens of the state. To this end they combined and confederated together, and filed a bill against petitioner and others on September 25, 1899, asking for injunction and relief, and secured an order restraining petitioner from unlawfully interfering with the gathering by such conspirators of oysters in the natural beds in the public streams or upon the banks of public marshes in Chatham county. The parties exercised the greatest haste, after procuring such a restraining order, in getting oysters from petitioner's leased beds, and were engaged night and day in depleting these beds. They will continue these depredations, and petitioner would be unable to protect his rights accorded him by the state, and would be compelled to institute many prosecutions and civil suits to maintain his rights, and possibly forced to use violent measures to protect his rights, his property, and his person, unless a court of equity afforded some immediate and substantial relief. Petitioner further alleged that the defendants had already taken over 4,000 bushels of oysters from his beds; that they were thoroughly and hopelessly insolvent, and had already damaged his property in the sum of $2,000, and would inflict infinitely more damage if permitted to continue their depredations. Petitioner prayed, among other things, that an injunction issue restraining the defendants from coming upon and into said territory, taking oysters therefrom, or in any manner interfering with said territory leased by your petitioner. Judgment was also prayed for the sum of $5,000 damages inflicted by the defendants, and there was a prayer for general relief. Upon this petition a temporary restraining order was granted, and at the June term, 1899, to wit, November 11, 1899, a hearing was had on said application. To the petition a demurrer was filed upon several grounds of a special nature to the effect that the petition was insufficient in law, in that it did not set up sufficient facts to show that the oyster leases, if they ever existed, were still in force; it did not show that the lands covered by the waters of Oyster Creek and Shad river did not contain natural oyster-beds, and did not show any lawful transfer of the leases covering said grounds, etc. This demurrer was met by an amendment to the petition, which amendment set forth particularly the acts of the legislature under which petitioner was claiming a lease to the land; that these leases were duly executed and delivered by the commissioners of Chatham county, in accordance with the act; that the leases made to Kayton and others were by them duly transferred to the Oemler Oyster Company, a corporation under the laws of this state; that this company executed to Karow and Gordon a deed of trust covering all its property; that the company and these trustees conveyed to the Merchants' National Bank and the Savannah Bank & Trust Company all of the property of the Oemler Oyster Company, including the territory in said county leased from the state in Oyster creek and Shad river, and the oysters planted thereon; and on May 1, 1899, the Merchants' National Bank and the Savannah Bank & Trust Company conveyed to petitioner, Augustus Oemler, by deed, the property in question held under the oyster leases. Petitioner further alleges that more than 100 bushels of dead shells per acre were deposited upon each of the five-acre tracts that were originally leased within the first year; and, in addition, there have been planted over 100 bushels of oysters to every acre of planting ground at the rate of one acre or more each year until each lot of five acres was planted. The amendment further alleges in detail that the conditions of the leases have been strictly conformed to by him and his predecessors in title. To the petition defendants filed an answer, alleging, in substance, that the plaintiff had no rights to the oyster beds in question; that all the beds in Oyster Creek and Shad river which are now there were there at the time of taking out the oyster leases, and have since remained, and still are, natural oyster beds, and as such free and open to all citizens of the state. They further denied that the leases were taken out under any act of the legislature, because the acts especially excepted from their operation the natural oyster beds within the waters of this state. In short, the answer contained a specific denial of plaintiff's right to the oyster beds in question, and asserted the right of defendants, in common with the public generally, to get oysters from the beds. After hearing the evidence, his honor, Robert Falligant, judge of the court below, on December 2, 1899, filed his judgment granting a temporary injunction against the defendants, as prayed for. To this judgment the plaintiffs in error except, and assign error thereon in their bill of exceptions.

1. There can be no question but that a state owns the beds of all tide waters within its jurisdiction. It has as absolute jurisdiction and control over such territory as it has over any other property it may own. It has the same power to grant, sell, or lease such beds, or any portion thereof, to any of its citizens, upon any terms or conditions which its legislature might prescribe, to the same extent that it would have the right to dispose of its wild or other lands which it owns. This fundamental principle, inherent in a sovereign power, we may say has been universally recognized by the learned law writers and the courts, state and federal, of this country. We deem it, in this connection, important only to cite a few of these numerous authorities on the subject. In the case of *McCready v. Virginia*, 94 U.S. 391, 24 L.Ed. 248, this doctrine of the ownership by a state, and its power over the beds of its tide waters, is clearly recognized. Mr. Chief Justice Waite, in

delivering the opinion in that case, on page 396, 94 U.S., and page 249, 24 L.Ed., says: "The planting of oysters in the soil covered by water owned in common by the people of the state is not different in principle from that of planting corn upon dry land held in the same way. Both are for the purposes of cultivation and profit; and if the state, in regulation of its public domain, can grant to its own citizens the exclusive use of dry lands, we see no reason why it may not do the same thing in respect to such as are covered by water. And as all concede that a state may grant to one of its citizens the exclusive use of a part of the common property, the conclusion would seem to follow that it might, by appropriate legislation, confine the use of the whole to its own people alone." The case of *Alabama v. Harrub (Ala.)* 10 So. 752, 15 L.R.A. 761, was a prosecution under a statute prohibiting the shipment out of the state of oysters taken in the public waters of the state, and also prohibited the taking of such oysters by any person who was not a resident of the state. It was held that the act was not unconstitutional as a regulation of interstate commerce. Coleman, J., delivering the opinion (on page 753, 10 South., and pages 763, 764, 15 L.R.A.), makes this observation: "We think it further settled that the people of Alabama, through its legislature, alone have the power to dispose of their property rights in their oyster beds and oysters, and, if they see proper, may dispose of them to their own people only. It is further settled that the legislature has ample authority to adopt all precautions and regulations deemed desirable or necessary for the preservation and increased production of its fisheries." It appears that in the state of Connecticut there are statutes regulating the designation of oyster beds to any person for oyster culture. Under a statute of 1875, selectmen of East Haven could designate for oyster culture ground in a certain part of the navigable waters within the town limits. East Haven, at its town meeting, appointed a committee to designate the grounds within its limits, and the committee assumed to designate lands not within the part specified in section 1 of the act. In 1877 a statute was enacted validating and confirming all designations made by authority of the town through its selectmen or oyster-ground committee. In the case of *State v. Bassett (Conn.)* 29 A. 471, the supreme court of Connecticut decided that the section under the latter act healed not only the errors of the committee, but the defects in their appointment. That case was a prosecution of one for trespassing upon certain territory which had been designated and allotted to a Mrs. Foote, of the town of East Haven, for the purpose of planting oysters thereon; and in the designation the committee made a mistake, as above indicated. The territory came by regular conveyances to the purchaser from the original grantee, who was in possession when the alleged trespass was made. One of the grounds of defense was that the designation was void, as originally made, because the place was a natural clam and oyster ground, which seems to have been excepted from being designated by the statute. The action for trespass was sustained upon the theory that the error was remedied by subsequent legislation. It was in the exercise of just such powers that the state of Georgia, through its legislature, in 1889, passed a law repealing various acts of the state with relation to oysters within its domain, and adopted a new system in reference to this matter. See Acts 1889, p. 143. The title of that act recited the various acts which it repealed, "and in lien and place thereof substituting an act providing in what manner, at what seasons, and for what purposes oysters may be caught in the state of Georgia; the method of lease of public domain within the state of Georgia, for oyster planting, propagation and cultivation; the revenue to be paid therefor; the penalties for violations of this act, and for other purposes therein mentioned." Under the provisions of section 7 of this act, the county commissioners of each of the counties named, which includes Chatham county, or, where there is no board of county commissioners, the ordinary of such county, upon the application of any person for certain territory in any of the navigable waters of this state, and within a distance of 1,000 feet from the shore at ordinary mean low tide, upon satisfactory proof on hearing had before the county commissioners or ordinary that such territory has been duly staked off at the line of ordinary mean low tide for a period of 30 days before the hearing of such application, shall execute a lease for 20 years, with a privilege of renewal for 30 years more, to such applicant as may first apply for such territory not already appropriated, where there are no natural public beds which have, prior to the applications, been resorted to by the public, for the purpose of procuring oysters with the use of tongs for consumption or sale. It was under the provisions of this act, as amended by the act approved September 22, 1891 (1 Acts 1890-91, p. 214), to which particular reference will be hereinafter made, that leases during the year 1892 were duly granted by the proper authorities of Chatham county to divers parties. Under these leases the petitioner asserts his rights in this case.

2. Upon the hearing, one ground of defense made in behalf of the plaintiffs in error was that the act of 1889, above cited, is unconstitutional, because "it contains matter different from what is expressed in the title thereof, and refers to more than one subject-matter." Nothing appears in the record specifying or intimating in what particular the act contains any matter different from what is expressed in its title, nor in what particular it refers to more than one subject-matter. Section 5527 of the Civil Code requires that the bill of exceptions shall specify plainly the decision complained of. Under this rule of law, the bill of exceptions is manifestly too vague and indefinite for any consideration by this court of the constitutional question it undertakes to raise. Not even in the argument of the case was any defect in the act pointed out upon which it was claimed that it was unconstitutional for the reasons stated. We will add, in this connection, upon reading the title of the act itself, our conclusion is that it refers to but one subject-matter within the meaning of the constitution; and that it contains nothing different from what is covered by the very voluminous title thereto.

Another ground of defense alleged in the bill of exceptions is as follows: "That section 6 of the act of 1891 is invalid, and unconstitutional;" and, further: "That contracts based upon said section 6 of the Acts of 1890-91, so far as the same makes Drake's chart conclusive evidence of the existence or nonexistence of oyster beds, are void, and do not vest any rights in the lessee." There is no specification whatever of what particular provision in the constitution the section in question violates. No light has been given upon this subject, not even in the argument; nor can we conceive what constitutional requirement has been violated by the provisions of that section. For the reason above indicated, this court cannot consider the vague and indefinite ground of complaint.

3. The evidence introduced upon the trial fully sustains the allegations of the position touching the leases in question. After the passage of the act of September 22, 1891 (Acts 1890–91, p. 214), the record shows that applications were made by several parties to the proper authorities in Chatham county for leases, which were made to these parties by such authorities, after proceedings were had in conformity to the act in question. These beds thus leased were located in Oyster Creek and Shad river, in Chatham county, and leases of them were made to various parties in January and February, 1892; and the lessees, about May 22, 1892, transferred the leases to the Oemler Oyster Company. On June 30, 1892, this company conveyed its property to Karow and Gordon, as trustees, and on August 4, 1893, the Oemler Oyster Company and these trustees conveyed the territory to the Merchants' National Bank and the Savannah Bank & Trust Company. On May 1, 1899, these banks conveyed the territory in question to August Oemler, the defendant in error in this case. Under section 8 of the act of 1889 (page 146), it is provided that a lessee should have no authority to sublet or to assign his lease until after the expiration of five years from the date of his entry thereunder; and it was further provided that, in the event the lessee should fail to comply with the requirements of this section as to the cultivation of said territory, he shall forfeit so much of said territory as had not been cultivated as hereinbefore required; and if the lessee shall at any time during the term of his lease abandon said territory, and cease to cultivate oysters thereon, for the space of one year, the lease shall be void, and the territory shall revert to the state. The evidence is uncontradicted that within the time and according to the terms prescribed by the act the lessees or their assigns had fully complied with the conditions named in this section of the act, and in fact had deposited the requisite bushels of shells and oysters, not

only on every acre of planting ground, but upon the entire ground, within a year from the granting of the lease. It is contended by counsel for plaintiffs in error that under the provisions of that
section of the act, it was the duty of the lessee to comply with these conditions before he had any
power of transferring the property acquired by the lease. This provision of the act of 1889 prohibiting the original lessee from subletting or assigning his lease until after the expiration of five
years from the date of his entry was expressly repealed by section 3 of the act of 1891 by striking
out the words, "provided further, that said lessee shall have no authority to sublet or assign his
lease until after the expiration of five years from the date of his entry thereunder." Section 8 of
the act of 1891 provides that nothing in the act "shall be construed to affect the titles of the lessees
of oyster territory which has heretofore been leased by county commissioners or ordinaries, and
all leases executed by them before the passage of this act, or any assignments which have been
made of the leases of five (5) acre tracts are hereby confirmed and validated." By the repeal,
therefore, of the restriction upon leases embodied in the act of 1889, it is quite manifest that the
legislature intended the original lessees should have the power to sublet the property they have
acquired at any time they saw proper; and this provision would necessarily apply to all transactions of this nature, where the original leases, as in this case, were made after the passage of this
amendatory statute. The act itself goes further in the last section cited by ratifying and confirming transfers made by original lessees even prior to the passage of the act, which the legislature
clearly had a right to do. There is nothing in the record to indicate that any of the original lessees
in this case had not done everything required of them by law, including the payment of the purchase money for their property to the state, up to the time they sublet or conveyed what they had
acquired or were in possession of other parties. Anything else that remained to be done necessarily devolved upon the successors in title; and, if these successors failed to comply with the conditions prescribed by law that would work a forfeiture, their right to the leases would terminate. As
above stated, the evidence shows there has been no such failure on the part of the assignees or the
petitioner in this case. There can be no question about the general principle of law that the right
of disposal of any interest which one has acquired is an incident annexed to all property, and this
right of alienation necessarily exists although the title may be of an inchoate nature, unless there
is some provision in the instrument or contract creating the title that especially and specifically
prohibits such alienation. In the case of *Winter v. Jones,* 10 Ga. 191 (13), it was decided: "A purchaser of land from the state, by the act of entry and payment, acquires an inchoate legal title,
which may descend, be aliened or devested, in the same manner as any other legal title." Lumpkin,
J., delivering the opinion, says: "The act of purchase transfers an immediate right of possession,
and the title thus acquired is sufficient to enable the purchaser to arrest an intruder by due course
of law; and the certificate which was required to be given until the patent could issue, where the
money was all paid, was evidence that the purchase was complete; and, by act of entry and payment of the purchase money, the purchaser of land from the United States (and, I add, from each
of the individual states) acquires an inchoate legal title, which would descend and might be
aliened or sold under execution, as any other legal title." We therefore conclude that there is absolutely nothing in this record to authorize the inference that there has been any forfeiture whatever
of these leases which were granted by the state. But, suppose there had been, who has the power
to institute proceedings to declare such a forfeiture? We think manifestly in this case no one but
the state itself. In the case of *Mortgage Co. v. Tennille,* 87 Ga. 28, 29, 13 S.E. 158, 12 L.R.A. 529,
it was decided that "under the act of February 28, 1877, providing that the state of Georgia will
not consent to foreign corporations owning 5,000 or more acres of land in this state unless they
shall become incorporated under the laws of Georgia, the state alone can make the question as to
the right of such corporations to hold said lands." See the able opinion of Justice Lumpkin (now
presiding justice) in that case, and authorities cited. In the case of *Norris v. Milner,* 20 Ga. 563, it
was decided that a condition does not defeat the estate, although it be broken, until entry by the

grantor or his heirs; and that a stranger cannot take advantage of the breach of a condition in a deed. In the case of Edmondson v. Leach, 56 Ga. 461, it was decided: "An estate forfeited by breach of condition subsequent is not revested in the grantor until after entry or action brought by him or his heirs." In the case of *Van Wyck v. Knevals*, 106 U.S. 360, 1 S.Ct. 336, 27 L.Ed. 201, the question arose with reference to the forfeiture of a grant by the United States to a company in consequence of the failure of the company to comply with its obligations under the contract. It was there decided that such a forfeiture could be enforced only by the United States through judicial proceedings, or the action of congress, and that a third party cannot set it up to validate his title. See, also, *Land Co. v. Griffey*, 143 U.S. 32, 12 S.Ct. 362, 36 L.Ed. 64. Under these principles, then, we think it is quite manifest that even if there had been any failure on the part of petitioner in this case, or of any one of his predecessors in title under whom he claims, to comply with conditions or obligations which might work a forfeiture of the lease, only the state, acting in its sovereign capacity, as the representative of the entire people, could institute proceedings to bring about such a result and to reclaim its lands. Certainly, where such holders of an estate for years are in the quiet and peaceable possession of it under grant from the state, by the state's permission and consent, and after the state had accepted compensation therefor, and expended it for the public good (in this case the fund paid for the leases was appropriated for the purposes of public education), individuals could not trespass upon such property, and, in a suit, brought against them by one thus holding under the state to prevent such trespass, could not defend by alleging that the plaintiff had forfeited his lease to the state.

4. By virtue of section 16 of the above-cited act of 1889, provision is made that the rights of any citizen of this state to enter upon and take from any public beds oysters, by the use of such implements as may have been heretofore in general use in this state, shall not be abridged or interfered with. One evident purpose of the act was to allow the public to use what were known as the "natural oyster beds," that they had been in the habit of using, and that the leases should not cover territory of this character. The defendants below claimed their right of entry upon the ground in question by virtue of such provision. We are not aware of any statute of the state, prior to the passage of the act of 1889 that specifically preserves such right; and in fact the defendants below rely upon that provision in the act of 1889, also recognized in the act of 1891. It is a little singular that they seek to have an act declared unconstitutional, by virtue of the terms of which they justify their conduct complained of by the petitioner in this case. It will be observed that this act of 1889 did not provide any plan or scheme by which natural oyster beds could be definitely distinguished from other territory covered by water, known as "vacant lands," which the state had provided for leasing. Consequently, great confusion seems to have originated, and great doubt and uncertainty necessarily existed, at times when leases were applied for on certain territory, as to whether or not there were located thereon the natural oyster beds referred to by the statute. Under the act as it stood, such questions necessarily were to be decided and passed upon by the county authorities granting the leases; and it would seem that their adjudication should be considered as final, especially as to any contest of the kind brought before them. In consequence of this confusion, uncertainty, and the frequent contests which were likely to arise touching the rights of one who had acquired privileges in certain territory for planting and cultivating oysters, and the rights of the public in such territory, after the passage of the act of 1889, the same legislature adopted a resolution to the effect that the governor be requested to ask the federal government to assist the state of Georgia in making a survey, as was done by the federal government for the state of North Carolina, and to detail one or more officers of the United States navy to make a physical examination of the waters of the state of Georgia, and prepare a chart showing in detail the productive and unproductive areas, as well as the natural oyster beds. See Acts 1889, p. 1429. A survey for Georgia was accordingly made in compliance with this resolution, which included the territory involved in this litigation. Plats of these surveys were made, and a report of the

result thereof was made by J. C. Drake, ensign in the United States navy, and assistant United States coast and geodetic surveyor. This report was introduced in evidence, and was known as "Bulletin No. 19." In it is a plat embracing the surveys about Savannah, in which was clearly indicated upon what portion of the territory were located natural oyster beds, and what portion was vacant; the latter, under the act of the legislature, being clearly subject to lease to private parties, but the former not. By virtue of the act of September 22, 1891 (1 Acts 1890–91, p. 214), embodied in section 1700 of the Political Code, it was provided: "The natural oyster beds of the state shall forever remain the property of this state, open to all her citizens for the procuring of oysters for consumption, sale, seed or propagating purposes; and for the better securing of this purpose, the charts made and published in consequence of a resolution passed by the legislature of this state, by United States geodetic survey known as 'Bulletin No. 19,' shall be conclusive evidence of the location of such natural oyster-beds and of vacant ground: provided, that wher-ever beds, shown by said 'bulletin No. 19' to be natural oyster-beds, shall as a matter of fact not extend below low-water mark, then the territory below low-water mark shall nevertheless be open to lease. Except as herein stated, it shall not be lawful for the county commissioners or ordinary to grant leases to any grounds shown on said 'bulletin No. 19' to contain a natural bed, and it shall be lawful for them to grant leases on any or all territory indicated on said 'bulletin No. 19' as vacant." The testimony in the record was uncontradicted that all the territory embraced by the leases involved in this contest was indicated on said bulletin No. 19, introduced in evidence, as being vacant land, no part of it containing natural oyster beds. We do not think there is any question about the power of the legislature to enact such legislation. Under the facts disclosed by this record, it was clearly a wise act to settle by legislation definitely in some way exactly what lands the county commissioners of Chatham county could lease, and exactly what they could not. A mere specification that vacant lands would be leased, and that natural oyster lands should not be, had given rise to confusion and uncertainty. The purpose of having these lands thoroughly surveyed by expert officials entirely competent to determine the distinction and the location of the two classes of territory mentioned was certainly a proper course to pursue under the circum-stances. The object of this survey, and the object of the legislature in afterwards adopting it as conclusive evidence touching the character of the territory embraced therein, were intended no more for the protection of the lessees of these lands than for the protection of the public as to their rights in these waters; and we think there can be no question about the right of the legisla-ture to make such evidence conclusive. This principle was decided in *White v. Petty (Conn.)* 18 A. 253, by the supreme court of Connecticut, where it was held: In a suit to restrain defendants from interfering with complainant's oyster grounds, the report of a committee appointed, under a certain law of Connecticut, to locate and describe all natural oyster beds in a certain town, which did not include grounds in the town that had been designated to complainant, was admissible to rebut evidence that the grounds were in 1886, and for 10 years before, natural oyster grounds, as the statute provides that such report, when accepted and recorded, shall be conclusive, as to the extent of natural oyster beds, at its date. The same principle is substantially announced in the case of *State v. Nash (Conn.)* 25 A. 451, where it appears that a certain statute designated certain locations of natural oyster beds. In a prosecution against a party for trespassing upon places that had been designated to private parties, it was held that the only evidence which could avail would be that the property in question was embraced in the locations and descriptions contained in the statute, and enumerated as natural oyster beds. This is exactly the effect of our statute in question; for it adopts the United States survey made of this territory, states it is conclusive evidence, and thereby, by special statute, recognizes certain specific and definitely located property as natu-ral oyster beds, and certain other property as vacant land subject to be leased. *State v. Bassett (Conn.)* 29 A. 471; *Rollins v. Wright,* 93 Cal. 395, 29 P. 58, deciding, in effect, that an act of the legislature providing that certain matters in a deed are conclusive evidence is valid and effective.

See, also, the case of *Abbott v. Lindenbower,* 42 Mo. 162, where it was held that an act making the recitals in certain tax deeds conclusive evidence was valid and constitutional. To the same effect, see *Marx v. Hanthorn (C. C.)* 30 F. 579; *Ensign v. Barse,* 107 N.Y. 329, 14 N.E. 400, 15 N.E. 401; *De Treville v. Smalls,* 98 U.S. 517, 25 L.Ed. 174.

It was contended, however, by plaintiffs in error that the provision in the act of 1891 making bulletin No. 19 conclusive evidence of the character of these lands is no longer of any force, for the reason that by the act of 1898 (Acts 1898, p. 48), the act of 1891 was amended by striking therefrom the word "conclusive." There is nothing in the act making this amendment to indicate that it was intended to be retroactive in its effects. When such is the case, the usual rule of construction is to give such acts a prospective effect only. We think, therefore, it was the intention of the legislature that, whatever effect striking the word "conclusive" may have been intended to have, it should apply only to leases or other contracts with reference to oyster beds made and entered into after the passage of the act, and was not intended to affect in any way leases made by the state to individuals before the passage of this amendment, and after the passage of the act of 1891. Were the intention otherwise, then we think the act would be obnoxious to that provision in the constitution, embodied in section 5730 of the Civil Code, in which it is declared that no ex post facto law, retroactive law, or law impairing the obligation of contracts or making irrevocable grants of special privileges or immunities, shall be passed. See, also, the provision in the constitution, embodied in section 5936 of the Civil Code, where it is declared, in effect, that all rights, privileges, etc., which may have accrued to any person or corporation in his or their own right, or in any fiduciary capacity, under and by virtue of any act of the general assembly, or any judgment, decree, or order, or other proceeding, of any court of competent jurisdiction in this state, heretofore rendered, shall be held inviolate by all courts before which they may be brought in question, unless attacked for fraud. It is true that acts purely of a remedial nature, and even acts amending the law upon the subject of evidence and the competency of witnesses, etc., may often constitutionally have a retroactive effect, but this rule is made applicable only in such cases where such legislation does not have the effect of impairing any substantial right which had accrued and vested in a person by virtue of a valid contract made before the passage of the act. The leases in this case were granted by the state several years before the passage of the act of 1898. They were founded upon the provisions of the act of 1891; and the parties are presumed to have taken them and to have spent their money on their faith in the guaranty of the state that the survey made by its direction should forever be conclusive evidence of the fact that the lands conveyed under the leases were vacant lands, within the meaning of the act.

The provision in the act of 1891 to which we refer necessarily constituted a part of the contract between the state and its lessees, and it could not have been more effectual or binding upon the state itself than it would have been had a grant been made with a plat attached, accurately locating the land, and definitely describing it by metes and bounds. In fact the provision of the act of 1891 does something more than simply lay down a rule of evidence. In the last sentence of section 1700 of the Political Code, embodying this provision, it is expressly declared that, while the county commissioners or ordinary shall not grant leases to any grounds shown on bulletin No. 19 to contain a natural bed, it shall be lawful for them to grant leases on any and all territory indicated on said bulletin as vacant. The state, therefore, by this provision adopted lands indicated on the chart to be vacant as the identical property which the county commissioners or ordinary had a right to lease; and the statute would not have been stronger in its effect, had a special law been passed, describing this territory minutely and accurately, as it was in the chart or survey, and declaring to the county commissioners of Chatham county that they had authority to lease that land. Of course, such a lease would have been binding upon the state, whether in point of fact there were any natural oyster beds upon the land or not. Such an express provision would necessarily have been a change or modification of any general law that might have been in existence, reserving any natural oyster beds for public use; and the lessee and his assigns would have taken the property free from any such restriction, or free from any rights of individuals or the public to enter thereon. Now, to so construe the act of 1898 as to hold that it was retroactive in its effect, and applied to the leases in this case, and thus permit a party to go behind the chart, and show that in point of fact some of the territory was not vacant, but contained natural oyster-beds, would be a direct violation and impairment of a solemn contract entered into between the state and some of its citizens. It would be allowing a party to show by parol testimony that the action of the county commissioners in their judicial capacity was absolutely void, notwithstanding that when they acted they proceeded in direct obedience to a positive mandate of the state legislature, and that mandate being expressed touching a matter over which the legislative department of the government had absolute control. We dare say that no respectable authority can be found in this country for the proposition that such a law, whether called a remedial statute, or a statute upon the admissibility of testimony in causes, could be retroactive, and could be so applied as to defeat the vested rights of a citizen. This principle is practically decided in the case of *Waters v. Manufacturing Co.,* 106 Ga. 592, 32 S.E. 636, where it was held, "When the lien of a material man has, under the terms of the statute, become fixed and secured, such lien is then a vested right, and no subsequent repeal or modification of the act under which it became fixed can destroy or modify such right." Certainly, the lien of the material man in that case was not any more fixed and not any more securely vested than was the title to this land when the same was leased by authority of the state.

It really does not appear, however, from this record, what the ruling of the judge below was upon the subject of the effect and weight as evidence of this chart made by the United States survey, and we fail to find in the bill of exceptions any exception to any views entertained or judgment that may have been made by the court on this question. It is simply stated "that the law of 1898, amending the law of 1891, making Drake's chart conclusive evidence, is the law which regulates procedure in the courts of this state in cases arising subsequent to the passage of said act of 1898, and the only recourse which the plaintiff has in a court must be had under the existing law regulating procedure." Nothing appears in the record indicating that there was any ruling by the judge to the contrary of this contention of counsel for plaintiffs in error. On the contrary, it appears from the brief of evidence that testimony was introduced in behalf of defendants below tending to show that some of the lands claimed by petitioner under the leases had upon them natural oyster beds. There was also evidence in behalf of petitioner proving the contrary, and the truth is that the only material conflict that seems to have existed in the testimony of the witnesses was in relation to the quantity of oysters that was upon the land in question at and before the granting of these leases. We think that the preponderance of evidence was decidedly with the petitioner in establishing the fact that while there may have been, years before this lease, natural oyster-beds, of advantage to the public in obtaining oysters, yet at the time of this lease, and for several years before, these beds had become so utterly depleted that even the public stopped resorting thereto for the purpose of getting oysters, either for consumption or sale, and that in point of fact, properly speaking, there are no natural oyster-beds anywhere upon these grounds. Simply because there have been at a certain place natural oyster beds, which have become entirely depleted, it does not follow that such territory, which has only a few scattering oyster shells about in places, is still an oyster-bed. In the case of *State v. Willis (N. C.)* 10 S.E. 764, the supreme court of North Carolina decided: "A natural, as distinguished from an artificial, oyster bed, is one not planted by man, and is any shoal, reef, or bottom where oysters are to be found growing, not sparsely or at intervals, but in a mass or stratum, and in sufficient quantities to be valuable to the public." Shepherd, J., in delivering the opinion in that case, enters into quite an able and thorough discussion of what constitutes a natural oyster bed, reaching the conclusion above quoted from the unanimous decision in the case by the court. Even, therefore, if we are in error about the conclusiveness of the evidence furnished by Drake's chart, we cannot say that there was not sufficient evidence, simply treating the chart and the action of the county commissioners as only prima facie evidence, to authorize the judge below in granting this temporary injunction.

5. It is further contended in behalf of plaintiffs in error that, even conceding that the conduct of the defendants in this case was a trespass upon the property of petitioner, it then involved the violation of a penal offense, and a court of equity has no jurisdiction to restrain a party from committing a crime. The authorities relied on to sustain this position refer simply to such criminal acts as involve only the public interests, and equity will not interfere to restrain their commission unless some private right of property is being endangered, and serious injury threatened thereto, for which the owner has no adequate relief at law. The fact that a trespass upon the property rights of others involves also the violation of a public penal statute does, of course, not prevent the owner of the property from appealing to the courts for such redress as law or equity may give him. The acts proven by the testimony in this case clearly show that the plaintiffs in error were trespassers upon the property of petitioner. Even at common law such an action could be maintained to protect planted oysters. In the court of appeals of New York in the case of *Post v. Kreischer (N. Y.)* 8 N.E. 365, it was decided: "Oysters planted in a bed clearly marked out in the tide waters of a bay or arm of the sea, where there are no oysters growing spontaneously at the time, are the property of the person who plants them; and the taking of them by another person is a trespass, for which an action lies." See, also, *Grace v. Willets (N. J. Err. & App.)* 14 A. 559, decided by the court of appeals of New Jersey, in which it appeared that plaintiffs deposited in a certain river a boat load of oyster shells. To these shells the germs of oysters, floating in these waters, attached themselves, and in about two years developed into marketable oysters. It was held that these oysters belonged to plaintiffs, and that they could maintain an action against the defendants for the removal and conversion of them. Certainly, then, under the facts of this case, this defendant in error had such an interest in the property trespassed upon as to entitle him to maintain an action for trespass on account of the depredations which the evidence shows were made by these plaintiffs in error on his oyster beds. It appears from the pleadings and the evidence that these defendants below were all insolvent; that they had combined in considerable numbers, and were committing continuous and repeated invasions upon petitioner's property, and before the filing of the petition had actually destroyed about $2,000 worth of property, by taking several thousand bushels of oysters from the beds of defendant in error; that this injury would continue unless the parties were restrained, and it was evidently irreparable in damages, on account of the insolvency of the defendants below. Section 4916 of the Civil Code declares, "Equity will not interfere to restrain a trespass unless the injury is irreparable in damages, or the trespasser is insolvent, or there exists other circumstances which, in the discretion of the court, render the interposition of this writ necessary and proper, among which shall be the avoidance of circuitry and multiplicity of actions." In the case of *Cottle v. Harrold,* 72 Ga. 830, it was decided, "Equity will restrain a trespass when the threatened injury is irreparable in damages, or when the trespasser is insolvent, or when there exist other circumstances which, in the discretion of the court, render the interposition of the writ necessary and proper." See, also, *Peterson v. Orr,* 12 Ga. 464; *McGinnis v. Justices, etc.,* 30 Ga. 47; *Graham v. Mining Co.,* 71 Ga. 297; *Smith v. Smith,* 105 Ga. 106, 31 S.E. 135; *English v. Jones (Ga.)* 34 S.E. 122.

6. It is unnecessary to make further comment upon the character of the evidence introduced on the hearing of this application for a temporary injunction. Every material allegation of fact in plaintiff's petition was substantially sustained by the overwhelming weight of testimony introduced on the trial. Under the evidence disclosed by the record, and in view of the principles of law herein enunciated, there was no error in granting the injunction. Judgment affirmed.

Another Georgia case, *Fraser v. State*,[88] stresses the disturbance of what is termed "oyster-marks," equivalent to disturbing property monumentation. It also discusses the use of a geodetic chart, a standard reference to separate a natural oyster bed from a private bed. Deeds with descriptions, maps, and charts are all items of helpful evidence to anyone understanding and interpreting submerged areas and how they are controlled by the governing statutes.

The plaintiff in error was convicted of violating section 588 of the Penal Code, which reads as follows: "If any person shall, without authority from the owner, take or catch any oysters from any private bed, or remove or deface any oyster-marks, he shall be guilty of a misdemeanor." He made a motion for a new trial, which was overruled, and he excepted. The material questions thus presented, and the facts necessary to an understanding of our rulings thereon, are stated and dealt with below:

1. The indictment alleged that the offense was committed on the 28th day of March, 1900. Augustus Oemler, the only witness sworn in behalf of the state, testified that on September 20, 1899, he saw the accused taking oysters from a private bed belonging to the witness, and that he did not authorize or consent to such taking. Counsel for the accused thereupon moved for "a postponement of the case," stating "that they were surprised, and not prepared to try an accusation based on a taking occurring in September, and they had prepared their case to meet the charge and date set out in the indictment." Complaint is made of the refusal to grant the motion to postpone. We cannot hold that the court erred in this respect. The accused was chargeable by law with knowledge that it would be perfectly competent for the state to prove that the alleged misdemeanor was committed at any time within the two years next preceding the finding of the indictment, and it was incumbent upon him and his counsel to bear this in mind in preparing his defense. Doubtless, it is within the discretion of a trial judge, even after the parties have announced "Ready" in a criminal case, and the trial has accordingly been begun, to grant, at the request of the accused, a postponement, if it should be made to appear that the ends of justice require it. But nothing of the sort was made to appear in the present instance; for, taking the showing made for the postponement at its best, it entirely fails to disclose that the accused, even had he been granted further time, would have been able to overcome the evidence relied on by the state, or to produce any evidence whatever which would benefit him in the least. Clearly, the showing for a postponement made in this case would have been entirely without merit had the same, even before the accused announced "Ready," been made the basis of a motion to continue.

2. One ground of the motion for a new trial was as follows: "Because the court erred in ruling out the question, 'Did they make any claim that they had the right to take oysters there?'--the evidence sought to be introduced being that this defendant bona fide claimed the right to take and catch oysters at the place where he was taking them in the day testified to (September 20, 1899); this testimony being material for the jury's consideration in determining the good faith of defendant, and his freedom of any criminal intent." It is fairly inferable, though as to this matter the motion is certainly indefinite, that the question referred to in this ground was propounded to Oemler; and, in view of the scope covered by the examination of this witness, it is probable that the pronoun "they," as used in this question, was intended to have reference to the accused and others participating with him in committing the act upon which the indictment was predicated. Still, we are unable to perceive how this ground of the motion furnishes cause for a new trial. Even if the expected answer to the question had shown conclusively that the defendant did bona fide claim the right to take and catch oysters at the place where he was taking them on September 20, 1899, there is nothing to indicate at what time this claim was set up. What the accused claimed could in no possible view be admissible unless the claim was made at the very time he was actually taking the oysters, and for this reason constituted a part of the res gestae of the act for which he was indicted. Certainly a claim of good

---

[88] 112 Ga. 13, 37 S.E. 114 (Ga. 1900).

faith made afterwards, at any other time or place, would be nothing more than a declaration by the accused in his own favor, and clearly inadmissible. It does not, therefore, appear that, in declining to allow the question to be answered, the court excluded proof which it was competent for the accused to introduce in his defense.

3. In another ground of the motion, error is assigned upon the refusal of the court to allow the accused to introduce in evidence "the record in the case of Wm. Jones et al. v. Augustus Oemler." Doubtless, the record referred to was that of the case between the parties named, which this court passed upon at the last term (35 S.E. 375); but the motion does not affirmatively so disclose. It is therein stated that this record was offered for the purpose of showing certain facts, but the record itself is not set forth, either literally or in substance. Indeed, not a word of its contents appears. Accordingly, we are unable to determine, from anything appearing in the motion itself, whether the document in question would or would not have shown what the movant in his motion alleges it was offered to show. In this connection, see *Petty v. Railway Co.,* 109 Ga. 666, 674, 35 S.E. 82, and cases cited.

4. The state introduced in evidence a lease from the commissioners of Chatham county, Ga., to Augustus Oemler, purporting to cover a "tract of land under water, in the county and state aforesaid." The other descriptive terms employed in the lease were as follows: "Commencing at low-water mark at the lower mouth of Big creek; thence along the line of low water, 967.9 feet; thence to middle of the river, 225 feet; thence along the middle of the river, 967.9 feet; thence 225 feet to starting point." It is, in the motion for a new trial, complained that the court erred in admitting the lease in evidence "over the objection of the prisoner at the bar"; but the motion does not disclose what, if any, specific objection was made to the admissibility of the document when it was tendered. Though it is, in the motion, alleged that "the description was too vague and uncertain," it does not appear that this objection to it was presented and ruled upon at the trial, and accordingly we cannot undertake to pass upon the admissibility of the document. If, however, the description was too uncertain to identify any particular tract of land, the state failed to make out a case; for Oemler, after testifying that this lease covered the premises, in effect admitted that his claim to this land was based exclusively on this lease. We cannot say that the description in the lease was insufficient to identify the locus in quo. It was urged in the argument here that it could not be determined in what direction the first line mentioned in the lease ran – whether up the coast or down the coast – and that consequently the descriptive terms employed would apply equally to at least two entirely separate and distinct parcels of realty. This contention is not well founded. The "commencing" point was "low-water mark at the lower mouth of Big creek." It cannot be assumed, as matter of law, that this point is incapable of identification. This being so, and the first line therefrom being "the line of low water," it necessarily ran from the beginning corner down the coast; for it surely could not have been contemplated that it was to extend across the mouth of Big creek, which would have resulted from running it up the coast. This being settled, it does not affirmatively appear that there was any practical difficulty in applying the description embraced in the lease to the premises in dispute. It is true that the instrument fails to state at what angles the two lines of 225 feet each were, respectively, to intersect the first line; but, for aught that appears, it may be that the courses of the third line, "along the middle of the river," and the length of that line, were such as to make certain at what angles the second and fourth lines must intersect the first in order to give them a length of exactly 225 feet each. A lease cannot, as matter of law, be held to be void for uncertainty as to description unless it in some way appears that it is too indefinite to be applied to its subject matter. No such thing was shown in the present instance. On the contrary, there was positive testimony that the lease did cover the very land upon which the offense charged against the accused was alleged to have been committed.

5. The foregoing disposes of all the questions presented by the motion for a new trial which are of sufficient importance to be specially noticed, save one; and upon it the case, in our judgment,

really turns. Besides proving that Oemler, under the lease above mentioned, claimed and held the land from which the accused took the oysters, the state introduced in evidence "Drake's chart, known as 'Bulletin No. 19,' United States geodetic survey"; and it was admitted that according to this chart the land leased by Oemler was "Vacant of natural oyster beds." Taking into view the entire brief of the evidence, and the statement of the accused, the jury were fully warranted in finding that at the very time he took the oysters he knew of the geodetic survey and of Drake's chart; also, that he was aware that the land upon which he was fishing was claimed by Oemler under a lease, and that according to the chart this land was not a natural oyster bed. The real defense was that, notwithstanding all this, the accused had a right to take the oysters if the land was in fact a natural oyster bed, and had been erroneously marked as "vacant" on the chart. Accordingly, complaint is made in the motion for a new trial that the court erred in rejecting certain evidence tending to show that the land in dispute was, as matter of fact, and notwithstanding what was shown by the chart, a natural oyster bed. Granting that the rejected evidence would have gone to this extent, there was no error in refusing to admit it. Under the law as it stood when the lease was granted to Oemler, Orake's chart was conclusive on the question at issue. *Jones v. Oemler*, cited supra. The accused did not make any mistake of fact, but, conceding that he acted in good faith, he did judge erroneously with respect to the law. This cannot, of course, excuse him. *Levar v. State,* 103 Ga. 42, 29 S.E. 467. It is true that, at the time Fraser committed the act for which he was indicted, this court had not decided the Jones Case; but the accused was nevertheless bound to know what the law was. It would never do to hold that one who did an act which the law makes criminal could be excused because he honestly misapprehended the true meaning of a statute. The question in such a case is not one of good faith or of diligence in endeavoring to find out what the law was. It is, simply, did the accused knowingly do the forbidden act? Judgment affirmed.

In Mississippi, a Gulf Coast state, similar issues have been addressed. The decision in *Parks v. Simpson*,[89] is about dredging and selling oyster shells by a third party.

W. L. Parks and Radcliff Gravel Co., Inc., filed their bill of complaint against W. G. Simpson, and others, comprising the members of the Mississippi Marine Conservation Commission, and Jahncke Service, Inc., to cancel a purported contract between the Commission and Jahncke, and to enjoin the removal of any materials from the waters of the Mississippi Sound pursuant to such contract.

It was alleged that Parks is a resident citizen and taxpayer of Harrison County; that he filed this action for himself, the general public, and all persons interested; and that he expressly invited such persons to join in the litigation. It was further alleged that, prior to instituting the action, he solicited both the Attorney-General of the State and the District Attorney of the Second Judicial District of Mississippi to bring the suit or permit the same to be brought in their names, but that said officers declined to do either. Radcliff Gravel Company is an Alabama corporation, qualified to do business in this state, and is engaged, among other things, in the business of dredging and selling shells.

It was further charged that the defendant Commission, on August 23, 1960, entered into the purported contract with Jahncke Service, Inc., for the sale of shells from the waters of Mississippi Sound. A copy of the contract and supplementary agreements was attached. Under the contract, the Commission granted, bargained and sold to Jahncke oyster shells and shell deposits, meaning the shells of dead oysters and other shellfish, which had accumulated over a long period of time and which were commonly referred to as "dead reef shells," "reef shells," or "cay shells" in the territory known as Mississippi Sound, with rights of ingress and egress for the purpose of mining, dredging, taking, and removing the shells. The compensation was 12 cents a cubic yard, with

---

[89] 242 Miss. 894, 137 So.2d 136 (Miss. 1962).

a minimum annual payment of $16,000, thus assuring the collection of revenue from that source of at least that amount annually. This contract was to run for a period of ten years, with certain rights of renewal. At the expiration of the period, Jahncke ceased to have any rights whatever in the shells. It is unnecessary to refer to other provisions and restrictions in the contract.

The answers of the defendants were separate but of substantially the same effect. They denied that the contract and its supplementary modifications was a nullity. It was averred that Chapter 173, Laws of 1960, constituted a complete revision of the seafood laws of the state; that Section 20 thereof repealed 81 other sections of such laws, and, at the same time, specifically authorized and empowered the Commission to enter into contracts for the sale of, and the right to dredge for and remove, shells from the Mississippi tidewater bottoms, and to make contracts such as are involved in this case; that the seafood industry, giving employment to thousands of persons, and their employment was threatened because of a decline in the efficiency of the industry; that the Legislature declared the public policy in respect to this industry under Section 1 of the Chapter and granted far-reaching powers to the Commission; that it had been the established policy of the state for the Commission, having jurisdiction of its seafoods, to be self-supporting; that, upon investigation, it was found that the revenue available from taxes levied on the industry were inadequate to carry out the duties of the Commission, and that an increase thereof would make it impossible for Mississippi citizens to compete in the national market; that under Subsection (4), Section 6, Chapter 173, Laws of 1960, as one of the powers of the Commission, the Legislature provided: "For the purpose of growing oysters, may acquire and dispose of shell seed oysters and other materials; * * *"; that the shells in question had accumulated over centuries, and, in some of the banks, they are over 100 feet thick from top to bottom; that in former years, the Seafood Commission had let contracts of this nature, but, because of non-performance, the revenue therefrom had been negligible; and that the contract here involved is an advantageous one because it insured the availability of at least $16,000 each year.

It was further charged that, in Subsection (4), of Section 6, supra, a comma appears to have been omitted, evidently a clerical error, between the words "shell" and "seed," but that, under established rules of statutory construction, a comma should be inserted so as to make the provision read: "For the purpose of growing oysters, may acquire and dispose of shell, seed oysters and other materials; * * *"; that seed oysters are very small, grow in thick beds, and are useless unless removed from the beds and scattered or planted where more food is available; that subsection (4), unless construed to mean both a sale of shells and seed oysters, would be meaningless because the sale of dead reef shells is the most important source of funds with which to rebuild and revitalize the seafood industry.

Consequently, the defendants averred that the court should construe said Subsection (4) and hold that it was the actual purpose and intention of the Legislature that the subsection should provide as follows: "(4) For the purpose of growing oysters, may acquire and dispose of shell, seed oysters and other materials; * * *"; that the Commission was authorized to make sales of shells and shell deposits from the water bottoms in question, and that the contract in question was valid.

Subsequently, the complainants moved the court to strike the answers of the defendants on the ground that they were insufficient in law to constitute a defense to the bill.

The motion to strike was overruled. The court recognized that the sole purpose of the motion to strike was to present the question as to whether or not the Commission had the requisite authority in law to enter upon the particular contract here involved. Hence, the chancellor granted the complainants an appeal to this Court.

The appellees quote the rule with reference to striking an answer as laid down in Griffith's Mississippi Chancery Practice, Section 367, pp. 351–352, showing that it is "not favored in our practice, and will be allowed only where there is no other available method and when the justice of it is so clear as to be fairly indisputable." They also correctly state that it is in the nature of a demurrer and with like effect. Consequently, they argue that, if evidence had been presented, they would have been able to show clearly the ambiguity in Subsection (4), supra. Griffith's Mississippi Chancery Practice, Section 288, pp. 271–272.

The parties agree that the question on this appeal is whether or not under chapter 173, Laws of 1960, the Commission had authority to sell the dead reef shells, as heretofore described, and thus enter into the contract about which this controversy arose.

The appellants say that these deposits, being property held by the state as trustee for the people, are public property and can be sold only by legislative authority and in the manner provided by law. Too, since the Legislature expressly repealed Section 25, Chapter 195, Laws of 1958, which gave the old Seafood Commission the power and authority to sell the deposits in accordance with the method there mentioned, and since there is no language in the new Act, Chapter 173, Laws of 1960, which purports to grant such authority as was contained in the 1958 Act, it follows that the Legislature ordained the repeal intentionally and with a purpose. Besides, it is vain to contend that the Legislature in omitting Section 25 of the old Act, and inserting nothing to take its place, in fact made no change in the substance of the law. This is not a correction of a mere clerical error or an inadvertent omission. The Legislature should not be accused of forgetting to insert those three paragraphs. Obviously, the courts have no authority to write into the statute something which the Legislature itself did not intend to write therein. They also say that shell, as used in Subsection (4) of Section 6 of the 1960 Act, when read with the balance of that subsection, means shells taken from the oysters processed in the factories.

On the other hand, the appellees say that the Legislature evinced its purpose and intention of bringing about a revitalization of the seafood law by its declaration of policy in Section 1 of the 1960 Act. State policy had always contemplated that the Commission should be self-supporting. The sale of such deposits was the practical and feasible way to get money for operation. Consequently, the Legislature could not have had it in mind to eliminate this source of revenue. There is an ambiguity in the bill. The proper construction necessitates extrinsic aids. There is no such thing, for instance, as "shell seed oysters." Subsection (4) should be construed to the effect that the Commission had authority to sell shells and seed oysters; otherwise the language is meaningless. Consequently, a comma should be placed between "shell" and "seed oysters" in order to effect a sensible construction. To show that this Court has corrected ambiguities, they cite *Adams v. Yazoo & M. V. R. Co.,* 75 Miss. 275, 22 So. 824; *Ott v. State ex rel. Lowery,* 78 Miss. 487, 29 So. 520; *Dukate v. Adams,* 101 Miss. 433, 58 So. 475; *Roseberry v. Norsworthy,* 135 Miss. 845, 100 So. 514; *Gandy v. Public Service Corporation of Mississippi,* 163 Miss. 187, 140 So. 687; *Mississippi Cottonseed Products Co. v. Stone et al.,* 184 Miss. 409, 184 So. 428; *Martin v. State,* 190 Miss. 32, 199 So. 98.

This Court, by an unbroken line of decisions over 100 years, has declared that the State "is the owner of the lands in the beds of all its shores, inlets, and adjacent to the islands, over which the tides of the sea ebb and flow, and that it holds title as trustee for the people of the *State." Giles v. City of Biloxi,* 237 Miss. 65, 112 So.2d 815, 823, 113 So.2d 544; *Xidis v. City of Gulfport,* 221 Miss. 79, 72 So.2d 153; *Crary v. State Highway Commission,* 219 Miss. 284, 68 So.2d 468; *State ex rel. Rice v. Stewart,* 184 Miss. 202, 184 So. 44, 185 So. 247; *Rouse v. Saucier's Heirs,* 166 Miss. 704, 146 So. 291; *Money v. Wood,* 152 Miss. 17, 118 So. 357; *Martin v. O'Brien,* 34 Miss. 21.

Manifestly, the shells, undertaken to be sold under the contract here in question, constitute state property – a part of the land. Section 95 of the Constitution prohibits the donation of land belonging to, or under the control of the state. It also contains this admonition: "Nor shall such land be sold to corporations or associations for a less price than that for which it is subject to sale to individuals." Besides "the granting of lands under the control of the State must be by general law." *Giles v. City of Biloxi*, supra.

By section 2, chapter 195, Laws of 1958, the Legislature recognized that such shells are the property of the State; and, by section 25 thereof, it was provided as follows:

"Shells are property of state.--All of the shells of dead oysters, clams and other shell fish; and all of the oyster shells, clam shells, mussel shells, and cay shells, being upon the bottoms of, or under the tide waters within the territorial jurisdiction of the State of Mississippi, and all beds, banks and accumulation of such shells within such territorial jurisdiction or upon the beaches or bottoms of such waters, or surrounded by such waters, are hereby declared to be the property of the State of Mississippi, and the title to same shall not be divested except in the manner herein provided.

'The seafood commission of the State of Mississippi shall have and it is hereby granted full authority and control over all of such shells, beds, banks and accumulation of shells and may, in its discretion, sell and dispose of same or of any part thereof, at private or public sale; the proceeds of any such sale or sales to be paid into the state treasury into the sea food fund to be used for any purpose for which sea food funds are authorized to be expended.

'The commission shall have the authority, in its discretion, to sell such shells, beds, banks or accumulation of shells by the yard or barrel or may grant the right to remove the same upon a contract under which a percentage of the shells so removed will be delivered to the seafood commission to be used in developing public oyster reefs."

By said section, quoted above, it was again declared that such shells are the property of the State and that title thereto could not be divested except in the manner therein provided. It is seen that alternative methods of sale and disposal were set up. The Legislature was evidently aware of the provisions of the Constitution and was making an effort to comply therewith. This section contained 249 words. It was evidently not omitted from Chapter 173, Laws of 1960, by inadvertence or mistake. Actually it was expressly repealed by Section 20 of the 1960 Act. Consequently there must have been a reason and purpose, although the motivation therefor does not appear.

Section 3, chapter 173, Laws of 1960, provides as follows:

"All sea foods the property of the state, until. All sea foods existing or living in said waters not held in private ownership legally acquired, and all beds and bottoms of rivers, streams, bayous, lagoons, lakes, bays, sounds and inlets bordering on or connecting with the Gulf of Mexico or Mississippi Sound within said territorial jurisdiction, including all oysters and other shell fish and parts thereof grown thereon, either naturally or cultivated, shall be, continue, and remain the property of the State of Mississippi, to be held in trust for the people thereof until title thereto shall be legally divested in the manner and form hereinafter authorized, and the same shall be under the exclusive control of the commission until the right of private ownership shall vest therein as hereinafter provided." (Emphasis supplied by author.)

Thus the Legislature again recognized that such shells are the property of the State and will remain so until title shall be legally divested in the manner and form "hereinafter authorized."

In section 6 of chapter 173, with reference to jurisdiction, authority and duties of the Commission, it is said:

"* * * In connection with its jurisdiction and authority, the commission:

* * *

'(4) For the purpose of growing oysters, may acquire and dispose of shell seed oysters and other materials; provided, however, fifty per cent (50%) of all the oyster shells produced from oysters taken from the public reefs of the State of Mississippi and fifty per cent (50%) of all oyster shells produced from the oysters processed within the State of Mississippi are hereby declared to be the nontransferable property of the State of Mississippi and all persons, firms or corporations dealing in or canning oysters taken from the public reefs of the state and all persons, firms or corporations processing oysters within the state shall deliver to the commission fifty per cent (50%) of the oyster shells taken or processed by said person, firm or corporation, delivery of same shall be at the place of business of the oyster processor, dealer or factory."

Actually the above Subsection (4) is the only basis on which the appellees claim that they have authority to make such sales.

But it must be borne in mind that, even if a comma is placed between the words "shell" and "seed oysters," in the above quotation, the shells, which were being referred to in that context, were the oyster shells "produced from oysters taken from the public reefs of the state" and "oyster shells produced from the oysters processed within the state." Such shells have no connection whatever with "dead reef shells" or "reef shells," or "cay shells," such as are involved in the present controversy.

Section 11 of the 1960 Act grants authority to the Commission to lease, not sell, bottoms. This applies to the gathering of seafood for a term of one year and of course has no application to the sale of shells as are under consideration in the present controversy.

The Court has been unable, after a careful check of the 1960 Act, to find that the Commission is clothed with authority to make sales of shells of the kind referred to in the contract here under consideration. In *United Gas Pipe Line Company v. Mississippi Public Service Commission, Miss.,* 133 So.2d 521, the Court had under consideration not whether the State had the power to regulate the kind of sales there involved but whether the Legislature granted to the Commission the power to do so, and it held that such power had not been granted. This is the question now before the Court--not whether the Legislature had the power to authorize this to be done, but whether in fact they actually did so.

After the Legislature omitted Section 25 of the Act of 1958, expressly repealed the section, and did not re-enact another section of like character and effect, it is too late to ask this Court, by a strained and indefensible construction, under the label of inadvertence, clerical error, or some other euphemistic phrase to stultify itself by holding that the Legislature, after all, did, in the first seventeen words of Subsection (4), empower the Commission to sell and dispose of such shells as are involved in this case, and, especially so, when those words are wholly out of context for such a construction. Besides, in addition to the opprobrium which the Court would heap upon itself, the announcement of such a construction would also erect a permanent monument of legislative stultification.

It follows that the Commission had no authority to enter upon the contract with Jahncke. Consequently, the contract was invalid and of no effect. The cause is therefore reversed and a decree will be entered here invalidating the contract and enjoining its performance.

Oyster shells belong to the state under the shellfish statute."

## REAL PROPERTY VS. PERSONAL PROPERTY

A fundamental consideration of ownership and of exercising rights is whether property is real or personal. Oysters, oyster-beds and farms, and oyster shells are distinguishable.

On the Pacific Coast, the California decision of *Darbee and Immel Oyster and Land Company v. Pacific Oyster Company,*[90] concerns property rights in the form of inheritance.

Plaintiff, a corporation, brings this action for a partition of certain premises which it avers to be "real property," and avers that it has an "estate of inheritance" therein. A demurrer to the complaint was filed by some of the defendants. The demurrer was general and special, and was sustained in the court below, and judgment was rendered in favor of defendants. From this judgment plaintiff appeals.

The action is based on section 752 *of the Code of Civil Procedure.* That section is under the head of "Actions for the partition of real property," and it provides that an action for partition may be brought by one or more cotenants of real property, in which one or more of them "have

---

[90] 150 Cal. 392, 88 P. 1090 (1907).

an estate of inheritance, or for life or lives, or for years." The court below held that the complaint shows that neither plaintiff nor any of the defendants had an estate of inheritance, or for life, or for years, in any real property, and upon that ground sustained the demurrer; and as we think that this conclusion was right, we need not examine any other question raised in the case.

In the complaint the alleged real property in which plaintiff is averred to have an estate of inheritance is described as follows: "The right to the exclusive use and occupation thereof for the purposes of laying down and planting oysters and taking up and carrying off the same in and from said real property, in accordance with and as provided by the terms of an act of the legislature of the state of California, entitled 'An act to encourage the planting and cultivation of oysters,' approved March 30, 1874." Those parts of said act of March 30, 1874, (Stats. of 1873–74, p. 940,) which are material here, are as follows: the title of the act is "An act to encourage the planting and cultivation of oysters," and the first section is as follows:

"Any citizen of the United States may lay down and plant oysters in any of the bays, rivers or public waters of this State; and the ownership of and the exclusive right to take up and carry off the same shall be continued and remain in such person or persons who shall have laid down and planted the same." The next sections, down to and including section 8, provide that the person desiring to use the right must define the limits of his claim by stakes, etc., must maintain thereon a sign on which must be painted the words "Oyster Beds," and must record a description of his bed or beds of oysters in the county recorder's office. (In the case at bar it does not appear that appellant complied with these provisions; but we will not consider that matter.) It is further provided that, after he has complied with these provisions, any person who enters thereon and carries off oysters or removes therefrom marks designating boundaries shall be guilty of a misdemeanor. Section 9 is as follows:

"This act shall not apply to any tide lands which the state may have sold to private parties; *provided, further*, that nothing herein shall be so construed as to interfere with the right of the state to sell and dispose of any of the tide lands, nor to affect in any manner the rights of purchasers at any sale of tide lands by the state."

In the property described in the complaint, there is no element of an estate of inheritance, or, as described in section 761 *of the Civil Code*, a "perpetual" estate. The privilege extended to all citizens by said act to temporarily use the unsold tide lands belonging to the state, if it can be considered as an estate at all in lands, is certainly of no higher dignity than an estate at will. But, in our opinion, it is really nothing more than a mere personal license. That was the view taken of similar statutes by the highest court of Maryland, where the oyster business is a very large one. In *Phipps v. State,* 22 Md. 380, [5 Am. Dec. 654], the court was dealing with statutes like ours and it said: "It abundantly appears from the nature of the privilege in dispute, as well as from the terms in which it was conferred, that no transfer of the state's title to lands covered by navigable water was contemplated. Permission to use given areas covered by navigable water for a particular purpose seems to be all that the legislature intended, and we think the language of its assent to that use should be construed, not as a grant binding the state, but as a conditional license, revocable at the pleasure of the legislature." Again, in *Hess v. Muir,* 65 Md. 586, [5 A. 540, 6 A. 673], Alvey, C. J., said: "These statutes, the better to promote the growth and to increase the supply of oysters in the waters of the state, provide that any of the citizens of the state may locate one lot, and only one, of five acres, in any unappropriated ground covered by the tide, and plant the same with oysters, and thereupon he is given exclusive control thereof. This, however, is not a grant of an indefeasible right or estate in the lot thus authorized to be located and planted with oysters. It is simply a conditional or qualified license or franchise, revocable at the will and pleasure of the state. (*Phipps v. State*, 22 Md. 380, [5 Am. Dec. 654].) It is neither inheritable nor transferable, but is purely a personal privilege in the party locating the lot."

**Author's Comment:** Note that harvesting oysters on public land is a license not an easement.

219 Or. 588 (Or. 1959)
348 P.2d 39
**COOS BAY OYSTER COOPERATIVE, an Oregon corporation, Appellant,**
v.
**STATE of Oregon, by and through its State Highway**
**Commission, composed of Charles H. Reynolds, M. K.**
**McIver, and Robert B. Chessman, Respondent.**
**Supreme Court of Oregon, Department 1.**
**December 31, 1959**
**Page 589**

The order of dismissal resulted from plaintiff's refusal to plead further after the trial court sustained defendant's demurrer to each of the seven causes of action alleged in plaintiff's amended complaint.

In 1952 the highway commission prosecuted to judgment 18 condemnation actions to acquire title to those lands in Silver Point No. 1 which were within the right-of-way boundaries of the new highway. Oysters were then growing on all the 296 tracts of Silver Point No. 1.

In each of the 18 condemnation cases it was necessary to determine the 'larger tract' involved in order to compute the damages to the remainder after the area within the right-of-way was taken. The larger tracts in each instance consisted of the total number of oyster tracts or fractions thereof contiguous, one to the other, and in common ownership as respects the fee.

In the instant case plaintiff's first cause of action consists of a claim for the alleged 'taking' by the state of the oysters growing on all 296 of the 1-acre tracts in Silver Point No. 1, which were not included in the 18 condemnation actions above mentioned. The oysters on each and all of the said 296 tracts were owned by plaintiff at the time the highway was constructed.

The second cause of action is based upon the alleged taking of a certain parcel of land consisting of several tracts located within Silver Point No. 1 and owned in fee by the plaintiff. Each of the remaining five causes of action is likewise for the taking of a separate and different parcel of land constituting a part of said Silver Point No. 1, the owner in fee in the case of each of said five parcels having assigned his claim to plaintiff.

Plaintiff asserts three assignments of error on this appeal. The first two assignments of error concern the correctness of the court's ruling in striking from plaintiff's original complaint portions thereof. The third assignment of error deals with the principal question raised on this appeal and relates to the alleged error of the court in sustaining defendant's demurrer to plaintiff's amended complaint.

In its original complaint plaintiff alleged in detail the manner and method of propagating oysters and the conditions essential for their growth. The trial court ordered all such averments stricken, on the ground that they constitute, at best, the pleading of evidence, if that. Plaintiff in its first assignment of error contends that the trial court erred in ordering said allegations stricken. In an inverse condemnation suit such as this the correctness of the trial court's ruling in striking such matter from the complaint as being irrelevant and immaterial requires no citation of authority.

Plaintiff in its original complaint alleged its ownership of all the oysters growing on the 296 1-acre tracts located within Silver Point No. 1 at the time the highway fill was constructed. In assignment of error No. 2 plaintiff challenges the court's ruling in striking from plaintiff's original complaint an allegation to the effect that the state had not paid any damages for the taking of the oysters, except as thereafter enumerated, listing 58 separate tracts of the 296 on which damages had been paid.

The 58 enumerated tracts consisted of those upon which payment previously had been made in the 18 prior condemnation actions filed by the state and brought to judgment. Those 18 cases were resolved pursuant to a stipulated compromise. The fact the state previously saw fit to make payments for damages to certain tracts is in no way material to any question properly an issue in this inverse condemnation action, and the allegations with reference thereto appearing in plaintiff's original complaint were properly stricken.

The defendant demurred to plaintiff's first cause of suit upon the following two grounds, to-wit:

'1. That the face of the amended complaint demonstrates that plaintiff's purported First cause does not state facts sufficient to constitute a cause of action.

'2. * * * upon the ground that said amended complaint discloses upon its face that the same calls for numerous causes of action improperly united, in that the several parcels of land and all the appurtenances a part of the land, including claimed crops of oysters growing thereon, are in diverse ownerships and thus result in a pleading where the claim of any one person will not affect all the parties to the action.'

In addition to the above grounds, defendant demurred to plaintiff's remaining six causes of action upon the further ground, to-wit:

'3. Defendant further demurs to plaintiff's purported Second cause of action upon the ground that it appears from the face of the amended complaint that the court has no jurisdiction of the person of the defendant in that said amended complaint does not allege that defendant has consented to be sued.'

The third ground of said demurrer hereinabove referred to was not seriously urged upon argument or in the brief of defendant, and we conclude that said third ground is without merit.

While the wording of the first two grounds of defendant's demurrer is not identical as to each of plaintiff's seven causes of action, the variation therein as to each cause is not material so far as our consideration thereof on this appeal is concerned.

The trial court sustained plaintiff's demurrer upon the first ground, to-wit: that each cause fails to state facts sufficient to constitute a cause of action. It would appear that the trial court's ruling was predicated on the premise that an oyster crop is a part of the realty when the subject of an inverse condemnation action, and that each of the seven purported causes of action consists of a claim for the taking of a fractional interest only in real property. Cause of action No. 1 is for an alleged taking of a crop of growing oysters only, while the remaining six causes of action are for the alleged taking of certain enumerated tracts comprising a part of the land on which the oysters for which recovery is sought in the first cause are presently growing.

Defendant in its brief states its position with reference to defendant's demurrer to plaintiff's amended complaint upon said first ground as follows:

> * * * All we are contending at this point is that plaintiff cannot allege the taking of one particular interest in a piece of realty in one count (e. g., the oysters, Cause of Action No. 1) and the taking of another or the remaining interest in the same parcel of realty in another count (e. g., the land underlying the oysters, Causes of Action II through VII). Having arbitrarily divided the fundamental unit (realty) claimed to have been taken into 'oysters' and 'land,' the causes of action for each interest are insufficient for want of the other.

If planted growing oysters constitute an interest in real property, there is an improper splitting up of plaintiff's claim for the taking of the realty, with the result that all the causes of action are fractional and incomplete. Allowance of separate actions for fractional claims would inevitably result in the pyramiding, duplicating and overlapping of damage items, to the detriment of the condemnor. *State, By and Through State Highway Commission v. Burk,* 200 Or. 211, 265 P.2d 783; *Pape v. Linn County,* 135 Or. 430, 296 P. 65. For holdings in inverse condemnation actions in keeping with the holding in the Burk case see *Camden & Rockland Water Co. v. Ingraham,* 85 Me. 179, 27 A. 94; *Meacham v. Fitchburg R. Co.,* 4 Cush. 291, 58 Mass. 291; *Hastings &*

*G. I. R. Co. v. Ingalls,* 15 Neb. 123, 16 N.W. 762; *Ackerman v. State,* 199 Misc. 76, 102 N.Y.S.2d 536; and *Davis v. La Crosse & M. R. Co.,* 12 Wis. 16, 18.

It would thus appear that defendant's contention that plaintiff's complaint fails to state facts sufficient to constitute a cause of action stands or falls upon a finding that the oysters with which we are here concerned are to be regarded as a part of the realty. It is conceded by defendant that if the oysters are personalty, they can properly be made the basis of a cause of action separate and distinct from the cause of action for the underlying land. If, however, the growing oysters are a part of the realty, then under the holding of the Burk case their value cannot be claimed for and proved separately.

No case from any jurisdiction has been cited, nor have we been able to find, any, which considers the question of whether planted growing oysters are realty or personalty in either direct or indirect condemnation proceedings. Many cases can be found which deal with the subject of planted growing oysters. However, we know of no case in which planted growing oysters are held to be realty in any situation.

In 22 Am.Jur., Shellfish 670, § 5, it is said:

The difference between the locomotive powers of swimming fish and shellfish, such as oysters and claims, justifies the law in making a distinction as to their ownership. In their natural state, clams and oysters are classified as ferae naturae, and their ownership is vested in the state in its sovereign capacity, but where planted where they do not naturally grow, in locations marked by posts or otherwise, they partake rather of the nature of ferae domitae and are the subjects of private ownership, although their owner has no greater actual possession than is evidenced by their planting and staking. In the latter case, they may be the subject of larceny; and if one injures or converts such shellfish, he is liable to respond in damages.

In *People v. Morrison,* 194 N.Y. 175, 86 N.E. 1120, 1121, 128 Am.St.Rep. 552, it is held:
"* * * When clams or oysters are reclaimed from nature and transplanted to a bed where none grew naturally, and the bed is so marked out by stakes as to show that they are in the possession of a private owner, they are personal property and may become the subject of larceny."

Where oyster seed which owners had planted in tidelands drifted onto adjoining tidelands during a storm, the court in *Edison Oyster Co. v. Pioneer Oyster Co.,* 22 Wash.2d 616, 157 P.2d 302, 307, held:
'There is no question but that the oyster seed, and the oysters resulting therefrom, constituted personal property. We think it must also be conceded that appellant could have immediately gone upon the Nauman tracts and reclaimed the oysters, or, if such rights were denied them, they could have instituted an action of replevin to obtain such oysters, upon demand first being made.'

In holding that mussels in a stream or pools are not part of the realty giving a right of action for treble damages for damage to real estate, Mr. Justice Holmes, speaking for the court in *McKee v. Gratz,* 260 U.S. 127, 43 S.Ct. 16, 18, 67 L.Ed. 167, said:
'As to the rule of damages in case the plaintiff recovers, in the absence of a decision by the Supreme Court of the State we should not regard the mussels as part of the realty within the meaning of the statute relied upon in the second count * * *.'

See also *Robins Island Clam Co. v. Gaffga,* 170 Misc. 362, 8 N.Y.S.2d 969; *Vroom v. Tilly,* 99 A.D. 516, 91 N.Y.S. 51; *State v. Taylor, 3 Dutch.* 117, 27 N.J.L. 117, 72 Am.Dec. 347.

We hold that the planted crop of oysters growing on the real property referred to as Silver Point No. 1 are personal property and should not be regarded as a part of the realty in the within inverse condemnation proceeding. We are unable to agree with the contention of the defendant that in a condemnation proceeding, whether direct or indirect, planted growing oysters should be likened to growing crops. We are not unmindful that property that is personalty in one situation may very well be realty in another. *Jones v. Adams,* 37 Or. 473, 59 P. 811, 62

P. 16, 50 L.R.A. 388. However, it would indeed require an unrestrained excursion in mental gymnastics to arrive at the conclusion that a crop of oysters (animals) constitutes a part of the realty in any situation. Unlike shrubs and other plants which under certain circumstances have been held to be a part of the realty, oysters are not in any way rooted to the soil and indeed obtain their sustenance from the water surrounding them and not from the land. It follows that the defendant's demurrer upon the ground that each cause of action appearing in plaintiff's amended complaint fails to state facts sufficient to constitute a cause of action is not well taken and should have been overruled.

The Washington case of *Ouellette et al. v. Olympia Jim et al.*[91] relates to the interpretation of real vs. personal property dependent on the existence of a meander line.

Three actions, by Elizabeth Ouellette and others against Olympia Jim and wife and others, against Dick Jackson and wife and others, and against Harrie T. Korter, which were consolidated for trial. From a judgment for defendants, plaintiffs appeal. Affirmed.

These several actions were instituted to recover possession of and quiet title to parts of lot 4 of section 13, Tp. 19, N. R. 3 west, W. M., in Thurston county, and were consolidated for the purposes of trial. From a judgment in favor of the defendants the plaintiffs have appealed.

The Northern Pacific Railroad Company received a patent for lot 4, and other lands, under the act of Congress of July 2, 1864, c. 217, 13 Stat. 365, and the appellants claim title to lot 4 through mesne conveyances from that company. The parts of lot 4 in controversy lie below the line of ordinary high tide on Totten Inlet, and are claimed by the respondents, as oyster lands, under deeds and contracts from the state. *The only question we deem it necessary to determine on this appeal is the correct location of the meander line along the westerly side of fractional section 13, of which lot 4 forms a part. Fractional section 24 lies immediately south of fractional section 13, and there is no dispute between the parties as to the true location of the following monuments bounding these two sections, viz., the northwest corner of fractional section 13, the northeast corner of fractional section 13, the southeast corner of fractional section 13, which is also the northeast corner of fractional section 24, and the southeast corner of fractional section 24.* [Emphasis supplied by author.]

The principal question in controversy is the true location of the southwest corner of fractional section 24, for with that corner located the meander line can be readily traced from the field notes by courses and distances. In locating this corner and tracing the meander line, the appellants followed the field notes of the original survey on file in the office of the Surveyor General, by courses and distances, from the southeast corner of the section, and, if this method is to be adopted and followed, it is conceded that the lands in controversy are within the confines of lot 4, and are included in the patent under which the appellants claim. *Kneeland v. Korter,* 40 Wash. 359, 82 P. 608, 1 L. R. A. (N. S.) 745. This line we will hereafter refer to for convenience as the appellants' survey. The respondents on the other hand made a survey of the south line of section 24, and the west line of fractional sections 24 and 13, which we will refer to as the respondents' survey. By this latter survey the southwest corner of fractional section 24 was located, by reference to the bearing or witness trees described in the field notes, at a point 5.90 chains east of the corner as fixed by the appellants' survey. In other words, the distance along the south line of fractional section 24, according to the measurements given in the field notes, is 79.50 chains, whereas the distance is only 73.60 chains, if the southwest corner is located by reference to the bearing or witness trees mentioned in the field notes. The rule prescribed by the Department of the Interior for relocating missing or lost corners is the following: "The identification of mounds, pits, and witness trees, or other objects noted in the field notes of survey, affords the best means of relocating the missing corner in its original

---

position. If this cannot be done, clear and unquestioned testimony as to the locality it originally occupied should be taken, if such can be at all obtained. In any event, whether the locus of the corner be fixed by the one means or the other, such locus should always be tested and proven by measurements to known corners. No definite rule can be laid down as to what shall be sufficient evidence in such cases, and much must be left to the skill, fidelity, and good judgment of the surveyor in the performance of his work." 1 Land Decision, 676.

Many of the rules given for the ascertainment of lost boundaries are not inflexible; but in this case the bearings of the post at the southwest corner of section 24 are thus given in the field notes:

A fir, 20 ins. dia., N. 43~ E., 66 lks. dist.,
A fir, 14 ins. dia., N. 29~ W., 100 lks. dist.,
A fir, 18 ins. dia., S. 3~ E., 18 lks. dist.,

– and an error in the measurement of the distance across the section is far more likely to occur than an error in the measurement of the shorter distances to the several witness trees. Furthermore, the respondents' survey located the west center post of section 24, by reference to the bearing trees at that point, and this location corresponds with the southwest corner of the section, when located by reference to the bearing trees there, whereas, the appellants' survey located this same post approximately six chains farther west. Again, according to the field notes the surveyor "set a post on the S.E. beach of Totten Inlet, for cor. of fracl. sec. 13 and 24, and piled stones around it as per instructions." This corresponds with the respondents' survey at the same point, which located this corner approximately at the line of ordinary high tide, whereas, the appellants' survey located the same corner approximately 25 rods below the line of ordinary high tide, in upwards of 18 feet of water at extreme high tide. We are convinced that no corner was ever located at such a point, and that no meander line was ever surveyed or located along the line indicated by the appellants' survey; for such a line could only be run at extreme low tide, if at all. The record in this case abundantly shows that government surveying is not one of the exact sciences, but we are fully convinced that the respondents' survey is approximately correct, and that no part of the land in controversy lies within lot 4, as claimed by the appellants. The judgment is therefore affirmed.

## EXTENDED LITIGATION BASED ON EARLY PATENTS

While the foregoing decisions related to ocean resources and farming, the foundation of rights, leases and regulations is based on the ownership of the seabed. The basis of the title is traceable back to the sovereign, in many cases a foreign power. Title creation is necessarily very early, and many times demands an examination and interpretation of the original grant, patent, or treaty. The following cases are based on wording and interpretation of 17th-century patents. It is an important illustration of what is necessary in an investigative process in order to arrive at the correct conclusion. It also demonstrates the importance of knowing and understanding the history of an area to relate early documents to the circumstances and conditions at the time, a basic requirement for document interpretation.

### FISH FARMS

#### Fish Farming or Pisciculture

It involves raising fish commercially in tanks or enclosures such as fish ponds, usually for food. It is the principal form of aquaculture, while other methods may fall under mariculture. A facility that releases juvenile fish into the wild for recreational fishing or to supplement a species' natural numbers is generally referred to as a fish hatchery. Worldwide, the most important fish species produced in fish farming are carp, tilapia, salmon, and catfish.

## Mariculture

It is a specialized branch of aquaculture involving the cultivation of marine organisms for food and other products in the open ocean, an enclosed section of the ocean, or in tanks, ponds, or raceways which are filled with seawater. An example of the latter is the farming of marine fish, including finfish and shellfish like prawns, or oysters and seaweed in saltwater ponds. Non-food products produced by mariculture include: fish meal, nutrient agar, jewelry (e.g. cultured pearls), and\ cosmetics.

Farming of abalone began in the late 1950s and early 1960s in Japan and China. Since the mid-1990s, there have been many increasingly successful endeavors to commercially farm abalone for the purpose of consumption. Overfishing and poaching have reduced wild populations to such an extent that farmed abalone now supplies most of the abalone meat consumed. The principal abalone farming regions are China, Taiwan, Japan, and Korea. Abalone is also farmed in Australia, Canada, Chile, France, Iceland, Ireland, Mexico, Namibia, New Zealand, South Africa, Spain, Thailand, and the United States.

After trials in 2012, a commercial "sea ranch" was set up in Flinders Bay, Western Australia to raise abalone. The ranch is based on an artificial reef made up of 5000 (as of April 2016) separate concrete units called *abitats* (abalone habitats). The 900 kilograms (2,000 lb) habitats can host 400 abalone each. The reef is seeded with young abalone from an onshore hatchery.

The abalone feed on seaweed that has grown naturally on the habitats; with the ecosystem enrichment of the bay also resulting in growing numbers of dhufish, pink snapper, wrasse, Samson fish, among other species.

Brad Adams, from the company, has emphasized the similarity to wild abalone and the difference from shore-based aquaculture. "We're not aquaculture, we're ranching, because once they're in the water they look after themselves."

Fish farms are of two types: inland and offshore. The most commonly known fish farm in this part of the world is the salmon farm, but in Asia, where fish farming is a huge industry, farms contain shellfish and fin-fish, including oysters, shrimp, eel, tilapia. Some of the coastal states in the United States have shellfish farms or various types, some private, but mostly commercial. Whatever the type, they have a title, and of necessity, boundaries, whether general, or specific.

Inland, fish farming has a long history of raising catfish and carp, and more recently species such as tilapia, and barramundi.[92]

Like inland vegetable farms, fish farms may be leased, inland on dry land and offshore within a designated riparian zone. Where a person does not own the underlying fee (seabed or lake bed) a person may rent or lease space for growing a crop, subject to required permits and licenses. Popular in some areas are food and sport fishes on private lands, and colorful pet fish in backyard pools.

# DOCKOMINIUMS AND BOAT SLIPS

### Inland Waters vs. Tidal Waters

Since the basic definitions of navigability vary between inland waters and tidal waters, as well as between states, the prevailing rules must be consulted for any particular location. Border waters generally have between-state agreements as to what uses can be made jointly, or specifically by state. The same applies to those border waters between countries. The basic concept is that one rule does not fit all situations.

---

[92] Recently, it was reported that a successful sustainable salmon farm is based in Miami, Florida. An article by Sarah Cox, December 2020, in *The Narwhal* details the attributes and future of land-based aquaculture.

## REAL PROPERTY VS. PERSONAL PROPERTY

Docks permanently attached to the upland when both are owned by the same entity, like buildings and other permanent structures, are generally considered to be part of the real estate. Removable docks and temporary structures may be personal property, and have not connection, physically or legally, to the upland real estate. The rules are not the same for each.

## DEFINITION

A dockominium is a series of boat slips on a dock that can be sold individually. *Island Properties Associates v. Reaves Firm, Inc.*[93] They are found on tidewater with its set of rules, or on inland waters, governed by a different set of rules, which may be federal, state, local, or a combination, depending on jurisdiction.

A **dockominium** is the water-based version of a condominium; rather than owning an apartment in a building, one owns a boat slip on the water. In addition to the exclusive right to use the boat slip, ownership also provides one with the right to use the common elements of the marina, much the same as one would have the right to use the common areas in a residential condominium development. Also, unit owners may use, rent, or sell their unit at any time, subject to association approval.

Similar to a condominium, a management company manages the common areas and provides all required services such as maintenance, security, insurance, bookkeeping, legal, and overall management and supervision of the dockominium facility. A monthly fee is charged to cover these expenses. Typically, water is included, while electricity and cable, etc. are billed separately via the management association. Real estate taxes are separately assessed by the municipality and are the responsibility of the unit owner.

A dockominium is created when a marina converts, sells or leases individual slips to individual owners. Traditionally, marinas are in the business of renting or leasing space. A comparison would be the conversion of a rental apartment to a condominium. An association is created that monitors the maintenance and operation of the marina. Individual owners are responsible for paying their monthly, quarterly, or annual association dues and for paying their own property taxes assessed on the slip. Dockominium conversions are a popular trend taking place in the marina industry in high demand areas focusing on the luxury markets.

However, despite the advantages, whether or not dockominium sales are legal varies according to laws of each area. Few marina owners also own the land under the water, and most have only an easement to the property. Individual unit sales may violate law, thus following the legal concept of the public trust doctrine that provides that public trust lands, waters, and living resources in a state are held by the State in trust for the benefit of all of the people.

---

[93] 413 S.W.3d 392 (Tenn.Ct.App. 2013).

# Section III

---

*Locating Original Surveys
and Related Information*

# 8 Finding the Original Survey

Ask and it will be given to you; seek and ye shall find.

**Matthew 7: 7**

## COMBINATIONS OF METES & BOUNDS AND RECTANGULAR SURVEYS AND DESCRIPTIONS

One of the greatest insights to approaching a situation is analyzing exactly what one is dealing with. A simple survey, regardless of its magnitude, is one thing, however most problems are more complex, sometimes far more complex. Separation into the correct system(s) is paramount to understanding the nature of the situation, and especially where information may be found. During the process, it will sometimes be discovered that more than one category is involved, and each must be kept separated, for information may exist in several locations, and approaches to resolutions of problems may vary depending on the nature of the category.

Nationwide, in the colonial states and in the Public Land Survey System (PLSS), rectangular systems, once created as such, are frequently later broken down by "metes and bounds" creations as well as other systems, leaving behind a conglomeration of combinations of surveyed and non-surveyed descriptions. One of the biggest challenges is to sort through the variations, giving priority to beginning creations of title, followed by a chronology of events tracing titles forward to present day. This can be very time-consuming and frustrating for the less-than-very experienced researcher. Frequently, such a study demands one or more specialists versed in a variety of title and survey considerations. A step-by-step process from beginning to present day is a key to understanding the history of what took place with a particular tract of land.

## VARIATIONS

In the colonial states, metes and bounds descriptions are prevalent, but a host of rectangular townships and their divisions are found in many areas. Some of these have been further subdivided, or otherwise described, with metes-and-bounds methods. When Tennessee was separated from North Carolina, nearly the entire state as we know it today, was divided in rectangular fashion (Figure 8.1). Subsequently, even after some title transfers had taken place, the system fell with disfavor and people reverted to metes and bounds descriptions for their conveyancing. In the PLSS, a number of systems may be found, depending on the history of the various settlements. French systems are found in Louisiana, and the Midwestern and lake states, as well as elsewhere. Spanish systems are found in Florida, Texas, and the Southwest, along with large grants known as Ranchos in some areas, particularly in California. Special grants within the PLSS, many described by metes and bounds, occurred for special purposes under separate authority, such as mineral surveys, homestead entries, townsites, and a host of others. These, along with early grants in colonial states before rectangular lotting, are earlier titles requiring special treatment. A glaring example is the numerous Spanish land grants made in Florida prior to acquisition by the United States followed by the superimposition of its rectangular system. Some of these overlap and give rise to conflicting title concerns, many of which require litigation to resolve. Many of the very earliest ones are senior surveys, and therefore take priority and are controlling over subsequent activities.

DOI: 10.1201/9781003032557-16

# Table of Consanguinity

Showing degrees of relationship

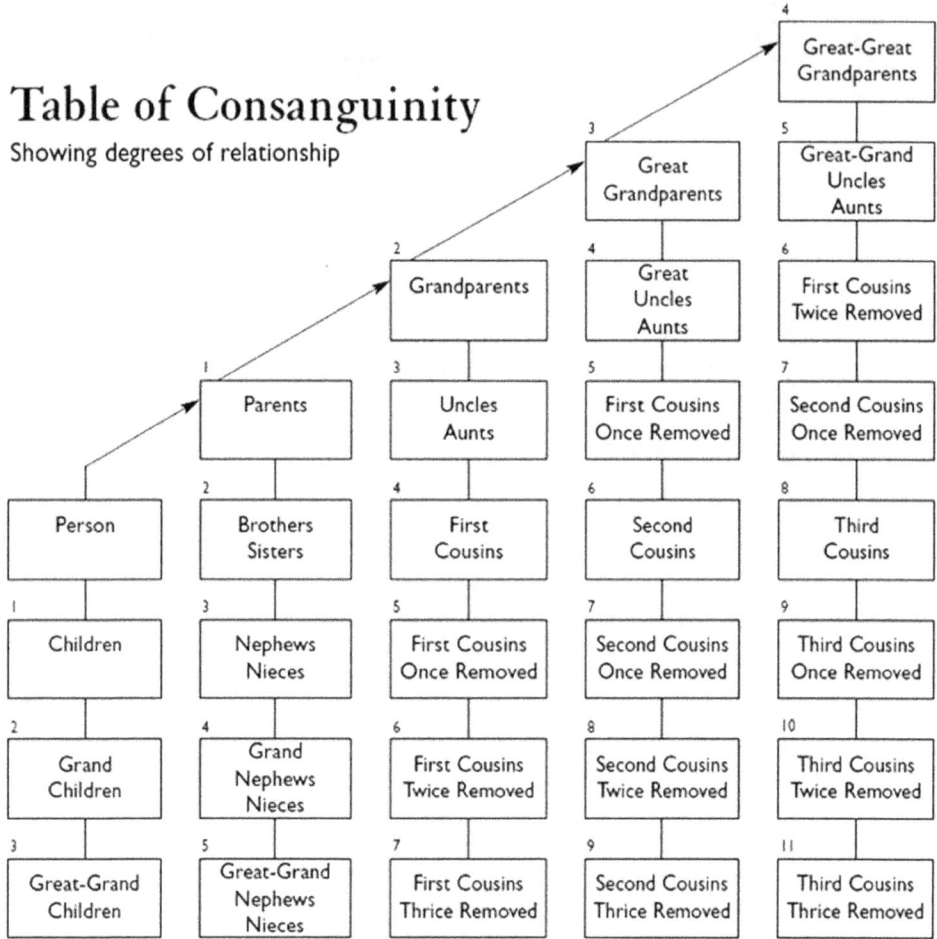

| | | | | Great-Great Grandparents (4) |
|---|---|---|---|---|
| | | | Great Grandparents (3) | Great-Grand Uncles Aunts (5) |
| | | Grandparents (2) | Great Uncles Aunts (4) | First Cousins Twice Removed (6) |
| | Parents (1) | Uncles Aunts (3) | First Cousins Once Removed (5) | Second Cousins Once Removed (7) |
| Person | Brothers Sisters (2) | First Cousins (4) | Second Cousins (6) | Third Cousins (8) |
| Children (1) | Nephews Nieces (3) | First Cousins Once Removed (5) | Second Cousins Once Removed (7) | Third Cousins Once Removed (9) |
| Grand Children (2) | Grand Nephews Nieces (4) | First Cousins Twice Removed (6) | Second Cousins Twice Removed (8) | Third Cousins Twice Removed (10) |
| Great-Grand Children (3) | Great-Grand Nephews Nieces (5) | First Cousins Thrice Removed (7) | Second Cousins Thrice Removed (9) | Third Cousins Thrice Removed (11) |

**FIGURE 8.1**   Refer to Figure 2.7. An original township division into lots in Maine (then Massachusetts). This town, Sapling Township, is easterly of and abuts Misery Township shown in Figure 2.8. (*Atlas of Somerset County, Maine.* Geo. N. Colby & Co. 1883.)

There are many more variations and considerations, far too numerous to be considered here. Mostly, many of them will be found to be unique, demanding a novel approach. However, a number of situations have been resolved by the courts, providing guidance. Otherwise, situations must be examined one at a time, as they arise, demanding special expertise to identify the particular problem(s), and seeking its ultimate resolution.

## HOW TO GET IT

**Considering that the court system requires, as a matter of law, that a retracement surveyor, as well as others, rely on the original survey.**[1] Where do we find the original survey? It depends

---

[1] There is no difference between private actions and the subdivision of public lands for which separation lines are created by the U.S. Congress and later marked on the ground by federal surveyors. The lines do not become permanent until the plat is appropriately filed, after which a title transfer takes place in some form of public grant, usually a patent. *Rivers v. Lozeau*, 539 So.2d 1147 (Fla. App. 5 Dist. 1989).

on the survey, why it was done and whether there is an official repository for it. Some are available, often found with difficulty, others are long gone and not recoverable. In such cases, the law provides for alternate means for location of property. We must undertake a diligent search and proceed if every original survey, or its equivalent, is recoverable. Only when there is absolutely no hope of finding original information may we resort to less desirable options. This is emphasized in a number of decisions, including *U.S. v. Doyle,*[2] which states, succinctly, "For corners to be lost, they must be so completely lost that they cannot be replaced by reference to *any existing data* or other sources of information, and before courses and distances can determine boundary, *all means* for ascertaining location of the lost monuments must first be exhausted." Emphasis by author.

The court system has dealt with missing and lost records for a very long time, and, over time, has established some reliable options. However, it should be emphasized that options are only valid and permissible, when better or prior evidence is forever absent.

**Author's Note:** Do not forget that, even though there are recognized alternative procedures, such as apportionment of excess and deficiency, and proportioning a lost corner, such procedures are only valid when better methods have failed – completely. Many a corner considered "lost" has been later recovered because someone exercised more perseverance, or sought to investigate evidence not considered before. Almost always, resorting to an alternate solution because of failure to follow the correct procedure, can only lead to exacerbating the initial problem, and likely creating additional concerns. *Suggestion:* Don't be the one to perpetuate a problem, or make one worse. Eventually it will be discovered, when land values demand it, or when investigated by some more experienced, more clever, or more stubborn than you.

## UNITED STATES PATENTS AND GRANTS

As previously noted several times, a federal patent does not become a legal entity until the survey has been completed and the appropriate paperwork filed in the required locations.

### ORIGINAL SURVEY RECORDS

Township plats furnish the basic data relating to the survey and the description of all areas in the particular township. All title records within the area of the former public domain are based upon a Government grant or patent, with description referred to an official plat. The lands are identified on the ground through the retracement, restoration, and maintenance of the official lines and corners.

The plats are developed from the field notes; both are permanently filed for reference purposes and are accessible to the public for examination or making of copies in the various offices listed in the supplement to the Manual of Instructions.[3]

Many supplemental plats have been prepared by protraction to show new or revised lottings within one or more sections; these supersede the lottings shown on the original township plat. There are also many plats of the survey of islands or other fragmentary areas of public land which were surveyed after the original survey of the township. These plats should be referred to governing the position and description of the subdivisions shown on them.[4]

### RESURVEY RECORDS

The plats and field notes of resurveys which become a part of the official record fall into two principal classes according to the type of resurvey, as follows:

---

[2] 468 F.2d 633 (C.A. Colo., 1972).

[3] *Restoration of Lost or Obliterated Corners and Subdivision of Sections*, USDI, BLM, various editions.

[4] *Restoration of Lost or Obliterated Corners and Subdivision of Sections* (1952, reprinted 1955).

The *dependent resurvey* is a restoration of the original survey according to the record of that survey, based upon the identified corners of the original survey and other acceptable points of control, and the restoration of lost corners in accordance with proportional measurement as described in the supplement to the Manual as noted above. Normally, the subdivisions shown on the plat of the original survey retained on the plat of the dependent resurvey, although new designations and areas for the public land subdivisions may be shown to reflect true areas.

The *independent resurvey* is designed to supersede the original survey and creates new subdivisions and lottings of the vacant public lands. Provision is made of the segregation of individual tracts of privately owned lands, entries, or claims that may be based upon the original plat, when necessary for their protections, or for their conformation, if feasible, to the regular subdivisions of the resurvey.[5]

## IMPORTANCE OF PLAT AND FIELD NOTES

The importance, or legal significance, of the plats and field notes is well stated in an opinion by the Department of the Interior (45 L.D. 330, 336) as follows:

It has been repeatedly held by both State and Federal Courts that plats and field notes referred to in patens may be resorted for the purpose of determining the limits of the area that passed under such patents. In the case of *Cragin v. Powell* (128 U.S. 691), the Supreme Court said:

> It is a well settled principle that when lands are granted according to an official plat of the survey of such lands, the plat, itself, with all its notes, lines, descriptions and landmarks, becomes as much a part of the grant or deed by which they were conveyed, and controls so far as limits are concerned, as if such descriptive features were written upon the face of the deed or the grant itself.[6]

## RECORDS TRANSFERRED TO STATES

In those states where the public land surveys are considered as having been completed, the field notes, plats, maps, and other papers relating to those surveys have been transferred to an appropriate State office for safekeeping as public records. No provision has been made for the transfer of the survey records to the State of Oklahoma, but in the other States the records are filed in the offices as listed in the supplement to the Manual.[7]

## GENERAL PRACTICES

The rules for the restoration of lost corners have remained substantially the same since 1883, when first published as such.[8] These rules are in harmony with the leading judicial opinions and the most approved surveying practices. They are applicable to the public land rectangular surveys, within the Public Land States, and to the retracements of those surveys, as distinguished from the running of property lines that may have legal authority only under State law, court decree, or agreement.[9]

---

[5] Ibid.

[6] Ibid.

[7] Ibid.

[8] See Wilson, *Boundary Retracement: Principles and Processes* for an extensive discussion of lost corners and a list of court decisions by state.

[9] "While these general rules [as reviewed in this decision] apparently have their origin in surveys reflecting government grants, such rules are equally applicable to private surveys." *Wood v. Starko*, 197 S.W.3d 255 (Tenn.App. 2006) quoting *Staub v. Hampton*, 117 Tenn. 706, 101 S.W. 776 (1907).

In the New England and Atlantic Coast States, except Florida, and in Pennsylvania, West Virginia, Kentucky, Tennessee, and Texas, jurisdiction over the vacant land remained in the States. The public land surveys were not extended in these States, and it follows that the practices outline herein are not applicable there, except as they reflect sound surveying methods.

The practices outlines are in accord with the related provisions of the Manual; they have been segregated for the convenience of the reader in order to separate them from the instructions pertaining only to the making of original surveys.[10]

## GENERAL RULES

The general rules followed by the Bureau of Land Management which are controlling upon the location of all public lands, are summarized in the following paragraphs:

*First: That the boundaries of the public lands, when approved and accepted, are unchangeable.*

*Second: That the original township, section, and quarter-section corners must stand as the true corners which they were intended to represent, whether in the place shown by the field notes or not.*

*Third: That quarter-quarter-section corners not established in the original survey shall be placed on the line connecting the section and quarter-section corners, and midway between them, except on the last half mile of section lines closing on the north and west boundaries of the township, or on the lines between fractional or irregular sections.*

*Fourth: That the center lines of a section are to be straight, running from the quarter-section corner on one boundary to the corresponding corner on the opposite boundary.*

*Fifth: That in a fractional section where no opposite corresponding quarter-section corner has been or can be established, the center line must be run from the proper quarter-section corner as nearly in a cardinal direction to the meander line, reservation, or other boundary of such fractional section, as due parallelism with the section boundaries will permit.*

From the foregoing it will be evident that corners established in the public land surveys remain fixed in position and are unchangeable; and that lost or obliterated corners of those surveys must be restored to their original locations from the best available evidence of the official survey in which such corners were established.

## RETRACEMENTS

After the original title and its accompanying survey are safely within the appropriate repositories, sometimes long after the placement, the next order of business, at some point, is its retracement. The rules are clear, the responsibilities spelled out, but the obstacles are many. Understanding the underlying laws, the sanctity of ownership (title) and knowing how to investigate, are the keys to overcoming those obstacles. It is quite simple once proper understanding and training are in place; in the federal system, it is as easy as it gets. Records are available in accessible locations, unlike earlier systems which are much older, with documentation either lost or in unknown locations. Two examples follow that illustrate the problems that were solved by the institution of an orderly system, despite its (relatively few) shortcomings.

Where the surveyor is called upon to retrace the lines of the rectangular public-land surveys, the problem requires a careful study of the record data. The first step is to assemble copies of

---

[10] *Restoration of Lost or Obliterated Corners and Subdivision of Sections* (1952, reprinted 1955).

the field notes and plats, and determine the names of the owners who will be concerned in the retracement and survey. A thorough search and inquiry with regard to the record of any additional surveys that have been made since the approval of the original survey should be made. The county surveyor, county clerk, register of deeds, practicing engineers and surveyors, landowners, and others who may furnish useful information should be consulted as to such features.

The matter of boundary disputes should be carefully reviewed, particularly as to whether claimants have based their locations upon evidence of the original survey and a proper application of surveying rules. If there has been a boundary suit, the record testimony and the court's opinion and decree should be carefully examined insofar as these may have a bearing upon the problem at hand.

The law requires that the position of original corners shall not be changed. There is a penalty for defacing corner marks, and for changing or removing a corner. The corner monuments afford the principal means for identification of the survey, and accordingly the courts attach the greatest weight to the evidence of their location. Discrepancies may be developed in the directions and length of lines, as compared with the original record, do not warrant any alteration of a corner position.

Obviously, on account of roadways or other improvements, it is necessary to preserve its position. Alterations of that kind are not regarded as changes in willful violation of the law, but rather as being in complete accord with the legal intent to safeguard the evidence.

Therefore, whatever the purpose of the retracement may be – if it calls for the recovery of the true lines of the original survey, or for the running of the sub-divisional lines of a section, the practices outlined require some or all of certain definite steps, as follows:

a. Secure a copy of the original plat and field notes;
b. Secure all available data regarding subsequent surveys;
c. Secure the names and contact the owners of the property adjacent to the lines that are involved in the retracement;
d. Find the corners that may be required;

First: by the remaining physical evidence;
Second: by collateral evidence, supplemental survey records, or testimony, if the original monument is regarded as obliterated, but not lost, or;
Third: by application of the rules for proportionate measurement, if lost;

e. Reconstruct the monuments as required, including the placing of reference markers where improvements of any kind might interfere, or if the site is such as to suggest the need for supplemental monumentation;
f. Note the procedure for the subdivision of sections where these lines are to be run; and,
g. Prepare and file a suitable record of what was found, the supplemental data that was employed, a description of the methods used, the direction and length of lines, the new markers, and any other facts regarded as important.

A knowledge of the practices and instructions in effect at the time of the original survey will be helpful. These should indicate what was required, and how it was intended that the original survey should be made.

The data used in connection with the retracements should not be limited to the section or sections under immediate consideration. It should also embrace the areas adjacent to those sections. The plats should be studied carefully; fractional parts of sections should be located on the ground as indicated on the plats.[11]

---

[11] Ibid.

## OTHER PATENTS

Prior to GLO (General Land Office) surveys (1785), and prior to the United States becoming a sovereign nation (1784), patents and other grants had been issued by other nations – England, France, Spain, Netherlands, Sweden, and Russia being the most prevalent. Original paperwork can be elusive, even impossible to find.

For example, one of the largest land development schemes ever was instituted by the Forbes Company in the southeastern part of the United States.

## AN UNSUCCESSFUL ATTEMPT AT LOCATING CRITICAL EVIDENCE: THE FORBES PURCHASE

According to one major source,[12] the Forbes Purchase of lands in Florida and beyond, has been called Florida's biggest land grab. It was a contemplated development of unprecedented proportions, at least from a private development perspective, second only to the Northeast speculations of William Bingham in the latter part of the 18th century.[13]

The Forbes Company, over more than 52 years (1783–1835), played a large and significant role in the development of lands in the Florida area. Florida is unique in that the area went through several regimes, resulting in a mixture of land grants and their descriptions. A brief history is important to understand the full impact of the various activities, culminating in a major court case postponed for lack of evidence.

Spain, having claimed the Territory of Florida based on discovery in 1513 by Ponce de Leon, made a number of land grants over the ensuing 250 years. In 1763 Spain exchanged Florida with Britain in return for control of Cuba, which had been captured by Britain. Under British control for 20 years, land grants were made to officers and soldiers who fought in the French and Indian War. In addition, settlers moved to Florida from other British colonies. Based on the Treaty of Paris in 1783, Britain was required to return Florida to Spain. Spain then encouraged settlement by offering headright grants. During this time, the boundary between the United States and Spain was surveyed and under the Treaty of San Lorenzo in 1795, the boundaries of east and west Florida defined. It was not until 1821, under the Adams-Onis Treaty, that Florida became part of the United States. It acquired statehood in 1845. The Forbes Company began acquiring lands in 1804, which inserted a wrinkle into the mix of land grants, titles, and resulting boundaries.

The Forbes saga began in 1765 with John Panton, a Scottish immigrant, coming to America and settling in Savannah, Georgia. By 1774, he became a partner in the trading firm of Moore & Panton, gaining experience as an Indian trader. The following year, he was appointed by the British as the official trader to the Creek Nation. Forming a new company with another Scotsman named Thomas Forbes; they called themselves Panton, Forbes & Company. Both were Tories, loyal to Britain, and the company relocated to St. Augustine, an area controlled by the Spanish at the outbreak of the Revolution. Properties they owned in South Carolina and Georgia were confiscated by the United States. Even so, the company continued to expand in Spanish Florida throughout the Revolutionary War.

Joined by three additional Scotsmen, forming Panton, Leslie & Company in 1783, the company had become well established. The Spanish, realizing that they were not successful Indian traders, gave the company a monopoly in East Florida. By 1785, the company expanded and moved its headquarters to Pensacola; both Forbes' and Leslie's younger brother joined the company and

---

[12] The Forbes Purchase; Florida's Biggest Land Grab. *Apalachicolabay.com/the-forbes-purchase-floridas-biggest-land-grab/*. Internet post, October 14, 2013.

[13] Approximately 1 million acres.

John Forbes opened a post in Mobile, Alabama. Within 10 years, the company had trading posts as far north as Memphis, westward to New Orleans, all along the Gulf Coast into central Florida, as well as the Bahamas and other Caribbean islands. The company had developed a working relationship with the native tribes, Creeks, Seminoles, Chickasaws, Choctaws, and Cherokees. By the late 1790's, the company had acquired 3 million acres of land in what later became the states of Mississippi and Alabama.

From extensive credit afforded the Native American partners, along with two robberies of the St. Marks trading post, the Creek and Seminole Tribes, in 1804, ceded to the company a large tract of land in North Florida that the Spanish recognized as being in trial ownership. Seven years later, to pay off additional debts, three additional tracts were transferred, who, at this time, changed its name to John Forbes and Company. The total area of all the grants was about 1.4 million acres, which included all of two counties and sizeable portions of three others. The company had Asa Hartfield of South Carolina survey their lands, first locating the northern boundary of their entire holdings, then inspecting the entire tract for potential areas suitable for settlement and cultivation.

In 1819, the Forbes Company sold the entire tract to Colin Mitchel, a Spanish merchant. Two large surveys were made, and the areas subdivided. After the territory of Florida was acquired by the United States in 1821, the private ownership of the Forbes Purchase tract was challenged, based on the well-established policy of not recognizing land transactions between Native Americans and private individuals. The Board of Commissioners disbanded and left to Congress the task of deciding whether the Forbes Purchase was valid. Congress failed to act, ignoring the problem for seven years, so Mitchel filed suit in territorial court, where a judge ruled against its validity. In 1831, Mitchel filed an appeal to the U.S. Supreme Court, which ultimately settled the dispute in Mitchel's favor, stating that the previous case of *Johnson v. M'Intosh*[14] did not apply since the Indians willingly settled their debts by transferring lands, and the Spanish government both approved and confirmed the Forbes Purchase, as required.[15]

Several lessons may be derived from a study of the Forbes Purchase. First is a search for records turns out to be less than productive. A large portion of the Purchase had already been subdivided into townships and ranges by the Public Land Survey between 1824 and 1826 before the legal dispute began. Also, the interior portion of the purchase not covered by the two surveys commissioned by Mitchel was also subdivided into townships and ranges. No contracts have been found for this work, so it is not clear whether the work was authorized by the federal government. For one of the surveys, no field notes have been found, although a plat without dimensions is available. There are field notes and a plat available for the other survey, but the plat has no measurements, and the field notes are silent as to whether corners were set.

Perhaps of even greater concern is the fact that during the preparation of the Mitchel suit, U.S. attorneys attempted to acquire supporting documents from Cuba. After four more years of delay, the court refused any further continuances, and heard arguments in 1835, ruling in a matter of three months. The ruling the last involving Chief Justice John Marshall who had heard and decided the previous suits, held that since the Spanish rulers approved the transfer, it would violate the U.S.-Spanish treaty if the transactions were not recognized as valid by the United States government.

Besides these inherent problems, a lingering question concerns grants made by the Forbes Company in areas outside of East and West Florida, since they had additional offices in what is now Alabama and Mississippi, Memphis, westward to New Orleans, along with the Bahamas and other islands in the Caribbean.

Some of these areas have also been subdivided in PLSS fashion, and without extensive chains of title back to the origin of title and survey, it cannot be known if there are overlapping titles,

---

[14] 21 U.S. 543 (1823).
[15] *Mitchel v. United States*, 34 U.S. 711 (1835).

and if so, who has superior title after so many years of occupancy, use, and trespass or if there are idle parcels of land with overlapping records, and no use because they either could not be found, descriptions when available were inadequate and ambiguous, or the land, even if identifiable, was not suitable for any desired use.

The frustrating part of this exercise was the fact that the trial was delayed for four years while a search was made for the pertinent records. They were not found, the trial proceeded regardless, with an outcome. There is no telling whether results may have been different had the records been found. In addition, surveying since the trial has been handicapped due to the lack of the records. In the future, if the records are ultimately found, what their impact may produce as a further result.

## A SUCCESSFUL ATTEMPT AT LOCATING CRITICAL EVIDENCE: THE POPHAM COLONY

Historically important is one of the first English colonies established in the New World. In June of 1606, King James I granted a charter to the Virginia Company to establish a colony in North America. Two separate factions of the Virginia Company – one based in London and one in Plymouth – immediately began planning expeditions to the coast of North America. The London Company was tasked with colonizing the more southerly section that include the modern day states of Virginia and North Carolina, while the Plymouth Company was given reign over a swath of what is now New England. In mid-May, the London Company landed first locating settlers in Virginia and establishing the Jamestown Colony. A few weeks later, the Plymouth Company set sail, headed by George Popham, with his second-in-command, Raleigh Gilbert, a nephew of Sir Walter Raleigh. The expedition arrived on the coast of Maine in August of 1607, selecting a site for their settlement. They erected numerous houses, a church, a storehouse, and a fort. A map of the settlement would later be drawn by John Hunt. The Popham residents also build a ship from local timber and sails and iron they had brought from England. The 50-foot pinnace, later christened the *Virginia of Sagadahoc*, was the first British vessel constructed in North America.

George Popham died of unknown causes in February of 1608, leaving Raleigh Gilbert in charge of the remaining settlers, a number of the originals having returned to England on the last voyage back in December. The settlement carried on until it received word that Raleigh Gilbert's brother had died, leaving Raleigh heir to the family estate. Thus the leader returned to England and his inheritance, taking the remaining settlers with him. Of the 45 settlers abandoning the settlement, some left on the ship that brought the latest news and supplies, while the remainder sailed home in the ship they had constructed.

Since the 1960s, the site of the colony and the fort have been investigated and excavated by archaeologists, uncovering an abundance of evidence. However, in 1888, during research for other projects, the only copy of the Hunt map, the sole documentation of the colony, was discovered in Simmancas, Spain. It had come into the hands of Spanish Ambassador Pedro de Zuniga and from him to Philip III of Spain. Zuniga was a well-known spymaster who kept a steady watch on events and information, with the Spanish government. England's attempts at colonization was a frequent topic. For nearly 300 years, the map, the only extant copy, had been held in good hands.

What now remained was to verify the authenticity of the map. This was done by accomplishing two things: comparing the known details with those on the map and, using the map, searching for previously unknown evidence and locations. Both procedures were successful. Between 1994 and 2005, additional excavations at the site were undertaken, locating the fort and several of its components. The most important piece of evidence documenting an entire early settlement was elusive for nearly three centuries, demonstrating that sometimes the answers are available, but located in an obscure place, taking a long time to surface.

The net result of these two examples, as well as others that could be examined is the stress on the importance of pertinent records. Once they are lost, there may not be any possibility of getting them back therefore resulting in incorrect decisions, and a difficult procedure at best. Many critical items were one-of-a-kind, sometimes created with multiple copies, totally without modern copying facilities. Preservation of evidence in the appropriate repository is very much a necessity.

## SYSTEMS OF GOVERNMENT ESTABLISHED THE FOUNDATION OF ORIGINAL RECORDS

To understand what and where the original title and survey documents are and their nature necessitates an understanding of the foundations of government and the structure of how land and land rights were once treated. As detailed in *Land Tenure, Boundary Surveys, and Cadastral Systems*,[16] the original colonies and their systems of government were of three types.

**Royal Colonies** were ruled directly by the English monarchy and administered by a royal governor appointed by the British crown with a representative assembly elected by the people. At the time of the Revolutionary War, the Royal Colonies were New Hampshire, New York, New Jersey, Virginia, North Carolina, South Carolina, and Georgia.

**Charter Colonies** were generally self-governing, and their charters granted to the colonists as opposed to proprietors. They were controlled by written contracts between the reigning king and the American colonists, defining the share that each should have in the government, which could not be changed without the consent of both parties. A charter bestowed certain rights, privileges, or franchises from the sovereign power to the colony. The Charter Colonies were all located in New England. At the onset of the Revolutionary War, the Charter Colonies were Massachusetts (later changing to Royal), Connecticut, and Rhode Island. The first Charter in North America was granted to Sir Walter Raleigh by Queen Elizabeth I. Raleigh sent his first colonization mission to the island of Roanoke, which is in present-day North Carolina.

**Proprietary Colonies** were territories granted by the English Crown to one or more proprietors, similar to a corporation, and often made up of speculators, many of whom never left Europe, who had full governing rights. These were the types of settlements in British Colonial America established between 1660 and 1690. King Charles II established the first Proprietary Colonies in order to repay debts and favors. The British Crown awarded their most loyal supporters with large tracts of land in colonial New York, New Jersey, Pennsylvania, Maryland, Delaware, and the Carolinas. These proprietors were given authority to supervise and develop the colonies into successful and profitable enterprises. The lands were quickly colonized with British subjects at the expense of the proprietors. By the time of the Revolutionary War, there were three Proprietary Colonies: Pennsylvania, Maryland, and Delaware.

These three types of colonies and systems of government in the North American British Colonies would sometimes change in status depending on political and economic changes in Great Britain, or with changes in controlling European power. The government of a self-governing charter Colony or a private Proprietary Colony therefore could become a Royal Colony when the power reverted to the king. The colonies of New York, New Jersey, North Carolina, and South Carolina began as proprietary colonies but later became royal colonies.

Any changes in status likely affect existing laws, rules, and procedures depending on who was in power. Therefore a system may be in place for awhile, then become changed, thereby affecting the survey process and the record-keeping procedures.

---

[16] George M. Cole and Donald A. Wilson. *Land Tenure, Boundary Surveys, and Cadastral Systems*. Boca Raton: CRC Press, 2017.

**Author's Note:** While not often considered by the average practitioner, early governments and their land claims and grants often can be of great significance to the land surveyor. These first plantings went hand in hand with the first surveys, done either immediately prior to the grant, or soon after.[17] Land records systems were established, sometimes experimentally and the first laws enacted, many of which were refined, some being abolished. Some think in terms of today's practices and regulations, but with early records, it is essential to think in terms of conditions at the particular time frame of the record. This is a fundamental principle of the court system. Such provides us a beginning, based on original activity.

**Example:** Value of ancient records:

The Mitchell Map is a map made by John Mitchell (1711–1768), which was reprinted several times during the second half of the 18th century. The Mitchell Map was used as a primary map source during the Treaty of Paris (1783) for defining the boundaries of the newly independent United States.

The map remained important for resolving border disputes between the United States and Canada as recently as the 1980s dispute over the Gulf of Maine fisheries. It likely remains relevant and useful today for resolving future controversies.

## NON-FEDERAL GRANTS

One of the biggest obstacles to overcome when dealing with original governmental records created by European claimants and original states, is that too often the records are not stored where the land lies, but are frequently discovered in a location relating to where they originated. The Popham Colony map described earlier is a prime example. The map went missing for over 200 years, yet still existed, and held the secrets of the Colony.

Some obvious examples are the Connecticut Firelands in Ohio, granted by Connecticut to survivors of early fires in Hartford and New Haven. For years, the original records were kept in Connecticut. North Carolina executed many grants in what is now Tennessee, as well as granting lands in South Carolina. Virginia granted lands in what is now Kentucky and West Virginia. Vermont contains original township grants made by New Hampshire, New York, and Massachusetts, all claimants at one time or another. Florida is historically known as a "Public Lands State," yet prior to the PLSS in Florida, Spain made approximately 120 small grants, all containing metes and bounds descriptions. There is an abundance of similar examples, most of which can be found almost at any place in the entire country. Spanish grants in the Southwest, particularly California, Russian grants in Alaska, Dutch Grants in New York, and French grants in Louisiana are further examples to be aware of.

Many hours have been spent, some wasted time, searching in the wrong places for original documents relating to title and survey, where if the searcher had understood the history of the grant and the law surrounding its creation, more success at greater efficiency would have been achieved. Field notes, some of the most important of all of the information, must be filed in the PLSS, but no such requirement exists in the Colonial states. Field notes generally remain the personal property of the individual surveyor, or survey firm, and catalogues of their locations can be invaluable in accessing original survey information. Many original lotting surveys and plats list the surveyor of record, but locating those records can be a formidable task.

---

[17] A pitch is a type of land grant location, defined as the selection of the general location of a claim or share by entry and occupation. See *Garland v. Rollins*, 36 N.H. 349 (1858). After formal lotting and granting, if a grantee was short-changed on acreage, or if a potential grantee appeared later, they could be allotted land in a common area, which they would selected a parcel on the ground, make measurements or have it surveyed, then make a return (description) in a separate book known as the "pitch book." These would be equivalent to original surveys and original descriptions, since they would establish a title and a location.

## SYSTEMS OF LAND TENURE

Most colonial states, as we know them today, whether a colony or otherwise, had a system of land grants which created original title sources (from the sovereign) with accompanying original location of parcels and their respective boundaries.

Colonies, which later became states in one form or another, each had one or more land systems that were unique to that government. Sometimes, when the government changed, a land system may also change accordingly.

Rectangular systems are found within a number of colonial (so-called metes and bounds) states, each unique by itself. Examples include most of Maine, New Hampshire, Vermont, and New York, Georgia with widespread rectangular lotting. Texas, with its own General Land Office, and Louisiana (considered a rectangular state) with their systems of longlots, and two variants – Tennessee lotted in rectangular fashion but later mostly abandoned in favor of traditional metes and bounds and Kentucky with a small section in the western part lotted in rectangular fashion, the remainder of the state consisting of traditional early colonial metes and bounds conveyancing.

## THE VALUE OF FIELD NOTES

Historically, and emphasized in court decisions, the importance of field notes cannot be over-stressed. In the PLSS, field notes are protected and archived, so are, mostly, readily available. In the colonial states, by contrast, no such filing requirements have been in place, leaving such records in the hands of the surveyors who transcribed them in the field. Often these were one-of-a-kind records, filed with the personal survey information, and frequently lost after the surveyor moved on. As the court stated in the *Outlaw*[18] case, and found in other cases, the work on the ground is paramount, field notes describe what was done on the ground, and the resulting plat is [supposed to be] a picture of the survey that was done. To find the evidence on the ground, field notes are an indispensable tool, as profusely illustrated in the case of *U.S. v. Champion Papers, Inc.*[19] Yet, they are not always treated with the respect they deserve. Plats, by comparison, being a product of field notes and memory, are inherently unreliable even though they are relied upon by users, and others, sometimes to their peril. If a plat, or other type of drawing, is the best, or only, remaining evidence, that is a fact which must be dealt with, but often that is not the case, and the field notes, as elusive as they might be, supersede the plat and are ordinarily the better evidence in the absence of the original points on the ground.

**Example:** A plan was produced based on historical research and a previous survey creating the lot as shown in Figure 8.2. The plan was challenged by the abutting owner, who had accomplished the same research of deeds and probate materials at the local court house. The abutter challenged the latest survey, citing that the previous survey did not close and the mathematics were unreliable. Their conclusion was that the location of the westerly (northwesterly) line was incorrect according to their conclusions. A copy of the field notes of the survey (Figure 8.3) were produced which stated that a [barbed] wire fence defined that line according to the previous (original) survey, and that the current survey for that parcel discovered the remains of the fence, and relied on its location to define the location of the line in question.

The second surveyor had not discovered the field notes, and had not seen a copy of the current survey, relying solely on the previous plan and accompanying deed, both of which contained secondary evidence in the form of measurements.

**Author's Comment:** There is a lesson to be learned from this case. Even though the fence was found and located during the original survey, whoever drafted the plan, supposedly from the field

---

[18] *Outlaw v. Gulf Oil Corp.*, 137 S.W.2d 787, (1940).
[19] 361 F.Supp. 481 (S.D.Tex.1973).

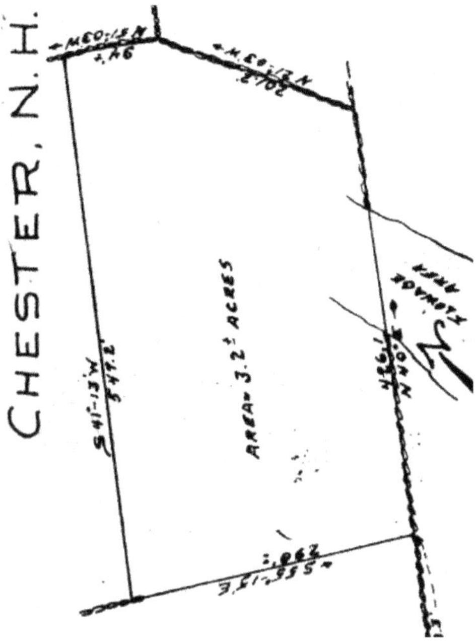

**FIGURE 8.2** Survey plat.

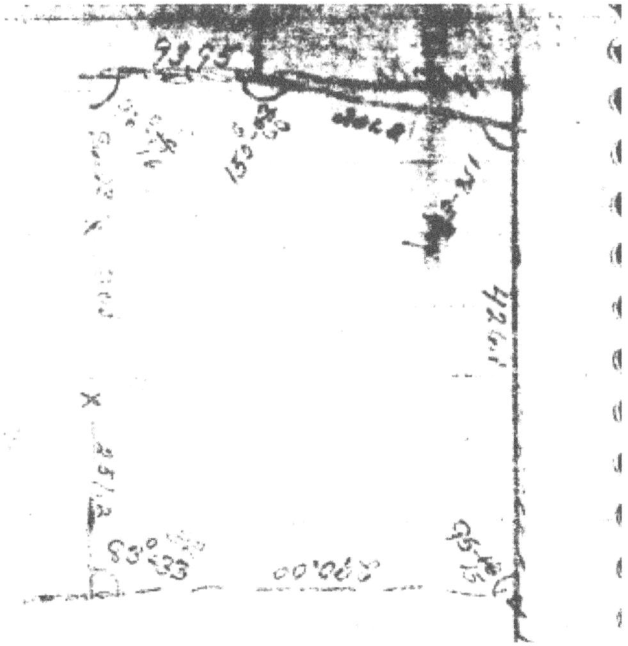

**FIGURE 8.3** Survey field note record. No wire fence shown on plat, but shown in the sketch in the field notes.

notes, neglected to show it on the plan. This was the source of the confusion, because strictly relying on the mathematics leads to the wrong conclusion.

**Example:** A survey was planned for Parcel 10 as shown on the assessor's map, Figure 8.4. The current deed contained a description by abutting owners dating back to 1890, stating an area of 40 acres. The only measurements provided were for the two outsales. The parent tract was traced back to the original survey of 1755, shown in Figure 8.5, with a plan and field notes describing the subject parcel and its abutting tracts. Deed research verified that the 40 acres was composed of 3 parcels as shown on the original survey, 32 acres, 4 acres, and 4 acres all acquired by Jeremiah Leavitt from 3 different owners. According to the original survey, the entire tract should be 40 rods on the road and extending back 160 rods, for an area of 40 acres. The deed in 1755, based on this partition survey, provided general directions and dimensions (based on the survey) of 40 by 160 rods. A retracement survey done in 1988 compared favorably.

This a classic example of relying on tax map information that can be far from reality. The true location of the boundary is derived from the other survey or the original title documents, not from someone's drawing for taxing purposes, based on far less than adequate information.

One of the more difficult tasks in the colonial states is finding the original field notes. Like early plats and plans, which were usually one copy, one-of-a-kind drawings, paperwork frequently gets lost, misplaced, or destroyed. People tend to focus on their difficulty in finding ancient marks and bounds, instead of the paper guides that often will aid their search for the physical evidence. Some states have gone to great lengths to preserve early records, and have

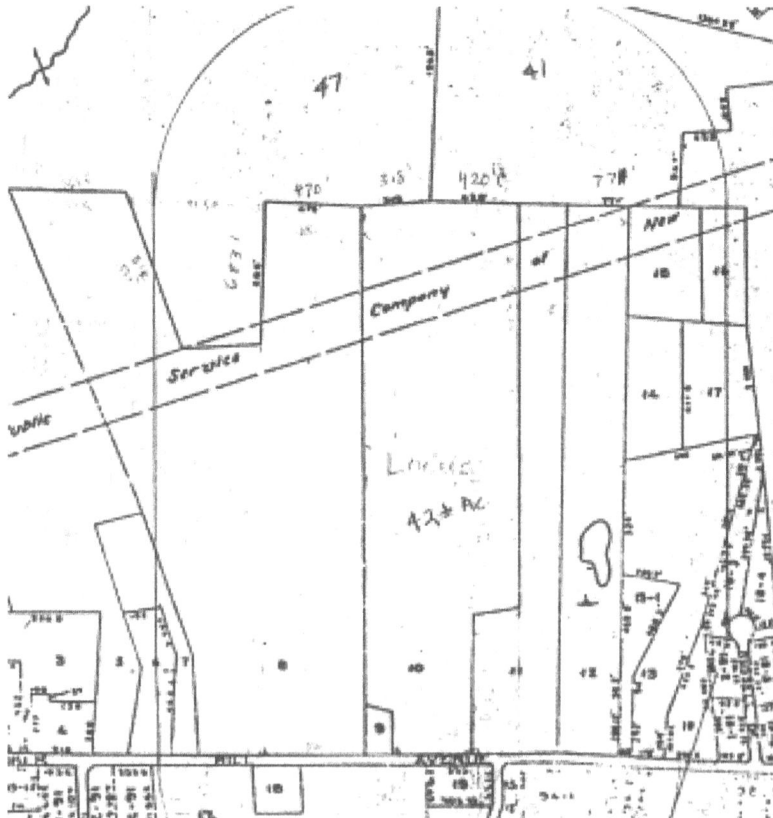

**FIGURE 8.4**  Assessor's map of parcel to be surveyed.

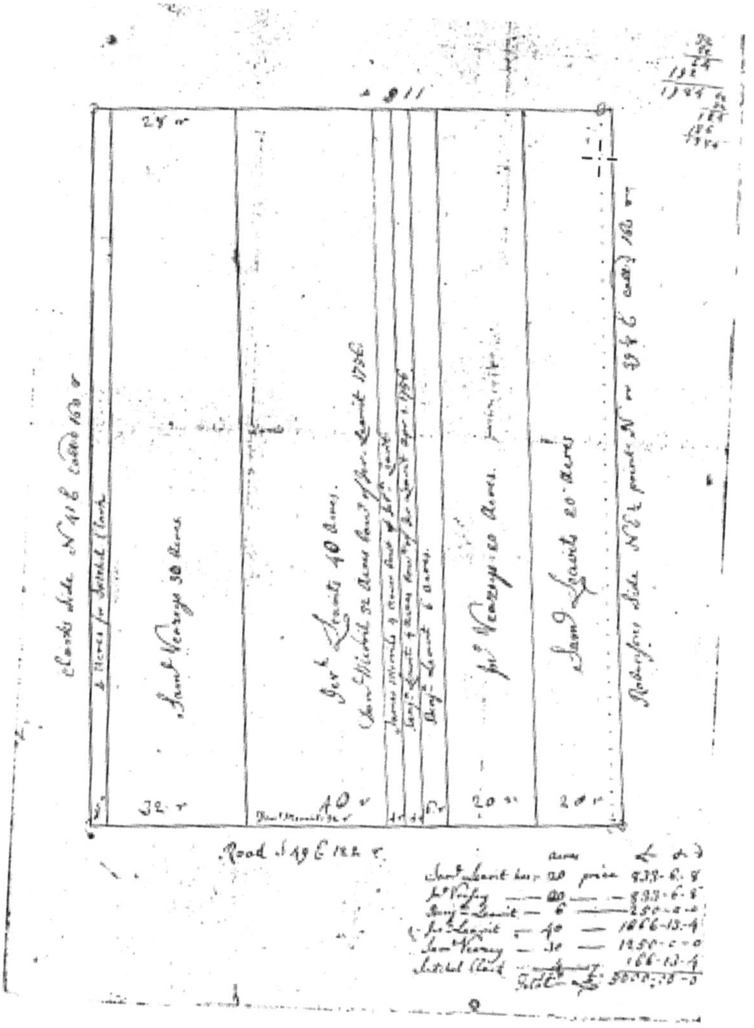

FIGURE 8.5   Original subdivision survey, 1755.

established archives to house them. Where available, these repositories should be searched for relevant information.

Even though the basic requirements for conducting a retracement, by law, are the same in the colonial states as in the PLSS, it can be a lot more challenging to acquire field notes in the former. Because of this difficulty, many surveyors sidestep this process and proceed with the information they have, acquired with much less difficulty. Such information typically appears in the forms of plats and deeds, both of which tend to be unreliable and misleading in many cases. Such practice can only result in situations illustrated by the two previous examples.

Details of, and information about, original surveys may come in a variety of forms, all of which is evidence. Depending on the form of the evidence, the questions become twofold: what is acceptable, and when two or more entities are found, which is the most reliable, or is controlling. Many times there is only one form to be found, so that the questions are moot, and one is left with no choice. Then it becomes a question of whether the evidence is reliable. That is one of the reasons that the work on the ground, and the corner evidence that is left, is controlling. As the court stated in the case of *Outlaw v. Gulf Oil Corporation*,[20] the survey is the substance. Numerous decisions have subsequently ruled that a map, the very thing that that most people want to rely on, is secondary, and yields to the original work on the ground. As several courts have stated, the map is a picture of the survey, and the picture may not be wholly accurate.

Several courts have also stated, the original markers control, *and have added*, "or the places where they once stood" (emphasis by author). The South Dakota case of *Titus v. Chapman*,[21] the original monuments, *or the places where they stood,* govern over unknown markers with no known basis and official maps with protracted, unsurveyed lines. The Maine court, through its decisions, came to the same conclusion. *In the case of Lloyd v. Benson,*[22] the court stated, "when a monument referenced in a deed is missing, it does not lose its significance as a monument if its original location can be determined. We have previously expressed this principle in both positive and negative terms. For example, in *Theriault v. Murray,* 588 A.2d 720, 722 (Me. 1991), we stated: 'The physical disappearance of a monument does not end its use in defining a boundary if its former location can be ascertained.'"

In contrast, in *Milligan v. Milligan,* 624 A.2d 474, 478 (Me. 1993), we stated, "[t]he physical disappearance of a monument terminates its status as a boundary marker *unless* its former location can be ascertained through extrinsic evidence." (Emphasis added.) We concluded in *Milligan* that "[b]ecause the unrebutted testimony in this case was that no pin … was ever located at [the terminus described in the deed], the pin could no longer be considered a monument. *Id.*"

Regardless of whether expressed in the positive or negative, the principle remains the same: the location of a monument that is described in a deed, but is missing from the face of the earth, can be established through extrinsic evidence. *See Hennessy,* 2002 ME 76, 796 A.2d at 48.

Once so established, the monument has the same legal significance as if it were not missing. *Theriault,* 588 A.2d at 722 (stating that if the locations of missing monuments can be determined, the "monuments as a matter of law must prevail over the deed's course and distance calls").

In the PLSS, it is required that the field notes of a survey be filed in an appropriate repository. With a government survey, two copies, one near the site and the other at BLM Headquarters. One is an original, the other a copy. Be wary of copies and transcriptions (of any kind), they often contain errors and therefore can be unreliable and misleading. They should be held as suspect until proven otherwise.

**Example:** A surveyor was retracing a political boundary, and attempted to follow a copy of the field notes, as the original set was not readily available. Too many things did not fit, so the surveyor returned to the record keeper and insisted on seeing the original notes. Comparing the two sets, he found *17 transcription errors*. It is incidents such as this that genealogists, for example, will not rely on a copy so long as an original is available. Sometimes the original is not available, but one must attempt to find it, because *any copy is suspect*. Emphasis by author.

Once again, remember what the court stated in the *Outlaw*[23] case:

A map is a picture of a survey; field notes a description thereof. The survey is the substance, and consists of the actual acts of the surveyor. If existing established monuments are on the ground evidencing these acts, such monuments control. Such established monuments control because same are the best evidence of what the surveyor actually did in making the survey. More, they are part at least of what the surveyor did. 9 Corpus Juris, 210.

---

[20] *Outlaw v. Gulf Oil Corp.*, 137 S.W.2d 787, (1940).
[21] 2004 SD 106, 687 N.W.2d 918 (S.D. 2004).
[22] 2006 ME 129, 910 A.2d 1048 (Me. 2006).
[23] *Outlaw v. Gulf Oil Corp.*, 137 S.W.2d 787, (1940).

## NOTEWORTHY DECISIONS REGARDING FIELD NOTES

Insofar as the field notes govern an area conveyed, that inquiry is not what the surveyor intended to do, but, if it can be determined, what he actually did. *Blackwell v. Coleman County,* supra; *Blake v. Pure Oil Co.,* 128 Tex. 536, 100 S.W.2d 1009.

Monuments are facts; the field notes and plats indicating courses, distances, and quantities are but descriptions which serve to assist in ascertaining those facts.[24]

Marks on the ground constitute the survey; courses and distances are only evidence of the survey. *Myrick v. Peet,* 180 P. 574 (Montana, 1919).

It is a familiar principle of our system, and one in reason applicable to this species of title, as well as any other, that it is the work on the ground, and not on the diagram returned,

which constitutes the survey, the latter being but evidence (and by no means conclusive) of the former................ *Wood v. Starko,* 197 S.W.3d 255 (Tenn.App. 2006)

With subsequent surveys, some field notes are filed with the official repository in the region. In the private sector, notes are filed there as well when required, otherwise with most private surveys, field notes are stored by the surveyor in the office files, or in storage. With official surveys, it is important to understand the entity responsible for the survey, some of which are difficult to track down, and may be found in extremely obscure, even unusual locations. Nothing is far-fetched, even the occasional encounter at a flea market, an estate sale, or an antique store. Hopefully private notes are preserved, rather than destroyed, although sometimes the latter is the ultimate fate. Fortunately, there exists less valuable information to be reviewed in the absence of primary evidence.

Some earlier surveys have proved extremely elusive to the practicing surveyor. Even though the usual locations are investigated, searchers often do not find the necessary records. Some long-standing examples are Spanish land grants in Florida, areas where the Forbes Purchase was active (records thought by some to be housed in Cuba[25]), and many of the larger early grants in the Colonial States. Although in most cases surveys were made, and records kept, if anything remains at all it is likely to be in the form of maps and written descriptions, the least important of evidence important in finding corner locations. To make matters worse, most descriptions have been derived from maps and plans, purportedly reflecting the actual survey, but often not.

## ELECTRONIC FIELD NOTES

We have now embarked on the era of electronic field notes. To the author's knowledge, no test of evidence integrity has been produced by the court system. Digital information tends to be suspect, since system failures and other anomalies may corrupt data, even result in a permanent loss. The suggestion is to print a hard copy, and include it as part of the filed record. Otherwise, the data may be permanently lost, or otherwise compromised.

## SUMMARY OF USUAL SOURCES OF VARIOUS TYPES OF INFORMATION

A few suggestions here speak to locations to search for original records. Genealogical researchers insist on the *original record* whenever it is possible to do so. However, under the right conditions, or rules, a copy (sometimes certification is necessary) can suffice. Both state and federal Rules of Evidence detail the requirements for the submission of evidence.

---

[24] In the absence of a monument, the point where it stood, if identifiable, will substitute for the missing monument and control as if the monument were still remaining.

[25] See discussion earlier in the chapter.

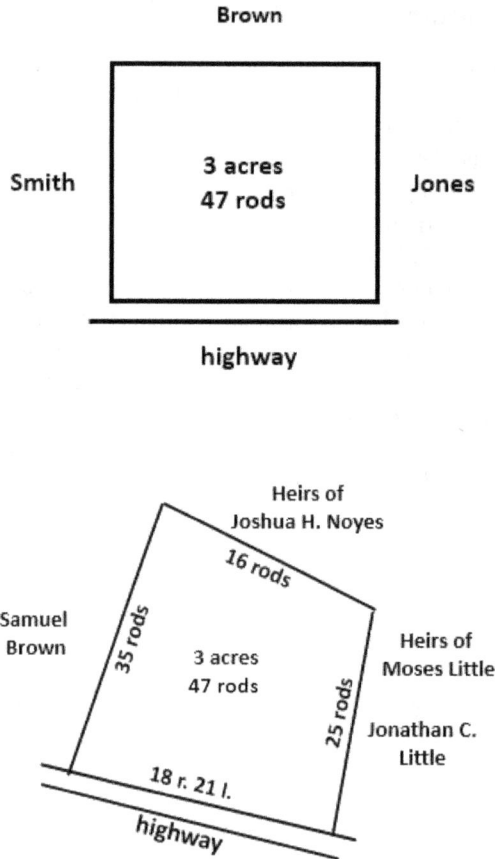

Brown

Smith    | 3 acres 47 rods |    Jones

highway

Heirs of
Joshua H. Noyes

16 rods

Samuel
Brown

35 rods

3 acres
47 rods

25 rods

Heirs of
Moses Little

Jonathan C.
Little

18 r. 21 l.

highway

**FIGURE 8.6** Current description reflects present or recent abutting parcels with a refined acreage, based on a survey (a). Tracing the description back in time reveals the measurements that gave rise to the computed acreage (b).

Understanding the origin and creation of the title is usually an important key to locating the original survey. Sometimes the survey is based on the title creation, and sometimes the title is based on the original survey. Knowing its origin will more than likely lead to the location of the information needed to find it and work with it (see Figures 8.6 and 8.7).

The Tennessee case of *Dillehay v. Gibbs*,[26] presents a modern example of the same scenario. In this case, all of the recent descriptions, back to 1878 were descriptions by adjoiners, with acreage estimates. Tracing back to 1878 produced the original survey by metes and bounds, which created the line separating two farms, both traceable to present day and parties to the litigation. The three surveyors involved in the case did not honor that survey and boundary creation, choosing instead to go by fences of unknown origin or otherwise trying to satisfy the acreage calls, which conclusions were far off from the called-areas. The Tennessee Court of Appeals refused to give any credence to the recent surveys, choosing to decide the case on other grounds.

Another refusal by the court system to accept so-called surveys submitted in evidence is found in the Wyoming decision of *Hagerman v. Thompson*,[27] wherein there were three conflicting sur-

---

[26] No. M2010-01750-COA-R3-CV; Court of Appeals of Tennessee, Nashville; June 16, 2011.
[27] 235 P.2d 750 (Wyo., 1951).

Lydia G. Bailey
to Ellen Giddings
1891

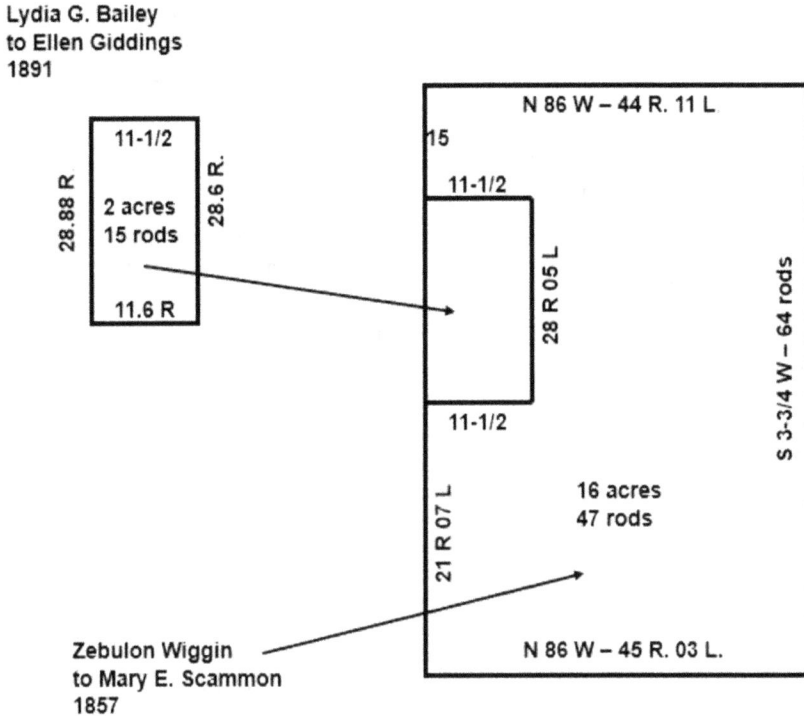

**FIGURE 8.7**   Example of improvement of description by checking earlier records. Later records sketchy, parcel actually made up of two surveyed tracts.

veys of a mining claim. Since the three independent opinions did not have a convincing basis, the court stated, "Three surveys in question in this case were resurveys, binding on no one, unless one of them perchance should ultimately in a proper proceeding be found to be correct."

## NOTES ACCORDING TO CATEGORY[28]

### TITLE BY TREATY

As with any description, follow the details of the description contained within the treaty. Be prepared for possible ambiguities, and interpretations of locations, depending on what map was used, or where information came from.

A logical place to begin searching for information on any particular treaty would be *Wikipedia*. In addition, many nation-states have a special bureau to keep track of activities along their borders, and to house and preserve historical records and descriptions, including surveys.

### TITLE BY PUBLIC GRANT (E.G., PATENT FROM THE UNITED STATES)

Lines do not become permanent until the plat is appropriately filed, after which a title transfer takes place in some form of public grant, usually a patent.[29]

---

[28] As listed in Chapter 2.
[29] *Rivers v. Lozeau*, 539 So.2d 1147 (Fla. App. 5 Dist. 1989).

With anything, there can only be one original. However, in the case of U.S. government surveys, and their field notes and plats, at the least, one copy is filed with the district, which may be the county, while the other copy is filed with headquarters. Make certain you have the *official* record.

## Title by Private Grant (Such as by Deed)

Since basic contract law requires, for a contract to be valid, the contract identify the subject matter. Sometimes a description in an instrument of conveyance is so insufficient in some way that a competent surveyor cannot locate it. Courts generally call such an attempted conveyance of rights as void, and therefore ineffectual. With any description, pay close attention to dates of transaction and date of recordation.

Generally, deeds are found in a county or town recording office (or, in a few cases, some other suitable repository). Sometimes a deed will not be part of the public record, never having been recorded. Many of these unrecorded documents are still extant, located in unusual places. Trunks in attics, shoeboxes at home, personal safes, and bank safety deposit boxes are all possible locations. People tend to hoard important papers, sometimes for extended periods of time.

**Author's Note:** While the local court house usually contains an abundance of information mostly in the form of documents, such study can be misleading. There are many critical sources that may be found elsewhere, or may not be of record any place due to their nature. Too many researchers believe that the extent of their research may be confined to the court house. This is a dangerous notion and, unfortunately, is found to be widespread and too common a practice. *Interpreting Land Records*[30] provides an extensive overview of critical items not expected to be found in regular repositories and where to search for necessary pieces of information.

## Title by Will (from the Decedent)

Wills are generally found filed with probate records at the local court house, although they may, under the right circumstances, be found elsewhere. Law offices and lawyers frequently have wills, or copies thereof, within their personal files if they were somehow involved with the estate.

A will may contain one or more line descriptions, although they may be somewhat sketchy. Occasionally a will may divide property, resulting in an original creation and description of a parcel.

Occasionally, filed with a will, there is an accompanying sketch of a parcel. This is like having a plat showing the parcel, although it is unlikely to be according to today's surveying and mapping standards.

## Title by Descent (Intestate Succession)

Divisions and partitions among heirs, along with dower descriptions may create new lines or new parcels. Some contain just a map, some written descriptions, other times a combination of both. They may also give rise to access (implied rights of way). Some lines may be coincident with previously created lines (from another survey).

A word of warning about a dower estate, which is a life estate usually given to a widow. Since it is a life estate, it terminates as an estate upon the death of the widow, with a provision within the probate documents as to its ultimate successor(s). Although the estate itself automatically disappears, survey information, including a monumentation set for location, generally persists. Such can be misleading, since it no longer has significance or priority. However, such things can

---

[30] Wilson, Donald A. (2nd edition, 2015).

be useful as means to other ends, such as tracking down survey data, or corner and parcel location. It is a perpetuation of information, so to speak.

Partitions and divisions are considered to be simultaneous creations, where all parcels have equal standing, and apportionment is often the rule in dealing with excess and deficiency. That means that the *entire* subdivision must be taken into account. Such divisions may be by map alone, by words alone, or by a combination of both.

Many partitions are the result of a probate court proceeding or equivalent, but they may also be found elsewhere, such as in a Legislative Act or in a deed. See following example of legislative act:

AN ACT FOR THE PARTITION OF CERTAIN REAL ESTATE WHEREOF JOHN INNES CLARK, AND JOSEPH NIGHTINGALE WERE POSSESSED AT THE TIME OF THE DECEASE OF THE SAID NIGHTINGALE.—

[Approved June 17, 1805. Original Acts, vol. 18, p. 88; recorded Acts, vol. 16, p. 78. See act of December 19, 1797. Laws of New Hampshire, vol. 6, p. 464.]

Whereas John Innes Clark of Providence, in the county of Providence, Esquire, late carrying on trade and commerce in company with Joseph Nightingale, of said Providence deceased, under the name and firm of Clark and Nightingale, and Lydia Clark, the wife of said John Innes Clark; Elizabeth Nightingale, widow and dowager of the said Joseph Nightingale, for herself and for William Nightingale, Joseph Nightingale and George C. Nightingale, sons of the said Joseph Nightingale deceased, and of the said Elizabeth, minors under the age of twenty one years, and to whose persons and estates the said Elizabeth is guardian, John Clark Nightingale, and Samuel W. Greene and Polly his wife, which said John Clark Nightingale and Polly Greene are children of the said Joseph Nightingale deceased, have petitioned the General Court setting forth that the said Joseph Nightingale deceased died intestate, and tenant in common or joint tenant with the said John Innes Clark, of divers parcels of land, and of divers tenements and hereditaments, or whereof one of said company was sole seized, in trust for the use and benefit of both, situate, lying and being in the several states of New Hampshire, Massachusetts, Rhode-Island, Connecticutt, Vermont, Ohio, and New York; that the said petitioners are desirous of making partition, and have agreed so to do, of the several parcels of land, and of the several tenements and hereditaments, herein after mentioned, between the said John Innes Clark of the one part, and the said widow as dowager and guardian as aforesaid, and the other heirs of the said Joseph deceased, on the other part, in manner and form following, that is to say; to the said John Innes; Clark, and to his heirs and assigns forever, are assigned and set off the lands, tenements and hereditaments, following, that is to say.

The key to finding original titles and surveys is dependent on an understanding of family relationships. Many times a professional genealogist must be consulted to connect family members by blood relationship. Thorough examination of family trees is indispensable.

## Degrees of Kindred

Under state intestacy laws, the amount remaining after the surviving spouse's share is deducted goes to the next of kin. The method most commonly used in the United States to determine the relatives who are most nearly related by blood is the civil law method. Under this method, each relationship to the decedent is assigned a number called a degree of kindred (see Figure 8.1). In general, the lowest numbered living kindred inherit the estate to the exclusion of all others. For example, second-degree kindred inherit only when there are no first-degree kindred alive. Similarly, third-degree kindred inherit only when there are no second- or first-degree kindred alive, and so forth.

The degree of kindred of a relative is calculated by counting upward from the decedent to the nearest common ancestor, then downward to the nearest relative. Each generation is called a degree. For example, parents and children of a decedent are related to the decedent in the first

AN ACT FOR THE PARTITION OF CERTAIN REAL ESTATE WHEREOF
JOHN INNES CLARK, AND JOSEPH NIGHTINGALE WERE POSSESSED
AT THE TIME OF THE DECEASE OF THE SAID NIGHTINGALE.—

[Approved June 17, 1805. Original Acts, vol. 18, p. 88; recorded Acts,
vol. 16, p. 78. See act of December 19, 1797, Laws of New Hampshire, vol.
6, p. 464.]

Whereas John Innes Clark of Providence, in the county of Providence, Esquire, late carrying on trade and commerce in company
with Joseph Nightingale, of said Providence deceased, under the
name and firm of Clark and Nightingale, and Lydia Clark, the
wife of said John Innes Clark; Elizabeth Nightingale, widow and
dowager of the said Joseph Nightingale, for herself and for William
Nightingale, Joseph Nightingale and George C. Nightingale, sons
of the said Joseph Nightingale deceased, and of her the said Elizabeth, minors under the age of twenty one years, and to whose persons and estates the said Elizabeth is guardian, John Clark Nightingale, and Samuel W. Greene and Polly his wife, which said John
Clark Nightingale and Polly Greene are children of the said Joseph
Nightingale deceased, have petitioned the General Court setting
forth that the said Joseph Nightingale deceased died intestate, and
tenant in common or joint tenant with the said John Innes Clark,
of divers parcels of land, and of divers tenements and hereditaments, or whereof one of said company was sole seized, in trust for
the use and benefit of both, situate, lying and being in the several
states of New Hampshire, Massachusetts, Rhode-Island, Connecticutt, Vermont, Ohio and New-York; that the said petitioners are
desirous of making partition, and have agreed so to do, of the several parcels of land, and of the several tenements and hereditaments, herein after mentioned, between the said John Innes Clark
of the one part, and the said widow as dowager and guardian as
aforesaid, and the other heirs of the said Joseph deceased, on the
other part, in manner and form following, that is to say; To the
said John Innes Clark, and to his heirs and assigns forever, are assigned and set off the lands, tenements and hereditaments, following, that is to say.

**FIGURE 8.8** New Hampshire legislative act dated 1805 authorizing the division of certain real estate.
Note that in the description, the subject lands lie in several states.

degree. Grandparent, grandchildren, brothers, and sisters are related to the decedent in the second degree. Uncles, aunts, nephews, nieces, and great-grandparents are third-degree relatives.
First cousins, great-uncles, great-aunts, and great-great-grandparents are fourth-degree relatives.
First cousins' children (first cousins once removed) are fifth-degree relatives, and first cousins'
grandchildren (first cousins twice removed) are sixth-degree relatives. Second cousins are persons who are related to each other by descending from the same great-grandparents and are also
sixth-degree relatives.

A few states follow the common law, or canon law, method of computing degrees of kinship.
Under this method, the degree of kinship is determined by counting the nearest common ancestor
down to the decedent, and then taking the longer of the two lines when they are unequal.

Within the degrees of kindred, certain priorities are recognized. For example, the decedent's
children receive preference over the decedent's parents, although they are both in the same degree.
Similarly, the decedent's brothers and sisters are favored over the decedent's grandparents.

## Consanguinity

*Definition: Blood relationship; the relation of people who descend from the same ancestor.*

Consanguinity is the basis of the laws that govern such matters as rules of Descent and Distribution of property, the degree of relation between which marriage is prohibited under the laws concerning Incest, and a basis for the determination of who may serve as a witness.

*Lineal consanguinity* is the relation in a direct line – such as between parent, child, and grandparent. It may be determined either upward – as in the case of son, father, grandfather – or downward – as in son, grandson, great-grandson.

*Collateral consanguinity* is a more remote relationship describing people who are related by a common ancestor but do not descend from each other – such as cousins who have the same grandparents.

Consanguinity is not the same as affinity, which is a close relation based on marriage rather than on common ancestry.[31]

**Consanguinity:** the relation subsisting among all the different persons descending from the same stock, or common ancestor. Vaughan, 322, 329; 2 Bl. Com. 202 Toull. Dr. Civ. Fr. liv. 3, t. 1, ch. n 115 2 Bouv. Inst. n. 1955, et seq.

2. Some portion of the blood of the common ancestor flows through the veins of all his descendants, and though mixed with the blood flowing from many other families, yet it constitutes the kindred or alliance by blood between any two of the individuals. This relation by blood is of two kinds, lineal and collateral.

3. Lineal consanguinity is that relation which exists among persons, where one is descended from the other, asbetween the son and the father, or the grandfather, and so upwards in a direct ascending line; and between the fatherand the son, or the grandson, and so downwards in a direct descending line. Every generation in this direct course males a degree, computing either in the ascending or descending line. This being the natural mode of computing the degrees of lineal, consanguinity, it has been adopted by the civil, the canon, and the common law.

4. Collateral consanguinity is the relation subsisting among persons who descend from the same common ancestor, but not from each other. It is essential to constitute this relation, that they spring from the same common root or stock, but in different branches. The mode of computing the degrees is to discover the common ancestor, to begin with him to reckon downwards, and the degree the two persons, or the more remote of them, is distant from the ancestor, is the degree of kindred subsisting between them. For instance, two brothers are related to each other in the first degree, because from the father to each of them is one degree. An uncle and a nephew are related to each other in the second degree, because the nephew is two degrees distant from the common ancestor, and the rule of computation is extended to the remotest degrees of collateral relationship. This is the mode of computation by the common and canon law. The method of computing by the civil law, is to begin at either of the persons in question and count up to the common ancestor, and then downwards to the, other person, calling it a degree for each person, both ascending and descending, and the degrees they stand from each other is the degree in which they stand related. Thus, from a nephew to his father, is one degree; to the grandfather, two degrees and then to the uncle, three; which points out the relationship.

**Author's Note:** Be mindful of the time frame. Inheritance laws have been modified significantly through time. Pay very close attention to the link in the chain of title and the date at which inheritance may have taken place.

## Title by Involuntary Alienation (Bankruptcy or Foreclosure)

This category likely does not have any significance other than being a link in the chain of title relating other conveyancing mechanisms together.

---

[31] *West's Encyclopedia of American Law*, Edition 2 (2008). The Gale Group, Inc.

## Title by Adverse Possession or Unwritten Agreement

Depending on the circumstances and nature of the possession, one or more new boundaries may be created. There may be informational paperwork available, such as survey plans depicting claims, or other descriptions providing color of title. Since this category is strictly a title question, it is usually not within the purview of the surveyor, other than a responsibility to recognize and report conflicts found on the ground with available title documents. During the investigation process however, a researcher should be aware of court actions and other procedures that present claims potentially affecting the subject title and/or its location and boundaries.

State statutes and court interpretations generally provide insight as to what may constitute "color of title." This is a study unto itself which varies widely.

## Title by Eminent Domain (Public Taking with Compensation)

A taking, under federal or state law, may involve an existing parcel where lines were created prior, or may be all or partly a unique tract, creating new line(s). Takings have both title and location considerations. Some takings appear within the records at the appropriate courthouse, but others may only be found at the appropriate agency, e.g., railroad, highway department, U.S. government agencies, and the like. The search for appropriate documentation and original survey data may take the investigator to some obscure locations, depending on the nature of the taking and the parties involved.

Takings may result in either an easement, or the fee. The common law principle is that absent a statute authorizing a fee taking, the taking results in an easement. This can be a complex area when dealing with railroads and similar categories.

## Title by Escheat (Property Reverting to the State)

When property, both real and personal, is left behind with no known owner, such as when an owner dies without a will or legal heirs, the property automatically transfers to the State under a common law process known as escheat. There generally is no paper trail to follow, but investigation into circumstances may reveal the necessary information from which to draw a conclusion. Since no new title or boundaries are created, it is of little consequence, other than providing a link in the chain of title in order to complete it.

## Title by Dedication (e.g., Easements for Public Use)

A dedication may create new parcel(s) with new boundaries; the resulting title may be easement or fee.

With four possible categories of any dedicated title, the area of dedication is one of the most complex that may be encountered. Dedication may be statutory (created according to statute law), common law, express or implied. In order for a dedication to become complete, there must be an acceptance by the appropriate authority, which is a discrete event, and it too may be either express or implied. When an entity, for example a road, public area, or the like exists, and no paper trail is readily found, dedication should be considered. However, it is sometimes difficult to distinguish dedication from creation through prescription or by custom, as the three create the same result, but in different manners.

Dedication may have recorded evidence to support it, such as when an area is shown on a recorded plan, or mentioned in a writing, but other forms may not be as obvious. Such entities may lead a searcher into some very obscure and difficult areas.

## TITLE WITH THE ELEMENT OF ESTOPPEL ENTERING

This category is likely of no consequence to the surveyor other than it is a legal question that may affect the chain of title. Unless there is a subsequent court action based on a claim, it is highly improbable that any kind of paper trail exists. One possible exception is the case where there are two deeds, or other sources of title, to the same piece of property, which deeds may or may not be on record, but are known. A party may be affected if they know about a previous claim. This is where notice could play an important part.

## TITLE THROUGH ACCRETION

Accretion, the gradual and imperceptible change of land by natural causes (water and wind) results in a new parcel attaching to an existing parcel(s). Title may be questionable, but the fact that the parcel exists, and has boundaries, is inescapable. The question is who owns it, strictly a matter of title, but a surveyor may locate and define it. Essentially, it is a new parcel of land, with new boundaries, but no written description until one is produced.

## TITLE BY PAROL GIFT (FOLLOWED BY ADVERSE POSSESSION OR ACTS OF PARTIES)

In some jurisdictions, and under certain circumstances, it is permissible to transfer land and land rights by parol (verbally), therefore no documentation. Again, since it is a matter of title, the only significance to a surveyor is one or more links to complete a chain of title. Boundaries have been created at some point in the past, and title has transferred.[32]

According to the court in *Ortmeyer v. Bruemmer*, 680 S.W.2d 384, (W.D. 1984), some states have also addressed this issue. In *Hurley v. Painter*, 180 Kan. 552, 306 P.2d 184 (1957), the court held that a parol gift of realty followed by possession and lasting and valuable improvements is valid, and may be enforced against the donor or anyone else. See also, *Hill v. Bowen*, 8 Ill.2d 527, 134 N.E.2d 769 (1956); *French v. French*, 125 Ariz. 12, 606 P.2d 830 (App.1980). The basis for this doctrine is estoppel and the prevention of fraud. *Root v. Mecom*, 542 S.W.2d 878, 879 (Tex.Civ.App.1976).

This may be in the category of Part Performance, which is strictly a legal consideration. The point is, just because there is no writing from one party to another and the Statute of Frauds states that contracts for the sale of land must be in writing to be enforceable, does not automatically mean there is no contract, or transfer of property rights.

## TITLE THROUGH OPERATION OF LAW N/A

Likely of little consequence to the surveyor except in a few cases, operation of law is a category whereby previous events have resulted in certain things taking place upon the completion of some conditions or set of conditions. An example is an estate for years. As soon as the designated number of years has elapsed, the estate terminates. Another common example is when an easement is termination, rights revert to some designated party, often a successor in title to the one who created the easement. Understanding the details and consequences of an event will usually lead to its ultimate fate.

## TITLE BY CUSTOM

Custom is the basis for a body of laws. In fact, Blackstone suggested that all laws, rules, and regulations owe their origin to some customary, accepted practice by members of society. Some

---

[32] For further information, insight, and examples from the court, consult two articles found in *American Law Reports.*

titles and rights arise through customary use over an extended period of time, sometimes genera-tions. It is somewhat important for the real estate professional to recognize as an alternative to the creation of easements by dedication or prescription. The ultimate result of all three is the same, however created under different conditions and with different requirements. There is no paper trail to follow, but there may be anecdotal evidence to consider.

## TITLE BY PRIOR APPROPRIATION

Prior appropriation is Blackstone's label for rights in the use of flowing water, which will even-tually attach to a property for a brief period of time, thereby giving rights to the landholder whose land it touches before it flows away and is replaced by the flowing water that follows it. Blackstone labelled it "transient property" and provide a category for it for identification as a property right. It is one of the forces of nature, falling under the broad category of natural law.

**While many of these operations establish a title, usually for the first time, not all of them establish a new, or original, boundary.**

## UNSURVEYED LANDS

The Florida court in *Hardee v. Horton*[33] stated:

> A map or plat which represents no survey, but is prepared by projecting lines of a prior erroneous government survey on paper over a space representing a large area of unsurveyed lands, purporting to represent section, township, and range lines according to the rectangular method of surveying, although adopted and referred to in deeds of conveyance as the official map of the grantor, when shown by competent testimony to be inaccurate and unreliable as an aid to locate the unsurveyed lands which are conveyed by description according to the rectangular method of describing lands, is insufficient as a survey of said lands.

> A complete title to unsurveyed public lands does not vest in the grantee until the lands conveyed have been identified by an authorized survey; and where unsurveyed lands are conveyed by description according to the rectangular method of describing lands, although the deed be a grant in praesenti, the title vests in the grantee upon delivery of the deed subject to the right and duty of the political authorities of the state to identify and separate by a survey the lands conveyed from the unsurveyed lands within which they are included.

> In the purchase of swamp and overflowed lands that have not been surveyed the vendees take them with notice that the lands described are to be located by an authorized survey, and that all property in the state is acquired and held subject to the due exercise by the state of its police power.

> Where purchasers of unsurveyed lands in large areas are made is given acreage described by sec-tions, townships, and ranges, which contemplate 640 acres to the section, and 36 sections to the township, the exact location and boundaries of the particular lands intended to be conveyed may be ascertained by an authorized survey; and if the lands so located are not all in fact precisely where the purchases supposed they would be, no harm is done the purchasers, even though the purchases were made with reference to a plat on which lines were merely protracted on the plat over the space representing the unsurveyed area, since an actual survey was contemplated and a particular acreage was intended to pass by description used.

> Where the boundaries mentioned in a deed of conveyance are inconsistent with each other, those are to be retained which best subserve the prevailing design manifested on the face of the deed, and the least certainty must yield to the greater certainty in the description.

---

[33] 90 Fla. 452, 108 So. 189 (Fla. 1925).

Where land is clearly and explicitly described in a deed, and a subsequent clause is added as a further description of it, but which is of doubtful import, or repugnant to the first clause, such latter description will be rejected.

Where an erroneous map, which represents no survey, is referred to in a deed conveying unsurveyed lands by a more particular description, and there is conflict between the map and the more particular description, the lands should be located by the more certain and definite description, and the erroneous map may be treated as surplusage.

Admission of Florida into Union by Act Cong. March 3, 1845, did not affect proprietary rights of United States in lands within state which had been ceded to United States by Spain by treaty of February 22, 1819, where such lands did not constitute beds or shores of navigable waters of state, or tidelands.

The allegation that the location of S.W. 1/4 of S.W. 1/4 of S.W. 1/4 of section 35, township 53 south, range 40 east, was changed in place by the survey cannot be sustained in law or fact. No one knew where section 35 was situated until township 53 was located by a survey and section 35 identified. It is obvious if the location of ten acres of a section be changed the location of every part of the section would be changed. A map of the character of the 1905–1907 map does not and cannot accurately locate land, and to hold that such a map is the measure of the land as well as the best evidence of its location, merely because it is referred to in a deed of conveyance which particularly describes the land and stating the acreage conveyed, would be the means of sanctioning the grossest errors and opening the door to intentional fraud.

The only method provided by law for an accurate identification of unsurveyed land, when described according to the rectangular method of describing land, is by a survey according to the rules established by law. The location of no part of township 53, and the location of no part of section 35, was changed by the survey. The lines established by the survey correspond with the lines of the map, except the south line of the township and section which was established by an actual measurement on the ground nearly half a mile north of where it erroneously appeared to be on the map. A fact established by an accurate survey is that the S.W. 1/4 of the S.W. 1/4 of the S.W. 1/4 of section 35, township 53 south, range 40 east, is not and never was included within the strip of land designated as lot No. 2 between townships 53 and 54.

## WHAT IF THE ORIGINAL SURVEY IS NO LONGER DISCERNIBLE?

*Cox v. Hart*[34] is a frequently cited decision dealing with such an anomaly.

Prior to the Act of March 28, 1908 (35 Stat. 52), unsurveyed lands, as well as surveyed lands, came within the purview of the Desert Land Laws (19 Stat. 377). That act, however, from and after its passage, restricted "the right to make entry of desert lands … to *surveyed* public lands," and expressly declared that "no such entries of unsurveyed lands shall be allowed or made of record." Then follows the proviso now being considered. The office of a proviso is well understood. It is to except something from the operative effect, or to qualify or restrain the generality, of the substantive enactment to which it is attached. *Minis v. United States,* 15 Pet. 423. Although it is sometimes misused to introduce independent pieces of legislation. *Georgia Railway & Banking Co. v. Smith,* 128 U.S. 174.; *White v. United States*, 191 U.S. 545. Here, however, the proviso is plainly employed in its primary character. The effect of the substantive enactment was to forbid the entry of unsurveyed lands. But the law theretofore had been otherwise, and one purpose of the proviso evidently was to exclude from the operative effect of the new rule cases

---

[34] 260 U.S. 427, 43 S.Ct. 154, 67 L.Ed. 332 (1922).

which might have arisen under the prior law – that is cases of persons who had taken possession of and undertaken to reclaim unsurveyed lands at a time when the law conferred the right to do so. Any such person, no less than one who acted subsequently, is within the words of the proviso. He is literally "a person who has, prior to survey, taken possession," etc. The proviso so construed impairs no vested right and brings into existence no new obligation which affects any private interest. No reason is perceived why the words employed should not be given their natural application and so applied the case of appellee is included. Indeed, this does not give the proviso a retroactive operation. The language in terms applies to one who at the time of the enactment occupied a particular status – *viz.,* the status of a person who has done the things enumerated. A statute is not made retroactive merely because it draws upon antecedent facts for its operation. *Regina v. Inhabitants of St. Mary, Whitechapel,* 12 Q.B. 120; *United States v. Trans-Missouri Freight Association,* 166 U.S. 290.

Passing this point, however, it is contended that the lands in question were in fact surveyed. It is true the lands had been surveyed in 1854–1856, but the lines of that survey by the year 1900 had disappeared to such a degree that, for practical purposes, they had become nonexistent. A survey of public lands does not ascertain boundaries; it creates them. *Robinson v. Forrest,* 29 Cal. 317; *Sawyer v. Gray,* 205 F. 160. Hence, the running of lines in the field and the laying out and platting of townships, sections and legal subdivisions are not alone sufficient to constitute a survey. Until all conditions as to filing in the proper land office and all requirements as to approval have been complied with, the lands are to be regarded as unsurveyed, and not subject to disposal as surveyed lands. *United States v. Morrison,* 240 U.S. 192; *United States v. Courtner,* 38 F. 1. It follows that, although the survey may have been physically made, if it be disapproved by the duly authorized administrative officers the lands which are the subject of the survey are still to be classified as unsurveyed. In other words, to justify the application of the term "surveyed" to a body of public land, something is required beyond the completion of the field work and the consequent laying out of the boundaries, and that something is the filing of the plat and the approval of the work of the surveyor. If, pending such approval, or, still more, if after disapproval of the survey, the lands in contemplation of law are unsurveyed, it is difficult to see why the same result may not follow when the survey originally approved and platted is subsequently annulled or abandoned, because the lines and marks established have become obliterated. A purpose to annul or abandon such survey we think may be disclosed by an act of Congress directing a resurvey plainly based upon the fact of such obliteration.

Turning now to the record in the case under consideration, it appears that the lines and marks of the original survey of 1854–1856 had for all practical purposes ceased to be. This is apparent not only from a consideration of the record, but is in accord with repeated declarations of the land department. *See* in re Peterson et al., 40 L.D. 562, where it is said not only that all the corners which had been established north of the fourth standard parallel were missing, but that the survey itself was "grossly inaccurate"; that, in making the resurvey, an attempt to retrace the old original survey had failed, "and it is now a physical impossibility to identify" the corners of Sections 16 and 36 "according to the original survey.... All vestiges of that survey have been wiped away." *See also Stephenson v. Pashgian,* 42 L.D. 113. In the land office contest between the parties hereto for the land in controversy, the Secretary of the Interior, although ruling in favor of appellant, stated that the description of the land by reference to the lines of the old survey "was an impossible condition." *Hart v. Cox,* 42 L.D. 592. Both courts below reached the same conclusion.

The district court said:

The evidence of the survey of 1856 upon the ground in that vast area covered by said act had become useless by reason of the fact that the lines of the survey were obliterated, and all that was left were some prominent monuments. This act [the resurvey act] recognized that the survey of 1856 was of no practical use, and that the lands were, for all practical purposes, unsurveyed lands. It was impracticable to dispose of these lands by congressional subdivision according to the survey of 1856.

The circuit court of appeals, after referring to the original survey and the fact that no settlement was made until nearly 50 years later, said: "It was then found that the marks of the survey had so far been obliterated that it was practically impossible to locate the lines thereof." *Cox v. Hart,* 270 F. 51.

From the foregoing it results, and we hold, that the Resurvey Act of 1902 was, in effect and intent, a legislative declaration that the lands therein described were, when the act was passed and for all purposes of settlement and sale, unsurveyed lands. With the disappearance of the physical evidences, the old survey survived only as an historical event. As a tangible, present fact, it ceased to exist, and a new survey became necessary to reestablish the status of the area over which it had extended as surveyed lands of the United States.

The decree below is

*Affirmed.*

**Author's Note:** These decisions illustrate the necessity of understanding the law at the time. As the court noted, a new survey is necessary, which, in such cases, would be a perfect example of the necessity of a resurvey.

One case referencing *Cox v. Hart*, 270 F. 51, the court stated:

The Desert Land Acts (19 Stat. 377; 26 Stat. 109 (Comp. St. Secs. 5029–5035)) do not require residence on the land. The principal requirements are that the claimant shall file a plat showing the mode of contemplated irrigation and shall expend a prescribed amount per acre in making permanent provision for such irrigation, and in addition thereto shall cultivate one-eighth of the land. In the Case of Virgil Patterson, 40 L.D. 264, it was held that the possession and improvement contemplated by the act of March 28, 1908–

> are not such as are required of a settler under the homestead law, but it is sufficient under that act if the possession and improvement conform to the requirements of the Desert Land Law and evidence the party's good faith under that law.

We are of the opinion that the court below properly held that at the time when the claims of the parties hereto were initiated the land was unsurveyed land, and that the act of March 28, 1908, was applicable thereto and conferred a preference right upon the appellee. The act of 1902 authorizing a resurvey was a legislative declaration that the lands were to be regarded as unsurveyed lands, that a new survey was to be substituted for the old, and that by the new survey the future disposition of the lands was to be regulated. By a proviso in that act protection was afforded as to then existing claims of occupants. The inference follows that as to all whose rights were initiated thereafter, the resurvey was intended to be controlling. This was the construction placed upon the act by the Land Department. By an order of the Commissioner of the Land Office of March 31, 1906, all entries in Imperial Valley were suspended after that date. About a year later the Commissioner directed that the order of suspension of said lands from entry be modified, so as to permit "the desert land entry thereof the same as though the lands were unsurveyed. "In *Nichols v. McCullom,* 169 Cal. 611, 614, 147 P. 271, 273 (L.R.A. 1915F, 638), the court said:

> The public lands are under the exclusive control of Congress, until title or a right to acquire title has, pursuant to some law, vested in some person or body corporate other than the United States. A person who claims no right in such public lands certainly cannot object to the act of Congress in abandoning a survey already made, and substituting another in its place.

The decree is affirmed.

An important decision from California is *Nichols, et al. v. McCullom, et al.*[35]:

This action was brought to recover possession of a tract of land in Imperial county, with damages for the withholding. There was a jury trial, which resulted in a verdict in favor of defendants. From the judgment entered pursuant to the verdict, the plaintiffs appeal, bringing up the evidence by means of a bill of exceptions.

The land in controversy is described as fractional section 16 of fractional township 17 south, range 15 east, San Bernardino base and meridian, containing 441.96 acres more or less. The land thus described was originally public land of the United States, and, being a sixteenth section, it passed from the United States to the state of California as a part of the school land grant. It is not disputed that the plaintiffs have acquired the state title as successors to the holders of a patent from the state, dated January 12, 1903. Nor is it questioned that the defendants are in possession of land which, according to the claim of plaintiffs, forms a part of the section granted by the state patent. The real point of contention is whether the plaintiffs have succeeded in establishing the identity of the land occupied by defendants with fractional section 16 of township 17 south, range 15 east, San Bernardino base and meridian. In other words, the defendants took at the trial, and now take, the position that, while the plaintiffs are unquestionably the owners of that section, the proof fails to show that the land claimed and withheld by the defendants is within the section thus described. The jury, by returning a verdict for the defendants, determined this issue in their favor. The appellants attack the sufficiency of the evidence to support the verdict, and assign as error certain instructions of the court.

The tract in controversy, with neighboring land, located in what is known as Imperial valley, was originally surveyed under the authority of the government of the United States in 1856. The survey was approved, and the approved plat filed in the Department of the Interior and in the local land office in 1857. It is a matter of common knowledge, and is, in fact, sufficiently shown by the record in this case, that the land embraced in this survey, or the greater part thereof, remained unoccupied for many years. When it was discovered that the apparently barren lands in Imperial valley could be rendered exceedingly productive by means of irrigation, an extensive settlement of the valley began. By that time the monuments referred to in the field notes of the survey of 1856 had, to a great extent, been obliterated, and it was difficult, and in many cases impossible, to trace upon the ground the boundaries of the governmental subdivisions as established by the survey. Moved, no doubt, by the knowledge of these conditions, the Congress of the United States passed an act, entitled "An act providing for the resurvey of certain townships in San Diego county, California." (The land is within the territory which has, since the passage of the act, been taken from San Diego County to [form the new county of Imperial].) By this act, which was approved July 1, 1902, it was enacted:

> That the Secretary of the Interior be, and he is hereby authorized to cause to be made a resurvey of the lands of San Diego county, in the state of California, embraced in and consisting of the tier of townships 13, 14, 15 and 16 south, of range 11, 12, 13, 14, 15 and 16 east, and the fractional township 17 south, of range 15 and 16 east, all of San Bernardino base and meridian; * * * provided, that nothing herein contained shall be so construed as to impair the present bona fide claim of any actual occupant of any of said lands to the lands so occupied.

Pursuant to this law resurveys of the townships described were ordered. Among others, instructions for surveying township 17 south, range 15 east, were issued to Legrand Friel, United States deputy surveyor. Mr. Friel made his survey, and filed his returns with the surveyor general. They were transmitted to the General Land Office for its approval. Such approval was refused for

---

[35] 169 Cal. 611, 147 P. 271 (1915).

reasons stated in a decision of the Commissioner of the General Land Office, who returned the papers to the surveyor general for California with directions to except from approval the establishment of the alleged school sections; the Commissioner stating that the sixteenth and thirty-sixth sections in each of the townships involved would be "platted and located in conformity with the claim lines of adjacent sections as shown by the resurvey." A corrected survey and plat were made, and were duly approved. The record contains a decision of the Commissioner, under date of November 4, 1910, adjudging that "tract 159" is a school section represented upon the plat of resurvey; i.e., section 16 of township 17. At the trial the defendants admitted that they were occupying "the tract of land on that map as section 16 by the new survey" – the land "marked as tract 159 on the plat."

It should be added that at the date of the approval of the act providing for a resurvey (July 1, 1902) the defendants had not gone into possession of the land, and had taken no steps toward acquiring any title thereto. They were not, therefore, within the scope of the proviso with which the enactment ends.

On the foregoing evidence, none of which is disputed, it seems clear that the plaintiffs made out a complete case of ownership and right of possession of the land in controversy, and that the verdict of the jury against their claim is unsupported.

The act of July 1, 1902, authorized the Secretary of the Interior to cause to be made a resurvey of certain townships. As we have already pointed out, the lines of the original survey could not with certainty be traced upon the ground. The purpose of the congressional enactment undoubtedly was that the new survey directed to be made should supersede the old one, which had become unavailable. It was intended that the dispositions of the public lands within the area affected should be regulated by the new survey, which, in theory at least, would reproduce the exact corners and lines of the old one. The Congress recognized, however, that some of these lands were held by occupants who claimed or might claim vested rights. Such claims were expressly saved from the effect of the act by the proviso at the end. But the very fact that this proviso was inserted lends added force to the view that, as to all persons not then in occupancy of any of the lands, the resurvey was intended to be binding. We see no constitutional objection to such an enactment. The public lands are under the exclusive control of Congress until title or a right to acquire title has, pursuant to some law, vested in some person or body corporate other than the United States. A person who claims no right in such public lands certainly cannot object to the act of Congress in abandoning a survey already made, and substituting another in its place. This would be true even if the monuments of the original survey were readily discoverable. The power of Congress to provide for a resurvey is much clearer where, as here, it has become impossible or exceedingly difficult to make a correct application of the original survey.

But, say the respondents, the sixteenth and thirty-sixth sections were identified by the survey of 1856, and, upon the approval of that survey, they at once passed to the state of California. Higgins v. Houghton, 25 Cal. 252; Sherman v. Buick, 45 Cal. 656. They then, it is argued, ceased to be public lands, and the power of the federal government to exercise any control over them, by survey or otherwise, ceased. But the location of these sections was necessarily dependent upon the location of the adjoining sections of public land still undisposed of. In delimiting the boundaries of the adjoining sections, it was necessary that the boundaries of the school sections should be marked out and determined. Furthermore, this contention, it seems to us, can hardly be urged against the appellants, who are accepting the resurvey as accurately describing the land owned by them. Such appellants might, perhaps, be in a position to attack the resurvey on the ground that its effect was to take from them land to which they already had title. But they are the only parties entitled to make this point. When they rely upon the resurvey as an accurate designation of the school section owned by them, the respondents, who do not claim under the state, and who were not making a claim of any kind when the statute authorizing the survey was passed, cannot object. If, according to the survey of 1856, the land in controversy was the sixteenth section, the appellants have a perfect title to it, and the respondents clearly have neither title nor right of possession. If it was not school land, it remained public land of the United States, and subject to the jurisdiction of Congress. In that aspect, the resurvey was unquestionably authorized, and the respondents can have no valid claim to land which was identified by the resurvey as school land which belonged to the state or its grantees, and was accordingly not open to entry.

The authorities cited by respondents to the effect that the Land Department has no authority by means of a resurvey or otherwise to affect the title to land after such title has passed from the government to its grantees (*Moore v. Robbins*, 96 U.S. 530, 24 L.Ed. 848; *Kean v. Roby*, 145 Ind. 221, 42 N.E. 1012; *Murphy v. Kirwan* [C. C.] 103 F. 104) have no application to the facts of this case. In none of these cases had there been congressional authorization to make a resurvey which should take the place (so far as claims to be thereafter initiated were concerned) of an existing survey.

If the act of 1902 has been rightly interpreted by us, there is no occasion to consider the argument of respondents to the effect that the decision of the Land Department of the government fixing the location of section 16 was erroneous, and that said section was placed by the new survey at a different place from that occupied by section 16 under the survey of 1856. The new survey superseded the old one. Parties going upon the land after the passage of the act, and with constructive knowledge of the fact that such survey was to be made, took the chance that such survey might establish that the land occupied by them was a school section, and not subject to entry as vacant lands of the United States.

The foregoing views make it unnecessary to consider specifically the various points made by appellants in attacking rulings on evidence and instructions to the jury. These questions can readily be solved by the application of the principles outlined in this opinion. Nor need we discuss the claim that the court erred in overruling plaintiffs' challenge to a juror. The same situation will not, in all likelihood, be presented on a new trial.

The judgment is reversed.

5. Public lands lose their status as "surveyed lands" and become "unsurveyed" when the lines and marks of the original survey have become obliterated for practical purposes and when, for that reason, a resurvey has been directed by an act of Congress. P. 436.

As discussed in detail in Chapter 2, "Title to land cannot exist without boundary."[36]

The following is from the Tennessee court:

> Title to land cannot exist without boundary, the plaintiff must show a marked boundary, or some proof of which boundary can be ascertained, before he can say to the defendant, even though he is a naked possessor: 'I have a better right to possess this particular piece of land than you have.' But then, in order to establish boundary, it is not indispensably necessary that some particular corner or marked line should be proven to exist. If it be proven to have existed, or any monument, corner, or marks from which the boundaries called for in a grant or deed can be satisfactorily ascertained, according to any easy and natural interpretation, it is sufficient. Nevertheless a grant can be lost for uncertainty in its boundary, there being no means by which its boundary can be defined. But not if by any reasonable means the intention of the contracting parties can be ascertained.

---

[36] 1 Meigs' Digest, p. 540. Quoted in *Moore v. Brennan*, 42 Tenn.App. 542, 304 S.W.2d 660 (Tenn.App. 1957).

# 9 Retracing and Locating Original Surveys

And that survey is the one we all go back to. When you find one of their original corners, it is like a handshake with the past.

**Andro Linklater**

In theory, theory and practice are the same. In practice, they ain't.

**Attributed to Yogi Berra, among others**

A brief overview of retracement principles is presented in this chapter. For detailed procedures and variations, consult the companion text, *Boundary Retracement: Processes and Procedures.*[1]

The Wyoming decision of *Hagerman v. Thompson*[2] succinctly states the function of a resurvey (retracement survey) and the sanctity of its control once established:

> The purpose of a resurvey is to ascertain lines of original survey and original boundaries and monuments as established and laid out by survey under which parties take title to land, and they cannot be bound by any resurvey not based on survey as originally made and monuments erected.

## DEFINITION OF RETRACEMENT

Two situations exist in the world of retracement – surveying an ancient description from its creation and following, and interpreting, subsequent so-called surveys by others. Either of these may appear as the subject parcel or as an abutting parcel.

The latter complicates the process in that (1) they must be examined for reliability, and (2) there may be a second set of records to search for and, if found, analyze. Unknowns will be inherent, like amount of traverse error, and what adjustments were made, if any. Without knowing what type of equipment was used, and something about the surveyors and chain people, unknowns such as measurement errors in several forms as well as compensating errors cannot be identified, leading to differences between actual measurements and those reported. Real measurements may be distorted, sometimes amounting to a significant difference.

If old enough, an intermediate survey may have been relied upon, and new boundaries created through unwritten means. One could not know if that is the case without making a comparison with the original information. There is an abundance of examples where an original survey was made, and either ignored or not found, and a later surveyor merely relied on evidence (in some case not even valid evidence, but merely physical items, particularly fences) that are easily found and conveniently "seem to fit" the description, or at least come close. Some of these approximate, some do not even come close, even to the point where the tract itself in question is grossly misplaced. When improvements are involved, they become catastrophes.

---

[1] See Wilson, Donald A. *Boundary Retracement: Processes and Procedures.* Boca Raton: CRC Press. 2017.
[2] 235 P.2d 750 (Wyo., 1951).

**Example:** A group of eight heirs, brothers and sisters, consulted their attorney concerning a parcel of land found in the father's estate. Their questions were all too common: where is this parcel located and does it have any value?

Researching the parcel and other parcels in the immediate vicinity resulted in a tax taking and resulting sale followed by a substantial improvement on the property by another party. Unfortunately, during the tax process, the wrong property was identified, none of the record title holders were notified of the taking or the proposed development. The improvement amounted to approximately a one million dollar facility. In short, it was a substantial factory constructed on the wrong lot. Action was brought against the factory owners, and subsequently the case settled, and the rightful owners received the money due them.

Even though the recent title was a tax taking, followed by a tax deed from the municipality making a difficult search, the parcel in question had a title history and an origin. Proper research uncovered the facts and the rightful owners.

## LINES OF AGREEMENT

Under proper circumstances, lines of agreement can establish a new boundary. They may be based on actual survey, referencing lines on a map, or merely putting selected words on paper. It is a new line, and therefore is, or is equivalent to, an original survey.

### PRIVATE

Much has been written, and litigated, concerning agreements between parties, commonly known as boundary line agreements. They should not be confused with boundary line adjustments, although the two terms are sometimes used interchangeably. However, they come under different requirements, may be according to different laws, and accomplish different purposes. They both create new lines, therefore are easily confused, not to mention persons being misled by the similarity in terminology.

Boundary line agreements are based on both statutory and common law requirements. One of the most important requirements that is often ignored is that a boundary between two ownerships, in order to be agreed upon, must be in dispute or otherwise unknown. Essentially there is no way to identify it, so that the rule is to allow the parties in interest on both sides of the line to locate it. Boundary line agreements are not intended for convenience, or for parties to shift a boundary to save the cost of a survey and transfer of property, as it would violate the Statute of Frauds.[3]

Boundary line adjustments are regulatory and zoning regulations that allow a relocation of an existing boundary by appropriate parties. It can be a convenient mechanism to change lot sizes to conform with zoning regulations, and to correct minor problems, such as a building design that violates setback requirements as well as improvements that cross the line. When the parties involved are willing to make accommodations, this process is often a convenient way to do it.

### TREATIES

Treaties fall into a select category that is a very complex and far-reaching area of international law. Treaties may be global in extent or local. International boundaries are the commonly known global situations, whereas Native Treaties (aka Indian treaties) are more often understood to be local, although they may affect tribes and their members in more than one country.[4]

---

[3] For an extensive discussion of boundary agreements, refer to Wilson, Donald A. *Boundary Retracement: Processes and Procedures.* Boca Raton: CRC Press. 2017.

[4] For an extensive list of treaties, types, and dates, consult Wikipedia.com.

One of the most important treaties affecting the United States is the Treaty of Paris (1783) between the United States and Great Britain, more locally establishing the boundary between the United States and Canada, among several other important items of consideration. They line was established by a committee of representatives from both countries based on the Mitchell map published in 1755, which was the most comprehensive map of this part of the world at the time. It remains useful today, along with historical significance.

Benjamin Franklin, a U.S. member of the committee, drew a red line on the map as a proposed boundary, which was essentially agreed upon. Tremendous difficulties ensued in locating the line on the surface of earth due to interpretation of landforms and conflicts with previous survey locations. It took 59 years to come to full agreement on the final line which was accomplished with the Webster-Ashburton Treaty as negotiated by Daniel Webster, Secretary of State (U.S.) and Lord Ashburton (William Baring-Great Britain) in 1842.

## WHAT IF THE ORIGINAL CORNERS ARE GONE?

As noted earlier, and stressed in the Chapter 8, several courts have stated that if the position of a corner, whether monumented or not, can be ascertained, it holds the same significance as if the original monument was still in place.

"The courses and distances in a deed always give way to the boundaries found upon the ground, or supplied by the proof of their former existence, where the marks or monuments are gone." *Cullacott v. Cash Gold & Silver Mining Co.,* 8 Colo. 179, 183, 6 P. 211, 214 (1885) (citing *Lodge v. Barnett,* 46 Pa. St. 477 (Pa. 1864)); 12 Am.Jur.2d Boundaries § 74.

*Lloyd v. Benson,* 2006 ME 129, 910 A.2d 1048 (Me. 2006):

When a monument referenced in a deed is missing, it does not lose its significance as a monument if its original location can be determined. We have previously expressed this principle in both positive and negative terms. For example, in *Theriault v. Murray,* 588 A.2d 720, 722 (Me. 1991), we stated: "The physical disappearance of a monument does not end its use in defining a boundary if its former location can be ascertained."

In contrast, in *Milligan v. Milligan,* 624 A.2d 474, 478 (Me. 1993), we stated, "[t]he physical disappearance of a monument terminates its status as a boundary marker *unless* its former location can be ascertained through extrinsic evidence." (Emphasis added.) We concluded in *Milligan* that "[b]ecause the unrebutted testimony in this case was that no pin ... was ever located at [the terminus described in the deed], the pin could no longer be considered a monument." *Id.*

Regardless of whether expressed in the positive or negative, the principle remains the same: the location of a monument that is described in a deed, but is missing from the face of the earth, can be established through extrinsic evidence. *See Hennessy,* 2002 ME 76, 796 A.2d at 48.

Once so established, the monument has the same legal significance as if it were not missing. *Theriault,* 588 A.2d at 722 (stating that if the locations of missing monuments can be determined, the "monuments as a matter of law must prevail over the deed's course and distance calls").

If monumentation is truly gone, then resort must, of necessity, be to appropriate procedures, only one of which is relying on courses and distances. This is probably the least desirable, since all measurements contain error, and there are many unknowns surrounding the procedures in making those measurements. The court in *Doyle* stated that means to be used to locate lost monuments or corners include collateral evidence such as fences that have been maintained, which should not be disregarded by surveyor, and artificial monuments such as roads, poles, and improvements may not be ignored; surveyor should also consider information from owners and former residents of property in the area.

However, court decisions state that in order for a corner to be considered lost, it must be so lost that it cannot be replaced with ANY evidence.[5] When results are not as anticipated, or desirable, resort is commonly had to apportionment procedures. This is a valid rule, but too often used as a shortcut to the required process. As Robert Griffin so aptly stated, apportionment may be used when retracement fails to uncover a corner.

He stated it thusly: "Apportionment is applied when retracement fails to yield sufficient evidence of the exact location of a lost property comer. When the whole length of the line between ascertained corners varies from the record length called for, and intermediate property corners are lost, it must be presumed in the absence of evidence indicating the contrary, that the variance arose from an imperfect measurement of the whole line. The variance then is distributed between the several subdivisions of the whole line in proportion to their respective lengths. This definition suggests the reason for the rule: 'it must be presumed in the absence of evidence indicating the contrary that the variance arose from an imperfect measurement of the whole line.'" Referencing, *Jones v. Kimble*, 19 Wis. 430, 432 (1865). This was an action concerned with the location of a lost quarter comer on the west side of Section 2 where there was a deficiency in the length of the west line of the section. Citing, *Moreland v. Page, Id.* See also: *Brooks v. Stanley*, 66 Neb. 826, 92 N.W. 1013 (1902); *Miller v. Topeka Land Co.*, 44 Kan. 354, 24 Pac. 420 (1890); *O'Brien v. McGrane*, 27 Wis. 446 (1871); 97 A.L.R. 1227.

**Author's Comment:** First, Griffin states, the corner must be lost, not difficult to find, or locate by appropriate legal rules. Second, the failure is the exact location of the corner, which may be as set, and not where later searchers think it should have been. What people intended to do, or what others thought they should have done, is irrelevant. It is what they actually did that matters, not matter how inconvenient or obnoxious it may be.

Continuing, Griffin states, "Where it is discovered that a gross blunder in the original survey causes the deficiency or surplus, the variance should not be distributed, but the correction applied at the location where the error was made, if its position can be determined. However, where small discrepancies are due to careless surveying and there are no circumstances suggesting the position of the error, the law of probabilities supports the appointment rule. Apportionment may be necessary to stabilize the location of intermediate parcels. When the variance occurs between two known corners recognized and accepted by the original survey, re-establishing the boundary lines of an intermediate parcel by using the record distance from one corner will produce a different location for the parcel than would re-establishment by using the record distance from the other corner. Stability of location is demanded in the public interest and can, very often, be obtained only by application of the rule of apportionment." (citing Skelton, note 4, §216.)[6]

**Author's Comment:** The caveat is that there is no gross blunder suggesting other measures. In addition, apportionment is based on the law of probabilities. A brief study of probabilities will quickly point out its inadequacies. Locating property corners based on chance should be taken seriously, not rushed into as a "quick-fix."

---

[5] For corners to be lost, they must be so completely lost that they cannot be replaced by reference to any existing data or other sources of information, and before courses and distances can determine boundary, all means for ascertaining location of the lost monuments must first be exhausted.

   *U.S. v. Doyle, 468 F.2d 633 (C.A. Colo., 1972).*

   Where the monuments and boundaries of the original government survey of a township were not so completely lost that they could not be retraced and relocated, the shortage in the township should not be apportioned equally to each of the six sections along the line; and "to be lost," when applied to section or township corners, means more than that they have been merely obliterated, tampered with, or changed, but they must be so completely lost that they cannot be replaced by reference to any existing data or other sources of information. *Mason v. Braught,* 146 NW 687 (S.D., 1914).

[6] Skelton, Ray Hamilton. *The Legal Elements of Boundaries and Adjacent Properties*, Indianapolis: The Bobbs-Merrill Company, Publishers.

To repeat Griffin's statement, apportionment is appropriate when retracement fails. When does retracement fail? Almost never. If a monument is truly gone, can the place where it stood be identified? In most cases, more than likely.

### *Lawson v. Viola Tp.*, **210 N.W. 979 (S.D., 1926):**

In boundary dispute, where location of original monument cannot be established by evidence, corner may be established by new survey made from points that can be determined and in accordance with field notes of original survey.

### *Reid v. Dunn*, **20 Cal.Rptr. 273, 201 Cal.App.2d 612:**

Existent corner; lost corner; obliterated corner.

## THINKING OUTSIDE THE BOX

The final thought if all the traditional procedures fail is get outside the box and think creatively. A much-used cliché, thinking outside the box has specific meaning. It means thinking differently, unconventionally, or from a new perspective. The phrase often refers to novel, creative and smart thinking, unimpeded by orthodox or conventional constraints.

Follow up on any idea, no matter how bizarre it may seem, and no matter how many people tell you that it will never work. Their opinion is not a fact. Their lack of adventure, creativity, or enthusiasm should not deter you from following your instinct, your curiosity, or the principles of law.

# 10 Failure to Find or Honor the Original Survey

You are free to make whatever choice you want,
   but you are not free from the consequences of the choice.

**Anonymous**

## FAILURE TO LOCATE ORIGINAL CORNERS, LINES, OR TITLES

A review of the court decisions from a variety of jurisdictions demonstrates that incorrect locations do not affect bona fide rights of previous purchasers. Or, in this author's words, "bad surveys do not destroy good titles." The crisis that arises is that once an incorrect "survey"[1] is produced, a number of potential problems arise:

- It puts a cloud on the title, which may or may not be recognizable in a standard title examination.
- It creates a problem for both the client and at least one neighbor, whether immediately discovered or not.
- It breeds needless, vexatious, possibly capricious litigation.
- It will not only involve the immediate parties, but also their lending institutions,[2] and possibly others.[3]
- Likely the problem will be discovered at a later date and be a burden on successors in title or other unsuspecting landowners and title holders.
- It will clutter the landscape with useless and meaningless markers.
- It will mislead innocent parties, landowners, inspectors, contractors, surveyors of other properties in the area.[4]
- It will influence the taxing process with misleading information.
- Once a plat is placed on record, it cannot be removed; at the least it is difficult to amend, or correct, it.

---

[1] Actually, it is not a survey at all, since it does not represent the truth, nor does it satisfy any definition of surveying. As one author stated, "it is an attempt at a survey that failed." An incorrect retracement does not satisfy its intent, and, in reality, may be either a case of breach of contract or a negligent act, depending on circumstances.

[2] This author was once involved in litigation representing a developer, who was being sued by an individual stating that the latter owned part of a subdivision wherein all the lots had been sold. There were ten lots in the subdivision, which meant there were ten defendants, along with not only their lawyers, but also ten title companies or banks, and their lawyers. In addition, also involved was the developer, along with his grantor in the event he lost and was seeking breach of warranty, which meant two more title companies. The courtroom that day was literally filled, with defendants, lawyers, surveyors, and consultants besides the parties themselves. The result was because the claim was bogus, the judge approved a motion for summary judgment, saving days of wasted court time and needless expense. Our lawyer estimated that the cost per hour during the time all the parties were present was about $10,000.

[3] Theoretically, the rule is that when an action is brought, it should involve everyone who potentially has a stake in it, that is, all parties who potentially have an interest.

[4] One example the author consulted on was a simple lot survey that took an inordinate amount of time, because of faulty locations and lack of discovery by intermediate surveyors. The surveyor was tempted to bring complaints against those previous surveyors – 18 of them.

DOI: 10.1201/9781003032557-18

## A DEPENDENT RESURVEY CANNOT CHANGE A BOUNDARY LINE

Precisely accurate resurvey cannot defeat ownership rights flowing from original grant and boundaries originally marked off.—*U.S. v. Doyle, 468 F.2d 633 (1972)*

The original survey as it was actually run on the ground controls. *United States v. State Investment Co.,* 264 U.S. 206, 44 S.Ct. 289, 68 L.Ed. 639; *Ashley v. Hill,* 150 Colo. 563, 375 P.2d 337. It does not matter that the boundary was incorrect as originally established. A precisely accurate resurvey cannot defeat ownership rights flowing from the original grant and the boundaries originally marked off. *United States v. Lane,* 260 U.S. 662, 43 S.Ct. 236, 67 L.Ed. 448; *Everett v. Lantz,* 126 Colo. 504, 252 P.2d 103, 108. The conclusiveness of an inaccurate original survey is not affected by the fact that it will set awry the shapes of sections and subdivisions. See *Vaught v. McClymond,* 116 Mont. 542, 155 P.2d 612, *Mason v. Braught,* 33 S.D. 559, 146 N.W. 687.

The actual location of a disputed boundary line is usually a question of fact. Gaines v. City of Sterling, 140 Colo. 63, 342 P.2d 651. "… [T]he generally accepted rule is that a subsequent resurvey is evidence, although not conclusive evidence, of the location of the original line." *United States v. Hudspeth,* 384 F.2d 683, 688 n. 7 (9th Cir.); accord, see *Ben Realty Co. v. Gothberg,* 56 Wyo. 294, 109 P.2d 455, 458, 459. And in its trespass action the burden of proving good title to the land rests on the *Government. Yakes v. Williams,* 129 Colo. 427, 270 P.2d 765; see also *Cone v. West Virginia Pulp & Paper Co.,* 330 U.S. 212, 67 S.Ct. 752, 91 L.Ed. 849.

After a tract of land has been surveyed and patented by the United States, its boundary cannot be affected, to the prejudice of the owner, by surveys and rulings of the Land Department. P. 212. 285 F. 128 affirmed.

When it has once made and approved a governmental survey of public lands and has disposed of them, the courts may protect the private rights acquired against interference by corrective surveys subsequently made by the Department. *Cragin v. Powell, supra,* p. 699. A resurvey by the United States after the issuance of a patent does not affect the rights of the patentee, the government, after conveyance of the lands, having "no jurisdiction to intermeddle with them in the form of a second survey." *Kean v. Canal Co.,* 190 U.S. 452. And although the United States, so long as it has not conveyed its land, may survey and resurvey what it owns, and establish and reestablish boundaries, what it thus does is "for its own information" and "cannot affect the rights of owners on the other side of the line already existing." *Lane v. Darlington,* 249 U.S. 331.

## A BOUNDARY LINE AGREEMENT CANNOT CHANGE A BOUNDARY LINE

Resurveys in no way affect titles taken under a prior survey. *Williams v. Barnett, 287 P.2d 789 (Cal. App., 1955)*

The doctrine of our law that coterminous owners may by agreement, implied from acquiescence, establish and fix their mutual boundary line, is applicable only where the true line is otherwise unknown or uncertain. There is no occasion for asserting that a boundary has been established by agreement, unless the description in the conveyance in reality designates a different boundary (*Mello v. Weaver,* 36 Cal.2d 456). "Of course, agreements of this type cannot be used where the true boundary line is known" (*Vowinckel v. N. Clark & Sons,* 217 Cal. 258, 261).

The owner of property must be presumed to have been familiar with the terms of the instrument which constituted his muniment of title. If he did not actually know the extent of his property and had the means of knowledge within reach, he would not be heard to say that a fence was located upon an accepted division line (*Janke v. McMahon,* 21 Cal.App. 781, 788).

Where there is an acquiescence in a wrong boundary, when the true boundary may be ascertained by the deed, it is treated both in law and equity as a mistake, and neither party is estopped from claiming the true line. The boundary is considered definite and certain when by survey it can be made certain from the deed (*Janke v. McMahon*, supra, 21 Cal.App. 781, 788).

A requisite to make applicable the doctrine of agreed boundary line is that the line acquiesced in by adjoining property owners must be specified, definite, and certain. The agreement is not controlling if that line is left indefinite, uncertain, or speculative (Garrett v. Cook, 89 Cal.App.2d 98).

Appellant and respondents are in agreement upon the principle that resurveys in no way affect titles taken under a prior survey.

Counsel for appellee seeks to justify the admission of the verbal testimony by the application thereto of the principle that owners of adjoining tracts may, by parol, settle and permanently establish a boundary line between their premises, which, if followed by possession according to such agreed line, may become conclusive upon such parties and their grantees with notice, as the true boundary line. This contention cannot be sustained. Adjoining proprietors, if the boundary line of their premises be in dispute or be unknown, may conclude themselves as to the true location of the line by an agreement establishing a line as the true line, and by holding possession and occupying their premises thereafter in accordance with such agreement. *Henderson v. Dennis*, 177 Ill. 547, 53 N.E. 65; *Duggan v. Uppendahl*, 197 Ill. 179, 64 N.E. 289. But this principle proceeds upon the ground that the extent of the ownership of such proprietors, only, may be so agreed upon and settled – not that title to land can be made to pass by parol agreements. It appears from the verbal testimony here under consideration that the true boundary lines were not in dispute or unascertained, but that there was a verbal agreement to convey other premises than those described in the deed of conveyance. The rule as to the establishment of a disputed or unascertained line between the lands of adjoining owners by verbal agreement, and possession in accordance with such agreement, has no application to the state of facts as disclosed by the testimony under consideration. *Grubbs v. Boon*, 201 Ill. 98, 66 N.E. 390 (Ill. 1903).

A survey establishing a line between adjacent landowners will not revive the right of an original owner against an established boundary, since all the survey does is establish the line and not the title. *Grell v. Ganser, 255 Wis. 381 (1949)*

## A FORM OF AGREEMENT, AN AGREED LINE THROUGH ACQUIESCENCE

The appellants brought this action claiming that they are the absolute and exclusive owners of all land connected with the S.W. forty up to the line marked by an ancient fence which was built some distance north of an old but now abandoned highway. Thus they claim that the land contribution to the old road came entirely from their premises, and, upon its abandonment, the land again became a part of their property. The appellants are insisting that the fence was placed and exists as a boundary fence.

The respondents disavow that fence line and base their claim to ownership upon a line fixed by a survey of the lines of the section.

It must be conceded that a line can be established by acquiescence of the parties that may differ from a surveyed or true line. *Wunnicke v. Dederich,* 160 Wis. 462, 152 N.W. 139, and cases cited therein; *Husted v. Willoughby,* 117 Mich. 56, 75 N.W. 279; *Pittsburgh & L. A., Iron Co. v. Lake Superior Iron Co.,* 118 Mich. 109, 76 N.W. 395; *White v. Peabody,* 106 Mich. 144, 64 N.W. 41; *Jones v. Pashby,* 67 Mich. 459, 35 N.W. 152, 11 Am.St.Rep. 589; Note, (1930) 69 A.L.R. 1491. When long continued possession of land up to a line has been acquiesced by all interested in the ownership of a contiguous piece of land, the boundary thereby established becomes the proper boundary irrespective of the operation of the principles which would otherwise fix and determine their location. A survey establishing a line between adjacent owners will not revive the right of an original owner against an established boundary since all that the survey does is to establish the line and not the title.

The court stated, "In Salmond, Jurisprudence, sec. 106, it is said: 'Possession * * * is the objective realization of ownership. It is in fact what ownership is in right. Possession is the de facto exercise of a claim; ownership [39 N.W.2d 399] is the de jure recognition of one. * * * The two things tend mutually to coincidence.' See, Readings on the History and System of the Common Law, Pound and Plucknett (3d ed.) 639. The doctrine has been followed by the courts not to disturb such long acquiescence in a boundary line. It seems to be the recognized policy of the law to encourage such settlements as a means of suppressing troublesome disputes. Jones v. Pashby, 67 Mich. 459, 35 N.W. 152, 11 Am.St.Rep. 589; 8 Am.Jur. 802."

## A RETRACEMENT SURVEY, OFTEN CALLED A RESURVEY, CANNOT CHANGE A BOUNDARY LINE

**A boundary is fixed by law, it is not fixed by survey. It is merely located by survey, and not always correctly.**

As stated in the 7th edition of *Brown's Boundary Control and Legal Principles*: "Retracing surveyors will encounter a minority of surveyors who, when finding an 'error' in the original survey, believe it is their responsibility to 'correct' the and make the original bearings and distances as they should be had they be surveyed correctly. These surveyors have no concept that, once the lines have been created, no subsequent surveyor has authority to re-create the original lines. When a creating surveyor indicates a distance or an angle, these are the original measurements, according to the creating surveyor's methodology and errors. By law, they are free from error."

In the case of *Hagerman et al. v. Hagerman et al.,*[5] the court stated "The purpose of a resurvey is to ascertain lines of original survey and original boundaries and monuments as established and laid out by survey under which parties take title to land, and they cannot be bound by any resurvey not based on survey as originally made and monuments erected.

All the three surveys in question here were resurveys, binding on no one, unless one of these perchance should ultimately in a proper proceeding be found to be correct."

> Purpose of resurvey is merely to ascertain lines of original survey, and original boundaries and monuments, and parties cannot be bound by any survey not based on that originally made and monuments erected thereunder.
>
> *Day v. Stenger, 274 P. 112 (Idaho, 1929)*

1. "In case of inability to establish monuments and lines of original survey, evidence of ancient fences and other improvements, though in its nature hearsay, is competent to prove original boundaries."

The court emphasized, "It is immaterial whether the original survey was accurate or not; when once made no subsequent survey can change such corners and lines. (4 R. C. L., secs. 41, 106; *Price v. De Reyes*, 161 Cal. 484, 119 P. 893; *Bayhouse v. Urquides*, supra; *Le Compte v. Lueders*, 90 Mich. 495, 30 Am. St. 450, 51 N.W. 542; *Bridenbaugh v. Bryant*, 79 Neb. 329, 112 N.W. 571.)"

The purpose of a resurvey is to ascertain the lines of the original survey and the original boundaries and monuments as established and laid out by the survey under which the parties took title (*Bayhouse v. Urquides,* 17 Idaho 286, 105 P. 1066; *Wing v. Wallace,* 42 Idaho 430, 246 P. 8). Parties cannot be bound by any survey not based upon the survey as originally made and monuments as erected. (*Idem.*)

---

[5] 68 Wyo. 515, 235 P.2d 750 (Wyo. 1951).

## FOOTSTEPS NOT FOUND, REMEDY

Where footsteps of the surveyor are not found, it is the court's duty to ascertain the surveyor's intention by his field notes and circumstances and conditions surrounding the survey.

*Howell v. Ellis,* 201 S.W. 1022 (1918)

Purpose of resurvey is to trace footsteps of original surveyor, and when marks of his footsteps are found, they control, but when they cannot be found, old use and occupancy and old recognition must suffice.

*Ballard v. Stanolind Oil & Gas Co.,* 80 F.2d 588 (Texas, 1935)

**Final Quote**

*There is no intellectual or emotional substitute
for the authentic, the original, the unique masterpiece.*

**Paul Mellon**

# Appendix A
# Original Survey, by State

## ALABAMA

The presumption must be indulged, that the government surveyor, in his original survey, adopted or erected monuments or marks to indicate that corner and line, unless the evidence shows that such presumption is contrary to the truth. The evidence does not show any thing irreconcilable with the truth of the presumption; and we shall assume, therefore, that the aforesaid disputed corner was established by the government surveyor, that his marks or monuments designating it have been wholly obliterated, that the aforesaid disputed line was also established by him, and that his marks indicating some two hundred yards of that part of it nearest to the township line have been obliterated.

If that assumption be correct, then it is for the jury to ascertain and settle at what precise point the disputed or lost corner was placed, and the disputed line marked, by the government surveyor in his original survey. And to enable the jury to perform that duty intelligently, any evidence, whether parol or written, may be submitted to them, which has any natural and reasonable tendency to show where that corner was placed or that line marked in the original survey. Recourse may be had to the unobliterated marks and corners of that survey, to the field-notes and plat, and to subsequent surveys made under their guidance. Such subsequent surveys cannot alter or control that survey; for, so far as it can be traced or proved, it must govern. But still they may aid the jury in ascertaining *the original position of its lost corner.* Their inaccuracies or errors may be so numerous, or so glaring, as to destroy all faith in them as evidence, on the part of the jury. On the other hand, their errors or inaccuracies may be such as are explainable by other evidence, or by the known imperfection of the means necessarily used in making all surveys; and may, therefore, not destroy all faith in them, nor render them worthless as evidence. With their aid, the jury may be enabled to ascertain with reasonable certainty where the lost corner was located in the original survey. Without their aid, the jury may not be able to ascertain that location. Their weight or influence as evidence must be determined by the jury. The mere fact that the party relying on them has not proved that they correspond *in all respects* with the original government survey, *does not authorize the court to instruct the jury to disregard them entirely in seeking the location of the lost corner or line.* The party cannot, in any case, prove such correspondence, without proving *every part* of that survey. In many cases, he cannot prove *a lost corner, or any other lost part of* that survey, without the aid of such subsequent surveys. And in all such cases, a rule which requires, as a condition for obtaining any influence for them, that the party relying on them should prove their correspondence *in all respects* with the original survey, would amount to a denial of the right to prove the location of a *lost corner or other lost part* of the original survey. There is no such rule. The court below erred, therefore, in the 1st charge given "at the instance of the defendant." The jury ought to consider such subsequent surveys, in connection with the other evidence in the cause; and if, after doing so, they believe that the original location of the lost corner was at a particular point designated with reasonable certainty by the evidence, such belief ought not to be disregarded in making up their verdict – McClintock v. Rogers, 11 Illinois Rep. 279; Doe, *ex dem. Miller v. Cullum,* 4 Ala. 679; *Bryan v. Beckley, Litt. Sel. Cases,* 91; *Wallace v. Maxwell, 1 J. J. Marsh.* 447.

If the original location of the lost corner is ascertained, there cannot be much difficulty in deciding as to the location of the disputed line. That line is marked from the known and established north-east corner of section 34, in a southern direction, to a point some two hundred yards distant from the admitted southern boundary of the section. As far as that line is marked, whether it be straight or not, it must be treated as established; and the presumption is, that from the point where all traces or proof of the marks fail, the government surveyor closed the survey, or completed the line, by running it straight to the disputed corner – *Thornberry v. Churchill*, 4 Monroe, 30; *Brown v. Hobson, 3 A. K. Marsh.* 382; *Wishart v. Cosby*, 1 *ib.* 382.

If the original location of the lost corner is not proved to have been at a point other than that at which the aforesaid marked line, if continued in its course, would intersect the said township line, then the intersection of these lines must be taken as the south-east corner of said section 34. The course of the line marked in part, must be followed the proper distance, that is, to the township line, unless there is or was an established corner to divert it, or unless there is something else deemed sufficient in law to control it – *Mercer v. Bate, 4 J. J. Marsh.* 334. See also authorities cited in the next preceding paragraph.

**Billingsley v. Bates, 30 Ala. 376 (Ala. 1857).**


## ALASKA

No decisions found respecting original surveys.


## ARKANSAS

The evidence disclosed that the government plat of date January 17, 1826, on file in Jefferson County, and the United States Land Office in Little Rock, only contained two lots numbered 1 and 2 in each quarter section; that all conveyances of said lands from the United State's Government down to the present owners described the lands in accordance with said plat; that after the government parted with its title to said lands under the description aforesaid, in accordance with the description in said plat, the original plat was amended by the United States Government on February 25, 1859, so as to divide each original lot 2 into lots 2 and 3 in each quarter section; that the amended plat was filed in the United States Land Office at Little Rock but never filed in Jefferson County, and no one in Jefferson County knew of the existence of the amended plat, and, for that reason, was not used for descriptions in conveying lands in that county; that the original plat is the only plat used by the people of Jefferson County in laying off and locating their lands; that all the engineers and surveyors refer to the original plat in making their surveys; and that a reference to any other plat would not be understood and would be misleading to the people of the county, and that the district in question was organized and outlined by the engineer by reference to the original plat.

In the organization of drainage districts, it is provided by section 1 of Act 279, Acts 1909, as follows: "The said engineer shall forthwith proceed to make a survey and ascertain the limits of the region which would be benefited by the proposed system of drainage; and such engineer shall file with the county clerk a report showing the territory which will be benefited by the proposed improvement, * * *." We are not familiar with any law requiring an engineer or surveyor to outline the boundaries of lands by reference to plats only. The boundaries to lands may be designated and described by reference to natural or artificial monuments. If the boundaries of the district or the boundaries of the lands contained therein are described so that the land owners and county court can understand where they are, such description would be sufficient. We think a survey or description referring to the only plat or map on file in the county, and in general use, is a sufficient description by which to designate and locate the boundary lines of a drainage district or the boundary lines of the lots of land contained therein. By reference to the original plat, made a part of the transcript in this case, it is definitely established that all the lands embraced within the district are contiguous.

*Hudson v. Quattlebaum*, **132 Ark. 613, 201 S.W. 1113 (Ark. 1918).**

## COLORADO

The original survey as it was actually run on the ground controls. *United States v. State Investment Co.,* 264 U.S. 206, 44 S.Ct. 289, 68 L.Ed. 639; *Ashley v. Hill,* 150 Colo. 563, 375 P.2d 337. It does not matter that the boundary was incorrect as originally established. A precisely accurate resurvey cannot defeat ownership rights flowing from the original grant and the boundaries originally marked off. *United States v. Lane,* 260 U.S. 662, 43 S.Ct. 236, 67 L.Ed. 448; *Everett v. Lantz,* 126 Colo. 504, 252 P.2d 103. The conclusiveness of an inaccurate original survey is not affected by the fact that it will set awry the shapes of sections and subdivisions. See *Vaught v. McClymond,* 116 Mont. 542, 155 P.2d 612; *Mason v. Braught,* 33 S.D. 559, 146 N.W. 687.

The actual location of a disputed boundary line is usually a question of fact. *Gaines v. City of Sterling,* 140 Colo. 63, 342 P.2d 651. "... [T]he generally accepted rule is that a subsequent resurvey is evidence, although not conclusive evidence, of the location of the original line." *United States v. Hudspeth,* 384 F.2d 683, 7 (9th Cir.); accord, see *Ben Realty Co. v. Gothberg,* 56 Wyo. 294, 109 P.2d 455, 458, 459. And in its trespass action the burden of proving good title to the land rests on the *Government. Yakes v. Williams,* 129 Colo. 427, 270 P.2d 765; see also *Cone v. West Virginia Pulp & Paper Co.,* 330 U.S. 212, 67 S.Ct. 752, 91 L.Ed. 849.

The procedures for restoration of lost or obliterated corners are well established. They are stated by the cases cited below and by the supplemental manual on Restoration of Lost or Obliterated Corners and Subdivisions of Sections of the Bureau of Land Management (1963 ed.). The supplemental manual sets forth practices and contains explanatory and advisory comments.

Practice 1 of the supplemental manual recognizes that an existent corner is one whose position can be identified by verifying evidence of the monument, the accessories, by reference to the field notes, or "where the point can be located by an acceptable supplemental survey record, some physical evidence, or testimony." Practice 2 recognizes that an obliterated corner is one at whose point there are no remaining traces of the monument, or its accessories, but whose location has been perpetuated, or the point for which may be recovered beyond a reasonable doubt, by the acts and testimony of the interested land owners, competent surveyors, or other qualified local authorities, or witnesses, or by some acceptable record evidence. Practice 3 states that a lost corner is one whose position cannot be determined, beyond reasonable doubt, either from traces of the original marks or from acceptable evidence or testimony bearing on the original position, and whose location can be restored only by reference to one or more interdependent corners.

The authorities recognize that for corners to be lost "[t]hey must be so completely lost that they cannot be replaced by reference to any existing data or other sources of information." *Mason v. Braught,* supra, 146 N.W. at 689, 690. Before courses and distances can determine the boundary, all means for ascertaining the location of the lost monuments must first be exhausted. *Buckley v. Laird.* 493 P.2d 1070, (Mont.); Clark, Surveying and Boundaries § 335, at 365 (Grimes ed. 1959); see advisory comments of the supplemental manual, supra at 10.

The means to be used include collateral evidence such as boundary fences that have been maintained, and they should not be disregarded by the surveyor. *Wilson v. Stork,* 171 Wis. 561, 177 N.W. 878, 880. Artificial monuments such as roads, poles, fences, and improvements may not be ignored. *Buckley v. Laird,* supra, 493 P.2d at 1073; *Dittrich v. Ubl,* 216 Minn. 396, 13 N.W.2d 384, 390. And the surveyor should consider information from owners and former residents of property in the area. See *Buckley v. Laird,* supra, 493 P.2d at 1073-1076. "It is so much more satisfactory to so locate the corner than regard it as 'lost' and locate by 'proportionate' measurement." Clark, supra § 335 at 365.

**United States v. Doyle, 468 F.2d 633 (10th Cir. 1972).**

"A lost or obliterated closing corner from which a standard parallel has been initiated or to which it has been directed will be reestablished in its original place by proportionate measurement from the corner used in the original survey to determine its position."

The corner in question is a closing corner, but not one from which a standard parallel has been initiated nor one to which a standard parallel has been directed; we do not see, therefore, that the rule relates to this case, but if it did we doubt that the corner can be regarded as lost or obliterated. It never existed, and so cannot, strictly speaking, be said to be lost or obliterated. If the monuments were lost or obliterated there would be some reason to attempt to relocate it, and perhaps the method prescribed in rule 47 is as good a way as any other, but when it is a myth, never on the ground, the natural, straightforward, and sensible way is to establish the corner at the place where the original surveyor ought to have put it, and that is where the north course of the east line of the section meets the correction line at right angles, and that is where the report puts it. Everybody knows that is where the section line ought to have closed, and where the original surveyor, honest or dishonest, meant to close it; that his duty required him to close it there, so that the inclosure of his lines might be a rectangle or nearly so. Why should courts be less reasonable than reasonable men?

**Lugon v. Crosier, 78 Colo. 141, 240 P. 462 (Colo. 1925).**

## CONNECTICUT

It appears from the finding that the land within the township of Stratford, including the premises in question, was included in deeds from the Indians; some being made to the inhabitants of the township, and others to the "townsmen" or their successors. Grants of this land were made to the proprietors of the town by the General Court in 1685 and 1703. In 1741 the proprietor's committee allotted to Zachariah Curtis, Jr., and Timothy Beach in 1741, three parcels of sedge on the Long Beach which is the foundation of the plaintiffs' claim. It may be assumed in favor of the plaintiffs' contention that Curtis and Beach acquired title to these three parcels which were contiguous to each other, and together contained 4 1/2 acres. They were thereafter conveyed to successive grantees as one parcel, and described as a sedge marsh. These parcels were not upon the beach proper, but were located between the beach and Mud creek which is a continuation of the Gut at its east end, and which penetrates extensive salt meadows, and a marsh connecting with the Housatonic river. These parcels were a mile east of the land now in dispute. No evidence of the Timothy Beach interest in this land thereafter is disclosed by the land records. The interest of Zachariah Curtis, Jr., passed by descent, devise, and conveyance by a description of a most general character, to one Jonas Hinman, who in 1797 conveyed his holding to Jonas McKinzee, describing it as 4 1/2 acres of land or sedge marsh, which deed referred to the original survey of Zachariah Curtis and Timothy Beach. In 1800 McKinzee conveyed the same parcel to Eli Tongue. No conveyance from Tongue appears of record, and the 4 1/2 acres in question were not made the subject of any subsequent conveyance of record, from the heirs, devisees, or representatives of Tongue. In 1835 one Abijah Ufford placed upon the land records of Stratford a deed, in which he purported to convey to one Timothy Risley a certain parcel of land or sand beach known as Long Beach Point, in Bridgeport Harbor at the mouth of the Gut, containing 10 acres, and bounded north on the Gut, east on land of David and Ira Curtis, south on Long Island Sound, and west on Bridgeport Harbor, "being property that I purchased of Eli Tongue, who purchased the same from James McKinzee." The grantor then refers by volume and page to the deed of McKinzee to Tongue, and further reference is made to deed of Jonas Hinman and to survey to Zachariah Curtis, as the basis of his title.

There is no other evidence that Ufford had ever acquired the westerly end of Long Beach. On the other hand he definitely refers as the sole basis of his title to this parcel of 4 1/2 acres, which was originally surveyed to Zachariah Curtis and Timothy Beach, and that parcel was situated near the other or easterly end of the beach (fully a mile to the eastward) and did not include any part of the beach proper, but was of sedge marsh below mean highwater. No evidence was offered other than this deed that Ufford ever in fact entered upon, took possession of, or in fact occupied any part of the easterly end of it, or any part of Long Beach. Risley did not enter upon, take possession of, or occupy any part of this parcel, unless upon the facts stated, such entry, possession, and occupation are facts necessarily presumed from the execution and record of the deeds to and from Risley.

*McMahon v. Town of Stratford*, **83 Conn. 386, 76 A. 983 (Conn. 1910).**

## DELAWARE

The problem seems to have originated when in a subdivision plan executed in 1967 the surveyor hired by Burke, who was the developer of Corner Ketch Farms, a residential area in New Castle County where the properties are located, consulted only his office calculations taken during a 1959 survey rather than conducting a field study or referring to actual field notes. The 1959 survey was the original survey for Corner Ketch Farms, containing 13 lots but reserving 5 acres, which held a farmhouse, barn, and other outbuildings, to Burke. When the surveyor at Burke's request executed the 1967 plan, in which Burke surrendered part of his retained acreage to create three new lots (Lots 14, 15, and 16) for development, the various buildings and their locations were not shown. Thus, the line between Burke's property and Lot 14 to the north, which was later acquired by the Collinses, was unwittingly drawn at a point on the plan where it actually bisected Burke's barn approximately two feet south of its northern wall, which faced Lot 14. This fact was not reflected on the 1967 subdivision plan, which was filed with the County and which served as the basis for the metes and bounds description prepared by the Collinses' attorney for the deed conveying Lot 14 from the Burkes to the Collinses.

*Collins v. Burke*, **418 A.2d 999 (Del. 1980).**

## FLORIDA

In the sale of lands in sections, or subdivisions thereof, including lots, according to the government survey, the survey as actually made controls. *Miller v. White*, 23 Fla. 301, 2 So. 614; *Liddon v. Hodnett*, 22 Fla. 442. It is the survey as it was actually run on the ground that governs, if the monuments, corners, or lines actually established can be located or proved. Courses and distances yield to such corners and lines so long as the latter can be located, and for the reason that the latter are the fact or truth of the survey as it was actually made, while the former are but descriptions of the act done, and, when inaccurate, they cannot change the fact. *McClintock v. Rogers*, 11 Ill. 279; *Yates v. Shaw*, 24 Ill. 367; *Bauer v. Gottmanhausen*, 65 Ill. 499; *Kincaid v. Dormey*, 47 Mo. 337; *Major's Heirs v. Rice*, 57 Mo. 384; *Willis v. Swartz*, 28 Pa. St. 413; *Riley v. Griffin*, 16 Ga. 141.

*Watrous v. Morrison*, **33 Fla. 261, 14 So. 805 (Fla. 1894)**

## GEORGIA

[4.] All lands are supposed to be actually surveyed; and the intention of the grant is, to convey the land according to that actual survey.

[14.] In ascertaining boundaries, the locations of the original surveyor, so far as they can be found, are to be resorted to; and where they vary from the proprietor's plan, the locations actually made, will control the plan.

**Riley v. Griffin, 16 Ga. 141 (Ga. 1854).**

## IDAHO

The purpose of a resurvey subsequent to the taking of title by purchasers and settlers is to ascertain the lines of the original survey and the original boundaries and monuments as established and laid out by the survey under which the parties originally procured their titles. (*Martz v. Williams*, 67 Ill. 306.) On such resurvey or re-established boundaries and monuments the question of the correctness of the original survey cannot enter into the matter at all, and is a matter that does not concern the surveyor, and is not a question to be ascertained by him. (*Diehl v. Zanger*, 39 Mich. 601; *Penry v. Richards*, 52 Cal. 672 (675); *Bullard v. Kempff*, 119 Cal. 9, 50 P. 780.)

**Bayhouse v. Urquides, 17 Idaho 286, 105 P. 1066 (Idaho 1909).**

## ILLINOIS

This court is impressed with the research and application of civil engineering procedures engaged in and followed by surveyor Harmon. For that reason and the further reason that we harbor doubts as to the value of Defendant's Exhibit D since it was concerned with laying out a roadway and not fixing boundaries, we are of the opinion that the trial court was correct in finding that Plaintiff's Exhibit 7 established the location of the division line between the quarter sections owned by plaintiff and defendant. We are, however, mindful of the fact that before final determination of such a question can be made it must be established that the procedures followed in arriving at the division line set forth in Exhibit 7 complied with the requisite legal principles.

Directing our attention to the applicable law we recognize that in our State it is well settled that monuments, when found, must control as against courses and distances given on field notes, when determining boundary lines. (*England v. Vandermark (1893)*, 147 Ill. 76, 35 N.E. 465; *Sawyer v. Cox (1872)*, 63 Ill. 130; *McClintock v. Rogers (1849)*, 11 Ill. 279.) In the instant case no monuments or stones were found along, in, or on the quarter section land between the property of the plaintiff and defendant. While such monuments may well have been placed in bygone days, they defied discovery when sought by the parties to this case and by the surveyors they employed.

There is no quarrel between the parties that monuments were found and relied upon by the surveyors for each party. Four monuments, being stones, were located along the half section line running through Sections 28, 29, and 30. These four stones were used by the surveyors in their efforts to establish the correct boundary line.

In the instant case the lines and corners of the quarter sections had become obliterated and no natural or artificial monuments could be found which by their very presence identified the boundary line, so surveyor Harmon resorted to field notes and plats of the original survey in order to locate the disputed division line. Such practice has long been approved of by our supreme court. See *Sawyer v. Cox (1872)*, 63 Ill. 130, and *McClintock v. Rogers (1849)*, 11 Ill. 279. The importance of original field notes and plats is recognized by a statutory provision which requires the county board of each county to acquire copies of them, and for the depositing of the same in the recorder's office. See Illinois Revised Statutes 1977, ch. 133, par. 5.

Mr. Harmon, the surveyor for the plaintiff, relied upon the original government survey plat and subsequent surveys and field notes concerned Sections 28, 29, and 30. Harmon's reliance upon field notes (Plaintiff's Exhibits 3, 4, and 5) and the 1917 plat of Horney (Plaintiff's Exhibit 6) convinced him that the quarter sections in question had originally been divided equally and that subsequent surveys continued to do so. A study of the exhibits compels a conclusion that Harmon's conviction was well founded, and that he was correct in adhering to the principle of prorating in order to determine the correct boundary line.

By contrast Greene, surveyor for the defendant, failed to follow the requisite standards and procedures but instead relied solely in preparation of Exhibit E on Defendant's Exhibit D, being a road plat on a half section line. It is difficult to conclude that Horney intended to establish section or quarter section lines on the road plat in light of the plat he prepared approximately nine years earlier, the sole purpose of which was to "fix boundary lines." In this earlier plat he treated the quarter sections as being equal.

*Pliske v. Yuskis*, **83 Ill.App.3d 89, 403 N.E.2d 710 (Ill.App. 3 Dist. 1980).**

## INDIANA

The land in controversy is situated in Lake county, in the extreme northwest corner of the State. It is described in the complaint as: "Lot number 5, in section 36, township 38 north; lots 8, 9, and 10, in section 1, township 37 north, and lots 5, 6, 7, and 8, in section 12, township 37 north, all in range 10 west, Lake county, Indiana, and containing 252.5 acres, more or less."

The appellant, in her cross-complaint, made claim to the same land, "in plaintiff's complaint described," and asked that her title thereto be quieted as against the plaintiff and all of her co-defendants; and that they be enjoined from setting up any claim thereto.

The appellees, who are in possession of the lands in controversy, claim title under the original government survey, by virtue of patents from the United States to the State of Indiana, and from the State to their remote grantors.

Townships 38 and 37 north, range 10 west, in which the lands are situated, were originally surveyed in 1834, under authority of the United States land department, as shown by the field notes and plats made a part of the record. These townships are both fractional, lying next to the State line dividing Indiana from Illinois, and are in part covered by a body of water known as Wolf lake.

The appellees claim that title to all of these lands passed from the general government to the State of Indiana by the Swamp land act of September 28, 1850, subject only to identification and selection by the State and approval thereof by the Secretary of the Interior. The patent from the United States to the State is dated March 24, 1853. In this patent it is recited that "The United States of America, in consideration of the premises and in conformity with the act of Congress aforesaid, have given and granted, and by these presents do give and grant, unto the said State of Indiana, in fee-simple, subject to the disposal of the legislature thereof, the tracts of land above described." The tracts so described include: "The whole of fractional sections 1, 12 * * all in township 37 north, of range 10 west; * * * also the whole of fractional section 36, in township 38 north, of range 10 west." The court found the title thus traced by appellees to be good, and held that they were entitled to continue in possession of the lands in dispute.

The appellant contends that the bed of Wolf lake, covering a part of the above described sections, as aforesaid, was not surveyed in the original survey of 1834; and shows that on representations to that effect, made to the land department of the United States, the commissioner of the general land office ordered a resurvey of the land within the meander lines of the lake, which re-survey was made in 1875. Appellant then claims that on such re-survey, by the land department, the lots in the lake bed became subject to entry and sale, and subject also to the right of appellant's remote grantors to locate "Sioux half-breed scrip" thereon; and she traces her title from patents issued for said lots to such remote grantors on such location of half-breed scrip.

Appellees, on the other hand, contend that all said fractional sections, including the bed of Wolf lake, were surveyed in 1834, and the lands and lots sold by the United States under such survey, so that, rightfully, the government had no such land to survey or sell when the order for the survey of 1875 was made and the lands in question attempted to be resold by the land department.

As the appellant must succeed, if at all, on the strength of her own title, it will be sufficient to decide the contention here made. If the lands in controversy were, in fact, surveyed in 1834, and sold by the United States under such survey, then it is clear enough that the government had no authority or power to re-survey the lands in 1875, or to sell them over again, and appellant's title must wholly fail.

The annexed plats show the original survey, in 1834, of section 36, in township 38, and of sections 1 and 12, in township 37; and also the re-survey, in 1875, of that part of the bed of Wolf lake, in the same townships and sections, being all that is necessary to indicate the location of the lands in dispute.

[SEE DRAWING IN ORIGINAL]

We are of opinion that in the original survey of 1834, both townships in question, and all of the sections were fully surveyed, including the bed of Wolf lake.

The plat, with the chains marked thereon, shows that the townships are both fractional. Township 38 consists only of fractional section 36, lying between the Illinois line and Lake Michigan. Township 37 is less than a mile in width, extending from the east township line west to the State line. The field notes show that the east section line, and also the west line, being the State line, were actually run; and that the section corners on said east section line of sections 1 and 12 were marked. The field notes also show that the east and west line between townships 38 and 37, being the north line of section 1, township 37, was actually run, except a short distance in the northeast corner, which extends into Lake Michigan; also that the east and west section line between sections 1 and 12, township 37, was run; also that the east and west section line on the south side of section 12, township 37, was run from the southeast corner of said section west to the lake, this line not being extended in the field across to the State line, or west line of the section. The field notes further show, as does the plat, that all the interior lines of section 36, in township 38, and the greater part of those in sections 1 and 12, in township 37, were actually run and the corners marked in the field.

Under the provisions of sections 2395 and 2396, R. S. U.S. the foregoing was a sufficient survey of the townships and sections named. By section 2395, R. S., *supra,* it is made a sufficient survey to run "each way, parallel lines at the end of every two miles; and by marking a corner on each of such lines, at the end of every mile." In the foregoing survey the east and west section lines, as actually run, are less than a mile apart. All the lines, interior and exterior, of section 36, township 38, are actually run; so also are all the lines, interior and exterior, of section 1, township 37. For section 12, township 37, the north line is actually run.

There remains only the south line of said section 12, a part only of which is run. For this defect, if it should be considered such, section 2396, U.S. R. S., *supra,* provides as follows: "The boundary-lines which have not been actually run and marked shall be ascertained, by running straight lines from the established corners to the opposite corresponding corners; but in those portions of the fractional townships where no such opposite corresponding corners have been or can be fixed, the boundary-lines shall be ascertained by running from the established corners due north and south, or east and west lines, as the case may be, to the watercourse, Indiana boundary-line, or other external boundary of such fractional township." Following this direction, the south line of section 12 will be found by "running from the established corner," at the southeast corner of said section, a "due east and west line, west to the State line, being the west external boundary of such fractional township."

It appears, therefore, that township 37, as well as township 38, was fully surveyed in the original survey of 1834; that the lines of the survey extend over both land and water, and that the United States conveyed to the State the fractional townships so surveyed.

It follows that all the territory in question, in the two townships, including that covered by Wolf lake, as well as the comparatively dry land on its borders, having passed to the State from the ownership of the United States by the Swamp land act of 1850, and the patents issued to the State therefor in 1853, no land in said sections remained, whether under Wolf lake or elsewhere, which could be re-surveyed or resold by the general government. The survey of 1875, and the sales thereunder, were nullities. Appellant, therefore, has no title to the lands claimed by her. See *Tolleston Club v. State,* 141 Ind. 197, 38 N.E. 214, where a similar question is considered, and a like decision reached.

Two objections made to this conclusion may be noticed:

First. It is said that the validity of the re-survey of 1875 having been affirmed by the United States land department, which department had full jurisdiction of the matter, the question is not reviewable in the courts. It is true, as said in *Hardin v. Jordan,* 140 U.S. 371, 35 L.Ed. 428, 11 S.Ct. 808, that the decision of the land department on matters of fact within its jurisdiction, made in the course of administration, cannot be called in question collaterally.

But, as said in the same case by the Supreme Court of the United States: "If the lands patented were not at the time public property, *having been previously disposed of,* * * * the department had no jurisdiction to transfer the lands, and their attempted conveyance by patent is inoperative and void." See also *Mitchell v. Smale,* 140 U.S. 406, 35 L.Ed. 442, 11 S.Ct. 819.

The other objection made is, that in the Beaver lake case, *State v. Portsmouth Sav. Bank,* 106 Ind. 435, 7 N.E. 379, this court decided that the sale of the border lots on Beaver lake, under the original government survey, did not carry title to the lands covered by the waters of the lake. The case is not in point, for the very good reason that in the Beaver lake case the lands under the lake were not surveyed; the survey ceased at the borders of the lake. In the case of Wolf lake, however, as we have seen, the lake itself was included in the survey; the section lines passed over and included the waters as well as the land. Had Beaver lake been actually surveyed, as Wolf lake was, the matter would be different; as it is, that case is not in point. The case at bar is ruled by the *Tolleston Club* case, *supra.*

**Kean v. Roby, 145 Ind. 221, 42 N.E. 1011 (Ind. 1896).**

## IOWA

Reference to the following plat will aid to an understanding of the case.

This is a plat of an addition to Des Moines known as Kauffman Place. The plaintiff is the owner of the north half of lot 22, and the defendants are the owners of the south half thereof. This lot faces east on Thirty-Sixth street. Thirty-Sixth street runs north and south. What appears as "A" street in the plat is referred to as Thirty-Seventh street in this record. From the southeast corner of lot 22 to the northeast corner of lot 14 is a distance of six hundred feet according to the plat, and according to the ground. This dimension also measures the distance between University avenue and Cottage Grove avenue as laid by the recorded plat. It is undisputed that lot 22, as platted, has a dimension of one hundred feet fronting east on Thirty-Sixth street, and that the parties hereto are entitled each to fifty feet thereof. The controversy is over the true location upon the ground of the north and south lines of such lot. Practically all the lots shown on the plat as fronting east on Thirty-Sixth street and numbered from 14 to 22, inclusive, are improved and occupied as residence properties. In locating and taking possession of their lots, the respective owners were guided by the presence of certain stakes which were supposed to represent the respective corners as fixed upon the ground by the original survey. These stakes were all consistent with each other; and the respective owners successively took possession in accord therewith, and each owner is in possession of his appropriate dimensions indicated upon the plat. The improvement and occupancy of these lots began about 1905. At that time neither Cottage Grove avenue nor University avenue nor Thirty-Sixth street had been improved. The original survey of that ground was made in 1902 by one Dickenson. This survey, however, laid open only six lots on this ground, giving to each a frontage of one hundred feet east on Thirty-Sixth street, and stakes were then set by the engineer one hundred feet apart, to indicate the boundaries of each of such lots. Such plat was not recorded in such form. Just when the plat was made in the above form does not appear. This later plat was filed and recorded in 1906, and after sales had been made therefrom. In April, 1907, the plaintiff purchased the north half of lot 22. He took possession in accordance with the stakes appearing upon the ground, and such possession was consistent with the claims of his neighbor on the north. The defendants also purchased in 1907 a few days prior to the purchase of the plaintiff. They also took possession of fifty feet south of plaintiff's assumed line. In the spring of 1909, after all the improvements above referred to had been made, except those of the defendants, a resurvey or measurement was had in pursuance of the call of the field notes of the original survey, and stakes were set in pursuance of this survey. The result of this survey was to disclose a discrepancy of approximately four feet between the call of the field notes and the stakes and lines which had been assumed and adopted by the respective owners. Under the call of the field notes every occupant was encroaching upon his neighbor to the south to the extent of approximately four feet, and was himself encroached upon in a likely manner by his neighbor on his north.

[SEE PLAT IN ORIGINAL]

In pursuance of this survey, the defendants claimed [**157 Iowa 240**] a four-foot strip of ground occupied by the plaintiff. It will be seen, therefore, that the controversy involves a possible readjustment of all the partition lines in the block.

The stakes which have been referred to were pointed out to plaintiff as the monuments fixing the boundaries of his proposed purchase, and he accepted them as such. If these stakes, or either of them, represented the monuments erected as a part of the original survey, then we have a case of conflict and discrepancy between the monuments upon the ground, on the one hand, and the field notes and plat as recorded, on the other. In such a case the law seems to be well settled that the survey upon the ground as ascertained by monuments then made to mark the boundaries of the lots is controlling, and the paper plat and field notes must give way thereto. *Root v. Town of Cincinnati,* 87 Iowa 202 at 204, 54 N.W. 206; *Bradstreet v. Dunham,* 65 Iowa 248; *Ufford v. Wilkins,* 33 Iowa 110; *McDaniel v. Mace,* 47 Iowa 509 at 510. To the same effect see *Olson v. City of Seattle,* 30 Wash. 687 (71 P. 201); *O'Farrel v. Harney,* 51 Cal. 125; *Holst v. Streitz,* 16 Neb. 249 (20 N.W. 307); *Flynn v. Glenny,* 51 Mich. 580 (17 N.W. 65); *Marsh v. Mitchell,* 25 Wis. 706; *Turnbull v. Schroeder,* 29 Minn. 49 (11 N.W. 147); *Burke v. McCowen,* 115 Cal. 481 (47 P. 367); *Morrow v. Whitney,* 95 U.S. 551 (24 L.Ed. 456).

The defendants do not controvert the legal propositions here involved. Their main contention is that the evidence fails to identify the stakes in question as being the monuments made upon the ground at the original survey. The case here is therefore made to turn upon this question of fact. The trial court held the evidence sufficient in that regard. From a careful reading of the evidence we also reach the conclusion that the identity was sufficiently proved. It is true that there is no specific identification by any witness who saw the stakes at the time of the original survey. But it is not legally necessary that the proof of the identity should be in that form. It is undisputed that the engineer Dickenson made the original survey upon the ground by setting stakes one hundred feet apart to indicate the boundaries of six one hundred-foot lots. In each case, the dividing line was to run due west from such indicated point, and parallel with the avenues. Dickenson also made the resurvey in 1909 from his original field notes, and thereby disclosed the discrepancy if discrepancy was there. He then saw the stakes upon which plaintiff and others relied. As a witness he would neither affirm nor deny whether such stakes were those that were set by him in 1902. If they were, they were not located where he intended to place them. In other words, they were not located in accordance with his field notes. In placing the stakes originally he intended to place them in accord with the notes and paper plat. If, in fact, he placed them otherwise, it was a mistake on his part. Of necessity he could not know that he made this mistake; otherwise he would not have made it. The fact, therefore, that the plaintiff failed to show the identity of the stakes, or the occurrence of a mistake by the testimony of this witness, is not very significant. This remark is not intended to reflect upon the witness. On the contrary, his testimony impresses us as entirely candid. One of the earliest persons to buy and improve in this locality was the witness Townsend. This was in 1905, and before the improvement of Thirty-Sixth street or either avenue. He then intended to buy upon the west side of Thirty-Sixth street. He looked at every lot on that side of the street, and ascertained its supposed boundaries. At the supposed southeast corner of lot 22 he found a stake, and a succession of stakes one hundred feet apart from there to University avenue. Due east from each one of these stakes on the east side of Thirty-Sixth street was a corresponding stake marking the boundaries on that side. He later bought a lot on the east side of the street and has occupied it ever since. He has been familiar ever since with the location of the stakes which he discovered in 1905. From 1905 down to the present time, the evidence of identity is abundant. The stakes and locations relied upon by plaintiff and others are the same as those ascertained by Townsend. They are located due west of similar stakes one hundred feet apart on the east side of Thirty-Sixth street. The fact that other stakes are found also which indicate smaller subdivisions of the original lots does not affect the question. Their location was determined by mere measurement from the original one hundred-foot points. These stakes were universally accepted by all parties in interest as representing the original survey until the survey of 1909. The record discloses no apparent advantage to be gained by any one by a shifting of the location of these stakes. So far as appears, every owner is in possession of the appropriate dimensions indicated by the plat, and this includes the defendants who are in possession of a little more than fifty feet. Their contention at this point, however, is that their possession is an encroachment of four feet upon Cottage Grove avenue, and that they hold such possession by sufferance, and not by right. The evidence shows that the south stake ascertained by Townsend purporting to be the southeast corner of lot 22 was actually located at the southeast corner of defendants' present location. It does appear that Cottage Grove avenue at this point is only approximately sixty-two feet wide, whereas it is supposed to be, according to the plat, sixty-six feet wide. The defendants' sidewalk apparently encroaches upon the platted street, approximately four feet beyond the ordinance provision. But such sidewalk as actually laid is nevertheless in a straight line with its extensions east and west. It is a somewhat inexplicable peculiarity of the situation that the sidewalk between Thirty-Fifth and Thirty-Seventh streets encroaches upon the width of Cottage Grove avenue, and that such encroachment is the result of keeping such sidewalk in a *straight line with its extensions east and west.* In other words, east of Thirty-Fifth street and west of Thirty-Seventh street, Cottage Grove avenue is sixty-six feet wide. The sidewalk is laid at such points one foot south of the lot line according to ordinance. Beginning, however, at any point in the sidewalk east of Thirty-Fifth street and extending the same west in a straight line, it encroaches upon the avenue as platted between Thirty-Fifth and Thirty-Seventh streets. The sidewalk as actually laid between Thirty-Sixth and Thirty-Seventh streets is laid in such straight line, and yet is five feet south of the lot line of 22 as claimed by defendants, or one foot south of such line as claimed by plaintiff. If the sidewalk were removed to its proper location as contended for by defendants, it would be four feet out of line with the sidewalk which is properly located east of Thirty-Fifth street and west of Thirty-Seventh. Such a change of location would give to Cottage Grove avenue its full width of sixty-six feet, but it would also throw its north boundary out of line for the two blocks mentioned. Cottage Grove avenue has been fully improved with paving, gutters, curb, and parking, and these improvements have been adapted to the encroachment, if such it is. If the monuments upon the ground are controlling as to the property owners, there is nothing in this record to indicate that they are not likewise controlling upon the city. In this view, the mistake or discrepancy, if any, has operated equitably upon all. It does not appear that any property owner has been deprived of any dimension or suffered in location. The sum of the whole trouble seems to be that there is a loss of width to Cottage Grove avenue, and a gain to University avenue.

We think the monuments or stakes upon the ground are sufficiently proved to have been a part of the original survey, and that they must accordingly control.

We reach the conclusion upon the whole case, therefore, that the decree of the trial court was right, and it is accordingly – *Affirmed.*

**Tomlinson v. Golden, 157 Iowa 237, 138 N.W. 448 (Iowa 1912).**

## KANSAS

S. H. BRUNT, county surveyor of Lincoln county, in February, 1887, made a survey of the township line in range 9 west, and between townships 12 and 13 south, which had been disputed. A certified report thereof is as follows:

"PLAT AND FIELD-NOTES of survey for Geo. Huhl and others of Lincoln county, state of Kansas, section 1 in township 13, south of base line, range 9, west of 6th principal meridian. George Huhl and Henry Lantz do hereby agree to record the north line of section 1, township 13, range 9 west, as surveyed February 2d and 3d, 1887, by S. H. Brunt, county surveyor.

TOWNSHIP 13 SOUTH, RANGE 9 WEST.

[SEE ILLUSTRATION IN ORIGINAL]

H. H. DAVIS, F. E. PERKINS, Chainmen.

LINCOLN, KANSAS, Feb. 5, 1887.

"In the matter of the survey of the township line in range 9 west, and between townships 12 and 13 south, which has been a disputed line: I made a survey of said township line, commencing February 2d, and closing February 3d, 1887. The manner in which I made my survey was as follows: Commencing at the township corner between ranges 8 and 9, and between townships 12 and 13, I ran a random line west to the township corner between ranges 9 and 10, all parties interested being satisfied that these township corners are the original government corners. I based my survey by these corners, and not by the government field-notes. In running my line west, I found no section corners established, but on closing on the township line between ranges 9 and 10, I found the township line to be six miles four chains and forty-two links in length. My closing was also one chain and ninety-four links south of the township corner. Then I retraced my line and established the quarter-corner on north line of section 1, township 13 south, in range 9 west, on a right line between the township corners. I also established the corner to sections one, two, thirty-five and thirty-six, in the same manner on a right line between the township corners, after making my connections. The length of township line by the government field-notes is 6 miles, 2 chains and 80 links. By actual measurement, I found it to be six miles, four chains and forty-two links, making a difference of one chain and sixty-two links. This overplus I divided equally in the first five miles, making these points equidistant, leaving 2 chains and 80 links in the last line between sections 6 and 36, as called for in the government notes.

"Between townships 12 and 13 S., range 9 W.: From the corner to townships 12 and 13 S., ranges 8 and 9 W., I ran west on a right line between sections 1 and 36.

Chains.

22.00. Cross a small draw; course north and south.

29.00. Top of low ridge; bears north and south.

33.00. Cross a small draw; course north, 3[degrees] west.

35.50. Top of low ridge; bears north, 3[degrees] west.

40.16. Set quarter-section corner, sandstone, 22x14x8, on east bank of creek.

40.30. Cross creek; course north, 29[degrees] west; timber on banks.

41.46. Cross large creek; course north and south; timber on banks. Enter bottom-land.

72.00. Leave bottom-land, and ascend gradually.

80.32 1/3. Set section corner to sections 1 and 2, 35 and 36, 24x16x10, limestone.

S. H. BRUNT, County Surveyor."

"STATE OF KANSAS, LINCOLN COUNTY, ss.: I, the undersigned, register of deeds in and for the county and state aforesaid, do hereby certify the above and foregoing to be a true, full and complete copy of certificate of survey for George Huhl and others by S. H. Brunt, county surveyor, as the same appears on file in my office.

"Witness my hand and official seal, this 26th day of March, 1887.

N. S. BRYANT, Register of Deeds."

Hedwig Reinert claimed that her interest was affected by the survey, and appealed from the report of the county surveyor to the district court of the county. Hon. S. O. HINDS, judge of the fourteenth judicial district, having been of counsel in the matters in controversy, the appeal was heard at the April term for 1887 by Hon. W. W. SMITH, who was duly elected and qualified as a judge pro tem. A general finding against the appellant and in favor of the appellee was made by the court; thereupon the court approved the report of the county surveyor, and adjudged that the cost of the survey should be taxed to the parties interested according to their respective interests therein. The appellant excepted to the rulings and judgment of the court, and brings the case here.

Judgment reversed and remanded.

**OPINION**

HORTON, C. J.:

This was a controversy in the court below over the correctness of a survey made by S. H. Brunt, the county surveyor of Lincoln county, of the township line between townships twelve and thirteen, in range nine west, in that county. Hedwig Reinert owns the south half of the southwest quarter of section thirty-six, and George Huhl owns the northwest quarter of section one. The other defendants in error own, also, lands north of this line, and have interest in common with Hedwig Reinert, the appellant in the court below. She claims that her southwest corner as located by the government survey is about thirteen rods south of where the survey made by S. H. Brunt, the county surveyor, established it; that the new survey makes a difference of twelve or thirteen acres in the quantity of land in her farm; that no attention was paid by him to any monuments other than the two township corners; that no attempt was made to find the lost corners, and that the field-notes were disregarded.

It appears that several years ago, W. Bishop, who has been a county surveyor of Lincoln county for ten years, surveyed the disputed line between sections thirty-six and one. Upon the trial, the principal contention seemed to be whether the lines and corners established by Bishop were in accordance with the government lines and monuments. On the one side it is claimed that the Bishop survey was a correct line; on the other, that the Bishop survey was erroneous, and therefore that the survey of Brunt was the true one. The great preponderance of the evidence tended to show that the true section corner of the northwest corner of section one was originally established at or near a prairie-dog hole, and that the quarter-corner was west of the creek. There was some evidence, however, tending to show that Bishop did not sufficiently examine or identify the stone at the dog-hole as a government monument; that the stone on the west side of the creek was a sandstone, and not the limestone described in the field-notes. Then, again, Robert Farnes testified that he knew of a stone as a corner, which had been broken off, on the east side of the creek. This evidence tended to contradict, in a slight way, the evidence given in support of the correctness of the survey of Bishop. It is probable that Farnes was mistaken, or did not fully understand the importance of the question. Therefore if the sole matter for our determination was whether the survey of Bishop was in all respects in accordance with the government monuments and corners, we would be compelled to follow the general finding of the trial court.

We think, however, that there is not sufficient evidence in the record to sustain the finding of the trial court in its approval of the report or survey of Brunt. Whatever may be said as to the corners testified to by Bishop, there is not sufficient evidence in the record to show that the survey of Brunt was correct. The report of the survey on its face shows that Brunt only followed township corners and disregarded the field notes. The report says that in running his line west, he found no section corners, but does not show he sufficiently attempted to search for the original or lost corners between the township corners. He found the township corners between ranges eight and nine, and between nine and ten, then ran a straight line from one to the other, and on this line placed the quarter and section corners between thirty-six and one. He did not take any evidence in the community or neighborhood as to the lost lines or corners, and paid no attention to hunting for monuments or corners, other than the two township corners. He says in his report that "I based my survey by these corners, and not by the government field-notes." In his testimony Brunt attempted to explain his disregard of the field notes by saying he merely disregarded the variations given in the notes.

While distances and bearings must be disregarded if the monuments on the ground for the corners as originally established can be found, or if lost, their original location can be ascertained, a surveyor should not disregard the field notes merely to make a straight line between township corners. The township line is not necessarily straight in all cases. (*McClintock v. Rogers*, 11 Ill. 279; *McAlpine v. Reicheneker*, 27 Kan. 257.) As affects the disputed corners, the field-notes of the government survey are as follows:

"From the corner of townships 13 and 13 S., R. 8 and 9 W., I run S. 89 [degrees] 54' W. on a true line between sections I and 36. Va. 11 [degrees] 40' E.

"33 chains – A creek 30 links wide runs N.E., narrow belt of timber on banks; enter creek bottom on left side.

"40 chains – Set limestone 18x15x4 for 1/4 sec. cor.

"75 chains – Leave bottom and enter upland; bears N.E. and S.W.

"80 chains – Set limestone 20x10x5 for cor. to secs. 1, 2, 35 and 36."

The field-notes mention the creek and describe the quarter-corner as west of it. The notes say that the course of the creek where the line crosses is northeast. Brunt says where he crossed the creek the course is northwest. Again, Brunt in his survey did not follow the directions stated in *Everett v. Lusk*, 19 Kan. 195. His survey was not as extensive as it should have been under the circumstances, in order to insure accuracy to a reasonable certainty. He should have reestablished all missing corners from all the nearest known original corners, in all directions, following section lines. (See also *McAlpine v. Reicheneker*, supra; Comp. Laws of 1885, ch. 25, §§ 161-165.)

Further, Brunt in his survey seems to have wholly disregarded the rule in *Tarpenning v. Cannon*, 28 Kan. 665, "that a boundary-line long recognized and acquiesced in is generally better evidence of where the real line should be than any survey made after the original monuments have disappeared." Also, 34 Kan. 595.

The judgment of the district court will be reversed, and the cause remanded for a new trial.

All the Justices concurring.

**Reinert v. Brunt, 42 Kan. 43, 21 P. 807 (Kan. 1889).**

**Author's Note:** This decision is included not so much as an original survey, but to illustrate the significance of proper procedure in collecting and analyzing evidence of original survey work.

Action by Johnson against Spawr, to recover a strip of land. Judgment for plaintiff. Defendant comes to this court. The opinion states the facts.

Judgment affirmed.

W. M. Glenn, for plaintiff in error:

When the plaintiff in error purchased this farm, certain lines and corners marked the boundaries to the land he was buying. These he recognized, and conformed his improvements thereto. Fences were built and roads laid out in conformity with these lines, which were and are the lines established by the government, by its last survey; and the land in dispute herein was a part of this purchase.

It is claimed by the defendant in error, that the purpose of the new survey was not to make new lines and corners, but to reestablish the old ones. Suppose this were true. Will it be claimed that if only one-fourth of the old section corners could be found by the Tweedale surveyors, that they may not establish the other three-fourths; and if they do or did establish them, and it afterward turns out that the corners so last established do not conform to the original corners, that therefore these last-established corners are not legally-established corners? Spawr testifies that he recognizes the Tweedale survey, for the reason, among others, that he has in his possession a letter, which is in evidence as "Exhibit B," from H. L. Muldron, dated "Department of the Interior, Washington, April 19, 1887."

In this letter from the acting secretary to the general land office, the secretary advises "recognizing the Tweedale survey as the most expedient and practicable method of settling this question," and then says: "You are therefore directed to recognize said surveys as the subsisting surveys of all the townships to which they relate," etc. This was accordingly done. Between plaintiff's and defendant's land the Tweedale survey puts the corners and line 13 rods farther east than the survey of 1871 put it. Will this direction or order of the acting secretary to the commissioner make this Tweedale line the legal and subsisting one? This is the claim of plaintiff in error. It is not so much a question of where this particular line is, as which is the legal and binding survey. We submit that the Tweedale survey must be recognized as the legal one, and where there, is a conflict between the survey of 1871 and it, that the Tweedale must stand; and if this be true, then the land in dispute belongs to plaintiff in error, and judgment of the lower court must be reversed.

Sankey & Campbell, for defendant in error:

We do not believe that any amount of "recognition" by plaintiff, Spawr, or by any number of citizens, can legalize an illegal survey; or could make that which was not the legally-established boundary lines and corners, the legal and "subsisting" lines and corners. Nor do we admit that any recognition by either the acting secretary of the interior or the commissioner of the general land office did or can legalize the Tweedale surveys in such a manner or to such an extent as to alter or change a single line or corner formerly established by the government.

The only question to be settled is: could the government, by the Tweedale surveys of 1884, change or alter the legal and subsisting corners? or, in other words, Had the government the power by this Tweedale survey to make new corners where the old ones existed, and at variance with them? The government was not asked by the petitioners to make new lines and corners, but to reestablish the old ones. The government could not change the established government corners after the title of the land had passed from the government and vested in the preemptors, the occupants. The government, even though the title had remained and was still in it in 1884, could not, without first declaring the survey of 1871 null and void, affect or change the corners by it established, by another survey. The new could not take the place of the old while the old still existed. The first must be abrogated, declared a nullity, by competent authority, before the second can have any legal force or authority.

The lands in western Harper county were surveyed under the authority of § 2396, Revised Statutes of the United States (2d ed.), 1878, as the then existing law, in 1871. Defendant Johnson settled on the land in dispute in 1880; received his certificate of purchase for same from the United States same year, and his patent a few months later. He testifies that it requires the strip of land 13 rods wide, and now in dispute, to make out the 160 acres which he purchased from the government. Three of the four corners of section 20, and two of section 21, of which the land in dispute forms a part, were distinctly marked by the survey of 1871, and the evidence shows were generally known and recognized by the early settlers, and were pointed out to Tweedale and his surveyors.

We maintain that nothing less than an act of congress would be sufficient to annul and render void these corners.

SIMPSON, C. All the Justices concurring.

**OPINION**

SIMPSON, C.:

The material facts are, that in 1880, William Johnson, the defendant in error, settled upon and preempted the southwest quarter of section No. 21, township 32, range 9 west of the sixth principal meridian, in Harper county. His settlement and preemption were made in accordance with lines, boundaries and corners as fixed by the original government survey, made in 1871. Spawr, the plaintiff in error, is the owner of the southeast quarter of section No. 20, in the same township and range. The subject-matter of the litigation is a strip of land 13 rods wide, east and west, and 160 rods long, north and south, which Johnson claims as a part of his quarter-section. According to the original government survey, this strip of 13 acres belongs to Johnson; but it is claimed that in the year 1884, the proper land officers ordered a resurvey of the land in that locality, and this survey is called the Tweedale survey, and by it this strip belongs to Spawr, the plaintiff in error.

The question is, Which of these two surveys shall govern – the original, or the survey made 13 years after by Tweedale? There is substantial agreement as to the facts. At the time of the first settlements all witnesses agree that many of the section and half-section corners established by the survey of 1871 were plainly visible. It is established that at the time Johnson took his land there was a government corner plainly visible at the southwest corner, one at the northwest and one at the southeast corner of the section in which his land is situated. His final certificate and patent were issued with reference to the lines, boundaries, and corners of the survey of 1871. This survey had been approved by proper authority and certified to the local land office. Under these circumstances, it seems too plain for argument that Johnson is entitled to the land. After the government parted with its title and it had been vested in the settlers, no officer of the land department had the right to order or approve a resurvey that changed the boundaries of the specific parcels of land to which the various settlers had title, in the absence of any claim or showing that fraud had been practiced by them or someone in their interest. Section 2396 of the revised statutes of the United States provides: "The boundary lines actually run and marked in the surveys returned by the surveyor general shall be established as the proper boundary lines of the sections or subdivisions for which they were intended." This section was in force at the time Johnson preempted his land and his patent was issued, and gives him a clear legal right to insist on the existing lines and boundaries as established by the original survey.

Another view can be taken, and that is that the location of the corner stones is a question of fact about which there was some conflict in the evidence, and the trial court having decided in favor of Johnson, and there being some evidence to sustain the finding, we will not disturb it. It seems to us from the recitations in the record that the weight of evidence was with Johnson at the trial.

We recommend that the judgment be affirmed.

By the Court: it is so ordered.

All the Justices concurring.

*Spawr v. Johnson*, **49 Kan. 788, 31 P. 664 (Kan. 1892).**


# KENTUCKY

Whether certain oil wells are embraced within the patent lines of two certain patents to Peter Phipps, one for 50 acres issued in 1850, and the other for 30 acres issued in 1851, is the question of fact presented by this appeal. The determination of the fact depended upon the correct location of the 50-acre patent and of one of the lines of the 30-acre patent. A large volume of evidence was taken. Several surveys of the land were had. Several maps were produced at the trial, showing the respective claims of the parties and theories of the surveyors. It was disclosed that neither survey was actually run out on the ground – at least at the points in dispute – when the land was taken up by Phipps. They were "call lines"; that is, taking certain known corners, the surveyors supposed that certain courses and distances would include the land intended to be patented. There is no doubt that the surveyors were mistaken in their assumptions. Still Peter Phipps and the adjacent owners supposed that the lines as run actually covered the land intended to be patented. Acting on that, they built division fences, and Phipps built a house, cleared fields, and lived upon his land for many years, and until his death, without his title being disputed. The corners now standing are in accordance with the claim and the settlement of Phipps. The facts bear out the theory of the plaintiff (appellee) that such was his intention, and that the lines as now located by the trial court were those actually intended in the original survey. The purpose of the survey was to appropriate the lands which Phipps intended to take up, and which he in fact settled upon and improved. The correct running of the lines, being made to yield their courses and distances so as to reach the actual corners adopted by the pre-emptor, is that found by the circuit court. Of the 30-acre patent, one line, the one is dispute, runs north, 76 poles, to a stake; thence to another known corner. Allowing the variation of two degrees for the 60 years since the survey was made, the correct running of the line is N. 2~ E., 76 poles; thence the same variation to the next known corner. That was done by the circuit court's judgment.

But it is insisted that there is old marked timber along that line on the due north course, which should control. If it were true that this timber was marked as a line of that survey, it would control. But it was not, we are morally sure. It was all a "called" survey. No corner was marked for it, unless it was the second. These hacked trees are along an old path or footway through the woods – probably the blazed way of the early settlers. There is no other accounting for them. They are not shown to have marks of the same age as the survey, which might have been done if it were the truth. Old people, acquainted with the lines when Peter Phipps lived there, testified that the course as adopted by the trial court's judgment was the true course as claimed by Phipps 40 or 50 years ago.

Copies of the plat, as made by the surveyors when the lands were first surveyed, are relied upon to show that the 50-acre survey was of a different shape from that adopted by the trial court. A "stake" patent, being made up in an office, and not on the ground, is very apt to be consistent in shape with its calls. It throws no light on how the conceded corners might affect its shape when run out on the ground. They merely confirm the mistakes of the surveyors. The duty of the court was to correct those mistakes by the true facts concerning the location of the lines and corners – known marked corners controlling all other evidence.

We perceive no error. Judgment affirmed.

***Ross Wetzell & Co. v. Mountain Oil Co.*, 141 Ky. 411, 132 S.W. 1040 (Ky.App. 1911).**

## LOUISIANA

This is a petitory action to recover sections 41 and 42 in township 12 south, range 1 west, Louisiana Meridian, as per survey of John W. Rhorer, made in May, 1905, containing about 230 acres, and rents for said property for the year 1908.

The defense is that the land sued for is a part of the S. 1/2 of section 33, township 11 south, range 1 west, according to the original survey made by Joseph Aborn in 1807, and a resurvey by W. H. R. Hangen in 1873, and is included in the chain of title of the defendants, emanating from the state of Louisiana.

There was judgment for the defendants, and the plaintiff has appealed.

There is no dispute as to titles, and the issue has been narrowed down to the question of the true boundary line between townships 11 and 12, and the solution of this question hinges on the proper location of the sixty-sixth milepost on the basis meridian line.

The basis meridian line south of the thirty-first degree of latitude was originally surveyed by Thomas Owing, in 1806, and he set posts at the end of each mile running south. Owing set a post at the end of the sixty-sixth mile south, which marked the southeast corner of township 11 south, range 1 west. The south boundary of the same township was originally surveyed by Joseph Aborn in 1807 "beginning at the sixty-sixth milepost on the basis meridian at the southeast corner of this township and section (36) and running due west."

The plaintiff contends that the Hangen survey of 1873 changed the south boundary of township 11 south, range 1 west by locating it *north* of the sixty-sixth milepost. Hangen commenced at the fifty-fifth milepost which he clearly identified, and, running south, re-established the mileposts, from 55 to 66, inclusive. Hangen was instructed to retrace the old basis meridian line as established by the previous surveys of Owing in 1806, Waile in 1808, and Phelps in 1856; and was further instructed to retrace the south boundary of township 11 south, range 1 west, by beginning at its southeast corner running due west, re-establishing the original corners where they could be clearly identified, and when they could not be so identified, to resurvey the line in conformity with the original field notes. Hangen resurveyed the northern boundary of township 12 south, range 1 west, which is the southern boundary of township 11, range 1 west, beginning at the sixty-sixth milepost as previously re-established by him. According to the resurvey of Waile made in 1808, the eastern boundary of township 11 south, range 1 west, measured 485.55 chains. The resurvey of Hangen in 1873 was from the same starting point – that is, the northeast corner of township 11 south, range 1 west – and agrees with the Waile survey in course and distance from mile to mile, and consequently in the total distance of 485.55 chains. The Owing field notes give no distances between the mileposts, except for the first and last mile. The difference between his survey and those of Waile and Hangen for the first mile is only 4 links, and the difference for the last mile is 2.30 chains. Owing's notes show that the last mile was run by him through a country partly inundated, and this fact, doubtless, was the cause of the discrepancy. The slight errors of Owing in his measurements were corrected by the Waile survey made two years later; and the correctness of his measurements was confirmed by the resurvey made by Hangen in 1873. There is no dispute as to the location of the sixty-sixth milepost by Hangen. The post is where it should be; that is, on the prolongation of the southern boundary of the township east of the basis meridian and directly opposite township 11 south, range 1 west.

Plaintiff's whole case hinges on the theory that the sixty-sixth milepost from which Aborn surveyed the south boundary of township 11 south, range 1 west, was 14 chains and a fraction south of the sixty-sixth milepost as re-established by Hangen in 1873. As Aborn took for his starting point the sixty-sixth milepost established by Owing in 1806, the plaintiff's theory is that the Owing sixty-sixth milepost was 14 chains and a fraction south of the sixty-sixth milepost, as re-established by Waile and Hangen. This theory is not supported by the field notes of Owing, and is exploded by the resurveys of Waile and Hangen.

It is unnecessary to consider surveys from starting points at long distances west of, and not prolonged to, the basis meridian, as such surveys cannot possibly change the location of mileposts established by measurement from the thirty-first degree of latitude, a mathematically certain starting point.

It is therefore ordered that the judgment below be affirmed.

**Elms v. Foote, 129 La. 975, 57 So. 306 (La. 1911).**

## MAINE

Two actions of trespass quare clausum fregit for cutting and removing timber from a lot in range D, in the town of Limington. The cases were tried together. The verdicts were for the defendant and are before this court on motions to set aside the verdicts as against law and evidence.

The cutting and removing of the timber was admitted and sought to be justified by denying that the title to the locus was in the plaintiffs, and by the claim that it was in the heirs of Luther Dole. The record shows that in 1855 there was conveyed to Luther Dole "the northerly half of lot No. 14, D range, containing fifty acres more or less"; that he continued to occupy the land until his death, February 17, 1892; that at the time of the acts complained of some of his heirs had conveyed their interest in the lot to the remaining heirs who owned it, and had possession of the northerly half of lot 14, sold the timber upon the locus in dispute, and authorized the acts complained of, and who defend these actions, claiming the cutting was upon the northerly half of lot 14, inherited by the heirs of Luther Dole.

Luther Dole's title came from the heirs of Henry Dole by four deeds, each of which described the land as "the northerly half of lot No. 14, range D, containing 50 acres more or less." Henry Dole obtained title to the property conveyed by his heirs, by warranty deed of Daniel Hodgdon to Henry Dole and Hosea Clark, February 9, 1829, and Hosea Clark to Henry Dole November 12, 1839, in which deeds the land was described as "one-half of lot No. 14 on D range, it being the northerly half of said lot, containing about 50 acres." Daniel Hodgdon obtained title to the premises by warranty deed of Joseph Hodgdon October 20, 1828; the premises being described as "it being one-half of lot No. 14 on D range, it being the northerly half of said lot, containing fifty acres except a small piece which I sold to Thomas Beal." Some of the other deeds reserve from the conveyance the piece of land sold to Thomas Beal. There is nothing in the record showing where the Beal land was located, but the plan drawn by the surveyor appointed by the court shows that a small parcel in the northerly corner of the northerly half of lot 14 was apparently taken from that lot, but the evidence does not refer to it, except in the deeds as above, and its location is not material in these cases.

From the above statement of title it is apparent that Luther Dole at his death owned the northerly half of lot 14, range D, and his predecessors in title had owned the same premises by deed so describing it at least from October, 1828.

The record shows that the plaintiffs in the two actions owned that part of lot 15, range D, adjoining lot 14 as called in the case the "Dole land," tracing their title back to deed of Robert Cole to Washington Ilsley, December 2, 1856, which deed and all other deeds of the premises, including the deed under which the plaintiffs claim, bound the plaintiffs' by land of Luther Dole.

At the trial the original survey or plan was not introduced, but a plan of lots 14, 15, and 16, range D, made by a surveyor appointed by the court to survey the premises and make a plan of them, was used. The accompanying sketch of the plan shows the claims of the parties:

(Image Omitted)

There is no dispute as to the range lines. It is claimed by the plaintiffs that the dividing line between lot 14, which is owned by the Dole heirs, and lot 15, the northerly part of which is owned by the plaintiffs, is a straight line as shown by the line from the post in the range line on the sand hill to a stone bound, halfway across the range, extending further to a stake on the range line in the swamp. The defendant claims the dividing line runs from the stone post in the range line to a stake halfway across the range, as shown by the dotted line, between which lines is the disputed lot upon which the cutting was done, containing about six acres.

It is objected by the defendant that in the writ of Ilsley et al. the plaintiffs' title in that portion of the northerly part of the disputed lot was obtained by a deed given in pursuance of a decree in a bill in equity brought to reform a deed given by Washington Ilsley in his lifetime to John Purrington, and that the Luther Dole heirs were not parties to the suit, and ought not to be bound by it.

The objection is without merit. The defendant, or the Dole heirs or their predecessors in title, were not necessary or proper parties to the bill in equity, and the decree or deed does not include any land owned by the defendant or Dole heirs, or Luther Dole, their predecessor in title, but corrects a mistake in a deed in which neither the defendant nor the Dole heirs or their predecessor in title were parties, so that the plaintiffs have title to land in range D, lot 15, that adjoins the Dole land in lot 14, and if the disputed lot was owned by Luther Dole no title to it passed to the plaintiffs by the deed, and, if the disputed lot was not owned by Luther Dole, then of course they were not injured by the decree. The only question in the case was the location of the division or check line between lots 14 and 15, because Luther Dole owned the northerly half of lot 14, and the plaintiffs' title is of that part of lot 15 which adjoins the land owned by Luther Dole in his lifetime.

The defendant did not prove the location of lot 14 by plan or admitted monuments upon the lot, or by measurements from admitted boundaries of other lots in range D, but claimed that the plaintiffs had not proved, as they should have done to entitle them to a verdict, that the disputed territory was in lot 15.

The record title of plaintiffs and the Dole heirs shows that the check line between lots 14 and 15 is the dividing line between their lands, and the burden was upon the plaintiffs to prove the original location of that line for the line run at the time the range and lots were originally located, for the boundaries are still the boundaries if their location can be found.

It is firmly established in this state that the survey must govern when its location can be shown; that when land is conveyed by lot, without further descriptions, the lot lines determine the boundaries of that lot when they can be located. *Bean v. Bachelder,* 78 Me. 184, 3 A. 279; *Stetson v. Adams,* 91 Me. 178, 39 A. 575; *Coleman v. Lord,* 96 Me. 192, 52 A. 645.

"It is the well-known practice of proprietors of townships in this state to have them surveyed and laid out in ranges and lots. *** They then sell by the number of the lot and the range, without a more particular description. And the purchaser is entitled to his lot according to" the original survey, "if that can be ascertained. *** Selling, as the proprietors do, by the number of the lot and of the range, the range and lot lines are referred to as monuments." *Warren v. Pierce, 6 Me. (6 Greenl.) 9,* 19 Am.Dec. 189.

The southerly half of lot 14 was conveyed to Jeremiah Gilpatric by warranty deed dated February 23, 1810, by Daniel Hodgdon, who was the owner of lot 14, and who conveyed the northerly half of the lot February 8, 1829, to the "Dole heirs" predecessors in title. The two deeds by Daniel Hodgdon show that lot 14 contained 100 acres, and that the range and check or division lines across the range were the boundaries.

The plaintiffs claim to have proved the location of the check line by monuments testified to, and admitted by adjoining owners. John Gilpatric, who was 86 years old at the time of his deposition, and whose father owned the southerly half of lot 14, having inherited it from his father, Jeremiah, who purchased it from Daniel Hodgdon in 1810, remembers the corners between the Gilpatric and Dole land for 70 years at least, and testified that the stone bound in the middle of the range was the boundary between the Dole lot and the Gilpatric lot on the check line between lots 14 and 15. He also remembers a stake in the swamp, where the evidence shows that the stake now stands, as the corners of lots 14 and 15, and also a pine tree as a monument on the sand hill at the opposite side of the range where there is now a post, and remembers the stone bound between the Dole and Gilpatric land upon the check line for many years. The father of John Gilpatric conveyed the southerly half of lot 14 in 1883 to James W. Foss, who now owns it. He was a witness for the defendant, and testified that the stone bound in the middle of the range was his corner, but did not think the stake in the swamp was the other corner, but by running the line from the stone bound to the range line in the swamp it is shown he was mistaken.

In October, 1888, Asa Libby, the then owner of the southwest one-quarter of lot 15 now owned by Chadbourne, showed the stone bound in the middle of the range as the corner of his land (the Gilpatric land), and the corner of the plaintiff Seavey's land, then owned by Robinson, and the Dole land; and Mr. Robinson, who owned the plaintiff Seavey's land for 21 years, was shown the stone monument by his predecessor in title, and, when he sold it in 1888 to Edgecomb, pointed out the stone bound in the middle of the range as the corner stone between the Gilpatric or Foss lots in lot 14 and the Dole lot in lot 14, the Libby and Chadbourne lot in lot 15, and the plaintiff Seavey's lot in lot 15.

Fourteen years before the trial, the Libby lot was purchased by Chadbourne, and a stake in the swamp afterwards replaced by a new stake in the same place by Mr. Chadbourne, and the stone bound in the middle of the range, which plaintiffs claim are on the check line, were pointed out to him as the line between lots 14 and 15, and his deed refers to them as the bounds, but Mr. Libby, who it is claimed pointed them out, does not remember that he pointed out the stake in the swamp, but he does not deny it. Several other witnesses identify the stone bound in the middle of the range and stake in the swamp as admitted corners between lots 14 and 15. Some 35 years before the trial, there being a dispute as to some of the division lines upon the lots, the county commissioners were called upon and went on the premises to establish the lines, and the point where the stone monument stands in the middle of the range was apparently by agreement used as the check line between lots 14 and 15.

The defendant admitted at the trial that the stone bound in the middle of the range was upon the check line, as claimed by Mr. Foss, the then owner, who testified that the stone represented the boundary of his land, as did all other witnesses, who testified to the corners of the lands in lots 14 and 15 one-half across the range where four different lots had cornered for so long that no one remembered when the bounds were first upon the land. During the examination of a witness as to the claim made on the lot by Mr. Foss while the owner of the south half of lot 14 adjoining the Dole land, counsel for defendant stated that "the James W. Foss land is not now, and has not been, in controversy so far as we know."

The owners of adjoining lands may agree as to the division line, and that agreement be binding upon them and those claiming under them; but it is not necessarily conclusive upon other owners whose lands are bounded by the same division line, but it is competent evidence, when the original monument cannot be found, as tending to prove, not a new boundary or corner, but that the line coincides with the original monument referred to in the deed. *Gilbert v. Curtis,* 37 Me. 45; *Gove v. Richardson, 4 Me. (4 Greenl.)* 327; *Loring v. Norton, 8 Me. (8 Greenl.)* 61.

As the range lines are not in dispute, if any part of the check line is proved, from that point a line in a straight and most direct course to the range line will be the check line between the lots (*Melcher v. Merryman,* 41 Me. 601), and a straight line so run from the stone bound in the middle of the range to the southerly range line runs to the stake in the swamp, and a line from the stone bound in the middle of the range to the northerly range line runs to the stake and stone on the sand hill and makes the line from range line to range line, as claimed by the plaintiffs, and locates all of the disputed tract in that part of lot 15 owned by the plaintiffs.

It is clearly proved that for at least 70 years the monuments now claimed by the plaintiffs, the stake in the swamp and a stone at the corner of the Gilpatric and Dole lands, or others in the same place destroyed by time, have been recognized as the true monuments by all the adjoining owners between lots 14 and 15 extending halfway at least across the range, which with the admission that the Foss (Gilpatric) corner is as claimed by the plaintiffs, and no evidence that before the 70 years there was ever any dispute as to the line, and the fact that the Dole lot was conveyed as one-half of lot 14 and the line as claimed by the plaintiffs, gives to the Dole lot one-half of the lot, and to accept the defendant's claim would give to the Dole lot six acres more than the Gilpatric lot, which was conveyed as one-half of the lot, and the fact that the line as shown by the monuments claimed by the plaintiffs runs in a straight course across the range, and the line as claimed by the defendant makes a jog of 176 feet, and makes the lot conveyed as one-half of lot 15 contain six acres less than the other part of lot 15, which was conveyed as one-half of the lot, raises a presumption that the line as shown by the monuments claimed by the plaintiffs was the line originally run between the lots sufficient, unless controverted by evidence, to prove the plaintiffs' title to the disputed lot.

The defendant claims that the check line between lots 15 and 14 begins at a stone on the north range line 176 feet distant from the post on the sand hill, as claimed by the plaintiffs, and extends halfway across the range to a stone post 176 feet from the stone bound claimed by the plaintiffs as on the cheek line at the corner of the Foss (Gilpatric) and Dole lands; but neither of the corners as claimed by the defendant are admitted by the adjoining owners, past or present, as the corners, and no measurements can be made from admitted corners of lots in 14, 15, or 16 that will even tend to show the lines as claimed by the defendant is the cheek line between lots 14 and 15; but, upon the contrary, every measurement from admitted lines and corners in the three lots places the check line between lots 14 and 15 as the line claimed by the plaintiffs that crosses the range from the post on the sand hill by the stone monument known as the Gilpatric corner to the stake in the swamp.

The fact that at some time Luther Dole fenced across the lot in dispute, as shown by the remains of an old fence, does not prove the title was in him. His line was a straight line across the range, and the plan shows that the old fence was not on any line, but was a crooked fence, and did not inclose all of the disputed lot, and the evidence clearly shows that it was a fence built to keep the cattle from the swamp and rye field for the convenience of the owners of the cattle, and not under a claim of ownership. There is evidence that Luther Dole claimed to own the fence, but evidence of a deceased owner making claims of title is not evidence of title, in such cases as these, unless he was upon the land pointing out the monuments at the time of the declaration. There is evidence that he did point out the corners of the fence as the corners of his land; it is testified to by men who helped build the fence and also in the deposition of two of his heirs (sons), but no witness testified he ever claimed to own the land except as a part of lot 14. That Luther Dole never claimed to own any part of lot 15 is the only conclusion that can be drawn from the testimony. Two of his sons, whose depositions have been referred to, testified that he said the corners now claimed by the defendant were the corners of his land; but that the sons understood him to refer to the corners of the northerly one-half of lot 14 is apparent from their conduct, for, after his death in 1892, they deeded to the other heirs their interest in the land and described it as the northerly one-half of lot 14, range D, and the heirs who defend these cases offer as evidence of their title to the disputed lot deeds in which all the heirs of Luther Dole are grantors or grantees describing the land claimed by them as the northerly one-half of lot 14.

The language of the court in *Brown v. Gay, 3 Me. (3 Greenl.)* 128, by changing the word "he" to "plaintiffs" and number "3" to "14" would exactly fit this case, where the court says:

"He is the owner of number 3, and he claims and defends the premises in dispute as a part of that lot. If they are no part of that lot, his claim is plainly founded in mistake. If the owner of a parcel of land, through inadvertency or ignorance of the dividing line, includes a part of an adjoining tract within his enclosure, this does not operate as a disseisin."

To allow the defendant's claim to include a part of lot 15 as in a deed of a part of lot 14 would, as held in *Robinson v. Miller,* 37 Me. 312, "be to contradict or vary the plain and unambiguous stipulations of his deed, and to have enlarged his grant, in a manner unauthorized by law."

That the check line as claimed by the plaintiffs is the true line is proved by applying the rule stated in Warren v. Pierce, 6 Me. (6 Greenl.) 11, 19 Am.Dec. 189, as follows:

"The burden of proof is doubtless upon the plaintiffs to make out their case; but when they show range lines between which their lot is bounded, and the side lines of the lot next below and next above theirs in number, they have located their lot, and made out their case, if it be not successfully controverted by opposing testimony."

There is no dispute as to the range lines, and the check line between lots 14 and 15 is the issue, and as lots 14 and 15 were originally conveyed as each containing 100 acres more or less, they being adjoining lots, by calling them one lot, and locating the check line between lots 13 and 14, and lots 15 and 16, the land in both lots will be shown as one lot; which, divided by a line extending through the middle across the range, will locate the check line between lots 14 and 15, and each lot will share, as it should, the surplus acreage. *Whitten v. Hanson,* 35 Me. 435.

The check line between lots 13 and 14 is not disputed, but claimed by all parties to be as shown upon the plan. The check line between lots 15 and 16 is not disputed, and is admitted by the abutting owners to be as shown upon the plan, and ancient monuments upon the land show the line to be as admitted. The territory between these two lines is lots 14 and 15, and, if that territory is divided in the middle by a line across the range, the line will run from the stake in the swamp by the stone bound known as the Gilpatric corner to the stake and stones on the sand hill, on the line as claimed by the plaintiffs, and locates all the disputed lot in that part of lot 15 owned by the plaintiffs, and gives to each lot 112.2 acres, while according to the plan and survey, the check line as claimed by the plaintiffs' lot 14 contains 112.8 acres, and lot 15 111.9 acres. The difference of nine-tenths of an acre on a lot more than 3,000 feet in length is so trifling in this case, where the land is wild and swampy, that it is immaterial, and that surplus is in lot 14, and therefore the defendant cannot complain.

By each of these two methods the plaintiffs have proved that the disputed lot is located in lot 15, that the title to it is in the plaintiffs, and the defendant has not by evidence successfully controverted their claim of title, and the record does not show evidence that authorized the jury to find that the title was in the Dole heirs, who defend the actions.

Motion sustained.

New trials granted.

### *Ilsley v. Kelley,* **113 Me. 497, 94 A. 939 (Me. 1915).**

The parties in this action dispute title to a 46.2-acre parcel of land lying west of Route 11 in Shapleigh. Plaintiff Ethiland Walker claims title to the property by a 1965 deed from the Hillside Fish Club to her late husband Vernon. The defendants, Joseph and Martha Provost, claim title to the disputed parcel by a 1971 warranty deed from Annette J. Dunnells. The Walkers commenced this action in 1981 seeking a judicial declaration that they are the owners in fee simple of the disputed parcel.[1] Shortly thereafter, the Provosts brought a third-party complaint against Dunnells seeking indemnification for any judgment entered against them.[2]

The parties agreed to submit the case to a referee, retired justice Donald W. Webber formerly of this court, who conducted a three-day hearing on the matter in October 1986. In his report, the referee concluded that Walker is the owner in fee simple of the disputed parcel. Over the objections of the Provosts and Joy, the Superior Court (York County, Brennan, J.) adopted the referee's findings. Finding no error, we affirm.

The Provosts and Joy contend that the referee erred in finding 1) that the deeds upon which Walker relies for her claim of title describe land containing the disputed parcel and 2) that a survey of the disputed parcel, introduced by Walker, is in any way based upon those deeds. The Superior Court is required to accept the referee's factual findings on these issues unless they are clearly erroneous. M.R.Civ.P. 53(e)(2). On appeal we will uphold the Superior Court's adoption of the referee's report as long as there is credible, probative evidence supporting the referee's factual findings, even though there may be evidence to support a contrary finding. See *Severance v. Choate,* 533 A.2d 1288, 1290 (Me.1987).

The record contains ample evidence to support the referee's conclusion that the deeds upon which Walker relies for her claim of title describe land containing the disputed parcel. The Walker chain of title can be traced directly back to the original conveyance of Lot 17, Range 5, in 1831. From 1831 to 1888, deeds described the property in various ways, but consistently defined a parcel containing 125 acres more or less, later reduced by certain out sales to 100 acres more or less. In 1888, one Clarissa Emery mortgaged this parcel, described as "containing 100 acres more or less." For the first time, the 1888 mortgage deed described the parcel with reference to the abutters on all four sides.

When Clarissa Emery died intestate in 1889, her son John Emery was appointed administrator. In 1893, John Emery conveyed his undivided half of all his mother's property to his sister, Lizzie, including a parcel described as containing 100 acres more or less bounded by essentially the same abutters, albeit at different compass points. An unremarkable chain of title runs from Lizzie Emery to the Walkers. The Walkers conveyed a portion of the parcel to the State in 1974, reserving for themselves all land "lying westerly of Route # 11."

The referee concluded that the property described in the 1888 mortgage deed is the same parcel described in the 1893 deed, finding that the scrivener erroneously rotated his compass directions 90 degrees when describing the abutters in 1888. The evidence supports this conclusion. Both the 1888 and 1893 deeds described identical quantities of land. See *Perkins v. Conary,* 295 A.2d 644, (Me.1972) (in determining a boundary, quantity of land is a valid control when not contradicted by monuments, courses or distances.) Further, John Emery in 1893 explicitly conveyed to Lizzie Emery an undivided half of all the real estate owned by their mother, Clarissa Emery. The referee committed no error in finding that the 1888 and 1893 deeds describe the same parcel. Accordingly, the Walkers had a valid, unbroken chain of title to the land when they conveyed a portion to the State, reserving for themselves the parcel now in dispute.

The Provosts and Joy assert that even if the 1888 and 1893 deeds describe the same "100 acre more or less" parcel, the referee erred in concluding that the survey introduced by plaintiff Walker accurately described the disputed 46.2-acre parcel. The Provosts and Joy maintain that there was no evidence linking the survey with Walker's source deeds.

The survey of the disputed parcel was prepared by Forrest Smart, the only witness providing relevant testimony on this issue. The referee found Smart to be a qualified and expert surveyor and gave his testimony considerable weight. See *Sargent v. Coolidge,* 399 A.2d 1333, (Me.1979) (weight given to the opinion of the surveyor is the prerogative of the referee). Smart testified that his primary responsibility in surveying the parcel was to establish the northerly and southerly boundaries of Lot 17, Range 5, in Shapleigh. Walker's chain of title can be traced back directly to the original conveyance of Lot 17 in Range 5 in 1831. Thus the referee had competent unrebutted expert testimony to support his conclusion that the survey accurately depicted the disputed parcel and was linked to Walker's source deeds.

No other challenge raised or intimated by the appellants merits discussion.

Judgment affirmed.

All concurring.

*Walker v. Provost,* **566 A.2d 749 (Me. 1989).**


## MARYLAND

The appellant applied to the commissioner of the land office for a patent to a triangular strip on the southern border of his land in Baltimore county, as a vacancy, and the appellees filed caveats on the ground that it had already been granted by the proprietary, and that there was therefore no vacancy which could be the subject of a new grant. The commissioner held with the caveators, and the caveatee appeals.

On November 10, 1695, a patent was issued to Walter Smith for a tract of land named "Bear Neck," of 500 acres, lying on the south side of Gunpowder Falls, and described as follows:

"Beginning at a bounded white oak standing on the south side of a high hill near to a place called Nen Port, thence down the said falls southeast and by south 250 perches to a bounded white oak standing on the south side of a branch called Cabbin branch, and running southwest and by west 320 perches, then northwest and by west 250 perches, thence straight to the first tree."

Four months later, on March 26, 1696, a patent was issued to Walter Burgess for another neighboring tract of 500 acres called "Cubhill," and described as follows:

"Beginning at two bounded white oaks standing on the south side of a great branch descending into the said falls, and running thence northwest and by north 150 perches to a bounded white oak by another great branch, it being a bounded tree of Walter Smith called Bear Neck, and running along the said Smith's line into the woods southwest and by west 534 perches, then southeast and by south 150 perches, then direct to the first tree."

The commissioner states, in an opinion filed with his decision, that the certificates of survey show that Thomas Richardson, surveyor of Baltimore county, made the original surveys for both of these patents. The certificates are not set out in the record.

It will be observed that the second line of each tract is to begin at a bounded white oak tree, and to run thence southwest by west, and that the line of Cubhill is also designated as "running along said Smith's line into the woods." The Cubhill line on that side was to extend 214 perches beyond Smith's line on the same side.

It appears from deeds and surveys of later date that Smith's second line has for an indefinite period in the past been run, not straight throughout, but with a turn to the north, somewhat less than one degree, at a point 94.3 perches from the beginning of that line; and that, on the other hand, surveys and deeds, fence lines, and a hedge line, of land opposite the extension of the Cubhill line, and also one deed of Cubhill land opposite Bear Neck, have taken this latter line to run straight from the common point of beginning, so that there has been a triangular strip left between the two lines thus separated. There is testimony that three county surveyors have found the Cubhill line to run straight and to leave this strip between the two tracts as now outlined. Evidence as to fences between the two in the past is in conflict. There has been no evidence offered to show title to the strip by adverse possession, the caveators relying, finally, on the descriptions in the patents alone. The commissioner was of opinion that the intention in the patents was clearly that the second lines of the two tracts should be one and the same to the full extent of the Bear Neck tract, and on that ground considered the land already patented, and so refused the patent now applied for. He considered the disputed area to have been included in the Cubhill tract. The caveatee objects that this conclusion must be erroneous because it is either inconsistent with the established Cubhill line to the west of the Bear Neck tract, or, if that is to be left as it stands, the Cubhill line in that part opposite Bear Neck must, under the commissioner's holding, be broken and advanced over the disputed strip, so that the northerly boundary of the whole Cubhill tract would be three lines instead of the one intended in the description.

It is our opinion that the original grants themselves made the two tracts contiguous to the full length of the Bear Neck tract, that there was no vacancy left, and that the state has no power now to grant the strip in dispute. The question is solely one of vacancy – whether the land remains in the ownership of the state. If there has been one grant of it, and the land has already passed from the state, then, until it shall have returned to the ownership of the state in some way, there can be no question of another grant. *Tolson's Lessee v. Lanham, 2 Har. & J.* 175. The patents described the adjacent liens of the two tracts as running from a common starting point, in the same direction, and as being contiguous for their joint length. The words "running along said Smith's line into the woods" mean nothing less than contiguous. Smith's line, is taken as the second line of Cubhill. In *Rogers v. Raborg, 2 Gill & J.* 54, 63, a tract running certain courses and distances from a beginning point, but without further calls, and described as "joining on Lunn's lot," was held not limited to any line of Lunn's lot in disregard of the courses and distances. The expression quoted was construed as no part of the description of the land itself, but as only indicating the general location of the tract. And in *Thomas v. Godfrey, 3 Gill & J.* 142, 149, a tract was described as "beginning at a bound hickory on the side of a hill, on the south side of the main falls of Patapsco, respecting to the west Chew's Resolution manor, and running with the said manor, south 53 degrees west, 200 perches to a bound hickory," and of this expression "running with the said manor," etc. Chief Judge Buchanan, for the court, said:

"It is not the case of a course and distance line of one tract of land, calling to, or to run with, or bind upon, a water course, or another tract of land, or a line of another tract, with no ulterior object called for, and looking only to the water course, or other tract, or line, as the definite object to be reached or run with, and to which the course and distance expressed, if not corresponding with it, is made to yield. But here there is a fixed ulterior object, a tree imperatively called for and designated as the boundary intended to be run to."

And upon this distinction the expression was held not meant to demand coincidence of lines of the two tracts.

The case now presented seems to stand exactly on the other side of Chief Judge Buchanan's distinction; it is one of a call to bind upon or run with the line of the other tract, wthout any ulterior object called for. A line called for is quite as controlling as any natural or artificial boundary. *Carroll v. Norwood, 5 Har. & J.* 155, 163.

We are not prepared to agree that the original survey, if it differed from the description in the patent, and inclosed a smaller area, would control the grant. *Tolson's Lessee v. Lanham, 2 Har. & J.* 175. But that question does not arise on the record before us, as there is no certificate of the original survey to show such a discrepancy. On the evidence before the court it would appear just as likely that a departure from the common boundary line occurred in subsequent surveys of one tract or the other; and that would not, of course, return the land left between the lines to the ownership of the state, so that it might grant it anew.

The appellant urges that there may at least be sufficient doubt or difficulty in deciding the point in dispute to render it desirable to issue the patent applied for and leave the parties to contest their rights in a suit at law. *Carswell v. Swindell,* 102 Md. 636, 640, 62 A. 956; *Railroad v. Hoye, 2 Bland,* 258, 263; *Chapman v. Hoskins, 2 Md. Ch.* 485, 495. But the question here seems to us to be peculiarly one of the construction of the original grants, which is a question for the court.

"That belongs exclusively to the court, whose peculiar province it is to expound patents, according to the intention to be collected from the terms or expressions used." *Thomas v. Godfrey, 3 Gill & J.* 142, 150.

And in our opinion the effect of the descriptions used in the two patents would not be so doubtful as to justify deferring the decision.

Order affirmed, with costs to the appellee.

***Ramsay v. Butler, Purdum & Co.*, 148 Md. 438, 129 A. 650 (Md. 1925).**

## MASSACHUSETTS

This was an action of trespass for breaking and entering two several closes of the plaintiff in Sheffield, and taking away two several parcels of fence thereon standing.

The cause was tried on the general issue, at the last May term in this county, before Jackson, J., from whose report of the trial it appears that one of the fences was on the dividing line between two lots of land owned by the parties respectively. The trespass complained of was the defendant's removing a part of that fence, which was made of rails. The defendant produced evidence to show that he built this same part of the fence about 23 years ago, and had ever since been accustomed to repair it, as his part of the division fence; and that, at the time in question, he removed the rails in order to make a stone-wall there; which he built accordingly in the following year, placing it rather nearer to his own land than the place where the rail fence had stood.

The judge instructed the jury that, if they believed the facts to be as above stated, they should find a verdict on this count for the defendant; and they returned a verdict for him accordingly.

The trespass complained of in the other count was the removing another parcel of rail-fence, standing on or near the dividing line between two other lots owned by the plaintiff and defendant respectively, and entering on a small piece or gore of land claimed by the plaintiff, and adjoining to said fence on the south side thereof. It appeared that the defendant had owned the north lot about 30 years. One Ashley previously owned the adjoining south lot, and devised it to one Dutchen, who owned it in 1802, and before the plaintiff had any title or claim to it. The fence between the two lots, which passed through a wood or swamp, did not run in a straight line, but was bent or inclined to the northward, so as to include in the south lot about two acres more than would have been included by a straight line.

The defendant produced evidence to prove that the said Ashley, in 1764, when he bought the south lot, caused it to be surveyed, and ran the dividing line straight, in the place where the defendant has now built the fence; that the said Ashley never claimed any land north of that line, but always declared that this was his boundary; and particularly in 1792, he caused the line to be run anew, in the same place, between him and the defendant, and then declared that he would not claim anything north of that line; also that the said Dutchen, after he owned the south lot, claimed to own to the same line, and agreed with the defendant, in 1802, to build a new fence on that line, to divide their lots; but the fence was never built by them.

The plaintiff did not deny that Ashley and Dutchen had agreed with the defendant upon the course and place of the division line as above stated; but produced evidence to prove that the defendant, in the years 1809, 1810, and 1811, had confessed or admitted that he had not a good title to the land in question. It was understood that the plaintiff had purchased the said south lot of the said Dutchen, but the deed was not produced. The plaintiff produced a deed of one Abraham Root, made in January, 1810, by which the said Root sold, released, and quitclaimed, a certain piece of land to the plaintiff, which, as he contended, included the premises in question.

The defendant produced a deed of the same A. Root, made in March, 1764, to the said Ashley, for the conveyance of about 250 acres of land, including, as the defendant contended, all that Root ever owned or claimed in or near the said south lot. And it appeared that Ashley immediately entered under the said deed, and always claimed to hold, in virtue thereof, all the lands south of the said lot, which was afterwards purchased by the defendant; and that the premises, so supposed to be conveyed by said Root to the plaintiff, had never been occupied by said Root after 1764; but had been included within the south lot, and occupied therewith, as before mentioned.

The judge left the whole evidence to the jury, and instructed them that, if it had been understood and agreed between the defendant and the said Ashley and Dutchen respectively, that the dividing line between their said lots was in the place now contended for by the defendant, and that the defendant owned all the said land, which was occupied by them on the north of said line – this was evidence of a seisin and possession thereof by the defendant; and that neither the said Ashley nor Dutchen could, under such circumstances, make a valid conveyance thereof to any other person; and that a boundary line, when agreed on without fraud or mistake by the owners of adjoining lots, would bind both parties, and their respective heirs or assigns, whether such boundary were marked by a fence on the line, or in any other manner; and, further, that, if the defendant was seised of the land in question, claiming it as his own, the plaintiff could not maintain his action on this count, and the jury should find a verdict thereon for the defendant.

The jury found a verdict accordingly for the defendant on this count also; and the plaintiff moved for a new trial, on account of the said opinions and directions of the judge.

Judgment according to the verdict.

Hulbert and C. Dewey for the plaintiff.

Ashmun and Whiting for the defendant.

**OPINION**

Parker, C. J.

There can be no question of the correctness of the verdict upon the first count. The action is trespass; and a possession, at the time the act supposed to be a trespass was done, must be proved, or the action cannot be maintained. Now, there was evidence of a possession by fence in the defendant for more than twenty years before the action was brought; and that the part of the fence, the removal of which is complained of, had been repaired by him during that time; also that the new fence or stone-wall was erected within the line on which the old fence stood.

The only question which could exist at the trial was, whether the facts there testified were true; and the jury having declared that they were, the verdict for the defendant was a necessary legal consequence.

There is nothing in the report from which an entry on the plaintiff's land can be inferred, unless such entry was necessary for the purpose of taking down the fence, in order to rebuild it, which would not be tortious. The part of the fence assigned to the defendant to keep in repair was his property, so far at least that the removal of it, for lawful purposes, could not make him a trespasser; and we do not think there was any joint-tenancy, or tenancy in common, of the materials of which the fence was composed.

There is not much more question upon the other count. Two borderers, having a known line between them, conformable to their title deeds, have, in consequence of the irregularity of the ground, acquiesced in a fence not corresponding with that line; but, nevertheless, neither party claims to own on the other side of the line, established by the deeds and by the express acknowledgment as late as 1802, of the person then in possession of the lot now held by the plaintiff, and under whom he derives his title, and in conformity with a survey made in 1764, by *Ashley*, the then owner, and again in 1792.

The line described in title deeds must always be considered the true original line between the owners of contiguous lots. This is generally ascertained by fence; but the fence is only evidence of the direction of the line, unless it is referred to in the deeds as a monument. Where parties have established a fence varying from the line described in the deeds, and each party has held and occupied up to his side of the fence, claiming to hold accordingly, for twenty years, neither can maintain a possessory action against the other. And, indeed, such act of ownership and acquiescence might be sufficient evidence of title in actions of a higher nature, especially if any doubt should exist as to the place and course of the line described in the deeds.

But where the parties have agreed upon a fence variant from the line, avowedly for convenience, and still have continued to claim according to the true line, neither party acquires a title, or even a right of possession, against the other, merely on account of the fence.

It was therefore a question of fact properly submitted to the jury, whether, under the circumstances of these parties, and the prior owners of the land, by whose acts the present owners are bound, the fence was to be considered as limiting the claims of the proprietors on either side, contrary to the boundary established by their supposed titles; and we have no doubt the jury have decided rightly upon the question.

*Judgment according to the verdict*
***Burrell v. Burrell*, 11 Mass. 294 (Mass. 1814).**


## MICHIGAN

COOLEY, J.

The bill in this case is filed to quiet the title of complainant to a small strip of land which defendant is endeavoring to recover from him. The parties own and occupy adjoining lots in the city of Marquette, and the dispute concerns the location of the boundary line. It appears that the original plat of the village of Marquette was laid out by the Cleveland Iron Company in the year 1854. It was crossed by Front, Third, and Fourth streets, running east and west, and by Fisher,

Rock, Superior, and Spring streets, crossing the others at right angles. Each block was divided into lots, which were intended to be parallelograms, of equal size. Subsequently, at several different times, adjoining territory was platted into blocks and lots, and the streets on the original plat, with the exception of Spring street, were extended by direct lines through the new plats.

In 1872 defendant purchased lot 9, fronting on Fisher street, between Third and Fourth, and complainant purchased the adjoining lot, numbered 8. Moriarty had previously become owner of lot 10, which was on the corner of Fisher and Fourth. The lots, according to the plat, were all 50 feet in width, and Moriarty had fenced off his lot of that width, beginning at the line of Fourth street where he and those who had previously bought lots upon it supposed it to be. When complainant purchased, he measured off 100 feet from the supposed line of Fourth street, and took that as the point of beginning. That left 50 feet between himself and Moriarty, and this was taken possession of by defendant as being lot 9. Complainant and defendant, immediately after their purchases, built upon their lots respectively, and joined in building a fence on what they supposed was the dividing line between their lots. The line of this fence is the one which complainant now insists upon and seeks to maintain. The other lots fronting on Fisher street in the same block were numbered from 1 to 7 inclusive, and these had all been sold off by the original proprietor. Lot 5 was purchased by one Steele in 1867, who was told by Peter White, the agent of the proprietor, that he would find stakes to indicate the boundary, and he did find stakes which located the lot with 100 feet for lots 6 and 7 between it and the premises now occupied and claimed by complainant as lot 8. One Neddeau purchased lot 4 the next year, and found it staked out in the same way as a 50-feet lot adjoining that of Steele. It is probable that these stakes were not the stakes planted at the time of the original platting, but there is nothing in the evidence to raise any suspicion that they had been planted without authority, or for any other purpose than to indicate the lot boundaries; and it is reasonable to infer that they were either the original stakes or others which had been planted in the same places when the original stakes had gone to decay. It seems very plain on the evidence that they were recognized by the original proprietor and its general agent as stated, which correctly designated the lot boundaries until all the lots fronting on Fisher street, on the block now in question, had been sold and taken possession of by purchasers.

Recently, and after these parties had been in possession of their respective **[17 N.W. 66]** lots for nearly 10 years, occupying with practical recognition of the fence they had jointly constructed as the dividing line, doubt has been thrown upon it by a survey which has been made of this part of the town, and which has led to the present litigation. A surveyor whose skill and competency is not questioned, taking for starting points certain monuments planted at the intersection of Front street with Spring street and Superior street, and also another at an intersection on one of the new plats, has demonstrated to his entire satisfaction, and by tests which in his opinion make his conclusions correct to a mathematical certainty, that the lot lines on this block are wrongly located; that Moriarty is some two feet over the line of Fourth street; that defendant encroaches on Moriarty to the same extent, and complainant upon defendant. Defendant insists upon locating the dividing lines in accordance with this survey, and complainant contends for the correctness of the practical location.

If it were certain that the monuments which are now found at the intersection of Front with Spring and Superior streets were planted at the time the original plat was laid out, and that the survey of the plat was made with those monuments as the guides, and was itself mathematically correct, the conclusions which have been reached on the survey just made would no doubt be reliable and accurate. But we have no evidence that the monuments were planted when the platting was done; we are only told that they are monuments which are generally accepted and relied upon. In the absence of any evidence of the time when or the person by whom they were located, it is quite as supposable that there may have been an error in locating Front street when the monuments were planted, as that there was an error in locating Fourth street when the lots upon it were purchased and taken possession of. It is pure assumption that there must necessarily be an error on Fourth street because the line as located does not conform to the line as it should be, measuring from the actual location of Front street.

It is also pure assumption that the original survey was mathematically correct. It is seldom or never that a town plat in a new country is made with perfect accuracy; and it is familiar knowledge in this state, if not elsewhere, that any attempt to make street lines and lot lines correspond with mathematical accuracy to the recorded plat after the lots have been occupied and improved, would disturb possessions in the most serious manner, and lead to infinite difficulty and litigation. Fortunately the rules of law do not admit of this. Purchasers of town lots have a right to locate them according to the stakes which they find planted and recognized, and no subsequent survey can be allowed to unsettle their lines. The question afterwards is not whether the stakes were where they should have been in order to make them correspond with the lot lines as they should be if the platting were done with absolute accuracy, but it is whether they were planted by authority, and the lots were purchased and taken possession of in reliance upon them. If such was the case, they must govern, notwithstanding any errors in locating them.

We have said on this general subject, in prior cases, all that we deem necessary, and shall content ourselves with a reference to those cases. *Joyce v. Williams,* 26 Mich. 332; *Stewart v. Carleton,* 31 Mich. 270; *Cronin v. Gore,* 38 Mich. 381; *Diehl v. Zanger,* 39 Mich. 601; *Dupont v. Starring,* 42 Mich. 492; [S.C. 4 N.W. 190.] The evidence is very strong that the parties to this litigation located the lines of their lots at the time they bought where they had been located in the original platting, and the evidence of the recent survey scarcely tends to cast a doubt upon it.

The decree of the circuit court was very clearly right, and must be affirmed, with costs.

(The other justices concurred.)

*Flynn v. Glenney,* **51 Mich. 580, 17 N.W. 65 (Mich. 1883).**

## MINNESOTA

The plaintiff has acquired the title of the patentee to the southeast quarter of section 31, township 127, range 28, and the defendant is the owner of the northeast quarter of the same section. The dispute in this case is over the boundary line. The location of the quarter-post on the east line of the section is agreed on by the parties, but they disagree as to the location of the quarter-corner on the west line of the section. In accordance with the acts of congress and rules of the land department, all subdivisional lines of a section must be straight lines running from the proper corner in one exterior line to the proper corner in the opposite exterior line of the section as established in the original survey. The boundaries, as established by the government surveyors and returned to and accepted by the government, are unchangeable, and control the description of lands patented. And it is well settled that mistakes in the surveys cannot be corrected by the judicial department of the government. *Cragin v. Powell,* 128 U.S. 691, (9 S.Ct. 203, 32 L.Ed. 566.) It is very clear, as remarked by the court in *Haydel v. Dufresne,* 58 U.S. 23, 17 HOW 23, 30, 15 L.Ed. 115, that "great confusion and much litigation would ensue if judicial tribunals, state and federal, were permitted to interfere and overthrow the public surveys." The purchaser from the United States, and those claiming under him, are bound by the descriptive calls, surveys, and plats designating what he buys. *Bates v. Illinois Cent. R. Co.,* 66 U.S. 204, 1 Black 204, 17 L.Ed. 158.

If, then, the quarter-corners as established by the government survey are ascertainable, a straight line running from the quarter-corner on the east line of the section above referred to, through the same to the quarter-corner so located on the west line, will establish the correct division line between the northeast and southeast quarter-sections now in dispute, whether the last-named quarter-corner was accurately located in the original survey, or not; that is to say, whether or not it is equidistant from the north and south section corners. The government plat indicates that the point of intersection of the east and west subdivision line with the west line of the section in question is in a lake or pond, and, in accordance with the rules governing the surveys of the public lands, a witness-corner appears to have been established on the section line on the northerly margin of the lake, and both the plat and field-notes show that this witness-corner, which consisted of a mound, pits, and a stake, is 14 chains and 85 links north of the quarter-corner as fixed by the surveyors. The call thus appearing by the records of the survey is controlling.

There is no doubt or dispute in respect to the witness-corner, which is clearly identified by the evidence, and the location of the quarter-corner in question is therefore a mere matter of measurement. Where no monuments are referred to in the description, and none are intended to be erected, the distance stated therein must control the location. *Negbauer v. Smith*, 44 N.J.L. 672. The end of the line measured south from the witness-corner on the section line, 14.85 chains in length, is the point fixed to designate the location of the quarter-section corner. *Wilson v. Hildreth*, 118 Mass. 578.

We do not see that it is material that in the survey of the adjacent township the description calls for a different location of the quarter-corner in the adjacent section. The survey of each section must stand upon its own records.

In answer to a suggestion of appellant's counsel, it may be assumed that if the survey appeared to be grossly or palpably erroneous, it would not have been approved by the land department until corrected; and if the call made the location of the corner impracticable, other methods would have to be resorted to in order to subdivide the section; and the law requires that corners of quarter-sections not marked shall be placed as nearly as possible equidistant between the two corners standing on the same line. The state statute (Gen. St. 1878, *c.* 8, § 235) is inoperative when it comes in conflict with the rules established by the acts of congress for subdividing sections. And when lines can be run and established in conformity with the government survey, as ascertained by fixed monuments, corners, and distances, they will control, and the rights of parties who purchase according to the government survey cannot be affected by the act of the legislature. *Neff v. Paddock*, 26 Wis. 546.

As there is substantially no conflict of evidence in this case, the court was right in directing a verdict for the plaintiff.

Order affirmed.

***Chan v. Brandt*, 45 Minn. 93, 47 N.W. 461 (Minn. 1890).**

## MISSISSIPPI

The intention of the parties will not control the plain and unambiguous terms of the deed as against third persons. 18 C. J. 278, and authorities there cited. The court takes judicial notice of the United States Government survey, and the legal subdivision therein. *Muse v. Richards,* 70 Miss. 581, 12 So. 821; *Shivers v. Insurance Company,* 55 So. 965. But complainant in the court below makes the government survey a part of his amended bill, brings it here as a part of the record, and it is agreed by both parties to this suit that it is correct, and that each and every section and subdivision thereof is correct and bounded as shown by said plat or survey.

Even the United States cannot by subsequent survey change the survey of a Spanish grant. *Doe v. King,* 3 H. 125. Still appellant comes to this court with the absurd proposition, that private parties, by mere misnomers and misdescriptions in their deeds, can change the original government survey and bind third parties thereby; but the most surprising thing of all is that appellant seems to think that such a ridiculous argument will have weight with an intelligent court.

It appears from the bill of complaint that when the territory now composing the county of George was acquired by the United States government a number of persons occupied various tracts of land therein, claiming ownership thereto, which claims were confirmed by the government. When the territory was surveyed and platted by the government, these tracts were set apart to the claimants, and each was designated on the township plats by a sectional number. Among such tracts of land was one situated in what is now George county, embracing the land now in controversy, and claimed by Daniel Green. The township in which this land is situated contained a number of such tracts, each of which is shown on the township plat as a separate section, resulting in the township being divided into 47 instead of 36 sections, the one set apart to Daniel Green being No. 42. It appears from the township plat, filed as an exhibit to the bill, that section 42 would have embraced and coincided with the Northwest quarter of section 27, the South half of section 22, and the Southwest quarter of section 23 if the township had been divided by the government surveyors into 36 regular sections. Sections 22, 23, and 27 were surveyed and designated on the government plat, but are only a fractional part of what they would have been had the township been divided into 36 regular sections.

Complainant deraigns his title through the following deeds: Daniel Green to Daniel McInnis; Daniel McInnis to John H. Sheppard; John H. Sheppard to H. C. Cochran and W. H. Goff; H. C. Cochran to W. H. Goff. The deed from McInnis to Sheppard, instead of describing the land as section 42, township 1, range 8 west, as it should have done, described it as:

"The northwest quarter of section twenty-seven (27) and the south half of section twenty-two (22) and the southwest quarter of section twenty-three (23) it being in township No. one (1) south and range No. eight (8) west, containing by estimate six hundred and forty (640) acres more or less."

The description in the deed from Sheppard to Cochran and Goff is:

"Northwest quarter section twenty-seven (27), the southwest quarter of southwest quarter and east half of west half of southwest quarter of section twenty-two, township one south, range eight west, containing five hundred and ninety acres."

The description in the deed from Cochran to Goff is as follows:

"Commencing at the one-half mile post between sections 22 and 21; thence run east 10 chains to place of beginning; thence continue due east along east and west center line 110.00 chains; thence south 26.66 chains; thence west 110.00 chains; thence north 26.66 chains to place of beginning containing 293 1/4 acres more or less, land being and lying in section 22 and 23 T. 1 S., R. 8 W."

The bill alleges that it was the custom of the land owners in that vicinity to describe the land situated in the irregular sections set apart by the government to the original claimants as hereinbefore set forth by what would have been the proper description thereof according to the government survey had the townships been divided into 36 regular sections, and that that was the plan adopted by the parties to these deeds, the grantors intending thereby to convey the land that would have been thereby described had the township been divided in 36 regular sections.

The land in controversy is not embraced in the description of the land conveyed in the deeds through which the complainant claims; for, where land is described in a conveyance by sectional numbers according to the government's survey thereof, the land thereby conveyed is that only which is situated within the designated sections as surveyed and platted by the government; the original corners and lines thereof established by the government being their true and only boundaries. That the parties to these deeds may have intended thereby to convey land that would have been included in the designated sections had they been of the retaliation size can avail the complainant nothing; for when the meaning of language is to be determined by the court the intent of the parties, expressed in the words they have used, must govern.

***Goff v. Avent*, 122 Miss. 86, 84 So. 134 (Miss. 1920).**

## MISSOURI

The circuit court, upon this evidence, directed a verdict for plaintiff, thus holding that the courses and distances, indicated upon the plat, should prevail over the lines actually surveyed and corners established.

I. When an authentic plat of a subdivided tract of land is referred to in a deed conveying a subdivision of such tract, the plat itself, in legal construction, becomes as much a part of the deed as if it had been fully incorporated in it. *Dolde v. Vodicka*, 49 Mo. 98; 2 Devlin on Deeds, sec. 1022.

II. While the deeds, under which plaintiff claims title to lot 1 in the subdivision, must be construed as describing the land conveyed as being of the full length shown by the plat, it does not follow that the particulars of the description contained in the plat are conclusive of the correctness of such description. The plat is only intended to be a representation of the actual survey as made upon the land itself. It is in the nature of a certified copy of an instrument which will be controlled by the original. So it is held, "where there are no express calls that determine a line with certainty, evidence *aliunde* is admissible to show where the line was actually run to which the deed refers, or to which it must have reference; and its location so fixed, by extrinsic evidence, will control the courses and distances named in the deed or in the survey. The right to prove the true line of the survey to which the deed refers, and which it follows, does not depend upon the rules applicable to ambiguities in written instruments. * * * It is not a question of construction but a question of fact." *Kronenberger v. Hoffner*, 44 Mo. 185. So in *Dolde v. Vodicka*, 49 Mo. 98, the court says: "Had this [lot] been so staked out in the original survey, there would be no difficulty, for the division of the lines of the lots would then have been actually located, and the location must govern."

It is a well-settled rule of construction that known and fixed monuments will control though they conflict with the courses and distances called for in the deed. *Myers v. St. Louis*, 82 Mo. 367; *Orrick v. Bower*, 29 Mo. 210; *Evans' Adm'r v. Temple*, 35 Mo. 494; 3 Wash. Real Prop. [4 Ed.] 405. While natural monuments are regarded of higher value in determining boundaries than artificial ones, the latter will also control courses and distances. "The order of applying descriptions of boundaries is, *first*, to natural objects; *second*, to artificial marks; *third*, to courses and distances given in the deed." 3 Wash. Real Prop. [4 Ed.] 405.

If the line between lots 1 and 4 was located, upon the land when surveyed and subdivided, and can now be ascertained and determined, that line will constitute the true division line between the lots though it conflicts with the description given in the plat.

Where the boundary line was actually located was a question for the jury, the evidence tending to prove a conflict between the calls in the deeds and plat, and the survey as located on the land. The court committed error in directing a verdict for plaintiff, and in refusing to submit the issue of fact to the jury.

**Whitehead v. Ragan, 106 Mo. 231, 17 S.W. 307 (Mo. 1891).**

The disputed corner was the northeast corner of section 7. The evidence tended to prove that the witness trees and other government monuments of this corner had been destroyed and obliterated and the corner lost prior to 1871; that the county surveyor, during that year, established such corner in the manner provided by the statute and again in 1889 another surveyor of the county reestablished the corner which corresponded with that previously established. On the other hand plaintiff offered evidence tending to prove that a stump of an original government witness tree to this corner was found by a surveyor of the county in 1887 and from that the true corner was found. These facts were testified to in chief by a witness, R. J. Stepp, who was county surveyor for eight years. The case was submitted to the court and taken under advisement.

If the corner was lost by reason of the destruction or obliteration of the government monuments, then the corner established would place the improvements of defendant on lot 8; if the government corner was known from the stump of the witness tree, then the proper line therefrom would place the improvements on the street as charged.

Afterwards, some days and during said term, on March 1, 1890, defendant filed in the cause his motion and affidavit asking to be permitted to introduce newly discovered, material evidence, which was denied him. One of the grounds for new trial assigned by defendant in his motion was the discovery since the trial of new and material evidence.

I. The declarations of law given at request of the plaintiffs of which complaint is made declared, in effect, that when a witness tree of a section corner remains, then the corner is not lost; and if witness Stepp, while county surveyor, when surveying the ground afterwards platted, and known as Allen and Burns' addition, found the stump of the witness tree at the corners of sections 5, 6, 7 and 8 and from said stump ran a line, the proper distance and at the proper angle as called for by the field notes of the original United States survey, and thereby located the said corner, and from said corner so located, established the section line between sections 7 and 8, by running a line from the southeast corner of section 7 to said corner so located, and that a line running parallel to said section line, and at a distance of seven feet to the west of the same, would lie west of the east line of the barn of Peter B. Burns, then the finding will be for the plaintiff.

"In establishing section lines, the witness trees and monuments of the original surveys made by the United States government, no matter how inaccurate they may be, control over more recent surveys, though the latter may clearly show that the former were grossly erroneous." It was admitted that the west line of the street was seven feet west of the east side of section 7.

This declaration of law correctly declares the well settled principle that corners which are shown to have been originally established by government surveyors are conclusive and shall be accepted as the true corners, however inaccurate the work by which they were originally established may have been. If witness trees or other government monuments are known or found and identified, the corners to which they refer must be ascertained by reference to them, and not by the regulations provided for establishing lost corners. *Climer v. Wallace*, 28 Mo. 556; *Knight v. Elliott*, 57 Mo. 317. The identity of these government corners, or of the monuments bearing witness to them, can be shown by the testimony of anyone having knowledge of the fact and does not depend for proof upon the record of a survey by a county surveyor. *Jacobs v. Moseley*, 91 Mo. 457, 4 S.W. 135; *Major v. Watson*, 73 Mo. 661.

**Mayor v. Burns, 114 Mo. 426, 19 S.W. 1107 (Mo. 1893).**

## MONTANA

This action involves a boundary line between the lands of the plaintiff and those of the defendant and cross-complainant in section 26, township 4 south, range 24 east, M. P. M. Carbon county.

Plaintiff, Dorothy Vaught, is the owner of the north half of the southwest quarter and the southwest quarter of the southwest quarter and cross-complainant, J. R. McClymond, is owner of the northwest quarter of section 26. The boundary line in dispute is the dividing line between the two quarters. The parties admit that plaintiff owns no part of the northwest quarter and that cross-complainant owns no part of the southwest quarter.

Cross-complainant contends that an east and west fence, which he built eight days before the commencement of this action, marks the boundary line between his quarter section and that of plaintiff, while plaintiff contends that the fence is located on her land and 309.54 feet south of the north boundary thereof.

The accurate location of the true boundary line between the northwest quarter and the southwest quarter of section 26, as *created* and established by the official United States survey thereof, will be determinative of the controversy.

The section was officially surveyed in 1890 and thereafter, by patents dated August 13, 1914 and March 22, 1915, respectively, the United States granted to Samantha E. Vaught certain described tracts, including among others all the lands in the west half of section 26 that are now owned by plaintiff and cross-complainant. In the patents, after the description of the lands granted, appear the words "according to the official plat of the survey of said lands returned to the General Land Office by the surveyor general." The patentee, Samantha E. Vaught, is the mother of plaintiff's husband, O. F. Vaught.

Commencing about 1914 the Vaught family farmed, as one unit, the land so granted including the west half of section 26 with no cross fence or other means of indicating the boundary line between the northwest quarter and the southwest quarter.

By deed dated October 22, 1940, there was conveyed to the cross-complainant, J. R. McClymond, the northwest quarter of section 26 "containing 160 acres more or less according to the United States government survey thereof."

About five months after he acquired title thereto, the cross-complainant engaged W. P. Burke, then county surveyor of Carbon county, to ascertain the boundaries of the northwest quarter and by survey, made on March 12, 1941, Mr. Burke located what he then "considered" the boundary lines of the quarter.

Thereafter, on March 31, 1941, cross-complainant informed plaintiff's husband, O. F. Vaught, that he intended to build a fence across the west half of section 26 on a line given him by Mr. Burke as the south boundary of the northwest quarter.

The proposed fence line ran through a tract of land which, according to the testimony of Mr. Vaught, had been farmed by him since 1914 and on which there was then growing a crop of winter wheat planted by him in the fall of 1940. Mr. Vaught protested, contending that Mr. Burke's survey was in error; that it erroneously placed the south boundary line of the northwest quarter a considerable distance to the south of the true dividing line between the two quarters, and that the proposed fence would enclose and take from plaintiff a strip of her ground lying in the north half of the southwest quarter.

In August 1941, W. P. Burke, at plaintiff's request, made a second survey for the purpose of locating the boundary line between the northwest quarter and the southwest quarter. After making allowance for an admitted mistake in the starting point and calculations made therefrom, Mr. Burke's second survey located the center of the section 309.54 feet north of where it had been placed by his first survey.

On September 17, 1941, cross-complainant built a fence running east and west on the line given him by Mr. Burke's first survey as the south boundary line of the northwest quarter. Cross-complainant also demanded of plaintiff, as the landowner's share, one-fourth of all the wheat harvested from the strip of ground lying north of such fence.

On September 25, 1941, being eight days after the fence was built, plaintiff commenced this suit for restoration of the strip of ground; for damages for withholding and for the rents and profits thereof alleging that the strip comprises the north part of the north half of the southwest quarter of which she is the owner.

By answer the defendant and cross-complainant, McClymond, admitted that he is the owner of the northwest quarter but denied the other allegations of the complaint and, by cross-complaint, alleged that about March 31, 1941, the plaintiff wrongfully entered into possession and ejected the cross-complainant from "The south 309.54 feet of the said Northwest Quarter of Section 26, Township 4 south, Range 24 east, M. P. M." The relief demanded was that plaintiff take nothing and that cross-complainant recover possession of the described strip of ground together with damages for the withholding and for the rents and profits thereof. Plaintiff's reply denied the allegations of the cross-complaint. The cause was tried to the court without a jury. Findings of fact and conclusions of law were made and filed and judgment rendered in favor of the cross-complainant, McClymond. By its findings and judgment, the trial court decreed the disputed strip of ground to be the south 309.54 feet of the northwest quarter and the boundary line between the northwest quarter and the southwest quarter to be the line indicated by Mr. Burke's first survey whereon cross-complainant had built his fence *after* this controversy arose.

The third finding of fact made by the trial court is to the effect, "That the true boundary line between said northwest quarter of said section 26 and said southwest quarter of said section 26 is on the line as alleged and contended by the defendant, J. R. McClymond, and as fenced by him on or about September 1, 1941."

Plaintiff has appealed from the judgment contending that the evidence fails to support the court's findings, conclusions and judgment as to the location of the boundary line and strip of ground in question.

The cross-complainant acquired and he holds the northwest quarter of section 26 "according to the United States government survey thereof." The deed of conveyance to him so specifies. When lands are granted according to an official plat of a survey, the plat itself, with all its notes, lines, descriptions and landmarks, becomes as much a part of the grant or deed by which they are conveyed, and controls so far as limits are concerned, as if such descriptive features were written out on the face of the deed or grant itself. *Pittsmont Copper Co. v. Vanina,* 71 Mont. 44, 227 P. 46; *Cragin v. Powell,* 128 U.S. 691, 9 S.Ct. 203, 32 L.Ed. 566; *Jefferis v. East Omaha Land Co.,* 134 U.S. 178, 10 S.Ct. 518, 33 L.Ed. 872; *United States v. Otley, 9 Cir.,* 127 F.2d 988; *Ohlson v. Batterton, Mo.,* 230 S.W. 110; *Read v. Bartlett,* 255 Ill. 76, 99 N.E. 345. The original corners and lines of section 26 created and established by the government constitute the true and only boundaries of the designated quarter section. *Goff v. Avent,* 122 Miss. 86, 84 So. 134; *Langle v. Brauch,* 193 Iowa 140, 185 N.W. 28.

Congress has provided a system for the survey of public lands, and the boundaries and limits of the several sections and subdivisions thereof, including quarter sections, must be ascertained in conformity with the principles laid down in the federal statutes. See Sections 751, 752, 43 U.S.C.A.

Certified copies of the official plat as well as the field notes of the official government survey were received in evidence. They disclose that in 1890 the United States government caused a survey to be made of section 26 and the township of which it is a part; that section 26 is regular both in size and form, containing 640 acres and subdivided into four quarter sections containing 160 acres each; that the four corners of the section and the four quarter section corners were located at the time of the original survey and sand stone monuments of survey with appropriate markings thereon were set to mark each of the corners so located.

"A survey of public lands does not ascertain boundaries; it *creates* them." *Cox v. Hart,* 260 U.S. 427, 43 S.Ct. 154, 157, 67 L.Ed. 332. "The quarter lines are not run upon the ground, but they exist, by law, the same as the section lines." *Keyser v. Sutherland,* 59 Mich. 455, 26 N.W. 865, 867. The location of corners and lines established by the government survey, when identified, are conclusive (*Hickerson v. Dillard, Mo.App.,* 247 S.W. 801) and the true corner of a government subdivision of a section is where the United States surveyors in fact established it, whether such location is right or wrong, as may be shown by a subsequent survey. *Beardsley v. Crane,* 52 Minn. 537, 54 N.W. 740. Original monuments of survey established during a government survey, when properly identified, control courses and distances (*Mitchell v. Hawkins,* 109 Neb. 9, 189 N.W. 175; Langle v. Brauch, supra) and field notes and an official plat of government surveys of record will control in ascertaining locations, even though the monuments established are gone. *Slovensky v. O'Reilly, Mo.,* 233 S.W. 478. In ascertaining the lines of land or in re-establishing the lines of a survey, the footsteps of the original surveyor, so far as discoverable on the ground, should be followed and it is immaterial if the lines actually run by the original surveyor are incorrect. *Ayers v. Watson,* 137 U.S. 584, 11 S.Ct. 201, 34 L.Ed. 803; *Galt v. Willingham,* 11 F.2d 757. "In surveying a tract of land according to a former plat or survey, the surveyor's only duty is to relocate, upon the *best evidence* obtainable, the courses and lines at the same place where originally located by the first surveyor on the ground. In making the resurvey, he has the right to use the field notes of the original survey. The object of a resurvey is to furnish proof of the location of the lost lines or monuments, not to dispute the correctness of or to control the original survey. The original survey in all cases must, whenever possible, be retraced, since it cannot be disregarded or needlessly altered after property rights have been acquired in reliance upon it. On [155 P.2d 617] a resurvey to establish lost boundaries, if the original corners can be found, the places where they were originally established are conclusive without regard to whether they were in fact correctly located." 8 Am.Jur.Boundaries, Sec. 102, p. 819, Emphasis ours.

The cross-complainant relies upon the accuracy of the two Burke surveys. The accuracy of the location of the line on which cross-complainant built his fence is dependent upon the accuracy of the line run by the first Burke survey. That the strip of ground in controversy is 309.54 feet in width rests wholly upon the accuracy of the lines run and measurements ascertained by the second Burke survey and, if the second survey be accurate, then such strip of ground lies south of the east-west center line of the section and is the property of plaintiff.

The first Burke survey commenced and ended at a point which he *assumed* to be located on the west boundary line of section 26; his second survey began on the east township line and stopped at a point which he then "considered" to be the center of section 26, while the government survey of section 26 started from the corner common to sections 1, 2, 35 and 36 on the south boundary of the township thence north 0~ 1' west between section 35 and 36, variation 18~ 35' east, a distance of 80 chains, where was established and marked the southeast corner of section 26. In neither of the Burke surveys was any attempt made to follow the tracks of the government surveyors and the second Burke survey retraced none of the lines run by the first Burke survey. The Burke surveys fail to agree upon the location of the center of section 26 and obviously they neither coincide with each other nor with the government survey.

At the time he made his first (March 1941) survey, there was a fence running north and south along the westerly side of section 26 which Mr. Burke *assumed* was located on the west boundary of section 26. However he did not verify this *assumption* by checking with the government plat or field notes of the original survey and on such *assumed* line he found no original corners or monuments of survey as established by the government surveyors. The starting point of the survey was a stone set along the above fence which Mr. Burke *assumed* indicated the quarter corner and which he *assumed* "was on the center line of the east and west center line of section 26, on the west–north and south line."

The field notes of the original survey show that on the west exterior boundary line of the section the government surveyors "Set a sandstone 20 X 10 X 5 ins. 15 ins. in the ground for 1/4 sec. cor. marked 1/4 on W. face; dug pits 18 X 18 X 12 ins. N. and S. of stone 5 1/2 ft. distant and raised a mound of earth 1 1/2 ft. high 3 1/2 ft. base alongside" and that on the east exterior boundary line of the section the government surveyors "Set a sandstone 20 X 12 X 4 ins. 15 ins. in the ground for 1/4 sec. cor. marked 1/4 on W. face; dug pits 18 X 18 X 12 ins. N. & S. of stone 5 1/2 ft. dist. and raised a mound of earth 1 1/2 ft. high 3 1/2 ft. base, alongside." Mr. Burke located neither of the above monuments of survey set to mark the quarter section corners established by the government surveyors nor did he at any time attempt to locate the east and west center line of the section by running a straight line from the quarter section corner on the west boundary to the quarter section corner on the east boundary of the section.

Thus, without checking with any of the monuments of survey established by the government surveyors and without running any lines to locate either of the center lines of section 26, Mr. Burke commenced his survey by *assuming* at the outset that the stone which he found along the west fence, and which does not purport to be a government monument of survey, marked the quarter section corner.

Mr. Burke testified: "Well, we found a stone set along the fence and that was marked, and it was on the center line of the east and west center line of section 26, on the west–north and south line. Then we measured a quarter of a mile north and found another corner. We continued there another quarter mile and found the northeast corner stone representing the northwest corner of section 26. There was three stones set there in that half a mile."

Mr. Burke also testified: "The line was established there by three corners being set from the center of section 26, on the west line, and the quarter corner 1,320 feet north; and another one set–another, 1320 feet north, northwest corner of section 26. Those corners were in–set and we checked the monuments."

Again Mr. Burke testified: "I surveyed the west line of the northwest quarter of section 26 beginning at the quarter corner and going north to the northwest section corner and then running east a half mile, and then running south a half mile, and then running west a half mile. I run out to find the center of section 26, and then chained back from the center of section 26 to the point of beginning to be sure I had the full acreage staked out."

To find the common corner of quarter sections or the legal center of a section of land, straight lines must be run from the quarter section corners on the boundary of the section to the opposite quarter corners, the point of intersection constituting the legal center, and the boundary line between two quarters cannot be legally established by measuring along one side of the section 160 rods, as Mr. Burke attempted to do in this case. Sec. 752, 43 U.S.C.A.; *Poleske v. Jones,* 192 Iowa 1015, 185 N.W. 917; *Gerke v. Lucas,* 92 Iowa 79, 60 N.W. 538; *Phillips v. Hink,* 21 S.D. 561, 114 N.W. 699.

As a starting point for his second survey made in August 1941, Mr. Burke used a brass topped marker set on the east township line of township 4 south, range 24 east, M. P. M. In making this survey Mr. Burke erroneously *assumed* that the "brass top" marker had been set to indicate, on the township line, the corner common to the southeast corner of section 25 and the northeast corner of section 36. Some time after completing his survey, he discovered that his starting point was 309.54 feet south of the true corner common to sections 25 and 36. He testified: "We found another marker at the northeast corner of 36, 309.54 feet north of the brass top. *** It was designated by 5 marks on the south side. *** We came back to the brass top corner 309 feet south of the stone corner and turned 90 degrees from that north and south line, using the brass top from there south as a sight, and run a mile and a half west. I tied to the marker on the southeast corner of section 36. *** We established a point a mile and a half west, and sighted back on the line on which we had just come, and turned a 90 degree angle north, and run another half mile. *** We made the mark a half mile from the line we had run from the east to the west and run a line down a half mile north." At this point Mr. Burke drove a lath in the ground, concerning which he testified: "Well, we will call it the center of section 26, *if the starting point from which we started was correct.*" Mr. Burke proceeded no further with his second survey but stopped at the point which he thus marked with the lath and which he stated he then "considered" to be the center of section 26.

Mr. Burke testified that at no time did he make a complete survey of township 4 south of range 24 east, M. P. M., nor did he at any time make a complete survey of section 26 thereof. He testified that in making his two surveys he did not consult the plat or field notes of the original government survey nor did he find or identify any monuments of survey, section corners or quarter section corners of section 26 originally established by the government surveyors.

The evidence shows that instead of searching for the original corners as established by the government surveyors Mr. Burke's first survey is tied into and largely dependent, for its accuracy, upon the accuracy of what purports to be a private survey of sections 22 and 27 and portions of sections 23 and 26 alleged to have been made in August 1911 by one A. B. Cooley. As evidence of the Cooley survey, the cross-complainant introduced in evidence as his exhibit B a purported original blueprint of the alleged survey. Mr. Burke testified that Mr. Cooley was formerly the county surveyor of Carbon county; that he considered him a competent surveyor and that at the time of the trial Mr. Cooley was dead. The witness was asked by the district judge as to his knowledge of whether Mr. Cooley had made the original blueprint, exhibit B, for the county or for private persons and the witness answered, "Apparently it was made for private purposes, because there is no record of it in the office." No field notes of the Cooley survey were introduced in evidence and there is nothing to show on what data or information the draftsman relied in making exhibit B.

Although Mr. Burke had no connection with, and personally knew nothing about, the making by Mr. Cooley of the alleged 1911 survey, yet he was permitted to testify as to his interpretation of the lines and points shown on exhibit B. Such testimony is valueless as evidence. "A surveyor's testimony is inadmissible except in connection with data from which he surveys, and testimony as to his location of a boundary line is incompetent where he [155 P.2d 619] has no means of verifying his survey." 11 C.J.S., Boundaries, § 107, p. 702.

Exhibit B affirmatively shows on its face that Mr. Cooley failed to find the monuments of survey set by the government for the northeast and southwest corners of section 26 and that he did not relocate either of such corners; that Mr. Cooley disregarded and declined to follow or abide by the government marker for the northwest corner of section 26 found set on the east and west section line at a point 189 feet west of what he *assumed* was the north and south line between sections 22 and 27 on the west and sections 23 and 26 on the east; that Mr. Cooley disregarded and declined to follow or abide by the government marker for the southeast corner of section 26 found at a point 489 feet west of what Mr. Cooley *considered* to be the east boundary line of the section and 400 feet north of what he *considered* to be the south boundary line of section 26; that Mr. Cooley did not locate any government monuments of survey for any quarter corners of section 26 and that, without following the method prescribed by statute, he *assumed* to relocate a quarter corner on what he *assumed* to be the west boundary line of section 26, without first locating the true section corners as established by the government survey and without locating or setting any quarter corner markers on the north, east or south exterior boundary lines of section 26. It is from the "quarter corner" so set by Mr. Cooley on what he *assumed* to be the west boundary line of section 26 that Mr. Burke commenced his first survey. "An obliterated quarter section corner should be fixed as originally located. Lost quarter section corners, in regular interior sections should, under the generally accepted rules, be relocated on a straight line between the section corners and equidistant therefrom, ***." 11 C.J.S., Boundaries, § 13, p. 556.

The cross-complainant owns land situate in section 27 which adjoins section 26 on the west and testimony was received tending to show that the line given him by Mr. Burke's first survey as the boundary line between the two west quarters of section 26 is in line with a fence running east and west through section 27. Where McClymond or others owning lands in section 27 built their fences could not possibly affect the location *according to government survey* of the true division line between the northwest quarter and the southwest quarter of section 26. *Fairfield v. Barrette*, 73 Wis. 463, 41 N.W. 624.

The Burke and Cooley surveys not only fail to conform to the official United States government survey of the lands in question but they also fail to comply with the laws and established rules governing the making of surveys to determine the boundaries and limits of lands conveyed and described according to the official government plat and survey thereof. Sections 751, 752, 43 U.S.C.A.; Sections 4837 and 4842, Revised Codes 1935.

It affirmatively appears that the private surveys upon which both plaintiff and cross-complainant rely were not made in accordance with the official plat, field notes, monuments of survey, lines, descriptions and landmarks made and established by the official government survey of the two quarter sections. For such reasons these surveys cannot and they do not furnish evidence of the location of the true boundary line between the northwest quarter and the southwest quarter of section 26 "according to the United States government survey thereof." The government survey conclusively fixes and controls the limits and boundaries of the two quarter sections involved and such limits, boundaries and corners, so established, may not be changed by the A. B. Cooley survey of 1911, by the W. P. Burke surveys of 1941, or by any other subsequent surveys. *Fair v. Ida County,* 204 Iowa 1046, 216 N.W. 952; *Thompson v. Darr,* 174 Ark. 807, 298 S.W. 1. The points where the government survey established the corners and set the monuments of survey for section 26 must prevail over both course and distance and these points the Cooley survey and the Burke surveys failed to locate.

"Original corners, as established by the government surveyors, if they can be found, or the places where they were originally established, if they can be definitely determined, are conclusive on all persons owning or holding with reference thereto, without regard to whether they were located correctly or not, and must remain the true corners or monuments by which to determine the boundaries. Errors of location cannot be corrected by the courts, or by a surveyor called on to locate government corners and lines. A change in the location of a township corner by state, county, or other surveyors will not affect the location of a government quarter corner, in the absence of evidence that such quarter corner was actually established at some other point by the government surveyor, and even though the new township corner was also marked by the government surveyor, the quarter corner would not be changed in the absence of evidence showing the establishment of a quarter corner consistent with such new corner. The fact that the location of the corner in accordance with an inaccurate government survey will set awry the shapes of the sections and the subdivisions affected thereby does not affect the conclusiveness of the survey." 11 C.J.S., Boundaries, § 11, p. 552.

"But the government surveys are, as a matter of law, the best evidence; *and, if the boundaries of land are clearly established thereby, other evidence is superfluous and may be excluded; the* best evidence *is the corners actually fixed upon the ground by the government surveyor, in default of which the field notes and plats come next, unless satisfactory evidence is produced that the corner was actually located upon the ground at a point different from that stated in the field notes."* Emphasis ours. *Halley v. Harriman,* 106 Neb. 377, 183 N.W. 665, 668. *See also Lawler v. Rice County,* 147 Minn. 234, 180 N.W. 37; *Littlejohn v. Fink,* 109 Neb. 282, 190 N.W. 1020; *Sala v. Crane,* 38 Idaho 402, 221 P. 556.

"Any section corner or quarter corner that is identified as having been established by an official survey of the United States government must stand as being correctly located, however plain it may appear that the location is wrong; because the government surveys cannot be changed in an action at law between individuals." *Houston Ice & Brewery Co. v. Murray Oil Co.,* 149 La. 228, 88 So. 802, 805. See also *Thompson v. Darr,* supra; *Fair v. Ida County,* supra; Section 752, 43 U.S.C.A.

Subsequent to the trial of this cause but some three months prior to the date of the filing of the court's findings, conclusions and judgment, the plaintiff served and filed a motion for the trial court to direct "the county surveyor of Carbon County, Montana, to make a survey of the boundaries between the lands of the plaintiff and defendant, said survey to be made according to the United States maps and field notes introduced as evidence at the trial of the above-entitled action and in accordance with such Government monuments as such surveyor may find which have been identified by the evidence given in this case." The motion was submitted upon briefs of counsel for the respective parties and by the trial court denied.

It is quite evident that, since the record contains no evidence showing that the line on which cross-complainant built his fence is situate on the true boundary line between the northwest quarter and the southwest quarter *according to and as created by the United States government survey thereof,* another survey, such as was contemplated by the above motion, must be made. See *Houston Ice & Brewing Co. v. Murray Oil Co.,* 149 La. 228, 88 So. 802.

The cross-complainant is entitled to all the lands embraced in the northwest quarter and the plaintiff to those embraced in the southwest quarter of section 26 according to and as shown by the United States survey thereof. "But to establish a practical location which is to divest one of a clear and conceded title by deed, the extent of which is free from ambiguity or doubt, the evidence establishing such location should be clear, positive, and unequivocal." *Beardsley v. Crane,* 52 Minn. 537, 54 N.W. 740, 742.

In view of the state of the record in this cause, we have concluded to remand the cause to have another survey made either by a surveyor or surveyors to be agreed upon by the parties, or by a surveyor or surveyors whom the district court may see fit to appoint. In running the boundary line between the northwest quarter and the southwest quarter, the government survey and official plat must control the action by any surveyor employed by the parties or appointed by the court to establish such dividing line. *Stewart v. Boyd,* 15 La.Ann. 171. The survey should be made pursuant to the statutes prescribing the rules regulating the survey of government lands and in accordance with instructions and regulations issued thereunder by the general land office for the guidance of county surveyors and others in ascertaining boundary lines created by the government survey and in relocating corners or monuments theretofore established which may have become lost or obliterated.

The judgment appealed from is set aside and it is ordered that this cause be remanded to the district court for the making of a new survey, or of new surveys, and for further proceedings not inconsistent with the foregoing opinion, each party to bear his own costs on this appeal.

*Vaught v. McClymond,* **116 Mont. 542, 155 P.2d 612 (Mont. 1945).**

## NEBRASKA

The plaintiff brought an action against the defendant, in the district court of Hamilton county, to recover possession of a parcel of land described as follows: Commencing at a point forty rods north from the quarter section line corner between sections four and nine, in township ten north, of range six west, thence north forty rods, thence west twenty-one rods, thence south forty rods, thence east twenty-one rods to the place of beginning, the plaintiff claiming that said land is a portion of the south-west quarter of section four, in said township and range. The defendant, in his answer, admits that the plaintiff is the owner of the south-west quarter of section four in the above described township and range, but denies that said land in dispute, or any portion of it, is in the south-west quarter of said section. Upon the trial of the cause the jury found in favor of the defendant. Judgment having been rendered on the verdict, dismissing the action, the plaintiff brings the case into this court by petition in error.

The errors assigned are that the verdict is not supported by the evidence, and is against the weight thereof, and that the court erred in overruling the motion for a new trial.

The only question to be determined is, does the testimony sustain the finding of the jury as to the proper location of the quarter section corner on the south line of section four in said township?

This is not a case where the government mounds have been obliterated and their location is to be determined by secondary evidence. All the testimony tends to show that the corner in dispute, as now designated, was established by the government surveyors when the original survey was made. The plaintiff himself, soon after settling on the land in question, measured from the south-west corner of said section four, one hundred and sixty rods east on the south line of said section, and drove a stake in what he seems at the time to have regarded as the quarter section corner; and it is proved by other testimony that this corner has been regarded as the quarter section corner from the time of the first settlement of that portion of the county, and it is clearly proved that the distance between the south-west corner of said section and the corner in dispute is one hundred and sixty rods and two links. The plaintiff therefore has a full quarter section of land, and has no cause of complaint, although the south-east quarter of the section may contain an amount in excess of one hundred and sixty acres. The mounds established by the government surveyors will control course and distance whenever they can be identified. If they have been destroyed, or their existence is in dispute, their location may be determined by secondary evidence, and course and distance may be considered for the purpose of re-establishing them. No unvarying rule can be laid down, as each case must be determined according to the circumstances surrounding it. It is clear that justice has been done in the case at bar, and that, under the testimony, no other judgment could be sustained. The judgment is therefore affirmed.

*Johnson v. Preston*, **9 Neb. 474, 4 N.W. 83 (Neb. 1880).**

This is an action of ejectment brought by the plaintiff against the defendant to recover the possession of "about fifty rods in width off of the east side across the north-east quarter of section 30, township 2 north, range 6 east, in Gage county." The defendant is the owner of the north-west quarter of the north-west quarter of section 29, township 2 north, range 6 east (lot 2). The question involved is the location of the section line between sections 29 and 30. An opinion was filed in this case in 1884, the judgment being reversed, and is reported in 20 N.W. 254.

A rehearing was granted, and the cause is again submitted.

The land in controversy is situated on what was formerly the Otoe reservation, and was surveyed in the year 1873. A stone purporting to be the corner stone at the north-east corner of section 30, is 52 rods east of the true line between sections 29 and 30, as shown by the section corner on the south line of said section 30, and by the section line on the north, so that the section line from the corner contended for by the plaintiff runs nearly south-west to the south line of the section. No witness swears to the location of the corner by the government surveyors at the point indicated, nor did any witness testify to seeing the stone there prior to 1876 or 1877. The court below found the issues in favor of the defendant, and rendered a decree quieting his title. We approve of the points stated in the syllabus in this case in the opinion heretofore filed, and also in *Johnson v. Preston*, 9 Neb. 474, 4 N.W. 83, and adhere to those decisions. And when it is shown that a corner was established by the government surveyors, its location cannot be changed by evidence showing that it is incorrect. But the question here is, where was the government corner located? That is the only question in the case. To determine this the court may examine all the facts tending to show the original location. Thus, the quantity of land returned as being contained in the several subdivisions of sections 29 and 30 is evidence tending to show the correctness of the survey and location of the lines. It is true this evidence would not prevail against proof of the location of a corner, but in the absence of such proof it may be considered. Thus, in this case we find that the plaintiff purchased and has title to one hundred and twenty acres of land; that the defendant purchased lot 2 in the N. W. 1/4 of section 29, which contains 41 20/100 acres. If the plaintiff should obtain the land she is seeking to recover, she would obtain nearly 140 acres, while the defendant would obtain but little more than one-half of the amount purchased and paid for by him.

The field notes also may be considered where a government corner is destroyed or its existence is in dispute; so with plats of the government survey. All these are fingerboards, as it were, which point in the direction of the original corner as located by the surveyors. The court will then weigh carefully all the evidence presented, and determine the fact. In this case we think that the court below was fully justified from the evidence in finding, as it must have done, that the stone 52 rods east of the true line was not placed there by the government surveyors, and in proceeding thereupon to determine from the evidence the true location.

There are strong equities in favor of the defendant, and justice as well as law approves the judgment of the court below, which is affirmed.

*Morrison v. Neff*, **18 Neb. 133, 24 N.W. 555 (Neb. 1885).**

## NEVADA

The only material question involved in this appeal is whether the evidence supports the court's finding as to the proper location of the north line of section 13, T. 21 S., R. 61 E., M.D.M., in Clark County, Nevada, and, more particularly, the location of the N 1/4 corner of said section approximately at the center of a straight line connecting the NW corner with the NE corner of said section. Our conclusion is that it does. Conversely stated, did the evidence justify the court's rejection of appellant's contention that a certain monument, consisting of an unmarked 4X 4 post in a mound of rocks, set some distance from the quarter corner point described, was the actual original N 1/4 corner. We are satisfied that the court was justified by the evidence in rejecting such contention. Other points are determined in the following opinion.

We refer to the respondents Arthur Fort and Ione Katherine Fort as Fort and to appellant Walter L. Gardner as Gardner. Gardner and Fort's predecessor were adjoining owners of so-called five acre tracts. Fort's predecessor, Humphrey, owned the tract to the north and Gardner the tract to the south. Fort acquired from his predecessor, Humphrey, approximately the south 273 feet of Humphrey's five acre parcel. The boundary line between Fort and Gardner is in dispute, Fort having sued Gardner to establish the line contended for by the former. The court so established it and Gardner has appealed.

The patent to Fort's predecessor Humphrey was for W 1/2 NW 1/4 NE 1/4 NW 1/4, containing five acres according to the official plat of the Bureau of Land Management. Patent to Gardner was for W 1/2 SW 1/4 NE 1/4 NW 1/4, containing five acres according to the official plat of the Bureau of Land Management. The original survey by the United States Land Office was in 1882 or 1886. This was followed by a survey of the General Land Office in 1942. The survey under which the plat was made by the Bureau of Land Management (which took over functions of this nature from the General Land Office) was made in 1952.

From 1950 to 1952 F. M. Eaton, a duly registered Civil Engineer and Land Surveyor, apparently surveyed the N 1/2 of section 13 for a number of people who desired to make entries upon five acre tracts and he recorded such survey in 1954. In making his survey for this purpose he found the NW corner of section 13 and the NE corner of section 13. The brass markers identifying these two corners had been placed there by the General Land Office in 1942 or 1943. In searching for the N 1/4 corner of the section, he found no such corner in place on the northerly section line of section 13, but did discover a 4 X 4 post in a mound of rocks 27.6 feet to the north and 20.6 feet to the east of the approximate center of the north line of section 13. From inquiries made of a neighboring landowner and from the latter's report that the point of diversion in an application to the state engineer for permission to appropriate underground water had been tied in to this monument as the N 1/4 corner of section 13 (which he verified from the state engineer's published notice), he accepted it as such and designated it on his map as such. His entire survey was based upon the acceptance of this monument as the N 1/4 corner of said section. It had no markings on it, official or otherwise. The original plat of the government survey of 1882 or 1886, while designating the NW corner and the NE corner and describing the markings thereon, did not, so far as disclosed by the evidence, indicate that any monument had been placed for the N 1/4 corner.

The survey made by the Bureau of Land Management in November, 1952 resulted in placing the N quarter corner of section 13 on the north boundary of that section, approximately halfway between the NW corner and the NE corner. A subsequent survey and grading work by the state department of highways in constructing a road along the north boundary of section 13 destroyed or covered the monument that Eaton had found. The patent issued to Humphrey in August, 1954 and the patent issued to Gardner in June, 1954, with respective descriptions as above set forth, were in accordance with the Bureau of Land Management survey. There was no evidence as to the respective dates of the entries (proper filings in the Bureau of Land Management and issue of certificates of entry and payment of fees) which resulted in the two respective patents.

Defendant asserts in his appeal (1) that the quarter corner found by Eaton is a monument and marker of the original survey; (2) that, if so, the survey by the Bureau of Land Management in 1952 was a resurvey; and (3), being such, it could not affect the rights of entrymen that had attached. Section 772, Title 43, U.S.C.A.; *Barringer v. Davis,* 141 Iowa 419, 120 N.W. 65.

There is no substantial factual support of these contentions.

Fort contends in addition that Gardner's present reliance upon the location of the monument found by his surveyor Eaton is a collateral attack upon the patent issued to Gardner as well as upon the patent issued to Humphrey, and that such collateral attack may not be made. *Cragin v. Powell,* 128 U.S. 691, 9 S.Ct. 203, 32 L.Ed. 566; *Earl v. Morrison,* 39 Nev. 120, 154 P. 75. This contention too we believe to be well founded.

Technically the judgment quieted plaintiffs' title in the metes and bounds description of the approximate south 273 feet of the five acre tract patented to their predecessor as described in his patent, and ordered the defendant to remove a fence placed by the defendant on plaintiffs' said tract and enjoined defendant from trespassing on plaintiffs' tract. The findings are amply supported and the conclusions drawn are warranted by said findings.

**Gardner v. Fort, 72 Nev. 160, 298 P.2d 468 (Nev. 1956).**

## NEW HAMPSHIRE

In the description of a line, what is most material and certain controls what is less material and uncertain. Boundaries marked on the land govern courses and distances. If the plan, or the line, described in a deed or charter, and the monuments made by an original survey of a tract or township of land, do not correspond, the monuments determine the true location; and the marks on the ground of an old survey, indicating the lines originally run, are the best evidence of the true location of that survey.

Where land is conveyed by a deed referring to a plan or to a charter line, between which and the actual original survey, as shown by fixed monuments upon the ground, there is a difference in the courses and distances, or in the location of lines and monuments, the lines and monuments, as originally located and marked on the ground, are to govern, however they may differ from those represented on the plan, or described in the charter.

No parol evidence of the understanding or intention of the parties can control the well settled construction and legal effect of the clear language of a deed; but, if that language contain a latent ambiguity, so that it may be applicable to several persons, to several parcels of land, to several species of goods, to several monuments, boundaries or lines; or, if the terms employed be vague and general, or susceptible of divers meanings, parol evidence is admissible of any extrinsic circumstances tending to show what persons or things were intended, or the true meaning of the terms used.

It is not improper for a witness to state that, among other indications of the existence of an established line, he noticed a ridge of land apparently marking the interval between the occupation for tillage or other purposes, on the one side and the other; what such ridge indicates, being almost as much a matter of fact and observation on the ground, as its existence.

WRIT OF ENTRY, for a tract of land in Manchester, described in part as bounded by "Derry old line." The case was tried upon the general issue. The plaintiffs introduced as evidence a copy of the charter of Londonderry, bearing date of 1792, in which said line was described as running from a certain monument north north-east, three miles. They also introduced a plan of Londonderry, admitted to be genuine, in which the same line is described as running from a tree, known as the Griffin tree, admitted to be the one shown to the jury on their view in this case, thence north north-east, three miles, to a red oak, marked, & c.

The plaintiff offered evidence, by the declarations of owners of land by said line, tending to show that a pine tree which formerly stood at a certain point 919 rods about north north easterly of said Griffin tree, was the true terminus of said line, and contended that the defendant could not be allowed to prove that said line was not a straight line.

The defendant contended that the said pine was not at said terminus, nor upon the Derry old line; it being neither in the course described in the charter and plan, nor at the distance named from said Griffin tree, and offered evidence tending to show that there are now standing ancient monuments at several points along the general course of said line, which were shown to the jury, and which the defendant contends made the "old Derry line," as originally located.

Judge Stark, who was called as a witness for the plaintiff, testified, on cross-examination by the defendant, that he had, as a surveyor, run many lots on and near said "old Derry line," and described the bounds which had been shown the jury as they appeared when he first saw them, and stated that the wood was cut off up to certain lines which he described.

He was inquired of by the defendant whether he saw a line visible on the land, near the premises in dispute, and where the defendant claims that the old Derry line was located, and said that he did see a ridge showing the line of occupation, being the same shown to the jury. To this inquiry and answer as to the line, the plaintiff objected, and the court overruled the objection, and the plaintiff excepted.

The plaintiff claims under a deed from Jane Young, and in the deed by which she derived title the disputed line is described as north 19 1/2 east by said "Derry old line."

The court ruled--

*First* – That if Derry line, described as straight in the charter, was in fact located and marked by monuments by parties to the charter, such location will govern, and is the true line, though the line as located is not straight.

*Second* – If the descriptions in the deeds under which both parties claim title were understood by the parties to refer to "Derry old line," as a known located line, marked by monuments, such located line defines their grants, whether it was or was not the true line of Derry.

A verdict having been returned for the defendant under these rulings, it was ordered that the questions arising upon the case be transferred to the Supreme Judicial Court for determination.

Morrison & Stanley, and Wm. Stark, for the plaintiffs.

Clark & Smith, for the defendant.

FOWLER, J.

The first position ruled by the court below was in strict accordance with the general principle repeatedly recognized in this State, that, in the description of a line, what is most material and certain shall control that which is less material and uncertain; that boundaries marked on the land, as being most material and certain, are to govern courses and distances; that if the plan, or the line described in a deed or charter, and the monuments made by an original survey of a tract or township of land, do not correspond, the monuments are always to determine the true location, and that the marks on the ground, of an old survey, indicating the lines originally run, are the best evidence of the true location of that survey. *Bowman v. Farmer,* 8 N.H. 402; *Griffin v. Bixby & al.,* 12 N.H. 454; *Hanson v. Russell,* 28 N.H. (8 Foster) 111; *Day v. Enfield,* 11 N.H. 525; *Prescott v. Hawkins,* 12 N.H. 267, and many other cases.

The same general rule has often been recognized in other jurisdictions, and it may be regarded as well settled, that where land is conveyed by a deed referring to a plan or to a charter line, between which and the actual original survey, as shown by fixed monuments upon the ground, there is a difference in the courses and distances, or in the location of lines and monuments, the lines and monuments, as originally located and marked on the ground, are to govern, however they may differ from those represented on the plan, or described in the charter. *Kellogg v. Smith & als.,* 7 Cushing 375; *Missouri v. Iowa,* 7 Howard 660; *Ripley v. Berry,* 5 Greenl. 24; *Brown v. Gay,* 3 Greenl. 126; *Esmond v. Tarbox,* 7 Greenl. 61; *Blasdell v. Bissell,* 6 Barr 258; *Thompson v. McFarland,* 6 Barr 478; *Cox v. Couch,* 8 Barr 147; *Mills v. Buchanan,* 14 Pa. State (2 Harris) 59; *Gratz v. Hoover,* 16 Pa. (4 Harris) 232. See, also, 1 U.S. Dig., Boundaries, I. and II., secs. 1-99.

To the second ruling of the court below, rightly understood, we can discover no valid objection, although it was apparently unnecessary, save to meet an assumption of the plaintiffs. The located line would seem very clearly to have been the true original line of Londonderry; and if it were so, whether the parties understood the deed to refer to "Derry old line," as a located line or not, was wholly immaterial, unless the language of the deed contained a latent ambiguity; for that language having named "Derry old line" as the boundary of the lot, when "Derry old line," or the original line of Londonderry, was shown to be an actually located line, that located line must be the boundary of the parties, whatever their understanding or intention might have been. No parol evidence of the understanding or intention of the parties could be received to control the well settled construction and legal effect of the language of the deed itself, unless it contained some latent ambiguity. Such construction is to be given to every instrument as will carry out the intention of the parties, whenever it is legally possible, consistently with the language of the instrument itself. *Chamberlain v. Crane,* 1 N.H. 64.

In the present case, the language of the deed was, "Derry old line." If these words naturally and necessarily meant the true, original line of Londonderry, whenever it was shown that such a known located line existed, the legal construction of the language of the deed would make it the boundary, such located line being legally the true original line of Londonderry, without regard to the actual intention of the parties. But if it appeared, as the plaintiffs by their evidence attempted to show in this case, that a certain located line was not the true original line of Londonderry, although called "Derry old line," then the expression, "Derry old line," might well be regarded as so ambiguous as to admit of parol testimony to prove what line was intended to be designated thereby, in which case it would be clearly competent for the parties to show how they understood and with what particular meaning the expression was used. *Hanson v. Russell,* 28 N.H. (8 Foster) 111. All, therefore, which the court can be understood to have intended by their second ruling is, that if "Derry old line," as located, had been shown, as the plaintiffs contended the fact to be, not to have been the true original line of Londonderry, then it was competent for the jury to consider and find, from the evidence before them, what line was understood and intended, by the parties to the deed, to be designated by that expression; and if they found it to have indicated a certain located line, marked by monuments, other than the true original line of Londonderry, but known as "Derry old line," the intention of the parties should govern. In other words, the court intended to rule, that if the "Derry old line" proved by the defendant, were not, as the plaintiffs contended it could not be, the true original line of Londonderry, but a certain located line marked by monuments, and known by that name, then the language of the deed was so ambiguous that parol testimony might be introduced and considered to determine what line was intended by it.

As all written instruments are to be interpreted according to their subject matter, and such construction given to them as will carry out the intention of the parties, whenever it is legally possible to do so, consistently with the language of the instruments themselves, parol or verbal testimony may be resorted to, to ascertain the nature and qualities of the subject matter of those instruments, to explain the circumstances surrounding the parties, and to explain the instruments themselves by showing the situation of the parties in all their relations to persons and things around them. Thus, if the language of the instrument is applicable to several persons, to several parcels of land, to several species of goods, to several monuments, boundaries or lines, to several writings, or the terms be vague and general, or have divers meanings, in all these and the like cases parol evidence is admissible of any extrinsic circumstances tending to show what person or persons, or what things, were intended by the party, or to ascertain his meaning in any other respect; and this without any infringement of the general rule, which only excludes parol evidence of other language, declaring the meaning of the parties, than that which is contained in the instrument itself. 1 Greenl. Ev., secs. 286, 287, 288, and authorities cited. *Chamberlain v. Crane,* 1 N.H. 64.

The second ruling of the court below is perfectly consistent with this general rule. The language of the deed, "Derry old line," was susceptible of various meanings, as the original line of Londonderry, or any other line marked by monuments and called by that name. It was, therefore, entirely competent to show what line was intended by this phraseology, by having the location of the land to which it was a boundary, and the understanding of the parties themselves as to the line actually designed to be designated by the terms employed by them.

We do not perceive any valid objection to the question and answer of the witness in regard to the visible marks of an existing line near the premises in dispute, to which exception was taken by the plaintiffs. It was proper for the witness to be asked, and to state, what indications of a marked line he noticed at the point in controversy; whether those indications consisted of spotted trees, walls, fences, or a ridge of land apparently marking the interval between the occupation for tillage on the one side and the other. The jury, composed as a large majority of it undoubtedly was, of practical farmers, could hardly fail to understand the witness correctly, or to appreciate the weight and importance of his testimony in this particular.

Nor is there any just exception to the answer as giving the opinion of the witness, when he describes the ridge which attracted his attention, as showing the line of occupation on the one side and the other. From the nature of the case, it was almost as much a matter of fact and observation upon the ground, whether this ridge indicated the line of occupation, as whether it existed at all; and as the jury had themselves viewed the premises, there could be hardly a possibility that the testimony could have misled them, or in any way injuriously have operated against the plaintiffs' interests. After the view there would seem to have been little occasion to offer the evidence to which exception was taken, but as it was competent, the defendant had a right to its admission, if he deemed it material.

As the rulings and instructions of the court below were correct, or at least sufficiently favorable to the plaintiffs, the exceptions thereto are overruled, and there must be *Judgment on the verdict*.

**Hall v. Davis, 36 N.H. 569 (N.H. 1858).**


## NEW JERSEY

No suitable case found.


## NEW MEXICO

The defendant in error, George W. Stoneroad, brought an action of ejectment against the plaintiff in error, James P. Stoneroad, to recover possession of two pieces or tracts of land, alleged to be within and constituting a portion of a larger tract known as the "Preston Beck Grant," situated in San Miguel county. The two pieces of land occupied by defendant at the date of the commencement of this suit, and held adversely by him, lie within the boundary calls of the Preston Beck grant, but without the limits of a survey thereof, made by the government without notice to the owners, and in which the claimants have never acquiesced.

Upon the trial the defendant in error introduced in evidence a true copy of the original and official translation of the original *espediente* of said grant, and of all proceedings had, and papers and documents filed, before and with the United States surveyor general for New Mexico, and the decision of the surveyor general approving the grant; a map (Exhibit B) showing the location of the objects given as boundary calls, and the location thereon of the two pieces so held by the plaintiff in error. He also introduced oral evidence, showing that the land sued for was within the boundary calls of said grant. The plaintiff in error admitted that he was in possession. He also admitted and stipulated to the correctness of the location of the "Pecos river," a boundary call in the grant, and of the pieces of land held adversely by him, as shown by Exhibit B; and that defendant in error had an undivided one-third of all the right and title of the original grantee in said grant; that said grant was duly confirmed by congress under the act of June 21, 1860; and that the same had been surveyed without notice to the claimants, and at the time of such survey some of the said claimants were minors and married women, and had not acquiesced in the same. Plaintiff in error offered nothing in evidence.

On the sixth of December, 1823, Juan Estevan Pino petitioned the governor of the province of New Mexico for a grant of lands in this territory, now a portion of San Miguel county, bounded as follows: On the north by the landmarks of the farm or land of Don Antonio Ortiz, and the table-land of the Aguage de la Yequa; on the south by the Pecos river; on the east by the table-land of Pajarito; and on the west by the point of the table-land of the Chupaines; and on the twenty-third of the same month and year the political chief or acting governor, by direction of the most excellent deputation, made the grant, reciting in the granting decree the same boundaries as set out in the petition of Pino, and directing that the said Pino, as the grantee, be put into possession by the alcalde, which was accordingly in due form done. Pino occupied the land until his death, and subsequently thereto his heirs sold to Preston Beck, who presented the claim to the surveyor general of the territory on May 10, 1855, for investigation and decision, as required by the eighth section of the act of congress of July 22, 1854.

After taking testimony, and after a due consideration of the case, the surveyor general recommended that congress confirm the grant to Preston Beck, Jr.; and by act of congress, approved June 21, 1860, the same was confirmed according to the report of the surveyor general as No. 1. The surveyor general had the grant surveyed, and the said survey approved by him, November, 1860. No notice, however, of this survey was given to the claimants, so far as the records show. The act of congress of June 21, 1860, confirmed the grant as reported.

The single question, under the facts stipulated and proven in this case, is, does the survey, made after the grant was confirmed, conclude the court from determining the extent and validity of the grant made by the Mexican government by specific boundary calls, and afterwards, under the act of congress of 1854, duly examined into and approved by the surveyor general, approved by the land department of the United States, and confirmed by act of congress. That the plaintiff in error was in possession and holding two parcels of land embraced within the limits of the Preston Beck grant, as confirmed by congress, if the boundary calls are to govern as to extent, is admitted. It is also admitted that these lands so held by him are without the limits of the survey, but near the line. It is further shown that the pieces so held, as shown by the map, lie some 10 or 12 miles within the limits of the grant, provided the grant is to be upheld as petitioned for and granted by the Mexican government, and of which he was put into judicial possession by the duly-authorized officer of the Mexican government. The surveyor general pursued the same description in his report, and the act of congress confirmed it as reported.

It is conceded by both parties that, where lands are to be surveyed by our government, the rules adopted by the interior department in surveys of grants like the one in question, where boundary calls are given, is to draw a straight line north and south, or east and west, through such point, on the side of the tract of which it constitutes such boundary, to the intersection of the boundaries on the other sides; and where the call is a meanderable object, such as a river, *mesa*, mountain, or arroyo, it should be meandered, to the extent that it constitutes such boundary, within the projected lines of the other sides. See *Ortez Mine Grant*, 2 Copp, Pub. Land Laws, 1276; *U. S. v. Soto*, 1 Hoff. L. Cas. 68; Tyler, L. Bound. 29, 187, 188.

It is not shown in the record that Preston Beck, or any person holding under him, ever applied to the government to have a survey made, or for a patent. It has not been urged that there is any act of congress making it obligatory upon the confirmance of a grantor, or those holding under him, to have the grant surveyed, or to call for a patent; nor is it insisted that there was any regulation of the executive department, having the force of law, making it obligatory on the surveyor general to survey such grants after confirmation, in operation at the date of the survey, in 1860.

If we apply the rule above stated, governing the survey of a Mexican or Spanish grant, it is obvious that the plaintiff in error was in possession of lands belonging in part to the defendant in error at the date of the institution of this suit.

The contention on behalf of plaintiff in error is that, after the survey, the defendant in error, and all other claimants under the grant, are concluded by the survey; and that, in an action of ejectment, the survey estops the plaintiff from asserting title to a greater quantity of land than is included in it. In support of this position the case of *Gallagher v. Riley*, 49 Cal. 473, and *Boyle v. Hinds*, 2 Sawy. 527, are cited.

In *Boyle v. Hinds*, the Mexican government, in 1839, granted a *rancho*, called "Estero Americano," to Edward Manuel McIntosh. The grant was for two square leagues, within certain designated boundaries, embracing six or more square leagues. It contained the usual provision for measuring the land, and leaving the surplus to the nation. The grant was approved by the departmental assembly. Afterwards, judicial possession was given by the alcalde of the jurisdiction within which the land was situated. The judicial possession was regular in all respects. This possession embraced six instead of ten square leagues. Prior to 1850, McIntosh conveyed his grant to O'Farrell, who, in 1852, presented the grant for confirmation, and it was subsequently confirmed to the extent of two leagues only. It was surveyed and patented for about two leagues. O'Farrell took his patent without objection, and placed it on record in the recorder's office of the county in which the lands were situated. The lands in question lie within the judicial possession given to McIntosh, but without the limits of the final survey and the patent issued to O'Farrell. The patent was issued to O'Farrell after the final decree of confirmation. After that decree the government patented the lands in controversy to Hinds, as a pre-emptor, who was in possession when the suit was brought. In delivering the opinion, SAWYER, J., said: "The plaintiff claims that McIntosh had a perfect Mexican title, and that it was unnecessary for his grantee, O'Farrell, to present his claim for confirmation; that, his title being perfect, congress had no constitutional power to deprive him of his land in case of his failure to present his claim under the act of 1851. Under the view I take, it may be conceded that it was unnecessary to present the claim; but the claimant did present his grant, and submit it to examination, and asked its confirmation under the act of congress. The question as to the genuineness and extent of the grant were litigated between the government and the claimant before a tribunal having jurisdiction to determine them. The grant was confirmed to the extent of two square leagues, and no more. The juridical possession was put in evidence, and the extent of the land to which the claimant was entitled in fact determined. The claimant did not appeal, and the determination became final. He had a right of appeal under the act, and could have gone from court to court, and ultimately had the question directly adjudicated by the supreme court of the United States in that proceeding, whether he had a title to the full extent of the juridical possession or not. The same court would then have passed upon his title in a direct proceeding to establish his claim to the whole that are now called upon to determine the same question collaterally. The law afforded him tribunals to determine this very question between him and the United States. The owner of the grant availed himself of the right afforded by the act of congress, and the question between him and the United States was litigated and determined. If the claimant chose to accept the decision of the inferior tribunal, he is bound by it." It was also held that, by the fifteenth section of the act of 1851, the adjudication between the government and the claimant became final, and must, for that reason, be considered as *res adjudicata*.

The proceedings in the case, ending in a decree limiting the confirmance to two leagues, was clearly judicial, and was final and binding on the parties. In that case the survey and patent but carried out the decree of confirmation. It was for these reasons held that the patent was the final authentic record of the proceedings, and was conclusive evidence between the parties of the extent of the grant, and the correctness of the location. There is nothing in this case to guide or control us in reaching a correct conclusion on the facts presented in the transcript before us. In the case above quoted the owner of the grant submitted his claim to a court of competent jurisdiction, and actually litigated with the United States upon the very question of the extent of his grant under the title derived from Mexico, and the act of juridical possession. The way was open to him, upon an adverse decision, to appeal finally, if necessary, to the supreme court of the United States. He submitted to an adverse decision without appeal, and thereby waived or lost the right to dispute with the government or its grantee the title to any portion of the lands lying without the terms of the decree, survey, and patent.

*Gallagher v. Riley, supra,* goes no further than to declare that in Mexican grants, where no juridical possession is given, and the grant is of a quantity of land as distinguished from a particular piece, by specific boundaries, a survey after confirmation by congress is necessary, and is made so by the thirteenth section of the act creating the board of land commissioners, (1851). WALLACE, C.J., in delivering the opinion, said: "In excluding the proffered evidence, the court below erred. If the survey had become final, as the plaintiff proposed to prove, it was conclusive of the boundaries of the Rancho Laguna de San Antonio. The court below was mistaken in supposing that the confirmation to Bartolome Bojorques was for a 'specific piece of land.' The quantity confirmed was six leagues in superficial area. The land confirmed was to be hereafter surveyed, located, and set apart to the confirmee by the proper authorities of the United States; and if the survey had become final, as the plaintiff offered to show, it was conclusive evidence in the action of the boundaries of the tract confirmed." In the latter case it was agreed that the defendant had possession of lands lying within the *diseno* or map describing the extent of the grant, but without the survey.

The quantity confirmed in the California grant was six leagues in superficial area, "being the same on which the said petitioner resides, and bordering towards the north-east on lands known in November, A. D. 1845, as the lands of Juan Martin, towards the north-west on Los Dos Piedros; towards the south-west on Los Tomales; and towards the north-east on lands known, at the date of the last mentioned grant, as the lands of Juan Miranda,–the said premises being of the extent of six square leagues." The plaintiff offered to prove that the survey, plat, and field-notes made by the surveyor general of the state of California of said Rancho Laguna de San Antonio, was filed in the office of said surveyor general on the tenth day of March, 1859, and approved by him on said day; that said surveyor general immediately thereafter, to-wit, on the fifteenth day of March, 1859, gave due notice by publication, according to law, that said claim had been surveyed, and a plat thereof made and filed as aforesaid; that no objection had been made; and further offered to show that said survey had become final. The court simply said that if the plaintiff could have proved that the survey was made as required by the statute in force in that state at that time, after due and proper notice to all persons interested in the grant, and that no objection had been made, and the survey was final, it conclusively settled the boundaries of the grant. It will be noticed that the decision was rendered after the act of 1851 was passed, and the parties in interest had submitted to the decision of a tribunal clothed with ample authority to finally settle disputed claims between grant owners and the government of the United States, and as between conflicting or adversary claimants. After the land-board had rendered judgment in cases like the one considered in *Gallagher v. Riley*, it became necessary, and was a part of the system adopted, to have a survey made, and assign the quantity to which the claimant was entitled, thereby segregating his claim from the public domain. The final survey became a necessary part of the proceedings in cases of quantity grants, in order to accurately locate the boundaries. Notice of such survey was, however, necessary. The act expressly conferred the power upon the surveyor general to do what he did in that case. His act, done in pursuance of such expressed authority, and upon notice, ought to have bound the claimants. There was no expressed duty imposed by the terms of the act of congress of 1854, creating the office, and defining the duties of the surveyor general of this territory. There is no evidence before us that any notice was ever given by the surveyor general of the territory of his survey in 1860. The stipulation on file shows that no such notice was ever given.

The grant in the case before us was a perfect one, made in 1823, by the Mexican government. This is admitted by plaintiff in error; or, at all events, he does not offer any objection to it on account of not conforming to some requirement of the laws of Mexico. It covers all the lands lying within certain named and well-known boundaries. The grant or patent made by Bartholome Baca, acting superior political chief of the province of New Mexico, after reciting the power conferred on him, and the previous consent of the most excellent provincial deputation of the territory, proceeds as follows: "I have granted, and by this patent do grant, in the name of the supreme government of the nation, to Don Juan Estevan Pino, the land he solicits on the Gallinis river, which shall be called the 'Hacienda of San Juan Baptista del Ojito del Rio de las Gallinis,' recognizing the boundaries: On the north, the land-marks of the farm of Don Antonio Ortiz, and the table-land of the Aguage de la Yequa; on the south, the Pecos river; on the east, the table-land of Parjarito; and on the west, the point of the table-land of the Chupaines, –on which fixed points you will place formal and well-constructed boundaries, so that in all times to come the dividing line of the land granted to him may be known." The alcalde, in obedience to directions from superior authority, and in strict conformity to law, put the grantee into the juridical possession of the lands embraced in the grant, as lying within the boundaries given. The surveyor general reported this grant for confirmation, following the description contained in the petition of Pino, and the grant from the Mexican government. The act of congress confirms the grant as reported.

In *U.S. v. Halleck*, 1 Wall., at page 455, FIELD, J., in passing upon a question of survey in that case, said: "The material question for determination is whether the survey approved conforms to the decree of confirmation. There must be a reasonable conformity between them, or the survey cannot be sustained." The survey here is not, by many miles, in conformity with the boundary calls of the grant and act of confirmation. It is at one point some 10 miles within the line of the grant, when drawn as prescribed by the rules of the interior department. There is no reasonable conformity, and the result is that the survey cannot be sustained, unless for reasons hereafter stated.

In *U.S. v. Pico*, 5 Wall. 536, the court said: "Were there any doubt of the intention of the governor to cede all the land contained within the boundaries designated by him, it would be removed by the juridical possession delivered to the grantees. This proceeding involved an ascertainment and settlement of the boundaries of the lands granted by the appropriate officers of the government, specially designated for that purpose, and has all the force and efficacy of a judicial determination. It bound the former government, and is equally binding upon the officers of our government." Such is the purport of the recent decision in the case of *Graham v. U.S.*, 4 Wall. 260. See, also, *Leese v. Clark*, 18 Cal. 567; *Page v. Scheibel*, 11 Mo. 187.

In *Graham v. U.S.*, 4 Wall. 259, Mr. Justice FIELD, in speaking of the effect of the juridical possession, used this language: "By this proceeding, called, in the language of the country, the delivery of the juridical possession, the land granted was separated from the public domain, and what was previously a grant of quantity became a grant of a specified tract. The record of the proceeding of this nature must necessarily control the action of the officers of the United States in surveying land claimed under a confirmed Mexican grant."

In the grant to Pino, juridical possession was delivered to him by the description and boundary calls, above stated. The south line was the Pecos river. The survey does not follow the specific and well-defined natural objects named in the grant and act of juridical possession as boundary lines and calls. If the record made by an officer, after delivering possession, is binding upon the officers of our government, it follows that the act of delivering such juridical possession, by such clearly and distinctly defined boundary lines as in this case, ought to be equally obligatory and binding upon the officers of our government. Had it been uncertain what lands the government of Mexico intended to convey to Pino, or the grant had been of a quantity within defined exterior limits, it would have been necessary for the officer to have measured the lands and fixed boundaries, and to have made some record of the act; otherwise a new survey by our government would have been necessary to separate the granted lands from the public domain. The boundaries given, and the rule for measurement ascertained, it is mathematically certain what lands were granted.

Mr. Justice DAVIS, in *Ryan v. Carter*, 93 U.S. 82, in speaking of the act of congress confirming lots and lands in the territory of Missouri to the inhabitants of certain towns and villages, under the act of 1812, said: "It does not require the production of proofs before any commission or other tribunal established for that special purpose, but confirms, *proprio viyore*, the right, title, and claims to the lands embraced within it, and operates as a grant to all intents and purposes. Repeated decisions of this court have declared that such statute passes the title of the United States as effectually as if it contained, in terms, a grant *de novo*, and that a grant may be made by law as well as by a patent pursuant to law."

In *Tameling v. United States Freehold, etc., Co.*, 93 U.S. 644, Mr. Justice DAVIS, in passing upon the effect of an act of congress confirming a Mexican grant, said: "Congress acted upon the claim as recommended for confirmation by the surveyor general. The confirmation being absolute and unconditional, without any limitation as to quantity, we must regard it as effectual and operative for the entire tract." The court reaffirmed the doctrine that an act of confirmation passes the title of the United States as effectually as if it contained in terms a *grant de novo*, and that a grant may be made by law as well as by a patent pursuant to law.

The act of June 21, 1860, confirmed to the claimant, Preston Beck, Jr., the entire tract, as recommended by the surveyor general, without any limitation as to quantity, or other condition whatever.

Plaintiff in error cites *U.S. v. Flint*, 4 Sawy. 61, as supporting his position in favor of the validity of the survey. In concluding the opinion, Justice FIELD of the supreme court said: "The courts can only examine into the correctness of a survey when, in a controversy between parties, it is alleged that the survey made infringes on the prior rights of one of them, and can then look into it only so far as may be necessary to protect such right." The right here alleged is a prior one, and we are asked to look into the survey only so far as to protect such prior right. The attorney of record for the plaintiff in error filed no brief. The only authorities presented were submitted by the United States attorney. The facts stated in the record were stipulated and filed in the court below. We have had our attention called to no case of a similar character arising under the act of July 22, 1854. The decision of the courts referred to under the California acts of 1851 and 1860 afford us very little aid in construing the powers of the surveyor general of New Mexico under the act of 1854. We have kept in view the fact that, at the date of the survey, November, 1860, some of the claimants were minors, while others were married women, who would not be bound by the act of the surveyor general to their prejudice, unless it was the evident and very clear intention on the part of congress of the United States to give the survey the effect claimed here. We do not think it was the intention of congress to confer any such power by the enactment of that statute.

### *Stoneroad v. Stoneroad*, 4 N.M. 181, 12 P. 736 (N.M. 1887).

(after stating the facts). It will thus be seen that the only question in this case is as to the true location on the ground of the section line between sections 10 and 15. Defendant claims the corner common to sections 10, 11, 14, and 15 to be at a point about 150 feet south of the coal mine, and that consequently the mine is in section 10, and the railroad company have no interest in it. Intervener claims the corner in question to be 960 feet north of the point claimed by defendant, and that consequently the mine is on section 15, and intervener is entitled to recover. The township was originally surveyed and subdivided in 1881. The United States deputy surveyor, who subdivided the township, is not produced as a witness, and there is no direct testimony as to where the corner in question was originally placed on the ground. Intervener produces evidence to the effect that in 1896 a monument, properly marked for the corner in question, is found in place at a point 960 feet north of the corner, as claimed by the defendant, but offers no proof, except the plat and field notes, which will be presently considered, to show that the corner was originally placed where the monument is now found. He also shows that a monument, properly marked for a quarter corner, is now found one-half mile east of the monument just mentioned, and a monument marked for the southeast corner of section 11 is now found at a point 710 feet north of where it should be, to be in harmony with the other corners of the township. Defendant meets this proof by several witnesses, who testify that they have in former years seen the monument of the corner in question in place and properly marked at the point claimed by defendant, some putting it prior to 1885. Various witnesses, by means of the official field notes, located the various monuments of adjoining corners, and found them on the ground in place, and demonstrate that the proper place for the corner in question, if it is to harmonize with the rest of the township, is where defendant claims it is. Defendant further shows that the monuments now found by intervener, and claimed by him to mark the true corner, bear evidence of having been recently placed in their present position, in that they are not set in the ground, and grass is found underneath them. Defendant further shows that the quarter corner between sections 11 and 14 now stands 412 feet north of where it formerly stood, and where it should stand if it is to be in harmony with the rest of the township. Looking at the testimony in the case, aside from the official plat and field notes, we feel convinced that the monument marking the corner in question was originally placed on the ground by the United States deputy surveyor at the point contended for by the defendant.

But it appears from the official field notes of the survey that the surveyor, in running the line between sections 14 and 15, started at the southeast corner of section 15, and ran north, and at 23 chains encountered a wagon road, at 28.50 chains encountered the Rio Puerco, at 29.50 chains encountered the Atlantic & Pacific Railroad, at 40 chains set quarter corner, at 64.50 chains encountered an arroyo, and at 80 chains set the corner in question. These topographical features appear upon the face of the official plat. But it appears from the testimony that the surveyor, in running this line, made a mistake in his measurements, and that, as a matter of fact, at 40 chains he had only crossed the wagon road, and the quarter corner should have been placed between the wagon road and the Rio Puerco, at least south of the railroad. In other words, it appears that the plat puts the wagon road, Rio Puerco, and railroad too near the starting point, and in this particular contradicts the field notes in the calls for distance. On the other hand, the field notes and plat put the corner in question at practically the same distance north of the wagon road, Rio Puerco, and railroad as claimed by intervener, and where the same is now found. In other words, starting at a point on the line between sections 14 and 15, at the railroad, for instance, the plat and field notes put the south corner of sections 14 and 15 nearer, and the corner in question further than we find from the evidence they were originally placed by the government surveyor. It will thus be seen that the only real difficulty in the case arises from the fact that the topographical features encountered in running this line have been placed in the wrong place on the line, both by the field notes and the plat. Counsel for intervener earnestly contends that the plat and field notes being the highest evidence of the location of the corner in question, and they, taking into consideration the calls made for natural objects in the field notes and appearing on the face of the plat, putting the corner in question where it now stands, this evidence must control, and establish the corner at that point, regardless of where it was originally put upon the ground by the surveyor. To this contention we cannot agree. It is, of course, a familiar rule that monuments, either natural or artificial, control courses and distances. The monument for the corner in question having been, as we find from the evidence, originally placed by the surveyor who subdivided the township at the place contended for by the defendant, the distance called for in the field notes for the line to run north of the wagon road, Rio Puerco, railroad, and quarter corner must yield to the monument placed at the end of the line. If this were a contest between the owner of the south half and the owner of the north half of section 15, it might be that the line would have to be extended north so as to include the natural objects called for in the field notes, and thus make the south half of the section much larger than the north half. But we know of no reason or rule of law requiring an error to be protracted beyond the point which will include the natural or artificial monuments called for in the field notes.

Even if the location of the original monument in question were not satisfactorily established – and we may say that it is not entirely so in this case – it is equally true that its original location at the point claimed by the intervener is not satisfactorily shown. And assuming, arguendo, that no monument, either at the place contended for by defendant or by intervener, is satisfactorily identified, our conclusion must still be the same; for, starting at the southeast corner of section 15, and running north, after having passed the natural and artificial monuments called for, course and distance from the starting point again takes control of the line, and thereby the same would terminate at the corner claimed by defendant.

But it is contended by counsel for intervener that the official plat of this township, as returned by the surveyor general, shows the corner in question to be where claimed by him, and, the plat being the highest evidence, it cannot be contradicted. He cites *Cragin v. Powell,* 128 U.S. 691, 9 S.Ct. 203, 32 L.Ed. 566; *Beaty v. Robertson (Ind. Sup.)* 30 N.E. 706. In Cragin v. Powell a controversy arose in Louisiana over the boundary of lands of contiguous owners in certain sections. The survey and plat placed upon the lands of the first entryman certain bayous, the lands upon either side of which were valuable as sugar lands. The second entryman bought by the same plat adjoining portions of the same sections. The latter claimed that the original survey of the government was erroneous, in that the eastern tier of sections of the township were not full size, and that, giving them the proper size, the said bayous would fall in the lands covered by his entry. The plat and field notes are not shown in the report of the case, nor are the locations of the monuments in controversy made to appear. But we do not understand the supreme court to be committed by this opinion to the proposition that the plat, by reason of topographical features appearing upon its face, can place a survey at a different place upon the ground from where it was actually placed by the surveyor. Thus, in *Horne v. Smith,* 159 U.S. 40, 15 S.Ct. 988, 40 L.Ed. 68, the plat showed that the lands in controversy bordered on the Indian river, while it appeared that in fact they were surveyed only to a bayou a mile or more east of the river. The court held that the true boundary was the bayou, and not the river. In Beaty v. Robertson, supra, it appears that section 6 had an Indian boundary running diagonally through it from northwest to southeast. The township was fractional, and the surveyor erroneously surveyed the south line of the section from west to east, thus throwing the excess land in the south half of the section on the eastern and interior subdivision, instead of the exterior subdivision, as should have been done. It further appears that the surveyor general made a note of this error on the field notes returned to him, and in issuing the official plat of the township corrected the error in accordance with his note on the field notes. This was within the scope of his authority, and was, in effect, a new survey, before the lands were sold by the plat made in accordance therewith. And the court held that the plat, representing this change in the field notes, controlled the location of the line between the subdivision involved. We see nothing in this case antagonistic to our views herein expressed. Authorities from state courts are abundant that monuments control both plat and field notes, some of which are cited. *Martin v. Carlin,* 19 Wis. 454; *McClintock v. Rogers,* 11 Ill. 279; *Ogilvie v. Copeland,* 145 Ill. 98, 33 N.E. 1085; *Robinson v. Laurer (Or.)* 40 P. 1012; *Bauer v. Gottmanhausen,* 65 Ill. 499; *Woods v. West (Neb.)* 58 N.W. 938; *Peterson v. Skjelver (Neb.)* 62 N.W. 43; *Cadeau v. Elliott (Wash.)* 34 P. 916. This conclusion renders it unnecessary to consider the other assignments of error, which relate largely to matters of practice and procedure. The decree of the lower court will be affirmed; and it is so ordered.

*Canavan v. Dugan,* **10 N.M. 316, 62 P. 971 (N.M. 1900).**

## NEW YORK

No suitable case found.

## NORTH CAROLINA

On a controversy as to boundaries, a parol agreement as to the location of the boundaries, made after the execution of the deed under which one party claims, could not alter the lines as fixed in the deed.

The controversy grew out of the question whether the plaintiffs' line should be run from a point designated as Figure 1 on the plat to 2, and thence to 12 and to 14, so as to include the disputed territory, included within the lines, 2, 12, 11, 2, or whether it should be located from 1 to 2, and thence north to 11, and thence to 12, as contended by the defendant, so as to cover no part of the defendant's possession. The call which gave rise to the dispute was "thence [from Figure 1] with the line of said tract [admitted to be the mountain field tract, 1, 2, 3, 4, 5, 1] to a stake on the west bank of the branch." In the beginning of the descriptive clause in plaintiffs' deed, the lands are described as lying on the head of Meadow's branch, on Bee Tree of Bull creek. The location of Bee Tree branch is shown on the plat attached, and which was agreed upon by the parties to be used in the statement; but the location of Figure 11 will be fixed further south, as indicated by an X mark on the line 10 to 11. The other material facts are stated in the opinion of the court.

The controversy hinges upon the location of a boundary line, and the facts are so far ascertained and admitted as to narrow it down to the legal question, what is the true line of the deed from Ramsey and others, executors, to Stephen Ammons? Beginning at Figure 14 on the plat, it is agreed that the line runs along a ridge to 15 and 16, and with the ridge to a Spanish oak opposite to Stephen Ammons' own corner of his mountain field tract, and thence to said corner at 4. From 4 the call of the deed is thence "with that line" (admitted to be the line of the mountain field tract) 127 poles to a stake, which both parties agree is at 5, and from 5 the call is thence with that line to the beginning corner, which is conceded to be located at Figure 1. The main question grows out of the location of the next line, and depends for its solution, in view of the undisputed facts, upon the construction which the court will give to the language of the next call, "thence with that line to a stake on the west bank of the branch." The land was described in the deed as "lying on the head of Meadow's branch, on the Bee Tree of Bull's creek." The nearest point of Bee Tree creek is at 17, which is 18 poles from 2, where the corner of the mountain field tract is located, and on a prolongation of the line 5, 1, 2; but the next succeeding line, of the mountain field tract (from 2 to 3) crosses Bee Tree branch at 11, and, according to the testimony of some of the witnesses, Cove branch, one of its tributaries, at 10. The point next called for being on Dan knob, at 12, the plaintiff contends that the proper location of the call which gives rise to the controversy is from 1, with the line of the mountain field tract, to 2, (where it diverges to the north,) thence to the nearest point on the west bank of Bee Tree, at 17, so as to include the disputed territory, where the defendant's possession lies, (within the boundaries 2, 12, 11, 2.) The defendant insists that the disputed line is to be so located as to run with the boundaries of the mountain field tract till they reach the west bank of Bee Tree branch at 11, and thus reconcile and fulfill the two descriptions by following the lines of the tract called for to the west bank of the branch. The general rule, which must be always observed, if possible, is that different descriptions of a boundary line should be reconciled so as to give effect to every expression of the grantor's intent. *Proctor v. Pool, 4 Dev.* 370; *Shaffer v. Hahn,* 111 N.C. 1, 15 S.E. Rep. 1033; *Shultz v. Young, 3 Ired.* 385.

It is settled that where there is direct proof of the actual running of a line at the time of the original survey or the location of a corner called for, as the line of another tract, or on the bank of a creek, the party claiming under the deed holds to such line or corner, notwithstanding the fact that some mistaken description may be inconsistent with such a location. *Cherry v. Slade, 3 Murph.* 82. It is equally true that where a point on the bank of a stream or in a given boundary is described in such a way as, with the aid of extrinsic proof, to fix its actual location, such description will prevail over and control a conflicting call for following the line of another tract. If the call here had been for a natural object, such as a certain rock on the creek or a stake corner of another tract, the location of which could be determined by running the boundaries of such tract, the line must have been run direct to such natural object, disregarding the call for following the line of another tract, if inconsistent with the more certain description. Cherry v. Slade, supra. In a conflict of descriptions, that one must always be adopted which is the more certain; and a known objective point, or one that can be made certain by proof, is, under that rule, to be preferred to one that can be fixed at a point on an extended line or surface only by groping along another line or lines to reach it. But no particular point on Bee Tree branch, known or ascertainable otherwise than by the use of the chain and compass, is designated as the terminus of the call. Indeed, the corner is by the terms of the description at an imaginary point – a stake. What did the grantor mean by the words, "thence with that line?" Did he mean the boundary line generally, or a particular line of the mountain field tract? The lines of the plaintiffs' deed, starting from the beginning corner at 14, first intersected with the 100-acre tract at 4, and the next line, which is described as running "thence with that line 127 poles to a stake," it is admitted, extends from 4 to 5. In running the call from 5, the language, "thence with that line to the beginning corner of said mountain field (100 acres) tract," is by common consent interpreted to mean, not the line which immediately preceded it in the deed, or a prolongation of it, but the next line of the same tract, running eastwardly at a right angle. That particular line of the mountain field tract terminates at 1; but the next call continues in the same direction to 2, falling short 17 poles of reaching the west bank of the creek by a direct line. It is evident that the description, "with that line," was properly construed by the parties, in running up to Figure 1, as meaning "with the boundaries of the mountain field tract," though it became necessary to change the course from that just previously pursued in order to conform to and follow its different lines. By continuing to place the same construction upon the identical language previously used after passing the beginning corner at 1, the two descriptions are met and reconciled. The boundary line of the mountain field tract is followed up to the very point fixed by its intersection with Bee Tree branch. Had it been impracticable to reach the west bank of Bee Tree branch by following, regardless of course, the boundaries of the other tract, both descriptions might have been met by diverging at 2 and running to the nearest point on the west bank of the branch. *Campbell v. Branch, 4 Jones, (N. C.)* 313. If a known point had been called for on the branch, course and distance, as determined by the compass or by known lines, might have been disregarded in order to reach it. *Redmond v. Stepp,* 100 N.C. 213, 6 S.E. Rep. 727.

The second line of defense contended for by the plaintiffs, in case they should fail in holding their first position, was that, conceding the correctness of the principle that both descriptions must be reconciled by following the lines, the first and nearest intersection with the waters of Bee Tree branch was at the west bank of Cove creek, (a tributary of Bee Tree,) at Figure 10, and by locating the corner at that point, and running thence to Dan knob, at 12, a part of the defendant's possession would still be covered by plaintiffs' deed.

The evidence was conflicting as to whether there was in fact any such stream as Cove branch crossing the line at 10, and, considering the language of the deed, the plaintiff had no ground to complain when the court told the jury that the boundary line of the mountain field tract must be followed to the branch; and it was for the jury to determine what branch was intended by the makers of the deed.

It was in evidence that the plaintiff and defendant had agreed by parol that the true boundary ran from 1 to 2, from 2 to 10, and thence to 12, and that they accordingly marked it. The judge very properly refused to charge the jury that by such a verbal agreement, made after the boundaries were fixed at the execution of the deed, they could be altered, and the parties estopped from disputing the new location so agreed upon. Shaffer v. Hahn, supra; *Caraway v. Chancy, 6 Jones, (N. C.)* 361. If the court below erred, it was in allowing the jury to take such running and marking into consideration at all, as tending to show the location of the branch called for, since the whole controversy depended upon ascertaining the point of intersection of the line with it, and neither an agreement subsequent to the execution of the deed, nor the running in accordance with it, was evidence tending to locate the lines run and marked contemporaneously with the execution of the original conveyance. Caraway v. Chancy, supra. We think that there was no error in the rulings or charge of the court of which the plaintiffs had just ground to complain.

### *Buckner v. Anderson*, 111 N.C. 572, 16 S.E. 424 (N.C. 1892).

The court refused the request of defendants' counsel, made in apt time, to instruct the jury, in effect, that, in fixing the location of the lower line of the Moravian deed, the proper and lawful method of conducting the survey was to run with the calls of the deed from an admitted corner, or from one which the jury believed was located by the testimony, instead of reversing the calls from such points. A corner admitted or ascertained by the usual marks, or established by the testimony to the satisfaction of the jury, is to be considered by them (as was said by Pearson, J., in *Safret v. Hartman, 7 Jones [N. C.]* 203) "a fact incorporated into the deed so as to make it a part of the description." If the principle contended for by the defendants' counsel is correct, it was immaterial whether the plaintiff admitted or the jury found from the testimony that the beech tree on Moravian creek was a corner of the Moravian tract. Whether established by proof or admission, it being in evidence that the line was reversed, when, by running forward, a different result would have been attained, it was error to refuse to instruct the jury that the location made by running, as the deed was originally run from a known corner or one established by proof, was to be adopted, rather than one ascertained by running in the opposite direction. It is a fact of which the courts must take and have taken notice that the measurements of boundary lines in making the original surveys for deeds and grants are often, if not always, inaccurate. Those discrepancies between the distance called for and the actual measurement occur much more frequently, too, in an undulating or mountainous section, because, as is a matter of general knowledge, it often happens that, in the original surveys of grants, only two or three lines of a square or parallelogram were actually run, and that the earlier surveyors, at least, universally adopted surface measurement. In running long lines from the top of one high and precipitous mountain to that of another, the area or acreage sold by the state to its citizens would have appeared much less than it actually was if the level measurement had been adopted in laying off large grants. It is therefore a well-known fact that, owing to inaccuracies in measurement, different results will follow from adopting one or the other of the two methods of surveying where many of the old monuments have perished or been removed. In determining which is correct, the courts proceed upon the idea that the object of legal investigation and inquiry is to find the lines, corners, and monuments which were agreed upon by the parties to the original conveyance, and that, in order to attain that object, the lines should be run in the direction and order adopted by them. *Harry v. Graham, 1 Dev. & B.* 78, 79; *Norwood v. Crawford,* 114 N.C. 519, 19 S.E. 349. There are some exceptional instances, in which it is manifest that reversing a line is a more certain means of ascertaining the location of a prior line than the description of such prior line given in the deed, but such cases are the rare exceptions to a well-established general rule. *Harry v. Graham,* supra, and *Norwood v. Crawford,* 114 N. C., at page 521, 19 S.E. 349; *Safret v. Hartman,* supra. The general rule is an established law of evidence, adopted as best calculated to ascertain what was intended to be conveyed, and it is incumbent on a party asking the courts to depart from it to show facts which bring the particular case within the exception to the rule.

We have rarely, if ever, had occasion to review a more confused statement of a case on appeal; but, construing all parts of it together, we think the defendants Denny and Cowles are entitled to the benefit of this assignment of error. The appeal as to the other defendant was dismissed for failure to print the record. The defendants Denny and Cowles are entitled to a new trial upon this ground. But we deem it proper, as a guide to the court below, to add that if it be true, as we understand the statement of the evidence and the exceptions, that the court below allowed the plaintiff, in the face of objection, to use the simple denial of the defendants in their answers that they were wrongfully and unlawfully in possession as evidence that the defendants were exercising such dominion over a virgin forest as to subject themselves to a possessory action (*Hamilton v. Icard,* 114 N.C. 532, 19 S.E. 607, and same case, decided at this term, 23 S.E. 354), that ruling was also erroneous. It is true that, under the late act of the legislature (Laws 1893, c. 6), a plaintiff may maintain an action to remove a cloud from his title without showing that the defendant is an occupant or any more than a claimant of the land in controversy. But, where he alleges an occupation as the cause of action, not only must the allegation and proof correspond, but the testimony offered to show possession is open to objection and exception on the ground of competency. It seems, however, that this action was brought before the 31st day of January, 1893, and the plaintiff can only recover upon a cause of action then existing.

*Duncan v. Hall*, **117 N.C. 443, 23 S.E. 362 (N.C. 1895).**

## NORTH DAKOTA

This action was brought to determine whether or not a certain triangular tract of land formed a portion of section 36, township 144, range 50, in Traill County, or a portion of section 35, same township. If the former, it belongs to plaintiff; if the latter, it belongs to defendant Agnes C. Walker. A trial to a jury resulted in a verdict for plaintiff. A motion for a new trial was denied, and judgment rendered on the verdict. There was an alleged irregularity about the motion for a new trial, which appellant seems to fear will prevent a review of the evidence in this court. But respondent makes no such claim. On the contrary, by his course he concedes that the evidence is before us. We need not, therefore, further notice the irregularity, if any there be; connected with the motion for a new trial. The claims of the respective parties will be more readily comprehended by a diagram:

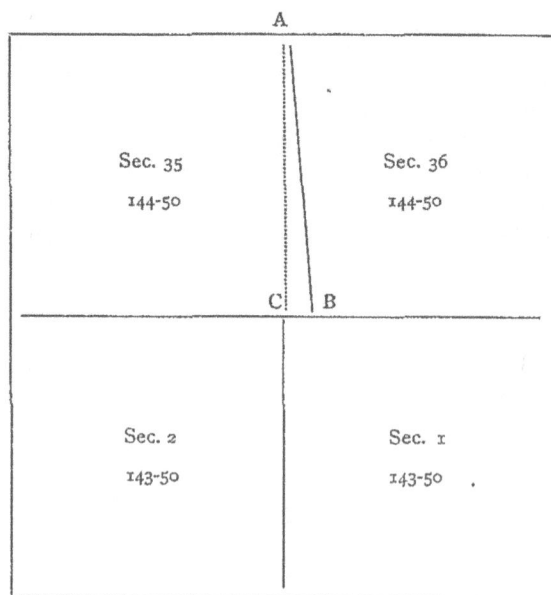

Sections 36 and 35 are, of course, on the south line of the township. It happens also, in this instance, that they are on the south line of the county; sections 1 and 2 directly south being in Cass County. It is conceded that point A is the original corner between said two sections on their North line, established by government survey. The contention is over the location of the original corner on their south line, both parties conceding that the original corner must govern. Plaintiff, who owns the west half of section 36, claims that the corner is at point C, while defendant Agnes C. Walker, who owns section 35, claims that it is point B. Defendants have occupied and cultivated the land up to line A– B, while plaintiff claims to the line A–C. The distance from B to C is 7 3/4 rods. The contest is over the triangular tract bounded by those lines.

Some errors are assigned upon the admission of testimony, but the rulings were clearly right, and the points raised involve nothing of general interest. The chief reliance of appellant is upon the assignment which raises the question of the sufficiency of the evidence to support the verdict. The controlling points in the evidence may be stated very briefly: Several parties testified that they had at different times assisted in measuring the distance along the south side of said section 36, commencing at the southeast corner (about the location of which there is no dispute,) and that the distance to point C was 320 rods, and that the south line of section 35, measuring from point B to west line, was 7 3/4 rods more than a mile. The original field notes of the government survey were introduced, showing the survey between townships 143 and 144 north, range 50 west, and from which it appears that, commencing at the point where the line between townships 143 and 144 intersects the line between ranges 50 and 49 (which would be the southeast corner of said section 36) the surveyor then ran west "on the true line between sections one and thirty-six" 80 chains (320 rods,) and "drove charred stake, and set post in mound, for corner to sections 1, 2, 35, and 36," and continuing west on line between sections 2 and 34, at the distance of 80 chains further, a like corner was established between sections 2, 3, 34, and 35. It was also shown that the highway which should be on the section line between sections 35 and 36 ran just east of the line A–B, but when it reached the south line of section 36 it turned directly west, and followed the east and west road to point C, and then ran south on the line between sections 1 and 2, but that there was no such offset at the township line in the highways on the section lines east or west. This was the testimony upon which the plaintiff relied. For the defense the appellant Thomas M. Walker testified that when he took possession of section 35, in 1880, the corner stakes were standing, and clearly visible, and were at point B; that in breaking up the land he established a straight line by means of intermediate stakes between the original government stakes at the southeast and northeast corners of said section 36, and broke the land, and has since occupied it with his co-defendant only up to the line so established. Another witness for the defense (Mr. Kenyon) testified that at one time he owned the northeast quarter of said section 2; that he took possession of it in 1878, and at that time the section corner stake at the northeast corner of his land (which would be the southeast corner of defendants' land) was standing; that it remained standing for some years, and was at point B: that the east line of his cultivated land was practically identical with the east line of the defendants' land. The witness also testified that in 1882 he was digging a ditch along there. The stake had at that time been burned away by a prairie fire, but the mound was still there, and to preserve the corner he buried some stones where the mound was, marking one with a cross. Plaintiff admitted that under the instructions of this witness he had uncovered the stones at the point where witness said he had buried them. Appellants insist that, as no witness went upon the stand to contradict either Mr. Walker or Mr. Kenyon, and as they each swore positively that the original corner was at point B, that any finding of a jury to the contrary must be set aside as without support in the testimony. But that is by no means true. The most direct and positive testimony may be completely demolished by circumstantial evidence, or overcome in the minds of a jury or the mind of a court by the establishment of other facts inconsistent therewith. This case is a good illustration. If the jury believed that it was only 320 rods from the southeast corner of section 36 to point C (and a number of witnesses so testify, and no one assumes to contradict it,) and if the jury also believed that the original section corner between said sections 1, 2, 35, and 36, was located 320 rods west from the southeast corner of 36 (and the original field notes so declare, and no one assumed to dispute the recital,) then the jury were bound to believe that the original stake had been moved after the corner was established (the survey was made more than 10 years before either witness claimed to have seen the stake,) or that the witnesses mistook something else for the original stake and mound. The two states of facts could not co-exist. There was strong evidence of the existence of each. It was the province of the jury to say which-did in fact exist, and their finding cannot be disturbed.

The court submitted two questions to the jury: (1) Was the original corner stake at point B? (2) Was the original corner stake at point C? Error is assigned in thus limiting the jury to those two points. It is urged that the jury should have been permitted to say where the original stake was. But it was not claimed by any one that it was at any point other than B or C. True, there was a slight discrepancy as to the exact location of point C, but it was too slight for the law's notice. Had the jury found that the corner was at a point other than B. or C, their finding could not have stood.

**_Black v. Walker_, 7 N.D. 414, 75 N.W. 787 (N.D. 1898).**

The parties to this action own adjoining farms, both situate in the same section, plaintiff's being the southwest quarter and defendant's the southeast quarter. The real dispute is as to the proper location of the boundary line dividing their farms. The tract in controversy consists of a narrow strip about 19 feet wide, and in quantity about one-half acre. Both parties agree upon the location of the original government corners marking the four corners of the section in which their land is located. Plaintiff fixes the beginning of the dividing line between them at a point on the south section line equidistant from the southeast and southwest corners of said section. The defendant, on the other hand, contends that the true line begins at a point on the south section line, about 19 feet west of the equidistant point where he undertakes to identify the location of the original quarter section corner established by the government surveyors. The plaintiff, alleging ownership in himself, sues defendant for damages for trespassing upon this tract. Defendant answers, likewise asserting ownership, and counterclaims for damages for alleged trespass. The jury returned a verdict for plaintiff. Defendant filed exceptions to the Court's charge, which, together with the refusal of certain requests, all of which are embodied in a settled statement of the case, he now urges as errors upon his appeal from the judgment. At the trial plaintiff rested his case for support upon the oral evidence of the county surveyor of Cass county, whose testimony, so far as it goes to aid in determining the proper location of this disputed line, and after satisfactorily identifying the section corners, which we have observed were not in dispute, is all contained in the following language: "I then placed a stake on the south section line half way between the southeast and southwest corners. * * * I did not find any quarter stake on the south line of the section." The defendant offered evidence tending to establish the exact location of the original quarter section corner. Defendant requested the following instruction: "The best evidence of the quarter section line between the lands in controversy is the original quarter section post or mound as placed there by the United States surveyors. The mark governs whether the survey was right or wrong. Hence, if you believe from the evidence that the quarter section post or mound was in existence, and was located where the defendant claims it was, then you must be governed by that line in deciding the controversy." The refusal of the foregoing is assigned as error. To this we cannot agree. The Court did charge as follows: "Now it is conceded that the land on both sides of this line originally came from the United States government, and I charge you, gentlemen of the jury, as the law of this case, that in a resurvey of the land which originally belonged to the United States, and which it has caused to be surveyed under its authority, such resurvey must conform to the survey made under the authority of the government, if the mounds and corners of the original government survey can be identified. If the stakes and monuments placed by the government in making the survey to indicate the section corners and quarter section posts can be found, or the places where they were originally placed can be identified, they are to control in all cases. Further, the corners established by the original surveyors under the authority of the United States cannot be altered. Whether properly placed or not, no error in placing them can be corrected by any surveyor deriving his authority from the laws of the state." Not only does that portion of the charge given and just quoted contain the substance of defendant's request, but as a correct statement of well-settled law it has our approval, and requires no citation of authority to support it.

The following was also excepted to: "The question of fact then, gentlemen of the jury, to be submitted to you is as to where the line between these two contending parties rest, bearing in mind that the line as fixed by the surveyor at this time is presumptively correct, and the burden of proof falls on the defendant to show that it is incorrect, and not according to the government survey." Here in the Court erred. The correctness of the line fixed by the county surveyor, so far as it was based upon courses and distances, and in the absence of the original quarter section corner, was not in dispute. The real issue was the existence or nonexistence of the original quarter section corner. This was a pure question of fact for the jury, and upon its determination on that one point confessedly the one party or the other would prevail on the question of ownership of the tract in dispute. By the instruction given, the jury were directed, in effect, not to weigh his evidence for what it was worth to aid them in determining the point in issue – that is, the location of the disputed corner and line – but to take his determination on that point as presumptively correct. This was wrong. We refer to his testimony in the language of the South Dakota Supreme Court in *Arneson v. Spawn,* 2 S.D. 269, 49 N.W. 1066, a similar case: "It was simply the evidence of a witness. Its probative value was not fixed by the law, as in the case of his official return, but went to the jury, like any other evidence." It is well settled that bounds and starting points are questions of fact to be determined by testimony, and surveyors have no more authority than other men to determine them upon their own notions. *Cronin v. Gore,* 38 Mich. 381; *Case v. Trapp,* 49 Mich. 59, 12 N.W. 908. Campbell, J., speaking for that Court in *Stewart v. Carleton,* 31 Mich. 270, said: "It appears to have been supposed that surveyors are competent not only to testify to measurements and distances, but also to pass judgment themselves, and on information of their own choosing, upon the position of lines and starting points. * * * This is a very dangerous error. The law recognizes them as useful assistants in doing the mechanical work of measurement and calculation, and it also allows such credit to their judgment as belongs to any experience which may give it value in cases where better means of information do not exist. But the determination of facts belongs exclusively to Courts and juries. Where a section line or other starting point actually exists is always a question of fact, and not of theory, and cannot be left to the opinion of an expert for final decision." The same judge, in the later case of *Gregory v. Knight,* 50 Mich. 61, 14 N.W. 700, expressed himself in language equally strong: "No one can be disturbed in his estate without the right to have his rights determined by legal process. * * * The law does not contemplate any such absurdity as the submission of question of title to land to anything but the judicial courts. It is quite manifest that in this case there seems to have been an idea that the exparte action of surveyors and commissioners was entitled to credence and authority upon the true lines and the facts of encroachment. Neither of them can affect vested rights or settled controversies. They may be useful witnesses when they speak of matters with which they are familiar, but they have no greater weight than any one else to determine starting points or boundaries." The trial court without doubt was led to give the instruction complained of by the language found in section 2028, Rev. Codes, which after generally providing for whom and in what manner county surveyors shall make surveys, adds that "his surveys shall be held as presumptively correct." The instruction complained of cannot be upheld in this case by the statute referred to for three reasons: First. It does not appear that the survey relied upon was made in conformity to the requirements of sections 2028-2034, Rev. Codes, inclusive, which provide for the making of field notes and plats, and what they shall contain; also for transcribing them into record books, if the county commissioners shall so require, as well as other matters of detail and for guidance. This is necessary, in our judgment, before the survey of a county surveyor as such becomes presumptively correct under the statute. In holding this view we agree with the interpretation placed upon the same statute in *Arneson v. Spawn* (S. D.) 2 S.D. 269, 49 N.W. 1066, wherein the Court said: "When section 689 (our section 2028) says, 'and his survey shall be held presumptively correct,' we think it means his surveys made, authenticated, and proved as provided by the statute." Second. The instruction given made the line established, and not the survey made, presumptively correct. The survey, so far as it related to courses, distances, and measurements, was not in dispute. The line was. The instruction not only gave undue weight to his testimony, which was relatively unimportant on the questions at issue, but gave not to his evidence alone, but to his determination of the line itself, the presumption of accuracy. See *Cockrell v. M'Quinn,* 4 T.B. Mon. 61; *Ott v. Soulard,* 9 Mo. 581; *Robinson v. White,* 42 Me. 209. Finally, we are of the opinion that the presumption of correctness which is extended to a survey made by a county surveyor under section 2028, Rev. Codes, refers not only to an official survey, in the sense of being made in compliance with the statute, but means courses, distances, variations, mathematical computation, and such acts as are done pursuant to the methods of an exact science, or as are provided for in the sections referred to, including field notes and plats, but does not include a determination of disputed boundaries or corners. The last, as already noticed, belongs to the courts. Other errors are urged. No useful purpose will be served by discussing them. That already considered is controlling.

    *Radford v. Johnson*, 8 N.D. 182, 77 N.W. 601 (N.D. 1898).

## OHIO

The issue in this case involves the location of the section line between Section 16 and Section 21 in Township 5, Range 20, Marion Township, Pike County, Ohio. The parties own adjoining parcels of land. Appellee owns a tract of land in Section 21. Appellants own a tract of land in Section 16, directly to the north of plaintiff's parcel. The ownership of 8.81 acres hinges upon the location of the section line between Section 16 and Section 21.

    Ronald P. Henry, a registered professional surveyor, testified for appellee. In 1986, Henry surveyed the property for appellee. Henry stated that the disputed line, the northern boundary line of Section 21 and the southern boundary line of Section 16, follows the centerline of Salem Cave Road. Henry testified the 8.81-acre tract, which lies south of the bend in Salem Cave Road, lies in Section 21.

    Henry relied upon an 1839 survey and plat of Section 21 made by the Pike County Surveyor and filed in the Pike County Engineer's Office and the survey and plat of Township 5, Range 20 filed in the State Auditor's Land Office. He based his opinion as to the location of the line on the presence of the road, a very old fence line along the road, and the two documents. Henry testified that the documents indicate the western boundary line of Section 21 is longer than the western boundary line of Section 16 and that the acreage in Section 21 is greater than the acreage in Section 16. Henry used overlays of the old surveys to show how the surveys correspond to the present-day location of Salem Cave Road. Henry noted that a 1927 map contained the earliest reference to Salem Cave Road.

    During his direct examination, Henry recited from Clark, Surveying and Boundaries, as follows:

    "The original survey must govern if it can be retraced. It must not be disregarded. In making a resurvey, it is a surveyor's duty to relocate the original lines and corners of the place it's actually established and not to run independent new lines even though the original lines were full of errors. We have a simpler version of it in that the following surveyor walks in the steps of the original surveyor."

Dale Exline, a registered professional surveyor, testified for appellants. He stated that the 8.81-acre tract lies within Section 16. Exline testified that the line between Section 21 and Section 16 is a straight line which deviates from Salem Cave Road as the road angles north. Exline based his opinion on his review of county road maps, USGS topographical maps, a field survey he conducted, the iron pin sets he found in the ground on that line, and his observation that section lines are usually straight lines.

On May 20, 1991, the court conducted a bench trial. The court's July 23, 1991 judgment entry provides as follows:

"1. Count Two of plaintiff's complaint essentially demands quiet title to and seeks recovery of the real estate described as part of Exhibit 'A'; ie 8.81 acres. See Pl.Ex. 1.

"2. The parties agree that a primary issue of fact is the location of the section line in the disputed area, between Sections 21 and 16 of Township 5, Range 21, Marion Township, Pike County. Plaintiff also claims the disputed area by adverse possession.

"3. The two surveyors, Henry and Exline, disagree upon the location of the Section line.

"Henry, relying upon old plats and surveys, states that the Section line actually changes direction, falling along Salem Cave Road in the disputed area. See Pl.Ex. 4, County Engineer's Plat Records. He also points out that State Land Office records show Section 21 containing more acreage (652 + acres), Section 16 (588 + acres), thus reinforcing his opinion. See Pl.Ex. 3. The overlays of Pl.Ex. 5 (aerial photo) and 5A (geological survey map) illustrate his opinion.

"Exline states that the easterly projection of the section line between adjoining Sections 17 and 20 provides the location for the Section line in the disputed area. See Def.Ex. 1 and 2., and Pl.Ex. 5A. He points out that his line is straight and extends thru two iron pins he found in a fence line. Exline states that the Congress Lands were surveyed in uniform Sections whose boundaries were straight lines and that did not change direction.

" * * *

"The Court is persuaded by a preponderance of the evidence, that the Henry Survey (Pl.Ex. 1) accurately locates the boundary line as lying along the centerline of Salem Cave Road in the disputed area. The Court is not persuaded by a preponderance of the evidence that plaintiff has established her claim of adverse possession of the disputed area.

"Therefore, it is the order of the Court that title to the disputed area shown as 8.1 acres on Pl.Ex. 1 (see attached) be quieted in plaintiff; and plaintiff is hereby awarded the right to peaceful possession, use and enjoyment of said property to the exclusion of the above named defendants. Upon request of plaintiff, the Clerk shall cause a writ of possession to issue to the Sheriff of Pike County, Ohio, to remove defendants and place plaintiff in possession of the described property.

"Upon motion of defendants made at the close of plaintiff's case-in-chief, the Court dismissed plaintiff's claim for damages."

Appellant filed a timely notice of appeal.

We note that on April 17, 1992, appellants Elmer Dillow, Louise Dillow, Gary Swanson, and Joyce Swanson filed a document captioned "exhibit." The "exhibit" states as follows:

"The Defendants feel their Surveyor, Dale A. Exline, has made the correct survey and plat of Section 16 and 21 in the disputed area.

"However, if the Appeal Court feel they don't have enough information to over-rule the Judgment in this case, they feel the case should be sent back to the Court of Common Pleas, Pike County, Ohio, to the Judge in said case. If not, they feel it should be sent to the County Commissioners to have them direct the County Engineer to locate the section lines as prescribed under O.R.C. 315.28-5553.06.

"The Defendants feel that if this is accomplished, the problems that they are experiencing will then be rectified."

Attached to the "exhibit" is a June 24, 1991 letter addressed to Elmer Dillow from Charles J. Neff, P.E., P.S., Chairman of the Ohio State Board of Registration for Professional Engineers and Surveyors. The letter states:

"At the June 13, 1991 Board meeting, the Board determined that with the information available to them at this time they could not determine if either surveyor had made a mistake.

"The Board recommends that you contact the Judge on this case or the County Commissioners and have them direct that the County Engineer locate the section lines as prescribed under the O.R.C. Section numbers 315.28, 5553.06. The Board feels that if this is accomplished, the problems that you are experiencing could then be rectified."

App.R. 9(A) and 12(A)(1)(b) limit a reviewing court's consideration to "original papers and exhibits thereto filed in the trial court, the transcript of proceedings, if any, including exhibits, and a certified copy of the docket and journal entries prepared by **[621 N.E.2d 423]** the clerk of the trial court * * *." App.R. 9(A). App.R. 12 provides in pertinent part:

"(1) On an undismissed appeal from a trial court, a court of appeals shall do all of the following:

"(a) Review and affirm, modify, or reverse the judgment or final order appealed;

"(b) Determine the appeal on its merits on the assignments of error set forth in the briefs under App.R. 16, the record on Appeal under App.R. 9 and, unless waived, the oral argument under App.R. 21."

In *State v. Johnston* (1990), 64 Ohio App.3d 238, 246, 580 N.E.2d 1162, 1168, quoting *State v. Ishmail* (1978), 54 Ohio St.2d 402, 8 O.O.3d 405, 377 N.E.2d 500, the court wrote:

"It is well established that '[a] reviewing court cannot add matter to the record before it, which was not a part of the trial court's proceedings, and then decide the appeal on the basis of the new matter.'" See, also, *State v. Callihan* (1992), 80 Ohio App.3d 184, 608 N.E.2d 1136.

We again note appellants filed this material on April 17, 1992. Obviously, the trial court did not consider this information prior to issuing its July 23, 1991 judgment entry. Therefore, because an appellate court cannot consider material which was not part of the trial court's proceeding, we may not consider this material as it is outside the record.

Because appellant's first and second assignments of error raise related issues, we will consider them jointly. In their first assignment of error, appellants assert the iron pin sets found by Exline constitute monuments and, therefore, should prevail as a boundary marker over Salem Cave Road. Appellee argues that Salem Cave Road, which was relied upon by appellee's surveyor, is also a monument. Appellee maintains that when determining the section line, the road and the iron pin sets are equal in significance. In their second assignment of error, appellants assert the trial court's finding is against the manifest weight of the evidence.

The Ohio Supreme Court has stated that when determining boundary disputes:

"It is well settled that monuments are of prime importance in settling boundary disputes. The general rule is well stated in 6 Thompson on Real Property (Perm.Ed.), 519 Section 3327, as follows:

"'A "monument" is a tangible landmark, and monuments, as a general rule, prevail over courses and distances for the purpose of determining the location of a boundary, even though this means either the shortening or lengthening of distance * * *. Generally, in determining boundaries, natural and permanent monuments are the most satisfactory evidence and control all other means of description, in the absence of which the following calls are resorted to, and generally in the order stated: First, natural boundaries; second, artificial marks; third, adjacent boundaries; fourth, course and distance, course controlling distance, or distance course, according to circumstances. Area is the weakest of all means of description. The ground of the rule is that mistakes are deemed more likely to occur with regard to courses and distances than in regard to objects which are visible and permanent. The reason assigned for this rule is that monuments are considered more reliable evidence than courses and distances.'

"* * *

"It is also settled that a public street or road or the boundary line of other property may be used as a monument." *Broadsword v. Kauer* (1954), 161 Ohio St. 524, 53 O.O. 395, 120 N.E.2d 111. See, also, *Haverstick v. Beaver (App.*1941), 34 Ohio Law Abs. 363, 37 N.E.2d 650; *Kramp v. Toledo Edison Co. (*1961), 114 Ohio App. 9, 18-19, 18 O.O.2d 298, 180 N.E.2d 172, 178; *Cardosi v. Wise (*1933), 44 Ohio App. 205, 209, 14 Ohio Law Abs. 698, 184 N.E. 863, 864; *Briggs v. Jeffers* (1965), 7 Ohio Misc. 44, 36 O.O.2d 36, 215 N.E.2d 622; and 2 McDermott, Ohio Real Property Law and Practice (3 Ed.1966) 126, Section 3-21G.

In *Sellman v. Schaaf (*1971), 26 Ohio App.2d 35, 55 O.O.2d 69, 269 N.E.2d 60, the court discussed resurveys and the function of the second surveyor:

"* * * When an original survey has been made, it is not the plat or the metes and bounds description that is primary. The primary function of the second surveyor is to find first where the boundaries were established by the first surveyor. Only where this becomes impossible of accomplishment does the second survey turn to the courses, distances, and still-existent monuments to determine the boundaries. The essential rule governing the resurvey is to follow the steps of the first surveyor.

"McDermott, supra, at 139, paragraph 3-26A states:

"'Conveyances are presumed to be made according to a prior actual survey. It is said that the primary purpose of construction is to follow the footsteps of the surveyor on the ground.'

"* * *

"In Clark, Surveying and Boundaries (2d Ed.1939), it is said at page 727, Section 665:

"'The original survey must govern if it can be retraced. It must not be disregarded. So, too, the places where the corners were located, right or wrong, govern, if they can be found. * * * In making a resurvey it is the surveyor's duty to relocate the original lines and corners at the places actually established and not to run independent new lines, even though the original lines were full of errors.'"

In Moffitt & Bouchard, Surveying (8 Ed.1987), Sections 18-16, 18-19, and 18-24, the authors discussed the methods used in the early surveys:

"The marking of the survey lines on the ground was supposed to be done in such a manner as to perpetuate the lines. This was done by monumenting the regular corners, by recording in the field notes all natural topographic features, and, whenever living timber was encountered, by means of blazing and hack marks on the trees.

"* * *

"Unfortunately for the present-day surveyor, the monuments used on the early surveys were not permanent in character and forest fires and settlers have destroyed most of the timber, so that the retracing of the old lines is often difficult and in many instances impossible.

"* * *

"The law provides that the original corners established during the progress of the survey shall forever remain fixed in position, and that even evident errors in the execution of the survey must be disregarded where these errors were undetected prior to the sale of the lands. The original monuments thus assume extreme importance in the location of land boundaries. Unfortunately, most of the public lands were surveyed before the present day regulations relative to the character of monuments went into effect, and as a consequence most of the monuments used were of a very perishable nature. Their disappearance or destruction has rendered the relocation of old lines a very difficult task.

"* * *

"In the original survey it was presumed that permanent monuments were being carefully established and witnessed, so that long lines of monuments would be perfect guides to the place of any one that chanced to be missing. Unfortunately, the "monuments" were often nothing but green sticks driven into the ground, lines were carelessly run, monuments were inaccurately placed, witnesses were wanting in permanence, and recorded courses and lengths were incorrect. As the early settlers made little effort to perpetuate either the corner monuments or the witnesses, the task of the present-day surveyor in reconciling much conflicting evidence (or in even finding any kind of evidence) is often an extremely difficult task. * * * It must be remembered that no matter how erroneously the original work is done, lines and monuments that can be identified still govern the land boundaries."

We note that in 2 History of the State of Ohio (1942) 136, the author discussed the difficulties relating to inaccurate land descriptions:

"The principal reason for disputes over land titles was the inaccurate measurement or description of land on the part of the early surveyors. No state displays a greater variety of surveying methods than does Ohio; it was the proving ground for the varied modifications of the rectangular system, and within the Virginia Military District the so-called 'indiscriminate survey' reached its ultimate absurdity, although Kentucky might lodge a counter-claim for this doubtful distinction."

When reviewing evidence presented at trial, an appellate court must not reweigh the evidence. In *C.E. Morris Co. v. Foley Constr. Co.* (1978), 54 Ohio St.2d 279, 8 O.O.3d 261, 376 N.E.2d 578, syllabus, the Ohio Supreme Court held:

"Judgments supported by some competent, credible evidence going to all the essential elements of the case will not be reversed by a reviewing court as being against the manifest weight of the evidence." See, also, *Vogel v. Wells (1991)*, 57 Ohio St.3d 91, 566 N.E.2d 154; *Ross v. Ross (1980)*, 64 Ohio St.2d 203, 18 O.O.3d 414, 414 N.E.2d 426.

An appellate court should not substitute its judgment for that of the trial court when there exists competent, credible evidence going to all the essential elements of the case. In *Seasons Coal Co. v. Cleveland (1984)*, 10 Ohio St.3d 77, 10 OBR 408, 410, 461 N.E.2d 1273, the court wrote:

"The underlying rationale of giving deference to the findings of the trial court rests with the knowledge that the trial judge is best able to view the witnesses and observe their demeanor, gestures and voice inflections, and use these observations in weighing the credibility of the proffered testimony."

In the instant case, after hearing both expert witnesses testify and after reviewing the exhibits, the trial court determined that Henry's survey accurately depicts the location of the section line between Section 21 and Section 16 as it appears in the original survey. In its decision, the trial court emphasized that Henry's use of the old plats, surveys, and the State Land Office Records were persuasive.

After a review of the evidence presented below, we find sufficient competent, credible evidence to support the trial court's judgment. The primary function of the second surveyor is to find the boundaries established by the first surveyor. The trial court found, by a preponderance of the evidence, that appellee's surveyor located the original section line. We will not substitute our judgment for that of the trial court.

Appellants assert the iron pins relied upon by Exline constitute artificial monuments and require the conclusion that Exline's survey accurately depicts the location of the section line. We disagree. Although we agree with appellants' proposition of law regarding the general rule that monuments are of prime importance in settling boundary disputes, we find no link established between the iron pins found by Exline and the original survey. Exline testified that section lines are usually straight lines and that, in his opinion, the location of the disputed section line coincides with the projection of the section line of adjoining sections. Again, we find no evidence the iron pins were used in the original survey.

We also note that appellee contends Salem Cave Road constitutes an artificial monument. We disagree with appellee. The evidence reveals that Salem Cave Road did not exist at the time of the original survey. Henry testified the earliest reference to the road is contained in a 1927 map. Thus, if the road did not exist at the time of the original survey, we question whether the road can be deemed to be a monument for purposes of the original survey. Henry's testimony did establish, however, that the disputed section line is located at the present location of Salem Cave Road.

**Sanders v. Webb, 85 Ohio App.3d 674, 621 N.E.2d 420 (Ohio App. 4 Dist. 1993).**

## OKLAHOMA

This action was brought by Plaintiffs-Appellants pursuant to 11 O.S.1971, § 530, to correct the plat of Carr Acres Third Addition (Carr Third) to the Town of Newcastle, McClain County, Oklahoma, which plat Appellants contended contained an error in the legal description. The Appellants all own lots within the addition, and the Appellees own all of the remaining lots in said addition, except the Appellee, Town of Newcastle, which was made a party pursuant to 11 O.S.1971, § 531, since the correction would affect a public street, and except the Appellees, J. L. Carr, Jr., individually and as Administrator of the Estate of James L. Carr, deceased, and of Norma F. Carr, deceased, Bruce Carr and Margaret Griffiths, who are the heirs at law and legal representatives of the original platters of said addition, who are now both deceased.

Appellee, Town of Newcastle, filed its answer admitting the allegations of the Appellants' petition and joined the Appellants in requesting that the plat of Carr Third be corrected.

Appellees, J. L. Carr, Jr., individually and as Administrator of the Estate of James L. Carr, deceased, and of Norma F. Carr, deceased, Bruce Carr and Margaret Griffiths, filed an answer neither admitting nor denying the ownership of lots within Carr Third (Appellees later in Answers to Request for Admissions specifically admitted such ownership), and specifically denying that there was error in the plat of Carr Third.

In addition, Appellees filed a Cross-Petition containing three causes of action. The first cause of action was for reasonable damages against the owners of Lots 15 through 28, Carr Third, for wrongful possession and taking of lands belonging to Appellees. The second and third causes of action sought to cancel and/or reform a certain correction deed executed and delivered to Appellee, Eugene C. Dobbs, also known as Gene Dobbs.

Appellants' Reply and Answer to the Cross-Petition of Appellees alleged that the titles of the Appellants and Appellee-owners were all deraigned from warranty deeds given by James L. Carr and Norma F. Carr, and that by reason of the warranties of title contained in said deeds the Appellees were legally and equitably estopped to claim any interest in the property of the Appellants and other Appellee-owners.

The following facts appear to be undisputed. On or about April 23, 1971, James L. Carr and Norma F. Carr, both now deceased, dedicated and platted Carr Third, as shown by the recorded plat thereof in Book 368, page 413, of the public records of McClain County, Oklahoma. The lots and streets of Carr Third, as constructed, developed and laid out on the surface, do not correspond to the location shown on the recorded plat. There is no vacant 160-foot strip of ground between Carr Third and Carr Acres Second Addition, notwithstanding that the recorded plats reflect existence of such a strip. The plat of Carr Third overlaps the plat of Carr Acres Fourth Addition. There is no satisfactory way to correct the physical location of the lots and streets of Carr Third to correspond to the recorded plat thereof since there is no vacant land available on either the east or west side thereof.

Prior to trial, Appellants filed a Motion for Judgment on Uncontroverted Facts, supported with an Affidavit of Roger Mayes, the County Surveyor of McClain County, stating he had resurveyed the area covered by the plat of Carr Third and determined the recorded plat did not correspond to the location of the addition as actually developed and laid out on the surface of the ground. He further stated that the plat of Carr Third described the point of beginning at a point 661.75

feet South and 1160 feet West of the Northeast Corner of the NW 1/4 of Section 15, Township 9 North, Range 4 West, but, in truth and in fact, the point of beginning of said addition is located at a point 661.75 feet South and 1000 feet West of the Northeast corner of the NE 1/4 of Section 15. Affiant stated that in the process of his resurvey he was able to locate some surveying pins that were already in place and which corresponded to his findings, and at the point of 1000 feet section line is located a surveying pin which had a red flag upon it. Based upon the color of the flag, he stated it had been there for some period of time, but he had no personal knowledge as to who placed it there, other than it was a surveying pin. Affiant stated that if the court would order the recorded plat corrected by this 160 feet it would correspond to the location of the addition as developed and constructed on the surface and that the plat would then not overlap Carr Acres Fourth Addition. It was Affiant's opinion that the erroneous call in the plat resulted from either a scrivener's error or from a mismeasurement of the actual distance on the ground.

Both parties submitted briefs to the court. The day prior to trial, November 3, 1976, the Trial Court overruled Appellants' Motion for Judgment on Uncontroverted Facts.

The matter came on for trial November 4, 1976. Appellants offered the testimony of three witnesses, including Roger Mayes, whose testimony was not completely taken, but was stipulated to by the parties and followed generally the statements made in his affidavit.

Loretta Fancher, Town Clerk of the Town of Newcastle, Oklahoma, testified she was familiar with the location of the houses in Carr Acres Second and Third Additions and that there was no vacant 160-foot strip between the additions as laid out on the surface. The same testimony was given by Bobby Joe Scarberry, owner of Lot 3 in Carr Acres Third Addition. Appellees offered no evidence, but demurred to the evidence of the Appellants. The Trial Court sustained the demurrer on the basis that it was not established that the plat was erroneously placed at 1160 feet, and stated:

"Alright, gentlemen, this is what I believe is going to happen in this matter. The plat will be reformed. There is no question in my mind but what that plat will be reformed. I don't believe that I have received enough evidence to do it either under that statute or in equity. I believe that enough evidence is available to do it in equity, and I don't believe that any of you all have got the witnesses here to establish an equity case today, and, I don't believe any of you have got witnesses here to establish values and that is why I don't believe you are prepared to proceed on the damages issue and I know that I am not prepared."

The Trial Court made no ruling on the Cross-Petition, but suggested that to expedite the matter on appeal the case should be tried independently of the statute, 11 O.S.1971, § 530, infra. No motion for new trial was filed. This appeal followed.

Subsequent to the filing of Appellants' Petition in Error in this Court, several of the Appellees, Billy J. Stow and Mary R. Stow, husband and wife, and Appellee, Town of Newcastle, Oklahoma, filed Confessions of Error in this Court, confessing error by the Trial Court and consenting to a reversal by this Court and entry of a judgment correcting the plat of Carr Third, or remand for a new trial.

On appeal, Appellants urge that the Trial Court erred in overruling their Motion for Judgment on Uncontroverted Facts, erred in sustaining Appellees' demurrer to the evidence of the Appellants, and that the Trial Court's judgment is contrary to law.

Appellants brought their action pursuant to 11 O.S.1971, § 530, which provides as follows:

"The district court shall have the jurisdiction to correct city and town plats and plats of additions thereto. The owner or owners of any lot or lots within the portion of the plat sought to be corrected of any city or town plat or the plat of an addition thereto, may file his petition in the district court of the county where the land is situated to correct such plat, or any portion thereof, when the same has been erroneously made by duplicating lot numbers in any block or incorrectly describing the distances on said plat or when the same is, in any manner, incorrect in description or otherwise. And the court in correcting the same, may correct the description of property in any conveyance of a lot or lots, where the plat is corrected, which may be necessary for the purpose of making a complete and correct descriptive chain of title to the lot, or lots, the description of which is corrected."

This statute was enacted in 1915 and has never been interpreted by this Court.

It appears the Trial Court adopted the theory that because Appellants were relying on the above quoted statute it was to be strictly construed against them; that the burden of proof upon Appellants was that they must prove it was the intention of the platters to develop the real property at the location where it was actually laid out and constructed on the surface of the ground, rather than 160 feet west of that location, as reflected by the recorded plat. On appeal, Appellants contend the only burden imposed by Section 530, supra, is to show that the dimension shown on the plat is not correct as it relates to the actual distance measured on the surface of the ground, unless Appellees can show some legal or equitable reason that the plat should not be corrected.

On the other hand, Appellees argue that Section 530, supra, operates only when an error has occurred in the plat and that nothing in the record of this case exclusively establishes this.

Appellants argue that it is the actual location of the boundary lines as established on the ground itself by the survey that is primary, that the plat is merely derivative and secondary, and in case of conflict between the two the existing monuments on the ground control. In support of this position, Appellants rely on *Roetzel v. Rusch,* 172 Okl. 465, 45 P.2d 518 (1935). Appellants do not contend *Roetzel,* supra, is directly applicable on the facts, but that the case manifests the proposition that the plat is secondary and the survey or existing monuments are primary and control. That case states in the fourth syllabus:

"4. Where a deed describes property by reference to plat or map, grantee takes title in accordance with the boundary so identified; but where discrepancy exists between call in deed for monuments and courses and distances shown by plat referred to, general rule that monuments control courses and distances will be followed."

Appellees attempt to distinguish Roetzel, supra, as involving a discrepancy between references to monuments within the written conveyance and a plat, urging that the discrepancy in the instant case is between the plat and whether construction was the fault of the platters or otherwise. However, it is to be noted that Appellees cite no authorities contrary to those relied on by Appellants.

While Appellants point out that they have been unable to find any specific Oklahoma authority, they urge that other jurisdictions have adopted the fundamental rule that where a survey as made and marked upon the ground conflicts with the plat, the survey must prevail, citing *Arnold v. Hanson,* 91 Cal.App.2d 15, 204 P.2d 97 (1949) and *Sellman v. Schaaf,* 26 Ohio App.2d 35, 269 N.E.2d 60 (1971).

In *Sellman,* supra, the Court said the subdivision is first done by the survey of the premises, establishing monuments, etc., so that a physical or semi-physical dividing of the land with the attendant markings takes place. Then a plat is made and the symbolic representation of what was done on the premises becomes the recorded documentation of the action taken and provision is made for the dedication and acceptance of public streets and ways. The Court posed the question: "What is the situation where the actual location of a corner by a surveyor in setting up a subdivision differs from the description or plat?"

The Court then discussed numerous authorities before arriving at the following conclusions set forth in syllabi 1 and 2:

"1. Where a survey of a parcel of land has been made by a surveyor, monuments placed or ascertained, and boundary lines established by such monuments and a plat is made thereafter and recorded which subdivides the land so surveyed into lots, the boundary lines of such lots as so established on the ground itself are primary. The plat is derivative and secondary.

"2. Where the original monuments as located by such surveyor are still ascertainable, the boundary lines determined by such monuments will determine the boundaries of the respective lots irrespective of deviation from the course or distance as set forth in the plat." (Emphasis ours.)

In Clark, *Surveying and Boundaries* (3rd Ed. 1959), it is said at page 534, Section 540:

"§ 540. Survey rather than plat fixes location of boundary. Where a plat delineates an actual survey, the survey rather than the plat fixes the location and the boundaries of the land. The plat is a picture, the survey the substance. In a conveyance referring to such plat, the lot bounded by the lines actually run upon the ground is the lot intended to be conveyed. The plat may be all wrong, but that does not matter if the actual survey can be shown. Thus, where there is a dispute as to the boundary line between a street and the abutting lots, the original survey will control the recorded plat. Where a surveyor of the land marks the division lines on the ground by monuments, such lines control calls and distances indicated on his map. * * *"

It is to be noted that the water line plan of Carr Acres Addition, Second and Third, prepared by the surveyor who made and executed the plat in question, and which purports to show the water line distribution system for the three additions, was admitted into evidence as being an official record of the City of Newcastle. It reflects that the three additions were to be located next to each other with no indication of a 160-foot strip between Carr Second and Third Additions.

In addition, the probate proceedings of the Estate of James L. Carr, Deceased, were offered into evidence and the Court took judicial notice thereof. This evidence consisted of an application by the Administrator of the estate and an order of the Trial Judge authorizing the granting of correction deeds covering tracts in Carr Fourth Addition, which was over-lapped by the plat of Carr Third. Appellants urge the granting of correction deeds further supports their position.

*Scarberry v. Carr*, **571 P.2d 458 (Okla. 1977).**

## OREGON

This is a suit in equity to determine the boundary lines between the lands of the plaintiff and the defendants, under the provisions of the act of 1887 relating to that subject. Code Or. § 506. The plaintiff is the owner in fee of the donation land claim of Solomon Eades, and claims that the eastern boundary of said claim is a straight line from its south-eastern corner to its north-eastern corner, as shown by the government survey, and that such boundary, as thus drawn and described, is the west boundary of the Chapman claim and the Lynch claim, now owned by the defendant Berry. The defendants claim that the line in dispute was surveyed and located in 1852 by the government, and that Eades and Chapman built a partition fence between their respective claims on said line, and that a continuation of said line southerly in the same course to the south-east corner of said Eades claim properly describes the line between the said Eades claim and the lands of the defendant Berry. A good deal of evidence was taken, and the court below, after hearing the respective arguments of counsel, made its findings, and rendered a decree thereon in favor of the plaintiff, and in accordance with the requirements of said act. To reverse this decree, and establish the boundary line as claimed by the defendants, is the object of this appeal. We shall not attempt to go into the details of the evidence, or to cite largely from it; for that would only incumber the record, without serving any useful purpose. As the case stands, it will probably be as well to consider first the grounds and the evidence upon which the defendants claim the true boundary line to be as alleged. It is admitted that the Eades donation claim is the older of these donation claims, and that those of the defendants are adjoining, and, necessarily, that the east line of the Eades donation claim must be the west line of the Chapman donation claim, and the land of Berry, as now owned by him. The contention of Mr. Chapman, one of the defendants, is based on his own evidence, and some other, corroborative in certain particulars, to the effect that after the first government survey, and when the corners were known, that he and Mr. Eades built a partition fence on the line thus located; and that, when the final survey was made by the government, the fence erected by them corresponded to the line then established as the east line of the Eades donation claim, and the west line of his donation claim. In 1862, Eades died, and there is nothing in the record to show that he was dissatisfied with the line, or that the final survey required any change of the fence built by them to be made. In 1864, Mr. Chapman testifies that he removed this partition fence 30 feet east on his land for the purpose of opening a road, and that it so remained until about 1876, when he moved it back to the original line upon which he and Mr. Eades had built the partition fence. His case therefore is that the fence, as removed in 1876, represents the true boundary line between his lands and the plaintiff's; and this line, protracted south to the south-east corner of the Eades donation claim, must be the true division line between the plaintiff and the defendant Bewley And he says, in respect to the building of this fence in 1858, that it was done by Mr. Eades and himself, and that at that time the stakes or monuments set up by the government were plainly visible, and that they set up a stake, with a flag upon it, at the north-east corner and the south-east corner; and, guided by these, Eades built the north end, and that he built the south end. These facts he claims are sufficient to indicate with certainty where the line between these adjacent claims is, when backed by other evidence of facts which preserve its identity. But Eades is dead, and there is no other person who testifies to the location of that fence, or knows whether it was correctly established on the division line between these claims. His wife is cognizant of the fact of the building of the fence in 1858, and testifies to his removal of it 30 feet east on his land in 1864, and its removal back to the original line in 1876, and that there never was any dispute between Mr. Eades and her husband in respect to it. And one of the sons of Mr. Eades testifies that he recollects the fact of the defendant Chapman's moving the fence east 30 feet in 1864 for the purpose of a road, and that the defendant Chapman and his father built the original fence from the section corners. There are also several witnesses who testify that they assisted in removing the fence in 1876, from the place where it was then located, to the original line, and that such line was distinctly traced on the ground by ground-chunks, rose-bushes, and briers, that still grew where the old fence originally stood. And Faulkener and Burden, who both became subsequently the owners of the Eades claim, corroborate, in some measure, the existence of such a line, although Faulkener says: "We just moved on the line where we supposed the line was." And, again: "It looked as if it was the same." And Burden says: "It was what was called the line;" and there were "some rose-bushes on the line," and a "few ground-chunks."

The fact that the fence was removed, in 1876, west, to a line that indicated by some ground-chunks and rose-bushes, the traces of an old fence is pretty well established; but whether this was the true division line or boundary, or whether it was the line on which Eades and Chapman had built their partition fence, is not clear, nor is it proved by these witnesses other than as before stated. That rests almost wholly upon the statement of the defendant Chapman; and the circumstances under which he narrates the building of that fence by Eades and himself are not favorable to exactness or accuracy in fixing the boundary between them, although it necessarily approximates to it, as the difference now, and as found by the surveyors, is not great. Taking this evidence as true, it only indicates that these neighbors, with adjoining lands, put up the fence as near as they could ascertain, or at least upon which they supposed was the east line of the Eades donation claim, or their division line. There was no disputed boundary between them, or disagreement of any character. They simply put up the fence on the land as already stated, and lived up to it, on the assumption that it was the true boundary; neither claiming anything from the other, and consequently nothing to acquiesce in. There are, however, some facts, independent of the evidence of several witnesses of the plaintiff to a state of facts in contradiction of the testimony of the defendant, that tend to establish quite strongly the correctness of the line as fixed by the court below. The Eades claim is the elder, and the east line of it, which is the boundary between these adjacent claims, must have been established by the survey before the Chapman claim may be said to have had an existence. Now, by the field-notes and patent, the east line of the Eades claim, which forms the boundary in question, is a straight line, while the line insisted upon by the evidence of the defendant would be deflected from a straight line. The field-notes put the south-east corner of the Eades donation claim in the line that divides or bounds the Sparks & Lynch donation claims, and the evidence shows that the corner, as fixed by the surveyor, Mr. Fenton, substantially agrees with the old fence on the line between these last-mentioned claims, which has stood for many years, and conforms to the line as called for by the survey, and makes the lines connect without deflection. Besides, it was said at the argument, and not contradicted, that to run the east line as now established, and according to the field-notes of the survey, gives, as it must, the proper quota of acres to the Eades and the Chapman donation claims; while to fix it as contended for by the defendants would make one of these claims less, and the other more, acreage than is called for. Moreover, when the situation and location of these adjacent claims are considered, the lines by which they are bounded as fixed by the government survey, and that the south-east corner of the Eades donation claim is a common point in the line between the Sparks and Lynch donation claims, it becomes apparent that to change this disputed line, or the south-east corner of the Eades donation claim, from the place as located in the decree, must disturb these lines, and result in the disarrangement of boundaries of several claims, and, perhaps, a confusion of titles. While it is true, in some particulars, there was some disagreement among the surveyors, they are for all practical purposes a unit as to the location of the south-west corner of the Eades donation claim; and this is in conformity, or substantially so, with the evidence of Lynch and Sparks, and there is much other corroborative testimony by several witnesses introduced by the plaintiff. It may be also said that there is considerable evidence contradicting the fact that Chapman ever moved the fence east, and which goes to negative his contention; yet much of the testimony introduced by Chapman in that particular is only valuable in an auxiliary way, and upon the assumption that the original fence, claimed to have been built by Eades and Chapman, was on the true division line. His evidence to this point is not convincing, in view of circumstances inherent in the case which overcome and outweigh it, although we may admit the building of such original fence as testified to, and the honest endeavor to fix it on the true line bounding these adjacent claims. To sum up the whole, the result of our consideration of the case is that the decree is in conformity with the evidence and the facts, and that this conclusion is in substantial conformity with the evidence of several witnesses whose long residence in the neighborhood, and opportunity to know, give additional weight and value to their testimony. We think the decree must be affirmed.

*Bewley v. Chapman*, **16 Or. 402, 18 P. 849 (Or. 1888)**

## PENNSYLVANIA

No suitable case found.

## RHODE ISLAND

No suitable case found.

## SOUTH CAROLINA

Where, in an action of trespass to try title, the Court thought the finding of the jury more conformable to the rules of location as applied to the facts of the case, than any other which they could prescribe, they refused to set aside the verdict.

The end arrived at by all the rules of location, is to establish for the plaintiff in an action to try the title, the lines run or described in the grant he produces, when such a muniment of title is produced elder and therefore paramount, so far as the evidence may permit, and no law or adverse proof may overrule; but when a plaintiff comes into Court with a grant and plat, which, though older, is surrounded by obscurity, arising either from the fact, that the surveyor platted by description or from any other carelessness, he is not entitled to demand that the Court shall push to a fanciful extent, or strain inordinately the idea, sometimes indulged, that the elder grant is to be more favorably located.

The case was narrowed into a question of location, which may, perhaps, be understood by reference to the annexed diagram. The small parcel on which the trespass was alleged to have been committed, is shown by D.

N

Deft's. 275.

L
E
K

B    Williams.    A    P
M                 Mathews.

R.

Kershaw.  F    C
G                H

Plaintiff's.

The plaintiff derived title from the Sumter grant, for 1500 acres; the defendant from the Burns grant, for 275 acres, which was of later date, and called for the Sumter land as a boundary.

These papers were in evidence:

The Sumter grant called for the Williams land, previously granted, as the boundary of one line, and for the Kershaw land as lying within the angle made by the two next lines on the West; the shape represented, being such as is made by lines A B H G with right angles at B and H; and no vacant land being called for between Williams and Kershaw, although the boundaries around every part of the Sumter grant are specially designated, and vacant land is called for in several places.

The Williams grant was introduced, and covers the triangle A B N. The Kershaw grant was not in evidence, but if there was such a grant, some or all the land embraced by it, must have been again embraced by the Burns grant. The plat annexed to the Sumter grant, represents the lines as if they had been run from A toward G, with the land on the left. It shows no interval between the Williams land and the Mathews land next on the east, such as is now really found between A and P. Beginning at A and running on the Williams line, the distance required by the Sumter grant stops at M. There is, however, no doubt that B is the true west corner of the Williams grant, and that G is a true pine corner of the Sumter grant. From some remote part of the Mathews land to G, the Sumter land was probably not actually surveyed before the grant, as open corners without any special designation are represented on the plat. The lines of the Burns plat are represented by B H G. From G toward F, are found marks on trees older than the Sumter grant, probably made for the Kershaw land. The representation which the Sumter grant makes, of the course of the line bounded by the Williams land, departs ten degrees from the true course; and there is the same difference between the course called for by the Sumter grant and the line G H; two and a half degrees on the other sides between the course called for and G C; B C and E F are parallel, and both on the course required by the Sumter grant; the distance of that line is shewn by F K or B R; and the distance on the line which terminates at G, is shewn by G F.

The surveyors of the plaintiff adverted to rules, which were understood to require that the senior grant should be located to the greatest advantage, and that, in connecting ascertained points where no marks are to be found, the boundaries which were called for should be followed to their termination, and then course and distance be conformed to as nearly as possible. Under these rules, (as the Kershaw boundary was not known) they thought it proper to begin at G, run on the course required (or rather the course of the old marks) to the termination of the distance at F, and thence on the required course of the next line toward E. One of them thought that he ought to stop at K, the end of the distance required, and thence run straight to B, the nearest point of the acknowledged boundary: – thus adding a line, and changing materially, the shape of the plat. The other thought, that from F he ought to run beyond K to the point E, where his line would be intersected by a line from B (run on the course which the Sumter plat lays down for the line, bounded in whole or in part by Williams,) and then from E to B: – thus, also, adding another line and departing somewhat (but less than the other) from the shape of the plat.

The surveyor for the defendant thought, that under the rules adverted to, a line from F to B, (which would divide the land in dispute) would be more proper than either two of the lines which the other surveyors respectively had adopted in place of one required. He, however, believed that B was a corner, which the Sumter plat represented as common to the Williams, Kershaw and Sumter grants; and taking that point and G as both fixed corners, he considered that the true mode of location was, to run on the required course from either, and establish the point of intersection as the immediate corner.

Adopting this mode (except that in following the marks between G and F, he departed two and a half degrees from the course) he established the lines A B C G, as the true boundary; and so showed that the land, whereupon the trespass was complained of, belonged to the defendant.

The Circuit Judge left it to the Jury to decide the location, instructing them to inquire whether B was a corner of the Sumter grant, and giving his opinion that, if it was, the mode adopted by the defendant's surveyor was correct, and the land in dispute belonged to the defendant.

His Honor thought that there were some indications that B, the acknowledged western corner of the Williams land, was represented on the Sumter plat as a corner, which he called to the consideration of the jury, viz:

1. No corner between A and that line of the Sumter land which runs west from A is laid down.

2. The distance from A to B is ten chains more than the distance required by the Sumter plat.

3. On that plat, the Williams land seems to be laid down as a boundary along the whole line running from A, and the Kershaw land as the boundary of the two next lines.

4. The shape of the plat, open corners, right angles, marks older than the Sumter grant, and other circumstances which influenced the opinion of the defendant's surveyor.

He thought F B, which would divide the land in dispute, was a nearer approach to the true location than F E B, or F K B; than either of these latter, he thought F L B would be better, made by a prolongation of the line A B to the point of intersection; but, as before said, he intimated his opinion in favor of G C B. He did not, however, say that the Sumter grant could not be located as the plaintiff's surveyors had done.

There was no other testimony which could affect the question, except, possibly, the fact proved that about 1795, one Trantham, (under whom the plaintiff claims) cleared a field, lying mostly east of the line B H which inched over the line C H a small distance, and over the line C F a less distance, so as in the whole, to embrace one fourth of an acre west of the line B H. This field was cultivated by Trantham ten years, and was turned out long ago, and has not been in cultivation for thirty years. Either party has had possession of his tract for thirty years and more, on some part distant from that now in dispute; the first actual possession under the Burns grant, having commenced about 1808.

The jury found for the defendant.

The plaintiff appealed and moved the Court of Appeals to set aside the verdict, and for a new trial, on the following grounds:

1. Because his Honor charged the jury, that if the surveyor of the Sumter grant intended to make the western corner of the Williams land a corner of the Sumter grant, the location made by defendant's surveyor was correct, and they should find for the defendant.

2. Because his Honor charged the jury, that there were indications on the plat of the Sumter grant, that the surveyor did intend to make the western corner of the Williams land, a corner of the Sumter grant.

3. Because his Honor charged the jury, that the Sumter grant could not be located to cover the whole of the trespass; and that, in his opinion, no part of the trespass was covered by that grant.

4. Because the verdict of the jury is contrary to the law and evidence of the case.

*Smart*, for the motion.

*Chesnut*, contra.

*Curia, per* WITHERS, J.

The end arrived at by all the rules of location, is to establish for the plaintiff, in an action to try the title, the lines run or described in the grant he produces, where, as in this case, such a muniment of title is produced, elder and therefore paramount. If the original surveyor made his tracks, the object is to retrace them; if he platted by description, the object is to fulfil the terms of it, so far as the evidence may permit and no law or adverse proof may overrule.

Since then the object is to follow the original surveyor, it is proper to begin where he did, if that point can be ascertained; at all events not to reverse the direction of his movements until, in pursuing them, we have reached a point beyond which his footprints are no longer to be discerned.

In regard to those lines that involve this contest, the direction of the original survey was from an easterly to a westwardly point. It may well be conjectured that the surveyor did not actually run the line along the Williams boundary, as the corners represented do not appear to have been marked. Yet the distance is given, 49 chains, and so is the course, S. 80 W. To establish that line, especially its termination westwardly, is all-important in this question of location. On that line the boundary called for is the Williams tract. Its beginning point is on the Mathews tract. That tract is represented on the re-survey. The Williams tract is satisfactorily located. Then there seems to be a propriety in beginning to trace that line either from the Mathews tract, or from the eastern corner of the Williams tract. It is not material in this case at which point, of these two, we may begin. The course of the plaintiff on that boundary is controlled by the line of Williams. Where shall he stop? His distance falls short of that of the Williams line; the latter terminates at B. Was the point B designed by the surveyor of plaintiff's plat to be his corner as well as that of Williams? If it was, then he must be confined to it. That question was distinctly submitted to the jury, and they have decided that B was plaintiff's corner. If there was evidence to lead to that conclusion, what is there in the testimony to overrule it? There was evidence enough to sustain such a finding by the jury, and we can discern none which shews that they were wrong. It is sustained by the facts, 1st that B is a corner of Williams' land; 2d the plaintiff's platt calls on that line for no other boundary, as to any portion of it; 3d more than his distance is secured by adopting that line, which is, so far, a favorable location for him. The plaintiff would go beyond B; what is to justify that? No adjacent boundary, no marked object, neither distance, nor quantity, nor the form of the plat, nor any necessary mode of closing with the next nearest ascertained point, can lead us beyond B. If the surveyor passes that, where shall he stop? We must, therefore, concur with the jury in fixing B as one corner for the plaintiff.

From that point how shall we proceed? The survey must be guided by such means as the plaintiff affords. His plat gives course and distance, and represents a corner between B and G. If G be given, and we have also the course from that, and there be nothing to control that course, the mode of closing is not at all doubtful; the lines must be run, by course, from the two ascertained points, and the place of intersection will show the intermediate corner. Now the point G is given; it is confessedly a corner for plaintiff. The result has been stated; the intermediate corner will be found at H, which is fatal to the plaintiff. If the line from G be conformed to certain old marked timber for a portion of the distance, the result to this case would be the same. In so closing from the ascertained points, B and G, there is no departure from the elements of location which the plaintiff's own plat affords, except in the particular of distance. In closing by any other mode from B to G, we should abandon the description of that plat, both as to course and distance, not to mention shape also, which would be necessarily incident to the other variations.

If the mode of closing between B and G was indicated by the plaintiff's platt to be by a single line instead of by two constituting an intermediate corner, the case of *Cain & Hodge* might be invoked. If distance be taken on the course from B, and the like be done on the course from G, and the points so made be closed by a straight line, while this would give three lines and two corners, instead of two lines and one corner, and be wholly unauthorized by any rule of location, the trespass would at the same time be excluded.

We shall not dwell upon the fact, that the grant under which the defendant claims, was located in the year and within a few months next following the date of the plaintiff's, so as to include the *locus in quo*. It called for land, using the language, "said to be General Sumter's," or some equivalent words. It is probable either that General Sumter's land was not located the year preceding, as the plaintiff claiming under him, now contends, or it was so carelessly and negligently designated, as to invite an intrusion by a subsequent grantee. When a plaintiff comes into Court with a grant and plat which, though older, is surrounded by obscurity, arising either from the fact that the survey has been made at the fire-place, rather than on the field, or from any other description of carelessness, he is not entitled to demand that we shall push, to a fanciful extent, or strain inordinately the idea, sometimes indulged, that the elder grant is to be more favorably located.

We think the finding of the jury in this case is more conformable to the rules of location as they apply to the facts adduced, than any other which we could prescribe. Unless we could dictate a better re-survey than that which is sanctioned by their verdict, it ought to stand; and, therefore, the motion must be refused.

*Faulkenberry v. Truesdell*, **36 S.C.L. 221 (S.C.App.Law 1850).**

## SOUTH DAKOTA

This is an action to recover the possession of about 15 acres of land, and for damages for its use and occupation. The case was tried to a jury, and, the verdict and judgment being in favor of the defendant, the plaintiff has appealed. The controversy in the case is as to the location of the original government corner, as established by the United States deputy surveyor by the original government survey, between the southwest and southeast corner of the two quarter sections. The plaintiff claimed the southwest quarter, and the defendant the southeast quarter, of the section.

It is contended by the appellant that the evidence was insufficient to justify the verdict of the jury, and that the court erred in its instruction to the jury. It seems to be the theory of counsel for the appellant that this court will review the evidence and determine as to its weight. This theory is clearly incorrect as applied to a case tried by a jury. In such cases, where the evidence is conflicting, this court will not weigh the evidence, or go further than to determine therefrom whether or not the successful party has given sufficient legal evidence to sustain the verdict, without regard to the evidence given on the part of the other party, except so far as such evidence tends to sustain his case. *Jeansch v. Lewis, 1 S. D.* 609, 48 N.W. 128. In a case tried to a jury, therefore, this court will not review the evidence with the view of determining its weight, but to simply ascertain whether or not there is sufficient legal evidence to support the verdict of the jury, assuming it to be uncontradicted, as the jury are the exclusive judges of the weight of the evidence and the credibility of the witnesses, subject to the exercise by the trial court of its sound judicial discretion in granting a new trial, where in its opinion the weight of the evidence is so strongly in favor of the losing party that justice requires the granting of a new trial; and the decision of the trial court in the exercise of its discretion in granting or refusing a new trial will not be disturbed, unless, in the opinion of this court, there has been a manifest abuse of such discretion. *Cannon et al. v. Deming, 3 S. D.* 421, 53 N.W. 863; *Walker v. McCaull et al.,* 13 S.D. 512, 83 N.W. 578; *Weiss v. Evans,* 13 S.D. 185, 82 N.W. 388; *Brewing Co. v. Mielenz, 5 Dak.* 136, 37 N.W. 728. A different rule is followed by this court in cases tried by the court without a jury. In such cases this court will review the evidence for the purpose of ascertaining whether or not the findings of the trial court are supported by the evidence, and, in case there is a clear preponderance of the same against the findings of the trial court, its decision will be reversed and a new trial ordered. In *Randall v. Burke Twp., 4 S. D.* 338, 57 N.W. 4, the rule is thus stated: "On such review of the evidence, this court will presume that the decision of the trial court or referee upon the weight of such evidence is correct; and it is only when this court is satisfied that there is a clear preponderance of the evidence against such decision that such presumption will be overcome, and the decision of the trial court or referee reversed." *Hill v. Whale Min. Co.,* 15 S.D. 574, 90 N.W. 853; *Krueger v. Dodge,* 15 S.D. 159, 87 N.W. 965. In the case at bar there was a sharp conflict in the evidence. It was claimed by the plaintiff, and there was evidence tending to prove, that the corner between the quarter section claimed by the plaintiff and the quarter section claimed by the defendant had been lost or destroyed, and that by a resurvey made by a competent surveyor, following the courses and distances given in the field notes of the original government survey, the line of plaintiff's land extended about 15 rods further to the east than the line as claimed by defendant. On the other hand, the defendant claimed and introduced evidence tending to prove that the line as it existed at the time the action was commenced was the true line as evidenced by the government corner established by the original survey, and which was in existence and was well known for many years after the two quarter sections were settled upon. The evidence on the part of the defendant that such corners existed and were seen and observed by the witnesses for many years after the settlement of that section of the country was direct and positive, and that the government corner was at the point now claimed by the defendant. If the jury believed this evidence – and it is apparent from their verdict that they did–their verdict was fully justified by the evidence, as this court has repeatedly held that the first object of a court and jury in the trial of cases of this class is to ascertain, if possible, the true points where the original government mounds were placed at the time of the original survey, and that, if these points can be ascertained from the evidence, the line must remain as established by the original government mounds. The rule is thus stated in the case of Randall v. Burke Twp., supra: "The rule is well settled that on a resurvey of land originally belonging to the United States, and which it has caused to be surveyed under its authority, such resurvey must follow the boundaries and monuments as run and made by the original government survey, if the monuments placed by the government, in making the survey, to indicate the section corners and quarter section posts, can be found, or the places where they were originally placed can be identified." *Tyler v. Haggart (S.D.)* 102 N.W. 682. There being sufficient evidence to justify the verdict of the jury, it will not be proper for us to examine the evidence with a view to determine its weight, and we do not therefore attempt its review in this opinion.

It is further contended that the court erred in giving to the jury the following instruction: "But it is the law that, if there is a discrepancy between the field notes and the actual location of the corner as made on the ground, the location prevails." This instruction of the court is clearly correct under the rule as laid down by this court in a number of cases. Randall v. Burke Twp., supra; *Dowdle v. Cornue, 9 S. D.* 126, 68 N.W. 194; Tyler v. Haggart, supra; *McGray v. Elevator Co.,* 16 S.D. 109, 91 N.W. 457.

**Unzelmann v. Shelton, 19 S.D. 389, 103 N.W. 646 (S.D. 1905).**

## TENNESSEE

This is an action of ejectment to recover a tract of land alleged to be located in Monroe and Polk counties. Complainants claim as the heirs of Francis W. Lea, who, in his lifetime, July, 7, 1858, obtained a grant for it from the state of Tennessee. On the trial of the cause in the court below, the chancellor held that complainants had failed to sustain the allegations of their bill, and dismissed the same; and they have appealed, and assigned as error the finding of the chancellor against their contention. The facts of the case, so far as they need to be stated, are as follows:

Prior to 1819, the United States government bought of the Cherokee Indians all their claim to the lands embraced in what was known as the "Hiwassee Land District"; having previously ceded the same to the state of Tennessee, subject to this Indian claim. After this purchase was made, the state of Tennessee appointed commissioners to lay off the district into ranges, townships, sections, and fractions of sections. In June, 1819, some disagreement having arisen with the Indians in regard to the boundary line of the district, R. Houston and James McIntosh, on the part of the United States, and "Squire" and "Crow," on the part of the Cherokee tribe of Indians, were appointed commissioners to resurvey and re-mark the line, which they proceeded to do, and on the 12th day of June, 1819, they agreed upon and signed their report at the forks of the Montagulee and Cause rivers. The Cherokees owned a section of county adjoining the Hiwassee purchase, lying south and east of it, which became known as the "Ocoee District." This was also purchased at a later date, and in 1836 the state of Tennessee directed a survey of this district into ranges, townships, sections, and fractions of sections. The line between the two districts ran in a direction north, 18~ west, from the beginning corner on the bank of the Hiwassee river, and the township lines were intended to close upon it, but could not do so at right angles, and the consequence was a great number of fractional sections and quarters were formed. Maps and surveys were filed in each district, showing the township sections and fractional divisions and lines, but the lengths of the fractional lines abutting on the boundary line were not given, but only the fractional areas.

On the 25th of February, 1854, the state of Tennessee passed an act authorizing the entry and grant of all unsurveyed lands in the Ocoee district, and in 1859 Francis W. Lea obtained a grant under this act for 8,270 acres of land. Complainants claim as his heirs, and their contention is that there is a strip of unsurveyed land lying between the Hiwassee and Ocoee districts, on the Ocoee side of the Indian boundary line. Defendants also hold under a chain of title from the state of Tennessee, by numerous grants. Their contention is that there is no unsurveyed, or vacant, unsectionized lands, between the two districts, but that the lines of the sections and fractional sections of surveyed lands in the two districts approach each other until they meet upon a common line, which is the original boundary line of 1819; and, second, that, as to a large portion of the land embraced in complainants' grant, the grant is void, because the land was, at the time the grant was issued, in the actual adverse possession and occupation of defendants and their predecessors, under older grants.

Complainants caused the lines called for by their grant to be run out by actual survey, using, in making the survey, not only the calls in complainants' grants, but also the calls of the fractional sections abutting on the district lines on both the Hiwassee and Ocoee sides, as obtained from their respective land offices; and upon completing such survey the surveyor came to the conclusion that there was an area of 3,610 acres of land not covered by defendants' title, or any other, and not sectionized, but laying like a gore between the two districts, and it is this strip or gore which complainants claim, disclaiming all right to any land covered by the calls of defendants' title papers as lying in the sectionized portion of the Ocoee district, and disclaiming all right to any lands in the Hiwassee district. It will thus be seen that the controversy narrows itself down to the question whether there is any vacant or unsectionized lands in the Ocoee district abutting on this Indian boundary line.

Defendants have also caused surveys to be made of the locality, and various tests to be applied; and the contention of each party is liberally illustrated by handsome maps covering all the territory, and fully sustaining the opposite contentions, and in each case showing the utter impossibility of the correctness of the opposing theory, from the standpoint taken.

The accuracy and competency of the surveyors is called in question. It is said that the defendants' surveyors are not experts; that they did not locate the Indian boundary line, nor attempt to do so, and that their testimony and conclusion are based upon the laws fixing the boundaries of the two districts, rather than upon any actual surveys and personal observations. On the other hand, it is shown that complainants' surveyor is incompetent; that he has had an experience of only five years in Georgia, and none in the mountains of East Tennessee; that he guessed at his beginning corner, and, in attempting to follow the Indian boundary line of 1819, he placed the variation of the compass needle on the wrong side of the meridian line; that he professed to find landmarks of the old line run in 1819 upon trees in a locality where it is shown the forest trees has been cleared away, and the land cultivated, for more than 20 years, and which had since been covered with a second or under growth for 30 years; that he also professed to find marked trees in other localities where it appeared that land had been cleared and in cultivation for 40 years. Upon an inspection of the title papers of the contesting parties, and a comparison of their maps, and the statements of the surveyors explaining the same, we are of opinion that complainants have failed to show satisfactorily that there is any vacant or unsectionized lands in the Ocoee district, bordering on this Indian line between the two districts covered by their grant. The testimony of Thomas, their surveyor, is wholly overthrown by that of Waring, Muller, Deakins, and McGuffey, the surveyors of defendants. These latter gentlemen appear to be competent, intelligent, experienced surveyors and engineers, of many years' standing and practice in their profession, with much experience in surveying mountain and other lands in the locality, and in locating old lines, including this Indian boundary line, which had been located by them in many places in the native forest after reaching Starr's mountain. Each of these surveyors states that from actual measurements and personal observations, and experiments with various fractional and sectional lines, it is impossible that there should be a vacant gore or strip between the two districts. They explain that while there is an apparent shortage in some of the fractional lines, and that others are too long, so as to cause, at some places, an interlap of the lines from the two districts, there is no discrepancy of any consequence at any place, and no more than would naturally and does usually result in surveys over rough mountains, and especially in the early history and settlement of the country; that while there is some confusion between lines and boundaries caused by the hasty and careless manner in which early surveys were made, there is no such discrepancy as cannot be cured and explained by the rules observed in establishing calls for old lines and natural objects; that the greatest difference observed did not exceed 40 poles in a line 6 miles in length, over very rough, mountainous grounds, while in other cases there were some lines that fell short small amounts.

It also appears that a part of the land covered by complainants' title papers has been in defendants' possession, under inclosure and in cultivation, for over 40 years, and other portions for more than 7 years, constituting what is known as the "Savannah Farm," which lies partly in each district, crossing the Indian boundary line, and lying on both sides of the Hiwassee river. The principal conflict seems to arise between a grant to Charles Seymour for 4,000 acres, made in 1874, and the calls of complainants' grant. Seymour, soon after obtaining his grant, conveyed the land to defendants, and they have been holding the same adversely, and in actual possession, ever since, as to considerable portions of the area embraced. In view of the evident purpose of the legislature to sectionize all the lands in each of these districts up to the Indian boundary line, and the surveys and plats to carry this purpose into effect, as manifested in the surveys and plats in the land offices of the two districts, and in view of the recognition by abutting landowners that they hold up to a common dividing line, and that there is no hiatus or gore left unsurveyed and unsectionized, we would not be disposed to adjudge that there was an unoccupied and vacant strip of land along the Indian boundary line, which extends across the state, as the dividing line between the two districts, unless the evidence was very plain and convincing. The act of 1854, under which this grant was obtained, is, at most, only a legislative recognition that there may be lands left vacant in the Ocoee district, but not an adjudication that there are such vacant or unsectionized lands.

Upon a careful examination of the entire case, we are convinced that complainants have failed to sustain their contention that there is a vacant gore of land between the two districts, a portion of which is covered by their grant, and the judgment of the court below is affirmed, with costs, and the bill dismissed.

*Hennegar v. Seymour*, **93 Tenn. 253, 23 S.W. 969 (Tenn. 1893).**

## TEXAS

A "map" is a picture of a survey, "field notes" constitute a description thereof, and the "survey" is the substance and consists of the actual acts of the surveyor, and, if existing established monuments are on the ground evidencing such acts, such monuments control because they are the best evidence of what surveyor actually did in making the survey and are part at least of what surveyor did.

*Outlaw v. Gulf Oil Corp.*, **137 S.W.2d 787 (Tex.Civ.App., 1940).**

In suit involving boundary question, search must be made for the footsteps of the original surveyor and, when found, the case is solved.

*Hart v. Greis*, **155 S.W.2d 997, Texas, 1941.**

The purpose of the inquiry in boundary dispute action is to locate and follow the footsteps of the original surveyor.

*U.S. v. Champion Papers, Inc.*, **361 F.Supp. 141 (D.C. Tex., 1973).**

## UTAH

The location of the boundary line between plaintiff's and defendant's premises is the matter in controversy in this action. The suit is one in ejectment. The particular strip in controversy is described in the complaint. It is alleged that the plaintiff is the owner; that the defendants are in possession of the same, had erected thereon fences and portions of a house, and are occupying the disputed strip without right or title, and against the will of the plaintiff. Judgment is sought for a restitution of the property, for damages, and a removal of the encroachments.

The crux of the defense is stated in the sixth paragraph of defendants' answer, as follows:

"That the line between the land of the plaintiff and the land of the defendants above described was marked and established many years ago, to wit, more than 30 years before the commencement of this action, by a certain fence; that the said fence remained intact for many years thereafter, and all improvements made upon said tracts by the then owners of the said parcels of land, the predecessors in interest of the plaintiff and defendants, respectively, were made with reference to the said fence; that the said fence remained intact until about 15 years ago, when it was moved, except that the marks of the old posts and some of the stumps of the same, showing the location of the said line, are still in existence upon the said line between the said tracts of land; that thereafter the grantors and predecessors in interest of both the plaintiff and defendants continued to occupy their said respective parcels of land up to the said line so formerly marked by the said fence, and the plaintiff and the defendants, since they have been the owners of the said tracts, respectively, have continued to occupy their said respective tracts up to the said line, and have made all improvements with reference to the same, and the plaintiff and the defendants, and their and each of their grantors and predecessors in interest, for more than 30 years before the commencement of this action, occupied their said respective tracts of land up to the said line, and have made all improvements upon said tracts of land respectively with reference to said line, and have recognized said line, and have acquiesced therein, as the true boundary line between their said respective tracts of land."

It is also alleged in the answer that the strip of land mentioned in the complaint lies to the east of the true boundary line formerly marked by said fence, and that such has been recognized as the true boundary line by the parties to the action and their predecessors in interest.

The court found the issues presented by the answer against the defendants, and entered its judgment awarding to the plaintiff the strip of land in controversy.

Findings of fact are stated in fifteen different paragraphs. Assignments of error assail each and all of the findings of the court. No assignment is made respecting the ruling of the court upon the admission or rejection of evidence, nor of the court's order denying the motion for a new trial.

The lands owned by plaintiff and defendants are contiguous, and constitute a part of lot 1, block 104, plat D, Salt Lake City survey. Defendants' lands are situated on the southeast corner of said block, with a south frontage on Seventh avenue and an east frontage on F street. Plaintiff's land lies immediately west of defendants' land and has a frontage on Seventh avenue of 2 1/2 rods, and a depth of approximately 100 feet. The strip in controversy is the border line of the west side of defendants' premises and the east side of plaintiff's premises. Plaintiff had her premises surveyed recently and the boundary lines of the same established. She introduced the testimony of the surveyor making the survey, and his testimony fixes the boundary line east of the line claimed by the defendants to be the true boundary line. The line established by this surveyor includes the premises in controversy in the description of plaintiff's land. A map was introduced made by this surveyor and his assistants from the notes of the survey. This map shows the location of the present fence and of the corner of defendants' house to be upon plaintiff's premises.

It seems that the original survey of Salt Lake City was made by one Jesse W. Fox, Jr. That survey is referred to in the record as the "Fox survey," and also as the "Old survey." The court found that the field notes of the Fox survey had been lost or destroyed. A later survey was made about the year 1890 or 1891 by one A. F. Doremus, who was at that time city engineer of Salt Lake City. This survey is referred to as to "Doremus survey," and also as the "new survey." The Doremus survey was authorized by the ordinances of Salt Lake City, and thereafter became, and was known as, the official survey. It is inferable from the testimony that the monuments established by these two surveys did not always coincide.

Property lines were established in the early settlement of Salt Lake City under what is known as the "Fox survey." The decisions of this court are to the effect that, when these property lines were marked upon the ground by fences and buildings open and visible, and the property owners had occupied the premises, and made improvements thereon, with reference to such survey, such boundaries would not be disturbed. *Holmes v. Judge*, 31 Utah 269, 87 P. 1009; *Young v. Hyland*, 37 Utah 229, 108 P. 1124; *Binford v. Eccles*, 41 Utah 453, 126 P. 333; *Tanner v. Stratton*, 44 Utah 253, 139 P. 940.

The findings of the court and the evidence take the instant case out of the rules announced in the above decisions. In other words, the testimony and the findings of the court are against the defendants' contention that an established boundary line existed and had been acquiesced in by the parties and their predecessors in interest.

The court found that the location monument on the southeast corner of said block 104, which is the tie point for the description of the property of both plaintiff and defendants, does not differ under the respective surveys, and also recites that the court does not find that the length of the block facing on Seventh avenue has varied, or that the frontage of defendants' said land has varied, or that there has been any variance in defendants' land, but on the contrary, does find that the true boundaries of said tract of land of defendants has not varied by reason of its location as surveyed under the Doremus survey. In other words, the findings of the court are to the effect that the location of defendants' land, measured and located under either the Fox or Doremus survey, is apparently identical.

There is much testimony in the record tending to establish the exact location of an old fence, claimed to have been the dividing line of the lands of plaintiff and defendants, in the rear of the lots. The effort to place the exact location of the fence was by certain old cedar fence posts. These posts were found by digging into the earth some 7 or 8 inches below the surface. There is testimony to support a finding that that fence was located at or near the present fence, but there was also testimony to support a finding that the same was located further east than the present fence. Witnesses who had known the premises for many years testified as to the location of the present boundary line of the property lying further west in the old block. A concrete wall located near the center of the block, running north from Seventh avenue, is particularly relied on by the defendants as supporting their contention that the east boundary line of plaintiff's property does not include the strip of ground in controversy. There is also testimony, respecting the location of other old fences on the block. The testimony, however, does not show that the boundaries of the lands of plaintiff and defendants were ever established or located by any reference to the cement wall or to the old fence in the block, nor that the location of this cement wall or other fences in the block in any way controlled or determined the boundary line between the lands of plaintiff and defendants. In fact, the testimony does not show that there was ever any recognized boundary line between plaintiff's and defendants' land acquiesced in by the owners of the respective properties.

The court after seeing and hearing the witnesses, and the plats submitted by the respective parties, found the issues against the defendants.

It is quite evident from the record that the court gave the matter careful consideration and considered the evidence, conflicting and otherwise, carefully before rendering its judgment. We find no sufficient reason in the record for disturbing the court's findings.

The court's finding that the defendants had failed to maintain their defense, or their affirmative claims to this property, is supported by the weight of the evidence, and, in our judgment, the great weight.

There is this to be added: The findings and judgment of the court give to each party the full frontage called for by the deeds of conveyance to them of their respective property.

The judgment is affirmed.

***Carlston v. Torkelson*, 69 Utah 261, 253 P. 904 (Utah 1926).**

## VERMONT

The action is trespass for breaking and entering plaintiffs' close described as lot 14, range 17, in the town of Lowell, with a count in trover for the conversion of certain saw logs. The defendant admitted entering the land in dispute and the removal of substantially the quantity of timber claimed by the plaintiffs, but claimed title in himself both to the land and timber. At the time of the entry complained of the plaintiffs owned lot 14, range 17, and the defendant owned lots 13 and 14, range 16, in said town. Range 17 lies immediately west of range 16 and the lots in each range are numbered consecutively from south to north; thus the plaintiffs' land abutted upon that of the defendant, the southeast corner of their lot being the southwest corner of defendant's lot 14, range 16. There is one more lot in range 16 than in range 17, owing to the irregular shape of the town. This lot is numbered 15, range 16, and was formerly owned by one Mulligan and referred to at the trial as the Mulligan lot. It will be seen that plaintiffs' land, which extended to the town line, abutted upon defendant's lot 14, range 16, and the Mulligan lot. The land in dispute is a rectangular tract 76 rods in width and extending across the entire width of lot 14, range 16. Both parties claimed title to the disputed tract, the plaintiffs as part of lot 14, range 17, and the defendant as part of lot 14, range 16; and the controversy was as to the location of the division line between said lots. The location of the land in controversy is shown by the annexed sketch and is the tract marked a, b, d, c. Both parties derive title to their respective lots from one Hanson E. Lewis who, as early as June 1, 1887, owned the tract embracing lots 12, 13 and 14, range 16, and lots 12, 13 and 14, range 17. The plaintiffs' chain of title is as follows:

Deed from Hanson E. Lewis to John R. Sullivan, Aug. 13, 1891;

Deed from John R. Sullivan to George E. Young. July 9, 1894;

Deed from George E. Young to Gilbert & Pope, Nov. 24, 1898;

Deed from Thomas Gilbert to W. H. Pope. Dec. 1, 1899;

Deed from W. H. Pope to J. H. Silsby & Co., July 22, 1912.

The defendant acquired title to his lots Sept. 30, 1911, by deed from Adelaide L. Lewis, Executrix of the will of Hanson E. Lewis.

The defendant's evidence tended to show that the true range line between ranges 16 and 17 is as shown in the above sketch and that the land in dispute is part of lot 14, range 16. The exceptions state, "The plaintiffs claimed, and the evidence tended to show, that the range line between ranges 16 and 17 as claimed by the defendant was run south of lot 12 and marked as testified by the defendant's surveyors, one hundred years ago; that said range line north of the south corner of 12 had no marks as old as the range line south of said corner; that there were a few trees near that line on lot 12 with marks older than the Webster survey of 1887 (hereinafter referred to); that on lot 13 there were no old marks; that across lot 14 in range 17 there were no marked trees indicating a range line west of the Mulligan line (also later referred to) * * * except the line run by the defendant since this controversy began." It appears from the exceptions that the defendant's evidence tended to show several old marked trees near his claimed range line on lot 12 and one old mark on lot 13, but no old marks on lot 14.

As we construe the exceptions the plaintiffs practically admitted that the original range line between ranges 16 and 17 south of lot 12 was marked upon the ground as claimed by the defendant. It also fairly appears that there was no testimony of existing marks upon the ground indicating that that line had been extended north between lots 14 in ranges 16 and 17 before this controversy arose. In this connection it appeared that plaintiffs owned lot 10, range 16, did not own the adjoining lot in range 17 and that they cut the timber on their lot to the range line claimed by the defendant.

It appeared that one Lawrence Mulligan owned lot 15, range 16, from March 4, 1885 to April 5, 1902; that in June, 1887, shortly after Hanson E. Lewis acquired title to said six lots, he sent John Webster, a land surveyor, and one Anderson to run out and mark the boundaries of his land; that pursuant to their directions they ran and marked the south lines of lots 12 in both ranges and the east line of lots 12, 13 and 14, range 16; that they also ran and marked as part of the north line of said lots the line from the northeast corner of lot 14, range 16, westerly to the so-called

"Mulligan corner," (marked d on the sketch), being the corner claimed by the plaintiffs as the northwest corner of lot 14, range 16, where they found a spruce tree marked as for a corner with new marks; that as the last work they did they ran and marked a line northerly from said corner to the Westfield line, being the line d-e on the sketch. This line the plaintiffs claim as part of the east line of their lot. The defendant's evidence tended to show that there was no marked line north from the Mulligan corner until the Webster survey, while the plaintiffs' evidence tended to show that it was run and marked earlier and that it "was marked when Mulligan purchased his lot in 1885." Whether we construe the exceptions to mean that the Mulligan line was run and marked at the time Mulligan purchased his lot, or that it had been marked at some previous time, it is evident that it was a comparatively recent marking.

In the fall of 1891, after he had purchased lot 14, range 17, John R. Sullivan caused a line to be run and marked south from the Mulligan corner a distance of 100 rods in the same course as the line that Webster had run in 1887 north from that corner, for the purpose of determining his south line. The line thus run by Sullivan extended across lot 14, range 16, and is the line plaintiffs claim as the division line between their lot and defendant's 14, range 16. It was unmarked before Sullivan caused the survey to be made. Referring to the Mulligan line and its extension south across lot 14, range 16, the exceptions state that the plaintiffs claimed that this was the true east line of their lot, whether, it was the true range line or not. The exceptions do not disclose any claim by plaintiffs that the original allotment of the town of Lowell located this line farther east than the line between ranges 16 and 17 south of lot 12. In view of what does appear we take the exceptions to mean that their claim in this regard was that this line had become the true east line of their lot by acquiescence or by adverse possession. The case is barren of any evidence that this line was ever established as a part of the range line, or as the true division line between lots 14 and 15, range 16, and lot 14, range 17, unless it had become such by one or other of these means. Plaintiffs say in their brief that whether their claimed line was the original range line was immaterial as the case stood. The court in its charge, which is before us, states the claims as follows: "The plaintiff and defendant disagree in regard to the lines of lot 14, range 17. The plaintiff claims that his lot, 14 in range 17, extends down east to what is called the Mulligan line. He claims that he has a right to have that extended down to the Mulligan line on account of having occupied to that line for so long a time that he can hold it by adverse possession, as it is called. He also claims that he has a right to that land down to the Mulligan line on the ground of what has been called here in the course of the trial as acquiescence to a line by the parties upon both sides, so it will be necessary for me to call your attention to these two claims and the grounds of them and the law bearing on them." Having thus stated plaintiffs' claim, it is to be presumed that the court has stated the whole claim. *Akeley v. Carpenter*, 87 Vt. 248, 251, 88 A. 897.

It is important to a proper understanding of the questions raised to consider first what, on the evidence, would be the location of the division line between lot 14, range 16, and lot 14, range 17, unaffected by acquiescence or adverse possession. It is a well-known fact that the original proprietors of townships usually had them laid out into lots which they and their successors have been accustomed to sell by number without more particular description. Described thus, the lot lines, if surveyed and marked upon the ground, serve as monuments in fixing the boundaries. It was held in *Spiller v. Scribner*, 36 Vt. 245, that the description of a lot by reference to its number is a description in legal effect according to the lines of such lot as surveyed and established in the original division of the town and is just as definite, though not as particular, as it would be if the lines were given; that the description should receive the same construction and have the same legal effect in one case as in the other; and that such description was controlling and determined the extent of the land in controversy. See *Warren v. Pierce*, 6 Me. 9, 19 Am. Dec. 189.

The actual location upon the ground of original lot lines will control, if capable of being ascertained; but, when such lines have never been surveyed or, if surveyed, their location upon the ground cannot be ascertained, resort may be had to the lines of adjacent lots to determine their location. A range line will be presumed, in the absence of marks upon the ground, to be a straight line. Plaintiffs' claim that the location of the Mulligan line by acquiescence determined the location of the range line between lots 14 depends upon this assumption. There being no evidence on the ground of the original range line across the disputed territory, its location is determined *prima facie* by extending the range line from a point where its actual location can be ascertained. It follows that on the case made the plaintiffs failed to show that the disputed land was a part of lot 14, range 17, as originally allotted; while the defendant made at least a *prima facie* showing that it was within the limits of lot 14, range 16.

On the questions of acquiescence and adverse possession the plaintiffs' evidence tended to show that the Mulligan line was run by Mr. Webster acting for Mr. Lewis; that he made notes of his survey for Lewis; that after this controversy arose Webster furnished plaintiffs a sketch of his surveys made from his notes which showed the Mulligan line as a continuation of the range line between ranges 16 and 17 south of lot 12; that Webster then supposed that the Mulligan line was a continuation of such range line and that he so believed until after this suit was instituted and he went onto the land; when he concluded it was not; that the Mulligan line has been claimed and recognized as the east line of lot 14, range 17, separating it from lot 15, range 16, since some time prior to 1885; and that the continuation of the Mulligan line south from the Mulligan corner has been claimed by the owners of lot 14, range 17, except George E. Young, as the east line of that lot ever since Lewis conveyed it to Sullivan. As to the claim and occupancy of said Young, it appeared that he never went onto the lot, never did nor caused to be done any acts of possession thereon and knew nothing about the boundaries thereof. It appeared that all the time Mulligan owned lot 15, range 16, he recognized the line claimed by the plaintiffs as the west line of his lot and never claimed nor occupied any land west thereof.

There was no evidence of actual occupation by Sullivan to the line run by him south from the Mulligan corner, except that he entered upon the land under his deed, caused the lines to be run and marked as before stated and cut timber on the land north of the Mulligan corner and between the Mulligan line and the range line. During all the time he owned the lot he claimed the line run by him south from the Mulligan corner as the east line of his lot. The fall Pope and Gilbert acquired title to the lot they erected a lumber camp on the disputed territory which they, and subsequently Pope alone, occupied while cutting the timber on their lot. Pope and Gilbert and Pope alone cut all the soft wood timber on lot 14, range 17 and down to the line run by Sullivan claiming that as the east line of the lot. Pope finished cutting in 1902. Thereafter he went onto the lot four or five times a year hunting and to see that no trespassing was committed. The lumber camp remained on the land and was claimed by Pope as his. So far as appears the next cutting on the disputed tract was in the fall in 1912 when the plaintiffs cut the logs sued for. In addition to cutting said logs the plaintiffs built roads on said lot and built a lumber camp on the disputed land. They also looked after the property to see that no trespassing was done. The exceptions state that there was no further evidence of occupation or possession of lot 14, range 17, and no evidence that any one ever entered upon, claimed to own, or exercised any rights in the territory in dispute adverse to the plaintiffs and their grantors or questioned the line run by Sullivan as the true line of lot 14, range 17, prior to the time defendant entered and removed the timber sued for, which was in the winter of 1912–1913, except that during all the time after Lewis conveyed the lot to Sullivan the defendant and his grantors held the record title and claimed to own lots 13 and 14, range 16.

On this evidence the plaintiffs failed to establish their claimed line across lot 14, range 16, by agreement or acquiescence. Their evidence tended to show that the line north from the Mulligan corner across lot 15, range 16, had been recognized by Mulligan and his grantees under circumstances and for a sufficient length of time to establish that line as the division line between their lot and the Mulligan lot. But acquiescence by Mulligan and his grantees, who are not in defendant's chain of title, cannot affect the defendant's rights in the adjoining lot. The most that could be claimed for the evidence of acquiescence in the Mulligan line would be that it had a tendency to prove that the claimed line coincided with original range line. We are not called upon to decide whether it had that tendency, for, as we have seen, the plaintiffs were not relying upon the claim that the Mulligan line and the line run by Sullivan coincided with the original range line. At the most the Mulligan line was shown to be a new boundary established by acquiescence and not the original boundary of plaintiffs' lot. There was no evidence that Lewis knew anything of the line run by Sullivan. It follows that he could not have recognized it as the west line of his lot 14, range 16. Mutual recognition is requisite to establish a line by acquiescence. *Sawyer v. Coolidge*, 34 Vt. 303. And so are all our cases. Knowledge is the *sine qua non* of acquiescence. It was said in *Pence v. Langdon*, 99 U.S. 578, 581, 25 L.Ed. 420, that there cannot be acquiescence without knowledge, though one cannot wilfully shut his eyes to what he might readily and ought to have known; that there must be knowledge of facts which will enable the party to take effectual action – nothing short of this will do. While there might be circumstances in which the court would impute knowledge, there is nothing in the evidence before us to justify such action. Plaintiffs' counsel admits that there was no evidence of acquiescence in the line on lot 14, range 16, unless Lewis' failure to object to the cutting by Pope and Gilbert and Pope alone furnishes a basis for such an inference. Without evidence that Lewis knew of the cutting, and there is none, this would not warrant the inference. Lewis lived at St. Albans and, so far as appears, was never on the land. Besides, if we were to presume that he knew of such cutting, the necessary time had not elapsed to establish the line by acquiescence.

The theory of the plaintiffs on this branch of the case was that Lewis, being the owner of the whole tract, by surveying and marking the line north from the Mulligan corner, established the line between lot 14, range 16, and lot 14, range 17; that recognition by Lewis of the Mulligan line as the west line of lot 15, range 16, was a recognition of that line extended south as the west line of lot 14, range 16; and that silence on his part for 15 years coupled with the claim of Sullivan and his grantees to the Mulligan line and its extension south across lot 14, range 16, established that line by acquiescence. The court below seems to have adopted this view. The law establishing boundaries by acquiescence does not go to this length. Carried to its logical conclusion it would compel a holding that if, by mistake Lewis recognized the Mulligan line as part of the range line, he, by that means, recognized its extension south across his whole tract, though not surveyed and marked, as the range line. It would follow that his grantee of lot 12, range 17, could, without Lewis' knowledge, by marking a line across lot 12, range 16, in range with the Mulligan line and claiming it as his east line for 15 years make the new line the division line by acquiescence, despite the fact that the true range line was plainly marked upon the ground and may in fact have been relied upon by Lewis as the division line. The doctrine of acquiescence in boundary lines of necessity implies what amounts to actual knowledge of their existence upon the ground as marking the limits of one's ownership. See *Palmer v. Osborne*, 115 Iowa 714, 87 N.W. 712; *Connell v. Clifford*, 39 Colo. 121, 88 P. 850; *Fitzimons v. Atherton*, 162 Cal. 630, 124 P. 250.

Nor do the plaintiffs make a case of adverse possession. Pope and Gilbert succeeded by Pope alone were the first in the plaintiffs' chain of title whose occupation of the disputed territory was of the character to ripen into title by prescription. Pope and Gilbert acquired title Nov. 24, 1898, and the writ was served May 22, 1913; so their occupancy tacked to that of the plaintiffs would fall short of the necessary 15 years. Their predecessor Young was never in actual possession of any part of the lot and knew nothing of its boundaries. His deed gave him constructive possession only to the true range line. Plaintiffs' claim is that Sullivan had constructive possession of all the land within the spotted line indicating the east boundary of lot 14, range 17; that this possession went with his deed to Young and continued down to the plaintiffs, because there was no evidence of an abandonment of that possession and because their constructive possession under color of title was never disturbed, interfered with nor questioned by anyone. The assumption that Sullivan had constructive possession of the disputed territory is not supported by the evidence. While·he claimed to the Mulligan line and the line run by him south, which was plainly marked, and was in actual possession of the land thus enclosed outside the disputed tract, he did not have constructive possession of the disputed tract. The defendant's grantor was claiming the whole of lot 14, range 16, under his deed and so had constructive possession thereof. Having the title, he would not be disseized by anything short of actual adverse possession. *Hodges v. Eddy*, 38 Vt. 327; *Jakeway v. Barrett*, 38 Vt. 316; *Rice v. Chase*, 74 Vt. 362. 52 A. 967, *Davenport v. Newton*, 71 Vt. 11, 42 A. 1087; *Chase v. Eddy*, 88 Vt. 235, 92 A. 99. While one may retain possession once acquired by intention merely, if uninterrupted, intention or claim of ownership is not sufficient for the acquisition of possession. *Webb v. Richardson*, 42 Vt. 465. By causing the line south from the Mulligan corner to be run and marked Sullivan did not take actual possession of the disputed tract. Such act would not interrupt the constructive possession of defendant's grantor. *Oatman v. Fowler*, 43 Vt. 462; *Kidder v. Kennedy*, 43 Vt. 717; *Paine & Slocum v. Hutchins*, 49 Vt. 314; *Rice v. Chase*, 74 Vt. 362, 367, 52 A. 967. Moreover, Sullivan was holding under paper title which limited his constructive possession to the bounds of lot 14, range 17. *Chase v. Eddy*, 88 Vt. 235, 92 A. 99; *Rice v. Chase*, 74 Vt. 362, 366, 52 A. 967. And besides, if all the elements of adverse possession in Sullivan had been present, it was interrupted during the time Young owned the lot and could not be resorted to help out the possession of Pope and Gilbert and their grantees.

These holdings dispose of all the questions raised by the exceptions to the charge and exceptions to the court's refusal to charge as requested that are of sufficient importance to merit attention. We do not deem it necessary to consider them *seriatim*.

**J. H. Silsby & Co. v. C. A. Kinsley, 89 Vt. 263, 95 A. 634 (Vt. 1915).**

## VIRGINIA

The court is of opinion, that in a controversy concerning the boundary or locality of a tract of land granted by the commonwealth, pursuant to a survey, the calls and description of a survey made by the same surveyor, about the same time, or recently thereafter, of a coterminous or neighbouring tract, upon which last mentioned survey, a grant has also issued from the commonwealth, whether to a party to the controversy, or a stranger, is proper evidence upon such question of boundary or locality, unless plainly irrelevant; to have such weight with the jury as under all the circumstances, they may consider it entitled to. And, therefore, that the said circuit court erred in rejecting the surveys for Thomas and Lewis, mentioned in the second branch of the demandants' first bill of exceptions.

The court is further of opinion, that in a controversy concerning the boundary or locality of a tract of land granted by the commonwealth, upon a survey made by a duly authorized surveyor, evidence is properly admissible of declarations by such surveyor, or by chaincarriers who assisted him in making such survey, or by other persons present at such survey, of the acts done by, or under the authority of, such surveyor, in making such survey, unless plainly irrelevant; and as to the persons present at such survey, to have such weight as the jury, under all the circumstances, may consider it entitled to: provided however, that such declarations were not made post litam motam, and are not in contradiction of such surveyor's official report of such survey; and provided that the person or persons who made such declarations, be dead at the time of the trial: and, therefore, that there is no error in the decision of the said court, admitting the evidence of the declarations of the surveyor William Pettyjohn, mentioned in the tenant's first bill of exceptions: but that said court erred in rejecting the evidence of the declarations of Major Powers, mentioned in the first branch of the demandants' first bill of exceptions.

The court is further of opinion, that on the trial of a writ of right, upon the mise joined on the mere right, the tenant is entitled to the opening and conclusion of the cause before the jury: and, therefore, that there is no error in the decision of the said circuit court stated in the demandants' second bill of exceptions.

The court is further of opinion, that where the land in controversy is embraced by conflicting grants from the commonwealth, to different persons; and the junior patentee enters thereupon, and takes and holds actual possession of any part thereof, claiming title to the whole under his grant; that such adversary possession of part of the land in controversy, is an adversary possession of the whole, to the extent of the limits of the younger patent; and to that extent is an ouster of the seisin or possession of the older patentee, if the latter has had no actual possession of any part of the land within the limits of his grant. But that if the older patentee, at the time of such entry of the younger patentee, is in the actual possession of any part of the land in controversy; then that the latter can gain no adversary possession, beyond the limits of his mere enclosure, without an actual ouster of the older patentee, from the whole of the land in controversy. And, moreover, that upon the question of adversary possession, it is immaterial whether the land in controversy, be embraced by one, or several coterminous grants of the older patentee; or one or several coterminous grants of the younger patentee: in either case, the lands granted to the same person by several patents, must be regarded as forming one entire tract. Wherefore, the court, so understanding the instruction of the said circuit court to the jury, stated in the demandants' third bill of exceptions, is of opinion that there is no error therein, so far as it conforms to the opinions above expressed.

The remaining question arising on the third instruction, is, whether the actual possession under the elder patent, that is to say, the pedis positio, or actual occupancy by building, clearing, cultivating, or enclosure, which would limit the adversary possession under the junior patent, to that part of the interlock or lap, within his enclosure, must be of a part of the land within the lap; and whether such actual possession under the elder patent, of part of the land not within the lap, would thus limit the adversary possession under the junior patent to his enclosure. On this question no judgment is pronounced, because of the diversity of the opinions of the judges thereon; and because of the high probability that whatever that judgment might be, it would have no practical effect in this particular case.

The court is further of opinion, that where lands have been granted by the commonwealth to different persons, by conflicting patents, the junior patentee cannot, under any circumstances, disseize or oust the older patentee from, or acquire an adversary possession of, the land in controversy, but by the actual occupation of some part thereof, or the use or enjoyment of some part thereof, by acts of ownership equivalent to such actual occupation: and that while such patented lands remain completely in a state of nature, they are not susceptible of a disseisin or ouster of, or adversary possession against, the older patentee, unless by acts of ownership effecting a change in their condition. And the court, so understanding the instruction stated in the demandants' fourth bill of exceptions, as asked for by the demandants' counsel, is of opinion, that the said court erred in not giving the same as asked for, without the addition thereto of the clause, "or by the open exercise of acts of ownership over the same"; whereby a disseisin or ouster of, or adversary possession against, the older patentee, was recognized as effected by acts of ownership falling short of such actual occupation, use or enjoyment as aforesaid.

*Overton's Heirs v. Davisson*, 42 Va. 211 (Va. 1844).

# WASHINGTON

This leaves but one question to be determined, namely, what is the proper method of determining the location upon the ground of the dividing line between the subdivisions above mentioned of the land formerly owned by Howard Carr? It is claimed by the respondents that this line was correctly ascertained and established, when the addition to the city was surveyed and platted, where respondents and their grantors have always claimed it to be, and that it cannot now be changed, after the lapse of so many years. In making the survey of the five-acre tract conveyed to Job Carr the surveyor located the northeast corner of the northwest quarter of the northeast quarter of section 31 – the starting point mentioned in the deed and also in that of the appellant – at a point on the north line of the section 1,320 feet west of the northeast corner thereof as established by the government survey, on the theory that, as the patent to Howard Carr stated that that quarter section contained 160 acres of land, the point sought for must of necessity be that distance-one-fourth of a mile-from the corner post. From the point thus established he proceeded to lay off the land covered by the deed by following the courses therein specified, and it is insisted that the north and south line thus located is the true line between the premises in dispute. This is controverted by the appellant, and it is urged on her behalf that the starting point called for in the deeds from Howard Carr lies midway between the northeast and the northwest corners of that quarter section. On this theory the appellant's survey was made. The surveyor who did the work testified that he ascertained this starting point by first platting the entire section, according to the government survey as shown by the original government posts, all of which were found, as well as the witness trees to those posts; that he carefully measured the four boundaries of the northeast quarter, and established the middle points of those boundaries, and then struck lines across, and quartered the northeast quarter; that in fact all of the boundaries were less than half a mile in length, although represented as being half a mile in the government field notes. In other words, he fixed his starting point exactly in the middle of the line between the northeast corner of the section and the half-mile post west on the north line, according to actual measurement upon the ground; and we think that the point so established was the true starting point called for in the deeds, and that a line drawn south from that point to the middle point of the south boundary is the true line of division between the premises in controversy. If the land in dispute lies west of that line, then the appellant is entitled to recover its possession, but otherwise if it lies on the east side thereof; and this will be the only question to be determined upon a new trial. The judgment is reversed, and the cause remanded to the court below for a new trial in accordance with this opinion.

*Packscher v. Fuller*, 6 Wash. 534, 33 P. 875 (Wash. 1893).

# WEST VIRGINIA

The second cause of error complained of is that the complainant was bound to prove the identity and location of the survey claimed by him, and upon the proofs he offered the court should have dismissed his bill, or at least should have directed an issue to try the question of identity. The identity of the 2,000 acre survey claimed by the claimant depended upon the identity of one or more other surveys in the same connection, because Sherwood, the only witness examined to prove the identity of it, says that he did not think that he could have identified this survey had it not been for the connection of surveys laid down on the plat made by him and marked "A." This witness made a survey of some of the lots of the connection of surveys in which the 2,000 acre tract claimed by the complainant is supposed to lie, and of some coterminous surveys, and in respect to their identity he says that in making said surveys he found "original work" sufficient to satisfy him that said surveys are located and exist as laid down in said plat marked "A," according to the calls of the patents for said land, and this is all the evidence in the case tending to show that the 2,000 acre survey is located as claimed by the complainant. This witness does not mention a single fact which would enable the court to determine the location of the land in question, but says that he found "original work" sufficient to satisfy him that the surveys are located as claimed.

The purpose of evidence is to satisfy the court and not the witness. It seems to me that this evidence was clearly insufficient to locate the land in question, as it amounted to nothing more than the opinion of the witness. 5 Graft., 120. It is claimed that the certificates of the auditor filed in the case show that the 2,000 acre tract claimed by the complainant was forfeited for omission and liable to entry. It is incumbent on him who alleges a forfeiture to show it. S. P. Moore, the patentee of the 2,000 acre tract claimed by the complainant, had upon the books of the commissioners of the revenue of the county of Harrison, for all of the years for which it might have been forfeited for omissions, a tract or tracts of land which from quantity and local description may have included this tract, and if it might have been included it must be presumed that it was until the contrary appears.

It is claimed here that the complainant has no title to the 2,000 acres claimed by him because the deed to him from Lamar Moore is not properly authenticated. The deed is acknowledged before a notary public in the republic of Texas, on the 14th day of December, 1844. This authentication I think is clearly insufficient. The authentication of the power of attorney from Lamar Moore, under which the partition of Samuel P. Moore was made, is insufficient in the same manner. Upon the whole case I think the decree complained of ought to be reversed, with costs to the appellant, and the cause remanded to the circuit court of Doddridge county for further proceedings to be had therein, with leave to the complainant to show, if he can, by an issue, or in any other appropriate manner, that the 2,000 acre tract of land claimed by him is located as claimed by him; and with leave, also, to the complainant to show, if he can, that the power of attorney from Lamar Moore to P. M. Adams, by virtue of which the lands belonging to the heirs of S. P. Moore were divided and the deed from Lamar Moore to the complainant for the 2,000 acre tract of land claimed by him were properly executed; and with leave to the defendant to show, if he can, that the said tract of land was forfeited and liable to entry.

Caldwell, J.

The bill of the complainant, in the court below, represents that the defendant, Randolph, by fraud and stratagem, procured a patent from the commonwealth of Virginia, about the year 1851, for 135 acres of land, the greater portion of which, if not all, lying in the county of Doddridge, and alleging further that the defendant, for fear of detection, went to the county of Harrison to obtain his patent for said land and has by various means undertaken to force possession of the same, he well knowing at the same time that the complainant had title to said land by and through S. P. Moore, to whom it had been patented many years before.

These allegations of the complainant's bill it is sufficient to notice in considering the first point made in the argument of the cause.

The defendant by his answer admits that he did obtain a grant for 135 acres of land, but says it is not true that the grant was obtained by fraud and stratagem, nor that defendant went to Harrison county to make his entry. That it was made in Doddridge county where the greater portion of the land lies; that a survey thereof was regularly and fairly made, and that the defendant fairly and *bona fide* deposited his plat and certificate of survey with the register of the land office, and obtained from the commonwealth a grant for the 135 acres, according to law. The defendant denies that he had by various means undertaken to force possession of said land, except that he had brought suit at law setting up his title to the land and for the purpose of obtaining possession of the same, and asserting that the law tribunal is fully competent to decide and settle the question of right between the complainant and himself to said land.

It will be seen from the transcript of the record that the allegations of fraud and stratagem in the procurement of the grant, as set up in complainant's bill, are fully and completely refuted by the answer and the exhibits filed in support thereof. This is conclusive of this part of the case. But then it is alleged in the last paragraph of the bill, "that from the premises stated, the complainant believes said grant of 135 acres of land is void or voidable, because it was issued contrary to law and to his prejudice." I remark first, the statute provides that a grant may be repealed where it is shown to have been issued "to the prejudice of the equitable rights of a party." The allegation of complainant that this grant was issued to his prejudice amounts to nothing in a question of the repeal of a grant. And secondly, that the grant here is either a void or a valid grant. It cannot be considered in any measure a voidable grant. True, it is absolutely void if issued contrary to law, as claimed by the complainant.

Now the question arises, what is the purport and effect of the terms in the statute "contrary to law?" Grants for land in Virginia were obtained by making due entry of the land proposed to be taken up, obtaining a plat and certificate of survey thereof within the time prescribed by law, and returning the same to the register of the land office in the manner required by the statute. If these steps have not been fully taken the grant would be adjudged as having issued contrary to law. But I do not understand that where a grant of land may conflict with the claim of any other land proprietor, such grant having been obtained regularly, all the requirements of the statute being fully complied with, it should, therefore, be determined to have been issued contrary to law in the sense properly to be given, in my opinion, to the terms of the statute under consideration. Upon the application of a party, Virginia land warrants are issued, requiring the surveyor of any county in the State to enter on his land book and to survey for the applicant "any vacant and unappropriated land or lands liable to entry," as he might designate. When the entry has been duly made, and a plat and certificate of survey of the land returned, within the time prescribed by law to the land office, a grant will issue, unless restrained by a caveat proceeding. Such grant, I maintain, will not have issued contrary to law, though it may be alleged to conflict with a grant previously issued. The commonwealth for a consideration awards a grant, assuring, to some extent, to the patentee, title to the land embraced by such grant. This being so, there can be no inquiry in a court of equity as to any conflict between various grantees touching their respective claims to land granted to them by the commonwealth. The jurisdiction of such courts is to adjudge whether in the procurement of a grant of land, fraud has been perpetrated, or the steps required to be taken towards its issue have been pursued, or whether it was obtained to the prejudice of another party's equitable rights, and to this extent only.

In my opinion the court below erred in not dismissing the complainant's bill, and the decree should therefore be reversed.

*Randolph v. Adams*, **2 W.Va. 519 (W.Va. 1868).**

## WISCONSIN

This is the common case of contest over the boundaries of land by adjoining owners. The evidence tended to show that both parties purchased their lands and went into the possession and occupancy thereof about the same time, and about thirty years ago, and that the defendant built line-fences on the lines denoted by the monuments then standing of the original government survey between his land and that of the plaintiff, and that afterwards the plaintiff, when he came into the possession of his land, joined with the defendant in building fences on the same lines, and they divided the fences between them as their respective shares or parts to be kept up and maintained in the same place and on the same lines by the parties. At one time the plaintiff set out a hedge on his portion of the line as a fence, and at another time he built a board fence on the same line. Both parties have occupied their lands according to the lines or boundaries indicated by such fences, and improved and cultivated the same, and acquiesced in such boundary lines, and believed that they were the true lines for a long time, and until a new survey was made by the county surveyor, in 1872, at the request of the plaintiff, and a new line established by such resurvey, and, even after that, and near the time of the commencement of this suit, the defendant had never been notified that the old lines had been changed by that survey. The parties have continued to occupy and enjoy their respective premises according to the old lines indicated by such fences nearly to the time of the commencement of this suit without question or controversy. By the survey of 1872, and a survey agreeing therewith made recently, there are two narrow and uneven strips, of less than one acre in all, of the land so occupied by the defendant, on the north and west of the plaintiff's land, which the plaintiff seeks to recover in this suit. In addition to such practical location of the division lines between the parties, there was evidence tending to show that the fences built on such lines by them were placed upon the lines indicated by existing monuments of the government survey. This of itself might have justified the jury in rendering a verdict for the defendant, because, according to many cases in this court, such monuments must govern in such cases of subsequent controversy. But this question was not specially submitted to the jury by instructions.

The answer, after denying that the plaintiff was seized of these strips of land, or had any estate or interest in them, or that the defendant unlawfully withholds them, or that the plaintiff is entitled to the possession of them, sets up the defense of twenty years adverse possession and the statute of limitations. We have examined carefully the instructions of the court to the jury, and those asked by the plaintiff, and it seems to us that those given were unusually clear and correct, and embraced all of those asked by the plaintiff which ought to have been given. They were all in respect to the questions of adverse possession and the statute of limitations, ignoring all other defenses. Most of the instructions asked on behalf of the plaintiff were inapplicable to the case, there being no evidence to which they could apply. The instructions asked were to the effect that if the jury find that at any time since 1865 the defendant recognized the plaintiff's title to these strips of land by trying to purchase them from him, or that if the defendant negotiated for their purchase, then he did not hold adversely, and they must find for the plaintiff. There was some evidence by the son of the plaintiff that the defendant offered to buy these strips of land in 1882 of the plaintiff, but this the defendant denied, and there was some evidence that after this there was some talk between the parties about arbitrating the question of the lines. To make such an instruction proper on this evidence, the inference to be drawn from it must have been that the defendant thereby submitted to the title of the plaintiff, and yielded his claim to the land in dispute.

But such an inference would be most violent and unreasonable. Then, too, there was threatened litigation by the plaintiff, and the natural inference would be that the defendant wished to avoid such litigation, and made the offer for that purpose, if he made it at all. This is rendered the more probable by the subsequent talk of arbitration between them. It is very questionable whether this evidence should have affected the verdict one way or the other, and we do not think it error in the refusal of the court to so charge the jury, in view of the other evidence in the case which so clearly and conclusively showed that the defendant had held adversely for over twenty years, and that the plaintiff had not been seized or in the possession of the land in dispute for the same length of time or within the period of limitation of the action.

As to the main question of adverse possession, the authorities are uniform that if two adjacent or coterminous proprietors agree upon and establish a dividing line between their premises, and actually claim and occupy the land on each side of that line continuously for twenty years, such possession will be *adverse* and confer a *title* by prescription. *Bader v. Zeise,* 44 Wis. 96, and the authorities cited in the opinion. The learned counsel has cited numerous authorities from various states to the same effect, but there being no conflict of authority upon the question, I will not swell this opinion by their citation.

The only other remaining question discussed in the briefs of counsel is whether, conceding that these strips of land belonged to the plaintiff at any time within the last twenty years, he had such a title to it all that time that he could have brought ejectment to oust the defendant therefrom. The land was sold by the state as a part of the school lands, June 1, 1850, to one George Fancher, who received a certificate therefor, which certificate was assigned to the plaintiff November 8, 1854, and the plaintiff obtained a patent thereon December 19, 1865. So, for a part of the last twenty years' possession of the defendant, the plaintiff held the land by this certificate. It is settled by the statute, as well as by the decisions of this court, that the plaintiff could have brought ejectment while holding only by this certificate, at any time. Sec. 20, ch. 24, R. S. 1849; sec. 51, ch. 28, R. S. 1858; sec. 220, R. S. 1878. These sections provide that the holder of the certificate has "the same rights of possession, enjoyment, descent, transmission, and alienation of the land therein described, and the same remedies for the protection of said rights as against all persons except the state, that he would possess if he were the owner thereof in fee." In *Gunderson v. Cook*, 33 Wis. 551, it is held under these statutes that the holder of a school land certificate may bring eject-ment against a person in possession, even under a subsequent patent; and it was first so held that ejectment would lie, in *McGlone v. Prosser*, 21 Wis. 273. These authorities are conclusive, both as to the plaintiff's estate and interest in the land under his certificate and his right of action, and it would be a waste of time and space to cite other authorities on a point so plain by the statute.

We can find no error in the record, and we cannot doubt but that the jury rendered the proper verdict on the evidence. There is not likely to be much merit in such an action, between two old settlers and neighbors, who entered upon the possession, improvement, and enjoyment of contiguous wild land, for homes and farms, and did the best they then could to ascertain their boundary lines by the monuments then existing, and fixed and agreed upon the lines and built fences thereon and continuously occupied, cultivated, and improved the same accordingly for from thirty to forty years, by mutual acquiescence, and then one of them obtains a new survey and thereby finds his neighbor in possession on his side of the old fence of a little narrow strip of half an acre. In such a case new surveys, after all the ancient monuments except the old fences are gone, are not to be encouraged, and they are very seldom correct.

**Tobey v. Secor, 60 Wis. 310, 19 N.W. 99 (Wis. 1884).**

## WYOMING

The law is well settled that boundaries of land entered upon under an original survey cannot be changed by a resurvey to affect title acquired under the original survey. *Weaver v. Howatt (Cal.)* 152 P. 926; *Russell v. Maxwell*, 158 U.S. 256; 39 L.Ed. 971; *Bentley v. Jenne*, 33 Wyo. 1; *Porter v. Carstensen (Wyo.)* 274 P. 1072; *Cragin v. Powell*, 128 U.S. 693. The point for decision is the determination of the lands fixed by the original survey. Plaintiff did not acquire title to the dis-puted land by adverse possession.

"Although the power to correct surveys of the public land belongs to the political department of the Government and the Land Department has jurisdiction to decide as to such matters while the land is subject to its supervision and before it takes final action * * * this power of supervision and correction by the Department is 'subject to the necessary and decided limitation' that when it has once made and approved a governmental survey of public lands and has disposed of them, the courts may protect the private rights acquired against interference by corrective surveys sub-sequently made by the Department * * *. A resurvey by the United States after the issuance of a patent does not affect the rights of the patentee; the government after conveyance of the lands, having no 'jurisdiction to intermeddle with them in form of a second survey.' * * * And although the United States, so long as it has conveyed its land, may survey and resurvey what it owns, and establish and re-establish boundaries, what it thus does is 'for its own information,' and 'cannot affect the rights of owners on the other side of the line already existing.'"

*In Washington Rock Co. v. Young*, 29 Utah 108, 80 P. 382, it was held that where an original government survey of land was made before the township line was established, the fact that a retracing of such survey placed the corner of a section east of the township line as subsequently established, and in another township, could not injuriously affect the rights of a party holding under a government patent based on the original survey, and that the original survey is controlling. In *Harrington v. Boehmer,* 134 Cal. 196, 66 P. 214, 489, it is stated that "a government township lies just where the government surveyor lines it out on the face of the earth." In *Galt v. Willingham,* 11 F.2d 757, the court stated: "The only thing on which appellants reasonably can rely is the failure of the government surveyor to run his range line due north and south. But in re-establishing the lines of the survey the footsteps of the original surveyor should be followed, and it is immaterial that the lines actually run by him are not correct." If that is true of a range line, it must be equally true of the lines of a standard parallel. The only distinctive feature in this case is that it is sought, under the provisions of rule 378 supra, to deprive a true government corner, which is the only one in that particular place, or in the neighborhood thereof, which marked the north boundary of the township and the closing corner of two sections, of the character as such corner. We find no justification therefor under the decisions, which are uniform to the effect that government corners mark the true boundaries. The trial court in *Galt v. Willingham,* 300 F. 761, said: "Granted that the rules governing surveyors of government land are required to run range lines due north and south, yet if the surveyor does not do this, as I understand the law, when it comes to re-establishing the lines, they are to be run as the surveyor ran them at the time of making his survey, and not what he ought to have done. And so strict is this rule that not even the government can change the lines to the detriment of private interests." So in this case, granted that the surveyors of 1883 should have established the closing corner here in question 218.01 feet north of where they did, yet since they did not do so, the rights of the patentees here in question and those of their successors in interest cannot be prejudiced thereby.

**See Ben Realty Co. v. Gothberg, 56 Wyo. 294, 109 P.2d 455 (Wyo. 1941).**

# Appendix B
# Notice Requirements, by State

The following is a list of decisions regarding the notice requirements, selected from the various states. While there are many more, if notice, in any form, becomes an issue, it is suggested that the individual undertakes a thorough study of, at the very least, the requirements in the state(s) where the question(s) arose.

**Author's Note:** It is suggested that the reader or researcher read this entire list to understand the significance of the notice requirements in the various states and jurisdictions. It is obvious that the courts view notice very seriously, and that any professional, or investigator, should constantly be on guard as to the significance of even a subtle piece of evidence, in whatever form it may appear. Even the most seemingly insignificant item can result in failure, or a later result giving a different, even opposing, conclusion.

## ALABAMA

"[If a purchaser has] sufficient information to stimulate inquiry which would lead to knowledge of adverse or hostile and superior claim or title, and fails therein, the injury is the result of his own folly he is wanting in good faith, an indispensable element of a purchaser (for value) without notice and a court of equity will not protect such reckless purchaser." *Leslie v. Click*, 221 Ala. 163, 128 So. 170, (1930).

In this jurisdiction the rule is, whatever is sufficient to put one on his guard and call for inquiry is notice of everything to which due inquiry would lead. *Wittmeir v. Leonard*, 219 Ala. 314, 122 So. 330 (Ala. 1929).

## ALASKA

"A person is alerted to the fact that he or she has a potential cause of action when he or she discovers, or reasonably should have discovered, the existence of all elements essential to his or her cause of action. It is settled law that when a reasonable person is so alerted, the person is put on inquiry notice. This means that the person is "deemed to have notice of all facts which reasonable inquiry would disclose." *Mine Safety Appliances Co. v. Stiles*, 756 P.2d 288, (Alaska 1988) (quoting *Vigil v. Spokane County*, 42 Wash.App. 796, 714 P.2d 692, (1986)). Referenced in *Pedersen v. Zielski*, 822 P.2d 903 (Alaska 1991).

## ARIZONA

"Reasonable diligence requires that a person make inquiry when it is reasonable and prudent to do so, and a person is charged with notice of all facts that a reasonable inquiry would reveal." *Alameda Materials, Inc. v. Capital Title Agency, Inc.*, 1 CA-CV 11-0203 (AZAPP1) (2011).

## ARKANSAS

"Actual notice of the facts upon which an action may be sustained is not necessary to put the statute in motion," as was held in the late case of *Bland v. Fleeman*, 58 Ark. 84. As was said by the United State circuit court of appeals for the eighth circuit, in the case of *Percy v. Cockrill*, in the 53 F. Rep. 872, "Notice of facts and circumstances which would put a man of ordinary intelligence and prudence on inquiry is, in the eye of the law, equivalent to knowledge of all the facts a reasonably diligent inquiry would disclose. Whatever is notice enough to excite attention, and put the party on his guard, and call for inquiry, is notice of everything to which such inquiry might have led. Where a person has sufficient information to lead him to a fact, he shall be deemed conversant with it." *Thomas v. Sypert*, 33 S.W. 1059, 61 Ark. 575 (Ark. 1896).

## CALIFORNIA

"Notice of facts and circumstances which would put a man of ordinary prudence and intelligence on inquiry is.......equivalent to knowledge of all of the facts a reasonably diligent inquiry would disclose." *Schneider v. Henley*, 61 Cal.App. 758, 215 P. 1036 (1923).

## COLORADO

"We have traditionally recognized three forms of notice: actual notice, constructive notice, and inquiry notice. Actual notice occurs when a party has actual knowledge of a title defect. While both constructive notice and inquiry notice operate to impute knowledge to a party under certain specific conditions, we recognize them as separate inquiries. Constructive notice arises where a search of the title records would have revealed a defect. Inquiry notice arises when a party becomes aware or should have become aware of certain facts which, if investigated, would reveal the claim of another." *Martinez v. Affordable Housing Network, Inc.*, 123 P.3d 1201 (Colo. 2005).

## CONNECTICUT

A purchaser who has sufficient information to put him on inquiry is chargeable in equity with full notice. *Boswell v. Goodwin*, 31 Conn. 84; *Sigourney v. Munn*, 7 Id., 333; *Peters v. Goodrich*, 3 Id., 146; *Booth v. Barnum*, 9 Id., 286; *Curtis v. Mundy*, 3 Met., 405; *Blackwood v. Jones*, 4 Jones' Eq., 654; 1 Story's Eq. Jur., § 400; Willard's Eq., 249; *Berry v. Mutual Ins. Co.*, 2 Johns. Ch., 612.

Notice of a deed is notice of its contents. *Mertins v. Joliffe*, Ambler, 311; Hilliard on Vend. and Pur., 403; *Patten v. Moore*, 32 N.H. 382; *Powell v. Devlin*, 2 Ball & B. 416; *Dunham v. Dey*, 15 Johns., 555; (overruling Dey v. Dunham, 2 Johns. Ch., 182) *Berry v. Mut. Ins. Co.*, 2 Johns. Ch., 611; *Cuyler v. Brady*, 2 Caines Cas., 326.

Notice to a man's attorney, agent, or counsel is sufficient notice to him. Powell on Mortgages, 588, 591; Hilliard on Vend. and Pur., 408; 1 Story's Eq. Jur., § 408; Newland on Cont. in Eq., 505; *Hargrave v. Rothwell*, 2 Keen, 154. Such notice is not constructive but actual. *Lenehan v. McCabe*, 2 Irish Eq. R., 342.

Men of ordinary prudence will use all reasonable means to ascertain the state and condition of their own titles. Hence, we may lay it down as a rule, founded upon the experience of mankind, that one who has knowledge of the existence of a deed, to which he has access, and which affects the title to property in which he is interested, will, in equity, be presumed to have knowledge of the contents of the deed. And generally, when a purchaser cannot make out a title but by a deed which leads him to another fact, he shall be presumed to have knowledge of that fact. 1 Story's Eq. Jur., § 400, and cases there cited. Under our recording system a deed duly recorded is constructive notice to all the world; and the law conclusively presumes that every person interested has knowledge, not only of the deed, but of its precise language, where that is material. These principles apply in full force to this case. If a man will, under certain circumstances, be presumed to have knowledge of the contents of the deed of another, how much more reasonable is it to presume that he has knowledge of the contents of his own deed. Occasional hardships may result from the application of this rule; but it is believed to be founded in sound policy, and that, in a large majority of cases, it will tend to prevent fraud and promote the cause of justice. *Hamilton v. Nutt*, 34 Conn. 501, (1868).

## DELAWARE

In circumstances where blameless ignorance is present, the limitations period will not begin to run until the "plaintiff is objectively aware of the facts giving rise to the wrong, i.e., when the plaintiff is on inquiry notice." Inquiry notice can be found when a plaintiff is aware of the "'existence of facts sufficient to put a person of ordinary intelligence and prudence on inquiry, which, if pursued, would lead to the discovery' of such facts." Inquiry notice is satisfied once a plaintiff is in possession of facts sufficient to make a reasonable person suspicious of the alleged wrongdoing. *Rich Realty, Inc. v. Meyerson & O'Neill*, C. A. N13C-06-270 MMJ.

Plaintiff alleges the statute of limitations was tolled by (i) Defendants' fraudulent concealment of the facts; (ii) principles of equitable tolling; or (iii) the inherently unknowable nature of the Ceiling Cracks' cause. All those tolling theories, however, even if applicable to the facts of this case, only apply until a person is on inquiry notice of the facts supporting a claim. "Inquiry notice" means a person is aware of facts "sufficient to put a person of ordinary intelligence and prudence on inquiry which, if pursued, would lead to the discovery." Once a person is on inquiry notice of an injury, the statute begins to run even if that person is unaware of the cause of the injury. *Armstrong v. Council of Devon*, N16C-09-026AML 2018.

## FLORIDA

Florida recognizes three types of notice in cases concerning covenants running with the land-constructive notice, actual notice, or implied actual notice.

"'Constructive notice' has been defined as notice imputed to a person not having actual notice." *Sapp v. Warner*, 105 Fla. 245, 141 So. 124, (1932). "The usual instance of constructive notice, is of course, restrictions in a recorded deed or plat. And the authorities are practically unanimous in holding that the recorded deed containing such restriction is not necessarily the immediate deed by which the instant owner takes or has taken title; it may be in an antecedent deed, even the deed from the original common grantor." *Hagan v. Sabal Palms, Inc.*, 186 So.2d 302, (Fla. 3d DCA 1966).

*Park Avenue BBQ* is an actual notice case. The barbeque restaurant had actual knowledge of the sports bar's exclusivity provision; the restaurant's president "hand wrote language into his lease allowing [the restaurant] to televise sporting events because he was concerned about [the sports bar's] exclusivity provision." 746 So.2d at 482. Actual or "express" notice is based on "direct information" leading to "actual knowledge of the fact in question." *Sapp*, 141 So. at 127; *McCausland v. Davis*, 204 So.2d 334, (Fla. 2d DCA 1967).

Implied actual notice includes "notice inferred from the fact that the person had means of knowledge, which it was his duty to use and which he did not use." *Sapp*, 105 Fla. at 255. If a person has information that would lead a reasonable man to make further inquiry for his own protection, but fails to further investigate and learn what the inquiry would reasonably have uncovered, the person "must suffer the consequence of his neglect." *Flanigan's Enters., Inc. v. Shoppes at 18th & Commercial, Inc.*, No. 4D06-1276, 2007 WL 1342543 (Fla. 4th DCA May 9, 2007) (quoting *Chatlos v. McPherson*, 95 So.2d 506, (Fla.1957)); *Rafkind v. Beer*, 426 So.2d 1097, (Fla. 3d DCA 1983); see also *Citgo Petroleum Corp. v. Fla. E. Coast Ry. Co.*, 706 So.2d 383, (Fla. 4th DCA 1998) (citing *Chatlos*). Implied actual notice is based upon the principle that a person has no right to shut his eyes or ears to avoid information, and then say that he has no notice; that it will not suffice the law to remain willfully ignorant of a thing readily ascertainable by whatever party puts him on inquiry, when the means of knowledge is at hand.

*Sapp*, 141 So. at 127 (quoting *Knapp v. Bailey*, 9 A. 122 (Me. 1887)). *Winn-Dixie Stores, Inc. v. Dolgencorp, Inc.*, 4D06-1585 2007.

## GEORGIA

The law in Georgia is well-settled that in the purchase and sale of real estate there is an underlying principle of law that one cannot be permitted to claim that he has been deceived by false representations about which he could have learned the truth of the matter and could have avoided damage. When the means of knowledge are at hand and equally available to both parties if the purchaser does not avail himself of these means he will not be heard to say, in impeachment of the contract, that he was deceived by the representations of the seller.

*BPP069, LLC v. Lindfield Holdings, LLC*, A18A0072 (GACA) (2018).

OCGA § 23-1-17 provides: "Notice sufficient to excite attention and put a party on inquiry shall be notice of everything to which it is afterwards found that such inquiry might have led. Ignorance of a fact due to negligence shall be the equivalent to knowledge in fixing the rights of parties." See also *Reidling v. Holcomb*, 225 Ga.App. 229, (1) (483 S.E.2d 624) (1997). "For more than a century, it has been recognized that a purchaser of land in this state is charged with notice of every fact shown by the records, and is presumed to know every other fact which an examination suggested by the records would have disclosed." (Citations and punctuation omitted.) *Deljoo v. SunTrust Mortgage*, 284 Ga. 438, (668 S.E.2d 245) (2008). Indeed, Georgia law provides for "broad constructive notice upon recording." *Id.* at 440. Accordingly, "[a] purchaser of land is charged with constructive notice of the contents of a recorded instrument within its chain of title." *Virginia Highland Civic Assn. v. Paces Properties*, 250 Ga.App. 72, (550 S.E.2d 128) (2001). "Chain of title includes all recorded instruments pertaining to the property that are executed by an entity holding a recorded interest in the property at the time of the execution of the instrument." (Citation and punctuation omitted.) *Suntrust Bank v. Equity Bank*, 312 Ga.App. 644, (719 S.E.2d 539) (2011). Finally, we have held that an incorrect property description in a deed will not necessarily render the deed "invalid and take it out of the chain of title." *Id.*

It is only when a description of property is manifestly too meager, imperfect, or uncertain to serve as adequate means of identification that the court can adjudge the description insufficient as a matter of law. Indeed, while the description may be inaccurate, meager, or erroneous, yet if it is expressed in such a manner or connected with such attendant circumstances as that a purchaser should be deemed to be put upon inquiry, if he fails to prosecute this inquiry he is chargeable with all the notice he might have obtained had he done so.

## HAWAII

***American Sec. Bank v. Read Realty, Inc.*, 1 Haw.App. 161, 616 P.2d 237 (Hawai'i App. 1980):**
Constructive notice is a legal inference or presumption by which, for promotion of sound policy or purpose, the legal [1 Haw.App. 165] rights and interests of parties are treated as though they had actual notice or knowledge. *Black v. Public Service Elec. & Gas Co.*, 98 N.J.Super. 366, 237 A.2d 495 (1968), *Merriman v. Cash-Way Inc.*, 35 Wis.2d 112, 150 N.W.2d 472 (1967).

***Corbett v. Association of Apartment Owners of Wailua Bayview Apartments*, 70 Haw. 415, 772 P.2d 693 (Hawai'i 1989):**

"Actual notice" means that the defendant actually knew the condition was unreasonably dangerous. "Constructive notice" of an unreasonably dangerous condition arises by operation of law and may be inferred if the facts indicate a reasonably prudent person would have discovered the dangerous condition in time to remedy it or give warning before the injury occurred.

## IDAHO

Whatever is notice enough to excite the attention of a man of ordinary prudence and prompt him to further inquiry, amounts to notice of all such facts as a reasonable investigation would disclose. *Farrell v. Brown*, 111 Idaho 1027, 729 P.2d 1090 (1986).

## ILLINOIS

Whatever is sufficient to put a purchaser upon inquiry is notice of whatever the inquiry would disclose: *Bent v. Coleman*, 87 Ill. 364; *U.S. Mortgage Co. v. Gross*, 93 Ill. 483; *Watt v. Schofield*, 76 Ill. 261; 1 Story's Eq. Jur. § 483; 2 White & Tudor's Lead. Cas. 117. *Battenhausen v. Bullock*, 11 Ill.App. 665, (1882).

## INDIANA

A lease for more than three years must be recorded in the recorder's office of the county where the land is situated. Ind.Code § 32-21-4-1(a) (2012). When such a lease is "properly acknowledged and placed on record" as required by statute, it is "constructive notice of [its] existence," and a subsequent grantee is charged with "notice of all that is shown by record, including recitals in instruments so recorded." *C. Callahan Co. v. Lafayette Consumers Co.*, 102 Ind.App. 319, 2 N.E.2d 994, 1000 (1936); *see also Bank of New York v. Nally*, 820 N.E.2d 644, 648 (Ind. 2005) (holding that a "purchaser of real estate is presumed to have examined the records" in the chain of title and "is charged with notice, actual or constructive," of all such properly recorded instruments).

Moreover, actual notice may be inferred from the fact that a person who is charged with a duty of searching the records of a particular property had the means of knowledge that he did not use. *Keybank Nat. Ass'n v. NBD Bank*, 699 N.E.2d 322, (Ind.Ct.App. 1998).

Whatever fairly puts a reasonable, prudent person on inquiry is sufficient notice to cause that person to be charged with actual notice, where the means of knowledge are at hand and he omits to make the inquiry from which he would have ascertained the existence of a deed or mortgage. *Altman v. Circle City Glass Corp.*, 484 N.E.2d 1296, (Ind.Ct.App. 1985), *trans. denied*. Thus, the means of knowledge combined with the duty to utilize that means equates with knowledge itself. *Id.* Whether knowledge of an adverse interest will be imputed in any given case is a question of fact to be determined objectively from the totality of the circumstances. *Id.* at 1299. *Wilson v. Huff*, 60 N.E.3d 294 (Ind.App. 2016).

## IOWA

As noted, Townsend also had to prove that Nickell had "express notice" of the prescriptive easement claim. *Johnson*, 637 N.W.2d at 180. This requirement ensures the landowner knows another's use of the property is claimed as a right hostile to the landowner's interest in the land. Otherwise, the landowner may incorrectly assume the other's use results merely from the landowner's willingness to accommodate the other's desire or need to use the land. *Larman v. State*, 552 N.W.2d 158, (Iowa 1996). The express notice must be actual or "from known facts of such nature as to impose a duty to make inquiry which would reveal the existence of an easement." *Johnson*, 637

N.W.2d at 180 (quoting *Collins Trust v. Allamakee County Bd. of Supervisors*, 599 N.W.2d 460, (Iowa 1999)). Based on the facts described above, we conclude there was sufficient evidence of Townsend's claim to Parcel C to place Nickell on inquiry notice of the existence of the easement. *Townsend v. Nickell*, 031109 IWCA, 08-1058.

## KANSAS

The recording act declares that conveyances of real estate not filed for record shall be void except as between the parties thereto and such persons as have actual notice thereof. The courts have uniformly held that "actual notice" as used in the act may be either express or implied. *Pope v. Nichols*, 61 Kan. 230, 59, 59 P. 257. Pac. 257; *Faris v. Finnup*, 84 Kan. 122, 113 P. 407. If the holder of an unrecorded deed is in the open, notorious, exclusive possession of the premises, a subsequent purchaser, and all others, have such implied notice of his rights as will amount to actual notice of his title. *Deetjen v. Richter*, 33 Kan. 410, 6 P. 595; *Stough v. Lumber Co.*, 70 Kan. 713, 79 P. 737; *Lang v. Adams*, 71 Kan. 309, 312, 80 P. 593.

The English rule is that possession by the tenant is notice simply of his tenancy, and never notice itself of the title of the lessor. And the English courts have held that it is not incumbent upon the purchaser to ask tenants found in possession to whom they pay their rent. *Hunt v. Luck* [1900], 1 Ch. 45, 70 L. J. Ch. 30; Note, 13 L. R. A., n. s., 99. There is great confusion and conflict among the American cases. Some follow the English rule; others hold that the possession of the tenant is the possession of the landlord, and that notice of the former is notice of the latter as fully as his own possession would be. *Glendenning v. Bell*, 70 Tex. 632, 8 S.W. 324. See, also, Note, 13 L. R. A., n. s., 100. We think the better rule, and the one which finds abundant authority for its support, is that although possession by a tenant is the landlord's possession, it is not itself notice of the landlord's title, but is, however, sufficient notice to put a person about to deal with the title upon inquiry. The law presumes that, having notice of the possession, he has either made inquiry and ascertained the extent of the occupant's right, or has been so negligent that he cannot be regarded as a bona fide purchaser. This presumption, however, is but a mere inference of fact which may be rebutted by proof that the purchaser made diligent inquiry and failed to discover the prior right. *Williamson v. Brown*, 15 N.Y. 354; *Betts v. Letcher et al.*, 1 S.D. 182, 46 N.W. 193; *Fair v. Stevenot*, 29 Cal. 486. Thus, in *Thompson v. Pioche*, 44 Cal. 508, it was held that although the possession of a tenant is not of itself notice of the landlord's title, yet such possession is sufficient to put a person dealing with the property upon inquiry, and is proof of notice, unless it is shown that the inquiry, having been prosecuted with due diligence, did not disclose the title of the landlord.

Possession is notice and necessitates inquiry, which must be prosecuted with due diligence. In *Deetjen v. Richter*, 33 Kan. 410, 6 P. 595, it was held under the facts in that case that due diligence required the purchaser to follow up the suggestion implied from information obtained from the tenant, by making further inquiry of the landlord. Through inquiry of the tenant the purchaser learned the name of the landlord; and it was said in the opinion, quoting from section 286 of Wade on Notice: "Inquiry cannot safely stop here; for the next step suggested by the circumstances would be to inquire of the landlord." *Penrose v. Cooper*, 86 Kan. 597, 121 P. 1103, (1912).

## KENTUCKY

Notice, generally, may be defined as that which imparts information of the fact to the one to be notified, and is divided by the law into several classes, such as actual, constructive, implied, and presumptive notice. *Burdine v. White*, 173 Ky. 158, 190 S.W. 687 (Ky.App. 1917).

## LOUISIANA

"Actual notice is knowledge of dangerous defects or conditions by a corporate office or employee of the public entity having a duty either to keep the property involved in good repair or to report defects and dangerous conditions to the proper authorities." *Boddie v. State,* 27,313, p. 6 (La.App. 2 Cir. 9/27/95), 661 So.2d 617, 622. Constructive notice, according to La.R.S. 9:2800(D), means "the existence of facts which infer actual knowledge." *Cowboy Connection, Inc. v. State,* 060607 LACA3, 07-68.

## MAINE

"Actual notice and actual knowledge are not necessarily synonymous expressions. Actual notice is that which gives actual knowledge, or the means to such knowledge. It is a warning brought directly home to one whom it concerns to know. Actual notice may be either express or implied. It is express when established by direct proof. It is implied when inferable as a fact by proof of circumstances. 'Express actual notice' is, perhaps, its own best definition. Implied actual notice is that which one who is put on a trail is in duty bound to seek to know, even though the track or scent lead to knowledge of unpleasant and unwelcome facts." *Hopkins v. McCarthy,* 121 Me. 27, 115 A. 513, (1921).

Elaborating more particularly on the concept of "implied actual notice," we earlier said in *Knapp v. Bailey,* 79 Me. 195, 9 A. 122, (1887), that:

"The doctrine of actual notice implied by circumstances (actual notice in the second degree) necessarily involves the rule that a purchaser before buying, should clear up the doubts which apparently hang upon the title, by making due inquiry and investigation. If a party has knowledge of such facts as would lead a fair and prudent man, using ordinary caution, to make further inquiries, and he avoids the inquiry, he is chargeable with notice of the facts which by ordinary diligence he would have ascertained. He has no right to shut his eyes against the light before him. He does a wrong not to heed the 'signs and signals' seen by him. It may be well concluded that he is avoiding notice of that which he in reality believes or knows. Actual notice of facts which, to the mind of a prudent man, indicate notice is proof of notice." *Gagner v. Kittery Water Dist.,* 385 A.2d 206 (Me. 1978).

There is a conflict in the cases and among writers as to what is actual notice. Much of the difference is said to be verbal only – more apparent than real. Certain propositions, however, are quite well agreed upon by a majority of the authorities. Notice does not mean knowledge; actual knowledge is not required. Mr. Wade describes the modes of proving actual notice as of two kinds. One he denominates express notice, and the other implied. "Implied, which imputes knowledge to the party because he is shown to be anxious of having the means of knowledge, though he does not use them; in other words, where he chooses to remain voluntarily ignorant of the fact, or is grossly negligent in not following up the inquiry which the known facts suggest." Wade, Notice, (2d Ed.) § 5. Some writers use the word "implied" as meaning constructive, and would regard what is here described to be implied actual notice as constructive notice merely. As applicable to actual notice, such as is required by the sections of the statute under consideration, we think the classification of the author whom we quote is satisfactory. The author further explains the distinction by adding that "notice by implication differs from constructive notice, with which it is frequently confounded, and which it greatly resembles, with respect to the character of the inference upon which it rests; constructive notice being the creature of positive law, or resting upon strictly legal inference, while implied notice arises from inference of fact." It amounts substantially to this: that actual notice may be proved by direct evidence, or it may be inferred or implied (that is, proved) as a fact from indirect evidence – by circumstantial evidence.

A man may have notice or its legal equivalent. He may be so situated as to be estopped to deny that he had actual notice. We are speaking of the statutory notice required under the conveyances act. A higher grade of evidence may be necessary to prove actual notice appertaining to commercial paper. *Kellogg v. Curtis,* 69 Me. 212. The same facts may sometimes be such as to prove both constructive and actual notice; that is, a court might infer constructive notice, and a jury infer actual notice, from the facts. There may be cases where the facts show actual, when they do not warrant the inference of constructive, notice; as where a deed is not regularly recorded, and not giving constructive notice, but a second purchaser sees it on the records, thereby receiving actual notice. *Hastings v. Cutler,* 24 N.H. 481.

Mr. Pomeroy (2 Eq. Jur. 596, note) summarizes the effect of the American cases on the point under discussion in the following words: "In a few of the states the courts have interpreted the intention of the legislature as demanding that the personal information of the unrecorded instrument should be proved by the direct evidence, and as excluding all instances of actual notice established by circumstantial evidence. In most of the states, however, where this statutory clause is found, the courts have defined the 'actual notice' required by the legislature as embracing all instances of that species in contradistinction from constructive notice; that is, all kinds of actual notice, whether proved by direct evidence or inferred as a legitimate conclusion from circumstances."

The doctrine of actual notice implied by circumstances (actual notice in the second degree) necessarily involves the rule that a purchaser, before buying, should clear up the doubts which apparently hang upon the title, by making due inquiry and investigation. If a party has knowledge of such facts as would lead a fair and prudent man, using ordinary caution, to make further inquiries, and he avoids the inquiry, he is chargeable with notice of the facts which by ordinary diligence he would have ascertained. He has no right to shut his eyes against the light before him. He does a wrong not to heed the "signs and signals" seen by him. It may be well concluded that he is avoiding notice of that which he in reality believes or knows. Actual notice of facts which to the mind of a prudent man indicate notice is proof of notice. 3 Washb. Real Prop. (3d Ed.) 335. *Knapp v. Bailey,* 79 Me. 195, 9 A. 122 (Me. 1887).

## MARYLAND

Notice is critical to the discovery rule. Before an action can accrue under the discovery rule, "a plaintiff must have notice of the nature and cause of his or her injury." *Frederick Rd.,* 360 Md. at 96, 756 A.2d at 973. There are two types of notice: actual and constructive. *Poffenberger,* 290 Md. at 636-37, 431 A.2d at 680. Actual notice is either express or implied. *Id.* at 636, 431 A.2d at 680. As the name suggests, express notice "is established by direct evidence" and "embraces not only knowledge, but also that which is communicated by direct information, either written or oral, from those who are cognizant of the fact communicated." *Id.* at 636–637, 431 A.2d at 680 (citation and internal quotation marks omitted). Implied notice, also known as "inquiry notice," is notice implied from "knowledge of circumstances which ought to have put a person of ordinary prudence on inquiry (thus, charging the individual) with notice of all facts which such an investigation would in all probability have disclosed if it had been properly pursued." *Id.* at 637, 431 A.2d at 681 (citation and internal quotation marks omitted). Stated simply, inquiry notice is "circumstantial evidence from which notice may be inferred." *Id.* at 637, 431 A.2d at 680 (citation and internal quotation marks omitted). Constructive notice is notice presumed as a matter of law. *Id.* at 636, 431 A.2d at 680. Unlike inquiry notice, constructive notice does not trigger the running of the statute of limitations under the discovery rule. *Id.* at 637, 431 A.2d at 681. *Windesheim v. Larocca,* 443 Md. 312, 116 A.3d 954 (Md. 2015).

## MASSACHUSETTS

### *Ryan v. Gardner Franco-America Federal Credit Union*, 18-P-1414 (MACA):

"[T]he statute of limitations starts when the plaintiff discovers, or reasonably should have discovered, 'that [he] has been harmed or may have been harmed by the defendant's conduct.'" *Id.*, quoting *Bowen v. Eli Lilly & Co.*, 408 Mass. 204, (1990). "Reasonable notice that ... a particular act of another person may have been a cause of harm to a plaintiff creates a duty of inquiry and starts the running of the statute of limitations." *Bowen*, supra at 210.

## MICHIGAN

### *Antilla v. Cbocs Properties, Inc.*, 342924 (MICA):

Whether a condition is open and obvious depends on "whether an average user with ordinary intelligence would have been able to discover the danger and the risk presented upon casual inspection." *Id.* (quotation marks, alteration, and citation omitted). This is an objective test, "and the inquiry is whether a reasonable person in the plaintiff's position would have foreseen the danger, not whether the particular plaintiff knew or should have known that the condition was hazardous."

## MINNESOTA

.......asserts that inquiry or implied notice may prove actual knowledge. This court has held that "[i]mplied notice differs from constructive notice in that, [it] relates to what one can learn by reasonable inquiry. *It arises from actual notice of the circumstances*, and not from constructive notice." *Elfelt*, 756 N.W.2d at 508 n.5 (second alteration in original) (quotation omitted) (emphasis added). Moreover, "[i]mplied notice charges a person with notice of everything that he could have learned by inquiry where there is sufficient actual notice to put him on guard and excite attention." *Id. Ariola v. City of Stillwater*, 889 N.W.2d 340 (Minn.App. 2017).

## MISSISSIPPI

"[A]ctual notice" is defined as notice expressly and actually given ... while "constructive notice" is defined as information or knowledge of a fact imputed by law to a person (although he may not actually have it), because he could have discovered the fact by proper diligence, and his situation was such as to cast upon him the duty of inquiring into it. *Backstrom v. Briar Hill Baptist Church, Inc.*, 184 So.3d 323 (Miss.App. 2016).

## MISSOURI

Whatever is enough to excite attention, put on guard, or call for inquiry, is notice of everything to which such inquiry might lead. *Harbison v. James*, 90 Mo. 411, 2 S.W. 292 (Mo. 1886).

[H]e was put upon inquiry as to the facts; and being put upon inquiry, the well-established principle of law is, that he stands affected with the knowledge of every fact which the prosecution of that inquiry would have disclosed. *Mathias v. O'Neill*, 94 Mo. 520, 6 S.W. 253 (Mo. 1887).

**Note – Capable of ascertainment:**

The Missouri Supreme Court defined "capable of ascertainment" as when "the evidence [is] such to place a reasonably prudent person on notice of a potentially actionable injury." *Powel v. Chaminade Coll. Preparatory, Inc.*, 197 S.W.3d 576, (Mo. 2006) (emphasis removed). This "objective" test is from the standpoint of a "reasonable person in [plaintiff's] situation." Id. at 584. Both the "character of the condition … and its cause" must be capable of ascertainment. *Elmore v. Owens-Illinois, Inc.*, 673 S.W.2d 434, (Mo. 1984). Cited in *Levitt v. Merck & Co., Inc.*, 914 F.3d 1169 (8th Cir. 2019).

## MONTANA

"The plat gave notice to [] that the property was being conveyed pursuant to a particular recorded document. It also put her on inquiry notice of the easement……Knowledge of the existence of a claim will be imputed to a party who has sufficient information to put it on inquiry notice of that claim." *Pearson v. Virginia City Ranches Ass'n*, 298 Mont. 52, 993 P.2d 688 (Mont. 2000).

## NEBRASKA

Whatever fairly puts a person on inquiry is sufficient notice, where the means of knowledge are at hand; and if he omits to inquire, he is then chargeable with all the facts which, by a proper inquiry, he might have ascertained. *League v. Vanice*, 221 Neb. 34, 374 N.W.2d 849 (1985). *In re Estate of Seidler*, 241 Neb. 402, 490 N.W.2d 453 (Neb. 1992).

The rule is:

> "Whatever fairly puts a person on inquiry is sufficient notice, where the means of knowledge are at hand; and if he omits to inquire, he is then chargeable with all the facts which, by a proper inquiry, he might have ascertained. This, in effect, means that notice of facts which would lead an ordinarily prudent man to make an examination which, if made, would disclose the existence of other facts is sufficient notice of such other facts." 20 R.C.L. 346, § 7.

*Baxter v. National Mortg. Loan Co.*, 128 Neb. 537, 259 N.W. 630 (Neb. 1935).

## NEVADA

"While difficult to define in concrete terms, a person is put on 'inquiry notice' when he or she should have known of facts that 'would lead an ordinary prudent person to investigate the matter further.'" Citing Black's Law Dictionary (2009). *Winn v. Sunrise Hosp. & Medical Center*, 128 Nev. Adv. Op. 23, 277 P.3d 458 (Nev. 2012).

## NEW HAMPSHIRE

Notice may be of three types: (1) actual; (2) record; and (3) inquiry.

"Actual notice" refers to "information concerning the fact – as, for example, concerning the prior interest, claim, or right – directly and personally communicated to the party."

"Record notice" is "notice arising solely from the record." "Record notice" is sometimes referred to as "constructive notice." "A subsequent purchaser … is on constructive notice of all claims revealed by the record, regardless of whether [the subsequent purchaser] ever looks at the record or ever sees the information contained there." "Record notice" is "notice inferred from the record."

"Inquiry notice" is notice that arises from a legal inference. "Inquiry notice" is notice of a fact that is "sufficiently 'curious' or 'suspicious,' according to normal human experience, that the purchaser should, as a matter of law, make an investigation into it." "If upon making the investigation into this first fact a second fact, namely that another person has a claim to the title of the property, is revealed, then the purchaser is considered to have inquiry notice of the claim itself." "In other words, because of the nature of the first fact, of which the purchaser has actual or constructive notice, a rebuttable inference is made that the purchaser has notice of the second fact. However, if a reasonable inquiry would not reveal the second fact, then the inference is rebutted." *Bilden Properties, LLC & a. v. S. Gerald Birin & a.*, 165 N.H. 253 (2013).

## NEW JERSEY

>........it will be sufficient if it is shown that he had the means of knowledge, that is, that he had notice of such facts as would have led a prudent man to further inquiry, which inquiry, if pursued, would have disclosed the infirmity of the paper. *Hamilton v. Vought,* 34 N.J.L. 187.

*The National Bank of Republic, of New York v. Young*, 41 N.J.Eq. 531, 7 A. 488 (N.J.Eq. 1886).

## NEW MEXICO

"A person who purchases an estate in the possession of another than his vendor is, in equity, that is, in good faith, bound to inquire of such possessor what right he has in the estate. If he fails to make such inquiry, which ordinary good faith requires of him, equity charges him with notice of all facts that such inquiry would disclose." *McBee v. O'Connell*, 19 N.M. 565, 145 P. 123 (N.M. 1914).

## NEW YORK

These known, conceded facts, assuming that of themselves they are not sufficient to constitute that full knowledge, action upon which will constitute a ratification, are at least abundantly sufficient notice to have put him upon inquiry to find out all the details that could be ascertained. *Kelly v. Connecticut Mut. Life Ins. Co.*, 50 N.Y.S. 139, 27 A.D. 336 (1898).

Notice is of two kinds: actual and constructive. Actual notice embraces all degrees and grades of evidence, from the most direct and positive proof to the slightest circumstance from which a jury would be warranted in inferring notice. It is a mere question of fact, and is open to every species of legitimate evidence which may tend to strengthen or impair the conclusion. Constructive notice, on the other hand, is a legal inference from established facts; and like other legal presumptions, does not admit of dispute. "Constructive notice," says Judge Story, "is in its nature no more than evidence of notice, the presumption of which is so violent that the court will not even allow of its being controverted." Story's Eq. Juris., § 399.

A recorded deed is an instance of constructive notice. It is of no consequence whether the second purchaser has actual notice of the prior deed or not. He is bound to take, and is presumed to have, the requisite notice. So, too, notice to an agent is constructive notice to the principal; and it would not in the least avail the latter to show that the agent had neglected to communicate the fact. In such cases, the law imputes notice to the party whether he has it or not. Legal or implied notice, therefore, is the same as constructive notice, and cannot be controverted by proof.

But it will be found, on looking into the cases, that there is much want of precision in the use of these terms. They have been not unfrequently applied to degrees of evidence barely sufficient to warrant a jury in inferring actual notice, and which the slightest opposing proof would repel, instead of being confined to those legal presumptions of notice which no proof can overthrow.

The phraseology uniformly used, as descriptive of the kind of notice in question, "sufficient to put the party upon inquiry," would seem to imply that if the party is faithful in making inquiries, but fails to discover the conveyance, he will be protected. The import of the terms is that it becomes the duty of the party to inquire. If, then, he performs that duty is he still to be bound, without any actual notice? The presumption of notice which arises from proof of that degree of knowledge which will put a party upon inquiry is, I apprehend, not a presumption of law, but of fact, and may, therefore, be controverted by evidence. *Williamson v. Brown*, 15 N.Y. 354 (1857).

## NORTH CAROLINA

The principle of constructive notice arises out of the duty of any would-be purchaser to reasonably and in common prudence see that his vendor has a prima facie good title, and because of this duty the purchaser will be affected with notice of the provisions of such deeds and other documents as are necessary to show the vendor's title. *Allen v. Allen*, 121 N.C. 328, 28 S.E. 513 (N.C. 1897).

Where observable defects in a highway have existed for a time so long that they ought to have been observed, notice of them is implied, and is imputed to those whose duty it is to repair them; in other words, they are presumed to have notice of such defects as they might have discovered by the exercise of reasonable diligence. *Willis v. City of New Bern*, 191 N.C. 507, 132 S.E. 286 (N.C. 1926).

Essential notice arises from: (1) actual notice or knowledge directly imparted to the proper officials of the municipality; (2) implied, constructive, or imputed notice. The principle creating and governing implied, constructive, or imputed notice is thus stated in *Shearman & Redfield on the Law of Negligence* (6th Ed.) vol. 2, § 369. *Pickett v. Carolina & N.W. Ry.*, 200 N.C. 750, 158 S.E. 398 (N.C. 1931).

## NORTH DAKOTA

Whatever is notice enough to excite attention, and put the party on guard, and call for inquiry, is notice of everything to which such inquiry might have led. When a person has sufficient information to lead him to a fact, he shall be deemed conversant with it. *Wetzler v. Nichols,* 53 Wash. 285, 132 Am. St. Rep. 1075, 101 P. 867; *Shauer v. Alterton,* 151 U.S. 607, 38 L. ed. 286, 14 S.Ct. 442; *Johnson v. Erlandson,* 14 N.D. 521, 105 N.W. 722; *Hingtgen v. Thackery,* 23 S.D. 329, 121 N.W. 839; *Webb v. John Hancock Mut. L. Ins. Co.*, 162 Ind. 616, 66 L.R.A. 632, 69 N.E. 1006. *McDowall v. Herbert*, 31 N.D. 217, 153 N.W. 464 (N.D. 1915).

## OHIO

This distinction becomes apparent upon examination of the various sources purporting to define actual and constructive notice. Certainly, there is some confusion among these sources and reference is often made to "actual notice," "implied actual notice," "inquiry notice," or "constructive notice" in describing the same circumstances. Nevertheless, regardless of the terminology used, the overwhelming majority of sources describe actual or constructive notice in terms which would attribute knowledge of conditions on a property only to a purchaser who has either (1) actual knowledge of the condition or at the very least, or (2) actual knowledge of some circumstance which would reasonably incite inquiry.

Note, for example, the following excerpts from Black's Law Dictionary (5 Ed.1979) 957–958:

"Actual notice. * * * In this sense [express and implied] actual notice is such notice as is positively proved to have been given to a party directly and personally, or such as he is presumed to have received personally because the evidence within his knowledge was sufficient to put him on inquiry" (Emphasis added).

"Constructive notice. Constructive notice is information or knowledge of a fact imputed by law to a person (although he may not actually have it), because he could have discovered the fact by proper diligence, and his situation was such as to cast upon him the duty of inquiring into it" (Emphasis added).

"Implied notice. * * * [Implied notice] is notice inferred or imputed to a party by reason of his knowledge of facts or circumstances collateral to the main fact, of such a character as to put him on inquiry, and which, if the inquiry were followed up with due diligence, would lead him definitely to the knowledge of the main fact. * * *" (Emphasis added.)

In addition, note the following passages from 58 American Jurisprudence 2d (1971), 488–493, Notice, Sections 5–8:

"Actual notice has sometimes been said to be of two kinds, namely, (1) express, which includes what might be called direct information, and (2) implied, which is said to include notice inferred from the fact that the person had means of knowledge, which he did not use. Notice of the latter kind has frequently been classified by the courts as constructive notice * * *." Id. at 488–489. (Emphasis added.)

"Means of knowledge and knowledge itself are, in legal effect, the same thing where there is enough to put a party on inquiry. When a party has information or knowledge of certain extraneous facts which of themselves do not amount to or tend to show, an actual notice, but which are sufficient to put a reasonably prudent man upon inquiry * * *." Id. at 491. (Emphasis added.)

"It has also been said that there are two kinds of constructive notice, one of which is notice which results from a record or which is imputed by the recording statutes, and the other of which is notice which is presumed because of the fact that a person has knowledge of certain facts which should impart to him, or lead him to knowledge of the ultimate fact. * * *." Id. at 490. (Emphasis added.)

Finally, we find the language immediately following the above definitions from American Jurisprudence 2d (1971), 492–493, Notice, Sections 8 and 9, to be particularly instructive to the case before us:

"* * * This does not mean, however, that mere want of caution amounts to notice of what would have been discovered by the use of caution, where there is no element of knowledge of facts suggesting inquiry. The question is not only whether an inquiry would have revealed the fact, but also whether, when acting as an ordinarily prudent person would have done, the person to be charged was called upon, under the circumstances, to make the inquiry." *John W. Johnston v. Faith Baptist Church, Inc.*, 89-LW-1385, 1-87-14 (1989).

The principal purpose of the recording statute is to protect a bona fide purchaser, who does not have actual notice at the time of his purchase, against legal claims under unrecorded conveyances and encumbrances. R.C. 5301.25. Constructive notice is in legal effect the equivalent of actual knowledge. 54 Ohio Jur. 2d, Vendor and Purchaser, See 87. Under the recording laws, all persons dealing with the land in question are chargeable with constructive notice of properly recorded instruments in the chain of title. Statements and references contained in instruments in his chain of title bind the owner, and he is charged with knowledge he would have obtained from reasonable inquiry. *Arnoff v. Williams* (1916), 94 Ohio St. 145. Knowledge sufficient to put a person on inquiry which would disclose unrecorded facts is sometimes called constructive notice but is treated as actual notice. Actual notice may be inferred from the fact that means of knowledge is available. *In re Fahle's Estate* (1950), 90 Ohio App. 195. Also see 89 ALR 3d 901, Annotation: Recorded Real Property Instrument as charging third party with constructive notice of provisions of extrinsic instrument referred therein.

In *Davis v. Tebbs* (1886), 81 Va 600, the court held that references to the will in subsequent deeds which were links in the chain of title put the defendant upon notice of the provisions of the will. The defendant had obtained the property under a deed purporting to convey the entire tract from a third person who had succeeded to the remainder interest of 2/3 children surviving the life tenant of the property, as devised by the will. The plaintiff claimed through the surviving child under the will. The court concluded that the reference in subsequent deeds to the will put the defendant on notice of its provisions, whether the will was recorded or not.

A purchaser of land is chargeable with notice of recitals or disclosures in instruments in the chain of title, from which, in the exercise of reasonable diligence, he would discover the existence of outstanding rights in the property. 47 Ohio Jur. 2d Records and Recording, Sec. 38.

Patton on Land Titles, 2nd Ed. 1957 discusses the effect of constructive notice upon a purchaser to whom it is charged at Section 19:

(A) purchaser is chargeable with notice of all matters affecting the title which are expressly set forth in any instrument or proceeding forming an essential link in the title, and also of all other material facts which an inquiry thereby reasonably suggested would have disclosed. The recording acts are an extension to title records of the doctrine of notice, with the result that a purchaser bound thereby with constructive notice of a prior right takes subject thereto. Consequently, when the earlier conveyance is of record, there cannot be a subsequent purchaser who fulfills the statutory or judicial requirement of being "innocent' or "without notice.' The legal consequence of constructive notice afforded by the record is the same as that of actual notice. And the fact that a subsequent purchaser searched the records and failed to find an earlier conveyance properly recorded does not impair the notice afforded by it. He is charged with the notice created by the record, whether he searches the records or not; if he does not examine the records, he is charged with constructive notice of their contents; and, if he examines them, he has actual notice of their contents. *Judith Ferguson v. Paul Zimmerman*, 86-LW-4364, 9426 (C.P. 81-2490) (1986).

## OKLAHOMA

"Constructive notice" is a presumption of law, making it impossible for one to deny the matter concerning which notice is given, while "implied notice" is a presumption of fact, relating to what one can learn by reasonable inquiry, and arises from actual notice of circumstances and not from constructive notice.

Under Revised Laws Oklahoma 1910, § 2926, providing that every person having actual notice of circumstances sufficient to put a prudent man on inquiry as to a particular fact, and omitting to make inquiry with reasonable diligence, is deemed to have constructive notice of the fact, held, where a proposed mortgagee has notice from a record, and also actual notice of the fact that under a guardian's sale the title vested in the guardian's wife within 63 days from the date of the guardian's deed, it is incumbent upon said proposed purchaser to make inquiry with reasonable diligence, and where said purchaser makes a mere superficial inquiry, such inquiry is not sufficient to make out a defense of bona fide purchaser for value without notice.

Where a mortgagee, claiming to be an innocent purchaser for value, was charged with notice of suspicious circumstances sufficient to put a prudent man on inquiry, and failed to submit said title to a lawyer for examination, and merely took form affidavits setting up formal statements, without disclosing essential facts relating to payment or amount of consideration, and where the complete result of said inquiry does not satisfactorily appear in the testimony, held, that the finding of fact of the lower court that said mortgagee had failed to sustain its defense of bona fide purchaser for value without notice, is not against the clear weight of the testimony and must be sustained; held further that examination of the record in this case fails to disclose the diligent inquiry required by law of a prudent purchaser. *Gay v. Williams*, 219 P. 906, (1923).

3. Notice–Actual notice of circumstances essential to put one on inquiry.

Under Rev. Laws Okl. 1910, § 2926, providing that every person having actual notice of circumstances sufficient to put a prudent man on inquiry as to a particular fact, and omitting to make inquiry with reasonable diligence, is deemed to have constructive notice of the fact, one is not put on inquiry by constructive notice contained in a record, but only by actual notice of circumstances sufficient to do so.

4. Records–Recorded deed is notice of what appears on its face.

Under Rev. Laws Okl. 1910, § 1155, a recorded deed is constructive notice of what appears on its face.

5. Notice–'Constructive notice' and 'implied notice' distinguished.

'Constructive notice' is a presumption of law, making it impossible for one to deny the matter concerning which notice is given, while 'implied notice' is a presumption of fact, relating to what one can learn by reasonable inquiry, and arises from actual notice of circumstances, and not from constructive notice.

10. Notice–Test is what prudent man should do, and not what extremely cautious person would do.

The test as to notice is not what an extremely cautious man might do, because of the circumstances of which he has notice. *Gay v. Williams*, 219 P. 906, (1923).

## REFERENCE

Probably few subjects in the law have been so prolific of litigation as that of notice affecting the rights of third parties. Oklahoma has by statute sought to make certain the doctrine of notice actual and constructive. The Revised Laws of Oklahoma of 1910 provide as follows:

"2923. Notice.– Notice is either actual or constructive.

"2924. Actual Notice.– Actual notice consists in express information of a fact.

"2925. Constructive Notice.– Constructive notice is notice imputed by the law to a person not having actual notice.

"2926. Presumption of Constructive Notice.– Every person who has actual notice of circumstances sufficient to put a prudent man upon inquiry as to a particular fact, and who omits to make such inquiry with reasonable diligence, is deemed to have constructive notice of the fact itself."

It is to be noted that this section relates to putting those on inquiry who have actual notice of circumstances, etc.; that, before being charged with constructive notice of what inquiry might develop, the party must have express information of circumstances. He is not put upon inquiry be constructive notice contained in a record. *Charles v. Roxana Petroleum Corp.*, 282 F. 983, (Cir. 1922).

While the general doctrines of notice, as enunciated in the decisions of most of the courts, do not go as far as the Oklahoma statute on the subject of constructive notice, yet the doctrine of implied notice from knowledge of facts sufficient to put a reasonably prudent man on inquiry in its results reaches the same end and accomplishes practically the same purpose. We refer to some cases on this subject: In *Wood v. Carpenter*, 101 U.S. 135, 141 (25 L.Ed. 807), the court quotes with approval the language of *Kennedy v. Greene*, 3 Myl. & K. 722, as follows:

"Whatever is notice enough to excite attention, and put the party on his guard, and call for inquiry, is notice of everything to which such inquiry might have led. When a person has sufficient information to lead him to a fact, he shall be deemed conversant of it."

This doctrine charges a person with notice of everything that he could have learned by inquiry where there is sufficient notice to put him on guard and excite attention. True, it does not make such notice constructive; but, if a party actually has notice, the only difference between constructive or implied notice would be that one could be rebutted and the other could not. The same doctrine is announced in *Weniger v. Success Mining Co. et al.* (this circuit) 227 F. 548, 557, 142 C.C.A. 180, 189. Referring to the question of notice the court says:

"For ignorance which is the effect of inexcusable negligence is no excuse for laches, and knowledge of facts and circumstances which would put a person of ordinary prudence and diligence on inquiry is, in the eyes of the law, equivalent to a knowledge of all the facts which a reasonably diligent inquiry would disclose."

The rule is well stated in 23 Am. & Eng. Enc. of Law (2d Ed.) 495, as follows:

"It is a well-settled rule that, where a purchaser has knowledge or information of facts which are sufficient to put an ordinarily prudent man upon inquiry, and the inquiry, if followed with reasonable diligence, would lead to the discovery of defects in the title or of equitable rights of others affecting the property in question, the purchaser will be held chargeable with knowledge thereof, and will not be heard to say that he did not actually know of them. In other words, knowledge of facts sufficient to excite inquiry is constructive notice of all that the inquiry would have disclosed."

Mr. Pomeroy, in his work on Equity Jurisprudence (volume 2, par. 600), say:

"What facts are sufficient to put the party upon an inquiry, so that he may thereby be charged with the actual notice inferred from circumstantial evidence? Among the facts to which, as evidence, such force has been attributed are close relationship, personal intimacy, or business connections existing between the purchaser and the party with whom he is dealing, or between him and the holder of the adverse claim; great inadequacy of the price, which might arouse the purchaser's suspicion, and put him upon inquiry as to the reasons for selling the property at less than its apparent value; the sight or knowledge of visible material objects upon or connected with the subject-matter, which may reasonably suggest the existence of some easement or other similar right."

The rule is laid down in *Coder v. McPherson*, 152 F. 951, 953, 82 C.C.A. 99, 101, by this court as follows:

"Notice of facts which would incite a man of ordinary prudence to an inquiry under similar circumstances is notice of all the facts which a reasonably diligent inquiry would disclose."

The California statutes on this subject are similar to the Oklahoma statutes, and in *Prouty v. Devin*, 118 Cal. 258, 50 P. 380, the court discusses this question and says:

"The question for determination is whether there was actual notice upon the part of Bates of circumstances sufficient to put him as a prudent man upon inquiry. If so, then, in law he is chargeable with constructive notice and knowledge of the fact, etc."

In *Gaines v. Saunders*, 50 Ark. 322, 7 S.W. 301, the rule laid down by the Supreme Court of Arkansas has a measure of clearness and simplicity:

"All persons acquiring an interest in lands 'take with constructive notice of whatever appears in the conveyances constituting their chain of title'; and whatever appears in such conveyances sufficient to put a prudent man on inquiry, 'which, if prosecuted with ordinary diligence, would lead to actual notice of some right or title in conflict with that they are about to acquire, it is their duty to make inquiry, and if they do not make it' the law 'will charge them with the actual notice they would have received if they had made it."

*Charles v. Roxana Petroleum Corp.*, 282 F. 983, (Cir. 1922).

## OREGON

"The law has been well settled from early times in this state that a subsequent purchaser with notice of an outstanding unrecorded title, encumbrance or interest in real property takes title subject to the outstanding unrecorded title. If the subsequent purchaser has information sufficient to put him upon inquiry as to such outstanding unrecorded title and reasonable inquiry would lead to full information regarding such unrecorded interest and he neglects to make such inquiry, he takes title to the land charged with such outstanding title or interest." "The general doctrine is, that whatever is sufficient to direct the attention of a purchaser to the prior rights and equities of third persons, and to enable him to ascertain their nature by inquiry, will operate as notice." *Greene v. Legacy Emanuel Hosp. and Health Care Center*, 335 Or. 115, 60 P.3d 535 (Or. 2002).

## PENNSYLVANIA

Whatever puts a party upon inquiry amounts to notice, provided that inquiry became a duty, and would lead to the knowledge of the requisite fact by the exercise of ordinary diligence and understanding. *Lodge v. Simonton*, 23 Am.Dec. 36, 2 Pen. & W. 439, (1831).

The law presumes that a purchaser of real estate will not trust merely to the title papers and records, but will inquire of the person in possession whether he claims title to the land: if the possession is distinct and unequivocal, it is sufficient to put the purchaser on inquiry, and amounts to constructive notice. *Sailor v. Hertzog*, 4 Whart. 259, (1839).

## RHODE ISLAND

Implied notice has been defined as a factual inference of the possession of knowledge, inferred from the availability of a means of acquiring such knowledge, when the party charged therewith had the duty of inquiry. *McCausland v. Davis*, 204 So.2d 334 (Fla.App.1967). In short, where one acquires knowledge of facts that are reasonably informative of the existence of an ultimate fact, a reasonably cautious person would thereby be led to the ultimate fact, and the courts will imply that such person had actual notice and, therefore, is chargeable with the knowledge of the ultimate fact. *Hardy v. Zoning Bd. of Review of Town of Coventry*, 113 R.I. 375, 321 A.2d 289, (1974).

## SOUTH CAROLINA

The law presumes he was aware of the location of the property lines the deed created. *See Binkley v. Rabon Creek Watershed Conservation Dist. of Fountain Inn*, 348 S.C. 58, 558 S.E.2d 902, (Ct. App.2001) ("Notice of a deed is notice of its whole contents ... and it is also notice of whatever matters one would have learned by any inquiry which the recitals of the instrument made it one's duty to pursue" (quoting 66 C.J.S. *Notice* § 19 [1998])); [391 S.C. 538] *see also S.C. Dep't of Transp. v. Horry County*, 391 S.C. 76, 705 S.E.2d 21 (2011) (Shearouse Adv. Sh. No. 2 at 18–19) (citing *Binkley* for the proposition that onstructive notice of recorded instruments forecloses a claim of lack of knowledge). Nothing was presented by either side to overcome this presumption. *Pendarvis v. Cook*, 391 S.C. 528, 706 S.E.2d 520 (S.C.App. 2011.)

## SOUTH DAKOTA

"Actual notice consists in express information of a fact." SDCL 17-1-2. "Constructive notice is notice imputed by the law to a person not having actual notice." SDCL 17-1-3. "One having actual notice of circumstances sufficient to put a prudent person on inquiry about 'a particular fact, and who omits to make such inquiry with reasonable diligence, is deemed to have constructive notice of the fact itself.'" *Strassburg*, 1998 S.D. 72, ¶ 10, 581 N.W.2d at 514 (quoting SDCL 17-1-4). *East Side Lutheran Church of Sioux Falls v. Next, Inc.*, 2014 SD 59, 852 N.W.2d 434 (S.D. 2014).

Notice of an easement is generally imputed to a purchaser where the easement is of such character that a purchaser acting with ordinary diligence would know or learn of its existence. Thus, where the easement is open and visible, the purchaser will be charged with notice even though the easement was created by a grant which was not then recorded. The grantee is bound where a reasonably careful inspection of the premises would disclose the existence of the easement or where the grantee has knowledge of facts sufficient to put a prudent buyer on inquiry. *Peterson v. Beck*, 537 N.W.2d 375 (S.D. 1995).

## TENNESSEE

"Constructive notice" is defined as "information or knowledge of a fact imputed by law to a person (although he may not actually have it) because he could have discovered the fact by proper diligence, and his situation was such as to cast upon him the duty of inquiring into it." *Hawks v. City of Westmoreland,* 960 S.W.2d 10, (Tenn. 1997) (quoting *Kirby v. Macon Cnty.,* 892 S.W.2d 403, [Tenn. 1994]).

## TEXAS

.......our courts of last resort have placed the construction that, if the purchaser for valuable consideration had knowledge of such facts as would excite the suspicion of a man of ordinary prudence, and put him upon inquiry as to the fraudulent intent, knowledge of such facts is constructive notice, and is equivalent to actual notice. *Scheuber v. Wheeler,* 15 S.W. 503 (Tex.App. 1891).

## UTAH

"Whatever is notice enough to excite attention and put the party on his guard and call for inquiry is notice of everything to which such inquiry might have led. When a person has sufficient information to lead him to a fact, he shall be deemed conversant of it." *First American Title Ins. Co. v. J.B. Ranch, Inc.,* 966 P.2d 834 (Utah 1998).

"The inquiry notice maxim is that the means of knowledge is equivalent to knowledge." "Whatever is notice enough to excite attention and put the party on his guard and call for inquiry is notice of everything to which such inquiry might have led. When a person has sufficient information to lead him to a fact, he shall be deemed conversant of it." *McBroom v. Chile,* 2016 UT 38, 392 P.3d 835 (Utah 2016).

## VERMONT

Upon the discovery of facts constituting the basis of the cause of action *or* the existence of facts sufficient to put a person of ordinary intelligence and prudence on inquiry which, if pursued, would lead to the discovery. Thus, the statute of limitation begins to run when the plaintiff has notice of information that would put a reasonable person on inquiry, and the plaintiff is ultimately chargeable with notice of all the facts that could have been obtained by the exercise of reasonable diligence in prosecuting the inquiry. *Kaplan v. Morgan Stanley & Co.,* 2009 VT 78, ¶ 7, 186 Vt. 605, 987 A.2d 258 (mem.) (quotation omitted); see also *Turner v. Roman Catholic Diocese of Burlington, Vermont,* 2009 VT 101, ¶ 50, 186 Vt. 396, 987 A.2d 960 (applying "discovery rule" standard). *In re Estate of Alden,* 190 Vt. 401, 35 A.3d 950 (Vt. 2011).

Equity recognizes both actual and implied, or inquiry, notice. Thus, if a party has "sufficient facts concerning [another's] interest in the property to call upon him to inquire, he is charged with notice of such facts as diligent inquiry would disclose." *Black River Assocs. v. Koehler,* 126 Vt. 394, 233 A.2d 175, (1967); see also *Fed. Land Bank v. Pollender,* 137 Vt. 42, 399 A.2d 512, (1979). Here, defendants' title to the now-landlocked land was of record, and a reasonable inquiry would have shown that its only road access was through the mortgaged property. Indeed, the obligation of the mortgagee to inquire and learn of the potential way of necessity is implied in our decision in *Traders, Inc.* See *William Dahm Realty Corp. v. Cardel*, 128 N.J. Eq. 222, 16 A.2d 69, (Ch.1940) (subsequent grantee of land burdened by easement of necessity charged with notice thereof because examination of land records and inspection of area would have revealed landlocked character of land in favor of which easement was implied). *Myers v. LaCasse*, 176 Vt. 29, 838 A.2d 50 (Vt. 2003).

## VIRGINIA

"It is a general rule that whatever puts a person on inquiry amounts in judgment of law to notice, provided the inquiry becomes a duty, and would lead to a knowledge of the facts by the exercise of ordinary intelligence and understanding. A person who has sufficient information to lead him to a fact is deemed conversant with it, and a person who has notice of facts which would cause a reasonably prudent person to inquire as to further facts is chargeable with notice of the further facts discoverable by proper inquiry." *Chavis v. Gibbs*, 198 Va. 379, 94 S.E.2d 195 (Va. 1956), quoting 66 C.J.S., Notice, § 11.

## WASHINGTON

The minimum notice necessary to defeat a buyer's claim of bona fide purchase is inquiry notice, defined as "such information as would excite apprehension in an ordinary mind and prompt a person of average prudence to make inquiry," *Glaser v. Holdorf*, 56 Wn.2d 204, 352 P.2d 212 (1960), or "a visible state of things as is inconsistent with a perfect right in him who proposes to sell," *Paganelli v. Swendsen*, 50 Wn.2d 304, 311 P.2d 676 (1957) (citation omitted). Our Supreme Court has elaborated on this requirement:

[Notice] need not be actual, nor amount to full knowledge, but it should be such "information from whatever source derived, which would excite apprehension in an ordinary mind and prompt a person of average prudence to make inquiry." [(Quoting] *Bryant v. Booze*, 55 Ga. 438 [(1875))]. It follows, then, that it is not enough to say that diligent inquiry would have led to a discovery, but it must be shown that the purchaser had, or should have had, knowledge of some fact or circumstance which would raise a duty to inquire.... "There must appear to be, in the nature of the case, such a connection between the facts discovered and the further facts to be discovered, that the former may be said to furnish a clue-a reasonable and natural clue-to the latter." [Quoting *Birdsall v. Russell*, 29 N.Y.220, (1864).] *Daly v. Rizzutto*, 59 Wash. 62, 109 P. 276 (1910). *Gervais v. Miederhoff*, 47852-8-II.

We stated in *Daly v. Rizzutto*, 1910, 59 Wash. 62, 109 P. 276, 29 L.R.A.,N.S., 467, that

It [notice] need not be actual, nor amount to full knowledge, but it should be such 'information from whatever source derived, which would excite apprehension in an ordinary mind and prompt a person of average prudence to make inquiry.' [Citing cases.] It follows, then, that it is not enough to say that diligent inquiry would have led to a discovery, but it must be shown that the purchaser had, or should have had, knowledge of some fact or circumstance which would raise a duty to inquire." (Italics ours.)
*Paganelli v. Swendsen*, 50 Wn.2d 304, 311 P.2d 676, (1957).

## WEST VIRGINIA

"Whatever is sufficient to direct the attention of a purchaser to prior rights and equities of third parties, so as to put him on inquiry into ascertaining their nature, will operate as notice." This Court continued in syllabus point three of *Pocahontas*: "That which fairly puts a party on inquiry is regarded as sufficient notice, if the means of knowledge are at hand; and a purchaser, having sufficient knowledge to put him on inquiry, or being informed of circumstances which ought to lead to such inquiry, is deemed to be sufficiently notified to deprive him of the character of an innocent purchaser." *Id.* at 60 S.E. at 890, syl. pt. 3. In syllabus point four of *Clark v. Lambert,* 55 W.Va. 512, 47 S.E. 312 (1904), this Court also addressed this issue of notice and held: "One who claims the protection of a court of equity as a bona fide purchaser must show that he had acquired the legal title before notice or knowledge of facts equivalent to notice." *Kourt Security Partners, LLC v. Judy's Locksmiths, Inc.*, 239 W.Va. 757, 806 S.E.2d 188 (W.Va. 2017).

The law, widely recognized, and controlling herein, is most concisely stated by an eminent authority (Warvelle on Vendors, § 263) as follows: "Whatever fairly puts a party on inquiry is regarded as sufficient notice where the means of knowledge are at hand; and a purchaser, whenever he has sufficient knowledge to put him on inquiry, or where he has been informed of circumstances which ought to have led to such inquiry, is deemed to have been sufficiently notified to deprive him of the character of an innocent purchaser. It is the duty of every person who may have knowledge or information of facts sufficient to put a prudent man on inquiry, as to the existence of some right or title in conflict with that he is about to purchase, to prosecute the same, and to ascertain the extent of such prior right; and if he wholly neglects to make inquiry, or, having begun it, fails to prosecute it in a reasonable manner, the law will charge him with knowledge of all facts that such inquiry would have afforded. A purchaser is bound to take notice of all recitals in the deed through which the title is derived, and is affected with notice of every matter or thing stated in the several conveyances constituting his chain of title. All such statements and recitals are sufficient to raise an inquiry, and the corresponding duty is thrust upon the purchaser to investigate and fully explore everything to which his attention is thereby directed." This authority also tells us: "Notice, to bind a purchaser, need not consist of positive information, for any fact that would put an ordinarily prudent man on inquiry will suffice; nor is it essential that notice of an equitable interest should come from the interested party or his agent, for such notice may be imparted aliunde, provided it is of a character likely to gain credit. Vague rumor or mere surmises are insufficient in themselves; but where parties assume to speak from knowledge, and particularly when such parties stand in situations which may reasonably be presumed to afford them the means of knowledge, the purchaser cannot disregard the information so obtained." Further, it says: "It has been held that where a party has heard of a sale of land before he purchased, and from a source entitled to reasonable credit, and under circumstances not likely to be forgotten, a duty would devolve upon him of tracing out the matter and ascertaining its truth."

These principles are so universally accepted that it seems useless to cite any of the many decisions approving them. *Pocahontas Tanning Co. v. St. Lawrence Boom & Mfg. Co.*, 63 W.Va. 685, 60 S.E. 890 (W.Va. 1908).

## WISCONSIN

The violation of the statute, therefore, appears to be only in the omission of the proper index to each volume. The statute does not make this index necessary to the record of an instrument, and it has never been held by this court essential to the constructive notice of the record. On the contrary, the proper entries of an instrument in the general index have been uniformly held to be constructive notice as far as they go, and constructive notice of the whole instrument until recorded *in extenso*. *Shove v. Larsen*, 22 Wis. 142; *Hay v. Hill*, 24 Wis. 235; *Pringle v. Dunn*, supra. This volume index appears to be a convenience which it is the statutory duty of the register to keep. But the failure to keep it does not, in the terms of the statute or in reason, affect the registry of an instrument or its constructive notice. *The Oconto Company v. Jerrard*, 46 Wis. 317, 50 N.W. 591 (Wis. 1879).

When the information brought home to the aggrieved party is such as to indicate where the facts constituting the fraud can be effectually discovered upon diligent inquiry, it is the duty of such party to make the inquiry, and if he fails to do so within a reasonable time, he is, neverthe-less, chargeable with notice of all facts to which such inquiry might have led. But here the dis-covery of the two deeds, standing by themselves and unconnected with the other facts, gave no indication of any of the facts constituting the fraud; nor did they point out where any of those facts could, upon inquiry, be ascertained. In the language of the court in *Kuhn's Appeal,* 87 Pa. 100: "The foulness was not on paper; it was in the assurances and concealments of him who was bound to disclose everything to his ward [here, client] which she ought to know." We may well conceive how it might have been regarded as an honest transaction. "The discovery by the aggrieved party of the facts constituting the fraud" is an actual discovery, and not a mere constructive discovery. *McMahon v. McGraw,* 26 Wis. 614. The fact that the plaintiff knew of the deeds in July, 1876, simply charged her with notice of the contents of the deeds, but did not charge her with notice of any facts transpiring prior to the deeds, and to which no reference was made in either of the deeds. *Godbold v. Lambert,* 8 Rich. Eq. 155.

In a case like this, the statute of limitations will not begin to run against the aggrieved party until such party has discovered the facts constituting the fraud, or has information of such a nature as would, upon diligent inquiry, lead to the discovery of such facts. Of course, no person is bound to commence an action on mere suspicion or rumor. The discovery, or the information which, upon diligent inquiry, would lead to the discovery, of facts sufficient to set the statute of limitations in motion, must be such facts as would impress a reasonable person with the belief that the transaction was, in fact, fraudulent. The views above expressed are not in conflict, but in harmony, with the opinion of the learned judge in *Carr v. Hilton,* 1 Curt. 390, 5 F. Cas. 137, cited and relied upon by counsel for the defendants, and, moreover, they seem to be supported by abundance of authority. *Kennedy v. Green,* 3 Mylne & K. 699; *Hovenden v. Lord Annesley,*

2 Schoales & L. 607; *Martin v. Smith,* 1 Dill. C. C. 85; *Bailey v. Glover,* 21 Wall. 342, 22 L.Ed. 636; *First Mass. Turnpike Co. v. Field,* 3 Am. Dec. 124; *Homer v. Fish,* 1 Pick. 435; *Rice v. Burt,* 4 Cush. 208; *Kane v. Bloodgood,* 7 Johns. Ch. 90; *S. C.* 11 Am. Dec. 417; *App v. Dreisbach,* 2 Rawle 287; *S. C.* 21 Am. Dec. 447; *Reeves v. Dougherty,* 27 Am. Dec. 496; *Haynie v. Hall's Ex'r,* 42 Am. Dec. 427; *Kuhn's Appeal,* 87 Pa. 100. *O'Dell v. Burnham*, 61 Wis. 562, 21 N.W. 635 (Wis. 1884).

The jury in finding that defendant had "information" obviously found and intended to find that defendant had knowledge of the facts referred to. Information is defined: "Knowledge acquired, derived, or inculcated, as by observation, or by reading or study, or in conversation." Stand. Dict. 924. And notice is held to mean information by whatever means communicated; knowledge given or received. *U.S. v. Foote,* 13 Blatchf. 418, 25 F. Cas. 1140. Prof. Pomeroy says that actual notice "is information concerning the fact ... directly and personally communicated to the party." 2 Pom. Eq. Jur. (3d ed.) § 595. Actual notice is also said to be "hat which consists in express infor-mation of a fact." *Prouty v. Devin,* 118 Cal. 258, 50 P. 380; 5 Words & Phrases, 4840. So we think that under the finding of the jury the defendant was chargeable with notice or knowledge of the fact that some person other than plaintiff was interested in the property and was not justified in relying upon the representations of the plaintiff in that regard. The defendant was bound to use ordinary care and prudence and avail itself of the information at hand. Ordinarily, such notice as would put a person of ordinary prudence upon inquiry is sufficient to charge the person with knowledge of such facts as he might by proper inquiry ascertain. *Thomas v. Flint,* 123 Mich. 10, 81 N.W. 936; *Osborne v. Ala. S. & W. Co.* 135 Ala. 571, 33 So. 687; *U. S. v. Sliney,* 21 F. 894; *Walker v. Neil,* 117 Ga. 733, 45 S.E. 387; *Martel v. Somers,* 26 Tex. 551; *Collins v. Davis,* 132 N.C. 106, 43 S.E. 579; *Bolles v. Chauncey,* 8 Conn. 389. So we think the defendant under the findings of the jury and the evidence must be charged with having acted upon its own judgment and not upon the representations of plaintiff. *Metcalf v. Mutual Fire Insurance Company of Towns of Albany*, 132 Wis. 67, 112 N.W. 22 (Wis. 1907).

## WYOMING

Inquiry notice, as used herein, arises from a legal inference. A subsequent purchaser may have actual (or even constructive) notice of some fact. However, that fact is not that another person has a claim to the property. Instead, the notice of some apparently extraneous fact. However, it is a fact that is sufficiently "curious" or "suspicious," according to normal human experience, that the purchaser should, as a matter of law, make an investigation into it. If, upon making the investigation into this first fact a second fact, namely that another person has a claim to the title of the property, is revealed, then the purchaser is considered to have inquiry notice of the claim itself. In other words, because of the nature of the first fact, of which the purchaser has actual or constructive notice, a rebuttable inference is made that the purchaser has notice of the second fact. However, if a reasonable inquiry would not reveal the second fact, then the inference is rebutted.

Because this notice is based upon a legal inference, the purchaser cannot void notice of the second fact by merely failing to make the inquiry. The rule is often stated that when the facts suffice to impose the duty of investigation, the purchaser is charged with notice of what a proper investigation would have revealed, whether the investigation is made or not. Ignoring this principle quickly renders the doctrine of inquiry useless. However, if the search, even though not conducted, was certain to be futile, no notice should be imputed. This statement is consistent with the prior principle, because the fact that the search would be futile indicates that a reasonable inquiry would have revealed no information that could be charged to the purchaser. Perhaps a better title for this form of notice, as defined here, would be *inferred notice,* or more specifically, *notice inferred from secondary facts* (Emphasis in original).

The last two subsections noted that a person who has actual or constructive notice of a particular claim may have an obligation to inquire into the evidence to ascertain the validity of the claim. In contrast, a person with inquiry notice or inferred notice does not have notice of the claim itself; she merely has notice of some extraneous or secondary fact. Inquiry is necessary to disclose whether any claim exists. If she ascertains that a claim exists, then the purchaser is obligated to further investigate the evidence to determine the validity of the claim. This two-step process of inquiry is quite different from the one-step inquiry into the validity of a claim that may be necessary for actual or constructive notice.

With inquiry notice, the purchaser must watch for those "curious" or "suspicious" facts that give her notice that she should inquire further. Exactly how inquisitive or distrustful a purchaser must be has been the subject of considerable debate. Professor Philbrick has asserted, in an extensive argument, that the standard should be reasonableness from the perspective of the ordinary person purchasing real estate in the purchaser's situation at the time of the purchase. This means that a court should find an obligation to inquire only from those extraneous or secondary facts that, by ordinary human experience, suggest a genuine potential for a defect. In addition, the extent of the inquiry necessary to satisfy the burden of investigation must be governed by some standard of reasonableness. A more exacting standard would give unwarranted protection to prior unrecorded interests at the expense of subsequent purchasers, contrary to the general purpose of the recording acts to protect a subsequent purchaser from an unrecorded prior interest. 14 Powell on Real Property, § 82.02[1][d][iii] (footnotes omitted). Inquiry notice, then, has two parts: (1) extraneous or secondary facts that suggest a genuine potential for a title defect; and (2) a reasonable inquiry based on the existence of those facts would disclose a claim by another person to the title of the property. However, notice will not be imputed to a purchaser if a reasonable inquiry would not disclose a claim to the property.

*Wyoming Bank and Trust v. Haught,* 2003 WY 111, 76 P.3d 301 (Wyo. 2003).

# References

Expand your references, and you will immediately expand your life.

**Tony Robbins**

Alexander, William DeWitt. *A Brief History of Land Titles in the Hawaiian Kingdom*. Honolulu: P.C. Advertiser, 1882.

Andrews, Evan. *The Lost Colony of Popham. History Stories*, 2017; updated 2018.

Batchellor, Albert Stillman, Ed. *Laws of New Hampshire. Volume 2: Province Period, 1702–1745*. Concord: Rumford Printing Company, 1913.

Bederman, David J. *Custom as a Source of Law*. Cambridge: Cambridge University Press, 2010.

Blackstone, Sir William. Commentaries on the Laws of England. Clarendon Press, 1765–1769.

Cadastral Survey Training Staff, Denver Service Center. *Glossaries of BLM Surveying and Mapping Terms*. 2nd Edition. Denver: U.S. Department of the Interior, Bureau of Land Management, 1980.

Cazier, Lola. Surveys and Surveyors of the Public Domain, 1785–1975. Washington, DC: U.S. Government Printing Office, 1976.

Clement, Donald B., Ed. Restoration of Lost or Obliterated Corners and Subdivision of Sections. A Supplement to the Manual of Surveying Instructions. United States Department of the Interior, Bureau of Land Management. Washington: U.S. Government Printing Office, 1952. Reprint with corrections, 1955.

Cole, George M. & Donald A. Wilson. *Land Tenure, Boundary Surveys, and Cadastral Systems*. Boca Raton, FL: CRC Press, 2017.

Getzler, Joshua. *A History of Water Rights at Common Law. Oxford Studies in Modern Legal History*. Oxford: Oxford University Press, 2006.

Gray, John Chipman. *The Nature and Sources of the Law*. 2nd Edition by Roland Gray. New York: The Macmillan Company, 1921.

Higgins, Pat. Popham Colony: A Slice of Time. The Maine Story, 2000, 2014.

History.com Eds. Jamestown Colony, 2010; updated 2019.

Lambden, David W. & Izaak de Rijcke. *Boundaries and Surveys*. Toronto, Canada: Carswell Company Limited, 1985.

Orn, Clayton L. Vanishing Footsteps of the Original Surveyor. *Baylor Law Review*, Vol. IV, No. 3. Spring. 1952.

Patton, Carroll G. & Rufford G. Patton. *Patton on Land Titles*. 2nd Edition. Volume 3. St. Paul: West Publishing Company, 1957.

Reed, Michael W. Shore and Sea Boundaries. Volume 3. U.S. Department of Commerce. National Oceanic and Atmospheric Administration, Coastal Services Center, Office of Coast Survey. Washington, DC: U.S. Government Printing Office, 2000.

Rose, Christine. *Military Bounty Land*, 1776–1855. San Jose, CA: CR Publications, 2011.

Shalowitz, Aaron L. Shore and Sea Boundaries. Volume 1. U.S. Department of Commerce. Coast and Geodetic Survey. Washington, DC: U.S. Government Printing Office, 1962.

United States Department of the Interior. Manual of Surveying Instructions: For the Survey of the Public Lands of the United States. Bureau of Land Management. Denver, CO: Government Printing Office, 2009.

Uzes, Francois D. *Chaining the Land*. Sacramento: Landmark Enterprises, 1977.

Watson, Blake A. Buying West Florida from the Indians: The Forbes Purchase and *Mitchel v. United States* (1835). *FIU Law Review*. Vol. 9, No. 2, Spring 2014.

Wikipedia. Benson Syndicate.

Wikipedia. Oregon Land Fraud Scandal.

Wikipedia. Pine Barrens Speculation.

Wikipedia. Yazoo Land Scandal.

Wilson, Donald A. Case Studies in Boundary Research. Professional Paper Presented to The Michigan Society of Registered Land Surveyors 38th Annual Meeting, February 7–9. Kalamazoo, Michigan. 1979.

Wilson, Donald A. *Boundary Retracement: Processes and Procedures*. Boca Raton, FL: CRC Press. 2017.

Wilson, Donald A. *Interpreting Land Records*. 2nd Edition. Hoboken: John Wiley & Sons, Inc. 2015.

# Index

## A

*Abbott v. Abbott*, 167
*Abbott v. Lindenbower*, 320
Accretion, title through, 18–19, 365
*Ackerman v. State*, 333
Acreage, 81–82
Acts of owners, 27
Actual notice, 101, 479–480, 481–482, 483, 484,
    485–486, 487, 488, 489–495, 499
*Adams v. Blodgett*, 171
*Adams v. United States*, 195
*Adams v. Yazoo & M. V. R. Co.*, 327
Adverse possession/unwritten agreement, title by, 17–18, 364
*Ainsa v. United States*, 147
Air rights, 195–196
Airspace
    ownership and use of, 194
        easements through air, 194–195
        glide path, 195
        undefined rights in, 195
*Akeley v. Carpenter*, 468
*Akin v. Godwin*, 113
Alabama
    notice requirements in, 479
    original survey in, 385–386
*Alabama v. Harrub*, 315
*Alameda Materials, Inc. v. Capital Title Agency, Inc.*, 479
Alaska, notice requirements in, 479
*Alden v. Pinney*, 146
Alexander, William DeWitt, 44
Alienated lands, tract survey and description of, 35
*Allen v. Allen*, 490
*Allen v. Bates*, 169
*Allen v. Duvall*, 178
*Allen v. Pulliam*, 243
*Alston et al. v. Limehouse, et al.*, 308
*Altman v. Circle City Glass Corp.*, 483
*American Ass'n, Limited, v. Innis, etc.*, 243
*American Dock and Improvement Co. v. Trustees of
    Public Schools*, 243, 271
*American Sec. Bank v. Read Realty, Inc.*, 482
*Andreu v. Watkins*, 146, 163
*Antilla v. Cbocs Properties, Inc.*, 487
*Appellant, v. Foster's Heirs*, 29
*App v. Dreisbach*, 499
Aquaculture, 222
    clam flats, 222, 223–259
    lobster wars/oyster wars, 222
    mussel farms, 259–262
    oyster bed, 263–265
        ownership of, 291–294
        survey of, 295–329
    oyster farms, 265–291
    oyster lands, 262
    real property vs. personal property, 329–335

*Archibald v. N.Y.C. & H. R. Ry. Co.*, 242
*Ard v. Brandon*, 268
*Ariola v. City of Stillwater*, 487
Arizona, notice requirements in, 479
Arkansas
    notice requirements in, 479–480
    original survey in, 386–387
*Armstrong v. Council of Devon*, 481
*Armstrong v. Du Bois*, 169
*Arneson v. Spawn*, 445, 446
*Arnoff v. Williams*, 491
*Arnold v. Hanson*, 453
*Arnsperger v. Crawford*, 298
*Ashley v. Hill*, 106, 380, 387
*Aurrecoechea v. Bangs*, 76, 77
*Aurrecoechea v. Sinclair*, 76
*Averill v. Hull*, 285
Avulsion, law of, 214
*Ayers v. Watson*, 420

## B

*Backstrom v. Briar Hill Baptist Church, Inc.*, 487
*Bader v. Zeise*, 476
*Bagley v. Morrill*, 135
*Bailey v. Glover*, 499
*Baker v. Roslyn Swim Club*, 28
*Baldwin v. Hinton*, 178
*Ballard v. Stanolind Oil & Gas Co.*, 383
*B. & H. Turnpike Co. v. Union R. Co.*, 296
*Bank of New York v. Nally*, 483
*Barden v. Northern Pacific Railroad Co.*, 276
*Bar Harbor Banking & Trust Co. v. Alexander*, 261
*Barnard's Heirs v. Ashley's Heirs*, 76
Barren bottom, 295
*Barringer v. Davis*, 427
*Barter v. Commonwealth*, 170
*Bartholomew v. Baker*, 184
*Bartlett Land and Lumber Company vs. Saunders*, 151
*Bass v. Rublee*, 131
*Bates v. Illinois Cent. R. Co.*, 414
*Battenhausen v. Bullock*, 483
*Bauer v. Gottmanhausen*, 113, 389, 438
*Baxter v. National Mortg. Loan Co.*, 488
*Bayhouse v. Urquides*, 110n66, 382, 390
*Beach v. Fay*, 134
*Bealmear v. Hutchins*, 242
*Bean v. Bachelder*, 50, 127, 404
*Bean v. Central Maine Power Co.*, 198
*Beardsley v. Crane*, 420, 425
*Beaty v. Robertson (Ind. Sup.)*, 438
*Beckley v. Bryan & c.*, 85, 137, 139n9, 140n13, 145
*Beckley v. Bryan and Ransdale*, 7
*Bell v. Morse*, 169
*Ben Realty Co. v. Gothberg*, 106, 109n65, 380, 387, 478
Benson, John A., 40

503

For Product Safety Concerns and Information please contact our EU
representative GPSR@taylorandfrancis.com
Taylor & Francis Verlag GmbH, Kaufingerstraße 24, 80331 München, Germany

www.ingramcontent.com/pod-product-compliance
Lightning Source LLC
Chambersburg PA
CBHW060953210326
41598CB00031B/4810